Piotr T. Chruściel
Helmut Friedrich

The Einstein Equations

and the

Large Scale Behavior of Gravitational Fields

50 Years of the Cauchy Problem
in General Relativity

Springer Basel AG

Editors:

Piotr T. Chruściel
Laboratoire de Mathématiques
et de Physique Théorique
CNRS UMR 6083
Avenue Grammont
37200 Tours
France
e-mail: Piotr.Chrusciel@lmpt.univ-tours.fr

Helmut Friedrich
Max-Planck-Institut für Gravitationsphysik
Am Mühlenberg 1
14467 Golm
Germany
e-mail: hef@aei-potsdam.mpg.de

2000 Mathematical Subject Classification 35J60, 35L45, 58J05, 58J45, 53C50, 83C05

A CIP catalogue record for this book is available from the
Library of Congress, Washington D.C., USA

Bibliographic information published by Die Deutsche Bibliothek
Die Deutsche Bibliothek lists this publication in the Deutsche Nationalbibliografie;
detailed bibliographic data is available in the Internet at <http://dnb.ddb.de>.

ISBN 978-3-0348-9634-4   ISBN 978-3-0348-7953-8 (eBook)
DOI 10.1007/978-3-0348-7953-8

© 2004 Springer Basel AG
Originally published by Birkhäuser Verlag ,Basel, Switzerland in 2004
Softcover reprint of the hardcover 1st edition 2004
Printed on acid-free paper produced of chlorine-free pulp. TCF ∞

ISBN 978-3-0348-9634-4

9 8 7 6 5 4 3 2 1

www.birkhauser-science.com

Contents

Preface vii

R. Bartnik and J. Isenberg
    The Constraint Equations 1

H.L. Bray and P.T. Chruściel
    The Penrose Inequality 39

L. Andersson
    The Global Existence Problem in General Relativity 71

H. Friedrich
    Smoothness at Null Infinity and the Structure of Initial Data 121

L. Lehner and O. Reula
    Status Quo and Open Problems in the Numerical Construction
    of Spacetimes 205

A.D. Rendall
    The Einstein-Vlasov System 231

Y. Choquet-Bruhat
    Future Complete U(1) Symmetric Einsteinian Spacetimes,
    the Unpolarized Case 251

L. Andersson and V. Moncrief
    Future Complete Vacuum Spacetimes 299

J.L. Friedman
    The Cauchy Problem on Spacetimes That Are Not
    Globally Hyperbolic 331

M.T. Anderson
    Cheeger-Gromov Theory and Applications to General Relativity 347

Contents

G.J. Galloway
Null Geometry and the Einstein Equations 379

T. Barbot and A. Zeghib
Group Actions on Lorentz Spaces, Mathematical Aspects:
A Survey 401

R. Geroch
Gauge, Diffeomorphisms, Initial-Value Formulation, Etc. 441

Index 479

On the cover the reader will find the general form of a toroidal Gowdy metric and the vacuum Einstein equations for that metric. The mountain landscape is a result of a numerical simulation of the time evolution of the function P appearing in a closely related metric. We are very grateful to Beverly Berger for making the results of her numerical simulations available to us.

The Einstein Equations and the Large Scale Behavior of Gravitational Fields

50 years of the Cauchy problem in general relativity

Preface

The essence of the theory of general relativity is encoded in the specific features of individual solutions and also in the structure of the set of all solutions to Einstein's equations. Extracting this information from the field equations has been a major effort in the field since the birth of general relativity and still much has to be done.

At early stages, constructing explicit solutions played a dominant role. The analysis and interpretation of some of these solutions revealed unexpected features of general relativity. Guided by explicit solutions, approximation methods were invented to describe weak gravitational fields. Using highly sophisticated versions of these techniques to interpret recent observations, one was able to establish the first indirect proof of the reality of gravitational radiation (see [1, 22]).

Strong gravitational fields and highly dynamical situations are not accessible by those methods. Techniques of abstract analysis are required to establish the general qualitative picture, to clarify conceptional questions, and to supply structural insights for numerical work. Numerical computer codes are needed to provide quantitative information for comparison with observational data.

Due to the intricacies of Einstein's equations, and the lack of mathematical tools, the initial investigation of the solution manifold proceeded at a slow pace (cf. [6, 20, 21, 16] for early work). A fundamental breakthrough in the field was achieved by Yvonne Choquet-Bruhat's article *Théorème d'existence pour certains systèmes d'équations aux dérivées partielles non linéaires* [8], published in 1952, where she proves local existence of solutions to the Cauchy problem for Einstein's field equations.

Independent strands of development, which emphasized the geometrical content of the theory, began in the late 1950's. Attempts to understand *gravitational radiation* and *gravitational collapse* brought new insights, highlighting in particular the important role of the global structure of space-time.

The concept of asymptotic flatness led to a general notion of *gravitational radiation field* which does not require mathematical approximations [2, 18] (compare [23]). These studies relied on the analysis of formal expansions and assumptions on the fall-off behavior of solutions at null infinity. The *singularity theorems* showed that geodesic incompleteness is a stable feature of Einstein's theory [19, 9, 10]. The proofs exploited, however, only certain general features of the field equations and supplied little information on the solutions.

Because the asymptotic behavior and the singularity structure of its solutions express the individuality of a quasi-linear hyperbolic equation in a most subtle and precise way, it is quite remarkable that these results could be obtained at all. In fact, if Einstein's equations are looked at purely from the PDE point of view, their general covariance poses even further difficulties by the need to invent and control gauge conditions and to determine the underlying manifold as part of the solution. The geometrical nature of the equations supplies, however, arguments of remarkable strength which help in various considerations.

The developments indicated above raised questions about the large scale structure of solutions to the field equations. The underlying assumptions needed justification, one had to face the possible failure of predictability or the occurrence of naked singularities. Attempts to resolve these questions by the original methods failed. Combined with physical insight, the traditional calculational tools and stability arguments of the physicist can be surprisingly powerful, but they cannot supply definitive, general, and detailed answers to global problems. This requires control of the long time behavior of solutions to Einstein's equations in terms of suitably prescribed Cauchy or other boundary data.

The existence results up to the early 1980's were local in time (see [3, 7] for a survey), but the situation changed drastically in the last twenty-five years. The analysis of the Cauchy problem evolved into a tool for performing large scale studies of classes of solutions with distinguished qualitative features. The questions indicated above played guiding roles in most of the new work.

In view of this progress it seemed appropriate to mark the fiftieth birthday of the local existence theorem by organizing a summer school in which younger participants would learn from leading experts about recent progress in the field, while their older colleagues would have an opportunity to meet, exchange ideas, and present their work. This is how the Cargèse Summer School "50 years of the Cauchy problem in general relativity" was born. The reader will find the audio, video, and supporting material from several lectures held there on the enclosed DVD.

After the school had taken place it became clear to us that a volume describing the state of the art in mathematical general relativity would be useful. This was not planned as a proceedings of the school: not all topics which we felt should be represented were covered during the school; moreover, the contents of some of the talks was already adequately covered elsewhere. This is how the idea of this book was conceived.

The reader will find here an overview of the current knowledge on global properties of solutions of Einstein equations, and on techniques which in recent years brought progress to the understanding of the subject. It should be recognized that there are three notable gaps: Firstly, the techniques developed in the the proof of the non-linear stability of Minkowski space-time are not represented at all. However, those techniques have already been presented in detail in monographs ([5, 11]) as well as in introductory reviews ([4, 12]), which have recently been complemented by [13]. Secondly, the theory of low-regularity solutions of the Cauchy problem developed in [14, 15] is not presented. This aspect of the field is rapidly evolving, and it is useful to wait a few years before attempting to present a synthetic overview. Finally, an important event in the field, which took place after the structure of this book had been finalized, is a new proof of the stability of Minkowski space-time based on wave coordinates [17]. While the new proof provides less information about the solutions than the previous ones, it leads to dramatic simplifications of the global existence argument.

There does not appear to be a natural ordering of the articles in this book except the most uninspired one, namely that by alphabet. The current ordering arose from a mixture of logical and pedagogical arguments, together with the desire to guide a newcomer to the field through its physical and mathematical, theoretical and technical aspects. In any case, the readers should feel strongly encouraged to start reading wherever they like and to look at whatever raises their interest, and try to discover the intricate web of relations between the different contributions.

We are pleased that mathematical relativity is thriving, and that young people contribute new deep results, or solve long standing problems. We hope that this volume will help keep the field alive and inspire new developments.

Both this book, and the Cargèse Summer School, would not have been possible without the financial support of the European Community (under the Euro Summer Schools Program), the Centre National de Recherche Scientifique, the Clay Institute (Boston), the Tomalla Foundation (Zürich), la Collectivité Territoriale Corse, the Centre International de Mathématiques Pures et Appliquées (Nice), the International Association of Mathematical Physics, le Laboratoire de Mathématique Physique et Théorique de Tours, and Compaq. We wish to express our warmest thanks to all those Institutions.

Throughout her whole career Yvonne Choquet-Bruhat has kept contributing key results to a field which she helped grow and evolve. Her work has been a constant inspiration for scientists entering the field. It is a pleasure to dedicate this volume to her.

Piotr T. Chruściel, Helmut Friedrich, Tours and Golm, April 2004.

References

[1] L. Blanchet. Post-Newtonian Gravitational Radiation. In: B. Schmidt (ed.), *Einstein's Field Equations and Their Physical Implications*. Berlin, Springer, 2000.

[2] H. Bondi, M.G.J. van der Burg, A.W.K. Metzner. *Gravitational waves in general relativity VII. Waves from axi-symmetric isolated systems*. Proc. Roy. Soc. A 269 (1962) 21–52.

[3] Y. Choquet-Bruhat, J.W. York. *The Cauchy problem*. In: A. Held (ed.) *General Relativity and Gravitation*, Vol. 1. New York, Plenum, 1980.

[4] D. Christodoulou. The Global Initial Value Problem in General Relativity. In: V.G. Gurzadyan et al. (eds.) Proceedings of the 9th Marcel Grossmann Meeting World Scientific, New Jersey, 2002.

[5] D. Christodoulou, S. Klainerman. The Global Nonlinear Stability of the Minkowski Space. Princeton University Press, Princeton, 1993.

[6] M.G. Darmois. Les équations de la gravitation Einsteinienne. Mém. Sci. Mathématiques, Acad. Sci. Paris. Fasc. XXV (1927)

[7] Fischer, A.E., Marsden, J.E. (1979) The initial value problem and the dynamical formulation of general relativity. In: Hawking, S.W., Israel, W. (Eds.) General Relativity: an Einstein Centenary Survey. Cambridge University Press, Cambridge.

[8] Y. Fourès-Bruhat. Théorème d'existence pour certains systèmes d'équations aux dérivées partielles non linéaires. *Acta Math.* 88 (1952) 141–225.

[9] S.W. Hawking, R. Penrose. *The singularities of gravitational collapse and cosmology.* Proc. Roy. Soc. A **314** (1970) 529–548.

[10] S.W. Hawking, G.F.R. Ellis. *The Large Scale Structure of Space-Time.* Cambridge, Cambridge University Press, 1973.

[11] S. Klainerman, F. Nicolò. The Evolution Problem in General Relativity. Birkhäuser, Basel, 2003.

[12] S. Klainerman, F. Nicolò. On local and global aspects of the Cauchy problem in general relativity. *Class. Quantum Grav.* 16 (1999), R73–R157.

[13] S. Klainerman, F. Nicolò. Peeling properties of asymptotically flat solutions to the Einstein vacuum equations. *Class. Quantum Grav.* 20 (2003) 3215–3257.

[14] S. Klainerman, I. Rodnianski. Rough solution for the Einstein vacuum equations. Preprint (2001) math.ap/0109173.

[15] S. Klainerman, I. Rodnianski. The causal structure of microlocalized Einstein metrics. Preprint (2001) math.ap/0109174.

[16] A. Lichnerowicz. L'intégration des équations de la gravitation relativiste et le problème des n corps. *J. Math. Pures Appl.* 23 (1944) 37–63.

[17] H. Lindblad and I. Rodnianski. Global existence for the Einstein vacuum equations in wave coordinates. Preprint (2003) math.ap/0312479.

[18] R. Penrose. Asymptotic properties of fields and space-time. Phys. Rev. Lett., 10 (1963) 66–68.

[19] R. Penrose. Gravitational collapse and space-time singularities. Phys. Rev. Lett., **14** (1965) 57–59.

[20] K. Stellmacher. Zum Anfangswertproblem der Gravitationsgleichungen. *Math. Annalen* 115 (1938) 136–152.

[21] K. Stellmacher. Ausbreitungsgesetze für charakteristische Singularitäten der Gravitationsgleichungen. *Math. Annalen* 115 (1938) 740–783.

[22] J.H. Taylor. Binary pulsars and relativistic gravity. *Rev. Mod. Phys.* 66 (1994) 711–719.

[23] A. Trautman. Radiation and boundary conditions in the theory of gravitation. *Bull. Acad. Pol. Sci., Série sci. math., astr. et phys.* VI (1958) 407–412.

The Constraint Equations

Robert Bartnik and Jim Isenberg

Abstract. Initial data for solutions of Einstein's gravitational field equations cannot be chosen freely: the data must satisfy the four Einstein constraint equations. We first discuss the geometric origins of the Einstein constraints and the role the constraint equations play in generating solutions of the full system. We then discuss various ways of obtaining solutions of the Einstein constraint equations, and the nature of the space of solutions.

Mathematics Subject Classification (2000). 53C99, 83C57.

Keywords. constraint equations; Einstein equations; conformal method; quasi-spherical; thin sandwich.

1. Introduction

Yvonne Choquet-Bruhat's epic work of over 50 years ago shows that if a set of smooth initial data which satisfies the Einstein constraint equations is given, then we can always find a spacetime solution of the Einstein equations which contains an embedded hypersurface whose metric and second fundamental form agree with the chosen data. In the years since then, arguably the most important method for constructing and studying solutions of Einstein's equations has been the initial value (or Cauchy) formulation of the theory, which is based on Yvonne's result. Especially now, with intense efforts underway to model astrophysical events which produce detectable gravitational radiation, the Cauchy formulation and numerical efforts to implement it are of major interest to gravitational physicists.

To understand the Cauchy formulation of Einstein's theory of gravitation, we need to understand the constraint equations. Not only do the constraints restrict the allowable choices of initial data for solutions; they also effectively determine the function space of maximally globally hyperbolic solutions of the theory, and they play a role in generating the evolution of the initial data via their appearance in the Hamiltonian for Einstein's theory.

The goal of this review paper is to provide some measure of understanding of the Einstein constraints. We start in Section 2 by explaining the geometric origin of the constraint equations. To do this, we discuss $3 + 1$ foliations of spacetimes, the Gauss-Codazzi decompositions of the curvature, and the consequent $3 + 1$ projection of the spacetime Einstein equations. Next, in Section 3, we discuss the relationship between the constraints and evolution. Here, after first reviewing the proof of well-posedness, we discuss the ADM Hamiltonian formulation of Einstein's theory, noting the relationship between the Hamiltonian functional and the constraints. We also examine the evolution equations for the constraints, noting that if a set of data satisfies the constraints initially, it will continue to satisfy them for as long as the evolution continues.

We discuss methods for constructing solutions of the Einstein constraint equations in Section 4. We focus first on the most useful approach to date: the conformal method. After describing how the conformal method works, we discuss its success in parameterizing the set of all constant mean curvature (CMC) solutions of the constraints, the difficulties which arise when constructing solutions with non-constant mean curvature, and the issue of finding physically relevant sets of data which satisfy the constraints.

Closely related to the conformal method is the conformal thin sandwich approach. We describe how it relates to the standard conformal method, and its major advantages and disadvantages.

Instead of exploiting conformal variations to enforce the constraints, the original thin sandwich formulation [8, 21] varies the lapse (via an algebraic relation) and the shift vector. Under certain conditions this procedure leads to an elliptic system for the shift [17], and an implicit function theorem then shows that the system is solvable for all nearby data. This constructs an open set of solutions of the constraint equations, from unconstrained data. However, the restrictions arising from the surprising ellipticity condition, that π^{ij} be positive or negative definite, mean that the original thin sandwich approach is viable only for a limited range of geometries, and at this stage it must be considered more of a curiosity than a practical solution technique.

It is also possible to construct solutions of the Hamiltonian constraint by solving a semi-linear parabolic equation. Recent work [69, 71, 72] has shown that the essential feature of the original "quasi-spherical" construction [14] can be generalized beyond the quasi-spherical foliation condition to give a flexible technique for constructing exterior metrics of prescribed scalar curvature, satisfying geometric inner boundary conditions. Whilst the resulting parabolic method is restricted by the requirement that the background must be quasi-convex, it is able to handle several questions involving the constraints which cannot be addressed using the conformal method.

We finish our review of constraint construction techniques by discussing the idea of gluing, and how this idea has been implemented to date. Two approaches to gluing have been developed and applied to solutions of the constraint equations. Connected sum or IMP gluing [52] builds new solutions by adding a cylindrical bridge (or wormhole) connecting a pair of points either on a single given solution or on a pair of given solutions. The Corvino-Schoen technique [36, 37, 33] gives a local projection from approximate solutions to exact solutions of the constraints. This may then be applied to smoothly attach a finite region in a given asymptotically Euclidean solution to an exterior Schwarzschild or Kerr solution, with a transition region connecting the two regions. Both approaches have been very useful in answering a number of longstanding questions regarding solutions of the constraints.

We conclude this paper with a number of comments on important issues concerning the constraints which need to be addressed.

2. Deriving the constraints

The Cauchy formulation is used primarily to construct new solutions of the Einstein gravitational field equations from specified initial data. The best way to understand the origins of the constraint equations is to assume that we have a spacetime solution and to consider the induced data on spacelike hypersurfaces. We do this here, showing that the constraint equations necessarily must be satisfied by initial data (M^3, γ, K), where M^3 is a manifold, γ is a Riemannian metric on M^3 and K is a symmetric tensor on M^3, if this data is to be induced by a spacelike hypersurface in a spacetime solution of Einstein's equations.

2.1. $3+1$ foliations of spacetimes

Let V^4 be a smooth 4-dimensional Lorentzian spacetime, with the smooth[1] metric g having signature $(-1, 1, 1, 1)$. A *hypersurface* in V is an (embedded) submanifold $M^3 \hookrightarrow V^4$ of codimension 1. M is *spacelike* if the induced bilinear form $\gamma := i^* g$ is a Riemannian metric on M. Equivalently, M is spacelike if at each point $x \in M$ there is a timelike future unit normal vector n. We have the familiar diagram:

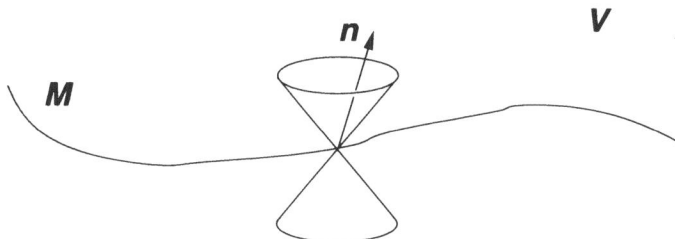

If X, Y are any vector fields tangent to M, then as a consequence of the embedding, we can consider them as vectors in V and we can decompose the V-covariant derivative $D_X Y$ into tangential and normal components,

$$D_X Y = \nabla_X Y + K(X, Y)n, \tag{1}$$

where ∇ is the induced connection on M (it is also the Levi-Civita connection of the induced Riemannian metric on M), and K is a bilinear form (rank 2 tensor)

---

[1] Although it is useful in addressing certain questions to consider lower regularity, for our present purposes it is most convenient to assume that the spacetime and the metric are C^∞.

on M, called the *second fundamental form*, or *extrinsic curvature* in the physics literature. From the relation

$$K(X,Y) = g(D_X n, Y) \qquad (2)$$

and from the fact that the Lie bracket of X and Y is tangent to M, $[X, Y] \in TM$, we find that K is a symmetric form, $K(X,Y) = K(Y,X)$.

A function $t \in C^1(V)$ is a *time function* if its gradient is everywhere timelike. We say that a time function t is *adapted to the hypersurface* M if M is a level set of t, in which case we can choose adapted local coordinates (x, t). In terms of such coordinates the normal vector field takes the familiar lapse-shift form

$$n = N^{-1}(\partial_t - X^i \partial_i) , \qquad (3)$$

where N is the *lapse function* and $X = X^i \partial_i$ is the *shift vector*. Equivalently,

$$\partial_t = Nn + X. \qquad (4)$$

This time evolution vector field ∂_t need not necessarily be timelike everywhere. (The shift vector $X \in TM$ is of course necessarily spacelike wherever it is non-vanishing.) However, if ∂_t is timelike then the $x = const.$ paths together make up a timelike congruence of spacetime observers.

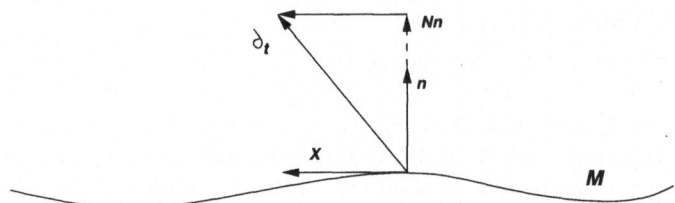

The spacetime metric can be expressed in terms of the lapse and the shift and the spatial metric in adapted coordinates by the formula[2]

$$g = -N^2 \, dt^2 + \gamma_{ij}(dx^i + X^i \, dt)(dx^j + X^j \, dt); \qquad (5)$$

and the second fundamental form is given by

$$K_{ij} = K(\partial_i, \partial_j) = \tfrac{1}{2} N^{-1}(\partial_t \gamma_{ij} - \mathcal{L}_X \gamma_{ij}) \qquad (6)$$

where

$$\mathcal{L}_X \gamma_{ij} = \nabla_i X_j + \nabla_j X_i \qquad (7)$$

is the Lie derivative in M of the spatial metric γ in the direction $X \in TM$. Thus we obtain an expression for the ∂_t evolution of the spatial metric,

$$\partial_t \gamma_{ij} = 2NK_{ij} + \mathcal{L}_X \gamma_{ij}. \qquad (8)$$

---

[2]Our index conventions here are: mid latin alphabet indices (i, j, etc.) run from 1 to 3 and correspond to directions tangent to the spacelike manifold M, while early latin indices (a, b, etc.) run from 0 to 3 and correspond to spacetime directions.

2.2. The Gauss and Codazzi equations and the constraints

Based on the orthogonal decomposition of the covariant derivative operator $D = \nabla + K$, and on the definition of the curvature on V, the Gauss and Codazzi equations relate certain spacetime curvature components to spatial curvature and other expressions formed solely from data intrinsic to the submanifold M. For example, starting from the expression for the curvature on V

$$\text{Riem}^V(X, Y, Z, W) = g((D_X D_Y - D_Y D_X - D_{[X,Y]})Z, W), \tag{9}$$

if X, Y, Z, W are all vector fields tangent to M, then we find by a simple computation that

$$\begin{aligned} &\text{Riem}^V(X, Y, Z, W) \\ &= \text{Riem}^M(X, Y, Z, W) + K(X, W)K(Y, Z) - K(X, Z)K(Y, W). \end{aligned} \tag{10}$$

This is the *Gauss equation*; it shows that the intrinsic curvature of M, measured by Riem^M, is determined by curvature components of the ambient spacetime V and the second fundamental form K. In the perhaps more familiar index notation, the Gauss equation takes the form

$$R^V_{ijkl} = R^M_{ijkl} + K_{il}K_{jk} - K_{ik}K_{jl}, \tag{11}$$

where the indices i, j, k, l refer to an (unspecified) basis of the spatial tangent space TM, which may be equally well determined either by a coordinate basis ∂_i (holonomic basis) or by an orthonormal basis e_i, $i = 1, \ldots, 3$.

A similar computation gives the *Codazzi identity*,

$$\begin{aligned} \text{Riem}^V(X, Y, n, Z) &= g((D_X D_Y - D_Y D_X - D_{[X,Y]})n, Z) \\ &= \nabla_X K(Y, Z) - \nabla_Y K(X, Z), \end{aligned} \tag{12}$$

where

$$\nabla_X K(Y, Z) = D_X(K(Y, Z)) - K(\nabla_X Y, Z) - K(Y, \nabla_X Z) \tag{13}$$

is the covariant derivative in M of the tensor K. In index notation the Codazzi identity takes the form

$$R^V_{ijnk} = K_{jk;i} - K_{ik;j}.$$

What about $\text{Riem}^V(n, Y, n, Z)$? A calculation similar to those used to derive the Gauss and Codazzi equations does not lead to an expression solely in terms of γ, K, and their spatial derivatives. If we introduce the Lie derivative

$$\mathcal{L}_n K(Y, Z) = D_n(K(Y, Z)) - K(\mathcal{L}_n Y, Z) - K(Y, \mathcal{L}_n Z),$$

along with the lapse and its spatial derivatives, then we have

$$\text{Riem}^V(Y, n, n, Z) = -\mathcal{L}_n K(Y, Z) + K^2(Y, Z) + N^{-1}\text{Hess}(N)(Y, Z), \tag{14}$$

where $K^2_{ij} := K^k_i K_{jk}$ and $\text{Hess}(N) = \nabla^2 N$ is the Hessian or second covariant derivative of the lapse function N. This is sometimes referred to as the *Mainardi equation*, and has the index form

$$R^V_{innj} = -\mathcal{L}_n K_{ij} + K^k_i K_{jk} + N^{-1}N_{;ij}. \tag{15}$$

The index expressions for the Gauss, Codazzi and Mainardi equations given above are useful for calculating the $3 + 1$ expressions for contractions of the Riemann tensor, such as the Ricci curvature tensor, the scalar curvature, and the Einstein tensor. For example, the Einstein tensor $G_{ab} = \text{Ric}_{ab}^V - \frac{1}{2} R^V g_{ab}$ satisfies

$$2G_{nn} = R^M + (\text{tr}_\gamma K)^2 - ||K||^2 \tag{16}$$

$$G_{in} = K_{i;j}^j - K_{j;i}^j. \tag{17}$$

We note that these expressions for G_{nn} and G_{in} involve just γ, K, and their spatial derivatives. Hence it follows from the Einstein field equations

$$G_{ab} = 8\pi T_{ab} \tag{18}$$

that the G_{nn} and G_{in} equations are constraints on the choice of the initial data γ and K. These are the *Einstein constraint equations*

$$R^M + (\text{tr}_\gamma K)^2 - ||K||^2 = 16\pi T_{nn}, \tag{19}$$

$$K_{i;j}^j - K_{j;i}^j = 8\pi T_{ni}. \tag{20}$$

Here T_{ab} is the stress-energy tensor and describes the matter content of the ambient spacetime. Although T_{ab} will itself generally satisfy some evolution equation, we will regard it as a prescribed field and often focus for definiteness on the case $T_{ab} = 0$ of vacuum (matter-free) spacetimes.

It is clear from our discussion here that the constraint equations are conditions on the data (M, γ, K) which are *necessary* for the data to arise from a spacelike hypersurface in a spacetime satisfying the Einstein field equations (18). That these conditions are also *sufficient* (for the existence of a spacetime satisfying the Einstein equations and which induces the data on some hypersurface) is the content of the fundamental theorem of Y. Choquet-Bruhat [43], which we discuss in the next section.

3. The constraints and evolution

The constraint equations comprise four of the ten Einstein equations. The remaining six equations

$$G_{jk} = 8\pi T_{jk} \tag{21}$$

describe how the data (γ, K) evolve in a spacetime solution. This can be seen by rewriting the curvature terms in (14), giving

$$\mathcal{L}_n K_{ij} = G_{ij} - \frac{1}{2} G_c^c g_{ij} + 2K_{ij}^2 - \text{tr}_\gamma K K_{ij} - \text{Ric}_{ij}^M + N^{-1} N_{;ij}. \tag{22}$$

Yet, as noted in the introduction, the constraints play a role in the evolution as well. We discuss this role here. We first show that the constraints are sufficient as well as necessary for a spacetime solution to evolve from a given set of data. We next explore the role of the constraints in the canonical, Hamiltonian, formulation of Einstein's theory. Finally, we discuss the evolution of the constraint functions under the evolution of the data generated by (21).

3.1. Well-posedness of the vacuum Einstein equations

The Cauchy formulation of a given field theory is well posed if (a) for any choice of initial data of specified regularity, there exists a solution which is consistent with that data, and (b) the map from the space of initial data to solutions is continuous. As shown by Choquet-Bruhat [43], the Cauchy formulation of Einstein's (vacuum) theory of gravitation, for smooth initial data, is well posed in coordinates of wave map gauge (harmonic) type. We now review the proof of this result [59, 43, 31, 40, 44, 74].

The vacuum Einstein equations take the form $\mathrm{Ric}^V(g) = 0$. If this were a hyperbolic system, well-posedness would follow from general results on quasi-linear hyperbolic systems (eg. [58, 74]). However, the system is not hyperbolic, which we verify by noting that by diffeomorphism invariance we have $\Phi^*(\mathrm{Ric}^V(g)) = \mathrm{Ric}^V(\Phi^*(g))$ for any diffeomorphism Φ of V, so the symbol (leading order terms in the linearisation) of $\mathrm{Ric}^V(g)$, considered as a partial differential equation in local coordinates, has a very large kernel.

To get around this difficulty, we consider the *reduced Einstein equations* [59, 39]

$$\rho(g) := \mathrm{Ric}^V(g) + \tfrac{1}{2}\mathcal{L}_W g = 0, \tag{23}$$

where $\mathcal{L}_W g_{ab} = D_a W_b + D_b W_a$ is the Lie derivative of the metric g in the direction of the vector field W, and W is chosen to have leading terms

$$W_a = -g^{cd}(\partial_c g_{da} - \tfrac{1}{2}\partial_a g_{cd}) + F(g) \tag{24}$$

where the terms $F(g)$ do not involve derivatives of g. A short calculation shows that $\rho(g)$ has symbol $\tfrac{1}{2}g^{ab}\xi_a\xi_b$, exactly that of the wave equation of the Lorentzian metric g itself. Thus $\rho(g) = 0$ forms a quasi-linear hyperbolic system,

$$g^{cd}\partial^2_{cd} g_{ab} = Q_{ab}(g, \partial g), \tag{25}$$

where Q_{ab} is quadratic in ∂g and depends also on the $F(g)$ terms in (24). As noted above, the initial value problem for $\rho(g) = 0$ is well posed.

However in general solutions of $\rho(g) = 0$ will not satisfy the vacuum Einstein equations. The idea now is to choose the initial conditions (and, perhaps, the precise form of the terms $F(g)$) in such a way that the vector field W can be shown to vanish identically.

To understand the meaning of W, recall the definition of the *tension field* $\omega(\phi)$ of a map $\phi : (V, g) \to (N, \hat{\nabla})$,

$$\omega^\alpha(\phi) = \hat{\nabla}^a D_a \phi^\alpha = g^{cd}\left(\partial^2_{cd}\phi^\alpha - \Gamma^e_{cd}\partial_e \phi^\alpha + \hat{\Gamma}^\alpha_{\gamma\delta}\partial_c \phi^\gamma \partial_d \phi^\delta\right), \tag{26}$$

where N is some chosen four-dimensional manifold and $\hat{\nabla}$ is an arbitrary connection on N. If we now assume ϕ is a diffeomorphism, which we use to identify the coordinates x^a on V with y^α on N by pull-back $x^a = \phi^*(y^a)$, then the tension becomes

$$\omega^a(\phi) = -g^{ab}g^{cd}(\hat{\nabla}_c g_{db} - \tfrac{1}{2}\hat{\nabla}_b g_{cd}), \tag{27}$$

which shows by comparison with (24) that W may be chosen as the metric dual of the push-forward by ϕ^{-1} of the tension,

$$W_a = g_{ab}\phi_*^{-1}(\omega^b(\phi)) \, . \tag{28}$$

Note that the reduction to a quasi-linear hyperbolic system holds for any choice of the target $(N, \hat{\nabla})$, where the connection $\hat{\nabla}$ need not be compatible with any metric on N. In the simplest case $\hat{\nabla}$ is the flat connection on $\mathbb{R}^4$ and the condition $W = 0$ translates into $\Box_g x^a = 0$ in local coordinates x^a on V and N. The gauge condition $W(\phi) = 0$ is known as the *wave map* or *Lorentz-harmonic* coordinate gauge. The common terminology "harmonic gauge" is inappropriate since solutions of the wave equation do not have the mean-value ("harmonic") property implied by the corresponding Euclidean Laplace equation.

Good existence theorems for the Cauchy initial value problem for the reduced Einstein system (25) are well known (see, for example, [74, 58, 65]), and the initial data $(g_{ab}(0), \partial_t g_{ab}(0))$ are freely prescribable, subject only to the condition that $g_{ab}(0)$ has Lorentz signature with M_0 spacelike. In particular, if the geometric data (γ_{ij}, K_{ij}) are given (subject only to the vacuum constraint equations (19)–20)), then for any choice of the lapse and shift g_{0a} we may recover $\partial_t g_{ij}$ via (8). Since $g_{ij} = \gamma_{ij}$, we see that initial data $(g_{ab}(0), \partial_t g_{ab}(0))$ may be chosen consistent with (γ, K) for any choice of $(g_{0a}(0), \partial_t g_{0a}(0))$.

Of course, it is not sufficient to just solve the reduced Einstein equations; in order to recover a solution of the full vacuum Einstein equations we must also ensure that the solution satisfies $W = 0$. However, it is an easy consequence of the second Bianchi identities and $\rho(g) = 0$ that W must satisfy

$$\Box_g W + \mathrm{Ric}^V(g)W = 0. \tag{29}$$

Thus we aim to construct the full initial data $(g_{ab}(0), \partial_t g_{ab}(0))$ for the evolution equation $\rho(g) = 0$ in such a way that $W(0) = 0$ and $\partial_t W(0) = 0$. If we can do this, then uniqueness for solutions of the Cauchy problem for the hyperbolic evolution equation (29) will imply that $W = 0$ everywhere.

To ensure that $W(0) = 0$ we may either choose the background metric h on N appropriately or we may expand (24) and choose the components $\partial_t g_{0a}(0)$ appropriately. As noted above, the choice of $\partial_t g_{0a}(0)$ does not affect the prescribed geometric data (γ, K).

To find the conditions necessary for choosing initial data with $\partial_t W(0)$ we combine the reduced Einstein equation (23) and the condition $W(0) = 0$ to obtain $-2\mathrm{Ric}_{ab}^V = \partial_a W_b + \partial_b W_a$ at $t = 0$, which may be rearranged into the form

$$g^{00}\partial_t W_a = -2G_a^0 + (\text{terms vanishing at } t = 0). \tag{30}$$

Then, noting that G_a^0 are the constraint functions, we see from (30) that the condition that the geometric data (γ, K) satisfy the constraint equations ensures that $\partial_t W(0) = 0$. Since solutions to (29) are uniquely determined by the initial data $(W(0), \partial_t W(0))$, the corresponding solution of the reduced equation (23) will have $W = 0$ everywhere and thus g will satisfy the vacuum Einstein equations with geometric data (γ, K). This shows that the constraint equations are also sufficient for the existence of a compatible vacuum spacetime, as claimed.

In this formulation of the Einstein equations, the lapse and shift are determined from $g_{0a}(t)$ by solving the reduced Einstein equations. Alternatively, we may ask whether it is possible to *prescribe* the lapse and shift as given functions of x and t (for example, $N(x,t) = 1$ and $X(x,t) = 0$ for all $x \in M$ and $t \geq 0$). As shown recently by Choquet-Bruhat [28], this is possible although in a weaker sense than the Einstein system in the wave map gauge described above. Specifically, the Einstein system for general lapse and shift is "non-strictly hyperbolic"; as a consequence, the system is well posed in Gevrey spaces, and so has local existence. We note as well that a number of researchers [31, 44] have formulated extended first-order versions of Einstein's vacuum equations and have verified symmetric hyperbolicity and therefore well-posedness in more standard spaces for these versions.

Because the initial lapse-shift variables $g_{0a}(0)$ (cf. (5)) are freely specifiable in the above construction, there remains the question of the uniqueness of the spacetime development of a given set of initial data. It turns out that solutions constructed using different choices of the lapse and the shift but the same geometric data (γ, K) are in fact equal in some neighborhood of the initial surface, after suitable coordinate changes [29]. This property is established by constructing a wave map ψ from one solution to the other, and using the identification of the initial data to ensure the wave map initially at least is a spacetime isometry. The two metrics can then be compared directly (via g_1 and $\psi^*(g_2)$), and uniqueness for the reduced Einstein equations leads to geometric uniqueness. This shows that the geometric data (γ, K) determine the resulting spacetime uniquely, at least locally.

3.2. Einstein's theory as a dynamical system and the ADM Hamiltonian formulation

If we fix a choice of the lapse and shift for which the system of Einstein's equations is well posed (see the discussion above), then Einstein's equations can be viewed formally as a dynamical system. More specifically, fixing a three-dimensional manifold M and letting $T\mathcal{M}$ denote the tangent space to the space $\mathcal{M}$ of smooth Riemannian metrics on M, we find that (with a fixed choice of lapse N and shift X) the Einstein evolution equations

$$\frac{\partial}{\partial t}\gamma_{ij} = 2NK_{ij} + \mathcal{L}_X \gamma_{ij}, \tag{31}$$

$$\frac{\partial}{\partial t}K_{ij} = \nabla^2_{ij}N + \mathcal{L}_X K_{ij} + N(\text{Ric}^V_{ij} - \text{Ric}^M_{ij} + 2K^k_i K_{jk} - \text{tr}_\gamma K \, K_{ij}) \tag{32}$$

specify a flow on $T\mathcal{M}$. The K_{ij} evolution equation follows from the Mainardi equation, cf. (14) and (22). For the vacuum Einstein case $\text{Ric}^V_{ij} = 0$; in the nonvacuum case, Ric^V_{ij} depends on the matter fields through the stress-energy tensor T_{ab}.

There is an important sense in which the constraints generate the flow, which we see by recasting Einstein's theory in the canonical Hamiltonian form. Originally motivated by attempts to obtain a theory of gravity consistent with quantum theory, and often labelled the "ADM" Hamiltonian formulation because of the early work on these ideas by Arnowitt, Deser and Misner [7], this is based on the analysis of the Einstein-Hilbert Lagrangian $\int_V R^V \, dv_g$.

One way to approach this analysis is via the second variation formula for spacelike hypersurfaces [70, 10]; for more traditional treatments see [31, 75, 57]. We let $F : (-\epsilon, \epsilon) \times M \to V$ be a 1-parameter family of embeddings $s : M \mapsto M_s = F(s, M)$, with variation vector $Y = F_*(\frac{\partial}{\partial s})$, with normal vector $n = n_s$ and with second fundamental form $K = K_s$ along M_s. If Y is everywhere transverse to M then $s \mapsto M_s$ is a foliation and we can regard Y, n, and K as fields on V; otherwise they live naturally on $(-\epsilon, \epsilon) \times M$ via pull-back. The first variation of the area $|M_s|$ is given by

$$D_Y(dv_\gamma) = -g(Y, n)\mathrm{tr}_\gamma K \, dv_\gamma \,, \tag{33}$$

and the second variation formula is

$$D_Y \mathrm{tr}_\gamma K = g(Y, n)(\|K\|^2 + \mathrm{Ric}^V(n, n)) + g(Y, \nabla\mathrm{tr}_\gamma K) - \Delta_M g(Y, n) \,, \tag{34}$$

where $D_Y = \frac{\partial}{\partial s}$ and the s-dependence of n and K is understood. This follows also from the Mainardi equation (15). Since from (10) and (12) we have

$$2\mathrm{Ric}^V(n, n) = -R^V + R^M - \|K\|^2 + (\mathrm{tr}_\gamma K)^2,$$

and since $-g(Y, n) = N$ is the lapse if $Y = \frac{\partial}{\partial s}$ is the evolution vector of a foliation by spacelike hypersurfaces, we find that (34) can be rewritten as

$$R^V = R^M + \|K\|^2 + (\mathrm{tr}_\gamma K)^2 - 2N^{-1}\Delta_M N + 2D_n(\mathrm{tr}_\gamma K) \,. \tag{35}$$

Integrating this quantity over the region bounded by two spacelike hypersurfaces $F(\{s_0\} \times M)$ and $F(\{s_1\} \times M)$, we obtain the Lagrangian

$$
\begin{aligned}
\mathcal{L}(g, K) &= \int_{F([s_0, s_1] \times M)} R^V \, dv_V \\
&= \int_{s_0}^{s_1} \int_{M_s} \left(R^M + \|K\|^2 - (\mathrm{tr}_\gamma K)^2\right) N - 2\Delta_M N \, dv_\gamma \, ds \\
&\quad + \int_{M_{s_1} - M_{s_0}} 2\mathrm{tr}_\gamma K \, dv_\gamma \,,
\end{aligned}
\tag{36}
$$

using (33). From this Lagrangian, we calculate π^{ij}, the momentum conjugate to γ_{ij} (see [75]) as

$$\pi^{ij} = \frac{\delta \mathcal{L}}{\delta \dot{\gamma}_{ij}} = \sqrt{\det \gamma} \left(K^{ij} - \mathrm{tr}_\gamma K \gamma^{ij}\right) \,. \tag{37}$$

Then ignoring boundary terms, we calculate the Hamiltonian $\mathcal{H} = \int_M \pi^{ij} \dot{\gamma}_{ij} - \mathcal{L}$ in ADM form:

$$\mathcal{H}_{ADM} = -\int_M \xi^a \Phi_a(g, \pi), \tag{38}$$

where $\xi^a = (N, X^i)$ is the four-vector consisting of the lapse and the shift, and where

$$\Phi_0 = R^M \det(\gamma)^{1/2} - \det(\gamma)^{-1/2}(||\pi||^2 - \tfrac{1}{2}(\mathrm{tr}_\gamma \pi)^2) \tag{39}$$

$$\Phi_i = -2\gamma_{ij}\nabla_k \pi^{jk} \tag{40}$$

are the constraint operators, written in densitized form, in terms of the canonical variables (γ, π) rather than (γ, K). It is easily verified that the equations $\Phi_a(\gamma, K) = 0$, $a = 0, \ldots, 3$ are equivalent to (19),(20) with $T_{0a} = 0$.

Given an explicit expression for the Hamiltonian such as (38), we can readily calculate the evolution equations for (γ, π) and for any functional of these quantities. In particular, we have the ADM form [7] of the evolution equations(31), (32)

$$\frac{d}{dt}\begin{bmatrix} \gamma \\ \pi \end{bmatrix} = \begin{bmatrix} 0 & 1 \\ -1 & 0 \end{bmatrix} D\Phi_a(\gamma, \pi)^* \xi^a, \tag{41}$$

where $D\Phi_a^*$ is the formal adjoint of the linearization (or functional derivative operator) $D\Phi_a$. Explicitly we have [42]:

$$D\Phi(\gamma, \pi)(h, p)$$
$$= \left(\left(\delta\delta h - \Delta \mathrm{tr}_\gamma h - h_{ij}\left(\mathrm{Ric}^{Mij} - \tfrac{1}{2}R^M \gamma^{ij}\right)\right)\sqrt{\gamma} \right.$$
$$+ h_{ij}\left(\mathrm{tr}_\gamma\pi\,\pi^{ij} - 2\pi^i_k \pi^{kj} + \tfrac{1}{2}|\pi|^2\gamma^{ij} - \tfrac{1}{4}(\mathrm{tr}_\gamma\pi)^2\gamma^{ij}\right)/\sqrt{\gamma}$$
$$+ p^{ij}\left(\mathrm{tr}_\gamma\pi\gamma_{ij} - 2\pi_{ij}\right)/\sqrt{\gamma} ,$$
$$\left. \pi^{jk}\left(2\nabla_j h_{ik} - \nabla_i h_{jk}\right) + 2h_{ij}\nabla_k\pi^{jk} + 2\gamma_{ik}\nabla_j p^{jk} \right), \tag{42}$$

$$D\Phi(\gamma, \pi)^* \xi$$
$$= \left(\left(\nabla^i\nabla^j N - \Delta N\gamma^{ij} - N\mathrm{Ric}^{Mij} + \tfrac{1}{2}NR^M\gamma^{ij}\right)\sqrt{\gamma} \right.$$
$$+ N\left(\mathrm{tr}_\gamma\pi\,\pi^{ij} - 2\pi^i_k \pi^{kj} + \tfrac{1}{2}|\pi|^2\gamma^{ij} - \tfrac{1}{4}(\mathrm{tr}_\gamma\pi)^2\gamma^{ij}\right)/\sqrt{\gamma}$$
$$+ \left(X^k\nabla_k\pi^{ij} + \nabla_k X^k\pi^{ij} - 2\nabla_k X^{(i}\pi^{j)k}\right) ,$$
$$\left. N\left(\mathrm{tr}_\gamma\pi\gamma_{ij} - 2\pi_{ij}\right)/\sqrt{\gamma} - 2p^{ij}\nabla_{(i}X_{j)} \right), \tag{43}$$

where $\delta\delta h = \nabla^i\nabla^j h_{ij}$ and $\xi = (N, X^i)$.

Note that in the literature there are a number of different explicit expressions for the time derivatives of γ and π [3]. These expressions can all be related by the addition of terms which vanish if the constraint equations are satisfied.

The adjoint used in (41) is formal, in that it ignores the contributions of boundary terms. For spatially compact spacetimes, there are of course no boundary terms. For asymptotically flat spacetimes, the resulting asymptotic boundary terms in the Hamiltonian are closely related to the ADM total energy-momentum [75], and they play a major role in the analysis of the physics of such spacetimes.

An important consequence of the form (41) of the evolution equations is the result of Moncrief [63], which shows that a vector field $\xi = (N, X^i)$ satisfies $D\Phi(\gamma, K)^*\xi = 0$ for sufficiently smooth vacuum initial data (γ, K) if and only if ξ is the restriction to M of a Killing field in the spacetime development of the initial data.

3.3. Preserving the constraints

If (γ_0, π_0) is initial data which satisfies the (vacuum) constraint equations and has vacuum evolution $(\gamma(t), \pi(t))$ corresponding to some choice of lapse-shift (N, X^i), it is natural to ask if the evolving data satisfies the constraints for all values of t. A first step in showing this is to calculate the time derivative of the constraint operators $H \equiv \Phi_0$ and $J_i \equiv \Phi_i$, cf. [2]:

$$\frac{\partial}{\partial t} H = \nabla_m(X^m H) + 2\gamma^{ij}\nabla_i N J_j + N\nabla_i J^i + \tfrac{1}{2}N\gamma^{-1/2}H\gamma_{ij}\pi^{ij} , \quad (44)$$

$$\frac{\partial}{\partial t} J_i = 2\nabla_i N H + N\nabla_i H + \nabla_k(X^k J_i) + \nabla_i X^k J_k . \quad (45)$$

These formulas follow directly from the conservation law $\nabla^b G_{ab} = 0$ and the fact that the evolution equation (32) implies $G_{ij} = 0$, $1 \leq i, j \leq 3$. The virtue of these expressions for the time derivatives of the constraints is that together they comprise a symmetric hyperbolic system. We may then use standard energy arguments (based on an energy quantity which is quadratic in H and J_i) to argue that if initially H and J_i vanish, then for all time consistent with a foliation of the spacetime development of the initial data, H and J_i vanish as well.

The formal Hamiltonian interpretation of the evolution equations (41) regards $(\gamma(t), \pi(t))$ as defining a trajectory in the cotangent bundle $T^*\mathcal{M}$ of the space $\mathcal{M}$ of Riemannian metrics γ_{ij} on M. It is straightforward to endow this phase space $\mathcal{F}$ with the structure of a Hilbert or Banach manifold, for example $(\gamma_{ij}, \pi^{ij}) \in H^s(M) \times H^{s-1}(M)$, $s \geq 2$ for compact M using the Sobolev spaces H^s, and then the evolution (41) is determined from a *densely defined* vector field on $\mathcal{F}$. The fact that the constraints are preserved shows that a trajectory of the evolution (41) lies in the constraint set $\mathcal{C}$, the subset of $\mathcal{F}$ satisfying the constraint equations, provided the initial data $(\gamma(0), \pi(0))$ lies in $\mathcal{C}$.

Note that while the constraints are preserved under the explicit, smooth, evolution generated by (41), if we evolve instead numerically (with the consequent unavoidable numerical errors), then the dynamical trajectory will leave the constraint set $\mathcal{C}$. Indeed, numerical tests suggest that such a trajectory will exponentially diverge from $\mathcal{C}$. This property, and how to control it, is under intensive study by numerical relativists; for a recent example see [60].

3.4. Linearization stability

The linearized Einstein equations govern the evolution of perturbations δg of a given background spacetime metric g_0 and have been extensively analyzed, in view of their importance in astrophysics and cosmology. It is then natural to ask whether linearized solutions actually correspond to solutions to the full nonlinear equations; that is, whether a given solution of the linearized equations is tangent to some curve of solutions of the full equations. Such a perturbation solution is called *integrable*. By the fundamental local existence theorem [43] of Yvonne Choquet-Bruhat, a perturbation is integrable if all solutions of the linearized constraint equations $D\Phi_{(\gamma,\pi)}(h,p) = 0$ correspond to full solutions of the constraints; i.e., if the constraint set $\mathcal{C}$ is a Hilbert submanifold. This is the question of *linearization stability* and it is somewhat surprising that this fails in certain situations.

The implicit function theorem (see, e.g., [1]) shows that $\mathcal{C}$ is a Hilbert submanifold of $\mathcal{F}$ provided the linearization $D\Phi_{(\gamma,\pi)}$ has closed range and is surjective. Now the range of $D\Phi_{(\gamma,\pi)}$ is $L^2(M)$-orthogonal to the kernel of the adjoint $D\Phi^*_{(\gamma,\pi)}$, which corresponds by work of Moncrief [63, 64] to space-time Killing vectors. Using this insight and elliptic theory for the operator $D\Phi_{(\gamma,\pi)}D\Phi^*_{(\gamma,\pi)}$, it has been shown by Fischer and Marsden [42, 41, 6] that if the data (γ,π) does not admit Killing vectors and if M is compact then $\mathcal{C}$ is locally a Hilbert manifold, at least near smoother data.

Moreover, at points of $\mathcal{C}$ corresponding to data for spacetimes with a Killing vector field, $\mathcal{C}$ is not a submanifold but instead instead has a cone-like singularity arising from a quadratic relation [41]. This condition arises from the second derivative of a curve in $\mathcal{C}$, $\lambda \mapsto \Phi(\gamma(\lambda), \pi(\lambda)$, about (γ_0, π_0) admitting a Killing field ξ. Integrating over M, discarding boundary terms and using $D\Phi^*_{(\gamma,\pi)}\xi = 0$ gives

$$\int_M \xi \cdot D^2\Phi_{(\gamma_0,\pi_0)}((h,p),(h,p))\, d^3x = 0, \qquad (46)$$

which is the required quadratic condition on the constrained variations (h,p). In [64] it is shown that this condition is equivalent to the vanishing of the Taub quantity, constructed from a solution of the linearized Einstein equations about a spacetime admitting a Killing vector.

The situation for asymptotically flat spacetimes differs significantly from the case of compact M. Using weighted Sobolev spaces (with local regularity $\gamma \in H^s$, $s = 2$) it is possible to show that the $\mathcal{C}$ is everywhere a Hilbert submanifold [9]; although Killing vectors such as rotations are possible, they do not satisfy the asymptotic conditions needed to prevent surjectivity of $D\Phi_{(\gamma,\pi)}$ and thus the quadratic conditions on constrained variations do not arise. This work [9] also shows that the asymptotically flat phase space $\mathcal{F} = (\mathring{g} + H^2_{-1/2}) \times H^1_{-3/2}$ provides a natural setting in which the ADM Hamiltonian and energy-momentum functionals become smooth functions, for example.

4. Solving the constraint equations

There are three reasons for seeking to construct and study solutions of the Einstein constraint equations. First, we would like to obtain initial data sets which model the initial states of physical systems. Evolving such data, we can model the gravitational physics of those physical systems. Second, we would like to understand the space of all globally hyperbolic solutions of Einstein's gravitational field equations. Since, as noted earlier, Einstein's equations are well-posed, the space of solutions of the constraints parametrizes the space of solutions of the spacetime field equations. Third, we would like to know enough about the nature of the space of solutions of the constraints and the Hamiltonian dynamics of the classical Einstein equations on this space to be able to consider descriptions of gravitational physics which are consistent with the quantum principle (i.e., to "quantize gravity").

The easiest way to find a set of initial data (M^3, γ, K) which satisfies the Einstein constraint equations is to first find an explicit solution (V^4, g) of the spacetime Einstein equations (e.g., the Schwarzschild or Kerr solutions), and then to choose a Cauchy surface $M^3 \hookrightarrow V^4$ in that spacetime solution: The induced Riemannian metric γ and second fundamental form K on M^3 together solve the constraints. Since the set of known spacetime solutions is very limited, this procedure is not especially useful for making progress towards the goals discussed above (although we shall see below that solutions of the constraints obtained from slices of known spacetime solutions can be very handy as building blocks for "gluing constructions" of solutions of the constraints, which are in turn potentially very useful for modelling physical systems).

Our focus here is on more comprehensive methods for finding and studying solutions of the constraints. We first discuss the conformal method. This is by far the most widely used approach, both for physical modelling and for mathematical analysis, and so it receives the bulk of our attention here. We next consider the thin sandwich and conformal thin sandwich ideas. The latter is closely related to the conformal method; we compare and contrast the two here. The fourth approach we consider is the quasi-spherical ansatz. Although not as comprehensive as the conformal method, the quasi-spherical ansatz is potentially useful for certain key applications, which we discuss below. We conclude by describing the recently developed procedures for gluing together known solutions of the constraints, thereby producing new ones.

For convenience, we work primarily in this paper with the vacuum constraint equations. We note, however, that most of what we discuss here generalizes to the Einstein-Maxwell, Einstein-Yang-Mills, Einstein-fluid, and other such nonvacuum field theories.

4.1. The conformal method

The vacuum Einstein constraint equations (19), (20) with $T_{nn} = 0$ and $T_{ni} = 0$ constitute an under-determined system of four equations to be solved for twelve unknowns γ and K. The idea of the conformal method is to divide the initial data on M^3 into two sets – the "Free (Conformal) Data", and the "Determined Data" – in such a way that, given a choice of the free data, the constraint equations become a *determined* elliptic PDE system to be solved for the Determined Data.

There are at least two ways to do this (see [35] for others). The sets of Free Data and Determined Data are the same for both procedures; we have

Free ("conformal") data:

λ_{ij} –   a Riemannian metric, specified up to conformal factor;

σ_{ij} –   a divergence-free[3] ($\nabla^i \sigma_{ij} = 0$), trace-free ($\lambda^{ij}\sigma_{ij} = 0$) symmetric tensor;

τ –   a scalar field,

Determined data:

ϕ –   a positive definite scalar field,

W^i –   a vector field.

The difference between the two procedures has to do with the form of the equations to be solved for the Determined Data, and the way in which the two sets of data are combined to obtain γ and K satisfying the constraints. In one of the procedures, which we label the "Semi-Decoupling Split" (historically called "Method A"), the equations for W and ϕ take the form

Semi-decoupling split

$$\nabla_i(LW)^i_j = \tfrac{2}{3}\phi^6 \nabla_j \tau \tag{47}$$

$$\Delta\phi = \tfrac{1}{8}R\phi - \tfrac{1}{8}(\sigma^{ij} + LW^{ij})(\sigma_{ij} + LW_{ij})\phi^{-7} + \tfrac{1}{12}\tau^2\phi^5, \tag{48}$$

where the Laplacian Δ and the scalar curvature R are based on the λ_{ij}-compatible covariant derivative ∇_i, where L is the corresponding conformal Killing operator, defined by

$$(LW)_{ij} \equiv \nabla_i W_j + \nabla_j W_i - \tfrac{2}{3}\lambda_{ij}\nabla_k W^k, \tag{49}$$

and we construct γ and K from the free and the determined data as follows

$$\gamma_{ij} = \phi^4 \lambda_{ij} \tag{50}$$

$$K_{ij} = \phi^{-2}(\sigma_{ij} + LW_{ij}) + \tfrac{1}{3}\phi^4\lambda_{ij}\tau. \tag{51}$$

Using the other procedure, which we label the "Conformally Covariant Split" (historically "Method B"), the equations for W and ϕ are

Conformally covariant split

$$\nabla_i(LW)^i_j = \tfrac{2}{3}\nabla_j\tau - 6(LW)^i_j \nabla_i \ln\phi \tag{52}$$

$$\Delta\phi = \tfrac{1}{8}R\phi - \tfrac{1}{8}\sigma^{ij}\sigma_{ij}\phi^{-7} - \tfrac{1}{4}\sigma^{ij}(LW)_{ij}\phi^{-1}$$
$$+ \tfrac{1}{12}(\tau^2 - (LW)^{ij}(LW)_{ij})\phi^5 \tag{53}$$

and the formulas for γ and K are

$$\gamma_{ij} = \phi^4 \lambda_{ij} \tag{54}$$

$$K_{ij} = \phi^{-2}\sigma_{ij} + \phi^4 LW_{ij} + \tfrac{1}{3}\phi^4\lambda_{ij}\tau. \tag{55}$$

Each of these two methods has certain advantages. For the semi-decoupling split, if one chooses the mean curvature τ to be constant, then the ϕ dependence drops from (47), and the focus of the analysis is on (48), the Lichnerowicz equation. This is not true for the conformally covariant split; however in this latter case, one

---

[3]In the free data, the divergence-free condition is defined using the Levi-Civita covariant derivative compatible with the conformal metric λ_{ij}.

finds that a solution exists for free data $(\lambda_{ij}, \sigma_{ij}, \tau)$ if and only if a solution exists for $(\theta^4 \lambda_{ij}, \theta^{-2} \sigma_{ij}, \tau)$. Far more is known mathematically about the semi-decoupling split, and this approach has been used much more in applications, so we focus on the semi-decoupling split in the rest of this paper. We do note, however, that numerical relativists have very recently begun to apply the conformally covariant split for certain studies [35].

Is it true that, for every choice of the free data $(\lambda_{ij}, \sigma_{ij}, \tau)$, one can always solve equations (47) and (48) for the determined data (ϕ, W) and thereby obtain a solution of the constraint equations? It is easy to see that this is not the case: Let us choose the manifold M^3 to be the three sphere, and on M^3 we choose a metric λ with non-negative scalar curvature, we choose σ to be zero everywhere, and we choose τ to be unity everywhere. One readily verifies that every solution to the equation $\nabla_i (LW)^i_j = 0$ has $LW_{ij} = 0$. The Lichnerowicz equation then becomes $\Delta \phi = R\phi + \phi^5$. Since the right-hand side of this equation is positive definite (recall the requirement that $\phi > 0$), it follows from the maximum principle on closed (compact without boundary) manifolds that there is no solution.

In light of this example, the main question is really the following: For which choices of the manifold M^3 and the free data can one solve (47) and (48)? It turns out that we know quite a bit about the answer to this question, yet still have quite a bit to learn as well. To describe what we know and do not know, it is useful to categorize the question using the following criteria:

Manifold and Asymptotic Conditions:
- Closed
- Asymptotically Euclidean
- Asymptotically hyperbolic
- Compact with boundary conditions
- Asymptotically Euclidean with interior boundary conditions

Regularity:
- Analytic
- Smooth
- Sobolev and Hölder classes
- Weak
- Asymptotic fall off conditions at infinity

Metric Conformal Classes:
- Yamabe positive
- Yamabe zero
- Yamabe negative

Mean Curvature:
- Constant ("CMC")
- Near Constant (small $\frac{\max |\nabla \tau|}{\min |\tau|}$)
- Non Constant

σ **Classes:**

- σ is zero everywhere on M^3
- σ not zero everywhere on M^3

The mean curvature of the data turns out to be the most important factor in separating those sets of free data for which we know whether or not a solution exists from those sets for which we do not. In fact, if the mean curvature is either constant or near constant, we can almost completely determine whether or not a solution exists (at least in those cases with no boundaries present). On the other hand, if the mean curvature is neither constant nor near constant, we know very little. For sufficiently smooth free data, we may summarize the known results as follows:

Constant Mean Curvature:

- *Closed*: Completely determined. For all but certain special cases (see example discussed above), solutions exist. See [31] and [50].
- *Asymptotically Euclidean*: Completely determined. For all but certain special cases (depending entirely on the conformal class), solutions exist. See [26] and [25].
- *Asymptotically Hyperbolic*: Completely determined. Solutions always exist. See [5] and [4]
- *Compact with Boundary*: Some cases determined, most unresolved.
- *Asymptotically Euclidean with Interior Boundary*: Some cases determined, some unresolved. See [61] and [38]

Near Constant Mean Curvature:

- *Closed*: Mostly determined. For all but certain cases, solutions exist. One special case known of nonexistence. A small number of special cases unresolved. See [54], [51] and [55]
- *Asymptotically Euclidean*: Mostly determined. No cases known of nonexistence. Some cases unresolved. See [30]
- *Asymptotically Hyperbolic*: Mostly determined. No cases known of nonexistence. A small number of cases unresolved. See [56]
- *Compact with Boundary*: Small number of cases determined, most unresolved.
- *Asymptotically Euclidean with Interior Boundary*: Nothing determined.

Non Constant Mean Curvature:

Very little known.

This summary is the compilation of results from a large number of works (note references listed above), including some not yet written up. The summary is fairly sketchy, since there is inadequate space here to include many of the details. We do, however, wish to give a flavor of what these results say more precisely, and how they are proven. To do this we shall discuss a few representative sub-cases.

We first consider the case in which the manifold is presumed closed, and in which we choose free data with constant τ. As noted above, as a consequence of the

CMC condition, the equation (47) for W is readily solved, and we have $LW = 0$. What remains is the Lichnerowicz equation, which takes the form

$$\Delta\phi = \tfrac{1}{8}R\phi - \tfrac{1}{8}\sigma^{ij}\sigma_{ij}\phi^{-7} + \tfrac{1}{12}\tau^2\phi^5. \qquad (56)$$

There are three key PDE analysis theorems which allow us to determine, for any set of free data consistent with these conditions, whether or not a solution to the Lichnerowicz equation exists. We state each of these theorems in a form most suited for this purpose (see [50] for much more detail):

1. *The Maximum Principle:* If we choose free data such that for any $\phi > 0$ the right-hand side of the Lichnerowicz equation (56) is either positive definite or negative definite, then (with M^3 presumed closed) for that free data there can be no solution.

2. *The Yamabe Theorem:* For any given Riemannian metric λ on a closed three manifold, there is a conformally related metric $\theta^4\lambda$ which has constant scalar curvature. For fixed λ, all such conformally-related constant scalar curvature metrics have the same sign for the scalar curvature. This allows one to partition the set of all Riemannian metrics on M^3 into three classes, depending on this sign. We label these *Yamabe classes* as follows: $\mathcal{Y}^+(M^3)$, $\mathcal{Y}^0(M^3)$, and $\mathcal{Y}^-(M^3)$.

3. *The Sub and Super Solution Theorem:* If there exist a pair of positive functions $\phi_+ > \phi_-$ such that

$$\Delta\phi_+ \le \tfrac{1}{8}R\phi_+ - \tfrac{1}{8}\sigma^{ij}\sigma_{ij}\phi_+^{-7} + \tfrac{1}{12}\tau^2\phi_+^5 \qquad (57)$$

and

$$\Delta\phi_- \ge \tfrac{1}{8}R\phi_- - \tfrac{1}{8}\sigma^{ij}\sigma_{ij}\phi_-^{-7} + \tfrac{1}{12}\tau^2\phi_-^5, \qquad (58)$$

then there exists a solution ϕ of the Lichnerowicz equation (56), with $\phi_+ \ge \phi \ge \phi_-$.

The Yamabe Theorem is very useful because the Lichnerowicz equation in the form (56) (with $LW = 0$) is conformally covariant in the sense that it has a solution for a set of free data (λ, σ, τ) if and only if some function $\psi > 0$ it has a solution for the free data $(\psi^4\lambda, \psi^{-2}\sigma, \tau)$. Hence, to check solubility for a given set of free data, we can always first perform a preliminary conformal transformation on the data and therefore work with data for which the scalar curvature is constant.

It is now straightforward to use the Maximum Principle (together with the Yamabe Theorem) to show that there is a collection of classes of CMC free data for which no solution exists. This holds for all of the following classes of free data:

$$(\lambda \in \mathcal{Y}^+(M^3), \sigma \equiv 0, \tau = 0), \qquad (\lambda \in \mathcal{Y}^+(M^3), \sigma \equiv 0, \tau \neq 0),$$
$$(\lambda \in \mathcal{Y}^0(M^3), \sigma \equiv 0, \tau \neq 0), \qquad (\lambda \in \mathcal{Y}^0(M^3), \sigma \neq 0, \tau = 0),$$
$$(\lambda \in \mathcal{Y}^-(M^3), \sigma \equiv 0, \tau = 0), \qquad (\lambda \in \mathcal{Y}^-(M^3), \sigma \neq 0, \tau \neq 0).$$

Note that here, "$\sigma \neq 0$" means that the tensor σ is not identically zero on M^3.

It is true, though not so simple to show, that for all other classes of CMC free data on a closed manifold, a solution does exist. We show this using the sub and super solution theorem. For free data of the type $(\lambda \in \mathcal{Y}^-(M^3), \sigma \equiv 0, \tau \neq 0)$, it is relatively easy to show that there are constant sub and super solutions. We merely need to find a constant ϕ_+ sufficiently large so that $-\phi_+ + \frac{1}{12}\tau^2\phi_+^5$ is everywhere positive, and a constant ϕ_- sufficiently small (yet positive) so that $-\phi_- + \frac{1}{12}\tau^2\phi_-^5$ is everywhere negative. Constant sub and super solutions are also readily found for free data of the type $(\lambda \in \mathcal{Y}^-(M^3), \sigma \neq 0, \tau \neq 0)$, or $(\lambda \in \mathcal{Y}^0(M^3), \sigma \equiv 0, \tau = 0)$. For the remaining types of free data, we need to find either a nonconstant sub solution or a nonconstant super solution. This takes a bit of work; we show how to do it in [50].

It is worth noting that the proof of the sub and super solution theorem (see [50]) is a constructive one: Starting with the super solution ϕ_+, one solves a sequence of linear equations, and one then shows that the monotonic sequence of solutions $\phi_+ \geq \phi_1 \geq \phi_2 \geq \cdots \geq \phi_-$ converges to a solution of the Lichnerowicz equation. This constructibility can be useful for numerical relativity, as has been shown in [45].

We next consider the case in which the free data is chosen to be asymptotically Euclidean (see, for example, [30] for a definition of this property), with $\tau \equiv 0$. For asymptotically Euclidean metrics, a Yamabe type theorem again plays a key role. Proven by Brill and Cantor [25], with a correction by Maxwell [61], this result says that if λ is asymptotically Euclidean, and if for every nonvanishing, compactly supported, smooth function f we have

$$\inf_{\{f \neq 0\}} \frac{\int_M (|\nabla f|^2 + Rf^2)\sqrt{\det \lambda}}{||f||_{L^{2*}}^2} > 0, \tag{59}$$

then for some conformal factor θ, the scalar curvature of $\theta^4 \lambda$ is zero. Moreover, if an asymptotically Euclidean metric fails to satisfy this condition, then there is no such transformation. Metrics which do satisfy this condition have been labeled (somewhat misleadingly) as "positive Yamabe metrics".

The positive Yamabe property just defined is exactly the condition which determines if, for a set of asymptotically Euclidean free data, the Lichnerowicz equation admits a solution. That is, as proven in [26], (56) admits a solution ϕ with suitable asymptotic properties if and only if the (asymptotically Euclidean) free data $(\lambda, \sigma, \tau = 0)$ has positive Yamabe metric λ. This is true regardless of whether σ vanishes or not. This result is not proven directly using a sub and super solution theorem, but the proof does involve a converging sequence of solutions of linear equations, and is therefore again constructive.

The last case we consider here is that of near constant mean curvature free data on closed manifolds. This case is more complicated than the CMC case, since now we must work with the coupled system (47)–(48). However, with sufficient control over the gradient of τ, we can handle the coupled system, and determine whether or not solutions exist for almost all sets of near CMC free data.

The key extra tool we use to carry out this analysis is the elliptic estimate for the "vector Laplacian" operator $\nabla_i L(\ )^i_j$ appearing on the left side of (47). Such

a result holds for any linear, elliptic, invertible operator [22] such as this one.[4]
One has the *Elliptic Estimate:* $||W^j||_{H_{k+2}} \leq C||\nabla_i(LW)^i_j||_{H_k}$, where H_k is the
Sobolev space of square integrable vector fields for which the first k derivatives
are square integrable as well, where $|| \ ||_{H_k}$ is the corresponding norm, and where
C is a constant depending on the chosen geometry (M^3, λ_{ij}). Based on this estimate, together with the Sobolev embedding theorem [22] and standard integration
inequalities, we obtain a *pointwise* estimate of the form

$$|LW| \leq \tilde{C} \max_{M^3} \phi^6 \max_{M^3} |\nabla \tau|, \tag{60}$$

where $\tilde{C}$ is also a constant depending only on the geometry (M^3, λ_{ij}). This pointwise inequality (60) is the crucial estimate which allows us to handle the coupled
system, for sufficiently small $|\nabla \tau|$.

To see how to prove that a solution exists in one of these near CMC cases,
let us consider free data with $\lambda \in \mathcal{Y}^-(M^3), \tau > 0$, and with $\frac{\max_{M^3}|\nabla\tau|}{\min_{M^3}|\tau|}$ sufficiently
small.[5] We claim that, for such free data, a solution exists. To show this, we
consider the sequence of semi-decoupled PDE systems

$$\nabla_i(LW_{(n)})^i_j = \tfrac{2}{3}\phi^6_{(n-1)}\nabla_j\tau \tag{61}$$

$$\Delta\phi_{(n)} = \tfrac{1}{8}R\phi_{(n)} - \tfrac{1}{8}(\sigma^{ij} + LW^{ij}_{(n)})(\sigma_{ij} + LW_{(n)ij})\phi^{-7}_{(n)} + \tfrac{1}{12}\tau^2\phi^5_{(n)}. \tag{62}$$

Choosing a suitable initializing value for ϕ_0, we first show that a sequence
$(\phi_{(n)}, W_{(n)})$ of solutions to (61–62) exists. For the decoupled equation (61), the
existence of a solution $W_{(n)}$ follows from the invertibility of the vector Laplacian.
For (62), existence follows from the sub and super solution theorem, since we readily find a sequence of constant sub and super solutions for this choice of free data.
We next show that there are (positive) uniform upper and lower bounds for the
sequence $(\phi_{(n)})$. This is where we first need to use the estimate (60) for $|LW_{(n)}|$:
Specifically, after arguing that there is a uniform (constant) upper bound for the
$\phi_{(n)}$'s if we can find a positive constant ζ such that

$$\zeta^3 \geq \frac{3}{2\min_{M^3}\tau^2}\zeta^2 + \frac{3}{\min_{M^3}\tau^2}\sigma^{ij}\sigma_{ij} + \frac{3}{\min_{M^3}\tau^2}LW^{ij}_{(n)}LW_{(n)ij}, \tag{63}$$

we use the $|LW_{(n)}|$ estimates to show that it is sufficient to find a positive constant
ζ satisfying

$$\zeta^3 \geq \frac{3}{2\min_{M^3}\tau^2}\zeta^2 + \frac{3}{\min_{M^3}\tau^2}\sigma^{ij}\sigma_{ij} + C\frac{\max_{M^3}|\nabla\tau|^2}{2\min_{M^3}\tau^2}\zeta^3 \tag{64}$$

---

[4]The operator $\nabla_i L(\ )^i_j$ is invertible only if λ has no conformal Killing field. However, the pointwise
estimate (60) we obtain for $|LW|$ can be derived even if λ has a conformal Killing field, using a
slightly more complicated elliptic estimate than the one presented here.

[5]The statement that "$\frac{\max_{M^3}|\nabla\tau|}{\min_{M^3}|\tau|}$ (is) sufficiently small" may seem nonsensical, since this quan-
tity is dimensional. However, as seen in the proof [54], the more precise statement of this condition
involves the constant C, which has dimensions as well.

We immediately see that if, for fixed λ and σ, we choose $C \frac{\max_{M^3} |\nabla\tau|^2}{2 \min_{M^3} \tau^2}$ sufficiently small, then such a ζ exists. The existence of the uniform upper bound for $\phi_{(n)}$ follows.

Establishing these uniform bounds is the crucial step in our proof that solutions exist for free data of the type we are discussing here, as well as for other classes of near CMC free data. Once we have these bounds, we can carry through a contraction mapping argument to show that the sequence $(\phi_{(n)}, W_{(n)})$ converges (a bit more squeezing of $|\nabla\tau|$ is needed to carry this out). We can then go on to use continuity arguments to show that the limit is a weak solution, and finally use bootstrap arguments to show that in fact the weak solution is a strong solution, of the desired smoothness. The details of this proof are presented in [54].

A very similar argument can be used to prove that equations (47)–(48) admit solutions for a number of classes of near CMC free data, including the following:

1. $\lambda \in \mathcal{Y}^-(M^3), \tau < 0$, and $\frac{\max_{M^3} |\nabla\tau|}{\min_{M^3} |\tau|}$ sufficiently small;

2. $\lambda \in \mathcal{Y}^+(M^3)$, σ not identically zero, and $\frac{\max_{M^3} |\nabla\tau|}{\max_{M^3} |\sigma|}$ sufficiently small;

3. $\lambda \in \mathcal{Y}^0(M^3)$, τ nowhere zero, σ not identically zero, and $\frac{\max_{M^3} |\nabla\tau|}{\max_{M^3} |\sigma| + \max_{M^3} |\nabla\tau|}$ sufficiently small.

What about the other near CMC cases on closed M^3? Just recently, with Niall O'Murchadha, we have found for the first time a class of near CMC free data for which a solution *does not* exist. The data is of the following type:

$$\lambda \in \mathcal{Y}^+(M^3) \cup \mathcal{Y}^0(M^3), \sigma \equiv 0, \tau \text{ nowhere } 0, \text{ and } \frac{\max_{M^3} |\nabla\tau|}{\min_{M^3} |\tau|} \text{ sufficiently small.}$$

The proof that no solution exists for free data of this type is a relatively simple application of the pointwise estimate (60) for $|LW|$ discussed above, and the maximum principle. The details are presented in [55]. As for near CMC data on a closed manifold which is not one of the types discussed here, nothing is yet known. Note that, in a rough sense, these remaining unresolved cases are not generic.

As noted in the summary above, there are a number of other classes of free data for which we know whether or not solutions to (47)–(48) exist. Methods similar to those discussed here resolve the existence question for most CMC or near CMC free data which are asymptotically Euclidean or asymptotically hyperbolic, as well for such data on closed manifolds. The situation for free data specified on compact manifolds with boundary is much less understood; however this is most likely a result of neglect rather than difficulty. Since numerical relativity has motivated a recent interest in this question, and since the methods discussed here are believed to work for data on manifolds with boundary, it is likely that in the next few years, the solvability of (47)–(48) for such data is likely to be relatively well understood. Indeed, this motivation has led to the very recent results of Maxwell [61] and Dain [38], who give sufficient conditions on asymptotically Euclidean free data with interior "apparent horizon" boundary conditions for solutions to exist.

The situation is very different for non-constant mean curvature data with no controls on $|\nabla\tau|$. Almost nothing is known, and there are no promising techniques known at this point. New ideas are needed.

It is worth noting that the underlying approach of specifying the mean extrinsic curvature τ is geometrically natural, since the converse problem, of finding a hypersurface of prescribed mean curvature in a given spacetime, leads to a quasi-linear elliptic equation bearing some similarities with the Euclidean minimal surface equation. It is known that solutions of the Lorentz mean curvature equation are strictly spacelike and smooth, at least to the extent permitted by the regularity of the ambient spacetime and boundary conditions, so such prescribed mean curvature hypersurfaces provide natural spacelike slices of the spacetime [27, 19, 46, 10, 11]. Although there are examples of spacetimes not admitting maximal ($\tau = 0$, [24]) or CMC ($\tau = const.$, [12], [34]) hypersurfaces, such non-existence behavior is driven by global causal topology, which allows area-maximizing sequences of hypersurfaces to become unbounded to the past or future; see [11, 12]. It is therefore not surprising that *a priori* control of the interior causal geometry is an essential ingredient in the proof of existence of entire maximal hypersurfaces in asymptotically flat spacetimes [10].

To wind up this discussion of the implementation of the conformal method to find solutions of the constraint equations, it is important to note two facts: 1) The map from sets of free data to solutions of the constraints is surjective. This follows from the observation that if (γ, K) is a solution of the constraints, then $\phi = 1, W = 0$ is clearly a solution of (47–48) for the free data $\lambda = \gamma, \sigma = K - \frac{1}{3}\lambda \mathrm{tr}_\gamma K, \tau = \mathrm{tr}_\gamma K$. 2) For all those sets of free data for which solutions of (47-48) are known to exist (excepting for the very special case of λ flat, $\sigma = 0$, and $\tau =$ on $M^3 = T^3$) the solution is unique.

As a consequence of these facts, those sets of free data for which solutions exist together parametrize the space of constraint-satisfying initial data for Einstein's equations. It follows that, once we determine exactly which free data sets map to solutions, we will have made significant progress toward understanding the "degrees of freedom" of Einstein's theory.

Also as a consequence of these facts, if we seek initial data invariant under an isometry group, it is sufficient to choose free data with this invariance.[6]

In view of the dominant role the conformal method has assumed as a tool for mathematical as well as numerical analysis of the constraints, it is important to point out its limitations. As noted above, the conformal method replaces the original under-determined PDE system of the constraint equations (four equations to be solved for twelve unknowns), by a determined system of elliptic character. This is very useful for a wide variety of studies (such as the parametrization question). However for a number of other problems, it is a bad idea. Consider for example a

---

[6]Note that the presence of symmetries generally introduces integrability conditions which must be satisfied by the initial data, and consequently by the free data. These conditions are relatively easy to handle.

given solution of the constraint equations on a finite radius ball. Can one smoothly extend the solution onto a neighborhood properly containing the ball? Can it be smoothly extended to a complete, asymptotically Euclidean solution? Might there be a smooth extension which is identically Schwarzschild or Kerr outside some (larger) ball? To study questions like these, casting the constraint equations into the form of a determined elliptic system is not at all useful. It is better to bypass the conformal method and work with the original under-determined system.

4.2. The thin sandwich construction

Some time ago, Wheeler [62, 8] discussed a possible alternative to the initial value formulation for determining a spacetime solution of Einstein's equations. Rather than specifying initial data in the form of a Riemannian metric γ and a symmetric tensor K which satisfy the Einstein constraint equations, he asked whether one might specify a pair of Riemannian metrics h_1 and h_2 and seek to find a (unique?) spacetime (V^4, g) in which h_1, h_2 arise as the induced metrics on disjoint Cauchy surfaces in (V^4, g), and which satisfies the Einstein equations, i.e., (V^4, g) is the spacetime solution "connecting" h_1 and h_2. This is the *sandwich conjecture*.

Considering the analogous question in electrodynamics, however, we find that the uniqueness part of the Maxwell version of the sandwich conjecture is false. Hence there is reason to believe that the uniqueness assertion in the sandwich conjecture for the Einstein equations may also be false. This led Wheeler to instead propose the *thin sandwich conjecture*, which postulates that given a freely chosen Riemannian metric γ and symmetric tensor J representing the time derivative of the metric evolution in a spacetime, then we can use the constraints to determine a lapse function and shift vector field so that indeed, with respect to a choice of foliation and coordinates compatible with that lapse and shift, J *is* the time derivative of the metric. More explicitly, having chosen γ and J, we seek a lapse N and shift X so that if we set

$$K_{ij} = -N^{-1}(\tfrac{1}{2}J_{ij} - X_{(i;j)}), \tag{65}$$

then (γ, K) satisfy the constraint equations (19,20) with source terms $\rho = 16\pi T_{nn}$, $S_i = 8\pi T_{ni}$. Regarding (γ, J, ρ, S) as *given* fields, the constraints give four equations for (N, X), the so-called *thin sandwich* equations.

It is clear that if (γ, K) satisfy the constraint equations, then J is the time derivative $\partial_t \gamma$ of the spatial metric in the corresponding spacetime evolution of the data (γ, K) with respect to coordinates with lapse-shift (N, X).

The Hamiltonian constraint is readily solved for the lapse, giving

$$N = \sqrt{\frac{(\Gamma^i_i)^2 - \Gamma^{ij}\Gamma_{ij}}{\rho - \mathrm{R}^M}}, \tag{66}$$

where we have set $\Gamma_{ij} = \tfrac{1}{2}J_{ij} - X_{(i;j)}$ for brevity. Substituting this expression for N into the momentum constraint (20) we obtain the *reduced thin sandwich equations* ("RTSE")

$$\nabla_j \left[\sqrt{\frac{\rho - \mathrm{R}^M}{(\Gamma_k^k)^2 - \Gamma^{kl}\Gamma_{kl}}} (\Gamma^{ij} - \Gamma_k^k \gamma^{ij}) \right] = S^i, \tag{67}$$

which we view as a system of partial differential equations for X^i, the components of the shift.

The linearization of (67) is shown to be elliptic in [17], provided that the conjugate momentum π^{ij} (recall the formula (37) for π) is either positive or negative definite. This rather surprising condition is guaranteed, so long as the condition

$$\rho - \mathrm{R}^M > 0 \tag{68}$$

is satisfied everywhere in M and provided that J is chosen in such a way that the formula (66) for the lapse function satisfies

$$N > 0. \tag{69}$$

An additional necessary condition for solvability of the RTSE is that the equation

$$\nabla_{(j} V_{i)} = \mu K_{ij} \tag{70}$$

only has solutions (V_i, μ) with $V_i = 0$.

An implicit function theorem argument [17] shows then that the RTSE is solvable for all data (γ, J, ρ, S) in an *open* neighborhood of a reference configuration $(\mathring{\gamma}, \mathring{J}, \mathring{\rho}, \mathring{S})$ satisfying the conditions (68, 69, 70). Here "open" is with respect to a Sobolev space in which the RTSE are posed.

The fact that the thin sandwich equations can be solved for an open class of data (γ, J, ρ, S) can be interpreted as showing that this reformulation of the constraints leads to "generic" solutions. However, (68) is rather restrictive – it excludes asymptotically flat data for example – so these conditions, taken together, limit the prospects of constructing most interesting spacetime initial data via the thin sandwich approach.

4.3. Conformal thin sandwich

An interesting hybrid of the conformal method and the thin sandwich approach has been suggested by York [76]. Before specifying how this *conformal thin sandwich* approach works, let us briefly compare the two approaches discussed so far:

1. Conformal Method: The free data consists of a conformal metric λ_{ij}, a divergence-free trace-free tensor field σ_{ij}, and a scalar field τ (8 free functions). Solving the constraints produces a vector field W and a scalar field ϕ (4 functions). We recompose to get a metric γ_{ij} and a symmetric tensor K_{ij} satisfying the constraints (12 functions). The lapse and shift are ignored.

2. Thin Sandwich: The free data consists of a metric γ_{ij} and a symmetric tensor J_{ij} (12 free functions). Solving the constraints produces a vector field X^i and a scalar field N (4 functions). We recompose to get a metric γ_{ij} and a symmetric tensor K_{ij} satisfying the constraints, plus the lapse N and the shift X^i (16 functions).

The idea of the conformal thin sandwich approach is to specify as free data a conformal metric λ_{ij}, a trace-free symmetric tensor U_{ij}, a scalar field τ, and another scalar field η. In terms of the loose function counting system used above, this amounts to 12 free functions. In the spirit of the thin sandwich, the tensor U_{ij} represents the time derivative of the conformal metric, while τ is the mean curvature and η represents the lapse function, up to a conformal factor. It follows from this interpretation, together with a choice of conformal scaling, that the second fundamental form K_{ij} is expressed as

$$K_{ij} = \psi^{-2}\tfrac{1}{2\eta}((LX)_{ij} - U_{ij}) + \tfrac{1}{3}\psi^4 \tau \lambda_{ij} \tag{71}$$

where X^i is the shift vector and ψ is the conformal factor, neither of which is known at this stage. To determine X^i and ψ, we use the constraint equations. These take the form

$$\nabla^j((2\eta)^{-1}(LX)_{ij}) = \nabla^j((2\eta)^{-1}U_{ij}) + \tfrac{2}{3}\psi^6 \nabla_i \tau, \tag{72}$$

and

$$8\Delta\psi - R(\gamma)\psi = -A_{ij}A^{ij}\psi^{-7} + \tfrac{2}{3}\tau^2\psi^5. \tag{73}$$

where we use the convenient short hand $A_{ij} = \tfrac{1}{2\eta}((LX)_{ij} - U_{ij})$. Thus, if we can solve (72-73) for X and ψ (4 functions), we have K (obtained from (71)) and

$$\gamma_{ij} = \psi^4 \lambda_{ij} \tag{74}$$

satisfying the constraints, plus a specification of the lapse $N = \psi^6\eta$ and the shift X^i.

Compared to the conformal method ("CM"), the conformal thin sandwich approach ("CTSA") has a number of virtues:

1. Unlike CM, specifying a set of CTSA free data does not require a projection to the divergence-free part of a symmetric tracefree 2-tensor.

2. The map from CTSA free data to a solution of the constraint equations is conformally covariant in the sense that if (γ_{ij}, K_{ij}) is a solution corresponding to a particular choice of free data $(\lambda_{ij}, U_{ij}, \tau, \eta)$, then it is also a solution for the CTSA data $(\theta^4\lambda_{ij}, \theta^{-2}U_{ij}, \tau, \theta^6\eta)$, for any positive function θ. This is true whether or not τ is constant, whereas CM data has this property only if τ is constant.

3. The mathematical form and hence the mathematical analysis of the CTSA equations (72)–(73) is very similar to that of the CM equations (47)–(48). Hence we have essentially the same existence and uniqueness results for the two sets of equations.

4. CTSA free data is arguably closer to the physics we wish to model, since it includes the time derivative of the conformal metric, while the CM data only has the divergence-free trace-free (transverse traceless) part of the second fundamental form.

Along with these virtues, CTSA has one troubling feature: Say we want to find a set of CMC initial data (γ_{ij}, K_{ij}) with the lapse function chosen so that the evolving data continues to have constant mean curvature. In the case of the conformal method, after solving (47) and (48) to obtain a solution of the constraints, one achieves this by solving a linear elliptic equation for the lapse function. This equation is not coupled to the constraint equations, and solutions are readily verified to exist. By contrast, in the CTSA the extra equation takes the form

$$\Delta(\psi^7\eta) = \tfrac{1}{8}\psi^7\eta R + \tfrac{5}{12}(\psi\eta)^{-1}(LX - U)^2 + \psi^5 X^i\nabla_i\tau - \psi^5, \tag{75}$$

which is coupled to the CTSA equations (72)–(73). This coupling cannot be removed by choosing τ constant. Indeed, whether the data is CMC or not, the mathematical analysis of the full system (72–75) is not very tractable. We emphasize, however, that if we do not require that the lapse be chosen to preserve the mean curvature, then the analysis of the CTSA equations is no more difficult than that of the conformal method equations. For further discussion of the conformal thin sandwich approach, both in theory and in practice, see [76] and [35].

4.4. Parabolic methods and the quasi-spherical ansatz

It was first shown in [14] that 3-metrics[7] of prescribed scalar curvature can be constructed by solving a certain *parabolic* equation on S^2. This leads to a method for constructing solutions of the Hamiltonian constraint which has found some interesting applications [14, 71, 69].

To describe the general construction [71], consider the second variation identity for a foliation $r \to \Sigma_r$ in a Riemannian 3-manifold with metric g,

$$R(g) = 2D_n H + 2K_G - H^2 - \|II\|^2 - 2u^{-1}\Delta u, \tag{76}$$

where n is the outward unit normal vector to the level sets Σ_r, II is the second fundamental form, $H = \mathrm{tr}_{\Sigma_r} II$ is the mean curvature, and K_G, $\Delta = r^{-2}\Delta_o$ are respectively the Gauss curvature and Laplacian of the metric $r^2\mathring{g}$ on the foliation 2-surfaces (which are usually assumed to be topological 2-spheres). This is the Riemannian version of the Lorentzian formula (35). The name "second variation" arises from the term $D_n H$, since the mean curvature measures the first variation of the area of Σ_r. This interpretation is not important for the present application, although it does reflect some deep relationship between mean curvature and minimal surfaces on the one hand, and positivity properties of mass on the other [67, 69].

Consider (76) when the surface variation $r \to \Sigma_r$ defines a foliation, so the metric takes the general form

$$g = u^2\,dr^2 + \mathring{g}_{AB}(rd\theta^A + \beta^A dr)(rd\theta^B + \beta^B dr), \tag{77}$$

where $u = u(r, \theta^A) > 0$, $\beta = (\beta^A(r, \theta^B))$ and the rescaled angular metric $\mathring{g}_{AB}(r, \theta^C)$ are arbitrary fields. The *quasi-spherical* (QS) case considered in [14] arises as the

---

[7]Although we restrict discussion here to 3-manifolds, the parabolic method generalizes to all dimensions.

special case where $\mathring{g} = d\vartheta^2 + \sin^2 \vartheta \, d\varphi^2$ is the standard 2-sphere metric. Generally, the mean curvature of the leaves Σ_r is

$$H = -\frac{1}{ru} \left(2 - \text{div}_o \beta + r\partial_r \log \sqrt{\det \mathring{g}} \right), \qquad (78)$$

where $\text{div}_o \beta$ is the divergence of the tangential vector field $\beta = \beta^A \partial_A$ in the metric $\mathring{g}$, and $n = u^{-1}(\partial_r - r^{-1}\beta^A \partial_A)$ is the exterior-directed unit normal vector. Assume throughout the *quasiconvexity condition*

$$h := 2 - \text{div}_o \beta + r\partial_r \log \sqrt{\det \mathring{g}} \; > \; 0, \qquad (79)$$

and let $n_o = un$, $H_o = uH = -h/r$, $II_o = u \, II$ be the corresponding quantities determined by the metric (77) with $u = 1$. The key observation is that substituting H from (78) into $D_n H$ gives a term $hu^{-3}r^{-1}\partial_r u$ which combines with the term $u^{-1}\Delta u$ to show that if the scalar curvature is specified then (76) may be read as a parabolic equation for the lapse u. Here β^A, $\mathring{g}_{AB}$ can be freely chosen, subject only to the quasiconvexity condition (79), or equivalently, $H < 0$. Explicitly we have

$$h(r\partial_r - \beta^A \partial_A)u \; - \; u^2 \Delta_o u \qquad (80)$$
$$= \; ru(r\partial_r - \beta^A \partial_A)H_o + \tfrac{1}{2}r^2 u(H_o^2 + |II_o|^2) - u^3(K(\mathring{g}) - \tfrac{1}{2}R(g)r^2),$$

which clearly shows the parabolic structure of the equation satisfied by u.

Thus if $R(g)$ is determined by the Hamiltonian constraint equation (19) and β, $\mathring{g}_{AB}$ are regarded as prescribed fields, then solving (80) leads to a 3-metric which satisfies the Hamiltonian constraint.

Global $(r \to \infty)$ existence theorems were established for (80) in the quasi-spherical case $\mathring{g} = d\vartheta^2 + \sin^2 \vartheta \, d\varphi^2$ [14], assuming β satisfies the quasiconvexity condition (79), which becomes $\text{div}_o \beta < 2$, and assuming the prescribed scalar curvature is not too positive ($r^2 R \leq 2$ is sufficient but not necessary). With suitable decay conditions on the prescribable fields β^A (for example, if β^A is compactly supported), these results give asymptotically flat solutions of the Hamiltonian constraint, with either black hole $H = 0$ or regular centre $r = 0$ inner boundary conditions. Setting $R(g) = 0$ gives a large class of solutions of the vacuum Hamiltonian constraint (19) in the *time-symmetric* case $K_{ij} = 0$, with two degrees of freedom corresponding to the choice of β^A. Analogous results have been established for general $\mathring{g}_{AB}$ [71, 72, 69], which demonstrate that this provides a flexible technique for extending solutions to be asymptotically flat, or asymptotically hyperboloidal if $R(g) = -6$.

The parabolic method enables us to construct metrics of prescribed scalar curvature which have properties not achievable by conformal methods. For example, any bounded domain (Ω, g) with smooth mean-convex ($H < 0$) boundary $\partial\Omega \simeq S^2$ and non-negative scalar curvature can be extended smoothly to an asymptotically flat manifold, also with $R \geq 0$; we simply extend the Gaussian normal foliation in a neighborhood of $\partial\Omega$ smoothly to a metric which is flat outside a neighborhood of Ω, with a mean-convex exterior foliation approaching the standard spherical foliation of $\mathbb{R}^3$. Solving the parabolic lapse equation for $R \geq 0$ with initial data $u = 1$ on a Gaussian level set $\Sigma_{-\epsilon} \subset \Omega$ produces the required 3-metric.

One consequence of this argument, first observed in [14], is the existence of $R = 0$ metrics on $\mathbb{R}^3$ which are asymptotically flat and non-flat, but which have a flat interior region.

The parabolic method also gives solutions satisfying "geometric boundary conditions" [16] $(r^2 \mathring{g}, H)$, which specify the boundary metric $r^2 \mathring{g}_{AB}$ and mean curvature H. This is clear from (78) since H is determined by specifying the well-posed initial condition $u(r_0)$ for the parabolic lapse equation. Note that this is not possible with conformal methods, since specifying both the boundary metric and mean curvature leads to simultaneous Dirichlet and Neumann boundary conditions on the conformal factor. These boundary conditions are ill posed for the elliptic equation imposed on the conformal factor, which will not then admit any solution in general. The problem of constructing 3-metrics with prescribed boundary metric and mean curvature arises naturally from the geometric definition of quasi-local mass [13] and provided the original motivation for the quasi-spherical construction.

Examples of 3-metrics satisfying the Geroch (inverse mean curvature flow) foliation condition [47, 48] can be constructed by choosing β^A, $\mathring{g}_{AB}$ satisfying $\mathrm{div}_o \beta = r \partial_r (\log \det \mathring{g})$ and then solving the parabolic lapse equation with suitably prescribed scalar curvature. This provides a large class of metrics for which the Geroch identity directly verifies the Penrose conjecture. Note that this does not help to solve the much harder problem considered in [48], of finding a Geroch foliation in a *given* metric.

Another advantage of the parabolic lapse method is that it is generally easier and cheaper numerically to solve a 2+1 parabolic equation than the 3D elliptic equations arising in the conformal methods, particularly when the solution is required near spatial infinity. Finally, in the original quasi-spherical gauge [14], the freely-specifiable fields $\beta^A, A = 1, 2$ give a rather explicit parameterization of the "true degrees of freedom" allowed by the Hamiltonian constraint, since fixing β removes all diffeomorphism freedom in the quasi-spherical gauge (at the linearized level at least).

Dual to the problem of *constructing* metrics having quasi-spherical form, is the harder problem of *finding* a quasi-spherical foliation in a given metric. Some simple observations suggest that QS foliations exist for general perturbations of a QS metric, but for interesting technical reasons a complete proof is not yet available.

Write the general QS 3-metric as $g = u^2 dr^2 + \Sigma_A (\beta_A dr + r \sigma_A)^2$ where σ_A, $A = 1, 2$ is an orthonormal coframe for the unit sphere S^2, and consider the effect of an infinitesimal diffeomorphism generated by $X = \zeta^A \tau_A + z(r \partial_r - \beta^A \tau_A)$, where τ_A is the dual frame to σ_A. The QS condition will be preserved under the metric variation δg_{ij} and infinitesimal diffeomorphism X exactly when $(\mathcal{L}_X g + \delta g)_{AB} = 0$, which gives the equation

$$\zeta_{(A|B)} + (\sigma_{AB} - \beta_{(A|B)})z + \tfrac{1}{2}\delta g_{AB} = 0. \tag{81}$$

Taking trace and trace-free projections and simplifying gives

$$\mathcal{D}_\beta \zeta \quad := \quad \zeta_{(A|B)} - \tfrac{1}{2}\mathrm{div}_o\zeta\sigma_{AB} + \mathcal{B}_{AB}\mathrm{div}_o\zeta$$

$$= \quad -\tfrac{1}{2}(\delta g_{AB} - \tfrac{1}{2}\delta g_C^C\sigma_{AB} + \mathcal{B}_{AB}\delta g_C^C), \tag{82}$$

where

$$\mathcal{B} := \frac{\beta_{(A|B)} - \tfrac{1}{2}\mathrm{div}_o\beta\sigma_{AB}}{2 - \mathrm{div}_o\beta} \quad \text{and} \quad z = -\frac{\mathrm{div}_o\zeta + \tfrac{1}{2}\delta g_A^A}{2 - \mathrm{div}_o\beta}.$$

The operator $\mathcal{D}_\beta\zeta$ in (82) is elliptic exactly when $|\mathcal{B}|^2 < 1$, and we can show that its adjoint $\mathcal{D}_\beta^\dagger y_{AB} = -y_{AB}{}^{|B} + (y_{BC}\mathcal{B}^{BC})_{|A}$ has trivial kernel when $|\mathcal{B}|^2 < 1/3$ (pointwise). This latter condition also guarantees that $\mathcal{D}_\beta$ is surjective, whereupon (82) can be solved for ζ_A, uniquely if the $L = 1$ spherical harmonic components of χ_A are specified. Hence existence of QS coordinates can be established at the linearized level, but the presence of $\mathrm{div}_o\zeta$ in the expression for z means that it is not possible to directly apply an implicit function theorem to conclude the existence for all metrics in an open neighborhood of g.

It is also interesting to consider the direct approach to finding quasi-spheres as graphs over S^2 in $\mathbb{R}^3$. From the second fundamental form of the graph $r = e^u$, $u : S^2 \to \mathbb{R}$,

$$II_{AB} = \frac{e^u}{\sqrt{1 + |Du|^2}}(u_{AB} - u_A u_B - \sigma_{AB})$$

where σ_{AB} is the S^2 metric and $|Du|^2 = \sigma^{AB}u_A u_B$, the Gauss and mean curvatures are given by

$$K(u) \quad = \quad \det II_{AB} / \det \bar\sigma_{AB} \tag{83}$$

$$H(u) \quad = \quad e^{-2u}\bar\sigma^{AB}II_{AB}, \tag{84}$$

where $\bar\sigma_{AB} = e^{2u}(\sigma_{AB} + u_A u_B)$ is the induced metric on the graph. It follows that the linearisations about the unit sphere with respect to infinitesimal changes $v = \delta u, h = \delta g$ are

$$-\delta K(v,h) \quad = \quad (\Delta + 2)v + \tfrac{1}{2}((\Delta + 1)(\sigma^{AB}h_{AB}) - h_{AB}{}^{|AB}) \tag{85}$$

$$H v \quad = \quad (\Delta + 2)v + h_{rr} + \sigma^{AB}(h_{AB} + \nabla_A h_{Br} - \tfrac{1}{2}\partial_r h_{AB}). \tag{86}$$

Now the equation $(\Delta + 2)\phi = f$ is solvable for ϕ exactly when $\oint_{S^2} fY_1 = 0$ for any $l = 1$ spherical harmonic Y_1, and it follows that for generic metric variations h, it is not possible to find a corresponding variation v of the graph which preserves the condition $H = -2$. This instability of constant mean curvature foliations can be overcome if the background metric has positive ADM mass [49]. In contrast, the equation $\delta K(v,h) = 0$ is solvable for v for any metric variation h, since

$$\oint_{S^2} (\Delta + 1)\sigma^{AB}h_{AB} - h_{AB}^{|AB})Y_1 = 0$$

for all Y_1 satisfying $\Delta Y_1 = -2Y_1$ and then ∇Y_1 is a conformal Killing vector of S^2 and $\nabla_{AB}^2 Y_1 = (\Delta Y_1 + Y_1)\sigma_{AB}$. Thus at the linearised level it is always possible to preserve the condition $K = 1$.

Further evidence supporting the conjecture that QS foliations exist for generic metrics near a given QS metric satisfying $|\mathcal{B}|^2 < 1/3$ comes from the thesis of Spillane [73], which establishes existence in the axially symmetric case. However, the general case remains open, as does the related problem of finding a QS foliation of a null hypersurface [15]. Note that this *foliation* existence question is irrelevant for the *metric* existence question, which is resolved by the proof in [14] of a large class of metrics in QS form.

The momentum constraint equations are not yet well understood in the parabolic method, although some results have been established [68]. In the quasi-spherical case, we may introduce the parameterization

$$K_{ij}\theta^i\theta^j = (\eta_{AB} + \tfrac{1}{2}\mu\delta_{AB})\theta^A\theta^B + \kappa_A(\theta^A\theta^3 + \theta^3\theta^A) + (\tau - \mu)\theta^3\theta^3, \qquad (87)$$

where $\theta^1 = \beta^1\,dr + r\,d\vartheta$, $\theta^2 = \beta^2\,dr + r\sin\vartheta\,d\varphi$, $\theta^3 = u\,dr$ is the QS orthonormal coframe. In terms of the parameters η_{AB}, μ, κ_A, $\tau = \mathrm{tr}_\gamma K$, the momentum constraint equations then take the form

$$
\begin{aligned}
8\pi T_{03}ru &= -(r\partial_r\mu - \beta^A\nabla_A\mu + (3 - \mathrm{div}\,\beta)\mu) \\
&\quad + u\,\mathrm{div}\,\kappa + \kappa^A\nabla_A u + \eta^{AB}\beta_{(A|B)} + (2 - \mathrm{div}\,\beta)\tau
\end{aligned}
\qquad (88)
$$

$$
\begin{aligned}
8\pi T_{0A}ru &= (u\eta_{AB})^{|B} + \mu\nabla_A u + \nabla_A(u(\tfrac{1}{2}\mu - \tau)) \\
&\quad + r\partial_r\kappa_A - \beta^B\kappa_{A|B} + ((3 - \mathrm{div}\,\beta)\delta_{AB} - \beta_{B|A})\kappa^B.
\end{aligned}
\qquad (89)
$$

The matter fields T_{00}, T_{0A}, T_{03} are prescribed and (88) provides either an equation for τ, by choosing μ, or vice-versa, choosing τ and solving for μ. Likewise, (89) can be regarded either as an equation for κ (with the symmetric traceless 2-tensor η_{AB} arbitrarily prescribable) or as an elliptic equation for η with κ_A freely prescribable.

Local existence in r for the momentum constraints with prescribed τ, κ_A has been established by Sharples [68], but it is not clear whether global results are possible without additional restrictions. Much work still remains to be done on these systems.

Finally we note that a characteristic version of the quasi-spherical gauge, where a foliation of outgoing ($\rho_{NP} < 0$) null hypersurfaces is assumed to admit a QS radial coordinate, has been described in [15]. In this case there is no parabolic equation. The resulting hypersurface Einstein equations are considerably simpler than those derived in the Bondi gauge [66, 23], and forms the basis for a 4th order numerical code [18], which heavily exploits the exact spherical geometry of the r-level surfaces.

4.5. Gluing solutions of the constraint equations

The conformal method, the conformal thin sandwich method, and the quasi-spherical ansatz are all procedures for generating solutions of the Einstein constraint equations from scratch. We now consider procedures for constructing new solutions of the constraints from existing solutions.

We first discuss a procedure for gluing connected sums of solutions. The idea of this "IMP Gluing" [52] is the following: Say we have two solutions of the constraint equations, (M_1, γ_1, K_1) and (M_2, γ_2, K_2). Let $p_1 \in M_1$ and $p_2 \in M_2$. Can we find a set of initial data $(M_{(1-2)}, \gamma_{(1-2)}, K_{(1-2)})$ such that 1) M_{1-2} is homotopic to the connected sum[8] $M_1 \# M_2$; 2) $(\gamma_{(1-2)}, K_{(1-2)})$ is a solution of the constraints everywhere on $M_{(1-2)}$; and 3) on that portion of $M_{(1-2)}$ which corresponds to $M_1 \setminus \{\text{ball around } p_1\}$, the data $(\gamma_{(1-2)}, K_{(1-2)})$ is isomorphic to (γ_1, K_1), with a corresponding property holding on that portion of $M_{(1-2)}$ which corresponds to $M_2 \setminus \{\text{ball around } p_2\}$? If so, we say that the sets of data admit IMP gluing.

IMP gluing can be carried out for quite general sets of initial data. The sets can be asymptotically Euclidean, asymptotically hyperbolic, specified on a closed manifold, or indeed anything else. The only condition the data sets must satisfy is that, in sufficiently small neighborhoods of each of the points at which the gluing is to be done, there do not exist nontrivial solutions ξ to the equation $D\Phi^*_{(\gamma, K)}\xi = 0$, where $D\Phi^*_{(\gamma, K)}$ is the linearization operator defined in (43) (with K replacing π). In [20] it is shown that this condition is indeed generic.

The proof that IMP gluing can be carried out to this degree of generality is detailed in [34], based on [52], [53], and [32]. We note here two features of it. First, the proof is constructive, in the sense that it outlines a systematic, step-by-step mathematical procedure for doing the gluing: One conformally blows up the balls surrounding p_1 and p_2 to produce two half cylinders extending from the original initial data sets; one joins the two half cylinders into a bridge, and splices together the data from each side using cutoff functions; one uses the local constant mean curvature to decouple the constraints in the neighborhood of the bridge; one uses tensor projection operators based on linear PDE solutions to find a new conformal K which solves the momentum constraint; one solves the Lichnerowicz equation (the argument that this can be done, and that the solution is very close to 1 away from the bridge, relies on the invertibility of the linearized equation and on a contraction mapping); one recomposes the data as in (50) and (51); and finally one does a nonconformal data perturbation away from the bridge to return the data there to what it was before the gluing. This procedure can be largely carried out numerically, although we note that it requires us to solve elliptic equations on topologically nontrivial manifolds.

The second feature we note regarding the proof is that it relies primarily on the conformal method, but it also uses a nonconformal deformation of the data at the end, to guarantee that the glued data is not just very close to the given data on regions away from the bridge, but is exactly equal to it.

While IMP gluing is not the most efficient tool for studying the complete set of solutions of the constraints, it has already proven to be very useful for a number of applications, including the following:

---

[8]The connected sum of these two manifolds is constructed as follows: First we remove a ball from each of the manifolds M_1 and M_2. We then use a cylindrical bridge $S^2 \times I$ (where I is an interval in R^1) to connect the resulting S^2 boundaries on each manifold.

1. *Multi-Black Hole Data Sets:* Given an asymptotically Euclidean solution of the constraints, IMP gluing allows a sequence of flat space initial data sets to be glued to it. The bridges that result from this gluing each contain a minimal surface, and consequently an apparent horizon. With a bit of care [32], one can do this in such a way that indeed the apparent horizons are disjoint, and therefore likely to lead to independent black holes.

2. *Adding a Black Hole to a Cosmological Spacetime:* Although there is no clear established definition for a black hole in a spatially compact solution of Einstein's equations, one can glue an asymptotically Euclidean solution of the constraints to a solution on a compact manifold, in such a way that there is an apparent horizon on the bridge. Studying the nature of these solutions of the constraints, and their evolution, could be useful in trying to understand what one might mean by a black hole in a cosmological spacetime.

3. *Adding a Wormhole to Your Spacetime:* While we have discussed IMP gluing as a procedure which builds solutions of the constraints with a bridge connecting two points on different manifolds, it can also be used to build a solution with a bridge connecting a pair of points on the *same* manifold. This allows one to do the following: If one has a globally hyperbolic spacetime solution of Einstein's equations, one can choose a Cauchy surface for that solution, choose a pair of points on that Cauchy surface, and glue the solution to itself via a bridge from one of these points to the other. If one now evolves this glued-together initial data into a spacetime, it will likely become singular very quickly because of the collapse of the bridge. Until the singularity develops, however, the solution is essentially as it was before the gluing, with the addition of an effective wormhole. Hence, this procedure can be used to glue a wormhole onto a generic spacetime solution.

4. *Removing Topological Obstructions for Constraint Solutions:* We know that every closed three-dimensional manifold M^3 admits a solution of the vacuum constraint equations. To show this, we use the fact that M^3 always admits a metric Γ of constant negative scalar curvature. One easily verifies that the data $(\gamma = \Gamma, K = \Gamma)$ is a CMC solution. Combining this result with IMP gluing, one can show that for every closed M^3, the manifold $M^3 \setminus \{p\}$ admits both an asymptotically Euclidean and an asymptotically hyperbolic solution of the vacuum constraint equations.

5. *Proving the Existence of Vacuum Solutions on Closed Manifolds with No CMC Cauchy Surface:* Based on the work of Bartnik [11, 12], one can show that if one has a set of initial data on the manifold $T^3 \# T^3$ with the metric components even across a central sphere and the components of K odd across that same central sphere, then the spacetime development of that data does not admit a CMC Cauchy surface. Using IMP gluing, one can show that indeed initial data sets of this sort exist.

4.6. The Corvino-Schoen method

There is another very useful form of gluing which has been applied recently to construct interesting solutions of the Einstein vacuum constraint equations. Developed by Corvino and Schoen [36, 37, 32], this method has the following remarkable application. Let (M^3, γ, K) be a smooth, asymptotically Euclidean solution of the constraint equations. If certain asymptotic conditions hold, then for any compact region $\Sigma^3 \subset M^3$ for which $M^3 \setminus \Sigma^3 = \mathbb{R}^3 \setminus B^3$ (where B^3 is a ball in $\mathbb{R}^3$), there is a smooth asymptotically Euclidean solution on M^3 which is identical to the original solution on $\Sigma^3 \subset M^3$, and is identical to Cauchy data for the Kerr solution on $M^3 \setminus \tilde{\Sigma}^3$ for some $\tilde{\Sigma}^3 \subset M^3$. In words, their technique allows us to smoothly glue any interior region of an asymptotically Euclidean solution to an exterior region of a slice of the Kerr solution. For asymptotically Euclidean solutions of the constraints with $\mathrm{tr}_\gamma K = 0$, this method glues any interior region to an exterior region of a slice of Schwarzschild.

Combining the Corvino-Schoen gluing techniques with some results of Friedrich [44], [32] showed that there is a large class of vacuum spacetime solutions of Einstein's equations which admit complete null infinity regions of the form "scri", as hypothesized by Penrose. The tools developed by Corvino and Schoen have also been used to strengthen the IMP gluing results [33].

The Corvino-Schoen method aims to solve the constraint equations through a projection using the linearized operator $D\Phi$ and its adjoint $D\Phi^*$. We sketch the method in the time-symmetric case $\pi = 0$, where $\Phi(\gamma, \pi)$ is replaced by the scalar curvature $R(\gamma)$ and the lapse-shift ξ is replaced by the lapse N. In this case the arguments are essentially the same while the calculations are considerably simpler.

We start with the observation that because DR^* has injective symbol, it satisfies an elliptic estimate on any domain Ω,

$$\|N\|_{H^2(\Omega)} \le C(\|DR^* N\|_{L^2(\Omega)} + \|N\|_{L^2(\Omega)}), \tag{90}$$

which importantly does not require any control on N at the boundary $\partial\Omega$. It follows easily that similar weighted estimates hold, with weight function $\rho \in C_c^\infty$ which is positive in Ω and vanishes to high order at $\partial\Omega$:

$$\|N\|_{H^2_\rho(\Omega)}^2 := \int_\Omega \rho(N^2 + |\nabla^2 N|^2)\, dv_\gamma \le C \int_\Omega \rho|DR^* N|^2 + N^2)\, dv_\gamma, \tag{91}$$

where the final term N^2 on the right can be removed if there are no Killing vectors. With this assumption, for all $f \in L^2_\rho(\Omega)$ we can solve $DR(\rho DR^* N) = \rho f$ for $N \in H^4_{\mathrm{loc}}$, which in particular produces a solution to the linearized constraint equation $DR\, h = \rho f$. An iteration argument is used to solve the nonlinear problem $R(\gamma_0 + h) = R(\gamma_0) + S$ for any sufficiently small S. This solution $h \in H^2_{\rho^{-1}}(\Omega)$ has the remarkable property that it vanishes to high order on $\partial\Omega$. Thus, for example, if $R(\gamma_0)$ is sufficiently small and supported in Ω then there is a perturbation h, also supported in Ω, such that $R(\gamma_0 + h) = 0$.

To use this method to glue a Schwarzschild exterior to an asymptotically flat $R(\gamma) = 0$ metric across an annulus $B_{2R} \setminus B_R$, $R \gg 1$, requires one more idea because the flat space kernel ker $DR_\delta^* = \mathrm{span}(1, x^1, x^2, x^3)$ is non-trivial. This implies that the linearized problem $DR_\delta h = \sigma$ is solvable if and only if σ satisfies the four conditions $\int_\Omega \sigma(1, x^i) d^3 x = 0$, and the nonlinear problem $R(\gamma_0 + h) = 0$ is similarly obstructed for γ_0 close to flat. By choosing the cutoff radius R sufficiently large and rescaling back to $\Omega = B_2 \backslash B_1$ produces exactly this close to flat situation. However, it is possible to solve the projected problem $R(\gamma_0 + h) \in K := \mathrm{span}(1, x^i)$ with uniform estimates on $h \in H^2_{\rho-1}(\Omega)$. Now the Schwarzschild exterior metric can be characterized by the mass and centre of mass parameters (m, c^i), defined by

$$\gamma_{\mathrm{Schw}} = (1 + \tfrac{m}{2|\mathbf{x}-\mathbf{c}|})^4 \delta_{ij}. \tag{92}$$

Some delicate estimates show that the map $(m, c^i) \mapsto K$ is continuous and has index 1, so there is a choice of parameters (m, c^i) mapping to $0 \in K$, which gives $R(\gamma_0 + h) = 0$ as required. The extension of these ideas, and the considerable details of the above arguments, are given in the original references [36, 37, 33].

5. Conclusion

A considerable amount is known concerning the solutions of the Einstein constraint equations. Using the conformal method or the conformal thin sandwich method, we know how to construct constant mean curvature solutions which are asymptotically Euclidean, asymptotically hyperbolic, or live on a closed manifold. We also know how to do the same for nearly constant mean curvature solutions. We can glue together quite general solutions of the constraints, producing new solutions of both mathematical and physical interest. And, for certain asymptotically Euclidean solutions, we know how to show that there are solutions which include any compact region of the solution in the interior, and which are exactly Kerr or Schwarzschild in the exterior.

Much remains to discover as well. We would like to know how to construct solutions with mean curvature neither constant nor nearly constant. We would like to know much more about constructing solutions of the constraints with prescribed boundary conditions. We would like to know to what extent we can construct solutions with low regularity. And we would like to know which solutions on compact regions can be smoothly extended to either asymptotically Euclidean or asymptotically hyperbolic solutions.

Besides these mathematical issues to resolve, there are important questions concerning solutions of the constraints and physical modelling. In view of the pressing need to model astrophysical events which produce detectable amounts of gravitational radiation, one of the crucial questions we need to answer is how to systematically find solutions of the constraint equations which serve as physically realistic initial data sets for such astrophysical models. Since these models are generally constructed numerically, an equally crucial question is the extent to which the constraint functions, initially zero, remain near zero as the spacetime is numerically evolved.

It is not clear how close we are to resolving these mathematical and physical questions regarding the Einstein constraint equations and their solutions. However, in view of the recent rapid progress that has been made in these studies, we are optimistic that many of them will be resolved soon as well.

Acknowledgments

We thank Yvonne Choquet-Bruhat, Rafe Mazzeo, Vincent Moncrief, Niall O'Murchadha, Daniel Pollack and James York for useful conversations. JI thanks the Caltech relativity group and the Kavli Institute for Theoretical Physics for hospitality while this survey article was being written, and acknowledges support for this research from NSF grant PHY 0099373 at the University of Oregon. The work of RB is supported in part by the Australian Research Council.

References

[1] R. Abraham, J.E. Marsden, and T. Ratiu. *Manifolds, tensor analysis, and applications.* Springer-Verlag, 1988.

[2] A. Anderson and J.W. York Jr. Hamiltonian time evolution for general relativity. *Phys. Rev. Lett.*, 81:1154–1157, 1998. grqc/9807041.

[3] _____ and J.W. York. Fixing Einstein's equations. *Phys. Rev. Lett.*, pages 82:4384–4387, 1999.

[4] L. Andersson and P.T. Chruściel. On asymptotic behavior of solutions of the constraint equations in general relativity with "hyperboloidal boundary conditions". *Dissert. Math.*, 355:1–100, 1996.

[5] _____, _____, and H. Friedrich. On the regularity of solutions to the Yamabe equations and the existence of smooth hyperboloidal initial data for Einstein's field equations. *Comm. Math. Phys.*, 149:587–612, 1992.

[6] J. Arms, J. E. Marsden, and V. Moncrief. The structure of the space of solutions of Einstein's equations II: Several Killing fields and the Einstein-Yang-Mills equations. *Annals Phys.*, 144(1):81–106, 1982.

[7] R. Arnowitt, S. Deser, and C. Misner. The dynamics of general relativity. In L. Witten, editor, *Gravitation*, pages 227–265. Wiley, N.Y., 1962.

[8] R.F. Baierlein, D.H. Sharp, and J.A. Wheeler. *Phys. Rev.*, 126:1, 1962.

[9] R. Bartnik. Phase space for the Einstein equations. gr-qc/0402070.

[10] _____. The existence of maximal hypersurfaces in asymptotically flat space-times. *Commun. Math. Phys.*, 94:155–175, 1984.

[11] _____. The regularity of variational maximal surfaces. *Acta Math.*, 161:145–181, 1988.

[12] _____. Remarks on cosmological spacetimes and constant mean curvature surfaces. *Commun. Math. Phys.*, 117:615–624, 1988.

[13] _____. New definition of quasilocal mass. *Phys. Rev. Lett.*, 62(20):2346–2348, May 1989.

[14] _____. Quasi-spherical metrics and prescribed scalar curvature. *J. Diff. Geom.*, 37:31–71, 1993.

[15] _____ . Einstein equations in the null quasi-spherical gauge. *Class. Quant. Gravity*, 14:2185–2194, 1997. gr-qc/9611045.

[16] _____ . Energy in general relativity. In Shing-Tung Yau, editor, *Tsing Hua Lectures on Analysis and Geometry*, pages 5–28. International Press, 1997.

[17] _____ and G. Fodor. On the restricted validity of the thin sandwich conjecture. *Phys. Rev. D*, 48:3596–3599, 1993. gr-qc/9304004.

[18] _____ and A. Norton. Numerical methods for the Einstein equations in null quasi-spherical coordinates. *SIAM J. Sci. Comp.*, 22:917–950, 2000. gr-qc/9904045.

[19] _____ and L. Simon. Spacelike hypersurfaces with prescribed boundary values and mean curvature. *Commun. Math. Phys.*, 87:131–152, 1982.

[20] R. Beig, P.T. Chruściel and R. Schoen. KIDs are non-generic. gr-qc/0403042.

[21] E.P. Belasco and H.C. Ohanian. Initial conditions in general relativity: lapse and shift formulation. *J. Math. Phys.*, 10:1503, 1969.

[22] A. Besse. *Einstein manifolds*. Springer, 1987.

[23] H. Bondi. Gravitational waves in general relativity. *Nature*, 186:535, 1960.

[24] D. Brill. On spacetimes without maximal surfaces. In H. Ning, editor, *Proc. Third Marcel Grossman meeting*. North Holland, 1982.

[25] _____ and M. Cantor. The Laplacian on asymptotically flat manifolds and the specification of scalar curvature. *Composit. Math.*, 43:317, 1981.

[26] M. Cantor. The existence of non-trivial asymptotically flat initial data for vacuum spacetimes. *Commun. Math. Phys.*, 57:83, 1977.

[27] Y. Choquet-Bruhat. Spacelike submanifolds with constant mean extrinsic curvature of a Lorentzian manifold. *Ann. d. Scuola Norm. Sup. Pisa*, 3:361–376, 1976.

[28] _____ . Non-strict and strict hyperbolic systems for the Einstein equations. In Ancon and J. Vaillant, editors, *Partial Differential Equations and Mathematical Physics*. Dekker, 2003.

[29] _____ and R. Geroch. Global aspects of the Cauchy problem in general relativity. *Commun. Math. Phys.*, 14:329–335, 1969.

[30] _____ , J. Isenberg, and J. York. Einstein constraints on asymptotically Euclidean manifolds. *Phys. Rev. D*, 61:084034, 2000.

[31] _____ and J.W. York. The Cauchy problem. In A. Held, editor, *General Relativity and Gravitation – the Einstein Centenary*, chapter 4, pages 99–160. Plenum, 1979.

[32] P.T. Chruściel and E. Delay. Existence of non-trivial vacuum asymptotically simple spacetimes. *Class. Quant. Grav.*, 19:L71–L79, 2002.

[33] _____ and E. Delay. On mapping properties of general relativistic constraints operators in weighted function spaces, with applications. Mémoires de la Société Mathématique de France 94:1–103, 2003. gr-qc/0301073.

[34] _____ , J. Isenberg and D. Pollack. Initial data engineering. gr-qc/0403066.

[35] G. Cook. Private communication. 2003.

[36] J. Corvino. Scalar curvature deformation and a gluing construction for the Einstein constraint equations. *Commun. Math. Phys.*, 214:137–189, 2000.

[37] _____ and R. Schoen. On the asymptotics for the vacuum constraint equations. gr-qc/0301071.

[38] S. Dain. Trapped surfaces as boundaries for the constraint equations, 2003. gr-qc/0308009.

[39] Th. de Donder. *Théorie des champs gravitiques.* Gauthier-Villars, Paris, 1926.

[40] D.M. DeTurck. The Cauchy problem for Lorentz metrics with prescribed Ricci curvature. *Comp. Math.*, 48:327–349, 1983.

[41] A.E. Fischer, J.E. Marsden, and V. Moncrief. The structure of the space of solutions of Einstein's equations I. One Killing field. *Ann. Inst. H. Poincaré*, 33:147–194, 1980.

[42] A.E. Fischer and J.E. Marsden. Topics in the dynamics of general relativity. In J. Ehlers, editor, *Structure of Isolated Gravitating Systems*, pages 322–395. 1979.

[43] Y. Fourés-Bruhat. Théorème d'existence pour certains systèmes d'équations aux dérivées partielles non linéaires. *Acta Math.*, 88:141–225, 1952.

[44] H. Friedrich. On the hyperbolicity of Einstein's and other gauge field equations. *Commun. Math. Phys.*, 100:525–543, 1985.

[45] D. Garfinkle. Numerical simulations of generic singularities. Unpublished. 2003.

[46] C. Gerhardt. H-surfaces in Lorentzian manifolds. *Commun. Math. Phys.*, 89:523–553, 1983.

[47] R. Geroch. Energy extraction. *Ann. N.Y. Acd. Sci.*, 224:108–117, 1973.

[48] G. Huisken and T. Ilmanen. The inverse mean curvature flow and the Riemannian Penrose inequality. *J. Diff. Geom.*, 59:353–438, 2001.

[49] _____ and S.-T. Yau. Definition of center of mass for isolated physical systems and unique foliations by stable spheres with constant mean curvature. *Invent. Math.*, 124:281–311, 1996.

[50] J. Isenberg. Constant mean curvature solutions of the Einstein constraint equations on closed manifolds. *Class. Quant. Grav.*, 12:2249, 1995.

[51] _____. Near constant mean curvature solutions of the Einstein constraint equations with non-negative Yamabe metrics. Unpublished. 2003.

[52] _____, R. Mazzeo, and D. Pollack. Gluing and wormholes for the Einstein constraint equations. *Comm. Math. Phys.*, 231:529–568, 2002.

[53] _____, _____ and _____. On the topology of vacuum spacetimes. *Ann. Inst. H. Poincaré*, 4:369–383, 2003. gr-qc/0206034.

[54] _____ and V. Moncrief. A set of nonconstant mean curvature solutions of the Einstein constraint equations on closed manifolds. *Class. Quant. Grav.*, 13:1819, 1996.

[55] _____ and N.Ó Murchadha. Non-CMC conformal data sets which do not produce solutions of the Einstein constraint equations. gr-qc/0311057.

[56] _____ and J. Park. Asymptotically hyperbolic non-constant mean curvature solutions of the Einstein constraint equations. *Class. Quant. Grav.*, 14:A189, 1997.

[57] V. Iyer and R.M. Wald. Some properties of Noether charge and a proposal for dynamical black hole entropy. *Phys. Rev.*, D50:846–864, 1994. gr-qc/9403028.

[58] F. John. *Partial Differential Equations.* Springer, 4th edition, 1982.

[59] C. Lanczos. Ein vereinfachendes Koordinatensystem für die Einsteinschen Gravitationsgleichungen. *Phys. Zeit.*, 23:537–539, 1922.

[60] L. Lindblom and M.A. Scheel. Dynamical gauge conditions for the Einstein evolution equations. *Phys. Rev.*, 7:124005, 2003. gr-qc/0301120.

[61] D. Maxwell. Solutions of the Einstein constraint equations with apparent horizon boundary, 2003. gr-qc/0307117.

[62] C.W. Misner, K.S. Thorne, and J.A. Wheeler. *Gravitation.* Freeman, 1973.

[63] V. Moncrief. Spacetime symmetries and linearization stability of the Einstein equations I. *J. Math. Phys.*, 16(3):493–498, March 1975.

[64] _____. Spacetime symmetries and linearization stability of the Einstein equations II. *J. Math. Phys.*, 17(10):1893–1902 1976.

[65] R. Racke. *Lectures on Nonlinear Evolution Equations.* Friedrich Vieweg, 1992.

[66] R.K. Sachs. Gravitational waves in general relativity, VIII. Waves in asymptotically flat space-time. *Proc. Roy. Soc. Lond. A*, A270:103–126, 1962.

[67] R. Schoen and S.-T. Yau. Proof of the positive mass theorem. *Comm. Math. Phys.*, 65:45–76, 1979.

[68] J. Sharples. *Spacetime initial data and quasi-spherical coordinates.* PhD thesis, University of Canberra, 2001.

[69] Y. Shi and L.F. Tam. Positive mass theorem and the boundary behaviors of compact manifolds with nonnegative scalar curvature. *J. Diff. Geom.*, 62:79–125, 2002. math.DG/0301047.

[70] L. Simon. *Seminar on Minimal Submanifolds*, chapter Survey Lectures on Minimal Submanifolds, pages 3–52. Annals of Math. Studies 103. Princeton UP, 1983.

[71] B. Smith and G. Weinstein. On the connectedness of the space of initial data for the Einstein equations. *Electron. Res. Announc. Amer. Math. Soc.*, 6:52–63, 2000.

[72] _____ and _____. Quasiconvex foliations and asymptotically flat metrics of nonnegative scalar curvature. *Comm. Analysis and Geom.*, 2004. to appear.

[73] M. Spillane. *The Einstein equations in the null quasi-spherical gauge.* PhD thesis, Australian National University, 1994.

[74] M.E. Taylor. *Partial Differential Equations III*, volume 117 of *Applied Mathematical Sciences.* Springer, 1996.

[75] R.M. Wald. *General Relativity.* University of Chicago Press, 1984.

[76] J.W. York. Conformal "Thin-Sandwich" data for the initial-value problem of general relativity. *Phys. Rev. Lett.*, 82:1350–1353, 1999.

Robert Bartnik
School of Mathematics and Statistics
University of Canberra
ACT 2601, Australia
e-mail: robert.bartnik@canberra.edu.au

Jim Isenberg
Department of Mathematics and Institute for Theoretical Science
University of Oregon
Eugene, OR 97403, USA
e-mail: jim@newton.uoregon.edu

The Penrose Inequality

Hubert L. Bray[1] and Piotr T. Chruściel[2]

Abstract. In 1973, R. Penrose presented an argument that the total mass of a space-time which contains black holes with event horizons of total area A should be at least $\sqrt{A/16\pi}$. An important special case of this physical statement translates into a very beautiful mathematical inequality in Riemannian geometry known as the Riemannian Penrose inequality. This inequality was first established by G. Huisken and T. Ilmanen in 1997 for a single black hole and then by one of the authors (HB) in 1999 for any number of black holes. The two approaches use two different geometric flow techniques and are described here. We further present some background material concerning the problem at hand, discuss some applications of Penrose-type inequalities, as well as the open questions remaining.

Mathematics Subject Classification (2000). 83C05, 53C80.

1. Introduction

1.1. What is the Penrose conjecture?

We will restrict our attention to statements about space-like slices (M^3, g, h) of a space-time, where g is the positive definite induced metric on M^3 and h is the second fundamental form of M^3 in the space-time. From the Einstein equation $G = 8\pi T$, where G is the Einstein curvature tensor and T is the stress-energy tensor, it follows from the Gauss and Codazzi equations that

$$\mu = \frac{1}{8\pi} G^{00} = \frac{1}{16\pi} [R - \sum_{i,j} h^{ij} h_{ij} + (\sum_i h_i{}^i)^2], \tag{1.1}$$

$$J^i = \frac{1}{8\pi} G^{0i} = \frac{1}{8\pi} \sum_j \nabla_j [h^{ij} - (\sum_k h_k{}^k) g^{ij}], \tag{1.2}$$

1) Research supported in part by NSF grants #DMS-0206483 and #DMS-9971960 and by the Erwin Schrödinger Institute, Vienna.
2) Partially supported by a Polish Research Committee grant 2 P03B 073 24, by the Erwin Schrödinger Institute, Vienna, and by a travel grant from the Vienna City Council.

where μ and J are respectively the energy density and the current vector density at each point of M^3. Then the physical assumption of nonnegative energy density everywhere in the space-time as measured by observers moving in all future-pointing, time-like directions (known as the dominant energy condition) implies that

$$\mu \geq |J| \tag{1.3}$$

everywhere on M^3. Hence, we will only consider Cauchy data (M^3, g, h) which satisfy inequality (1.3).

The final assumption we will make is that (M^3, g, h) is *asymptotically flat*, which will be discussed in more detail below. Typically, one assumes that M^3 consists of a compact set together with one or more asymptotically flat "ends", each diffeomorphic to the complement of a ball in $\mathbb{R}^3$. For example, $\mathbb{R}^3$ has one end, whereas $\mathbb{R}^3 \# \mathbb{R}^3$ has two ends.

Penrose's motivation for the Penrose conjecture [70] goes as follows: Suppose we begin with Cauchy data (M^3, g, h) which is asymptotically flat (so that total mass of a chosen end is defined) and satisfies $\mu \geq |J|$ everywhere. Using this as initial data, solve the Einstein equation forward in time, and suppose that the resulting space-time is asymptotically flat in null directions so that the Trautman-Bondi mass is defined for all retarded times. Suppose further that the space-time eventually settles down to a Kerr solution, so that the Trautman-Bondi mass asymptotes to the ADM mass of the relevant Kerr solution. By the Hawking Area Theorem [48] (compare [29]), the total area of the event horizons of any black holes does not decrease, while the total Trautman-Bondi mass of the system – which is expected to approach the ADM mass at very early advanced times – does not increase. Since Kerr solutions all have

$$m \geq \sqrt{A_e/16\pi}, \tag{1.4}$$

where m is total ADM mass [5,38] and A_e is the total area of the event horizons, we must have this same inequality for the original Cauchy data (M^3, g, h).

The reader will have noticed that the above argument makes a lot of global assumptions about the resulting space-times, and our current understanding of the associated mathematical problems is much too poor to be able to settle those one way or another. The conjecture that (all, or at least a few key ones of) the above global properties are satisfied is known under the name of *Penrose's cosmic censorship hypothesis*. We refer the reader to the article by Lars Andersson in this volume and references therein for more information about that problem.

A natural interpretation of the Penrose inequality is that the mass contributed by a collection of black holes is not less than $\sqrt{A/16\pi}$. More generally, the question "How much matter is in a given region of a space-time?" is still very much an open problem [23]. In this paper, we will discuss some of the qualitative aspects of mass in general relativity, look at examples which are informative, and describe the two very geometric proofs of the Riemannian Penrose inequality. The most general version of the Penrose inequality is still open and is discussed in Section 4.2. The notes here are partly based on one of the author's

(HB) lectures at the "Fifty Years of the Cauchy Problem in General Relativity" Summer School held in August 2002 in Cargèse (videos of lectures available at URL http://fanfreluche.math.univ-tours.fr, or on the DVD enclosed with this volume), and some sections draw substantially on his review paper [17], following a suggestion of the editors of this volume. The mathematically oriented reader with limited knowledge of the associated physics might find it useful to become acquainted with [17] before reading the current presentation.

1.2. Total mass in general relativity

Amongst the notions of mass which are well understood in general relativity are local energy density at a point, the total mass of an asymptotically flat space-time (whether at spacelike or at null infinity; the former is usually called the ADM mass while the latter the Trautman-Bondi mass), and the total mass of an asymptotically anti-de Sitter space-time (often called the Abbott-Deser mass). On the other hand, defining the mass of a region larger than a point but smaller than the entire universe is not very well understood at all. While we will return to this last question in Section 4.3 below, we start here with a discussion of the ADM mass.

Suppose (M^3, g) is a Riemannian 3-manifold isometrically embedded in a $(3 + 1)$-dimensional Lorentzian space-time N^4. Suppose that M^3 has zero second fundamental form in the space-time. (Recall that the second fundamental form is a measure of how much M^3 curves inside N^4. M^3 is also sometimes called "totally geodesic" since geodesics of N^4 which are tangent to M^3 at a point stay inside M^3 forever.) The Penrose inequality (which in its full generality allows for M^3 to have non-vanishing second fundamental form) is known as the *Riemannian Penrose inequality* when the second fundamental form is set to zero.[1]

In this work we will mainly consider (M^3, g) that are asymptotically flat at infinity, which means that for some compact set K, the "end" $M^3 \backslash K$ is diffeomorphic to $\mathbb{R}^3 \backslash B_1(0)$, where the metric g is asymptotically approaching (with the decay conditions (1.7) below) the standard flat metric δ_{ij} on $\mathbb{R}^3$ at infinity. The simplest example of an asymptotically flat manifold is $(\mathbb{R}^3, \delta_{ij})$ itself. Other good examples are the conformal metrics $(\mathbb{R}^3, u(x)^4 \delta_{ij})$, where $u(x)$ approaches a constant sufficiently rapidly at infinity. (Also, sometimes it is convenient to allow (M^3, g) to have multiple asymptotically flat ends, in which case each connected component of $M^3 \backslash K$ must have the property described above.) A qualitative picture of an asymptotically flat 3-manifold is shown in Fig 1.

The assumptions on the asymptotic behavior of (M^3, g) at infinity will be tailored to imply the existence of the limit

$$m = \frac{1}{16\pi} \lim_{\sigma \to \infty} \int_{S_\sigma} \sum_{i,j} (g_{ij,i} \nu_j - g_{ii,j} \nu_j) \, d\mu \qquad (1.5)$$

---

[1] This terminology is somewhat misleading, in the following sense: the results discussed below hold as soon as the scalar curvature is non-negative. This will certainly be the case if $h_{ij} = 0$ and $\mu \geq 0$ in (1.1), but, *e.g.*, $\sum_i h_i{}^i = 0$, or various other conditions in this spirit, suffice.

where S_σ is the coordinate sphere of radius σ, ν is the unit normal to S_σ, and $d\mu$ is the area element of S_σ in the coordinate chart. The quantity m is called the *total mass* (or ADM mass [5]) of (M^3, g). Equation (1.5) begs the question of the geometric character of the number m: the integrand contains partial derivatives of a tensor, which makes it coordinate dependent. For example, if $g = \delta$ is the flat metric in the standard orthogonal coordinates x^i, one clearly obtains zero. On the other hand, we can introduce a new coordinate system (ρ, θ, ϕ) by changing the radial variable r to

$$r = \rho + c\rho^{1-\alpha} , \tag{1.6}$$

with some constants $\alpha > 0$, $c \in \mathbb{R}$. In the associated asymptotically Euclidean coordinate system $y^i = \rho x^i / r$ the metric tensor approaches δ as $O(|y|^{-\alpha})$:

$$\delta_{ij} dx^i dx^j = g_{ij} dy^i dy^j ,$$

with

$$g_{ij} - \delta_{ij} = \mathcal{O}(|y|^{-\alpha}) , \quad \partial_k g_{ij} = \mathcal{O}(|y|^{-\alpha-1}) . \tag{1.7}$$

A short calculation gives

$$m = \begin{cases} \infty , & \alpha < 1/2 , \\ c^2/8 , & \alpha = 1/2 , \\ 0 , & \alpha > 1/2 . \end{cases}$$

Thus, the mass m of the flat metric in the coordinate system y^i is infinite if $\alpha < 1/2$, can have an arbitrary positive value depending upon c if $\alpha = 1/2$, and vanishes for $\alpha > 1/2$. (Negative values of m can also be obtained by deforming the slice $\{t = 0\}$ within Minkowski space-time [25] when the decay rate $\alpha = 1/2$ is allowed.) The lesson of this is that the mass appears to depend upon the coordinate system chosen, even within the class of coordinate systems in which the metric tends to a constant coefficients matrix as r tends to infinity. It can be shown that the decay rate $\alpha = 1/2$ is precisely the borderline for a well-defined mass: the mass is an invariant in the class of coordinate systems satisfying (1.7) with $\alpha > 1/2$ and with $R \in L^1(M)$ [6, 26][2]. We note that the above example is essentially due to Denisov and Solov'ev [37], and that the geometric character of m in a space-time setting is established in [27].

Going back to the example $(\mathbb{R}^3, u(x)^4 \delta_{ij})$, if we suppose that $u(x) > 0$ has the asymptotics at infinity

$$u(x) = a + b/|x| + \mathcal{O}(1/|x|^2) , \tag{1.8}$$

with the derivatives of the $\mathcal{O}(1/|x|^2)$ term being $\mathcal{O}(1/|x|^3)$, then the total mass of (M^3, g) is

$$m = 2ab. \tag{1.9}$$

Furthermore, suppose (M^3, g) is any metric whose "end" is isometric to $(\mathbb{R}^3 \backslash K, u(x)^4 \delta_{ij})$, where $u(x)$ is harmonic in the coordinate chart of the end $(\mathbb{R}^3 \backslash K, \delta_{ij})$ and goes to a constant at infinity. Then expanding $u(x)$ in terms of spherical harmonics demonstrates that $u(x)$ satisfies condition (1.8). We will call these Riemannian manifolds (M^3, g) *harmonically flat at infinity*, and we note that the total mass of these manifolds is also given by equation (1.9).

A very nice lemma by Schoen and Yau [73] is that, given any $\epsilon > 0$, it is always possible to perturb an asymptotically flat manifold to become harmonically flat at infinity such that the total mass changes less than ϵ and the metric changes less than ϵ pointwise, all while maintaining nonnegative scalar curvature (discussed in a moment). Hence, it happens that to prove the theorems in this paper, we only need to consider harmonically flat manifolds. Thus, we can use equation (1.9) as our definition of total mass. As an example (already pointed out), note that $(\mathbb{R}^3, \delta_{ij})$ has zero total mass. Also, note that, qualitatively, the total mass of an asymptotically flat or harmonically flat manifold is the $1/r$ rate at which the metric becomes flat at infinity.

A deep (and considerably more difficult to prove) result of Corvino [34] (compare [28, 35]) shows that if m is non zero, then one can always perturb an asymptotically flat manifold as above while maintaining zero scalar curvature and achieving (1.8) without any error term.

We finish this section by noting the following "isotropic coordinates" representation of the *exterior Schwarzschild space-time metric*

$$\left(\mathbb{R} \times \left(\mathbb{R}^3 \setminus B_{m/2}(0) \right), (1 + \frac{m}{2|x|})^4 (dx_1^2 + dx_2^2 + dx_3^2) - \left(\frac{1 - m/2|x|}{1 + m/2|x|} \right)^2 dt^2 \right) . \tag{1.10}$$

---

[2]Actually the results in [6] use weighted Sobolev conditions on two derivatives of the metric, suggesting that the right decay conditions in (1.7) are $o(|y|^{-1/2})$ for the metric and $o(|y|^{-3/2})$ for its derivatives. It can be checked that the argument in [26] generalizes, and gives the result under those conditions.

The $t = 0$ slice (which has zero second fundamental form) is the *exterior spacelike Schwarzschild metric*

$$\left(\mathbb{R}^3\backslash B_{m/2}(0), (1 + \frac{m}{2|x|})^4 \delta_{ij}\right). \tag{1.11}$$

According to equation (1.9), the parameter m is of course the total mass of this 3-manifold.

The above example also allows us to make a connection between what we have arbitrarily defined to be total mass and our more intuitive Newtonian notions of mass. Using the natural Lorentzian coordinate chart as a reference, one can compute that geodesics in the Schwarzschild space-time metric are curved when $m \neq 0$. Furthermore, if one interprets this curvature as acceleration due to a force coming from the central region of the manifold, one finds that this fictitious force yields a radial acceleration asymptotic to $-m/r^2$ for large r. Hence, a test particle left to drift along geodesics far out in the asymptotically flat end of the Schwarzschild spacetime "accelerates" according to Newtonian physics as if the total mass of the system were m.

1.3. Example using superharmonic functions in $\mathbb{R}^3$

Once again, let us return to the $(\mathbb{R}^3, u(x)^4 \delta_{ij})$ example. The formula for the scalar curvature is

$$R(x) = -8u(x)^{-5}\Delta u(x).$$

Hence, since the physical assumption of nonnegative energy density implies nonnegative scalar curvature, we see that $u(x) > 0$ must be superharmonic ($\Delta u \leq 0$). For simplicity, let us also assume that $u(x)$ is harmonic outside a bounded set so that we can expand $u(x)$ at infinity using spherical harmonics. Hence, $u(x)$ has the asymptotics of equation (1.8). By the maximum principle, it follows that the minimum value for $u(x)$ must be a, referring to equation (1.8). Hence, $b \geq 0$, which implies that $m \geq 0$. Thus we see that the assumption of nonnegative energy density at each point of $(\mathbb{R}^3, u(x)^4 \delta_{ij})$ implies that the total mass is also nonnegative, which is what one would hope.

1.4. The positive mass theorem

Suppose we have any asymptotically flat manifold with nonnegative scalar curvature, is it true that the total mass is also nonnegative? The answer is *yes*, and this fact is known as the positive mass theorem, first proved by Schoen and Yau [72] in 1979 using minimal surface techniques and then by Witten [80] in 1981 using spinors. (The mathematical details needed for Witten's argument have been worked out in [6,21,51,69].) In the zero second fundamental form case, also known as the *time-symmetric* case, the positive mass theorem is known as the Riemannian positive mass theorem and is stated below.

Theorem 1.1. *(Schoen, Yau [72]) Let (M^3, g) be any asymptotically flat, complete Riemannian manifold with nonnegative scalar curvature. Then the total mass $m \geq 0$, with equality if and only if (M^3, g) is isometric to $(\mathbb{R}^3, \delta)$.*

1.5. Apparent horizons

Given a surface in a space-time, suppose that it emits an outward shell of light. If the surface area of this shell of light is decreasing everywhere on the surface, then this is called a trapped surface.[3] The outermost boundary of these trapped surfaces is called the apparent horizon. Apparent horizons can be computed in terms of Cauchy data, and under appropriate global hypotheses an apparent horizon implies the existence of an event horizon outside of it [48, 78] in the time-symmetric case. The reader is referred to [24] for a review of what is known about apparent horizons; further recent results include [36, 67].

Now let us return to the case where (M^3, g) is a "$t = 0$" slice of a space-time with zero second fundamental form. Then apparent horizons of black holes intersected with M^3 correspond to the connected components of the outermost minimal surface Σ_0 of (M^3, g).

All of the surfaces we are considering in this paper will be required to be smooth boundaries of open bounded regions, so that outermost is well defined with respect to a chosen end of the manifold [15]. A minimal surface in (M^3, g) is a surface which is a critical point of the area function with respect to any smooth variation of the surface. The first variational calculation implies that minimal surfaces have zero mean curvature. The surface Σ_0 of (M^3, g) is defined as the boundary of the union of the open regions bounded by all of the minimal surfaces in (M^3, g). It turns out that Σ_0 also has to be a minimal surface, so we call Σ_0 the *outermost minimal surface*. A qualitative sketch of an outermost minimal surface of a 3-manifold is in Fig. 2.

FIGURE 2

---

[3]The reader is warned that several authors require the trapping of both outwards and inwards shells of light in the definition of trapped surface. The inwards null directions are irrelevant for our purposes, and they are therefore ignored in the definition here.

We will also define a surface to be *(strictly) outer minimizing* if every surface which encloses it has (strictly) greater area. Note that outermost minimal surfaces are strictly outer minimizing. Also, we define a *horizon* in our context to be any minimal surface which is the boundary of a bounded open region.

It also follows from a stability argument (using the Gauss-Bonnet theorem interestingly) that each component of an outermost minimal surface (in a 3-manifold with nonnegative scalar curvature) must have the topology of a sphere [68].

Penrose's argument [70], presented in Section 1.1, suggests that the mass contributed by the black holes (thought of as the connected components of Σ_0) should be at least $\sqrt{A_0/16\pi}$, where A_0 is the area of Σ_0. This leads to the following geometric statement:

The Riemannian Penrose Inequality *Let* (M^3, g) *be a complete, smooth, 3-manifold with nonnegative scalar curvature which is harmonically flat at infinity with total mass m and which has an outermost minimal surface Σ_0 of area A_0. Then*

$$m \geq \sqrt{\frac{A_0}{16\pi}}, \qquad (1.12)$$

with equality if and only if (M^3, g) *is isometric to the Schwarzschild metric* $(\mathbb{R}^3\backslash\{0\}, (1 + \frac{m}{2|x|})^4 \delta_{ij})$ *outside their respective outermost minimal surfaces.*

The above statement has been proved by one of us (HB) [15], and Huisken and Ilmanen [57] proved it when A_0 is defined instead to be the area of the largest connected component of Σ_0. In this paper we will discuss both approaches, which are very different, although they both involve flowing surfaces and/or metrics.

We also clarify that the above statement is with respect to a chosen end of (M^3, g), since both the total mass and the definition of outermost refer to a particular end. In fact, nothing very important is gained by considering manifolds with more than one end, since extra ends can always be compactified as follows: Given an extra asymptotically flat end, we can use a lemma of Schoen and Yau [73] to make the end harmonically flat outside a bounded region. By an extension of this result in the thesis of one of the authors (HB) [14], or using the Corvino-Schoen construction [34], we can make the end exactly Schwarzschild outside a bounded set while still keeping nonnegative scalar curvature. We then replace the interior Schwarzschild region by an object often referred to as "a bag of gold", one way of doing it proceeds as follows: Since we are now in the class of spherically symmetric manifolds, we can then "round the metric up" to be an extremely large spherical cylinder outside a bounded set. This can be done while keeping nonnegative scalar curvature since the Hawking mass increases during this procedure and since the rate of change of the Hawking mass has the same sign as the scalar curvature in the spherically symmetric case (as long as the areas of the spheres are increasing). Finally, the large cylinder can be capped off with a very large sphere to compactify the end.

Hence, we will typically consider manifolds with just one end. In the case that the manifold has multiple ends, we will require every surface (which could have multiple connected components) in this paper to enclose all of the ends of the manifold except the chosen end.

1.6. The Schwarzschild metric

The (spacelike) Schwarzschild metric $(\mathbb{R}^3\backslash\{0\}, (1 + \frac{m}{2|x|})^4\delta_{ij})$ (compare (1.11)), referred to in the above statement of the Riemannian Penrose Inequality, is a particularly important example to consider, and corresponds to a zero-second fundamental form, space-like slice of the usual $(3+1)$-dimensional Schwarzschild metric. The 3-dimensional Schwarzschild metrics with total mass $m > 0$ are characterised by being the only spherically symmetric, geodesically complete, zero scalar curvature 3-metrics, other than $(\mathbb{R}^3, \delta_{ij})$. Note that this flat metric on $\mathbb{R}^3$ may be interpreted as the $m = 0$ case of the Schwarzschild metric. Negative values of m also give Schwarzschild metrics, but these metrics are not geodesically complete since they have a curvature singularity at the coordinate sphere $r = -m/2$. If this singularity is smoothed out in a spherically symmetric way, the resulting metric has very concentrated negative energy density (and scalar curvature) in the smoothed out region, which violates the assumption of positive energy density used throughout this paper.

The 3-dimensional Schwarzschild metrics with total mass $m > 0$ can also be embedded in 4-dimensional Euclidean space (x, y, z, w) as the set of points satisfying $|(x, y, z)| = \frac{w^2}{8m} + 2m$, which is a parabola rotated around an S^2. This last picture allows us to see that the Schwarzschild metric, which has two ends, has a Z_2 symmetry which fixes the sphere with $w = 0$ and $|(x, y, z)| = 2m$, which is clearly minimal. Furthermore, the area of this sphere is $4\pi(2m)^2$, giving equality in the Riemannian Penrose Inequality.

1.7. A brief history of the problem

The Riemannian Penrose Inequality has a rich history spanning nearly three decades and has motivated much interesting mathematics and physics. In 1973, R. Penrose in effect conjectured an even more general version of inequality (1.12) using a very clever physical argument [70], described in Section 1.1. His observation was that a counterexample to inequality (1.12) would yield Cauchy data for solving the Einstein equations, the solution to which would likely violate the Cosmic Censor Conjecture (which says that singularities generically do not form in a space-time unless they are inside a black hole).

In 1977, Jang and Wald [58], extending ideas of Geroch [43], gave a heuristic proof of inequality (1.12) by defining a flow of 2-surfaces in (M^3, g) in which the surfaces flow in the outward normal direction at a rate equal to the inverse of their mean curvatures at each point. The Hawking mass of a surface (which is supposed to estimate the total amount of energy inside the surface) is defined to be

$$m_{\text{Hawking}}(\Sigma) = \sqrt{\frac{|\Sigma|}{16\pi}} \left(1 - \frac{1}{16\pi} \int_\Sigma H^2\right),$$

(where $|\Sigma|$ is the area of Σ and H is the mean curvature of Σ in (M^3, g)) and, amazingly, is nondecreasing under this "inverse mean curvature flow." This is seen by the fact that under inverse mean curvature flow, it follows from the Gauss equation and the second variation formula that

$$\frac{d}{dt}m_{\text{Hawking}}(\Sigma) = \sqrt{\frac{|\Sigma|}{16\pi}} \left[\frac{1}{2} + \frac{1}{16\pi} \int_\Sigma 2\frac{|\nabla_\Sigma H|^2}{H^2} + R - 2K + \frac{1}{2}(\lambda_1 - \lambda_2)^2\right] \quad (1.13)$$

when the flow is smooth, where R is the scalar curvature of (M^3, g), K is the Gauss curvature of the surface Σ, and λ_1 and λ_2 are the eigenvalues of the second fundamental form of Σ, or principal curvatures. Hence,

$$R \geq 0,$$

and

$$\int_\Sigma K \leq 4\pi \quad (1.14)$$

(which is true for any connected surface by the Gauss-Bonnet Theorem) imply

$$\frac{d}{dt}m_{\text{Hawking}}(\Sigma) \geq 0. \quad (1.15)$$

Furthermore,

$$m_{\text{Hawking}}(\Sigma_0) = \sqrt{\frac{|\Sigma_0|}{16\pi}}$$

since Σ_0 is a minimal surface and has zero mean curvature. In addition, the Hawking mass of sufficiently round spheres at infinity in the asymptotically flat end of (M^3, g) approaches the total mass m. Hence, if inverse mean curvature flow beginning with Σ_0 eventually flows to sufficiently round spheres at infinity, inequality (1.12) follows from inequality (1.15).

As noted by Jang and Wald, this argument only works when inverse mean curvature flow exists and is smooth, which is generally not expected to be the case. In fact, it is not hard to construct manifolds which do not admit a smooth inverse mean curvature flow. One of the main problems is that if the mean curvature of the evolving surface becomes zero or is negative, it is not clear how to define the flow.

For twenty years, this heuristic argument lay dormant until the work of Huisken and Ilmanen [57] in 1997. With a very clever new approach, Huisken and Ilmanen discovered how to reformulate inverse mean curvature flow using an energy minimization principle in such a way that the new generalized inverse mean curvature flow always exists. The added twist is that the surface sometimes jumps outward. However, when the flow is smooth, it equals the original inverse mean curvature flow, and the Hawking mass is still monotone. Hence, as will be described in the next section, their new flow produced the first complete proof of inequality (1.12) for a single black hole.

Coincidentally, one of the authors (HB) found another proof of inequality (1.12), submitted in 1999, which provides the correct inequality for any number of black holes. (When the outermost horizon is not-connected, the Huisken-Ilmanen proof bounds the mass in terms of the area of its largest component, while the new argument gives the full inequality, with the sum of areas of all components.) The approach involves flowing the original metric to a Schwarzschild metric (outside the horizon) in such a way that the area of the outermost minimal surface does not change and the total mass is nonincreasing. Then since the Schwarzschild metric gives equality in inequality (1.12), the inequality follows for the original metric.

Fortunately, the flow of metrics which is defined is relatively simple, and in fact stays inside the conformal class of the original metric. The outermost minimal surface flows outward in this conformal flow of metrics, and encloses any compact set (and hence all of the topology of the original metric) in a finite amount of time. Furthermore, this conformal flow of metrics preserves nonnegative scalar curvature. We will describe this approach later in the paper.

Other contributions to the Penrose conjecture have been made by O'Murchadha and Malec in spherical symmetry [65], by Herzlich [50,52] using the Dirac operator with spectral boundary conditions (compare [9,66]), by Gibbons in the special case of collapsing shells [44], by Tod [75] as it relates to the hoop conjecture, by Bartnik [7] for quasi-spherical metrics, by Jezierski [59,60] using adapted foliations, and by one of the authors (HB) using isoperimetric surfaces [14]. A proof of the Penrose inequality for conformally flat manifolds (but with suboptimal constant) has been given in [18]. We also mention work of Ludvigsen and Vickers [63] using spinors and Bergqvist [11], both concerning the Penrose inequality for null slices of a space-time.

Various space-time flows which could be used to prove the full Penrose inequality (see Section 4.2 below) have been proposed by Hayward [49], by Mars, Malec and Simon [64], and by Frauendiener [39]. It was independently observed by several researchers (HB, Hayward, Mars, Simon) that those are special cases of the same flow, namely flowing in the direction $\vec{I} + c(t)\vec{I'}$, where $\vec{I}$ is the inverse mean curvature vector $-\vec{H}/\langle \vec{H}, \vec{H} \rangle$ (which is required to be spacelike outward pointing), $\vec{I'}$ is the future pointing vector with the same length as $\vec{I}$ and orthogonal to $\vec{I}$ in the normal bundle to the surface, and $\vec{H}$ is the mean curvature vector of the surface in the spacetime. The function $c(t)$ is required to satisfy $-1 \leq c(t) \leq 1$ but is otherwise free, with its endpoint values corresponding to Hayward's null flows, $c(t) = 0$ corresponding to Frauendiener's flow, and $-1 \leq c(t) \leq 1$ yielding hypersurfaces satisfying the Mars, Malec, Simon condition which implies the monotonicity of the spacetime Hawking mass functional. The catch, however, is that this flow is not parabolic and therefore only exists for a positive amount of time under special circumstances. However, as observed by HB at the Penrose Inequalities Workshop in Vienna, July 2003, there does exist a way of defining what a weak solution to the above flow is using a max-min method analogous to the notion of weak solution to inverse mean curvature flow (which minimizes an energy functional) defined by Huisken and Ilmanen [57]. Finding ways of constructing solutions which exist for an infinite amount of time (analogous to the time-symmetric inverse mean curvature flow due to Huisken and Ilmanen) is a very interesting problem to consider.

2. Inverse mean curvature flow

Geometrically, Huisken and Ilmanen's idea can be described as follows. Let $\Sigma(t)$ be the surface resulting from inverse mean curvature flow for time t beginning with the minimal surface Σ_0. Define $\bar{\Sigma}(t)$ to be the outermost minimal area enclosure of $\Sigma(t)$. Typically, $\Sigma(t) = \bar{\Sigma}(t)$ in the flow, but in the case that the two surfaces are not equal, immediately replace $\Sigma(t)$ with $\bar{\Sigma}(t)$ and then continue flowing by inverse mean curvature.

An immediate consequence of this modified flow is that the mean curvature of $\bar{\Sigma}(t)$ is always nonnegative by the first variation formula, since otherwise $\bar{\Sigma}(t)$ would be enclosed by a surface with less area. This is because if we flow a surface Σ in the outward direction with speed η, the first variation of the area is $\int_\Sigma H\eta$, where H is the mean curvature of Σ.

Furthermore, by stability, it follows that in the regions where $\bar{\Sigma}(t)$ has zero mean curvature, it is always possible to flow the surface out slightly to have positive mean curvature, allowing inverse mean curvature flow to be defined, at least heuristically at this point.

It turns out that the Hawking mass is still monotone under this new modified flow. Notice that when $\Sigma(t)$ jumps outward to $\bar{\Sigma}(t)$,

$$\int_{\bar{\Sigma}(t)} H^2 \leq \int_{\Sigma(t)} H^2$$

since $\bar{\Sigma}(t)$ has zero mean curvature where the two surfaces do not touch. Furthermore,

$$|\bar{\Sigma}(t)| = |\Sigma(t)|$$

since (this is a neat argument) $|\bar{\Sigma}(t)| \leq |\Sigma(t)|$ (since $\bar{\Sigma}(t)$ is a minimal area enclosure of $\Sigma(t)$) and we can not have $|\bar{\Sigma}(t)| < |\Sigma(t)|$ since $\Sigma(t)$ would have jumped outward at some earlier time. This is only a heuristic argument, but we can then see that the Hawking mass is nondecreasing during a jump by the above two equations.

This new flow can be rigorously defined, always exists, and the Hawking mass is monotone, if the scalar curvature is positive. In [57], Huisken and Ilmanen define $\Sigma(t)$ to be the level sets of a scalar-valued function $u(x)$ defined on (M^3, g) such that $u(x) = 0$ on the original surface Σ_0 and satisfies

$$\text{div}\left(\frac{\nabla u}{|\nabla u|}\right) = |\nabla u| \tag{2.1}$$

in an appropriate weak sense. Since the left-hand side of the above equation is the mean curvature of the level sets of $u(x)$ and the right-hand side is the reciprocal of the flow rate, the above equation implies inverse mean curvature flow for the level sets of $u(x)$ when $|\nabla u(x)| \neq 0$.

Huisken and Ilmanen use an energy minimization principle to define weak solutions to equation (2.1). Equation (2.1) is said to be weakly satisfied in Ω by the locally Lipschitz function u if for all locally Lipschitz v with $\{v \neq u\} \subset\subset \Omega$,

$$J_u(u) \leq J_u(v) \qquad \text{where} \qquad J_u(v) := \int_\Omega |\nabla v| + v|\nabla u|.$$

It can then be seen that the Euler-Lagrange equation of the above energy functional yields equation (2.1).

In order to prove that a solution u exists to the above two equations, Huisken and Ilmanen regularize the degenerate elliptic equation (2.1) to the elliptic equation

$$\text{div}\left(\frac{\nabla u}{\sqrt{|\nabla u|^2 + \epsilon^2}}\right) = \sqrt{|\nabla u|^2 + \epsilon^2}.$$

Solutions to the above equation are then shown to exist using the existence of a subsolution, and then taking the limit as ϵ goes to zero yields a weak solution to equation (2.1). There are many details which we are skipping here, but these are the main ideas.

As it turns out, weak solutions $u(x)$ to equation (2.1) often have flat regions where $u(x)$ equals a constant. Hence, the levels sets $\Sigma(t)$ of $u(x)$ will be discontinuous in t in this case, which corresponds to the "jumping out" phenomenon referred to at the beginning of this section.

We also note that since the Hawking mass of the levels sets of $u(x)$ is monotone, this inverse mean curvature flow technique not only proves the Riemannian Penrose inequality, but also gives a new proof of the Positive Mass Theorem in dimension three. This is seen by letting the initial surface be a very small, round sphere (which will have approximately zero Hawking mass) and then flowing by inverse mean curvature, thereby proving $m \geq 0$.

The Huisken and Ilmanen inverse mean curvature flow also seems ideally suited for proving Penrose inequalities for 3-manifolds which have $R \geq -6$ and which are asymptotically hyperbolic; this is discussed in more detail in Section 4.1.

Because the monotonicity of the Hawking mass relies on the Gauss-Bonnet theorem, these arguments do not work in higher dimensions, at least so far. Also, because of the need for equation (1.14), inverse mean curvature flow only proves the Riemannian Penrose inequality for a single black hole. In the next section, we present a technique which proves the Riemannian Penrose inequality for any number of black holes, and which can likely be generalized to higher dimensions.

3. The conformal flow of metrics

Given any initial Riemannian manifold (M^3, g_0) which has nonnegative scalar curvature and which is harmonically flat at infinity, we will define a continuous, one parameter family of metrics (M^3, g_t), $0 \leq t < \infty$. This family of metrics will converge to a 3-dimensional Schwarzschild metric and will have other special properties which will allow us to prove the Riemannian Penrose inequality for the original metric (M^3, g_0).

In particular, let Σ_0 be the outermost minimal surface of (M^3, g_0) with area A_0. Then we will also define a family of surfaces $\Sigma(t)$ with $\Sigma(0) = \Sigma_0$ such that $\Sigma(t)$ is minimal in (M^3, g_t). This is natural since as the metric g_t changes, we expect that the location of the horizon $\Sigma(t)$ will also change. Then the interesting quantities to keep track of in this flow are $A(t)$, the total area of the horizon $\Sigma(t)$ in (M^3, g_t), and $m(t)$, the total mass of (M^3, g_t) in the chosen end.

In addition to all of the metrics g_t having nonnegative scalar curvature, we will also have the very nice properties that

$$A'(t) = 0, \qquad m'(t) \leq 0$$

for all $t \geq 0$. Then since (M^3, g_t) converges to a Schwarzschild metric (in an appropriate sense) which gives equality in the Riemannian Penrose inequality as described in the introduction,

$$m(0) \geq m(\infty) = \sqrt{\frac{A(\infty)}{16\pi}} = \sqrt{\frac{A(0)}{16\pi}} \tag{3.1}$$

which proves the Riemannian Penrose inequality for the original metric (M^3, g_0). The hard part, then, is to find a flow of metrics which preserves nonnegative scalar curvature and the area of the horizon, decreases total mass, and converges to a Schwarzschild metric as t goes to infinity. This proceeds as follows:

The metrics g_t will all be conformal to g_0. This conformal flow of metrics can be thought of as the solution to a first order o.d.e. in t defined by equations (3.2)–(3.5). Let

$$g_t = u_t(x)^4 g_0 \tag{3.2}$$

and $u_0(x) \equiv 1$. Given the metric g_t, define

$$\Sigma(t) = \text{the outermost minimal area enclosure of } \Sigma_0 \text{ in } (M^3, g_t) \tag{3.3}$$

where Σ_0 is the original outer minimizing horizon in (M^3, g_0). In the cases in which we are interested, $\Sigma(t)$ will not touch Σ_0, from which it follows that $\Sigma(t)$ is actually a strictly outer minimizing horizon of (M^3, g_t). Then given the horizon $\Sigma(t)$, define $v_t(x)$ such that

$$\begin{cases} \Delta_{g_0} v_t(x) & \equiv & 0 & \text{outside } \Sigma(t) \\ v_t(x) & = & 0 & \text{on } \Sigma(t) \\ \lim_{x \to \infty} v_t(x) & = & -e^{-t} \end{cases} \tag{3.4}$$

and $v_t(x) \equiv 0$ inside $\Sigma(t)$. Finally, given $v_t(x)$, define

$$u_t(x) = 1 + \int_0^t v_s(x)ds \qquad (3.5)$$

so that $u_t(x)$ is continuous in t and has $u_0(x) \equiv 1$.

Note that equation (3.5) implies that the first order rate of change of $u_t(x)$ is given by $v_t(x)$. Hence, the first order rate of change of g_t is a function of itself, g_0, and $v_t(x)$ which is a function of g_0, t, and $\Sigma(t)$ which is in turn a function of g_t and Σ_0. Thus, the first order rate of change of g_t is a function of t, g_t, g_0, and Σ_0. (All the results in this section are from [15].)

Theorem 3.1. *Taken together, equations (3.2)–(3.5) define a first order o.d.e. in t for $u_t(x)$ which has a solution which is Lipschitz in the t variable, C^1 in the x variable everywhere, and smooth in the x variable outside $\Sigma(t)$. Furthermore, $\Sigma(t)$ is a smooth, strictly outer minimizing horizon in (M^3, g_t) for all $t \geq 0$, and $\Sigma(t_2)$ encloses but does not touch $\Sigma(t_1)$ for all $t_2 > t_1 \geq 0$.*

Since $v_t(x)$ is a superharmonic function in (M^3, g_0) (harmonic everywhere except on $\Sigma(t)$, where it is weakly superharmonic), it follows that $u_t(x)$ is superharmonic as well. Thus, from equation (3.5) we see that $\lim_{x \to \infty} u_t(x) = e^{-t}$ and consequently that $u_t(x) > 0$ for all t by the maximum principle. Then since

$$R(g_t) = u_t(x)^{-5}(-8\Delta_{g_0} + R(g_0))u_t(x) \qquad (3.6)$$

it follows that (M^3, g_t) is an asymptotically flat manifold with nonnegative scalar curvature.

Even so, it still may not seem like g_t is particularly naturally defined since the rate of change of g_t appears to depend on t and the original metric g_0 in equation (3.4). We would prefer a flow where the rate of change of g_t can be defined purely as a function of g_t (and Σ_0 perhaps), and interestingly enough this actually does turn out to be the case. In [15] we prove this very important fact and define a new equivalence class of metrics called the harmonic conformal class. Then once we decide to find a flow of metrics which stays inside the harmonic conformal class of the original metric (outside the horizon) and keeps the area of the horizon $\Sigma(t)$ constant, then we are basically forced to choose the particular conformal flow of metrics defined above.

Theorem 3.2. *The function $A(t)$ is constant in t and $m(t)$ is non-increasing in t, for all $t \geq 0$.*

The fact that $A'(t) = 0$ follows from the fact that to first order the metric is not changing on $\Sigma(t)$ (since $v_t(x) = 0$ there) and from the fact that to first order the area of $\Sigma(t)$ does not change as it moves outward since $\Sigma(t)$ is a critical point for area in (M^3, g_t). Hence, the interesting part of theorem (3.2) is proving that $m'(t) \leq 0$. Curiously, this follows from a nice trick using the Riemannian positive mass theorem.

Another important aspect of this conformal flow of the metric is that outside the horizon $\Sigma(t)$, the manifold (M^3, g_t) becomes more and more spherically symmetric and "approaches" a Schwarzschild manifold $(\mathbb{R}^3 \backslash \{0\}, s)$ in the limit as t goes to ∞. More precisely,

Theorem 3.3. *For sufficiently large t, there exists a diffeomorphism ϕ_t between (M^3, g_t) outside the horizon $\Sigma(t)$ and a fixed Schwarzschild manifold $(\mathbb{R}^3 \backslash \{0\}, s)$ outside its horizon. Furthermore, for all $\epsilon > 0$, there exists a T such that for all $t > T$, the metrics g_t and $\phi_t^*(s)$ (when determining the lengths of unit vectors of (M^3, g_t)) are within ϵ of each other and the total masses of the two manifolds are within ϵ of each other. Hence,*

$$\lim_{t \to \infty} \frac{m(t)}{\sqrt{A(t)}} = \sqrt{\frac{1}{16\pi}}.$$

Theorem 3.3 is not that surprising really although a careful proof is reasonably long. However, if one is willing to believe that the flow of metrics converges to a spherically symmetric metric outside the horizon, then theorem 3.3 follows from two facts. The first fact is that the scalar curvature of (M^3, g_t) eventually becomes identically zero outside the horizon $\Sigma(t)$ (assuming (M^3, g_0) is harmonically flat). This follows from the facts that $\Sigma(t)$ encloses any compact set in a finite amount of time, that harmonically flat manifolds have zero scalar curvature outside a compact set, that $u_t(x)$ is harmonic outside $\Sigma(t)$, and equation (3.6). The second fact is that the Schwarzschild metrics are the only complete, spherically symmetric 3-manifolds with zero scalar curvature (except for the flat metric on R^3).

The Riemannian Penrose inequality, inequality (1.12), then follows from equation (3.1) using theorems 3.1, 3.2 and 3.3, for harmonically flat manifolds [15]. Since asymptotically flat manifolds can be approximated arbitrarily well by harmonically flat manifolds while changing the relevant quantities arbitrarily little, the asymptotically flat case also follows. Finally, the case of equality of the Penrose inequality follows from a more careful analysis of these same arguments.

We refer the reader to [16, 17, 20] for further review-type discussions of the results described above.

4. Open questions and applications

Now that the Riemannian Penrose conjecture has been proved, what are the next interesting directions? What applications can be found? Is this subject only of physical interest, or are there possibly broader applications to other problems in mathematics?

Clearly the most natural open problem is to find a way to prove the general Penrose conjecture (discussed in the next subsection) in which M^3 is allowed to have any second fundamental form in the space-time. There is good reason to think that this may follow from the Riemannian Penrose inequality, although this is a bit delicate. On the other hand, the general positive mass theorem followed from the Riemannian positive mass theorem as was originally shown by Schoen and Yau using an idea due to Jang [71,73]. For physicists this problem is definitely a top priority since most space-times do not even admit zero second fundamental form space-like slices. We note that the Riemannian Penrose inequality does give a result which applies to situations more general than time symmetric, as the condition $R \geq 0$ holds, *e.g.*, for maximal initial data sets $\text{tr}_g h = 0$, as well as in several other situations ("polar gauge", and so on). However, the general situation remains open.

Another interesting question is to ask these same questions in higher dimensions. One of us (HB) is currently working on a paper to prove the Riemannian Penrose inequality in dimensions less than 8. Dimension 8 and higher are harder because of the surprising fact that minimal hypersurfaces (and hence apparent horizons of black holes) can have codimension 7 singularities (points where the hypersurface is not smooth). This curious technicality is also the reason that the positive mass theorem in dimensions 8 and higher for manifolds which are not spin has only been announced very recently by Christ and Lohkamp [22], using a formidable singularity excision argument, and it is conceivable that this technique will allow one to extend the Riemannian Penrose inequality proof to all dimensions.

Naturally it is harder to tell what the applications of these techniques might be to other problems, but already there have been some. One application is to the famous Yamabe problem: Given a compact 3-manifold M^3, define $E(g) = \int_{M^3} R_g dV_g$ where g is scaled so that the total volume of (M^3, g) is one, R_g is the scalar curvature at each point, and dV_g is the volume form. An idea due to Yamabe was to try to construct canonical metrics on M^3 by finding critical points of this energy functional on the space of metrics. Define $C(g)$ to be the infimum of $E(\bar{g})$ over all metrics $\bar{g}$ conformal to g. Then the (smooth) Yamabe invariant of M^3, denoted here as $Y(M^3)$, is defined to be the supremum of $C(g)$ over all metrics g. $Y(S^3) = 6 \cdot (2\pi^2)^{2/3} \equiv Y_1$ is known to be the largest possible value for Yamabe invariants of 3-manifolds. It is also known that $Y(T^3) = 0$ and $Y(S^2 \times S^1) = Y_1 = Y(S^2 \tilde{\times} S^1)$, where $S^2 \tilde{\times} S^1$ is the non-orientable S^2 bundle over S^1.

One of the authors (HB) and Andre Neves, working on a problem suggested by Richard Schoen, were able to compute the Yamabe invariant of RP^3 using inverse mean curvature flow techniques [19] (see also [13, Lecture 2]) and found that $Y(RP^3) = Y_1/2^{2/3} \equiv Y_2$. A corollary is $Y(RP^2 \times S^1) = Y_2$ as well. These techniques also yield the surprisingly strong result that the only prime 3-manifolds with Yamabe invariant larger than RP^3 are S^3, $S^2 \times S^1$, and $S^2 \tilde{\times} S^1$. The Poincaré conjecture for 3-manifolds with Yamabe invariant greater than RP^3 is therefore a corollary. Furthermore, the problem of classifying 3-manifolds is known to reduce to the problem of classifying prime 3-manifolds. The Yamabe approach then would be to make a list of prime 3-manifolds ordered by Y. The first five prime 3-manifolds on this list are therefore S^3, $S^2 \times S^1$, $S^2 \tilde{\times} S^1$, RP^3, and $RP^2 \times S^1$.

4.1. The Riemannian Penrose conjecture on asymptotically hyperbolic manifolds

Another natural class of metrics that are of interest in general relativity consists of metrics which asymptote to the hyperbolic metric. Such metrics arise when considering solutions with a negative cosmological constant, or when considering "hyperboloidal hypersurfaces" in space-times which are asymptotically flat in isotropic directions (technically speaking, these are spacelike hypersurfaces which intersect $\mathscr{I}$ transversally in the conformally completed space-time). For instance, recall that in the presence of a cosmological constant Λ the scalar constraint equation reads

$$R = 16\pi\mu + |h|_g^2 - (\operatorname{tr}_g h)^2 + 2\Lambda .$$

Suppose that $h = \lambda g$, where λ is a constant; such an h solves the vector constraint equation. We then have

$$R = 16\pi\mu - 6\lambda^2 + 2\Lambda =: 16\pi\mu + 2\Theta . \tag{4.1}$$

The constant Θ equals thus Λ when $\lambda = 0$, or $-3\lambda^2$ when $\Lambda = 0$. The positive energy condition $\mu \geq 0$ is now equivalent to

$$R \geq 2\Theta .$$

For $\lambda = 0$ the associated model space-time metrics take the form

$$ds^2 = -(k - \frac{2m}{r} - \frac{\Lambda}{3}r^2)dt^2 + (k - \frac{2m}{r} - \frac{\Lambda}{3}r^2)^{-1}dr^2 + r^2 d\Omega_k^2 , \quad k = 0, \pm 1 , \tag{4.2}$$

where $d\Omega_k^2$ denotes a metric of constant Gauss curvature k on a two-dimensional compact manifold M^2. These are well known static solutions of the vacuum Einstein equation with a cosmological constant Λ; some subclasses of (4.2) have been discovered by de Sitter [74] ((4.2) with $m = 0$ and $k = 1$), by Kottler [62] (Equation (4.2) with an arbitrary m and $k = 1$). The parameter $m \in \mathbb{R}$ can be seen to be proportional to the total Hawking mass (cf. (4.5) below) of the foliation $t = $ const, $r = $ const. We will refer to those solutions as the generalized Kottler solutions. The constant Λ in (4.2) is an arbitrary real number, but in this section we will only consider $\Lambda < 0$.

From now on the overall approach resembles closely that for asymptotically flat space-times, as described earlier in this work. For instance, one considers manifolds which contain asymptotic ends diffeomorphic to $\mathbb{R}^+ \times M^2$. It is convenient to think of each of the sets "$\{r = \infty\} \times M^2$" as a connected component at infinity of a boundary at infinity, call it $\partial_\infty M^3$, of the initial data surface M^3. There is a well defined notion of mass for metrics which asymptote to the above model metrics in the asymptotic ends, somewhat similar to that in (1.5). In the hyperbolic case the boundary conditions are considerably more delicate to formulate as compared to the asymptotically flat one, and we refer the reader to [30–32,79] for details. In the case when M^3 arises from a space-times with negative cosmological constant Λ, the resulting mass is usually called the Abbott-Deser mass [1]; when $\Lambda = 0$ and M^3 is a hyperboloidal hypersurface the associated mass is called the Trautman-Bondi mass. (The latter notion of mass has often been referred to as "Bondi mass" in the literature, but the name "Trautman-Bondi mass" seems more appropriate, in view of the work in [77], which precedes [12] by four years; see also [76].) A large class of initial data sets with the desired asymptotic behavior has been constructed in [2, 3, 61], and the existence of the associated space-times has been established in [40, 41].

The monotonicity argument of Geroch [43], described in Section 1.7, has been extended by Gibbons [45] to accommodate for the negative cosmological constant; we follow the presentation in [33]: We assume that we are given a three-dimensional manifold (M^3, g) with connected minimal boundary ∂M^3 such that

$$R \geq 2\Theta \, ,$$

for some strictly negative constant Θ (compare (4.1)). We further assume that there exists a smooth, global solution of the inverse mean curvature flow without critical points, with u ranging from zero to infinity, vanishing on ∂M^3, with the level sets of u

$$\Sigma(s) = \{u(x) = s\}$$

being compact. Let A_s denote the area of $\Sigma(s)$, and define

$$\sigma(s) = \sqrt{A_s} \int_{\Sigma(s)} (^2\mathcal{R}_s - \frac{1}{2}H_s^2 - \frac{2}{3}\Theta)d^2\mu_s \, , \tag{4.3}$$

where $^2\mathcal{R}_s$ is the scalar curvature (equal twice the Gauss curvature) of the metric induced on $\Sigma(s)$, $d^2\mu_s$ is the Riemannian volume element associated to that same metric, and H_s is the mean curvature of $\Sigma(s)$. The hypothesis that du is nowhere vanishing implies that all the objects involved are smooth in s. At $s = 0$ we have $H_0 = 0$ and $A_0 = A_{\partial M^3}$ so that

$$\begin{aligned} \sigma(0) &= \sqrt{A_{\partial M^3}} \int_{\partial M^3} (^2\mathcal{R}_0 - \frac{2}{3}\Theta)d^2\mu_0 \\ &= \sqrt{A_{\partial M^3}} \left(8\pi(1 - g_{\partial M^3}) - \frac{2}{3}\Theta A_{\partial M^3} \right) \, . \end{aligned} \tag{4.4}$$

Generalizing a formula of Hawking [47], Gibbons [45, Equation (17)] assigns to the $\Sigma(s)$ foliation a *total mass* M_{Haw} via the formula

$$M_{Haw} \equiv \lim_{\epsilon \to 0} \frac{\sqrt{A_{1/\epsilon}}}{32\pi^{3/2}} \int_{\{u=1/\epsilon\}} (^2\mathcal{R}_s - \frac{1}{2}H_s^2 - \frac{2}{3}\Theta)d^2\mu_s \, , \tag{4.5}$$

where A_α is the area of the connected component under consideration of the level set $\{u = \alpha\}$. It follows that

$$\lim_{s \to \infty} \sigma(s) = 32\pi^{3/2}M_{Haw} \, ,$$

assuming the limit exists. The generalization in [45] of (1.13) establishes the inequality

$$\frac{\partial \sigma}{\partial s} \geq 0 \, . \tag{4.6}$$

This implies $\lim_{s\to\infty} \sigma(s) \geq \sigma(0)$, which gives

$$2M_{Haw} \geq (1 - g_{\partial M^3})\left(\frac{A_{\partial M^3}}{4\pi}\right)^{1/2} - \frac{\Theta}{3}\left(\frac{A_{\partial M^3}}{4\pi}\right)^{3/2}. \qquad (4.7)$$

Here $A_{\partial M^3}$ is the area of ∂M^3 and $g_{\partial M^3}$ is the genus thereof. Equation (4.7) is sharp – the inequality there becomes an equality for the generalized Kottler metrics (4.2).

The hypothesis above that du has no critical points together with our hypothesis on the geometry of the asymptotic ends forces ∂M^3 to be connected. It is not entirely clear what is the right generalization of this inequality to the case where several black holes occur, with one possibility being

$$2M_{Haw} \geq \sum_{i=1}^{k}\left((1 - g_{\partial_i M^3})\left(\frac{A_{\partial_i M^3}}{4\pi}\right)^{1/2} - \frac{\Theta}{3}\left(\frac{A_{\partial_i M^3}}{4\pi}\right)^{3/2}\right). \qquad (4.8)$$

Here the $\partial_i M^3$'s, $i = 1, \ldots, k$, are the connected components of ∂M^3, $A_{\partial_i M^3}$ is the area of $\partial_i M^3$, and $g_{\partial_i M^3}$ is the genus thereof. This would be the inequality one would obtain from the Geroch–Gibbons argument if it could be carried through for u's which are allowed to have critical points, on manifolds with $\partial_\infty M^3$ connected but ∂M^3 – not connected.

As in the asymptotically flat case, the naive monotonicity calculation of [43] breaks down at critical level sets of u, as those do not have to be smooth submanifolds. Nevertheless the existence of the appropriate function u (perhaps with critical points) should probably follow from the results in [55, 56]. The open questions here are 1) a proof of monotonicity at jumps of the flow, where topology change might occur, and 2) the proof that the Hawking mass (4.5) exists, and equals the mass of the end under consideration. We also note that in the hyperbolic context it is natural to consider not only boundaries ∂M^3 which are minimal, but also boundaries satisfying

$$H = \pm 2.$$

This is related to the discussion at the beginning of this section: if $\lambda = 0$, then an apparent horizon corresponds to $H = 0$; if $\Lambda = 0$ and $\lambda = -1$, then a future apparent horizon corresponds to $H = 2$, while a past apparent horizon corresponds to $H = -2$.

Let us discuss some of the consequences of the (hypothetical) inequality (4.8). In the current setting there are some genus-related ambiguities in the definition of mass (see [33] for a detailed discussion of various notions of mass for static asymptotically hyperbolic metrics), and it is convenient to introduce a mass parameter m defined as follows

$$m = \begin{cases} M_{Haw}, & \partial_\infty M^3 = S^2, \\ M_{Haw}, & \partial_\infty M^3 = T^2, \text{ with the normalization } A'_\infty = -12\pi/\Theta, \\ \dfrac{M_{Haw}}{|g_{\partial_\infty M^3} - 1|^{3/2}}, & g_{\partial_\infty M^3} > 1. \end{cases} \qquad (4.9)$$

Here A'_∞ is the area of $\partial_\infty M^3$ in the metric $d\Omega_k^2$ appearing in (4.2). For generalized Kottler metrics the mass m so defined coincides with the mass parameter appearing in (4.2) when u is the "radial" solution $u = u(r)$ of the inverse mean curvature flow.

Note, first, that if all connected components of the horizon have spherical or toroidal topology, then the lower bound (4.8) is strictly positive. For example, if $\partial M^3 = T^2$, and $\partial_\infty M^3 = T^2$ as well we obtain

$$2m \geq -\frac{\Lambda}{3}\left(\frac{A_{\partial M^3}}{4\pi}\right)^{3/2}.$$

On the other hand if $\partial M^3 = T^2$ but $g_{\partial_\infty M^3} > 1$ from Equation (4.8) one obtains

$$2m \geq -\frac{\Lambda}{3|g_\infty - 1|}\left(\frac{A_{\partial M^3}}{4\pi}\right)^{3/2}.$$

Recall that in a large class of space-times[4] the Galloway–Schleich–Witt–Woolgar inequality [42] holds:

$$\sum_{i=1}^{k} g_{\partial_i M^3} \leq g_\infty. \tag{4.10}$$

It implies that if $\partial_\infty M^3$ has spherical topology, then all connected components of the horizon must be spheres. Similarly, if $\partial_\infty M^3$ is a torus, then all components of the horizon are spheres, except perhaps for at most one which could be a torus. It follows that to have a component of the horizon which has genus higher than one we need $g_\infty > 1$ as well.

When some – or all – connected components of the horizon have genus higher than one, the right-hand side of Equation (4.8) might become negative. Minimizing the generalized Penrose inequality (4.8) with respect to the areas of the horizons gives the following interesting inequality

$$M_{Haw} \geq -\frac{1}{3\sqrt{-\Lambda}}\sum_i |g_{\partial_i M^3} - 1|^{3/2}, \tag{4.11}$$

where the sum is over those connected components $\partial_i M^3$ of ∂M^3 for which $g_{\partial_i M^3} \geq 1$. Equation (4.11), together with the elementary inequality

$$\sum_{i=1}^{N} |\lambda_i|^{3/2} \leq \left(\sum_{i=1}^{N} |\lambda_i|\right)^{3/2},$$

---

[4]The discussion that follows applies to all (M^3, g, h)'s that can be isometrically embedded into a globally hyperbolic space-time $\mathcal{M}$ (with timelike conformal boundary at infinity) in which the null convergence condition holds; further the closure of the image of M^3 should be a partial Cauchy surface in $\mathcal{M}$. Finally the intersection of the closure of M^3 with $\mathcal{I}$ should be compact. The global hyperbolicity here, and the notion of Cauchy surfaces, is understood in the sense of manifolds with boundary, see [42] for details.

lead to

$$m \geq -\frac{1}{3\sqrt{-\Lambda}} \, . \tag{4.12}$$

Similarly to the asymptotically flat case, the Geroch–Gibbons argument establishing the inequality (4.4) when a suitable u exists can also be carried through when $\partial M^3 = \emptyset$. In this case one still considers solutions u of the differential equation (2.1) associated with the inverse mean curvature flow, however the Dirichlet condition on u at ∂M^3 is replaced by a condition on the behavior of u near some chosen point $p_0 \in M^3$. If the level set of u around p_0 approach distance spheres centred at p_0 at a suitable rate, then $\sigma(s)$ tends to zero when the $\Sigma(s)$'s shrink to p_0, which together with the monotonicity of σ leads to the positive energy inequality:

$$M_{Haw} \geq 0 \, . \tag{4.13}$$

It should be emphasized that the Horowitz-Myers solutions [54] with negative mass show that this argument breaks down when $g_\infty = 1$.

When $\partial_\infty M^3 = S^2$ the inequality (4.13), with M_{Haw} replaced by the Hamiltonian mass (which might perhaps coincide with M_{Haw}, but this remains to be established), can be proved by Witten type techniques [30,31] (compare [4,46,79,81]). On the other hand it follows from [10] that when $\partial_\infty M^3 \neq S^2$ there exist no asymptotically covariantly constant spinors which can be used in the Witten argument. The Geroch–Gibbons argument has a lot of "ifs" attached in this context, in particular if $\partial_\infty M^3 \neq S^2$ then some level sets of u are necessarily critical and it is not clear what happens with σ at jumps of topology. We note that the area of the horizons does not occur in (4.12) which, when $g_{\partial_\infty M^3} > 1$, suggests that the correct inequality is actually (4.12) rather than (4.13), whether or not black holes are present.

We close this section by mentioning an application of the hyperbolic Penrose inequality to the uniqueness of static regular black holes with a negative cosmological constant, pointed out in [33]. It is proved in that last reference that for such connected black holes an inequality *inverse* to (4.7) holds, with equality if and only if the metric is the one in (4.2). Hence a proof of the Penrose inequality would imply equality in (4.7), and subsequently a uniqueness theorem for such black holes.

4.2. Precise formulations of the (full) Penrose conjecture

In the next two subsections we discuss formulations of the Penrose conjecture and possible applications of these statements to defining quasi-local mass functionals with good properties and to defining total mass in surprisingly large generality. This discussion is based on the third lecture [13] given by one of us (HB) in Cargèse in the summer of 2002. Besides discussing various formulations of the conjecture in this subsection, we point out the value of its possible applications in the next subsection, which greatly motivates trying to prove the conjecture.

We begin with the question, "Given Cauchy data, where is the event horizon, and what lower bounds on its area can we make?" Inequality (1.4) is the most general version of the Penrose conjecture, but there are more "local" versions of it which have the advantage of possibly being easier to prove. Recall, for instance, that the exact location of event horizons can not be determined from the Cauchy data (M^3, g, h) without solving the Einstein equations infinitely forward in time. On the other hand, apparent horizons Σ can be computed directly from the Cauchy data and are characterized by the equation

$$H_\Sigma = \mathrm{tr}_\Sigma(h), \qquad (4.14)$$

that is, the mean curvature H of Σ equals the trace of h along Σ. Note that in the $h = 0$ case, this is the assumption that $H = 0$, which is the Euler-Lagrange equation of a surface which locally minimizes area. This leads to the first formulation of the Penrose conjecture, which seems to be due to Gary Horowitz [53]:

Conjecture 4.1. *Let (M^3, g, h) be complete, asymptotically flat Cauchy data with $\mu \geq |J|$ and an apparent horizon satisfying equation (4.14). Then*

$$m \geq \sqrt{A/16\pi}, \qquad (4.15)$$

where m is the total mass and A is the minimum area required for a surface to enclose Σ.

The logic is that since apparent horizons imply the existence of an event horizon outside of it, and all surfaces enclosing Σ have at least area A, then inequality (1.4) implies the above conjecture.

An alternative possibility would be to replace A_e in (1.4) by the area of the apparent horizon. We do not know the answer to this, but a counterexample would not be terribly surprising (although it would be very interesting). The point is that the physical reasoning used by Penrose does not directly imply that such a conjecture should be true for apparent horizons. Hence, a counterexample to the area of the apparent horizon conjecture would be less interesting than a counterexample to Conjecture 4.1 or 4.2 since one of the latter counterexamples would imply that there was actually something wrong with Penrose's physical argument, which would be very important to understand.

There are also good reasons to consider a second formulation of the Penrose conjecture, due to one of the authors (HB), for (M^3, g, h) which have more than one end. We will choose one end to be special, and then note that large spheres S in the other asymptotically flat ends are actually "trapped," meaning that $H_S < \mathrm{tr}_S(h)$ (note that the mean curvatures of these spheres is actually negative when the outward direction is taken to be toward the special end and away from the other ends). We can conclude that these large spheres are trapped, if, for example, the mean curvatures of these large spheres is $-2/r$ to highest order and $|h|$ is decreasing like $1/r^2$ (or at least faster than $1/r$). Hence, this condition also allows us to conclude that there must be an event horizon enclosing all of the other ends. Thus, we conjecture

Conjecture 4.2. *Let* (M^3, g, h) *be complete, asymptotically flat Cauchy data with* $\mu \geq |J|$ *and more than one end. Choose one end to be special, and then define* A *to be the minimum area required to enclose all of the other ends. Then*

$$m \geq \sqrt{A/16\pi}, \qquad (4.16)$$

where m *is the total mass of the chosen end.*

We note that, more precisely, A in the above conjecture is the infimum of the boundary area of all smooth, open regions which contain all of the other ends (but not the special end). Equivalently (taking the complement), A is the infimum of the boundary area of all smooth, open regions which contain the special end (but none of the other ends). A smooth, compact, area-minimizing surface (possibly with multiple connected components) always exists and has zero mean curvature.

The advantage of this second formulation is that it removes equation (4.14) and the need to define apparent horizons. Also, preliminary thoughts by one of the authors (HB) lead him to believe that the above two formulations are equivalent via a reflection argument (although this still requires more consideration). In addition, this second formulation turns out to be most useful in the quasi-local mass and total mass definitions in the next subsection.

4.3. Applications to quasi-local mass and total mass

The ideas of this subsection are due to HB, and were greatly influenced by and in some cases are simply natural extensions of ideas due to Bartnik in [7, 8]. All of the surfaces we are considering in this subsection are required to be boundaries of regions which contain all of the other ends besides the chosen one. Given such a surface Σ in a (M^3, g, h) containing at least one asymptotically flat end, let I be the inside region (containing all of the other ends) and O be the outside region (containing a chosen end). Then we may consider "extensions" of (M^3, g, h) to be manifolds which result from replacing the *outside* region O in M with any other manifold and Cauchy data such that the resulting Cauchy data $(\tilde{M}^3, g, h)$ is smooth, asymptotically flat, and has $\mu \geq |J|$ everywhere (including along the surgery naturally). We define a "fill-in" of (M^3, g, h) to be manifolds which result from replacing the *inside* region I in M with any other manifold and Cauchy data such that the resulting Cauchy data $(\tilde{M}^3, g, h)$ is smooth, asymptotically flat, and has $\mu \geq |J|$ everywhere. Also, we say that a surface is "outer-minimizing" if any other surface which encloses it has at least as much area. Note that for "enclose" to make sense, we need to restrict our attention to surfaces which are the boundaries of regions as stated at the beginning of this paragraph. The notion of "outer-minimizing" surfaces turns out to be central to the following definitions.

Suppose Σ is outer-minimizing in (M^3, g, h). Define the *Bartnik outer mass* $m_{\text{outer}}(\Sigma)$ to be the infimum of the total mass over all extensions of (M^3, g, h) in which Σ remains outer-minimizing. Hence, what we are doing is fixing (M^3, g, h) inside Σ and then seeing how small we can make the total mass outside of Σ without violating $\mu \geq |J|$. Intuitively, whatever the total mass of this minimal mass extension outside Σ is can be interpreted as an upper bound for the mass contributed by the energy and momentum inside Σ.

The definition begs the question, why do we only consider extensions which keep Σ outer-minimizing? After all, we are attempting to find an extension with minimal mass, and one might naively think that the minimal mass extensions would naturally have this property anyway, and locally the minimal mass extensions we defined above probably usually do (if they exist). However, given any Σ, it is always possible to choose an extension which shrinks to a small neck outside Σ and then flattens out to an arbitrarily small mass Schwarzschild metric outside the small neck. Hence, without some restriction to rule out extensions with small necks, the infimum would always be zero. Bartnik's original solution to this problem was to not allow apparent horizons outside of Σ, and this works quite nicely. For technical reasons, however, we have chosen to preserve the "outer-minimizing" condition on Σ, which allows us to prove that $m_{\text{outer}}(\Sigma) \geq m_{\text{inner}}(\Sigma)$, defined in a moment. (Thus, the definition given here is not identical to that in [7], and we do not know whether or not it gives the same number as Bartnik's original definition, although this is a reasonable conjecture under many circumstances. We also note that the work of Huisken and Ilmanen [57] described above shows that, in the $h = 0$ case, the Hawking mass of Σ is a lower bound for the total mass if Σ is outer-minimizing. They also show that $m_{\text{outer}}(\Sigma) = m$ in the case that Σ is entirely outside the black hole of a time-symmetric slice of the Schwarzschild metric of total mass m. These results support considering the outer-minimizing condition in the current context.)

Suppose again that Σ is outer-minimizing in (M^3, g, h). Define the *inner mass* $m_{\text{inner}}(\Sigma)$ to be the supremum of $\sqrt{A/16\pi}$ over all fill-ins of (M^3, g, h), where A is the minimum area needed to enclose all of the other ends of the fill-in besides the chosen end. Hence, what we are doing is fixing (M^3, g, h) outside Σ (so that Σ automatically remains outer-minimizing) and then seeing how large we can make the area of the global area-minimizing surface (which encloses all of the other ends other than the chosen one). Intuitively, we are trying to fill-in Σ with the largest possible black hole, since the event horizon of the black hole will have to be at least A. If we think of $\sqrt{A/16\pi}$ as the mass of the black hole, then the inner mass gives a reasonable lower bound for the mass of Σ (since there is a fill-in in which it contains a black hole of that mass).

Theorem 4.3. *Suppose (M^3, g, h) is complete, asymptotically flat, and has $\mu \geq |J|$. Then Conjecture 4.2 implies that*

$$m_{\text{outer}}(\Sigma) \geq m_{\text{inner}}(\Sigma) \tag{4.17}$$

for all Σ which are outer-minimizing.

Sketch of proof. Consider any extension on the outside of Σ (which keeps Σ outer-minimizing) and any fill-in on the inside of Σ simultaneously and call the resulting manifold $\bar{M}$. Since Σ is outer-minimizing, there exists a globally area-minimizing surface of $\bar{M}$ which is enclosed by Σ (since going outside of Σ never decreases area). Thus, by Conjecture 4.2,

$$m \geq \sqrt{A/16\pi}, \tag{4.18}$$

for $\bar{M}$. Taking the infimum on the left side and the supremum on the right side of this inequality then proves the theorem since the total mass m is determined entirely by the extension and the global minimum area A is determined entirely by the fill-in. $\qquad\square$

Theorem 4.4. *Suppose (M^3, g, h) is complete, asymptotically flat, and has $\mu \geq |J|$. If Σ_2 encloses Σ_1 and both surfaces are outer-minimizing, then*

$$m_{\text{inner}}(\Sigma_2) \geq m_{\text{inner}}(\Sigma_1) \tag{4.19}$$

and

$$m_{\text{outer}}(\Sigma_2) \geq m_{\text{outer}}(\Sigma_1) \tag{4.20}$$

Sketch of proof. The first inequality is straightforward since every fill-in inside Σ_1 is also a fill-in inside Σ_2. The second inequality is almost as straightforward. It is true that any extension of Σ_2 (in which Σ_2 is still outer-minimizing) is also an extension of Σ_1, but it remains to be shown that such an extension preserves the outer-minimizing property of Σ_1. However, this fact follows from the fact that any surface enclosing Σ_1 which goes outside of Σ_2 can be made to have less or equal area by being entirely inside Σ_2 (by the outer-minimizing property of Σ_2). But since Σ_1 was outer-minimizing in the original manifold, any surface between Σ_1 and Σ_2 must have at least as much area as Σ_1. $\qquad\square$

The last three theorems inspire the definition of the *quasi-local mass* of a surface Σ in (M^3, g, h) to be the interval

$$m(\Sigma) \equiv [m_{\text{inner}}(\Sigma), m_{\text{outer}}(\Sigma)] \subset R. \tag{4.21}$$

That is, we are not defining the quasi-local mass of a surface to be a number, but instead to be an interval in the real number line. Both endpoints of this interval are increasing when we move outward to surfaces which enclose the original surface. If $\Sigma \subset (M^3, g, h)$ and (M^3, g, h) is Schwarzschild data, then this interval collapses to a point and equals the mass of the Schwarzschild data (assuming Conjecture 4.2). Conversely, if the quasilocal mass interval of Σ is a point, then we expect that Σ can be imbedded into a Schwarzschild spacetime in such a way that its Bartnik data (the metric, mean curvature vector in the normal bundle, and the connection on the normal bundle of Σ) is preserved, which is a nongeneric condition. Hence, we typically expect the quasi-local mass of a surface to be an interval of positive length. We also expect the quasi-local mass interval to be very close to a point in a "quasi-Newtonian" situation, where Σ is in the part of the space-time which is a perturbation of Minkowski space, for example. We point out that so far there are not any surfaces for which we can prove that the quasi-local mass is not a point. This is because there are very few instances in which the inner and outer masses of a surface can be computed at all. These questions will have to wait until a better understanding of the Penrose conjecture is found. This definition of quasi-local mass leads naturally to definitions of *total inner mass*, $m_{\text{inner}}^{\text{total}}$, and *total outer mass*, $m_{\text{outer}}^{\text{total}}$, where in both cases we simply take the supremum of inner mass and outer mass respectively over all Σ which are outer-minimizing.

Conjecture 4.5. *If (M^3, g, h) is asymptotically flat with total mass m_{ADM}, then*

$$m_{\text{inner}}^{\text{total}} = m_{\text{outer}}^{\text{total}} = m_{ADM}. \tag{4.22}$$

Consider (M^3, g, h) which is not assumed to have any asymptotics but still satisfies $\mu \geq |J|$. Then we will say that Σ (again, always assumed to be the boundary of a region in M^3) is "legal" if Σ is outer-minimizing in (M^3, g, h) and there exists an asymptotically flat extension with $\mu \geq |J|$ outside of Σ in which Σ remains outer-minimizing. Note that (M^3, g, h) is not assumed to have any asymptotics. We are simply defining the surfaces for which extensions with good asymptotics exist, and giving these surfaces the name "legal." Note also that both $m_{\text{inner}}(\Sigma)$ and $m_{\text{outer}}(\Sigma)$ are well defined for legal Σ. Thus, total inner mass and total outer mass are well defined as long as (M^3, g, h) has at least one legal Σ. Finally, theorems 4.3 and 4.4 are still true for legal surfaces even when (M^3, g, h) is not assumed to be asymptotically flat.

During the "50 Years" conference in Cargèse, Mark Aarons asked the question, "When are the total inner mass and the total outer mass different?" This is a very hard question, but it is such a good one that it deserves some speculation. If we define

$$m_{\text{total}} \equiv [m_{\text{inner}}^{\text{total}}, m_{\text{outer}}^{\text{total}}], \tag{4.23}$$

then total mass is very often well defined (as long as there is at least one legal Σ), but is not necessarily a single value. Mark's question is then equivalent to, "When is the total mass single-valued and therefore well defined as a real number?"

According to conjecture 4.5, we are not going to find an example of total inner mass $m_{\text{inner}}^{\text{total}} \neq m_{\text{outer}}^{\text{total}}$, the total outer mass, in the class of asymptotically flat manifolds. In fact, we are not aware of any examples of $m_{\text{inner}}^{\text{total}} \neq m_{\text{outer}}^{\text{total}}$, though one can give arguments to the effect that such situations could occur. On the other hand we believe that in reasonable situations this will not happen:

Conjecture 4.6. *Suppose (M^3, g, h) has $\mu \geq |J|$ and that there exists a nested sequence of connected, legal surfaces $\Sigma_i = \partial D_i \subset M$, $1 \leq i < \infty$, with $\bigcup_i D_i = M$ and $\lim |\Sigma_i| = \infty$. Then*

$$m_{\text{inner}}^{\text{total}} = m_{\text{outer}}^{\text{total}} \in \mathbb{R} \cup \{\infty\}. \tag{4.24}$$

At first this conjecture seems wildly optimistic considering it is suggesting that total mass is well defined in the extended real numbers practically all of the time, where the only assumptions we are making are along the lines of saying that the noncompact end must be "large" in some sense. Note, for example, we are ruling out cylindrical ends and certain types of cusp ends. However, the idea here is that most kinds of "crazy asymptotics" cause both $m_{\text{inner}}^{\text{total}}$ and $m_{\text{outer}}^{\text{total}}$ to diverge to infinity. Hence, the reason this conjecture (or one similar to it) has a decent chance of being true is the possibility that either m_{inner}^{total} or m_{outer}^{total} being finite is actually a very restrictive situation. For example, if either the total inner mass or the total outer mass is finite, then it might be true that this implies that (M^3, g, h) is asymptotic to data coming from a space-like slice of a Schwarzschild space-time (in some sense). In this case, one would expect that both the total inner and outer masses actually equal the mass of the Schwarzschild space-time and therefore are equal to each other. Certainly in the case that (M^3, g, h) is precisely a slice (even a very weird slice) of a Schwarzschild space-time, it is only natural to point out that total mass should be well defined. These definitions seem to be an approach to defining total mass in these more general settings. However, a complete understanding of these definitions clearly depends on making further progress studying the Penrose conjecture.

References

[1] L.F. Abbott and S. Deser, *Stability of gravity with a cosmological constant*, Nucl. Phys. **B195** (1982), 76–96.

[2] L. Andersson and P.T. Chruściel, *On asymptotic behavior of solutions of the constraint equations in general relativity with "hyperboloidal boundary conditions"*, Dissert. Math. **355** (1996), 1–100.

[3] L. Andersson, P.T. Chruściel, and H. Friedrich, *On the regularity of solutions to the Yamabe equation and the existence of smooth hyperboloidal initial data for Einsteins field equations*, Commun. Math. Phys. **149** (1992), 587–612.

[4] L. Andersson and M. Dahl, *Scalar curvature rigidity for asymptotically locally hyperbolic manifolds*, Annals of Global Anal. and Geom. **16** (1998), 1–27, dg-ga/9707017.

[5] R. Arnowitt, S. Deser, and C.W. Misner, *The dynamics of general relativity*, Gravitation (L. Witten, ed.), Wiley, N.Y., 1962, pp. 227–265, gr-qc/0405109.

[6] R. Bartnik, *The mass of an asymptotically flat manifold*, Comm. Pure Appl. Math. **39** (1986), 661–693.

[7] ———, *New definition of quasilocal mass*, Phys. Rev. Lett. **62** (1989), 2346–2348.

[8] R. Bartnik, *Energy in general relativity*, Tsing Hua Lectures on Geometry and Analysis (S.T. Yau, ed.), International Press, 1997, http://www.ise.canberra.edu.au/mathstat/StaffPages/Robert2.htm.

[9] R. Bartnik and P.T. Chruściel, *Boundary value problems for Dirac-type equations*, (2003), math.DG/0307278.

[10] H. Baum, *Complete Riemannian manifolds with imaginary Killing spinors*, Ann. Global Anal. Geom. **7** (1989), 205–226.

[11] G. Bergqvist, *On the Penrose inequality and the role of auxiliary spinor fields*, Class. Quantum Grav. 14 (1997), 2577–2583.

[12] H. Bondi, M.G.J. van der Burg, and A.W.K. Metzner, *Gravitational waves in general relativity VII: Waves from axi–symmetric isolated systems*, Proc. Roy. Soc. London A **269** (1962), 21–52.

[13] H.L. Bray, *Global inequalities*, lectures given at the Cargèse Summer School on 50 years of the Cauchy problem in general relativity, August 2002, online at `fanfreluche.math.univ-tours.fr`.

[14] _____, *The Penrose inequality in general relativity and volume comparison theorem involving scalar curvature*, Ph.D. thesis, Stanford University, 1997.

[15] _____, *Proof of the Riemannian Penrose conjecture using the positive mass theorem*, Jour. Diff. Geom. **59** (2001), 177–267, math.DG/9911173.

[16] _____, *Black holes and the Penrose inequality in general relativity*, Proceedings of the International Congress of Mathematicians, Vol. II (Beijing, 2002) (Beijing), Higher Ed. Press, 2002, pp. 257–271.

[17] _____, *Black Holes, Geometric Flows, and the Penrose Inequality in General Relativity*, Notices of the AMS **49** (2002), 1372–1381.

[18] H.L. Bray and K. Iga, *Superharmonic functions in $\mathbf{R}^n$ and the Penrose inequality in general relativity*, Comm. Anal. Geom. **10** (2002), 999–1016.

[19] H.L. Bray and A. Neves, *Classification of prime 3-manifolds with Yamabe invariant greater than RP^3*, Annals of Math. (2003), in press.

[20] H.L. Bray and R.M. Schoen, *Recent proofs of the Riemannian Penrose conjecture*, Current developments in mathematics, 1999 (Cambridge, MA), Int. Press, Somerville, MA, 1999, pp. 1–36.

[21] Y. Choquet-Bruhat, *Positive-energy theorems*, Relativity, groups and topology, II (Les Houches, 1983) (B.S. deWitt and R. Stora, eds.), North-Holland, Amsterdam, 1984, pp. 739–785.

[22] U. Christ and J. Lohkamp, in preparation (2003).

[23] D. Christodoulou and S.-T. Yau, *Some remarks on the quasi-local mass*, Cont. Math. **71** (1988), 9–14.

[24] P.T. Chruściel, *Black holes*, Proceedings of the Tübingen Workshop on the Conformal Structure of Space-times, H. Friedrich and J. Frauendiener, Eds., Springer Lecture Notes in Physics **604**, 61–102 (2002), gr-qc/0201053.

[25] _____, *A remark on the positive energy theorem*, Class. Quantum Grav. **33** (1986), L115–L121.

[26] _____, *Boundary conditions at spatial infinity from a Hamiltonian point of view*, Topological Properties and Global Structure of Space–Time (P. Bergmann and V. de Sabbata, eds.), Plenum Press, New York, 1986, pp. 49–59, URL `http://www.phys.univ-tours.fr/~piotr/scans`.

[27] _____, *On the invariant mass conjecture in general relativity*, Commun. Math. Phys. **120** (1988), 233–248.

[28] P.T. Chruściel and E. Delay, *On mapping properties of the general relativistic constraints operator in weighted function spaces, with applications*, Mém. Soc. Math. de France. **94** (2003), 1–103, gr-qc/0301073.

[29] P.T. Chruściel, E. Delay, G. Galloway, and R. Howard, *Regularity of horizons and the area theorem*, Annales Henri Poincaré **2** (2001), 109–178, gr-qc/0001003.

[30] P.T. Chruściel and M. Herzlich, *The mass of asymptotically hyperbolic Riemannian manifolds*, Pacific Jour. Math. **212** (2003), 231–264, dg-ga/0110035.

[31] P.T. Chruściel, J. Jezierski, and S. Łęski, *The Trautman-Bondi mass of hyperboloidal initial data sets*, (2003), gr-qc/0307109.

[32] P.T. Chruściel and G. Nagy, *The mass of spacelike hypersurfaces in asymptotically anti-de Sitter space-times*, Adv. Theor. Math. Phys. **5** (2002), 697–754, gr-qc/0110014.

[33] P.T. Chruściel and W. Simon, *Towards the classification of static vacuum spacetimes with negative cosmological constant*, Jour. Math. Phys. **42** (2001), 1779–1817, gr-qc/0004032.

[34] J. Corvino, *Scalar curvature deformation and a gluing construction for the Einstein constraint equations*, Commun. Math. Phys. **214** (2000), 137–189.

[35] J. Corvino and R. Schoen, *On the asymptotics for the vacuum Einstein constraint equations*, gr-qc/0301071, 2003.

[36] S. Dain, *Trapped surfaces as boundaries for the constraint equations*, (2003), gr-qc/0308009.

[37] V. I. Denisov and V. O. Solov'ev, *The energy determined in general relativity on the basis of the traditional Hamiltonian approach does not have physical meaning*, Theor. and Math. Phys. **56** (1983), 832–838, English translation, original pagination 301–314.

[38] P.A.M. Dirac, *The theory of gravitation in Hamiltonian form*, Proc. Roy. Soc. London **A246** (1958), 333–343.

[39] J. Frauendiener, *On the Penrose inequality*, Phys. Rev. Lett. **87** (2001), 101101, gr-qc/0105093.

[40] H. Friedrich, *Cauchy problem for the conformal vacuum field equations in general relativity*, Commun. Math. Phys. **91** (1983), 445–472.

[41] _____, *Einstein equations and conformal structure: Existence of anti-de-Sitter-type space-times*, Jour. Geom. and Phys. **17** (1995), 125–184.

[42] G.J. Galloway, K. Schleich, D.M. Witt, and E. Woolgar, *Topological censorship and higher genus black holes*, Phys. Rev. **D60** (1999), 104039, gr-qc/9902061.

[43] R. Geroch, *Energy extraction*, Ann. New York Acad. Sci. **224** (1973), 108–117.

[44] G.W. Gibbons, *Collapsing shells and the isoperimetric inequality for black holes*, Class. Quantum Grav. **14** (1997), 2905–2915, hep-th/9701049.

[45] _____, *Gravitational entropy and the inverse mean curvature flow*, Class. Quantum Grav. **16** (1999), 1677–1687.

[46] G.W. Gibbons, S.W. Hawking, G.T. Horowitz, and M.J. Perry, *Positive mass theorem for black holes*, Commun. Math. Phys. **88** (1983), 295–308.

[47] S.W. Hawking, *Gravitational radiation in an expanding universe*, Jour. Math. Phys. **9** (1968), 598–604.

[48] S.W. Hawking and G.F.R. Ellis, *The large scale structure of space-time*, Cambridge University Press, Cambridge, 1973.

[49] S.A. Hayward, *Quasilocalization of Bondi-Sachs energy loss*, Class. Quantum Grav. **11** (1994), 3037–3048, gr-qc/9405071.

[50] M. Herzlich, *A Penrose-like inequality for the mass of Riemannian asymptotically flat manifolds*, Commun. Math. Phys. **188** (1997), 121–133.

[51] _____, *The positive mass theorem for black holes revisited*, Jour. Geom. Phys. **26** (1998), 97–111.

[52] _____, *Minimal spheres, the Dirac operator and the Penrose inequality*, Séminaire Théorie Spectrale et Géométrie (Institut Fourier, Grenoble) **20** (2002), 9–16.

[53] G.T. Horowitz, *The positive energy theorem and its extensions*, Asymptotic behavior of mass and spacetime geometry (F. Flaherty, ed.), Springer Lecture Notes in Physics, vol. 202, Springer Verlag, New York, 1984.

[54] G.T. Horowitz and R.C. Myers, *The AdS/CFT correspondence and a new positive energy conjecture for general relativity*, Phys. Rev. **D59** (1999), 026005 (12 pp.).

[55] G. Huisken and T. Ilmanen, *A note on inverse mean curvature flow*, (1997), Proceedings of the Workshop on Nonlinear Partial Differential Equations (Saitama University, Sept. 1997), available from Saitama University.

[56] _____, *The Riemannian Penrose inequality*, Int. Math. Res. Not. **20** (1997), 1045–1058.

[57] _____, *The inverse mean curvature flow and the Riemannian Penrose inequality*, Jour. Diff. Geom. **59** (2001), 353–437, URL http://www.math.nwu.edu/~ilmanen.

[58] P.S. Jang and R.M. Wald, *The positive energy conjecture and the cosmic censor hypothesis*, J. Math. Phys. **18** (1977), 41–44.

[59] J. Jezierski, *Positivity of mass for certain space-times with horizons*, Class. Quantum Grav. **6** (1989), 1535–1539.

[60] _____, *Perturbation of initial data for spherically symmetric charged black hole and Penrose conjecture*, Acta Phys. Pol. B **25** (1994), 1413–1417.

[61] J. Kánnár, *Hyperboloidal initial data for the vacuum Einstein equations with cosmological constant*, Class. Quantum Grav. **13** (1996), 3075–3084.

[62] F. Kottler, *Über die physikalischen Grundlagen der Einsteinschen Gravitationstheorie*, Annalen der Physik **56** (1918), 401–462.

[63] M. Ludvigsen and J.A.G. Vickers, *An inequality relating the total mass and the area of a trapped surface in general relativity*, Jour. Phys. A: Math. Gen. **16** (1983), 3349–3353.

[64] E. Malec, M. Mars, and W. Simon, *On the Penrose inequality for general horizons*, Phys. Rev. Lett. **88** (2002), 121102, gr-qc/0201024.

[65] E. Malec and N. Ó Murchadha, *Trapped surfaces and the Penrose inequality in spherically symmetric geometries*, Phys. Rev. **D 49** (1994), 6931–6934.

[66] E. Malec and K. Roszkowski, *Comment on the Herzlich's proof of the Penrose inequality*, Acta Phys. Pol. **B29** (1998), 1975–1978, gr-qc/9806035.

[67] D. Maxwell, *Solutions of the Einstein constraint equations with apparent horizon boundary*, (2003), gr-qc/0307117.

[68] W.H. Meeks, III and S.T. Yau, *Topology of three-dimensional manifolds and the embedding problems in minimal surface theory*, Ann. of Math. (2) **112** (1980), 441–484.

[69] T. Parker and C. Taubes, *On Witten's proof of the positive energy theorem*, Commun. Math. Phys. **84** (1982), 223–238.

[70] R. Penrose, *Gravitational collapse – the role of general relativity*, Riv. del Nuovo Cim. (numero speziale) **1** (1969), 252–276.

[71] R. Schoen and S.-T. Yau, *Positivity of the total mass of a general space–time*, Phys. Rev. Lett. **43** (1979), 1457–1459.

[72] ———, *Proof of the positive mass theorem*, Comm. Math. Phys. **65** (1979), 45–76.

[73] ———, *Proof of the positive mass theorem II*, Comm. Math. Phys. **79** (1981), 231–260.

[74] W. De Sitter, *On the curvature of space*, Proc. Kon. Ned. Akad. Wet. **20** (1917), 229–243.

[75] K.P. Tod, *The hoop conjecture and the Gibbons-Penrose construction of trapped surfaces*, Class. Quantum Grav. **9** (1992), 1581–1591.

[76] A. Trautman, *King's College lecture notes on general relativity, May-June 1958*, mimeographed notes; reprinted in *Gen. Rel. Grav.* **34**, 721–762 (2002).

[77] ———, *Radiation and boundary conditions in the theory of gravitation*, Bull. Acad. Pol. Sci., Série sci. math., astr. et phys. **VI** (1958), 407–412.

[78] R.M. Wald, *General relativity*, University of Chicago Press, Chicago, 1984.

[79] X. Wang, *Mass for asymptotically hyperbolic manifolds*, Jour. Diff. Geom. **57** (2001), 273–299.

[80] E. Witten, *A simple proof of the positive energy theorem*, Commun. Math. Phys. **80** (1981), 381–402.

[81] X. Zhang, *A definition of total energy-momenta and the positive mass theorem on asymptotically hyperbolic 3 manifolds I*, (2001), preprint.

Hubert L. Bray
Mathematics Department, 2-179
Massachusetts Institute of Technology
77 Massachusetts Avenue
Cambridge, MA 02139, USA
e-mail: bray@math.mit.edu

Piotr T. Chruściel
Département de Mathématiques
Faculté des Sciences
Parc de Grandmont
F-37200 Tours, France
e-mail: piotr@gargan.math.univ-tours.fr
url: www.phys.univ-tours.fr/~piotr

The Global Existence Problem in General Relativity

Lars Andersson*

Abstract. We survey some known facts and open questions concerning the global properties of 3+1-dimensional spacetimes containing a compact Cauchy surface. We consider spacetimes with an ℓ-dimensional Lie algebra of space-like Killing fields. For each $\ell \leq 3$, we give some basic results and conjectures on global existence and cosmic censorship.

1. Introduction

In this review, we will describe some results and conjectures about the global structure of solutions to the Einstein equations in $3 + 1$ dimensions. We consider spacetimes $(\bar{M}, \bar{g})$ with an ℓ-dimensional Lie algebra of space-like Killing fields. We may say that such spacetimes have a (local) isometry group G of dimension ℓ with the action of G generated by space-like Killing fields.

For each value $\ell \leq 3$ of the dimension of the isometry group, we state the reduced field equations as well as attempt to give an overview of the most important results and conjectures. We will concentrate on the vacuum case.

In Section 2, we present the Einstein equations and give the $3 + 1$ decomposition into constraint and evolution equations, cf. Subsection 2.2. Due to the gauge freedom in the Einstein equations, questions on the global properties of solutions to the Einstein equations must be posed carefully. We introduce the notions of vacuum extension and maximal Cauchy development and state the uniqueness theorem of Choquet–Bruhat and Geroch, for maximal vacuum Cauchy developments. We also collect here a few basic facts about Killing fields on globally hyperbolic spacetimes. In Subsection 2.1 a version of the cosmic censorship conjecture appropriate for vacuum spacetimes, with compact Cauchy surface, is stated and in Subsection 2.3 we discuss a few gauge conditions which may be of use for the global

---

*) Supported in part by the Swedish Natural Sciences Research Council, contract no. F-FU 4873-307, and the NSF under contract no. DMS 0104402.

existence problem for the Einstein equations. Section 2 is ended by a discussion of a few PDE aspects of the Einstein equations which are relevant for the topic of this paper, cf. Subsection 2.4.

In the cases $\ell = 3$ (Bianchi, cf. Section 3) and a special case of $\ell = 2$ (polarized Gowdy, cf. Section 4), the global behavior of the Einstein equations is well understood, both with regard to global existence of the evolution equations and the cosmic censorship problem. For the general $\ell = 2$ case (local $U(1) \times U(1)$ symmetry, cf. Section 4), there are only partial results on the global existence problem and the cosmic censorship problem remains open, although conjectures supported by numerical evidence give a good indication of what the correct picture is.

In the cases $\ell = 1$ ($U(1)$ symmetry, cf. Section 5) and $\ell = 0$ (no symmetry, i.e., full $3 + 1$-dimensional Einstein equations, cf. Section 6), the large data global existence and cosmic censorship problems are open. In the $U(1)$ case conjectures supported by numerical evidence give a good idea of the generic behavior, and there is a small data semi-global existence result for the expanding direction due to Choquet–Bruhat and Moncrief [42, 39].

For $3+1$ Einstein gravity without symmetries the only global existence results known are the theorem on nonlinear stability of Minkowski space of Christodoulou and Klainerman, the semi-global existence theorem of Friedrich for the hyperboloidal initial value problem and the semi-global existence theorem for spatially compact spacetimes with Cauchy surface of hyperbolic type, due to Andersson and Moncrief. These are all small data results, see Section 6 for discussion.

Due to the high degree of complexity of the numerical solution of the Einstein equations in $3 + 1$ dimensions it is too early to draw any conclusions relevant to the asymptotic behavior at the singularity for the full $3 + 1$-dimensional Einstein equations, from the numerical studies being performed. However, an attractive scenario is given by the so-called BKL picture, cf. Section 7 for some remarks and references.

The Einstein equations are derived from a variational principle, and can be formulated as a Hamiltonian system (or time-dependent Hamiltonian system, depending on the gauge), and therefore the Hamiltonian aspect of the dynamics should not be ignored, see, e.g., the work by Fischer and Moncrief on the Hamiltonian reduction of the Einstein equations, [67] and references therein. In fact, the Hamiltonian point of view on the Einstein equations has played a vital role as motivation and guide in the development of the results discussed here. The notion that the Einstein evolution equations in terms of canonical variables can be viewed as the geodesic spray for a metric on the phase space (deWitt metric) modified by a curvature potential, is natural from the Hamiltonian point of view, and this picture has been relevant to the development of ideas on asymptotic velocity dominance, see Sections 4 and 5.

In this review, however, we will concentrate exclusively on the differential geometric and analytical point of view. Even with this restriction, many important topics are left out and we make no claim of complete coverage. See also [139] and [110] for related surveys.

2. The Einstein equations

Let $(\bar{M}, \bar{g})$ be a smooth 4-dimensional Lorentz manifold[1] of signature $-+++$. The Lorentzian metric $\bar{g}$ defines a causal structure on $\bar{M}$. For the convenience of the reader we give a quick review of the basic causality concepts in Appendix A, see [165, Chapter 8], [24, 84, 132] for details.

We will here consider only the vacuum case, i.e., the case when $\bar{g}$ is Ricci flat,

$$\bar{R}_{ab} = 0. \tag{2.1}$$

Let $M \subset \bar{M}$ be a space-like hypersurface, i.e., a hypersurface with time-like normal T. We let e_i be a frame on M and use indices i, j, k for the frame components. Let g, k be the induced metric and second fundamental form of $M \subset \bar{M}$, where $k_{ij} = \langle \bar{\nabla}_i e_j, T \rangle$. Let t be a time function on a neighborhood of M. Then we can introduce local coordinates $(t, x^i, i = 1, 2, 3)$ on $\bar{M}$ so that x^i are coordinates on the level sets M_t of t. This defines the coordinate vector field ∂_t of t. Alternatively we can let $M_t = \mathbf{i}(t, M)$ where $\mathbf{i} : \mathbb{R} \times M \to \bar{M}$ is a 1-parameter family of embeddings of an abstract 3-manifold M. Then $\partial_t = \mathbf{i}_* d/dt$ where d/dt is the coordinate derivative on $\mathbb{R}$.

Define the lapse N and shift X w.r.t. t by $\partial_t = NT + X$. Assume that T is future oriented so that $N > 0$. A $3+1$ split of equation (2.1) gives the Einstein vacuum constraint equations

$$R - |k|^2 + (\mathrm{tr}k)^2 = 0, \tag{2.2a}$$

$$\nabla_i \mathrm{tr}k - \nabla^j k_{ij} = 0, \tag{2.2b}$$

and the Einstein vacuum evolution equations

$$\mathcal{L}_{\partial_t} g_{ij} = -2N k_{ij} + \mathcal{L}_X g_{ij}, \tag{2.3a}$$

$$\mathcal{L}_{\partial_t} k_{ij} = -\nabla_i \nabla_j N + N(R_{ij} + \mathrm{tr}k k_{ij} - 2k_{im} k^m_{\ j}) + \mathcal{L}_X k_{ij}, \tag{2.3b}$$

where $\mathcal{L}_{\partial_t}$ denotes Lie-derivative[2] w.r.t. ∂_t. In case $[\partial_t, e_i] = 0$, $\mathcal{L}_{\partial_t}$ can be replaced by ∂_t.

A triple (M, g, k) consisting of a 3-manifold M, a Riemannian metric g on M and a symmetric covariant 2-tensor k is a **vacuum data set** for the Einstein equations if it solves (2.2).

Definition 2.1. Let (M, g, k) be a vacuum data set.

1. A vacuum spacetime $(\bar{M}, \bar{g})$ is called a **vacuum extension** of (M, g, k) if there there is an embedding $\mathbf{i}$ with time-like normal T of (M, g, k) into $(\bar{M}, \bar{g})$ so that $g = \mathbf{i}^* \bar{g}$ and $k = -\mathbf{i}^* (\bar{\nabla} T)$.

---

[1] We denote the covariant derivative and curvature tensors associated to $(\bar{M}, \bar{g})$ by $\bar{\nabla}, \bar{R}_{abcd}$ etc. All manifolds are assumed to be Hausdorff, second countable and C^∞, and all fields are assumed to be C^∞ unless otherwise stated.

[2] for a tensor b on M, with $b_{ij} = b(e_i, e_j)$, we have $\mathcal{L}_{\partial_t} b_{ij} = \partial_t(b_{ij}) - b([\partial_t, e_i], e_j) - b(e_i, [\partial_t, e_j])$.

2. *A globally hyperbolic vacuum spacetime* $(\bar{M}, \bar{g})$ *is called a* **vacuum Cauchy development** *of* (M, g, k) *if there is an embedding* **i** *with time-like normal T of* (M, g, k) *into* $(\bar{M}, \bar{g})$ *so that* $\mathbf{i}(M)$ *is a Cauchy surface in* $(\bar{M}, \bar{g})$, $g = \mathbf{i}^*\bar{g}$ *and* $k = -\mathbf{i}^*(\bar{\nabla}T)$. *If* $(\bar{M}, \bar{g})$ *is* maximal *in the class of vacuum Cauchy developments of* (M, g, k) *then* $(\bar{M}, \bar{g})$ *is called the* **maximal vacuum Cauchy development** *(MVCD) of* (M, g, k). *In the following, when convenient, we will identify M with* $\mathbf{i}(M)$.

The Einstein vacuum equations are not hyperbolic in any standard sense due to the coordinate invariance ("general covariance") of the equation $\bar{R}_{ab} = 0$. Nevertheless, the Cauchy problem for the Einstein vacuum equation is well posed in the following sense.

Theorem 2.2 (Choquet–Bruhat and Geroch [40]). *Let* (M, g, k) *be a vacuum data set. Then there is a unique, up to isometry, maximal vacuum Cauchy development (MVCD) of* (M, g, k). *If* $\phi : M \to M$ *is a diffeomorphism, the MVCD of* (M, ϕ^*g, ϕ^*k) *is isometric to the MVCD of* (M, g, k).

The proof relies on the fact that in spacetime harmonic coordinates, $\Box_{\bar{g}} x^\alpha = 0$, the Ricci tensor is of the form

$$\bar{R}^{(h)}_{\alpha\beta} = -\frac{1}{2}\Box_{\bar{g}}\bar{g}_{\alpha\beta} + S_{\alpha\beta}[\bar{g}, \partial\bar{g}], \tag{2.4}$$

where $\Box_{\bar{g}}$ is the scalar wave operator in $(\bar{M}, \bar{g})$. Hence the Einstein vacuum equations in spacetime harmonic coordinates is a quasi-linear hyperbolic system and therefore the Cauchy problem[3] for $\bar{R}^{(h)}_{\alpha\beta} = 0$ is well posed and standard results give local existence. One proves that if the constraints and gauge conditions are satisfied initially, they are preserved by the evolution. This together with a Zorn's lemma argument gives the existence of a MVCD. Uniqueness is proved using the field equations to get a contradiction to the Hausdorff property, given a pair of non-isometric vacuum Cauchy developments, which are both maximal w.r.t. the natural partial ordering on the class of Cauchy developments.

A spacetime $(\bar{M}, \bar{g})$ is said to satisfy the **time-like convergence condition** (or **strong energy condition**) if

$$\bar{R}_{ab}V^aV^b \geq 0, \quad \text{for all } V \text{ with } \bar{g}_{ab}V^aV^b \leq 0. \tag{2.5}$$

Globally hyperbolic spacetimes with compact Cauchy surface and satisfying the time-like convergence condition are often called "cosmological spacetimes" in the literature, following [23]. Here we will use the term spatially compact to refer to the existence of a compact Cauchy surface. A spacelike hypersurface (M, g) in $(\bar{M}, \bar{g})$ has constant mean curvature if $\nabla_i \mathrm{tr}_g k = 0$, cf. Subsection 2.3 below.

We end this subsection with by stating a few facts about Killing fields.

---

[3] Note that the Einstein equations in spacetime harmonic gauge should be viewed as an evolution equation for (g, k, N, X).

Proposition 2.3 ([68]). *Let M be a compact manifold and let (M, g, k) be a constant mean curvature vacuum data set on M with MVCD $(\bar{M}, \bar{g})$. Let Y be a Killing field on $(\bar{M}, \bar{g})$ and let $Y = Y_\perp T + Y_\parallel$ be the splitting of Y into its perpendicular and tangential parts at M. Then $(Y_\perp, Y_\parallel)$ satisfy the conditions*

1. *$Y_\perp = 0$, $\mathcal{L}_{Y_\parallel} g = 0$, $\mathcal{L}_{Y_\parallel} k = 0$, in case g is non-flat or $k \neq 0$.*
2. *$Y_\perp$ is constant and $\mathcal{L}_{Y_\parallel} g = 0$ if g is flat and $k = 0$.*

On the other hand, given $Y_\perp, Y_\parallel$ on M satisfying conditions 1, 2 above, there is a unique Killing field Y on $\bar{M}$, with $Y = Y_\perp T + Y_\parallel$ on M. □

Proposition 2.4. *Let $(\bar{M}, \bar{g})$ be a globally hyperbolic spacetime.*

1. *Assume that $(\bar{M}, \bar{g})$ satisfies the time-like convergence condition and contains a compact Cauchy surface M with constant mean curvature. Then either $(\bar{M}, \bar{g})$ is a metric product $M \times \mathbb{R}$ or any Killing field Y on $(\bar{M}, \bar{g})$ is tangent to M. In particular, if $(\bar{M}, \bar{g})$ is vacuum and has a nonzero Killing not tangent to M, then $(\bar{M}, \bar{g})$ is flat.*
2. *Assume a compact group G acts by isometries on $(\bar{M}, \bar{g})$. Then the action of G is generated by space-like Killing fields and $(\bar{M}, \bar{g})$ is foliated by Cauchy surfaces invariant under the action of G.*
3. *Assume that $(\bar{M}, \bar{g})$ is $3 + 1$-dimensional. Let M be a Cauchy surface in $\bar{M}$, let Y be a Killing field on $\bar{M}$ and assume Y is strictly spacelike, $\bar{g}(Y, Y) > 0$, on M. Then Y is strictly spacelike on $\bar{M}$.*

Proof. Point 1 is a well-known consequence of the uniqueness result for constant mean curvature hypersurfaces of Brill and Flaherty [35], cf. [124]. Point 2 is essentially [27, Lemma 1.1]. For the proof, note that as G is compact we can construct a G invariant time function on $\bar{M}$ by averaging any global time function t on $\bar{M}$ w.r.t. the G action, cf. the proof of [27, Lemma 1.1]. The level sets of the averaged time function are Cauchy surfaces and are invariant under the action of G. The result follows.

The following argument for point 3 is due to Alan Rendall[4]. Let $\mathcal{N} = \{p \in \bar{M} : \bar{g}(Y, Y) = 0\}$ and assume for a contradiction $\mathcal{N}$ is nonempty. Choose $p \in \mathcal{N}$ and a time function t on $\bar{M}$ so that $t(M) = 0$ and $t(p) > 0$. Let A denote the intersection of the past of p with the future of M and let $t_1 = \inf\{t(q) : q \in A \cap \mathcal{N}\}$. The set $\mathcal{N}$ is closed and by global hyperbolicity A is compact and hence $t_1 > 0$ and there is a $q \in A \cap \mathcal{N}$ with $t(q) = t_1$. If $Y(q)$ is nonzero and null, then using the equation $Y^a \nabla_a (Y^b Y_b) = 0$ which holds since Y is Killing, gives a null curve of points in $\mathcal{N}$ where Y is null. Following this into the past, shows that there is a $q \in \mathcal{N}$ with $t(q) < t_1$, which gives a contradiction. In case $Y(q) = 0$, the linearization of Y acts by isometries on $T_q \bar{M}$, and as the sphere of null directions in $T_q \bar{M}$ is two-dimensional it leaves a null direction fixed. Using the exponential map shows that the action of Y near q leaves a null geodesic invariant, along which Y must be zero or null. This leads to a contradiction as above. □

---

[4]Private communication, 1999.

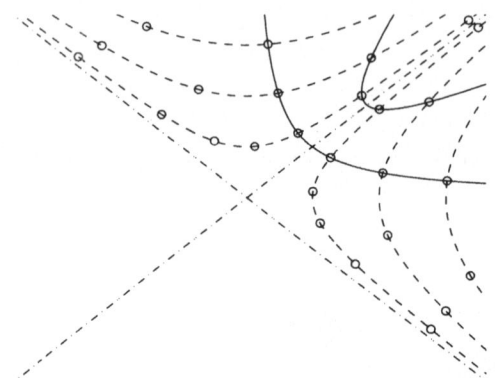

FIGURE 1. The $1+1$-dimensional Misner universe, showing a few
orbits of SO(1, 1) (dashed lines) and orbits of a discrete subgroup
$\Gamma \subset$ SO(1, 1) (circles). The boundary of a fundamental domain
for Γ is plotted with solid lines.

2.1. Cosmic censorship

Theorem 2.2 proves uniqueness of the MVCD of a given data set (M, g, k). However, examples show that the MVCD may fail to be maximal in the class of all
vacuum extensions, i.e., there exist examples of vacuum data sets (M, g, k) with
vacuum extensions $(\bar{M}, \bar{g})$ such that the MVCD of (M, g, k) is a strict subset of
$(\bar{M}, \bar{g})$.

Example 2.5. *Consider the $n + 1$-dimensional Minkowski space $\mathbb{R}^{n,1}$ with metric*
$\eta = -dt^2 + (dx^1)^2 + \cdots + (dx^n)^2$, *let $I^+(\{0\})$ be the interior of the future light cone.*
$I^+(\{0\})$ *is globally hyperbolic with the hyperboloids as Cauchy surfaces, and with*
the mantle of the light cone as Cauchy horizon. Let Γ be a cocompact discrete sub-
group of the Lorentz group SO(n, 1). Then the quotient space $\bar{M} = \Gamma \backslash I^+(\{0\})$ is a
globally hyperbolic, spatially compact spacetime. By choosing ρ to be the Lorentzian
distance from the origin, we get $\bar{g} = -d\rho^2 + \rho^2 \gamma$ where γ is the standard hyperbolic
metric on the compact quotient $M = \Gamma \backslash \mathbf{H}^n$. $(\bar{M}, \bar{g})$ is the MVCD of the vacuum
data set $(M, \gamma, -\gamma)$.

In case $n = 1$, $\mathbf{H}^1 = \mathbb{R}$, and a fundamental domain for Γ can be found
which intersects the null boundary of $I^+(\{0\})$ in an open interval. Therefore if
$n = 1$, there is a nontrivial extension of $\bar{M}$, which is still flat, but which fails to be
globally hyperbolic, cf. Figure 1. This spacetime is known as the Misner universe.
The maximal extension is unique in this case.

If $n > 1$, the ergodicity of the geodesic flow on (M, γ) can be shown to prevent
the existence of a spacetime extending $(\bar{M}, \bar{g})$ [104].

Higher-dimensional examples of flat globally hyperbolic spatially compact
spacetimes which admit nontrivial non-globally hyperbolic extensions can be con-
structed by taking products of the $n = 1$ Misner universe with the flat torus.

A maximal vacuum extension may be non-unique, as is shown by the Taub-NUT example, cf. [53]. If the MVCD of a vacuum data set is not a maximal vacuum extension, any extension of it must fail to satisfy the intuitively reasonable causality requirement of global hyperbolicity.

According to physical intuition, causality violations should be rare. This leads to the idea of cosmic censorship, essentially due to Penrose, see [133] for discussion. One way of stating this, relevant to the class of spacetimes we are concerned with here, is the following form of the strong cosmic censorship conjecture.

Conjecture 1 (Strong Cosmic Censorship). *Let M be a compact manifold of dimension 3. Then for generic vacuum data sets (M, g, k), the maximal vacuum Cauchy development of (M, g, k) is equal to the maximal vacuum extension of (M, g, k).*

In the case of asymptotically flat spacetimes (describing isolated systems in general relativity), the so-called weak cosmic censorship conjecture states that naked singularities (i.e., singularities which can be seen by an observer at infinity) should not occur generically, see the review paper by Wald [166] for a discussion of the status of the weak cosmic censorship conjecture. The work of Christodoulou, see [45] and references therein, see also the discussion in [166, §5], establishes weak cosmic censorship in the class of spherically symmetric Einstein-scalar field spacetimes, but also gives examples of initial data such that the Cauchy development has a naked singularity. For earlier surveys on the strong cosmic censorship conjecture, see [95] and [49].

The Penrose inequality, giving a lower bound on the ADM mass in terms of the area of a horizon in black hole spacetimes, was derived by a heuristic argument assuming the validity of the weak cosmic censorship conjecture. The proof of the Riemannian version of the Penrose inequality by Huisken and Ilmanen [91, 92], gives indirect support for the conjecture.

Let $(\bar{M}, \bar{g})$ be a spacetime and let $M \subset \bar{M}$ be a space-like hypersurface. The Cauchy horizon $H(M)$ is the boundary of the domain of dependence $D(M)$, cf. Appendix A for definition. If M is compact without boundary, then every point of $H^+(M)$ lies on a past inextendible null geodesic and every point of $H^-(M)$ lies on a future inextendible null geodesic, where $H^+(M)$ and $H^-(M)$ are the future and past components of $H(M)$, respectively.

Let $(\bar{M}, \bar{g})$ be a maximal vacuum extension of a vacuum data set (M, g, k) with M compact, and let $D(M) \subset \bar{M}$ be the MVCD of (M, g, k). If $D(M) \neq \bar{M}$, then the Cauchy horizon $H(M)$ is nonempty. One approach to SCC is to study the geometry of Cauchy horizons in vacuum spacetimes and to prove rigidity theorems as a consequence of extendibility of $D(M)$.

Isenberg and Moncrief proved for analytic vacuum or electrovac spacetimes, with analytic Cauchy horizon $H(M)$, that under the additional assumption that $H(M)$ is ruled by closed null geodesics, there is a nontrivial Killing field which extends to $D(M)$, see [100, 130]. This result was generalized to the C^∞ case by Friedrich et al. [72]. As spacetimes with Killing fields are non-generic, this may be viewed as supporting evidence for the SCC.

In the class of Bianchi spacetimes (i.e., spatially locally homogenous space-times, cf. Section 3), it has been proved by Chruściel and Rendall [55], generalizing work by Siklos [147] in the analytic case, that any C^∞ Bianchi spacetime which contains a compact locally homogeneous Cauchy horizon is a Taub spacetime, cf. Section 3, (3.7) for definition. This result may be viewed as a version of SCC in the class of Bianchi spacetimes. In this context, it is worth mentioning that work by Chruściel and Galloway [52] gives examples which indicate that Cauchy horizons may be non-differentiable, generically.

Conjecture 2 (Bartnik [23, Conjecture 2]**).** *Let* $(\bar{M}, \bar{g})$ *be a spatially compact glob-ally hyperbolic spacetime satisfying the time-like convergence condition* (2.5). *Then if* $(\bar{M}, \bar{g})$ *is time-like geodesically complete,* $(\bar{M}, \bar{g})$ *splits isometrically as a product* $(\mathbb{R} \times M, -dt^2 + g)$.

If the Bartnik conjecture 2 is true, then any vacuum, globally hyperbolic, spatially compact spacetime, is either flat and covered by $\mathbb{R} \times T^3$ or is has an inextendible time-like geodesic which ends after a finite proper time, i.e., it is time-like geodesically incomplete. A sequence of points approaching the "end" of a finite length inextendible geodesic is often thought of as approaching a singularity. See [76] for a discussion of the status of the Bartnik conjecture.

Inextendibility of $D(M)$ can be detected by monitoring the asymptotic be-havior of curvature invariants such as the Kretschmann scalar κ, defined by $\kappa = \bar{R}_{abcd}\bar{R}^{abcd}$. If κ blows up along causal geodesics, then $D(M)$ fails to be extendible, and therefore proving blowup for κ for generic spacetimes is an approach to prov-ing SCC. This is the method used in the proof of cosmic censorship for the class of polarized Gowdy spacetimes [54], cf. Section 4, and is likely to be important also in the cases with less symmetry. The structure of the horizon and extensions in the polarized Gowdy class can be very complicated as shown by the work of Chruściel et al., see [49] for discussion, see also [53]. It was proved by Ringström, cf. Theorem 3.1, that for vacuum Bianchi spacetimes of class A, either the spacetime is Taub, cf. Section 3, or κ blows up at the singularity.

2.2. The evolution equations

A solution to the vacuum Einstein evolution equations with initial data is a curve $t \mapsto (g(t), k(t), N(t), X(t))$ defined on some interval (T_0, T_1), satisfying (2.3).

Every regular solution (g, k, N, X) to the vacuum evolution equations (2.3) with initial data solving the vacuum constraint equations, gives a vacuum space-time. This is due to the fact that the constraint quantities

$$B = R + (\mathrm{tr}k)^2 - |k|^2$$

$$D_i = \nabla_i \mathrm{tr}k - \nabla^j k_{ij}$$

evolve according to a symmetric hyperbolic system and energy estimates together with an application of the Gronwall inequality allow one to show the constraints are satisfied during the time of existence of the solution curve. Now the fact that the Einstein vacuum equation is equivalent to the system of constraint and evolution equations shows that the spacetime $(\bar{M}, \bar{g})$ constructed from the curve (g, k, N, X) by letting $\bar{M} = (T_0, T_1) \times M$, and setting

$$\bar{g} = -N^2 dt^2 + g_{ij}(dx^i + X^i dt)(dx^j + X^j dt)$$

is a solution to the Einstein vacuum equations (2.1).

Note that in order for the solution to be well defined, it is necessary to specify the lapse and shift (N, X), either as functions on spacetime $M \times (T_0, T_1)$ or as functions of the data, $N = N[g, k], X = X[g, k]$; this may be viewed as a gauge fixing for the Einstein equations.

The choice of lapse and shift is crucial for the behavior of the solution curve. In particular, a foliation constructed for a particular choice of N, X may develop singularities which are not caused by any singular or irregular nature of the Cauchy development. Consider for example the Gauss foliation condition $N = 1, X = 0$. Then the hypersurface M flows in the direction of its unit normal and M_t is simply the level set of the Lorentzian distance function $t(p) = d(M, p)$. The foliation $\{M_t\}$ will develop singularities precisely at the focal set of M, which in general will be nonempty, even in Minkowski space.

Many authors have considered hyperbolic reformulations of the Einstein equations, see the paper by Friedrich [71] for discussion, see also [73, 93, 94] for related work. The development of singularities for hyperbolic systems presents a serious obstacle to the numerical treatment of the Einstein evolution equations using hyperbolic reformulations, see [1, 2] for discussion and examples. It is therefore necessary to consider also gauges which make (2.3) into an elliptic-hyperbolic system.

2.3. Constant mean curvature foliations

A particularly interesting choice of gauge condition for the lapse function is given by the constant mean curvature (CMC) condition

$$\nabla_i \operatorname{tr} k = 0,$$

i.e., the level sets M_t of the time function t are assumed to be hypersurfaces of constant mean curvature in $(\bar{M}, \bar{g})$. If $(\bar{M}, \bar{g})$ is globally hyperbolic, spatially compact and satisfies the time-like convergence condition (2.5), then for $\tau \in \mathbb{R}$, either there is at most one Cauchy surface with mean curvature $\operatorname{tr} k = \tau$ or $(\bar{M}, \bar{g})$ splits as a product, cf. [35].

This indicates that the mean curvature $\operatorname{tr} k$ may be useful as a time function on $(\bar{M}, \bar{g})$ in the spatially compact case. Setting $t = \operatorname{tr} k$ leads, using (2.2–2.3) to the lapse equation

$$-\Delta N + |k|^2 N = 1. \qquad (2.6)$$

Local well-posedness for the Einstein evolution equations in CMC and more general elliptic time gauges given by a lapse equation of the form $-\Delta N + |k|^2 N = NT(h)$ for given spacetime functions h has been proved by Choquet–Bruhat and York [44]. We may call this gauge the prescribed mean curvature (PMC) gauge. The proof is based on writing a wave equation for the second fundamental form, a technique which was also used in the work of Choquet-Bruhat and Ruggeri [43] on harmonic time gauge. The prescribed mean curvature condition, has been used in the $U(1) \times U(1)$ symmetric case by Henkel [85] who proved global existence in this gauge. An important advantage of using PMC gauge is that one avoids the problem of proving existence of a CMC Cauchy surface.

Equation (2.6) supplemented by an elliptic shift gauge makes the Einstein evolution equations (2.3) into an elliptic-hyperbolic system of evolution equations. Let $\hat{g}$ be a given smooth metric on M, and let $V^i = g^{mn}(\Gamma^i_{mn} - \hat{\Gamma}^i_{mn})$. Then $-V^i$ is the tension field of the identity map $\mathrm{Id} : (M, g) \to (M, \hat{g})$, and the CMC condition coupled with the spatially harmonic coordinate condition

$$V^i = 0$$

gives a new elliptic gauge condition for the Einstein equations, CMCSH gauge [10]. In [10], it is shown that the Einstein equations in CMCSH gauge forms an elliptic-hyperbolic system for the Einstein equations, which is well posed in $H^s \times H^{s-1}$, $s > n/2 + 1$.

The maximal slicing condition $\mathrm{tr}k = 0$ is of interest mainly for the asymptotically flat case. This was used in the proof of the nonlinear stability of Minkowski space by Christodoulou and Klainerman [46], cf. the discussion in Section 6. Due to the "collapse of the lapse" phenomenon, see [25], the maximal foliation is not expected to cover the whole MVCD except in the small data case. See [124] for a discussion of maximal slices. Asymptotically flat spacetimes satisfying certain restrictions on the causal structure are known to contain maximal hypersurfaces [21]. The existence of CMC hypersurfaces in asymptotically flat spacetimes was considered in [8].

The mean curvature operator satisfies a geometric maximum principle, see [7] for a proof of this under weak regularity. This allows one to use barriers to prove existence of constant mean curvature hypersurfaces. A spacetime is said to have **crushing singularities** if there are sequences of Cauchy surfaces with mean curvature $\mathrm{tr}k$ tending uniformly to $\pm\infty$. Gerhardt [78] proved, using a barrier argument, that any spacetime satisfying (2.5) with crushing singularities is globally foliated by CMC hypersurfaces. These facts indicate that the CMC foliation condition is an interesting time gauge for the Einstein evolution equation. See also [79, 65, 22, 23, 85] for results relevant to existence of CMC hypersurfaces.

Let $R[g]$ be the scalar curvature. A 3-manifold M is said to be of Yamabe type -1 if it admits no metric with $R[g] = 0$ (and hence no metric with nonnegative scalar curvature), of Yamabe type 0 if it admits a metric with $R[g] = 0$ but no metric with $R[g] = 1$ and of Yamabe type $+1$ if it admits a metric with $R[g] = 1$, cf. [67, Definition 9].

If the Cauchy surface M is of Yamabe type -1, it follows from the constraint equation that $(\bar{M}, \bar{g})$ cannot contain a maximal (i.e., $\mathrm{tr}k = 0$) Cauchy surface and therefore one expects that (if the dominant energy condition holds) the maximal time interval of existence for (2.3) in CMC time is of the form (after a time orientation) $(-\infty, 0)$ with $\tau \nearrow 0$ corresponding to infinite expansion.

If M is of Yamabe type 0, then one expects that either the maximal CMC time interval is $(-\infty, 0)$ (possibly after a change of time orientation) or $(\bar{M}, \bar{g})$ splits as a product, and therefore in the vacuum case is covered by $\mathbb{R} \times T^3$ with the flat metric. Finally in case M is of Yamabe type $+1$, one expects the maximal CMC time interval to be $(-\infty, \infty)$, i.e., the spacetime evolves from a "big bang" to a "big crunch". This is formalized in the "closed universe recollapse conjecture" of Barrow, Galloway and Tipler [20].

Conjecture 3 (Constant mean curvature foliations). *Let M be a compact 3-manifold and let (M, g, k) be a vacuum data set on M, with constant mean curvature. The Cauchy problem for the Einstein vacuum evolution equations with data (M, g, k) has global existence in the constant mean curvature time gauge, i.e., there is a CMC foliation in the MVCD $(\bar{M}, \bar{g})$ of (M, g, k), containing M, with mean curvature taking all values in $(-\infty, \infty)$ in case M has Yamabe type $+1$ and in case M has Yamabe type 0 or -1, taking all values in $(-\infty, 0)$ (possibly after a change of time orientation).*

Remark 2.1. *Conjecture 3 has been stated in essentially this form by Rendall [135, Conjecture 1], see also Eardley and Moncrief [129, Conjecture C2] for a closely related statement.*

Note that as $\tau \searrow -\infty$, the past focal distance of the (unique) CMC surface with mean curvature τ tends to zero, and hence the foliation exhausts the past of M. It follows that in case M has Yamabe type $+1$, then if Conjecture 3 is true, $(\bar{M}, \bar{g})$ is globally foliated by CMC hypersurfaces. In case M has Yamabe type 0 or -1 on the other hand, there is the possibility that the CMC foliation does not cover all of $\bar{M}$, due to the fact that as the mean curvature $\tau \nearrow 0$, the CMC hypersurfaces are expected to avoid black holes, by analogy with the behavior of CMC and maximal hypersurfaces in the Schwarzschild spacetime. See [135] for further remarks and conjectures related to this.

If one were able to prove Conjecture 3, then as remarked in [129], this would give the possibility of attacking the Cosmic Censorship Conjecture using PDE methods. There are no known counterexamples to Conjecture 3 for vacuum spacetimes. However, Isenberg and Rendall [103] give an example of dust spacetimes, not covered by a CMC foliation. Bartnik [23] gave an example of a spatially compact, globally hyperbolic spacetime satisfying the time-like convergence condition, which contains no CMC Cauchy surface. It is an interesting open question whether or not similar counter examples are possible in the vacuum case. The gluing results for solutions of the constraint equations, see [97, 98, 58] may lead to the construction of such examples.

The CMC conjecture 3 has been proved in a number of cases for spacetimes with symmetry, in the sense of the existence of a group G of isometries acting (locally) on $(\bar{M}, \bar{g})$ by isometries and with space-like orbits. In the case of Bianchi IX, cf. Section 3, which has Yamabe type $+1$, the closed universe recollapse conjecture and consequently Conjecture 3, was proved by Lin and Wald [121]. In the case of

2 + 1-dimensional vacuum spacetimes with cosmological constant, the conclusion of Conjecture 3 is valid [11]. In the 2+1 case, the CMC foliations are global.

The work of Rendall and Burnett, see [36] and references therein, proves under certain restrictions on the matter that a maximal, globally hyperbolic, spherically symmetric spacetime, which contains a CMC Cauchy surface diffeomorphic to $S^2 \times S^1$, is globally foliated by CMC hypersurfaces with mean curvature taking on all real values.

We end this subsection by mentioning the harmonic time gauge condition, $\Box_{\bar{g}} t = 0$ or

$$\bar{g}^{ab} \bar{\Gamma}^0_{ab} = 0$$

In case $X = 0$, this is equivalent to the condition $N = \sqrt{\det g}/\sqrt{\det e}$ where e_{ij} is some fixed Riemannian metric on M. The Einstein evolution equations with $X = 0$ were proved to be hyperbolic with this time gauge by Choquet–Bruhat and Ruggeri [43]. This time gauge also appears in the work of Gowdy and is used in the analysis of the Gowdy spacetimes as well as in the numerical work of Berger, Moncrief et al. on Gowdy and U(1) spacetimes, cf. Sections 4 and 5.

The papers of Smarr and York [148, 149] contain an interesting discussion of gauge conditions for the Einstein equations. See also Section 6 for a discussion of the spatial harmonic coordinate gauge and the survey of Klainerman and Nicolo [110] for further comments on gauges.

2.4. The Einstein equations as a system of quasi-linear PDE's

As mentioned above, the Einstein vacuum equations in spacetime harmonic coordinates form a quasi-linear hyperbolic system of the form

$$-\frac{1}{2}\Box_{\bar{g}} \bar{g}_{\alpha\beta} + S_{\alpha\beta}[\bar{g}, \partial\bar{g}] = 0. \tag{2.7}$$

The system (2.7) is a quasi-linear wave equation, quadratic in the first-order derivatives $\partial\bar{g}$ and with top-order symbol depending only on the field $\bar{g}$ itself. Standard results show that the Cauchy problem is well posed in Sobolev spaces $H^s \times H^{s-1}$, $s > n/2 + 1$. This was first proved for the Einstein equations by Hughes, Kato and Marsden [90]. It is also possible to prove this for elliptic-hyperbolic systems formed by the Einstein evolution equations together with the CMC-spatial harmonic coordinates gauge, see [10], see also Section 6 for a discussion of the spatial harmonic coordinates gauge.

Recent work using harmonic analysis methods by Bahouri and Chemin [18], Klainerman and Rodnianski [115, 114, 113] and Smith and Tataru [151], see also [157] has pushed the regularity needed for systems of the above type on $\mathbb{R}^{n,1}$ to $s > (n+1)/2$ for $n \geq 3$. In order to get well-posedness for s lower than the values given above, it is likely one needs to exploit some form of the null condition, [154, 152, 157]. The null condition for equations of the form $\Box_\eta u = F[u, \partial u]$ on Minkowski space states roughly that the symbol of the non-linearity F cancels null vectors. For a discussion of the null condition on a curved space background, see [153, 150]. Counter examples to well-posedness for quasi-linear wave equations with low regularity data are given by Lindblad, see [122].

The standard example of an equation which satisfies the null condition is $\Box u = \partial_\alpha u \partial_\beta u \eta^{\alpha\beta}$ where $\Box$ is the wave operator w.r.t. the Minkowski metric η on $\mathbb{R}^{n,1}$. This equation is well posed for data in $H^s \times H^{s-1}$ with $s > n/2$ [116], and has global existence for small data for $n \geq 3$. On the other hand, the equation $\Box u = (\partial_t u)^2$ which does not satisfy the null condition can be shown to have a finite time of existence for small data in the $3 + 1$-dimensional case.

For quasi-linear wave equations which satisfy an appropriate form of the null condition [89], global existence for small data is known in $3 + 1$ dimensions. The Einstein equations, however, are not known to satisfy the null condition in any gauge. In particular, it can be seen that in spacetime harmonic coordinates, the Einstein equations do not satisfy the null condition. However, the analysis by Blanchet and Damour [34] of the expansion of solutions of Einstein equations in perturbation series around Minkowski space indicates that the logarithmic terms in the gravitational field in spacetime harmonic coordinates, arising from the violation of the null condition, can be removed after a (nonlocal) gauge transformation to radiative coordinates where the coordinate change depends on the history of the field. A similar analysis can be done for the Yang–Mills (YM) equation in Lorentz gauge. It may further be argued that the small data, global existence proof of Christodoulou and Klainerman for the Einstein equations exploits properties of the Einstein equations related to the null condition. In a recent paper, Lindblad and Rodnianski [123] have introduced a weak form of the null condition and shown that the Einstein equations satisfy this.

Global existence is known for several of the classical field equations such as certain nonlinear Klein–Gordon (NLKG) equations and the YM equation on $\mathbb{R}^{3,1}$ (proved by Eardley and Moncrief, [63, 64]). The proofs for NLKG and the proof of Eardley and Moncrief for YM use light cone estimates to get a priori L^∞ bounds. The proof of Eardley and Moncrief used the special properties of YM in the radial gauge. This method was also used in the global existence proof for YM on $3 + 1$-dimensional, globally hyperbolic spacetimes by Chruściel and Shatah [56]. Klainerman and Machedon [109] were able to prove that the YM equations on $\mathbb{R}^{3,1}$ in Coloumb gauge satisfy a form of the null condition and are well posed in energy space $H^1 \times L^2$. They were then able to use the fact that the energy is conserved to prove global existence for YM. See also [108] for an overview of these ideas and some related conjectures.

An important open problem for the classical field equations is the global existence problem for the wave map equation (nonlinear σ-model, hyperbolic harmonic map equation). This is an equation for a map $\mathbb{R}^{n,1} \to N$, where N is some complete Riemannian manifold,

$$\Box_\eta u^A + \Gamma^A_{BC}(u)\partial_\alpha u^B \partial_\beta u^C \eta^{\alpha\beta} = 0. \tag{2.8}$$

Here Γ is the Christoffel symbol on N.

The wave map equation satisfies the null condition and hence we have small data global existence for $n \geq 3$. Small data global existence is also known for $n = 2$. Further, scaling arguments provide counterexamples to global existence for $n \geq 3$, whereas $n = 2$ is critical with respect to scaling. For $n \geq 2$, global existence for "large data" is known only for symmetric solutions, and in particular, the global existence problem for the wave map equation (2.8) is open for the case $n = 2$. For the case $n = 1$, global existence can be proved using energy estimates or light cone estimates. Tao proved [156] that the wave map equation in $2 + 1$ dimensions with the sphere as target is globally well posed for small energy. The proof is based on a new local well-posedness in $H^1 \times L^2$ and conservation of energy. This result has been generalized to the case of hyperbolic target by Krieger [119]. Tataru [158] has proved that the wave map equation in $n+1$ dimensions is locally well posed in $\dot{H}^{n/2} \times \dot{H}^{n/2-1}$ for general target embedded in $\mathbb{R}^m$. (Here $\dot{H}^s$ is the homogeneous Sobolev space. See [158] for the precise definition of well-posedness used.) See also [146, 159] for surveys.

The above discussion shows that the situation for the wave map equation is reminiscent of that for the Einstein equations, cf. Sections 4, 5. In particular, it is interesting to note that equations of the wave map type show up in the reduced vacuum Einstein equations for the Gowdy and U(1) problems.

3. Bianchi

Let $(\bar{M}, \bar{g})$ be a $3 + 1$-dimensional spacetime with 3-dimensional local isometry group G. Assume the action of G is generated by space-like Killing fields and that the orbits of G in the universal cover of $\bar{M}$ are 3-dimensional. This means there is a global foliation of $\bar{M}$ by space-like Cauchy surfaces M with locally homogeneous induced geometry. Such spacetimes are known as Bianchi spacetimes. The assumption of local homogeneity of the 3-dimensional Cauchy surfaces means that a classification of the universal cover is given by the classification of 3-dimensional Lie algebras.

Let e_a, $a = 0, \ldots, 3$ be an ON frame on $\bar{M}$, with $e_0 = u$, a unit time-like normal to the locally homogeneous Cauchy surfaces, let γ_{ab}^c be the commutators of the frame, $[e_a, e_b] = \gamma_{ab}^c e_c$. Let the indices i, j, k, l run over $1, 2, 3$. We may without loss of generality assume that $[e_a, \xi_i] = 0$ where $\{\xi_i\}_{i=1}^3$ is a basis for the Lie algebra $\mathfrak{g}$ of G.

Choose a time function t so that $t_{,a} u^a = 1$, i.e., the level sets of t coincide with the group orbits. Restricting to a level set M of t, the spatial part of the commutators γ_{ij}^k are the structure constants of $\mathfrak{g}$. These can be decomposed into a constant symmetric matrix n^{kl} and a vector a_i,

$$\gamma_{ij}^k = \epsilon_{ijl} n^{kl} + a_i \delta_j^k - a_j \delta_i^k.$$

We will briefly describe the classification used in the physics literature, cf. [66], [163, §1.5.1].

TABLE 1. Bianchi geometries

	Type	n_1	n_2	n_3			Type	n_1	n_2	n_3
	I	0	0	0			V	0	0	0
	II	0	0	+			IV	0	0	+
Class A	VI_0	0	-	+	Class B		VI_h	0	-	+
	VII_0	0	+	+			VII_h	0	+	+
	VIII	-	+	+						
	IX	+	+	+						

The Jacobi identity implies $n^{ij}a_j = 0$ and by choosing the frame $\{e_i\}$ to diagonalize n^{ij} and so that e_1 is proportional to the vector a_i, we get

$$n^{ij} = \mathrm{diag}(n_1, n_2, n_3), \qquad a_i = (\mathbf{a}, 0, 0). \tag{3.1}$$

The 3-dimensional Lie algebras are divided into two classes by the condition $\mathbf{a} = 0$ (class A) and $\mathbf{a} \neq 0$ (class B). The classes A and B correspond in mathematical terminology to the unimodular and non-unimodular Lie algebras. If $n_2 n_3 \neq 0$, let the scalar h be defined by

$$\mathbf{a}^2 = h n_2 n_3. \tag{3.2}$$

Table 1 gives the classification of Bianchi geometries. Note that the invariance of the Bianchi types under permutations and sign changes of the frame elements has been used to simplify the presentation. Here the notation VI_0, VII_0, VI_h, VII_h refers to the value of h defined by (3.2). In the list of Bianchi types I–IX, the missing type III is the same as VI_{-1}.

Due to the local homogeneity of the Cauchy surfaces M in a Bianchi space-time, the topologies of the spatially compact Bianchi spacetimes can be classified using the classification of compact manifolds admitting Thurston geometries.

The eight Thurston geometries S^3, E^3, H^3, $S^2 \times \mathbb{R}$, $H^2 \times \mathbb{R}$, Nil, $\widetilde{SL}(2, \mathbb{R})$, Sol, are the maximal geometric structures on compact 3-manifolds, see [145, 160] for background. Each compact 3-manifold with a Bianchi (minimal) geometry also admits a Thurston (maximal) geometry, and this leads to a classification of the topological types of compact 3-manifolds with Bianchi geometry, i.e., compact manifolds of the form X/Γ where X is a complete, simply connected 3-manifold with a Bianchi geometry and Γ is a cocompact subgroup of the isometry group of X. It is important to note that Γ is not always a subgroup of the 3-dimensional Bianchi group G.

The relation between the Bianchi types admitting a compact quotient and the Thurston geometries is given by Table 2. For each Bianchi type we give only the maximal Thurston geometries corresponding to it, see [74, 117, 118] for further details and references. We make the following remarks

Remark 3.1.

i. Let (M, g) be a 3-dimensional space form with sectional curvature $\kappa = -1, 0, +1$. A spacetime $(\bar{M}, \bar{g})$ with $\bar{M} = M \times (a, b)$ and a warped product metric $\bar{g} =$

$-dt^2 + w^2(t)g$, *satisfying the perfect fluid Einstein equations is called a (local) Friedmann–Robertson–Walker (FRW) spacetime. Specifying the equation of state for the matter in the Einstein equations leads to an ODE for w. The FRW spacetimes play a central role in the standard model of cosmology. In the vacuum case, only $\kappa = -1, 0$ are possible, and in this case, the spatially compact local FRW spacetimes are for $\kappa = 0$, the flat spacetimes covered by $T^3 \times \mathbf{E}^1$, a special case of Bianchi I, and for $\kappa = -1$, the local FRW spacetimes discussed in Example 2.5, which are Bianchi V.*

ii. *The Thurston geometry $S^2 \times \mathbb{R}$ admits no 3-dimensional group of isometries, and hence it does not correspond to a Bianchi geometry, but to a Kantowski–Sachs geometry [105] with symmetry group $\mathrm{SO}(3) \times \mathbb{R}$. The other cases with 4-dimensional symmetry group are vacuum LRS Bianchi I, II, III, VIII and IX, see [66, p. 133].*

iii. *The type of geometry depends on the subgroup Γ of the isometry group used to construct the compactification. The isometry group in turn depends on the Bianchi data. Note that $\widetilde{SL}(2, \mathbb{R})$ is both maximal and minimal, whereas the isometry group of $\mathbf{H}^2 \times \mathbb{R}$ has dimension 4 and is therefore not a minimal (Bianchi) group.*

iv. *The compactifications of Bianchi V and VII_h, $h \neq 0$ are both of the type discussed in Example 2.5, thus no anisotropy is allowed in the compactification of Bianchi type V and VII_h, $h \neq 0$.*

v. *The compactification of a Bianchi geometry, introduces new (moduli) degrees of freedom, in addition to the dynamical degrees of freedom, see [117] and references therein for discussion, see also [50]. The resulting picture is complicated and does not appear to have been given a definite treatment in the literature.*

In the rest of this section, we will concentrate on class A Bianchi spacetimes. We will also refrain from considering the moduli degrees of freedom introduced by the compactification, as it can be argued that these are not dynamical.

Let the expansion tensor θ_{ij} be given by

$$\theta_{ij} = \nabla_j u_i,$$

(i.e., $\theta_{ij} = -k_{ij}$ where k_{ij} is the second fundamental form). Decompose θ_{ij} as $\theta_{ij} = \sigma_{ij} + H\delta_{ij}$ where $H = \theta/3$, $\theta = g^{ij}\theta_{ij}/3$. Then H is the Hubble scalar. We are assuming (3.1) and in the vacuum class A case it follows that σ_{ij} is diagonal [163, p. 41].

Under these assumptions, $\gamma_{ab}^c = \gamma_{ab}^c(t)$ and we may describe the geometry of $(\bar{M}, \bar{g})$ completely in terms of γ_{ab}^c or equivalently in terms of the 3-dimensional commutators γ_{ij}^k and the expansion tensor given in terms of σ_{ij}, θ.

By a suitable choice of frame, σ_{ij} and n_{ij} can be assumed diagonal. Since σ_{ij} is traceless and diagonal, it can be described in terms of the variables

$$\sigma_+ = \frac{1}{2}(\sigma_{22} + \sigma_{33}), \quad \sigma_- = \frac{1}{2\sqrt{3}}(\sigma_{22} - \sigma_{33})$$

TABLE 2. Bianchi and Thurston geometries

Bianchi type	class	Thurston geometry	comments
I	A	$\mathbf{E}^3$	
II	A	Nil	
III $= \mathrm{VI}_{-1}$	B	$\mathbf{H}^2 \times \mathbf{E}^1$	cf. Remark 3.1 iii.
		$\widetilde{\mathrm{SL}}(2, \mathbb{R})$	
IV	—	—	no compact quotient
V	B	$\mathbf{H}^3$	cf. Remark 3.1 iv
VI_0	A	Sol	
$\mathrm{VI}_h, h \neq 0, -1$	—	—	no compact quotient
VII_0	A	$\mathbf{E}^3$	
$\mathrm{VII}_h, h \neq 0$	B	$\mathbf{H}^3$	cf. Remark 3.1 iv
VIII	A	$\widetilde{\mathrm{SL}}(2, \mathbb{R})$	
IX	A	S^3	

and similarly, n_{ij} can be represented by

$$(n_1, n_2, n_3) = \frac{1}{2\sqrt{3}}(n_{11}, n_{22}, n_{33}).$$

The curvature $b_{ij} = Ric(e_i, e_j)$ can be written in terms of n_{ij}, $b_{ij} = 2n_i{}^k n_{kj} - n_k{}^k n_{ij}$ and is therefore diagonal in the chosen frame. Decomposing b_{ij} into trace-free and trace parts s_{ij}, k we find that s_{ij} can be represented by two variables s_+, s_-.

Next we introduce the dimensionless variables (following Hsu and Wainwright [164], but normalizing with H instead of θ),

$$(\Sigma_+, \Sigma_-, N_1, N_2, N_3) = (\sigma_+, \sigma_-, n_1, n_2, n_3)/H$$

Similarly, we set

$$(S_+, S_-) = (s_+, s_-)/H^2$$

In addition to these choices we also define a new time τ by $e^\tau = \ell$, ℓ the length scale factor, or $dt/d\tau = 1/H$.

Clearly the equations are invariant under permutations $(\Sigma_i) \to P(\Sigma_i)$, $(N_i) \to P(N_i)$. Cyclic permutations of $(N_i), (\Sigma_i)$ correspond to rotations through $2\pi/3$ in the Σ_+, Σ_- plane.

We now specialize to the vacuum case. Then (for $H \neq 0$) the Einstein equations are equivalent to the following system of ODE's for Σ_+, Σ_-, N_1, N_2, N_3

(where the ′ denotes derivative w.r.t. the time coordinate τ):

$$N_1' = (q - 4\Sigma_+)N_1, \tag{3.3a}$$

$$N_2' = (q + 2\Sigma_+ + 2\sqrt{3}\Sigma_-)N_2, \tag{3.3b}$$

$$N_3' = (q + 2\Sigma_+ - 2\sqrt{3}\Sigma_-)N_3, \tag{3.3c}$$

$$\Sigma_+' = -(2 - q)\Sigma_+ - 3S_+, \tag{3.3d}$$

$$\Sigma_-' = -(2 - q)\Sigma_- - 3S_-, \tag{3.3e}$$

where

$$q = 2(\Sigma_+^2 + \Sigma_-^2),$$

$$S_+ = \frac{2}{3}[(N_2 - N_3)^2 - N_1(2N_1 - N_2 - N_3)],$$

$$S_- = \frac{2}{\sqrt{3}}(N_3 - N_2)(N_1 - N_2 - N_3).$$

The Hamiltonian constraint (2.2a) is in terms of these variables

$$\Sigma_+^2 + \Sigma_-^2 + [N_1^2 + N_2^2 + N_3^2 - 2(N_1 N_2 + N_2 N_3 + N_3 N_1)] = 1. \tag{3.4}$$

With our conventions, if (τ_-, τ_+) is the maximal time interval of existence for the solution to (3.3), then $\tau \to \tau_-$ corresponds to the direction of a singularity. For all non-flat vacuum Bianchi models except Bianchi IX, the spacetime undergoes an infinite expansion as $\tau \to \tau_+$ and is geodesically complete in the expanding direction, cf. [134, Theorem 2.1] which covers the vacuum case as a special case. For Bianchi IX, on the other hand, $\tau \to \tau_+$ corresponds to $H \to 0$, as follows from the proof of the closed universe recollapse conjecture for Bianchi IX by Lin and Wald [121], and hence to the dimensionless variables becoming ill defined.

We will now review the basic facts for Bianchi types I, II and IX.

I: Kasner. The Hamiltonian (3.4) constraint reads

$$\Sigma_+^2 + \Sigma_-^2 = 1, \tag{3.5}$$

and the induced metric on each time slice is flat. These spacetimes can be given metrics of the form

$$ds^2 = -dt^2 + \sum_i t^{2p_i} dx^i \otimes dx^i, \tag{3.6}$$

where $\sum_i p_i = 1$, $\sum_i p_i^2 = 1$, the Kasner relations, which correspond to the equation (3.5). The $\Sigma_\pm$ and p_i are related by $\Sigma_+ = \frac{3}{2}(p_2 + p_3) - 1$, $\Sigma_- = \frac{\sqrt{3}}{2}(p_2 - p_3)$. Clearly, the Kasner circle given by (3.5) consists of fixed points to the system (3.3). The points T_1, T_2, T_3, with coordinates $(-1, 0)$, $(1/2, \pm\sqrt{3}/2)$ correspond to flat spacetimes of type I or VII$_0$ (quotients of Minkowski space).

II: By using the permutation symmetry of the equations, we may assume $N_2 = N_3 = 0$. The solution curve which is a subset of the cylinder $\Sigma_+^2 + \Sigma_-^2 < 1$ has a past endpoint on the longer arc connecting T_2 and T_3 and future endpoint on the

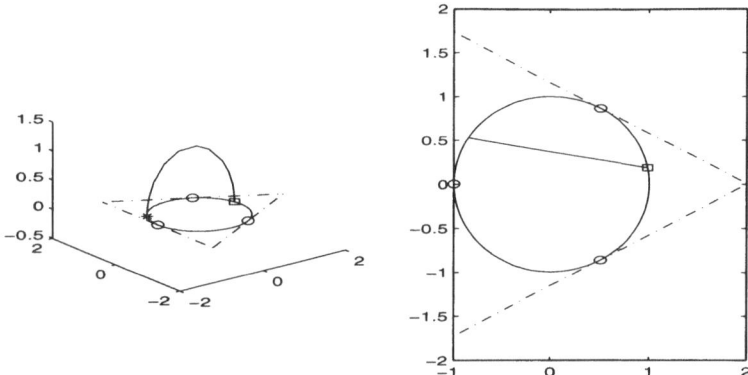

FIGURE 2. A type II solution, the Kasner circle and the triangle for the Kasner billiard in the (Σ_+, Σ_-)-plane are shown.

shorter arc connecting these points see Figure 2. This curve realizes the so-called Kasner map, cf. [163, §6.4.1]. We define the Bianchi II variety as the union of the three spheres in $\mathbb{R}^5$ given by the solutions to the constraint (3.4) with $(N_2, N_3) = 0$ and permutations thereof. The Kasner circle is the intersection of the spheres.

IX: Mixmaster, characterized by $N_i > 0, i = 1, 2, 3$. The heuristic picture is the following. The projection in the (Σ_+, Σ_-)-plane of a generic orbit in the direction $\tau \searrow \tau_-$ moves into the Kasner circle and stays there, undergoing an infinite sequence of bounces, which are approximately given by the Kasner billiard, cf. Figure 4. This picture is supported by numerical studies of the full Bianchi IX system, see, e.g., [30].

The Kasner billiard is the dynamical system given by mapping a (non-flat) point p on the Kasner circle to the point on the Kasner circle which is the end point of the type II orbit starting at p. This map can be described as follows. Let B be the nearest corner to p of the triangle shown in Figure 4. The ray starting at B through p intersects the Kasner circle in a point q, which is the image of p under the Kasner map, see also [163, Fig. 6.13]. Iterating this construction gives a sequence of points $\{p_i\}$ on the Kasner circle, which we may call the Kasner billiard.

The exceptional orbits which do not exhibit this infinite sequence of bounces are the Taub type IX solutions, cf. Figure 3. Up to a permutation these are given by the conditions

$$N_2 = N_3, \quad \Sigma_- = 0. \tag{3.7}$$

We call a Bianchi spacetime, satisfying (3.7) up to a permutation, a Taub space-time.

The past limit of the Taub type IX solution is the flat point $(-1, 0)$. The MVCD of Taub type IX data has a smooth Cauchy horizon, and is extendible, the extension being given by the so-called Taub–NUT spacetimes. As shown by

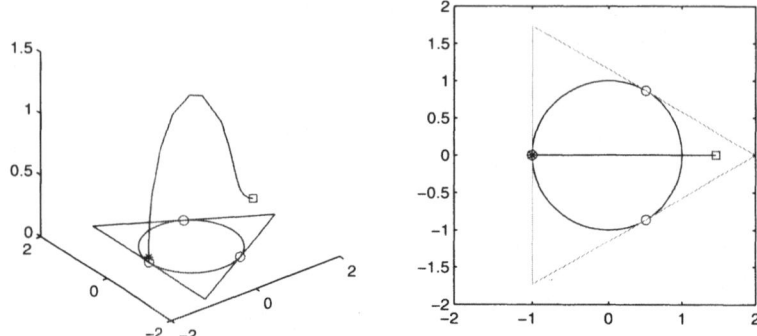

FIGURE 3. A Taub type IX solution, the Kasner circle and the triangle for the Kasner billiard in the (Σ_+, Σ_-)-plane are shown.

Chruściel and Rendall [55, Theorem 1.2], these are the only Bianchi IX spacetimes with a smooth Cauchy horizon, which gives a version of SCC for this class. See [55] for further details on the status of SCC in the locally homogeneous case. Chruściel and Isenberg proved that the MVCD of Taub type IX data has non-isometric maximal vacuum extensions, further emphasizing its pathological nature, cf. [53].

From the point of view of the cosmic censorship conjecture, the following theorem appears fundamental. A point which is a past limit point of $(\Sigma_+(\tau), \Sigma_-(\tau),$ $N_1(\tau), N_2(\tau), N_3(\tau))$ is called an α limit point. We say that the approach to the singularity is oscillatory, if the set of α limit points contains at least two points on the Kasner circle, at least one of which is distinct from the special points T_1, T_2, T_3, cf. [137].

Theorem 3.1 ([137, 141]). *A vacuum Bianchi spacetime of class A has exactly one of the properties*

1. *The Kretschmann scalar $\kappa = \bar{R}_{\alpha\beta\gamma\delta}\bar{R}^{\alpha\beta\gamma\delta}$ satisfies $\limsup_{\tau \searrow \tau_-} |\kappa| = \infty$*
2. *The MVCD has a smooth Cauchy horizon and the spacetime is a Taub spacetime.*

For non-Taub vacuum Bianchi VIII and IX spacetimes, the approach to the singularity is oscillatory.

Remark 3.2.

   i. *Rendall [137] proved the dichotomy in Theorem 3.1 for all Bianchi class A except VIII and IX. These cases as well as the oscillatory behavior for type VIII and IX were proved by Ringstrom [141]. Curvature blowup for non-Taub Bianchi A models with perfect fluid matter, including stiff fluid was proved by by Ringstrom [142].*
   ii. *In class B it is only the exceptional model Bianchi $VI_{-1/9}$ which exhibits oscillatory behavior as has been argued by Hewitt et al. [87].*

iii. *Theorem 3.1 shows that SCC holds in the class of vacuum Bianchi class A spacetimes, also with respect to C^2 extensions.*

iv. *See also Weaver [167] for a related result for Bianchi VI_0 with a magnetic field. Weaver proved that the singularity in magnetic Bianchi VI_0 is oscillatory and that curvature blows up as one approaches the singularity. It should be noted that vacuum Bianchi VI_0, on the other hand, is non-oscillatory.*

The dynamics of the Bianchi spacetimes has been studied for a long time, from the point of view of dynamical systems. In particular, it is believed that the Bianchi IX (mixmaster) solution is chaotic in some appropriate sense, see for example the paper by Hobill in [163] or the collection [88] as well as the work of Cornish and Levin [57], for various points of view. However, it does not yet appear to be clear which is the appropriate definition of chaos to be used, and no rigorous analysis exists for the full Bianchi IX system. In the course of the above-mentioned work, approximations to the Bianchi dynamics have been described and studied, such as the Kasner billiard (cf. Fig. 4, and the discussion above) and the BKL map, cf. [163, §11.2.3].

It has long been conjectured that the Bianchi II phase space is the attractor for the Bianchi VIII and IX dynamics, see [163]. The phase space for the Bianchi II model is given by the Hamiltonian constraint (3.4) together with the condition $N_1 N_2 = N_2 N_3 = N_1 N_3 = 0$. This defines a variety consisting of the union of three 2-spheres in $\mathbb{R}^5$. The Kasner circle is the intersection of these spheres. The conjecture has been proved in the Bianchi IX case.

Theorem 3.2 ([142]). *The Bianchi II variety is the asymptotic attractor for vacuum Bianchi IX.*

This result goes a goes a long way towards proving the chaotic nature of the Bianchi IX dynamics.

The questions of curvature blowup and oscillatory approach to the singularity can be studied also in the case of Bianchi class B models. In this case we have curvature blowup except for Bianchi III and V. However, in class B it is only the so-called exceptional Bianchi $VI_{-1/9}$ which, based on numerical work and qualitative analysis by Hewitt et al. [87] has an oscillatory singularity. The exceptional Bianchi $VI_{-1/9}$ has the same number of degrees of freedom as the most general class A models, Bianchi VIII and IX. Figure 3 shows an orbit of the Bianchi $VI_{-1/9}$ system.

4. G_2

In this section we consider the case when $(\bar{M}, \bar{g})$ is a $3 + 1$-dimensional, spatially compact, globally hyperbolic, vacuum spacetime, with a 2-dimensional local isometry group G_2 with the action of G_2 generated by space-like Killing fields. By passing to the universal cover, we see that the non-degenerate orbits of the G_2-action are 2-dimensional homogeneous spaces and hence the induced metric on the

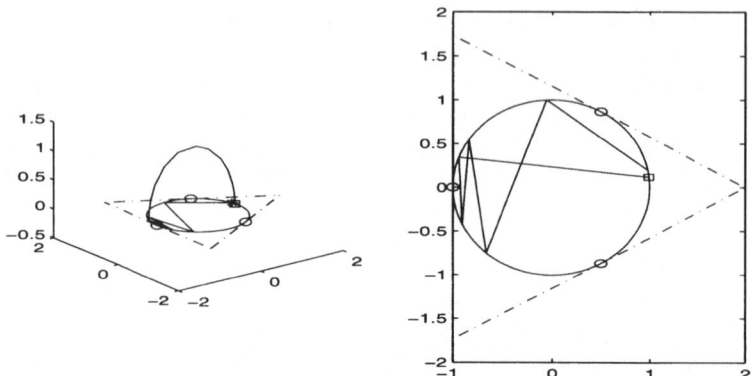

FIGURE 4. Bianchi IX orbit showing a few bounces. The vertical axis is N_1. The Kasner circle and the triangle for the Kasner billiard in the (Σ_+, Σ_-)-plane are shown.

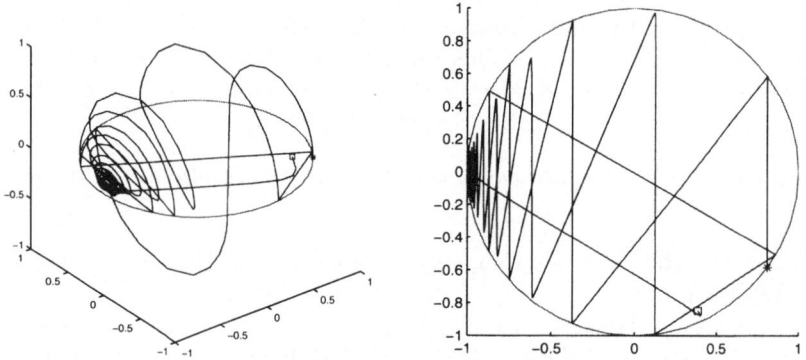

FIGURE 5. An orbit of the Bianchi $VI_{-1/9}$ system, showing a few bounces. Note that in contrast to the Bianchi IX system, the dynamics shows a combination of several types of bounces.

orbits must have constant curvature. The isometry group of the sphere S^2 has no two-dimensional subgroup and thus the orbits have geometry $\mathbf{E}^2$ or $\mathbf{H}^2$.

The special case when the group $U(1) \times U(1)$ itself acts on $\bar{M}$ was considered by Gowdy [80, 81]. In this case, it follows that orbits are compact (unless there is an extra Killing field), and M is covered by T^3, S^3 or $S^1 \times S^2$. Suppose p is a fix point for the action of $U(1) \times U(1)$ on M, then $U(1) \times U(1)$ acts by isometries on T_pM. As the isometry group $SO(3)$ of T_pM does not have a 2-dimensional subgroup, any degenerate orbit of G must be a closed curve.

Let $\bar{M}$ be a bundle over $S^1 \times \mathbb{R}$ with compact 2-dimensional fiber F and suppose that the orbits of the G-action on the universal cover of $\bar{M}$ cover the fibers F. If F has geometry $\mathbf{E}^2$ it follows that the Killing fields generating the G-action commute, and hence it is natural, following Rendall [136], to use the term local $U(1) \times U(1)$ symmetry for this situation.

Space-times with local $U(1) \times U(1)$ symmetry have also been considered by Tanimoto [155], who discussed the question of in which case a spacetime with local $U(1) \times U(1)$ symmetry can be considered as a dehomogenization of a Bianchi spacetime. It can be seen from the structure of the Bianchi groups that the only $U(1) \times U(1)$ symmetric spacetimes which have Bianchi (or Kantowski–Sachs) limits are covered by T^3 or S^3, and in the case of S^3 (Bianchi IX), it is only the Taub metrics that admit a $U(1) \times U(1)$ action by isometries. If we consider the case with local $U(1) \times U(1)$-symmetric case on the other hand, then all Bianchi models except Bianchi VIII and XI, and in case of Bianchi VIII and IX the Taub metrics, can be viewed as limits of locally $U(1) \times U(1)$-symmetric models, and therefore these serve as dehomogenization of the Bianchi models.

For simplicity, we concentrate in the rest of this section only the case of $U(1) \times U(1)$ symmetric spacetimes, and assume that the twist constants vanish, i.e., $\xi_1 \wedge \xi_2 \wedge d\xi_1 = \xi_1 \wedge \xi_2 \wedge d\xi_2 = 0$, where ξ_1, ξ_2 are one forms dual to the generators of the $U(1) \times U(1)$ action. This is a nontrivial restriction only in case $M \cong T^3$, see [81, p. 211]. Such spacetimes are known in the literature as Gowdy spacetimes. We further specialize to the case with Cauchy surface $M \cong T^3$

Let τ, x be coordinates on the $1+1$-dimensional Lorentzian orbit space $(U(1) \times U(1)) \backslash \bar{M}$ and let $A(\tau, \theta)$ be the area of the orbit. Gowdy showed that in the non-twisted case, with $M \cong T^3$, there are no degenerate orbits, i.e., $A \neq 0$, and further, the level sets of A in $(U(1) \times U(1)) \backslash \bar{M}$ are space-like. We may therefore choose coordinates so that $A = 4\pi^2 e^{-\tau}$, and choose the metric on the orbit as Ah where $h = h(\tau, x)$ is a unit determinant metric. $\bar{M}$ is causally incomplete in the direction $\tau \nearrow \infty$, which corresponds to a cosmological singularity.

By construction, the metric on the orbit, $h = h(\tau, x)$ is a unit determinant metric which is constant on each orbit. It therefore represents an element of the Teichmüller space $\mathcal{T}(T^2)$. The space $\mathcal{T}(T^2)$ with the Weil–Peterson metric is isometric to the hyperbolic plane $\mathbf{H}^2$. The identification of $\mathcal{T}(T^2)$ with $\mathbf{H}^2$ gives a map $u : (U(1) \times U(1)) \backslash \bar{M} \to \mathbf{H}^2$. This can be realized concretely for example by using the model for $\mathbf{H}^2$ with metric

$$dP^2 + e^{2P} dQ^2 \tag{4.1}$$

and letting $u = (P, Q)$ with

$$P = \ln(h_{11}), \qquad Q = e^{-P} h_{12}.$$

This is the parametrization that is used in the numerical work of Berger and collaborators. Thus the Gowdy (as well as the general $U(1) \times U(1)$ symmetric) Einstein equations on $M \cong T^3 \times \mathbb{R}$ can be viewed as equations for the evolution of a loop in $\mathbf{H}^2$. The velocity of a point $u(x, \tau)$ in $\mathbf{H}^2$ is given by $v_{\mathbf{H}^2}(x, \tau) = \sqrt{\langle \partial_\tau u, \partial_\tau u \rangle}$.

Define the asymptotic velocity $\hat{v}_{\mathbf{H}^2}(x)$ by $\hat{v}_{\mathbf{H}^2}(x) = \lim_{\tau \to \infty} v_{\mathbf{H}^2}(x, \tau)$, when the limit exists.

The spacetime metric may then be written in the form[5]

$$
\begin{aligned}
\ell_0^{-2} ds^2 = {} & e^{-\lambda/2 + \tau/2} (-e^{-2\tau} d\tau^2 + dx^2) \\
& + e^{-\tau} [e^P dy_2^2 + 2e^P Q dy_2 dy_3 + (e^P Q^2 + e^{-P}) dy_3^2]
\end{aligned}
\tag{4.2}
$$

where ℓ_0 is the unit of physical length. Here $(\tau, \theta) \in \mathbb{R} \times S^1$ are coordinates on the orbit space, $y^A, A = 2, 3$ are coordinates on the orbit, $0 \le y^A \le 2\pi$. The G invariance implies that we can assume all metric components depend on τ, θ only. Let $e = dx^2 + (dy^2)^2 + (dy^3)^2$. The lapse function N satisfies $N = \sqrt{\det(g)/\det(e)}$ and hence the time function τ is spacetime harmonic, cf. Subsection 2.3.

Let $\eta = -d\tau^2 + e^{2\tau} dx^2$. The Einstein evolution equations take the form

$$
\eta^{\alpha\beta} (\partial_{\alpha\beta} u^a + \Gamma^a_{bc}(u) \partial_\alpha u^b \partial_\beta u^c) = 0,
\tag{4.3}
$$

where Γ^a_{bc} are the Christoffel symbols on $\mathbf{H}^2$ with the metric (4.1). The system (4.3) is supplemented by a pair of equations for λ, which are implied by the Einstein constraint equations, and which are used to reconstruct the $3 + 1$ metric $\bar{g}$.

Equation (4.3) is a semilinear hyperbolic system, which resembles the wave-map equation, (2.8). Energy estimates or light cone estimates prove global existence on $(0, \infty) \times S^1$. The wave operator $-\partial_\tau^2 + e^{-2\tau} \partial_x^2$ degenerates as $\tau \nearrow \infty$, which corresponds to a singularity in the $3 + 1$ spacetime $(\bar{M}, \bar{g})$, since the area $e^{-\tau}$ of the orbit tends to zero.

The energy $E = \frac{1}{2} \int_{S^1} \langle \partial_\tau u, \partial_\tau u \rangle + e^{-2\tau} \langle \partial_x u, \partial_x u \rangle = E_K + E_V$, where E_K, E_V are the kinetic and potential energy terms, respectively, satisfies

$$
\partial_\tau E = -2E_V.
$$

Therefore E is monotone decreasing, with a rate determined by E_V. This shows immediately that there is a sequence of times (τ_k), $\lim_{k \to \infty} \tau_k = \infty$, so that $E_V(\tau_k) \to 0$ as $k \to \infty$, which indicates that the scale-free variables $(e^{-\tau} \partial_x P, e^{-\tau} \partial_x Q)$ become insignificant for the dynamics as $\tau \to \infty$. This leads to the idea that the Gowdy system behaves asymptotically as a dynamical system in the likewise scale-free variables $(\partial_\tau P, \partial_\tau Q)$. This heuristic is, as we shall see, supported by numerical work and some rigorous results.

We now briefly discuss some results and open problems for the Gowdy and $\mathrm{U}(1) \times \mathrm{U}(1)$ symmetric spacetimes. The following result proves Conjecture 3 for this class. The first global existence result for Gowdy spacetimes with topology $T^3 \times \mathbb{R}$ was due to Moncrief [125], who proved that vacuum Gowdy spacetimes with the stated topology are globally foliated by level sets of the area function, assuming the existence of a compact level set of the area function. This assumption was removed by Chruściel [47] who also studied global properties of the area function for Gowdy spacetimes on S^3 and $S^2 \times S^1$. A class of "nongeneric" metrics still remains to be studied, see [47]. The first result concerning global CMC foliations

---

[5]We have assumed here that the Killing fields are hypersurface orthogonal, see [47] for the most general form of the $\mathrm{U}(1) \times \mathrm{U}(1)$ symmetric metric on $T^3 \times \mathbb{R}$ with vanishing twist.

in Gowdy spacetimes was proved by Isenberg and Moncrief [99] for the case of vacuum Gowdy spacetimes. Recently the following was proved.

Theorem 4.1 (Andréasson, Rendall, Weaver [14]**).** *Nonflat spatially compact spacetimes with* $(\mathrm{U}(1) \times \mathrm{U}(1))$ *symmetry and Vlasov matter, are globally foliated by CMC hypersurfaces with mean curvature taking on all values in* $(-\infty, 0)$.

It was conjectured by Belinskii, Khalatnikov and Lifschitz [120] that in a generic spacetime with a cosmological singularity, spatial points will decouple as one approaches the singularity, and spatial derivatives become insignificant asymptotically. This leads to the idea that asymptotically near the singularity, the dynamics of the gravitational field should be explained by a family of ODE systems. This is a very rough idea. Yet, as it turns out, this principle appears to hold in the cases we have been able to study. The idea of BKL has been specialized and reformulated by among others Eardley, Liang and Sachs [62] and Isenberg and Moncrief [101], into the notion of asymptotically velocity term dominated (AVTD) singularities. Roughly, an AVTD solution approaches asymptotically, at generic spatial points, the solution to an ODE, the parameters of which depend on the spatial point. In particular, in an AVTD spacetime, locally near a fixed spatial point the spacetime approaches a Kasner limit, with parameters depending on the spatial point. The asymptotic model for an AVTD spacetime necessarily has a non-oscillatory singularity, which excludes generic spacetimes. The precise formulation of the BKL conjecture for generic spacetimes is more subtle, see [161, 13] for discussion of this problem.

The numerical studies referred to above indicate that for a generic Gowdy spacetime, as $\tau \nearrow \infty$, the velocity $v_{\mathbf{H}^2}$ is eventually forced to satisfy $0 \leq v_{\mathbf{H}^2} \leq 1$, except at isolated x-values, even if $v_{\mathbf{H}^2} > 1$ in some subsets of S^1 initially, and further that $v_{\mathbf{H}^2}$ has a limiting value $\hat{v}_{\mathbf{H}^2}(x)$ for each x as one moves toward the singularity.

The numerical solutions exhibit "spikes" at those x-values, where $v_{\mathbf{H}^2} \geq 1$ asymptotically, cf. Figure 6. The very sharp spikes seen (in Q) are coordinate effects, corresponding to a part of the solution loop approaching the point on the boundary of $\mathbf{H}^2$ sent to infinity by the transformation leading to the model with metric (4.1).

Gowdy spacetimes such that $\partial_x Q = 0$ are called polarized. Equation (4.3) then becomes linear. It was proved by Isenberg and Moncrief [101] that polarized Gowdy spacetimes are AVTD. Numerical studies by Berger et al., see [26] and references therein, see also [86, 29], support the idea that general Gowdy spacetimes are AVTD. It should be noted that there are polarized Gowdy spacetimes with $v > 1$ up to the singularity, but the above-mentioned work indicates that this behavior is non-generic. See also [48] for work on the case $v = 0$.

The equation (4.3) may, essentially under the restriction $0 < v < 1$, be written as a Fuchsian system

$$(t\partial t + E)\mathcal{U} = \mathcal{F}[\mathcal{U}],$$

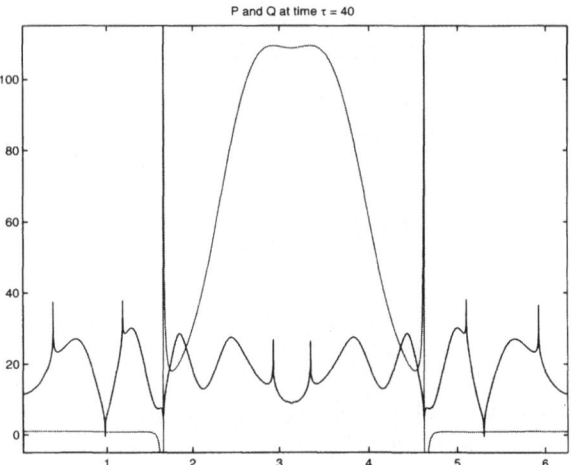

FIGURE 6. Spikes in P, Q. The very sharp spikes (in Q), so-called "false spikes" are coordinate effects.

where $t = e^{-\tau}$, and hence using a singular version of the Cauchy–Kowalewskaya theorem, cf. Kichenassamy and Rendall [107] and references therein, see also [19], AVTD solutions may be constructed given real analytic "data on the singularity". The solutions constructed are in terms of the time coordinate τ, of the form

$$P(\tau, x) = v(x)\tau + \phi(x) + e^{-\epsilon\tau}u(x, \tau) \tag{4.4a}$$

$$Q(\tau, x) = q(x) + e^{-2v(x)\tau}[\psi(x) + w(\tau, x)] \tag{4.4b}$$

where $\epsilon > 0$, $0 < v < 1$, and $w, u \to 0$ as $\tau \to \infty$. The Fuchsian method was generalized for the Gowdy case by Rendall [138] to the C^∞ case.

Rendall and Weaver [140] have constructed families of solutions with spikes, starting from a solution with a false spike at x_0 with $k = \hat{v}_{\mathbf{H}^2}(x_0) < 1$, by applying an explicit transformation, a new solution with $\hat{v}_{\mathbf{H}^2}(x_0) = 1 + k$. In the new solution $\hat{v}_{\mathbf{H}^2}$ has a discontinuity at x_0 with $\lim_{x \to x_0} \hat{v}_{\mathbf{H}^2}(x) = 1 - k$. By iterating this procedure spikes with arbitrarily high velocity can be constructed. The group formed by composing the transformations used by Rendall and Weaver has been termed the Geroch group [38]. Spikes with asymptotic velocity $\hat{v}_{\mathbf{H}^2} > 2$ correspond to higher-order zeros of $\partial_x Q$, and are therefore non-generic. Garfinkle and Weaver [77] have studied the dynamics of generic spikes with initially high velocity. Their analysis, based on a combination of heuristic arguments, related to the "method of consistent potentials", see [26] for discussion, and numerical work, shows that the velocity of these spikes is driven down into the interval $1 < v < 2$ by a sequence of bounces, the qualitative features of which can be explained in terms of the relative importance of the terms in the evolution equation.

Recently conditions on initial data (essentially a small energy condition together with restriction on the velocity) have been given by Ringstrom [144, 143], and Chae and Chruściel [38] under which the solution is of the form (4.4). This work shows that AVTD behavior holds on open and dense subsets of S^1, and makes earlier work [82] on formal expansions for the Gowdy field equations rigorous. The definition of AVTD solution used by Chae and Chruściel [38, Eq. (3.8–3.9)] is more general than (4.4) used by Ringstrom, allowing more general velocity v but with less precise control on the lower order terms.

In view of the above, it is reasonable to make the following conjecture.

Conjecture 4. *Generic vacuum, spatially compact* $U(1) \times U(1)$*-symmetric spacetimes with vanishing twist are AVTD at the singularity, in the complement of an at most countable closed subset E of S^1. The asymptotic velocity $\hat{v}_{\mathbf{H}^2}(x)$ exists for all $x \in S^1$ and satisfies $0 < \hat{v}(x) < 1$ for $x \in S^1 \setminus E$. $\hat{v}_{\mathbf{H}^2}$ is continuous on $S^1 \setminus E$.*

Remark 4.1.

1. *This is implicit in Grubišić and Moncrief [82].*
2. *Chae and Chruściel [38] have constructed solutions with an asymptotic velocity which is discontinuous on any closed set $F \subset S^1$, with nonempty interior. In particular F may have nonzero measure. If the above conjecture is correct, this behavior is nongeneric.*

The question of cosmic censorship for the Gowdy spacetimes may be studied by analyzing the behavior of the Kretschmann scalar κ as $\tau \to \infty$. For the class of polarized Gowdy spacetimes, this was done by Chruściel et al. [54]. It was proved by Kichenassamy and Rendall [107] that for generic AVTD spacetimes constructed using the Fuchsian algorithm, the Kretschmann scalar κ blows up at the singularity and hence generically, these spacetimes do not admit extensions, see also the discussion in [82, §3–4]. The same behavior was shown to hold also for the class of Gowdy spacetimes with asymptotic velocity $0 < \hat{v} < 1$ and bounds on the energy density, see [143, 38]. Further, the Gowdy spikes constructed by Rendall and Weaver also have κ blowing up, though at a different rate than nearby points. Therefore a reasonable approach to the SCC in the class of Gowdy spacetimes, is via conjecture 4.

It is relevant to mention here that the AVTD behavior for Gowdy symmetric spacetimes may be broken by the introduction of suitable matter, cf. [168], where numerical evidence for an oscillatory approach to the singularity is presented, for a locally $U(1) \times U(1)$ symmetric spacetime with magnetic field. Similarly, it is expected that scalar field or stiff fluid matter changes the oscillatory behavior of general $U(1) \times U(1)$ symmetric models to AVTD. However, even in the presence of a scalar field, one expects to see spikes forming.

A systematic way of deriving the system of ODE's governing the asymptotic behavior of the Gowdy spacetimes is to write (4.3) in first order form, using the scale invariant operators $\partial_\tau, e^{-\tau}\partial_x$, and then cancelling the terms involving $e^{-\tau}\partial_x$. Following this procedure and working in terms of H-normalized orthonormal frame

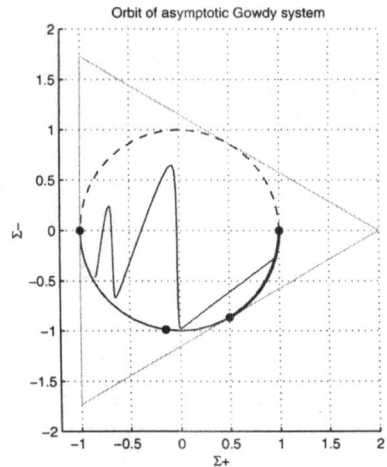

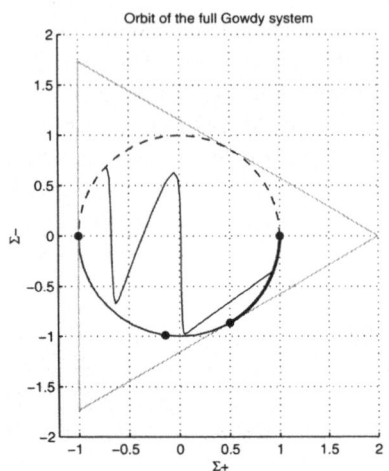

FIGURE 7.
Projection in
Kasner plane of
an orbit of the
asymptotic Gowdy
system (4.5–4.6).

FIGURE 8.
Projection in
Kasner plane of
the orbit of a
point of the full
Gowdy system.

variables adapted to the $U(1) \times U(1)$ orbits, constructed in a completely analogous way to those discussed in Section 3 gives, after cancelling the terms corresponding to $e^{-\tau}\partial_x$, the system

$$\mathcal{D}_0 \Sigma_+ = (q - 2)(1 + \Sigma_+) \tag{4.5a}$$

$$\mathcal{D}_0 \Sigma_- = (q - 2)\Sigma_- - 2\sqrt{3}\, N_-^2 + 2\sqrt{3}\, \Sigma_\times^2 \tag{4.5b}$$

$$\mathcal{D}_0 N_\times = (q + 2\Sigma_+)\, N_\times \tag{4.5c}$$

$$\mathcal{D}_0 \Sigma_\times = (q - 2 - 2\sqrt{3}\Sigma_-)\, \Sigma_\times - 2\sqrt{3}\, N_\times N_- \tag{4.5d}$$

$$\mathcal{D}_0 N_- = (q + 2\Sigma_+ + 2\sqrt{3}\Sigma_-)\, N_- + 2\sqrt{3}\, \Sigma_\times N_\times \ , \tag{4.5e}$$

where $\mathcal{D}_0 = (1 + \Sigma_+)\partial_\tau$, $q = 2(\Sigma_+^2 + \Sigma_-^2 + \Sigma_\times^2)$, subject to the constraint equation

$$1 = \Sigma_+^2 + \Sigma_-^2 + \Sigma_\times^2 + N_\times^2 + N_-^2. \tag{4.6}$$

See [13] for discussion of the system (4.5–4.6). This system contains the spatially homogeneous Bianchi class A models I–VII$_0$, but represented in terms of a non-Fermi-propagated frame.

The asymptotic behavior of the system (4.5–4.6) in the direction towards the singularity is easily analyzed, and one shows that all solutions of the system have limit points on the Kasner half-circle $\Sigma_+^2 + \Sigma_-^2 = 1$. Similar to the standard

Bianchi case, the Kasner circle consists of fixed points. For the Bianchi types II–VII$_0$, only part of the Kasner circle consists of stable fixed points. In the present case, only the arc with $\Sigma_- \leq 0$, $\Sigma_+ > \frac{1}{2}$ is stable, and a generic orbit of (4.5) ends on the stable arc. The stable arc corresponds to orbits with asymptotic velocity $0 < \hat{v} < 1$. See Figure 7 for an example of an orbit of the system (4.5–4.6).

According to the BKL proposal, a generic solution $u(t,x)$ of the full Gowdy system will have the property that it will asymptotically approach a solution of the Gowdy AVTD system (4.5–4.6). That this is indeed the case is indicated by Figure 8, which shows the projection in the Kasner plane of the orbit of one point for the full Gowdy system. We see in the figure the orbit of a point near a spike point. The points marked on the Kasner circle are (from the left) the points with velocity $v = \infty, 2, 1, 0$. The stable arc on the Kasner circle is plotted with a heavy line. It should be noted that the last part of the orbit which ends on the stable Kasner arc is approximately a Bianchi II orbit, with initial velocity in the interval $(1,2)$. This is the generic behavior, also close to spike points, see [13] for more details.

A class of AVTD polarized Gowdy spacetimes with non-vanishing twist has been constructed by Isenberg and Kichenassamy [96] using the Fuchsian algorithm. On the other hand for general non-polarised twisted $U(1) \times U(1)$ symmetric spacetimes, work of Berger et al. [31] supports the conjecture that these spacetimes show oscillating behavior as one approaches the singularity. This idea is also supported by the fact that the only Bianchi model with $U(1) \times U(1)$ symmetry and non-vanishing twist constants is the exceptional Bianchi VI$_{-1/9}$ which has been shown heuristically by Hewitt et al. [87] to have oscillatory behavior at the singularity.

5. U(1)

The U(1) symmetric vacuum $3+1$ Einstein equations is an important case which is of intermediate difficulty between the full $3+1$ Einstein equations and the highly symmetric Gowdy equations. In the presence of a hypersurface orthogonal space-like Killing field, the Einstein equations reduce to $2+1$ gravity coupled to wave map matter, the field equations and their reductions have been derived in [127, 126, 128], see also [37]. Choquet–Bruhat and Moncrief have proved that the Cauchy problem for the U(1) problem is well posed in H^2, see [42, 39].

In the above quoted papers, the spacetime is assumed to be a U(1) bundle over a spatially compact $2+1$ spacetime. The case of local U(1) symmetry does not appear to have been considered in connection with the Einstein equations, see however [131] for information about 3-manifolds with local U(1) action.

It is also possible to study the case when the reduced space is asymptotically flat. This case has been considered in work by Ashtekar and others, [17, 15], see also [16]. In these papers an analogue of the ADM mass at spatial infinity is introduced. It is proved that it is nonnegative and bounded from above. It is interesting to study the consequences of the presence of this conserved quantity for the $2+1$-dimensional Einstein-matter system given by the U(1) problem, as one expects that it gives a stronger bound on the fields than in the $3+1$ case. This appears to be a natural setting for a small data version of the U(1) problem.

In the following, we will consider the spatially compact case. Let $(\bar{M}, \bar{g})$ be a $3+1$-dimensional spacetime, assume $\bar{M} \cong B \times \mathbb{R}$, with $\pi : B \to \Sigma$ a principal U(1) bundle, Σ a compact surface. Further assume the group U(1) acts by isometries on $(\bar{M}, \bar{g})$ with the action generated by the Killing field J, which we assume to be space-like, $\langle J, J \rangle > 0$.

Let the function λ on $\Sigma \times \mathbb{R}$ be defined by $\pi^* \lambda = \frac{1}{2} \log(\langle J, J \rangle)$, and let $\theta = e^{-2\pi^* \lambda} J$. Then we can write the spacetime metric $\bar{g}$ in the form

$$\bar{g} = \pi^*(e^{-2\lambda} g) + e^{2\pi^* \lambda} \theta \otimes \theta,$$

where g is a Lorentzian metric on $\mathbb{R} \times \Sigma$.

Introduce a frame[6] e_α, $\alpha = 0, 1, 2, 3$ with $e_3 = e^{-\lambda} J$, and let the indices $a, b, c, \cdots = 0, 1, 2$. We may without loss of generality assume that $[J, e_a] = 0$. We have

$$dJ_{\alpha\beta} = \Theta_{\alpha\beta} + 2(e_\alpha(\lambda) J_\beta - e_\beta(\lambda) J_\alpha),$$

where $\Theta_{\alpha\beta} J^\beta = 0$. It follows that $\Theta_{\alpha\beta} = \pi^*(e^{2\lambda} F_{\alpha\beta})$ where $F_{\alpha\beta}$ is a 2-form on $\mathbb{R} \times \Sigma$. If $F_{\alpha\beta} = 0$, then J is hypersurface orthogonal.

To avoid cluttering up the notation we will in the following make no distinction between fields on the orbit space $\mathbb{R} \times B$ and their pull-backs by π.

The components of the $3 + 1$ Ricci tensor are

$$\bar{R}_{ab} = R_{ab} - 2\nabla_a \lambda \nabla_b \lambda + \nabla_c \nabla^c \lambda g_{ab} - \frac{1}{2} e^{4\lambda} F_{ac} F_a{}^c,$$

$$\bar{R}_{a3} = \frac{1}{2} e^{-\lambda} \nabla_c (e^{4\lambda} F_a{}^c),$$

$$\bar{R}_{33} = e^{2\lambda} [-g^{ab} \nabla_a \nabla_b \lambda + \frac{1}{4} e^{4\lambda} F_{ab} F^{ab}] \tag{5.1}$$

Let the one-form E on $\mathbb{R} \times \Sigma$ be given by $E = - \star_g (e^{4\lambda} F)$. One of the Einstein equations (5.1) implies that $dE = 0$.

Now the Einstein vacuum equations $\bar{R}_{\alpha\beta} = 0$ imply the system

$$R_{ab} = \frac{1}{2} (4\nabla_a \lambda \nabla_b \lambda + e^{-4\lambda} E_a E_b), \tag{5.2a}$$

$$\nabla^a \nabla_a \lambda + \frac{1}{2} e^{-4\lambda} E_a E_b g^{ab} = 0, \tag{5.2b}$$

$$\nabla^a \nabla_a \omega - 4\nabla^a \lambda E_a = 0. \tag{5.2c}$$

---

[6]In [41], a frame dx^α, θ^3 is used, where $\theta^3_\alpha = e^{-2\lambda} J_\alpha = e^{-\lambda} e^3_\alpha$. Furthermore, their $F_{\alpha\beta} = d\theta_{\alpha\beta} = e^{-2\lambda} \Theta_{\alpha\beta}$.

When $E_a = \nabla_a \omega$, we recognise (5.2) as the $2+1$-dimensional Einstein equations with wave map matter, for the wave map with components (λ, ω) with target hyperbolic space $\mathbf{H}^2$ with the constant curvature metric

$$2d\lambda^2 + \frac{1}{2}e^{-4\lambda}d\omega^2.$$

See Subsection 2.4 for some discussion of wave map equations.

We now specialize even further to the polarized case $E_a = 0$. This corresponds to assuming that the bundle $\pi : B \to \Sigma$ is trivial, and that the vector field J is hypersurface orthogonal. Then ω is constant and the equations (5.2) become

$$R_{ab} = \nabla_a \lambda \nabla_b \lambda \tag{5.3a}$$

$$\nabla^a \nabla_a \lambda = 0 \tag{5.3b}$$

which is precisely the $2 + 1$ Einstein equations coupled to a massless scalar field.

In the 2-dimensional case, the operator $k \to \operatorname{div} k$ is elliptic on symmetric 2-tensors with vanishing trace. Further, by the uniformization theorem, a compact 2-dimensional Riemannian manifold (Σ, h) is conformal to $(\Sigma, [h])$ where $[h]$ is a representative of the conformal class of g, i.e., a constant curvature metric. Therefore working in spatial harmonic gauge with respect to $(\Sigma, [h])$ (i.e., conformal spatial harmonic gauge) and CMC time gauge, the constraint equations form an elliptic system for (h_{ij}, k_{ij}). We get a representation of (h_{ij}, k_{ij}) in terms of $([h], \phi, k^{TT}, Y)$ where $[h]$ is the conformal class of h, corresponding to a point in Teichmüller space, ϕ a conformal factor determined by the Hamiltonian constraint equation (a nonlinear elliptic system for ϕ), k^{TT} a trace-free, divergence free 2-tensor on $(\Sigma, [h])$, corresponding to a quadratic differential, and finally, Y is a vector field determined from the momentum constraint equation. Note that $([h], k^{TT})$ represents a point in $T^*\mathcal{T}(\Sigma)$, the cotangent bundle of the Teichmüller space of Σ.

Due to the ellipticity of the constraint equations in the $2 + 1$-dimensional case, in the gauge as described above, it is possible to eliminate the Einstein equation (5.2a) from the system (5.2) and instead solve the elliptic-hyperbolic system consisting of the hyperbolic system (5.2b–5.2c), coupled to a nonlinear system which determines components of the spacetime metric $\bar{g}$ in terms of the data via the constraint and gauge fixing equations, as well as an ODE system determining the evolution of the Teichmüller degrees of freedom of the metric on Σ. In the special case $\Sigma = S^2$, Teichmüller space is a point, and further $H^1(S^2) = 0$, which means that $E = d\omega$. Therefore ignoring problems with gauge fixing, the reduced $U(1)$ system would consist of the wave equation coupled to the system which determines the components of $\bar{g}$.

A special case of the polarized $U(1)$ equations is given by setting $\lambda \equiv$ constant. Then the field equations (5.3) are just

$$R_{ab} = 0,$$

the $2 + 1$-dimensional vacuum equations. In this case, the spacetime is a 3-dimensional Lorentzian space-form. The dynamics of $2 + 1$-dimensional vacuum gravity has been studied by Andersson, Moncrief and Tromba [11] who proved global existence in CMC time for $2+1$-dimensional vacuum spacetimes, with cosmological constant, containing at least one CMC hypersurface. This proves Conjecture 3 for the class of $2+1$-dimensional vacuum spacetimes.

Andersson and Rendall [12] showed that the $3+1$-dimensional Einstein scalar field system can be formulated as a Fuchsian system. Using similar techniques, Damour et al. [61, §4] have shown that the polarized U(1) equations may be formulated as a Fuchsian system, and therefore AVTD solutions may be constructed using a singular version of the Cauchy-Kowalewskaya theorem, as was done by Kichenassamy and Rendall [107] for the Gowdy case. See also Isenberg and Moncrief [102] which deals with the "half-polarized" U(1) system in addition to the polarized case. This supports the following conjecture.

Conjecture 5. *Generic polarized* U(1) *spacetimes are AVTD at the singularity.*

Remark 5.1. *Conjecture 5 was essentially stated by Grubišić and Moncrief [83]. This is supported by numerical work of Berger and Moncrief [33]. Similarly to the case of Gowdy spacetimes, see Conjecture 4, one expects that "spiky features" will form at the singularity, and hence the AVTD property of the* U(1) *symmetric spacetimes should be understood to be generic also in the spatial sense.*

It seems reasonable to expect that polarized U(1) spacetimes which are AVTD at the singularity have a strong curvature singularity generically, and therefore that proving the AVTD conjecture for polarized U(1) would be a big step towards proving SCC for this class.

In contrast to polarized U(1), the generic U(1) spacetimes have sufficiently many degrees of freedom that one expects them to satisfy the BKL picture of an oscillatory approach to the singularity, as is also expected in the fully $3+1$-dimensional case, cf. the remarks in Section 7. This is supported by the numerical evidence so far, see [32].

In the expanding direction, on the other hand, a small data semi-global existence result holds [42, 39], similar to the one for the full $3+1$-dimensional case discussed below in Section 6. In this case, the notion of "small data" is taken to be data close to data for a background spacetime with space-like slice $M = \Sigma \times S^1$ where Σ is a Riemann surface of genus > 2 with metric σ of scalar curvature -1, and M has the product metric. In this case, the background spacetimes are products of flat spacetimes as in Example 2.5 with $n = 2$, with the circle (such spacetimes are of type Bianchi III). The proof uses energy estimates for a second order energy, which controls the $H^2 \times H^1$ norm of the data, for small data. The energy expression used is a combination of several terms. The first-order part is $\frac{\tau^2}{4}\mathrm{Vol}_g(\Sigma) + 2\pi\chi(\Sigma)$, which via the constraint equations may be related to the wave map energy. The second order term is defined in terms of the L^2 norm of the Laplacian of the wave map field, and the gradient of its time derivative. In order to use this method, one of the technical problems one has to deal with is controlling the spectral gap of the Laplacian. This is done by keeping control of the Teichmüller parameters during the evolution. For technical reasons, the assumption is made that the spectral gap Λ_0 of the Laplacian for the initial metric satisfies $\Lambda_0 > 1/8$. The proof shows that the Teichmüller parameter $\sigma(t)$ converges to a point in the interior of Teichmüller space and that the spacetime geometry converges in the expanding direction to one of the model spacetimes described above.

6. $3 + 1$

In the $3 + 1$-dimensional case with no symmetries, the only known facts on the global properties of spacetimes are Lorentzian geometry results such as the Hawking–Penrose singularity theorems [132] and the Lorentzian splitting theorem of Galloway [75], see [24] for a survey of Lorentzian geometry.

Here we are interested in results relevant to the SCC, Conjecture 1 and the CMC conjecture, Conjecture 3, i.e., results about the global behavior of solutions to the Cauchy problem for the evolution Einstein equations, in some suitable gauge. The Cauchy problem for the Einstein evolution equations has been discussed in Section 2.2.

With this limitation there are essentially 3 types of results known and all of these are small data results. The results are those of Friedrich on the "hyperboloidal Cauchy problem", of Christodoulou and Klainerman on the nonlinear stability of Minkowski space (generalized by Klainerman and Nicolò [111] to exterior domains) and the work of Andersson and Moncrief [9], on global existence to the future for data close to the data for certain spatially compact flat $\kappa = -1$ (local) FRW spacetimes, again a nonlinear stability result. We will briefly discuss the main features of these results.

The causal structure of a Lorentz space is a conformal invariant. This leads to the notion that the asymptotic behavior of spacetimes can be studied using conformal compactifications or blowup. The notion of isolated system in general relativity has been formalized by Penrose in terms of regularity properties of the boundary of a conformally related spacetime $(\tilde{M}, \tilde{g})$, with null boundary $\Im$, such that $\tilde{M}$ is a completion of $\bar{M}$, and $\bar{g} = \Phi^*(\Omega^{-2}\tilde{g})$, where $\Omega \in C^\infty(\tilde{M})$ is a conformal factor, $\Phi : \bar{M} \to \tilde{M}$ is a diffeomorphism of $\bar{M}$ to the interior of $\tilde{M}$. Given assumptions on the geometry of $(\tilde{M}, \tilde{g})$ at $\Im$, Penrose proved using the Bianchi identities that the components of the Weyl tensor of $(\bar{M}, \bar{g})$ decay at physically reasonable rates.

Friedrich derived a first-order symmetric hyperbolic system from the Einstein equations, the "regular conformal field equations". This system includes among its unknowns, components of the Weyl tensor, the conformal factor Ω and quantities derived from the conformally rescaled metric $\tilde{g}$. This system has the property that under the Penrose regularity conditions at $\Im$, the solution can be extended across $\Im$. The fact that the regular conformal field equations gives a well-posed evolution equation in the conformally compactified picture enabled Friedrich to prove small data global existence results by using the stability theorem for quasi-linear hyperbolic equations.

In [69, Theorem 3.5] Friedrich proved global existence to the future for data (M, g, k) close to the standard data on a hyperboloid in Minkowski space, satisfying asymptotic regularity conditions compatible with a Penrose type compactification (the hyperboloidal initial value problem). This was later generalized to Maxwell and Yang–Mills matter in [70]. Initial data for the hyperboloidal initial value problem were first constructed by Andersson, Chruściel and Friedrich [6], see also [4], [5].

The result of Friedrich is a semi-global existence result, in the sense that the maximal vacuum Cauchy development $D(M)$ of the data (M, g, k) is proved to be geodesically complete and therefore inextendible to the future, but not to the past. In fact typically $D(M)$ will be extendible to the past and (M, g, k) may be thought of as a partial Cauchy surface in a larger maximal globally hyperbolic spacetime $(\bar{M}, \bar{g})$, cf. Figure 9. In view of the inextendibility to the future of $D(M)$, the result of Friedrich may be viewed as supporting the cosmic censorship conjecture. In the case of the Einstein equations with positive cosmological constant, the method of Friedrich yields a global existence result for data close to the standard data on $M = S^3$ in deSitter space, cf. [69, Theorem 3.3].

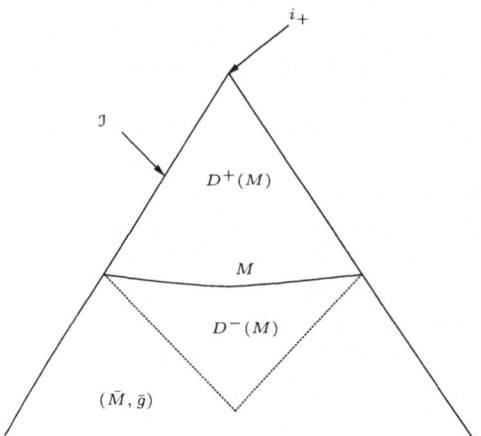

FIGURE 9. The semi-global existence theorem of H. Friedrich

The first true small data global existence result for the vacuum Einstein equations, was proved by D. Christodoulou and S. Klainerman [46]. They proved that for data (M, g, k) sufficiently close to standard data on a hyperplane in Minkowski space, with appropriate decay at spatial infinity, the MVCD $(\bar{M}, \bar{g})$ is geodesically complete and therefore inextendible, cf. Figure 10. The Christodoulou–Klainerman global existence theorem therefore supports the cosmic censorship conjecture.

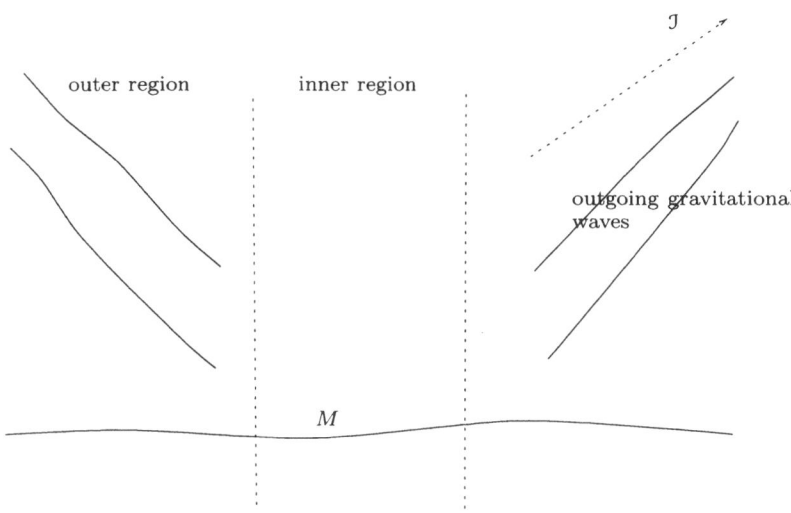

FIGURE 10. The global existence theorem of D. Christodoulou and S. Klainerman

Klainerman and Nicolò [110, 111] have proved a global existence result for data (M, g, k) which are close to standard flat Minkowski space data on an exterior region $M \setminus K$, where $K \subset M$ is compact so that $M \setminus K \cong \mathbb{R}^3 \setminus \mathrm{Ball}$, and with weaker asymptotic conditions compared to the Christodoulou–Klainerman theorem. The result of Klainerman and Nicoló states that for a vacuum data set (M, g, k) which is sufficiently close to the standard flat Minkowski data on $M \setminus K$, the outgoing null geodesics in the causal exterior region $D(M \setminus K)$ are complete and $D(M \setminus K)$ is covered by a double null foliation, with precise control over the asymptotics. It should be noted that the KN theorem therefore covers a more general class of spacetimes than the Christodoulou–Klainerman theorem and is not strictly a small data result. In particular, the maximal Cauchy vacuum development of (M, g, k) may be singular for data satisfying the assumptions of the KN theorem, cf. Figure 11. If the smallness assumption is extended also to the interior region, the Christodoulou–Klainerman theorem is recovered. The Einstein equations are quadratic in first derivatives, and therefore in $3 + 1$ dimensions, one needs something like a null condition in order to get sufficient decay for a global existence argument. The Einstein equations are not known to satisfy a null condition. However, by a detailed construction of approximate Killing fields and approximate conformal Killing fields, Christodoulou and Klainerman are able to control the behavior of components of Bel–Robinson tensors constructed from the Weyl tensor and its derivatives, and close a bootstrap argument which gives global existence for sufficiently small data. As part of this argument, it is necessary to get detailed control over the asymptotic behavior of light cones. Christodoulou and Klainerman also study the asymptotic behavior of components of the Weyl tensor and are able

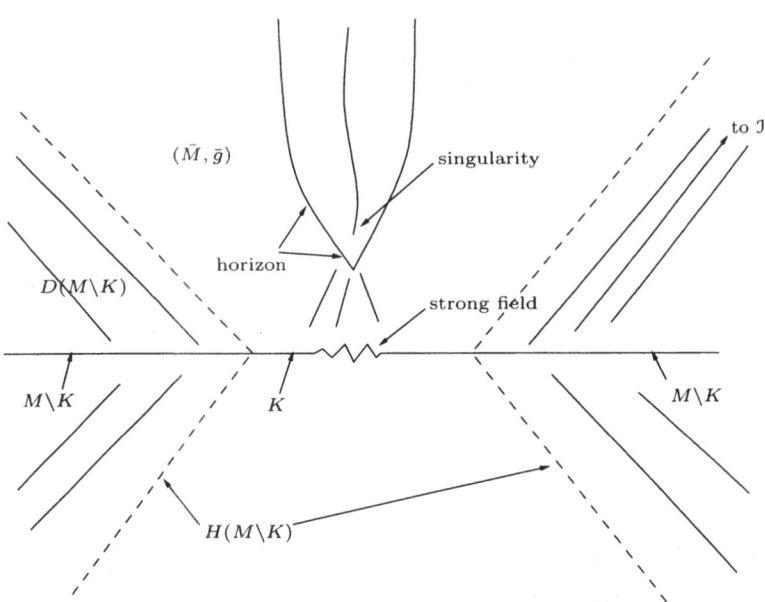

FIGURE 11. The exterior global existence theorem of S. Klainerman and F. Nicoló. The figure illustrates a situation which is covered by this theorem, with a singularity forming due to a strong
gravitational field in the interior region $I(K)$ while the exterior
region $D(M \setminus K)$ has complete outgoing null geodesics reaching ℐ.

to prove that some, but not all of these have the decay implied by the Penrose
conditions on ℐ.

The question about existence of vacuum asymptotically Minkowskian spacetimes with regular conformal completion, in the sense of Penrose, has been open
for a long time. Recently, Delay and Chruściel [51] were able to use a gluing construction of Corvino and Schoen [59] to construct vacuum spacetimes which are
exactly Schwarzschild near spatial infinity, and which are close enough to Minkowski space, so that the global existence result of Friedrich applies to show existence
of a global ℐ. Further, Klainerman and Nicoló [112] have been able to show that
a class of asymptotically Euclidean initial data with suitable decay at infinity has
a Cauchy development where the null components of the Weyl tensor have decay
corresponding to the peeling conditions of Penrose.

The work in [6] and [4] shows that generic hyperboloidal data do not satisfy
the required regularity at the conformal boundary, which therefore may be viewed
as an indication that Penrose regularity at ℐ is non-generic. It is still an open
question what general conditions on initial data on an asymptotically flat Cauchy
surface give a Cauchy development with regular conformal completion. H. Friedrich
has been developing a programme which approaches this problem by analyzing the

conformal structure at spatial infinite in detail, see [60] for references on this, and see also the paper [162] by Valiente-Kroon, which points out some new obstructions to regularity.

In contrast to the results of Friedrich and Christodoulou–Klainerman–Nicoló, the semi-global existence result of Andersson and Moncrief [9] deals with spatially compact vacuum spacetimes.

Let (M, γ) be a compact hyperbolic 3-manifold with metric γ of sectional curvature -1. Then the spacetime $\bar{M} = M \times \mathbb{R}$ with metric $\bar{\gamma} = -dt^2 + t^2 \gamma$ is a $\kappa = -1$ (local) FRW spacetime. A flat spacetime metric on $\bar{M}$ defines a geometric structure with group $SO(3, 1) \ltimes \mathbb{R}^4$. The moduli space of flat spacetimes with topology $M \times \mathbb{R}$ has the same dimension as the moduli space of flat conformal structures on M, see [3]. We say that M is **rigid** if the dimension of the moduli space of flat conformal structures on M is zero. This is equivalent to the condition that there are no nonzero traceless Codazzi tensors on (M, γ). Kapovich [106] has constructed examples of rigid compact hyperbolic 3-manifolds.

The global existence theorem proved in [9] states that if M is a rigid manifold of hyperbolic type, then for a vacuum data set (M, g, k), sufficiently close to the standard data in a spatially compact $\kappa = -1$ (local) FRW spacetime, the MVCD $(\bar{M}, \bar{g})$ is causally geodesically complete in the expanding direction. It is a consequence of the singularity theorem of Hawking and Penrose that $(\bar{M}, \bar{g})$ is singular, i.e., geodesics in the collapsing direction are incomplete, cf. Figure 12.

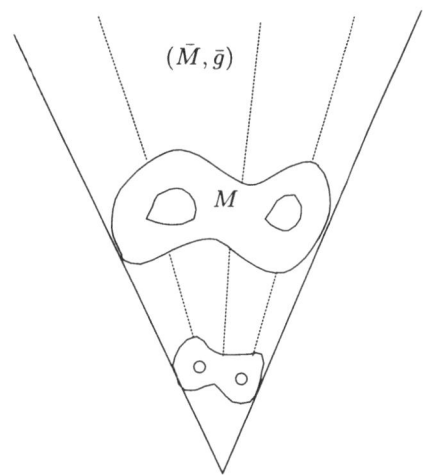

FIGURE 12. The semi-global existence theorem of L. Andersson and V. Moncrief.

The proof uses local well-posedness and a continuation principle for the Einstein equations in constant mean curvature and spatial harmonic coordinates (CM-CSH) gauge, cf. Subsection 2.3, together with energy estimates for a first-order Bel–Robinson type energy, defined in terms of the Weyl tensor and its covariant derivative w.r.t. T. Let W_{abcd} be the Weyl tensor and let ${}^*W_{abcd}$ be its dual. In a vacuum spacetime $W_{abcd} = R_{abcd}$. Define $E_{ij} = W_{iTjT}$ and $B_{ij} = {}^*W_{iTjT}$. Then E_{ij} and B_{ij} are the electric and magnetic parts of the Weyl tensor. These fields are traceless and symmetric and satisfy a Maxwell-like system. A computation shows that the zeroth-order Bel–Robinson energy

$$Q = \int |E|^2 + |B|^2$$

(which measures second derivatives of g in L^2), satisfies the equation

$$\partial_t = -3 \int_{M_t} N[(E \times E) \cdot k + (B \times B) \cdot k - \frac{1}{3}(|E|^2 + |B|^2)\mathrm{tr}k$$
$$- 2N^{-1}\nabla^i N(E \wedge B)_i]d\mu_{M_t}$$

where for symmetric traceless two-tensors A, B,

$$(A \times B)_{ij} = A_i{}^m B_{mj} + A_j{}^m B_{mi} - \frac{2}{3}A^{mn}B_{mn}g_{ij}$$

and

$$(A \wedge B)_i = \epsilon_i{}^{mn}A_m{}^k B_{kn},$$

Combining Q with the a term defined analogously in terms of $\bar{\nabla}_T W_{abcd}$ and extracting a scale invariant quantity, leads to an energy $\mathcal{E}$, which bounds the Cauchy data in $H^3 \times H^2$ for small data. This energy is shown to satisfy an energy inequality of the form

$$\partial_t \mathcal{E} \leq -2\mathcal{E} - C\mathcal{E}^{3/2}$$

Using this one may show that if $\mathcal{E}$ is small initially, then $\mathcal{E}$ will decrease, as long as local existence holds. A continuation principle for the Einstein equations in $H^3 \times H^2$ [10] completes the proof of global existence for small data.

7. Concluding remarks

The results and numerical studies so far can be argued to fit with the so-called BKL picture of cosmological singularities, which states that the generic singularity should be space-like, local and oscillatory, see [28] for a review. Roughly, one expects that the "points on the singularity" are causally separated and that locally at the singularity, the dynamics undergoes a chaotic sequence of curvature driven "bounces" interspersed with relatively uneventful "coasting" epochs, with Kasner like dynamics (i.e., the metric is, locally in space, approximately of the form given in (3.6)). The locality at the singularity, is easily checked for the non-vacuum Friedman-Robertson-Walker models of the standard model of cosmology, and is the cause of the so-called "horizon problem" in cosmology. Locality at the singularity can be proved for Gowdy and polarized U(1).

The picture with bounces and coasting is to a large extent inspired by the Mixmaster model (Bianchi IX, cf. Section 3) which is known to have an oscillatory singularity, cf. Theorem 3.1, and the above scenario may therefore be described by the slogan "generic singularities are spacelike and oscillatory". By an abuse of language we may say that the oscillatory behavior is of "mixmaster type".

An approach to understanding the asymptotic dynamics near singularities is to attempt to extract, in a systematic way, a dynamical system governing the asymptotic behavior of the Einstein equations. A promising approach is based on an analysis of the evolution equations written in Hubble-normalized scale-free orthonormal frame variables along the lines discussed for the Gowdy system in Section 4. The general situation has been discussed in [161]. In this setting the Einstein evolution equations appear in first-order form, and the natural version of the BKL conjecture is that $\partial_1 U \to 0$ in the direction of the singularity, where ∂_1 is a scalefree first-order operator analogous to the operator $e^{-\tau}\partial_x$ for the Gowdy system, and U denotes the scale free frame variables.

If this picture is correct, the asymptotic behavior near the singularity can be understood by analyzing the asymptotic behavior of orbits of the system of ODE's that one gets by cancelling the terms containing ∂_1 in the first-order form of the Einstein evolution equations. The possibilities that arise have been analyzed in [161]. It is now important to understand whether or not this version of the BKL conjecture is correct by doing numerical experiments. No complete results in this direction have been obtained so far.

The mixmaster type behavior may be prevented by the presence of certain types of matter such as a scalar field (stiff fluid). In fact work by Andersson and Rendall [12], shows that the $3 + 1$-dimensional Einstein-scalar field equations may be formulated as a Fuchsian system near the singularity, and AVTD solutions may be constructed using a singular version of the Cauchy–Kowalevskaya theorem. For the polarized $U(1)$ case, the reduced system is $2 + 1$-dimensional Einstein-scalar field equations. The Fuchsian method has been applied to this special case by Isenberg and Moncrief [102]. Damour et al. [61] have given a rather complete study of the application of the Fuchsian method to spacetimes of general dimension with matter.

It appears likely that some aspect of the picture sketched above will be relevant for the final analysis of the large data, global behavior of vacuum spacetimes.

Appendix A. Basic causality concepts

Here we introduce the basic causality concepts, see [165, Chapter 8], [24, 84, 132] for details.

A compact manifold $\bar{M}$ admits a Lorentz metric if and only if its Euler characteristic $\chi(\bar{M})$ vanishes, while all noncompact manifolds admit Lorentz metrics. A vector $V \in T\bar{M}$ is called space-like, null or time-like if $\bar{g}(V, V) > 0, = 0$ or < 0 respectively. V is called causal if it is null or time-like. A C^1 curve $c \subset \bar{M}$ is called time-like (causal) if $\dot{c}$ is time-like (causal). This extends naturally to continuous curves. A hypersurface $M \subset \bar{M}$ is called acausal if no causal curve meets M more than once, and is called space-like if the normal of M is time-like.

Given an acausal closed subset $S \subset \bar{M}$, the **domain of dependence** $D(S)$ is the set of points $p \in \bar{M}$ such that any inextendible causal curve containing p must intersect S. If $S \subset \bar{M}$ is a space-like hypersurface and $D(S) = \bar{M}$, S is called a **Cauchy surface** in $\bar{M}$. If $\bar{M}$ has a Cauchy surface M is called **globally hyperbolic**. A globally hyperbolic spacetime has a global time function, i.e., a function t on $\bar{M}$ so that $\bar{\nabla} t$ is time-like, with the level sets of t being Cauchy surfaces. In particular, $\bar{M} \cong M \times \mathbb{R}$ where M is any Cauchy surface.

$\bar{M}$ is called **time oriented** if there is a global time-like vector field on $\bar{M}$. A globally hyperbolic spacetime is time oriented by the above, and hence it makes sense to talk about the future and past domains of dependence $D^+(M)$ and $D^-(M)$ of a Cauchy surface M. In a time oriented space time, the chronological future $I^+(S)$ of $S \subset \bar{M}$ is the set of all points reached by future directed time-like curves, of nonzero length, starting on S. The time-like past $I^-(S)$ is defined analogously. The causal future and past $J^+(S)$ and $J^-(S)$ is defined analogously to $I^\pm(S)$ with causal curve replacing time-like curve (the causal curve is allowed to be trivial).

The variational problem for geodesics (w.r.t. Lorentzian length) is well behaved precisely when $(\bar{M}, \bar{g})$ is globally hyperbolic. Let $C(p, q)$ be the set of continuous causal curves between $p, q \in \bar{M}$. Then $C(p, q)$ is compact w.r.t. uniform convergence for all $p, q \in \bar{M}$, if $(\bar{M}, \bar{g})$ is globally hyperbolic. This makes global hyperbolicity a natural assumption in Lorentzian geometry, just as completeness is a natural assumption in Riemannian geometry.

Global hyperbolicity may also be characterized by strong causality and compactness of $C(p, q)$ for all $p, q \in \bar{M}$, or compactness of $J^+(p) \cap J^-(q)$ for all $p, q \in \bar{M}$, see [165, Chapter 8] for details.

Given a spacetime $(\bar{M}, \bar{g})$ and a space-like hypersurface $M \subset \bar{M}$, the future **Cauchy horizon** $H^+(M)$ is given by $H^+(M) = \overline{D^+(M)} \setminus I^-[D^+(M)]$, with the past Cauchy horizon $H^-(M)$ defined analogously. The set $H(M) = H^+(M) \cup H^-(M)$ is called the Cauchy horizon of M. It can be proved that $H(M) = \partial D(M)$, the boundary of $D(M)$.

Acknowledgements

I am grateful to Vince Moncrief for numerous conversations on the topics covered here and for detailed comments on an early version. Thanks are due to Håkan Andréasson, Piotr Chruściel, Jim Isenberg and Alan Rendall and others for helpful comments. I am happy to acknowledge the hospitality and support of the Institute of Theoretical Physics, UCSB, and Institut des Hautes Études Scientifiques, Bures-sur-Yvette, where part of the writing was done.

References

[1] Miguel Alcubierre, *Appearance of coordinate shocks in hyperbolic formalisms of general relativity*, Phys. Rev. D (3) **55** (1997), no. 10, 5981–5991.

[2] Miguel Alcubierre and Joan Massó, *Pathologies of hyperbolic gauges in general relativity and other field theories*, Phys. Rev. D (3) **57** (1998), no. 8, R4511–R4515.

[3] Lars Andersson, *Constant mean curvature foliations of flat space-times*, Comm. Anal. Geom. **10** (2002), no. 5, 1125–1150.

[4] Lars Andersson and Piotr T. Chruściel, *Hyperboloidal Cauchy data for vacuum Einstein equations and obstructions to smoothness of null infinity*, Phys. Rev. Lett. **70** (1993), no. 19, 2829–2832.

[5] _____, *On "hyperboloidal" Cauchy data for vacuum Einstein equations and obstructions to smoothness of scri*, Comm. Math. Phys. **161** (1994), no. 3, 533–568.

[6] Lars Andersson, Piotr T. Chruściel, and Helmut Friedrich, *On the regularity of solutions to the Yamabe equation and the existence of smooth hyperboloidal initial data for Einstein's field equations*, Comm. Math. Phys. **149** (1992), no. 3, 587–612.

[7] Lars Andersson, Gregory J. Galloway, and Ralph Howard, *A strong maximum principle for weak solutions of quasi-linear elliptic equations with applications to Lorentzian and Riemannian geometry*, Comm. Pure Appl. Math. **51** (1998), no. 6, 581–624.

[8] Lars Andersson and Mirta S. Iriondo, *Existence of constant mean curvature hypersurfaces in asymptotically flat spacetimes*, Ann. Global Anal. Geom. **17** (1999), no. 6, 503–538.

[9] Lars Andersson and Vincent Moncrief, *Future complete vacuum spacetimes*, gr-qc/0303045, in this volume.

[10] _____, *Elliptic-hyperbolic systems and the Einstein equations*, Ann. Henri Poincaré **4** (2003), no. 1, 1–34.

[11] Lars Andersson, Vincent Moncrief, and Anthony J. Tromba, *On the global evolution problem in $2 + 1$ gravity*, J. Geom. Phys. **23** (1997), no. 3–4, 191–205.

[12] Lars Andersson and Alan D. Rendall, *Quiescent cosmological singularities*, Comm. Math. Phys. **218** (2001), no. 3, 479–511.

[13] Lars Andersson, Henk van Elst, and Claes Uggla, *Gowdy phenomenology in scalefree variables*, Classical Quantum Gravity **21** (2004), S29–S57, *Spacetime Safari: Essays in Honor of Vincent Moncrief on the Classical Physics of Strong Gravitational Fields*, special issue of Classical and Quantum Gravity, eds. J. Isenberg and B. Berger.

[14] Håkan Andréasson, Alan D. Rendall, and Marsha Weaver, *Existence of CMC and constant areal time foliations in T^2 symmetric spacetimes with Vlasov matter*, Comm. Partial Differential Equations **29** (2004), no. 1–2, 237–262.

[15] Abhay Ashtekar, Jiří Bičák, and Bernd G. Schmidt, *Asymptotic structure of symmetry-reduced general relativity*, Phys. Rev. D (3) **55** (1997), no. 2, 669–686.

[16] _____, *Behavior of Einstein-Rosen waves at null infinity*, Phys. Rev. D (3) **55** (1997), no. 2, 687–694.

[17] Abhay Ashtekar and Madhavan Varadarajan, *Striking property of the gravitational Hamiltonian*, Phys. Rev. D (3) **50** (1994), no. 8, 4944–4956.

[18] Hajer Bahouri and Jean-Yves Chemin, *Équations d'ondes quasi-linéaires et estima-tions de Strichartz*, C. R. Acad. Sci. Paris Sér. I Math. **327** (1998), no. 9, 803–806.

[19] M. Salah Baouendi and Charles Goulaouic, *Remarks on the abstract form of non-linear Cauchy-Kovalevsky theorems*, Comm. Partial Differential Equations **2** (1977), no. 11, 1151–1162.

[20] John D. Barrow, Gregory J. Galloway, and Frank J. Tipler, *The closed-universe recollapse conjecture*, Mon. Not. R. Astron. Soc. **223** (1986), 835–844.

[21] Robert Bartnik, *Existence of maximal surfaces in asymptotically flat spacetimes*, Comm. Math. Phys. **94** (1984), no. 2, 155–175.

[22] _____, *Regularity of variational maximal surfaces*, Acta Math. **161** (1988), no. 3-4, 145–181.

[23] _____, *Remarks on cosmological spacetimes and constant mean curvature surfaces*, Comm. Math. Phys. **117** (1988), no. 4, 615–624.

[24] John K. Beem, Paul E. Ehrlich, and Kevin L. Easley, *Global Lorentzian geometry*, second ed., Marcel Dekker Inc., New York, 1996.

[25] R. Beig and N.Ó. Murchadha, *Late time behaviour of the maximal slicing of the Schwarzschild black hole*, Phys. Rev. D (3) **57** (1998), 4728–4737, gr-qc/9706046.

[26] B.K. Berger and D. Garfinkle, *Phenomenology of the Gowdy universe on $T^3 \times R$*, Phys. Rev. **D57** (1998), 4767–4777, gr-qc/9710102.

[27] Beverly K. Berger, Piotr T. Chruściel, James Isenberg, and Vincent Moncrief, *Global foliations of vacuum spacetimes with T^2 isometry*, Ann. Physics **260** (1997), no. 1, 117–148.

[28] Beverly K. Berger, David Garfinkle, James Isenberg, Vincent Moncrief, and Mar-sha Weaver, *The singularity in generic gravitational collapse is spacelike, local and oscillatory*, Modern Phys. Lett. A **13** (1998), no. 19, 1565–1573.

[29] Beverly K. Berger, David Garfinkle, and Vincent Moncrief, *Comment on "The Gowdy T^3 cosmologies revisited"*, gr-qc/9708050.

[30] Beverly K. Berger, David Garfinkle, and Eugene Strasser, *New algorithm for Mix-master dynamics*, Classical Quantum Gravity **14** (1997), no. 2, L29–L36.

[31] Beverly K. Berger, James Isenberg, and Marsha Weaver, *Oscillatory approach to the singularity in vacuum spacetimes with T^2 isometry*, Phys. Rev. D (3) **64** (2001), no. 8, 084006, 20.

[32] Beverly K. Berger and Vincent Moncrief, *Evidence for an oscillatory singularity in generic U(1) symmetric cosmologies on $T^3 \times \mathbf{R}$*, Phys. Rev. D (3) **58** (1998), 064023, gr-qc/9804085.

[33] _____, *Numerical evidence that the singularity in polarized U(1) symmetric cos-mologies on $T^3 \times \mathbf{R}$ is velocity dominated*, Phys. Rev. D (3) **57** (1998), no. 12, 7235–7240, gr-qc/9801078.

[34] Luc Blanchet and Thibault Damour, *Hereditary effects in gravitational radiation*, Phys. Rev. D (3) **46** (1992), no. 10, 4304–4319.

[35] Dieter Brill and Frank Flaherty, *Isolated maximal surfaces in spacetime*, Comm. Math. Phys. **50** (1976), no. 2, 157–165.

[36] Gregory A. Burnett and Alan D. Rendall, *Existence of maximal hypersurfaces in some spherically symmetric spacetimes*, Classical Quantum Gravity **13** (1996), no. 1, 111–123.

[37] John Cameron and Vincent Moncrief, *The reduction of Einstein's vacuum equations on spacetimes with spacelike U(1)-isometry groups*, Mathematical aspects of classical field theory (Seattle, WA, 1991), Amer. Math. Soc., Providence, RI, 1992, pp. 143–169.

[38] Myeongju Chae and Piotr T. Chruściel, *On the dynamics of Gowdy space-times*, Comm. Pure Appl. Math. **57** (2004), no. 8, 1015–1074.

[39] Yvonne Choquet-Bruhat, *Future complete Einsteinian space times with U(1) symmetry, the unpolarized case*, article in this volume, gr-qc/0305060.

[40] Yvonne Choquet-Bruhat and Robert Geroch, *Global aspects of the Cauchy problem in general relativity*, Comm. Math. Phys. **14** (1969), 329–335.

[41] Yvonne Choquet-Bruhat and Vincent Moncrief, *Existence theorem for solutions of Einstein's equations with 1 parameter spacelike isometry groups*, Quantization, nonlinear partial differential equations, and operator algebra (Cambridge, MA, 1994), Proc. Symp. Pure Math., vol. 59, Amer. Math. Soc., Providence, RI, 1996, pp. 67–80.

[42] _____, *Future global in time Einsteinian spacetimes with U(1) isometry group*, Ann. Henri Poincaré **2** (2001), no. 6, 1007–1064.

[43] Yvonne Choquet-Bruhat and Tommaso Ruggeri, *Hyperbolicity of the 3 + 1 system of Einstein equations*, Comm. Math. Phys. **89** (1983), no. 2, 269–275.

[44] Yvonne Choquet-Bruhat and James W. York, *Mixed elliptic and hyperbolic systems for the Einstein equations*, gr-qc/9601030.

[45] Demetrios Christodoulou, *The instability of naked singularities in the gravitational collapse of a scalar field*, Ann. of Math. (2) **149** (1999), no. 1, 183–217.

[46] Demetrios Christodoulou and Sergiu Klainerman, *The global nonlinear stability of the Minkowski space*, Princeton University Press, Princeton, NJ, 1993.

[47] Piotr T. Chruściel, *On space-times with U(1) × U(1) symmetric compact Cauchy surfaces*, Ann. Physics **202** (1990), no. 1, 100–150.

[48] _____, *On uniqueness in the large of solutions of Einstein's equations ("strong cosmic censorship")*, Proceedings of the Centre for Mathematics and its Applications, Australian National University, vol. 27, Australian National University Centre for Mathematics and its Applications, Canberra, 1991.

[49] _____, *On uniqueness in the large of solutions of Einstein's equations ("strong cosmic censorship")*, Mathematical aspects of classical field theory (Seattle, WA, 1991), Amer. Math. Soc., Providence, RI, 1992, pp. 235–273.

[50] _____, *On completeness of orbits of Killing vector fields*, Classical Quantum Gravity **10** (1993), no. 10, 2091–2101, ITP Preprint NSF-ITP-93-44.

[51] Piotr T. Chruściel and Erwann Delay, *Existence of non-trivial, vacuum, asymptotically simple spacetimes*, Classical Quantum Gravity **19** (2002), no. 9, L71–L79, Erratum, Classical Quantum Gravity **19** (2002), no. 12, 3389.

[52] Piotr T. Chruściel and Gregory J. Galloway, *Horizons non-differentiable on a dense set*, Comm. Math. Phys. **193** (1998), no. 2, 449–470.

[53] Piotr T. Chruściel and James Isenberg, *Nonisometric vacuum extensions of vacuum maximal globally hyperbolic spacetimes*, Phys. Rev. D (3) **48** (1993), no. 4, 1616–1628.

[54] Piotr T. Chruściel, James Isenberg, and Vincent Moncrief, *Strong cosmic censorship in polarised Gowdy spacetimes*, Classical Quantum Gravity **7** (1990), no. 10, 1671–1680.

[55] Piotr T. Chruściel and Alan D. Rendall, *Strong cosmic censorship in vacuum spacetimes with compact, locally homogeneous Cauchy surfaces*, Ann. Physics **242** (1995), no. 2, 349–385.

[56] Piotr T. Chruściel and Jalal Shatah, *Global existence of solutions of the Yang-Mills equations on globally hyperbolic four-dimensional Lorentzian manifolds*, Asian J. Math. **1** (1997), no. 3, 530–548.

[57] Neil J. Cornish and Janna Levin, *The mixmaster univese: A chaotic Farey tale*, Phys. Rev. D (3) **55** (1997), 7489–7510, gr-qc/9612066.

[58] Justin Corvino, *Scalar curvature deformation and a gluing construction for the Einstein constraint equations*, Comm. Math. Phys. **214** (2000), no. 1, 137–189.

[59] _____, *Scalar curvature deformation and a gluing construction for the Einstein constraint equations*, Comm. Math. Phys. **214** (2000), no. 1, 137–189.

[60] Sergio Dain and Helmut Friedrich, *Asymptotically flat initial data with prescribed regularity at infinity*, Comm. Math. Phys. **222** (2001), no. 3, 569–609.

[61] T. Damour, M. Henneaux, A.D. Rendall, and M. Weaver, *Kasner-like behaviour for subcritical Einstein-matter systems*, Ann. Henri Poincaré **3** (2002), no. 6, 1049–1111.

[62] Douglas M. Eardley, Edison Liang, and Rainer Kurt Sachs, *Velocity dominated singularities in irrotational dust cosmologies*, J. Math. Phys. **13** (1972), 99–107.

[63] Douglas M. Eardley and Vincent Moncrief, *The global existence of Yang– Mills–Higgs fields in 4-dimensional Minkowski space. I. Local existence and smoothness properties*, Comm. Math. Phys. **83** (1982), no. 2, 171–191.

[64] _____, *The global existence of Yang–Mills–Higgs fields in 4-dimensional Minkowski space. II. Completion of proof*, Comm. Math. Phys. **83** (1982), no. 2, 193–212.

[65] Klaus Ecker and Gerhard Huisken, *Parabolic methods for the construction of spacelike slices of prescribed mean curvature in cosmological spacetimes*, Comm. Math. Phys. **135** (1991), no. 3, 595–613.

[66] G.F.R. Ellis and M.A.H. MacCallum, *A class of homogeneous cosmological models*, Comm. Math. Phys. **12** (1969), 108–141.

[67] Arthur Fischer and Vincent Moncrief, *Hamiltonian reduction of Einstein's equations of general relativity*, Nuclear Physics B (Proc. Suppl) **57** (1997), 142–161.

[68] Arthur E. Fischer, Jerrold E. Marsden, and Vincent Moncrief, *The structure of the space of solutions of Einstein's equations. I. One Killing field*, Ann. Inst. H. Poincaré Sect. A (N.S.) **33** (1980), no. 2, 147–194.

[69] Helmut Friedrich, *On the existence of n-geodesically complete or future complete solutions of Einstein's field equations with smooth asymptotic structure*, Comm. Math. Phys. **107** (1986), no. 4, 587–609.

[70] _____ , *On the global existence and the asymptotic behavior of solutions to the Einstein-Maxwell-Yang-Mills equations*, J. Differential Geom. **34** (1991), no. 2, 275–345.

[71] _____ , *Hyperbolic reductions for Einstein's equations*, Classical Quantum Gravity **13** (1996), no. 6, 1451–1469.

[72] Helmut Friedrich, István Rácz, and Robert M. Wald, *On the rigidity theorem for spacetimes with a stationary event horizon or a compact Cauchy horizon*, gr-qc/9811021, 1998.

[73] Simonetta Frittelli and Oscar A. Reula, *First-order symmetric-hyperbolic Einstein equations with arbitrary fixed gauge*, Phys. Rev. Lett **76** (1996), 4667–4670, gr-qc/9605005.

[74] Yoshihisa Fujiwara, Hideki Ishihara, and Hideo Kodama, *Comments on closed Bianchi models*, Classical Quantum Gravity **10** (1993), no. 5, 859–867, gr-qc/9301019.

[75] Gregory J. Galloway, *The Lorentzian splitting theorem without the completeness assumption*, J. Differential Geom. **29** (1989), no. 2, 373–387.

[76] _____ , *Some rigidity results for spatially closed spacetimes*, Mathematics of gravitation, Part I (Warsaw, 1996), Polish Acad. Sci., Warsaw, 1997, pp. 21–34.

[77] David Garfinkle and Marsha Weaver, *High velocity spikes in Gowdy spacetimes*, Phys. Rev. D (3) **67** (2003), no. 12, 124009, 5.

[78] Claus Gerhardt, *H-surfaces in Lorentzian manifolds*, Comm. Math. Phys. **89** (1983), no. 4, 523–553.

[79] _____ , *Hypersurfaces of prescribed mean curvature in Lorentzian manifolds*, Math. Z. **235** (2000), no. 1, 83–97.

[80] Robert H. Gowdy, *Gravitational waves in closed universes*, Phys. Rev. Lett. **27** (1971), 826–829.

[81] _____ , *Vacuum spacetimes with two-parameter spacelike isometry groups and compact invariant hypersurfaces: topologies and boundary conditions*, Ann. Physics **83** (1974), 203–241.

[82] Boro Grubišić and Vincent Moncrief, *Asymptotic behavior of the $T^3 \times \mathbf{R}$ Gowdy space-times*, Phys. Rev. D (3) **47** (1993), no. 6, 2371–2382, gr-qc/9209006.

[83] _____ , *Mixmaster spacetime, Geroch's transformation, and constants of motion*, Phys. Rev. D (3) **49** (1994), no. 6, 2792–2800, gr-qc/9309007.

[84] S.W. Hawking and G.F.R. Ellis, *The large scale structure of space-time*, Cambridge University Press, London, 1973, Cambridge Monographs on Mathematical Physics, No. 1.

[85] Oliver Henkel, *Global prescribed mean curvature foliations in cosmological space-times. I, II*, J. Math. Phys. **43** (2002), no. 5, 2439–2465, 2466–2485.

[86] Simon D. Hern and John M. Stewart, *The Gowdy T^3 cosmologies revisited*, Classical Quantum Gravity **15** (1998), no. 6, 1581–1593.

[87] Conrad G. Hewitt, Joshua T. Horwood, and John Wainwright, *Asymptotic dynamics of the exceptional Bianchi cosmologies*, gr-qc/0211071, 2002.

[88] David Hobill, Adrian Burd, and Alan Coley (eds.), *Deterministic chaos in general relativity*, New York, Plenum Press, 1994.

[89] Lars Hörmander, *Lectures on nonlinear hyperbolic differential equations*, Springer-Verlag, Berlin, 1997.

[90] Thomas J.R. Hughes, Tosio Kato, and Jerrold E. Marsden, *Well-posed quasi-linear second-order hyperbolic systems with applications to nonlinear elastodynamics and general relativity*, Arch. Rational Mech. Anal. **63** (1976), no. 3, 273–294 (1977).

[91] Gerhard Huisken and Tom Ilmanen, *The Riemannian Penrose inequality*, Internat. Math. Res. Notices **1997**, no. 20, 1045–1058.

[92] _____, *The inverse mean curvature flow and the Riemannian Penrose inequality*, J. Differential Geom. **59** (2001), no. 3, 353–437.

[93] Mirta S. Iriondo, Enzo O. Leguizamón, and Oscar A. Reula, *Einstein's equations in Ashtekar's variables constitute a symmetric hyperbolic system*, Phys. Rev. Lett **79** (1997), 4732–4735, gr-qc/9710004.

[94] _____, *On the dynamics of Einstein's equations in the Ashtekar formulation*, Adv. Theor. Math. Phys. **2** (1998), no. 5, 1075–1103.

[95] James Isenberg, *Progress on strong cosmic censorship*, Mathematical aspects of classical field theory (Seattle, WA, 1991), Amer. Math. Soc., Providence, RI, 1992, pp. 403–418.

[96] James Isenberg and Satyanad Kichenassamy, *Asymptotic behavior in polarized T^2-symmetric vacuum space-times*, J. Math. Phys. **40** (1999), no. 1, 340–352.

[97] James Isenberg, Rafe Mazzeo, and Daniel Pollack, *Gluing and wormholes for the Einstein constraint equations*, Comm. Math. Phys. **231** (2002), no. 3, 529–568.

[98] _____, *On the topology of vacuum spacetimes*, Ann. Henri Poincaré 4 (2003), no. 2, 369–383, gr-qc/0206034.

[99] James Isenberg and Vincent Moncrief, *The existence of constant mean curvature foliations of Gowdy 3-torus spacetimes*, Comm. Math. Phys. **86** (1982), no. 4, 485–493.

[100] _____, *Symmetries of cosmological Cauchy horizons with exceptional orbits*, J. Math. Phys. **26** (1985), no. 5, 1024–1027.

[101] _____, *Asymptotic behavior of the gravitational field and the nature of singularities in Gowdy spacetimes*, Ann. Physics **199** (1990), no. 1, 84–122.

[102] _____, *Asymptotic behaviour in polarized and half-polarized U(1) symmetric vacuum spacetimes*, Classical Quantum Gravity **19** (2002), no. 21, 5361–5386.

[103] James Isenberg and Alan D. Rendall, *Cosmological spacetimes not covered by a constant mean curvature slicing*, Classical Quantum Gravity **15** (1998), no. 11, 3679–3688.

[104] Akihiro Ishibashi, Tatsuhiko Koike, Masaru Siino, and Sadayoshi Kojima, *Compact hyperbolic universe and singularities*, Phys. Rev. D (3) **54** (1996), no. 12, 7303–7310, gr-qc/9605041.

[105] Ronald Kantowski and Rainer Kurt Sachs, *Some spatially homogeneous anisotropic relativistic cosmological models*, J. Mathematical Phys. **7** (1966), 443–446.

[106] Michael Kapovich, *Deformations of representations of discrete subgroups of* SO(3, 1), Math. Ann. **299** (1994), no. 2, 341–354.

[107] Satyanad Kichenassamy and Alan D. Rendall, *Analytic description of singularities in Gowdy spacetimes*, Classical Quantum Gravity **15** (1998), no. 5, 1339–1355, preprint at http://www.aei-potsdam.mpg.de.

[108] S. Klainerman, *On the regularity of classical field theories in Minkowski space-time* $\mathbf{R}^{3+1}$, Nonlinear partial differential equations in geometry and physics (Knoxville, TN, 1995), Birkhäuser, Basel, 1997, pp. 29–69.

[109] Sergiu Klainerman and Matei Machedon, *Finite energy solutions of the Yang-Mills equations in* $\mathbf{R}^{3+1}$, Ann. of Math. (2) **142** (1995), no. 1, 39–119.

[110] Sergiu Klainerman and Francesco Nicolò, *On local and global aspects of the Cauchy problem in general relativity*, Classical Quantum Gravity **16** (1999), R73–R157.

[111] ———, *The evolution problem in general relativity*, Progress in Mathematical Physics, vol. 25, Birkhäuser Boston Inc., Boston, MA, 2003.

[112] ———, *Peeling properties of asymptotically flat solutions to the Einstein vacuum equations*, Classical and Quantum Gravity **20** (2003), no. 14, 3215–3257.

[113] Sergiu Klainerman and Igor Rodnianski, *The causal structure of microlocalized Einstein metrics*, math.AP/0109174, 2001.

[114] ———, *Rough solution for the Einstein vacuum equations*, math.AP/0109173, 2001.

[115] Sergiu Klainerman and Igor Rodnianski, *Ricci defects of microlocalized Einstein metrics*, J. Hyperbolic Differ. Equ. **1** (2004), no. 1, 85–113.

[116] Sergiu Klainerman and Sigmund Selberg, *Remark on the optimal regularity for equations of wave maps type*, Comm. Partial Differential Equations **22** (1997), no. 5-6, 901–918.

[117] Hideo Kodama, *Canonical structure of locally homogeneous systems on compact closed 3-manifolds of types* E^3, *Nil and Sol*, Progr. Theoret. Phys. **99** (1998), no. 2, 173–236.

[118] Tatsuhiko Koike, Masayuki Tanimoto, and Akio Hosoya, *Compact homogeneous universes*, J. Math. Phys. **35** (1994), no. 9, 4855–4888.

[119] Joachim Krieger, *Global regularity of wave maps from* $\mathbf{r}^{2+1}$ *to* $\mathbf{h}^2$, preprint.

[120] E.M. Lifshitz and I.M. Khalatnikov, *Investigations in relativistic cosmology*, Adv. in Phys. **12** (1963), 185–249.

[121] Xue-Feng Lin and Robert M. Wald, *Proof of the closed-universe-recollapse conjecture for diagonal Bianchi type-*IX *cosmologies*, Phys. Rev. D (3) **40** (1989), no. 10, 3280–3286.

[122] Hans Lindblad, *Counterexamples to local existence for quasilinear wave equations*, Math. Res. Lett. **5** (1998), no. 5, 605–622.

[123] Hans Lindblad and Igor Rodnianski, *The weak null condition for Einstein's equations*, C. R. Math. Acad. Sci. Paris **336** (2003), no. 11, 901–906.

[124] Jerrold E. Marsden and Frank J. Tipler, *Maximal hypersurfaces and foliations of constant mean curvature in general relativity*, Phys. Rep. **66** (1980), no. 3, 109–139.

[125] Vincent Moncrief, *Global properties of Gowdy spacetimes with* $T^3 \times \mathbf{R}$ *topology*, Ann. Physics **132** (1981), no. 1, 87–107.

[126] ———, *Reduction of Einstein's equations for vacuum space-times with spacelike* U(1) *isometry groups*, Ann. Physics **167** (1986), no. 1, 118–142.

[127] _____, *Reduction of Einstein's equations for cosmological spacetimes with spacelike* U(1)-*isometry groups*, Physique quantique et géométrie (Paris, 1986), Hermann, Paris, 1988, pp. 105–117.

[128] _____, *Reduction of the Einstein-Maxwell and Einstein-Maxwell-Higgs equations for cosmological spacetimes with spacelike* U(1) *isometry groups*, Classical Quantum Gravity **7** (1990), no. 3, 329–352.

[129] Vincent Moncrief and Douglas M. Eardley, *The global existence problem and cosmic censorship in general relativity*, Gen. Relativity Gravitation **13** (1981), no. 9, 887–892.

[130] Vincent Moncrief and James Isenberg, *Symmetries of cosmological Cauchy horizons*, Comm. Math. Phys. **89** (1983), no. 3, 387–413.

[131] Peter Orlik and Frank Raymond, *On 3-manifolds with local* SO(2) *action*, Quart. J. Math. Oxford Ser. (2) **20** (1969), 143–160.

[132] R. Penrose, *Techniques of differential topology in relativity*, SIAM, Philadelphia, PA., 1972.

[133] R. Penrose, *Some unsolved problems in classical general relativity*, Seminar on Differential Geometry, Princeton Univ. Press, Princeton, N.J., 1982, pp. 631–668.

[134] Alan D. Rendall, *Global properties of locally spatially homogeneous cosmological models with matter*, Math. Proc. Cambridge Philos. Soc. **118** (1995), no. 3, 511–526.

[135] _____, *Constant mean curvature foliations in cosmological space-times*, Helv. Phys. Acta **69** (1996), no. 4, 490–500, Journées Relativistes 96, Part II (Ascona, 1996), gr-qc/9606049.

[136] _____, *Existence of constant mean curvature foliations in spacetimes with two-dimensional local symmetry*, Comm. Math. Phys. **189** (1997), no. 1, 145–164.

[137] _____, *Global dynamics of the mixmaster model*, Classical Quantum Gravity **14** (1997), no. 8, 2341–2356.

[138] _____, *Fuchsian analysis of singularities in Gowdy spacetimes beyond analyticity*, Classical Quantum Gravity **17** (2000), no. 16, 3305–3316.

[139] _____, *Theorems on existence and global dynamics for the Einstein equations*, Living Rev. Relativ. **5** (2002), 2002–6, 62 pp. (electronic).

[140] Alan D. Rendall and Marsha Weaver, *Manufacture of Gowdy spacetimes with spikes*, Classical Quantum Gravity **18** (2001), no. 15, 2959–2975.

[141] Hans Ringström, *Curvature blow up in Bianchi VIII and IX vacuum spacetimes*, Classical Quantum Gravity **17** (2000), no. 4, 713–731.

[142] _____, *The Bianchi IX attractor*, Ann. Henri Poincaré **2** (2001), no. 3, 405–500.

[143] _____, *Asymptotic expansions close to the singularity in Gowdy spacetimes*, Classical Quantum Gravity **21** (2004), S305–S322.

[144] _____, *On Gowdy vacuum spacetimes*, Math. Proc. Cambridge Philos. Soc. **136** (2004), no. 2, 485–512.

[145] Peter Scott, *The geometries of 3-manifolds*, Bull. London Math. Soc. **15** (1983), no. 5, 401–487.

[146] Jalal Shatah, *The Cauchy problem for harmonic maps on Minkowski space*, Proceedings of the International Congress of Mathematicians, Vol. 1, 2 (Zürich, 1994) (Basel), Birkhäuser, 1995, pp. 1126–1132.

[147] Stephen T.C. Siklos, *Occurrence of whimper singularities*, Comm. Math. Phys. **58** (1978), no. 3, 255–272.

[148] Larry Smarr and Jr. York, James W., *Kinematical conditions in the construction of spacetime*, Phys. Rev. D (3) **17** (1978), no. 10, 2529–2551.

[149] _____, *Radiation gauge in general relativity*, Phys. Rev. D (3) **17** (1978), no. 8, 1945–1956.

[150] Hart F. Smith and Christopher D. Sogge, *On Strichartz and eigenfunction estimates for low regularity metrics*, Math. Res. Lett. **1** (1994), no. 6, 729–737.

[151] Hart F. Smith and Daniel Tataru, *Sharp local well-posedness results for the nonlinear wave equation*, http://www.math.berkeley.edu/~tataru/nlw.html, 2001.

[152] Chrisopher D. Sogge, *Lectures on nonlinear wave equations*, Monographs in Analysis, vol. II, International Press, Cambridge, MA, 1995.

[153] Christopher D. Sogge, *On local existence for nonlinear wave equations satisfying variable coefficient null conditions*, Comm. Partial Differential Equations **18** (1993), no. 11, 1795–1821.

[154] _____, *Fourier integral operators and nonlinear wave equations*, Mathematics of gravitation, Part I (Warsaw, 1996), Polish Acad. Sci., Warsaw, 1997, pp. 91–108.

[155] Masayuki Tanimoto, *New varieties of Gowdy space-times*, J. Math. Phys. **39** (1998), no. 9, 4891–4898.

[156] Terence Tao, *Global regularity of wave maps. II. Small energy in two dimensions*, Comm. Math. Phys. **224** (2001), no. 2, 443–544.

[157] Daniel Tataru, *Nonlinear wave equations*, Proceedings of the ICM, Beijing 2002, vol. 3, 2003, pp. 209–220, math.AP/0304397.

[158] _____, *Rough solutions for the wave-maps equation*, http://math.berkeley.edu/~tataru/nlw.html, 2003.

[159] Daniel Tataru, *The wave maps equation*, Bull. Amer. Math. Soc. (N.S.) **41** (2004), no. 2, 185–204 (electronic).

[160] William P. Thurston, *Three-dimensional geometry and topology. Vol. 1*, Princeton University Press, Princeton, NJ, 1997, Edited by Silvio Levy.

[161] Claes Uggla, Henk van Elst, John Wainwright, and George F.R. Ellis, *The past attractor in inhomogenous cosmology*, Phys. Rev. D **68** (2003), 103502.

[162] Juan Antonio Valiente Kroon, *A new class of obstructions to the smoothness of null infinity*, Comm. Math. Phys. **244** (2004), no. 1, 133–156.

[163] John Wainwright and George F.R. Ellis (eds.), *Dynamical systems in cosmology*, Cambridge University Press, Cambridge, 1997, Papers from the workshop held in Cape Town, June 27–July 2, 1994.

[164] John Wainwright and Lucas Hsu, *A dynamical systems approach to Bianchi cosmologies: orthogonal models of class A*, Classical Quantum Gravity **6** (1989), no. 10, 1409–1431.

[165] Robert M. Wald, *General relativity*, University of Chicago Press, Chicago, Ill., 1984.

[166] _____, *Gravitational collapse and cosmic censorship*, Black holes, gravitational radiation and the universe, Fund. Theories Phys., vol. 100, Kluwer Acad. Publ., Dordrecht, 1999, pp. 69–85.

[167] Marsha Weaver, *Dynamics of magnetic Bianchi VI$_0$ cosmologies*, Classical Quantum Gravity **17** (2000), no. 2, 421–434, gr-qc/9909043.

[168] Marsha Weaver, James Isenberg, and Beverly K. Berger, *Mixmaster behavior in inhomogeneous cosmological spacetimes*, Phys. Rev. Lett **80** (1998), 2984–2987, gr-qc/9712055.

Lars Andersson
Department of Mathematics
University of Miami
Coral Gables, FL 33124
USA
e-mail: `larsa@math.miami.edu`

Smoothness at Null Infinity and the Structure of Initial Data

Helmut Friedrich ·

Abstract. We describe our present understanding of the relations between the behavior of asymptotically flat Cauchy data for Einstein's vacuum field equations near space-like infinity and the asymptotic behavior of their evolution in time at null infinity.

1. Introduction

There are no doubts any longer that the idea of *gravitational radiation* refers to a real physical phenomenon. Framing, however, a precise underlying mathematical concept still poses problems. The work on gravitational radiation by Pirani [62], Trautman [69], Sachs [63], [64], Bondi [10], Newman and Penrose [58] and others, which was brought in a sense to a conclusion by Penrose [59], [60], is based on the idealization of an *isolated self-gravitating system*. It requires information on the long time evolution of gravitational fields which at the time could only be guessed. Ten years before these developments Y. Choquet-Bruhat had achieved a breakthrough in the mathematical analysis of the local Cauchy problem for Einstein's field equations [11]. However, the technical means to derive the fall-off behavior of gravitational fields at far distances and late times from 'basic principles' were not available in the 1960's. In the meantime there has been a considerable progress in controlling the asymptotic structure of solutions to Einstein's field equations but it is still not quite clear which 'basic principles' to assume here.

In the following we shall report on work which aims at closing various gaps in the study of gravitational radiation, the analysis of the Einstein equations, and the calculation of wave forms. Sections 2–5 present a fairly detailed discussion of the underlying analytical structures and of the recent results which led to the author's present understanding of the situation. To maintain the flow of the arguments, the reader is referred for derivations to the original literature. In Sections 6 and 7 will be given new results and detailed arguments.

Penrose's proposal to characterize far fields of *isolated systems* in terms of their conformal structure ([59], [60]) has been criticized over the years on several grounds; various variations, alternatives, etc. have been proposed (cf. [5], [13], [14], [17], [26], [66], [70], [73], and references given therein). Some authors consider the smoothness requirements on the conformal boundary as too restrictive and suggest generalizations (cf. [17], [70], [73]). Doubts have been raised as to whether non-trivial asymptotically simple solutions to the vacuum field equations exist at all ([14]) and it has been argued that the smoothness of the conformal boundary required in [59] excludes interesting physics ([13]). The wide range of opinions on the subject is illustrated by the curious contrast between this emphasis on subtleties of the asymptotic smoothness and claims that 'null infinity is too far away for modelling real physics' (cf. [26], [66]).

In [26] even the *asymptotically flat model* is abandoned and replaced by a *time-like cut model*. The latter introduces a spatially compact time-like hypersurface T which is chosen in an ad hoc fashion to cut off 'the system of interest' from the rest of the ambient universe. The idea then is to study the system which has thus been 'isolated' as an object of its own.

The usefulness of any such suggestion can only be demonstrated by analyzing its mathematical feasibility. This becomes clear when one tries to calculate wave forms numerically. Such calculations cannot be based on hand waving or physical intuition. The design of an effective numerical computer code requires a precise mathematical formulation.

The analysis of the time-like cut model reduces to a study of the initial boundary value problem for Einstein's field equations in which boundary data are prescribed on T and Cauchy data are given on a space-like hypersurface S which intersects T in the space-like surface ∂S. In [42] has been given a fairly complete analysis of this problem for Einstein's vacuum field equations. This study is only local in time, but it provides insights into the basic problem. So far, the time-like cut model raises many more questions than it appears able to answer.

How is T to be chosen? Physical considerations may lead to suggestions when the system of interest is 'sufficiently far' away from other systems. However, there is in general no preferred physical or geometrical choice for T. (It is instructive to compare this with the anti-de Sitter-type solutions, where the time-like boundary $\mathcal{J}$ at space-like and null infinity is determined geometrically and the boundary data can be prescribed in covariant form (cf. [36]).)

The boundary must be characterized by some implicit or explicit geometrical condition. A natural choice is to prescribe its mean extrinsic curvature. Its evolution in time is then defined implicitly by a quasi-linear wave equation which itself depends in a non-local way on the data given on S and T (cf. [42]). Long time calculations thus require an extra effort to control the regularity of the boundary.

The gauge is related on the time-like boundary T directly to the evolution process. It depends on the (implicit) choice of a time-like unit vector field tangent to T. While the data which are prescribed on the space-like hypersurface S allow one to analyze the local geometry near S at any desired order, the data which can be prescribed on the boundary $\mathcal{T}$ provide very little information on the local geometry near $\mathcal{T}$. All this makes it particularly difficult to show that the gauge and the constraints are preserved under the evolution in time.

These properties imply in general a non-covariance of the boundary conditions and data. Moreover, due to the fact that no causal direction is distinguished on $\mathcal{T}$ there does not seem to exist a natural 'no incoming radiation condition' and, in particular, no natural concept of 'outgoing radiation'. In fact, it appears difficult to associate with the initial boundary value problem any 'simple' quantities which characterize the system and its dynamics and which can be related to observational data.

While the discussion in [42] singles out data which are mathematically admissible, it is far from clear what should be prescribed on $\mathcal{T}$ from the physical point of view. The 'correct' data induced by the ambient universe will never be known. The information fed into 'the system' by the data prescribed on $\mathcal{T}$ can hardly be assessed. In long time calculations it may alter the character of the system drastically.

Because of these difficulties the time-like cut model appears not very promising. Nevertheless, it is of interest because of its similarity to the standard approach to numerical relativity, where an artificial time-like boundary is introduced to render the computational grid finite. It is expected here that the assumption of asymptotic flatness together with a judicious choice of the boundary will alleviate some of the difficulties pointed out above.

At present the only satisfactory solution to the gravitational radiation problem is based on the assumption of asymptotical flatness and the most elegant and geometrically natural definition of the latter is provided by the idea of the *conformal boundary at null infinity* introduced in [59]. While useful physical concepts can be associated with a conformal boundary which is sufficiently smooth (cf. [4], [46], [61] and the references given there), the possible degree of differentiability, which encodes the fall-off behavior of the gravitational field, still poses questions. This article deals with this particular issue and tries to disentangle its various aspects and difficulties.

Einstein's field equations admit certain conformal representations which in the following will be referred to as *conformal field equations*. These equations are 'regular' in the sense that they imply in a suitable gauge equations which are hyperbolic even at points of null infinity ([28], [29]). This fact has been used to show that the smoothness of the conformal boundary is preserved if it is guaranteed on the initial slice $\mathcal{S}$ of an hyperboloidal initial value problem ([31], [33], cf. also [38]). The subsequent analysis of hyperboloidal initial data ([3], [1], [2]) showed the existence of a large class of smooth hyperboloidal data for the conformal field equations. The construction of such data requires the 'free data' to satisfy a finite number of conditions at the space-like boundary $\partial \mathcal{S}$ at which the hyperboloidal slice $\mathcal{S}$ intersects future null infinity $\mathcal{J}^+$.

However, the work referred to above also shows the existence of a large class of hyperboloidal data which are smooth on $S \setminus \partial S$ but possess a non-trivial *poly-homogeneous expansion* at ∂S, i.e., an asymptotic expansion in terms of $x^k \log^j x$ where x is a defining function of the boundary ∂S, which vanishes on ∂S. Logarithmic terms can occur as a consequence of the constraint equations even if the free data extend smoothly to ∂S. Recently, it has been shown that certain hyperboloidal data which are polyhomogeneous at ∂S evolve into solutions to the conformal field equations which possess *generalized conformal boundaries* near the initial slice ([18], [57]). While the precise behavior of these solutions near that boundary still needs to be analyzed, the result shows that the use of the conformal field equations and the characterization of the edge of space-time in terms of its conformal structure are not restricted to asymptotically regular situations.

We conclude from these results that in the standard Cauchy problem the field equations decide on the degree of smoothness of the conformal boundary at null infinity in arbitrarily small neighborhoods of space-like infinity.

There are other reasons to study the region near space-like infinity. The hyperboloidal initial value problem is intrinsically time-asymmetric. To analyze in the same picture incoming radiation, a non-linear scattering process, and outgoing radiation, one needs to include space-like infinity (as pointed out already in [60]). Also, if the hyperboloidal data are not distinguished by special features as, for instance, the presence of a trapped surface, it is not clear which part of the imagined space-time is covered by their evolution. They could represent a hypersurface close to time-like infinity or close to a Cauchy hypersurface (a difficulty shared with the characteristic initial value problem and the initial boundary value problem).

This should not obscure the fact that numerical calculations of space-times from hyperboloidal data allow one to determine wave forms for many 'realistic' physical processes. So far the only semi-global calculations of space-times, including their radiation fields at null infinity, are based on hyperboloidal and characteristic initial value problems (cf. [27], [52], [53], and the article by L. Lehner and O. Reula, this volume).

We are thus left with the following task: (i) characterize the data which evolve near space-like infinity into solutions of prescribed smoothness at null infinity, (ii) analyze for which of these data physical concepts and requirements (linear and angular momentum at space-like and null infinity, reduction of the asymptotic gauge (BMS) group to a Poincaré group, ...) can be meaningfully introduced and a satisfactory physical picture can be established.

The first step is technically the most difficult one. It requires us to control under fairly general assumptions the effect of the quasi-linear, gauge hyperbolic field equations over infinite regions of space-time. Moreover, the asymptotic behavior of the solutions has to be determined with a precision which excludes any further refinement.

Once this step has been taken, many considerations of the second step will reduce to straightforward, though possibly quite lengthy, calculations. Nevertheless, the second step is of crucial importance. At this stage one has to observe that the notion of asymptotic flatness is not part of the general theory; it is an idealization which is chosen to serve a purpose. While it is suggested to us by important solutions such as those of the Kerr family, it is far from being determined by the equations alone. There remains a large freedom to decide on the asymptotic behavior of the fields.

To make one's choice, one needs to know the mathematical options and has to decide on the physical questions to be answered. A theorem which characterizes the most general Cauchy data on $\tilde{\mathcal{S}} = \mathbb{R}^3$ for which the maximal globally hyperbolic Einstein development is null geodesically complete and for which the Riemann tensor goes to zero at (null) infinity would be mathematically quite an achievement but, by itself, insufficient from the point of view of physics.

We are not interested here in discussing 'observations' in asymptotically flat solutions which refer to the roughness of the asymptotic structure as, for instance, in [70]. We rather wish to understand whether (i) the solution models a 'system of physical interest' and (ii) its far field and asymptotic structure allow one to extract information on the system which characterizes its physical nature and can be related to observational data.

This task is neither easy nor well defined. The studies of the last 40 years provide some understanding of the situations one may expect to observe (collapse to a black hole, mergers of black holes, ...). By exploring, however, the questions above in a general setting, new phenomena may be encountered (cf. [12], [49] for an example). But given that the interior is understood to some degree, what do we do about (ii)?

Recent results on the constraint equations exhibit possibilities to modify asymptotically flat vacuum data 'far out' without affecting the interior. The data can be made to agree near space-like infinity with exact Schwarzschild or Kerr data ([20], [21]), with even more general static resp. stationary data, or with data which are only *asymptotically static resp. stationary* ([16]) (cf. also the discussion in the article by R. Bartnik and J. Isenberg, this volume, for other techniques of modifying or extending solutions to the constraints).

These results have been used to settle a question which has been open for a long time. Since data which are static or stationary near space-like infinity evolve into solutions which possess a smooth conformal boundary at null infinity (cf. [23]), these solutions contain smooth hyperboloidal hypersurfaces. Recently P. Chruściel and E. Delay have shown the existence of families of Cauchy data on $\mathbb{R}^3$ which are static outside a fixed radius and have members of arbitrarily small ADM-mass. The corresponding solutions contain hyperboloidal hypersurfaces to which the results of [33] apply. This demonstrates the existence of non-trivial asymptotically simple solutions to Einstein's vacuum field equations with prescribed smoothness of the asymptotic structure ([15]).

More recently S. Klainerman and F. Nicolò revisited their work in [55] and showed ([56]) that their solutions will have the *Sachs peeling property* ([63], [64]) if the data are subject to certain asymptotic conditions. However, the class of data which meet these requirements still needs further analysis.

The new flexibility in constructing asymptotically flat initial data also allows one to illustrate some difficulties of the asymptotically flat space-time model. Let $(\mathcal{S}, d)$ denote the initial data where $\mathcal{S}$ is the hypersurface considered in our discussion of the time-like cut model and d indicates the fields induced on $\mathcal{S}$ by the cosmological model. Suppose that $(\mathcal{S}', d')$ is an asymptotically flat initial data set for which there exists an embedding $\phi : \mathcal{S} \to \mathcal{S}'$ such that the push forward of d by ϕ is in a suitable sense 'close' to d' on $\phi(\mathcal{S})$. The evolution in time of the data $(\mathcal{S}', d')$ can then be considered in some neighborhood of $\phi(\mathcal{S})$ as a good approximation of the evolution of d in the cosmological model.

If the set $\mathcal{S}$ is chosen large enough and close to the region where the system is undergoing a wave generation process, the main part of the wave signal will reach null infinity at a finite retarded time. The fact that for very late times the data on $\mathcal{S}' \setminus \phi(\mathcal{S})$ will create a deviation of our solution from the cosmological one is likely to be irrelevant in many interesting situations. From a pragmatical point of view it may be considered the main purpose of the asymptotically flat space-time extension beyond the domain of dependence of $\phi(\mathcal{S})$ to allow perturbations of the gravitational field generated near $\phi(\mathcal{S})$ to unfold into a clean wave signal which can be read off at null infinity.

Since changes near space-like infinity affect the field, however weakly, at all later times, they may have an important effect in the case of black hole solutions. One may envisage the collapse of pure gravitational radiation to a black hole as being modelled by vacuum solutions which arise from smooth asymptotically flat data on $\mathbb{R}^3$, admit smooth, complete (cf. [47]) conformal boundaries $\mathcal{J}^{\pm}$, and possess future event horizons while all past directed null geodesics require endpoints on $\mathcal{J}^-$. At present nothing is known about such solutions and they may not exist. Is it conceivable then that there exist solutions which show all the (suitably generalized) features listed above but have a rough conformal boundary? Could such boundaries allow for a 'higher radiation content'? If that were the case the restriction to smooth conformal boundaries might preclude the discussion of certain interesting physical phenomena.

Clearly, the large freedom in constructing asymptotically flat extensions should neither be used to import irrelevant information into the system nor to suppress important features. The replacement of an extension by one which is strictly Kerr (say) near space-like infinity introduces a transition zone on the initial hypersurface which mediates between the given and the Kerr data. The resulting 'wrinkles' in the solution are recorded in the radiation field at null infinity. Is this information physically insignificant or does it indicate that something important has been ironed out by forming the new extension?

This question points again to the need of understanding the detailed behavior of the fields near space-like infinity. In the standard conformal representation space-like infinity with respect to the solution space-time is represented by a point, usually denoted by i^0. With respect to an asymptotically flat Cauchy hypersurface $\tilde{S}$ space-like infinity is then also represented by a point, denoted by i, which becomes under conformal compactification a point in the extended 3-manifold $S = \tilde{S} \cup \{i\}$. Unfortunately, if $m_{ADM} \neq 0$ the conformal initial data are strongly singular at i. This is the basic stumbling block for analyzing the field near space-like infinity in terms of the standard conformal rescaling.

The constraint equations on the Cauchy hypersurface $\tilde{S}$ impose only weak restrictions on the asymptotic behavior of the data near space-like infinity. It is easy to construct data which at higher orders will become quite 'rough' near i and which can be expected to affect the smoothness of the fields near null infinity in a physically meaningless way. Thus one will have to make a reasonable choice and find a class of data which allows one to perform a sufficiently detailed analysis of their evolution in time while still being sufficiently general to recognize basic features of the asymptotic behavior at null infinity.

In the following it is assumed that the data on $\tilde{S}$ represent a space-like slice of time reflection symmetry, so that the second fundamental form vanishes, and that their conformal structure, represented by a conformal 3-metric h on S, extends smoothly to the point i. Only these conditions will be used in the following discussion, no a priori assumptions on the evolution in time will be made. We note that the assumptions are made to simplify the calculations, there exists a large space for generalizations.

Somewhat unexpectedly, the attempt to analyze for data as described above the evolution near space-like infinity i in the context of the conformal field equations led to a *finite regularization of the singularity at space-like infinity* ([37]).

In a certain conformal scaling of the conformal initial data the choice of the frame and the coordinates is combined with a blow-up of the point i to a sphere $\mathcal{I}^0$ such that the initial data and the gauge of the evolution system become smoothly extendable to $\mathcal{I}^0$ *in a different conformal scaling*. Moreover, the *general conformal field equations* imply in that scaling a system of hyperbolic reduced equations which also extends smoothly to $\mathcal{I}^0$ (Section 5.5). This allows one to define a *regular finite initial value problem at space-like infinity*.

Under the evolution defined by the extended reduced system the set $\mathcal{I}^0$ evolves into a set $\mathcal{I} =]-1, 1[\times \mathcal{I}^0$, which represents a boundary of the physical space-time. This *cylinder at space-like infinity* is defined solely in terms of conformal geometry and the general conformal field equations.

In the given coordinates, the sets $\mathcal{J}^\pm$ which represent near space-like infinity the conformal boundary at null infinity are at a finite location. They 'touch' the set $\mathcal{I}$ at certain *critical sets* $\mathcal{I}^\pm = \{\pm 1\} \times \mathcal{I}^0$, which can be regarded as the two components of the boundary of $\mathcal{I}$ and, simultaneously, as boundaries of $\mathcal{J}^\pm$. Due to the peculiarities of the construction the solution is determined on the closure $\bar{\mathcal{I}} \equiv \mathcal{I} \cup \mathcal{I}^- \cup \mathcal{I}^+ \simeq [-1, 1] \times \mathcal{I}^0$ of $\mathcal{I}$ uniquely by the data on $\tilde{S}$ and there is no freedom to prescribe boundary data on $\bar{\mathcal{I}}$.

This setting discloses the structure which decides on the asymptotic smoothness of the fields. At the critical sets $\mathcal{I}^\pm$ occurs a *break-down of the hyperbolicity of the reduced equations.* As explained in Section 5.3, a subtle interplay of this degeneracy with the structure of the initial data near $\mathcal{I}^0$, which is mediated by certain *transport equations* on $\mathcal{I}$, turns out critical for the smoothness of the conformal structure at null infinity. This peculiar situation is not suggested by general PDE theory, it is a specific feature of Einstein's theory, the geometric nature of the field equations, and general properties of conformal structures.

The transport equations, which are linear hyperbolic equations interior to $\mathcal{I}$, allow one to calculate the coefficients u^p of the Taylor series of the solution at $\mathcal{I}$ from data implied on $\mathcal{I}^0$ by the Cauchy data on $\mathcal{S}$ (cf. 5.82). Near $\mathcal{I}^\pm$ this series can be interpreted as an asymptotic expansion. The coefficients u^p are smooth functions on $\mathcal{I}$ which can be calculated order by order by following an algorithmic procedure.

It turns out that the coefficients u^p develop in general logarithmic singularities at the critical sets $\mathcal{I}^\pm$. This behavior foreshadows a possible non-smoothness of the conformal structure at null infinity. In the linearized setting it follows that the logarithmic singularities are transported along the generators of null infinity ([39]). In the non-linear case their effect on the conformal structure at null infinity is not under control yet, however, the solutions are unlikely to be better behaved than in the linear case.

The evidence obtained so far suggests cases which range from conformal structures of high differentiability to ones with low smoothness at null infinity. The non-smoothness may take the form of polylogarithmic expansions in terms of expressions $c\,(1-\tau)^k \log^j (1-\tau)$. Here τ is a coordinate with $\tau \to 1$ from below on $\mathcal{J}^+$, the coefficients $c = c(\rho, t)$ are smooth functions of a coordinate ρ along the null generators and suitable angular coordinates t on the set of null generators of $\mathcal{J}^+$, and k, j are non-negative integers. If k is small enough Sachs peeling fails and Penrose compactification results in weak differentiability.

Can one 'loose physics' if one insists on extensions which are smooth at null infinity? This certainly would be true if the coefficients c would encode important physical information. The discussion of the regularity conditions in Section 5.5 suggests that at low orders the coefficients are determined by the data in an arbitrarily small neighborhood of space-like infinity. By the results on the constraints referred to above these data seem to be rather arbitrary, only weakly related to 'the system' characterized by the data on $\mathcal{S}$, and thus of little relevance. As described in the following the situation is more complicated at higher orders, depends then in a more subtle way on the non-linearity of the equations, and requires further study.

Since the coefficients u^p can be calculated at arbitrary orders, we expect that this analysis will also allow us to describe in detail the relations between physical concepts defined at space-like infinity and concepts defined on null infinity (ADM resp. Bondi linear and angular momentum, etc.). The behavior of the fields at the sets $\mathcal{I}^\pm$ is also critical for the possibility to reduce the BMS group to the Poincaré group (cf. [41]). Thus, the precise understanding of the behavior of the fields near the critical sets should provide us with a rather complete physical picture.

Eventually one would like to make statements on the smoothness of the so-lution space-time at null infinity in terms of properties of the initial data. Thus one needs to control how the behavior of the asymptotic expansion at the critical sets depends on the structure of the initial data and to derive *regularity conditions* on the initial data which are necessary and sufficient for the smoothness of the coefficients u^p at the critical sets.

The information on the coefficients which is available so far has been used to derive conditions which are *necessary* for the regularity ([37]). The derivation of the complete condition is still difficult because of the algebraic complexities of the calculations involved.

Recently J.A. Valiente Kroon obtained with the help of an algebraic computer program complete and explicit expressions at higher order which are pointing at the possibility that *asymptotic staticity* (or, if the time reflection symmetry is dropped, *asymptotic stationarity*) may play a decisive role in deriving sufficient regularity conditions ([71], [72]).

We are thus led to revisit the static vacuum solutions (Sections 4.2, 6 and 7). Because of the loss of hyperbolicity of the conformal field equations at $\mathcal{I}^{\pm}$, it is not clear whether the smoothness of the conformal structure at null infinity observed for static and stationary vacuum solution can be understood as resulting from the possible regularity of the extended solutions at the critical sets. We show that for static solutions our setting is smooth, in fact real analytic, in a neighborhood of the set $\mathcal{I}^{-} \cup \mathcal{I} \cup \mathcal{I}^{+}$.

This narrows down the range in which the final regularity condition is to be found. We know that asymptotic staticity is sufficient and that the conditions found in [37], which are implied by asymptotic staticity, are necessary for the regularity of the asymptotic expansion on $\bar{\mathcal{I}}$. There is still the possibility that the final condition ' fizzles out' and depends on the specific data but we expect to arrive at the end at a definite, geometric condition.

To assess how restrictive such conditions would be, it is instructive to consider the results by Chruściel and Delay in [16]. They allow us to conclude that there exist large classes of solutions to the constraints, which are essentially arbitrary on given compact subsets of the initial hypersurface $\tilde{S}$, whose evolutions in time admit asymptotic expansions at $\mathcal{I}^{\pm}$ with coefficients which extend smoothly to the critical sets $\mathcal{I}^{\pm}$ up to a given or at all orders.

So far we ignored a question which is of central importance for gravitational wave astronomy: can the replacement of an asymptotically flat extension by an-other one result in a drastic change of the wave signal at null infinity? If that were the case, it would be hard to see how specific physical processes could be identified in the wave forms calculated at $\mathcal{J}^{+}$.

There should be characteristics of wave signals which are specific to 'the system' and which are stable under changes of the asymptotically flat extension if these extensions are restricted to 'reasonable' classes. This problem should be amenable to analytical and numerical investigations and we expect our analysis to contribute to its solution. In fact, it appears that with the regularity conditions mentioned above the field equations themselves hint at a first 'reasonable' class of asymptotically flat extensions.

2. Conformal field equations

Our analysis of the gravitational field near space-like and null infinity relies on a certain conformal representation of Einstein's vacuum field equations referred to as the *general conformal field equations*. We give a short introduction to these equations and point out various facts about the equations and the underlying mathematical structures which will be important in the following. For derivations, detailed arguments, and further background material such as the theory of *normal conformal Cartan connections*, which is the natural home of many of the concepts used in the following, we refer the reader to the original article [36] and the survey article [38].

The aim is to discuss a solution $(\tilde{M}, \tilde{g})$ to Einsteins vacuum field equation

$$R_{\mu\nu}[\tilde{g}] = 0, \tag{2.1}$$

in terms of a suitably chosen *conformal factor* Θ and the *conformal metric* $g = \Theta^2 \tilde{g}$. Denoting by ∇ the Levi-Civita connection of g, the transformation law of the Ricci tensor under the conformal rescaling above takes in four dimensions the form

$$R_{\nu\rho}[g] = R_{\nu\rho}[\tilde{g}] - \frac{2}{\Theta} \nabla_\nu \nabla_\rho \Theta - g_{\nu\rho} g^{\lambda\delta} \left(\frac{1}{\Theta} \nabla_\lambda \nabla_\delta \Theta - \frac{3}{\Theta^2} \nabla_\lambda \Theta \nabla_\delta \Theta \right). \tag{2.2}$$

If Θ is considered here as being given, equation (2.1) implies with (2.2) an equation for g with a similar principal part as (2.1).

As explained in the next section, we will mainly be interested in the behavior of the field in space-time domains where $\Theta \to 0$. Because the right-hand side of (2.2) is formally singular in this limit, an abstract discussion of the solutions near the sets $\{\Theta = 0\}$ becomes very delicate. It will be seen, however, that under suitable assumptions on the initial data for the field and with an appropriate behavior of the conformal factor the right-hand side of (2.2) can attain smooth limits. This result is obtained by a more sophisticated use of the behavior of the fields and the equations under transformations which preserve the conformal structure.

2.1. The general conformal field equations

In [28], [29] has been obtained a system of equations which is regular in the sense that there occur no factors Θ^{-1} on the right-hand sides or factors Θ in the principal part of the equations. Its unknowns are Θ, g, and fields derived from them such as the *rescaled conformal Weyl tensor* $W^i{}_{jkl} \equiv \Theta^{-1} C^i{}_{jkl}$. These have been used to derive various results about the asymptotic behavior of solutions to the Einstein equations. The specific behavior of the conformal fields near space-like infinity discussed in the next sections asks, however, for a particularly careful analysis of the equations and the gauge conditions. It turns out that this is considerably simplified by making use of the full freedom offered by conformal structures.

A *Weyl connection* for the conformal structure defined by g is a torsion free connection $\hat{\nabla}$ which satisfies

$$\hat{\nabla}_\rho \, g_{\mu\nu} = -2 \, f_\rho \, g_{\mu\nu}, \tag{2.3}$$

with some 1-form f_ρ. It is distinguished by the fact that it preserves the conformal structure of g (and thus of $\tilde{g}$). If a frame $\{e_k\}_{k=0,1,2,3}$ is conformal at a given point p in the sense that it satisfies there $g(e_j, e_k) = \Lambda^2 \, \eta_{jk}$ with some $\Lambda > 0$ and $\eta_{jk} = diag\,(1, -1, -1, -1)$, then it satisfies such a relation with a point dependent function Λ along a given curve γ through p if it is parallely transported along γ with respect to the connection $\hat{\nabla}$. In particular, if the 1-form f_ρ is closed the connection $\hat{\nabla}$ is locally the Levi-Civita connection of a metric in the conformal class.

Assuming under $\tilde{g} \to g = \Theta^2 \, \tilde{g}$ the transformation law

$$\tilde{f}_\rho \to f_\rho = \tilde{f}_\rho - \Theta^{-1} \tilde{\nabla}_\rho \Theta,$$

the defining property (2.3) is expressed in terms of the metric $\tilde{g}$ equivalently by $\hat{\nabla}_\rho \, \tilde{g}_{\mu\nu} = -2 \, \tilde{f}_\rho \, \tilde{g}_{\mu\nu}$ where $\tilde{\nabla}$ denotes the Levi-Civita connection of $\tilde{g}$. It follows from (2.3) that the connection $\hat{\nabla}$ defines with the connection ∇ the difference tensor $\hat{\nabla} - \nabla = S(f)$ given by the specific expression

$$S(f)_\mu{}^\rho{}_\nu \equiv \delta^\rho{}_\mu \, f_\nu + \delta^\rho{}_\nu \, f_\mu - g_{\mu\nu} \, g^{\rho\lambda} \, f_\lambda. \tag{2.4}$$

This, in turn, can be used to specify $\hat{\nabla}$ in terms of ∇ and the 1-form f_ρ. The three connections are related by

$$\hat{\nabla} - \tilde{\nabla} = S(\tilde{f}), \quad \nabla - \tilde{\nabla} = S(\Theta^{-1} d\,\Theta), \quad \hat{\nabla} - \nabla = S(f). \tag{2.5}$$

Important for us will also be the 1-form

$$d_\mu \equiv \Theta \, \tilde{f}_\mu = \Theta \, f_\mu + \nabla_\mu \Theta. \tag{2.6}$$

The decomposition

$$R^\mu{}_{\nu\lambda\rho} = 2 \, \{g^\mu{}_{[\lambda} \, L_{\rho]\nu} - g_{\nu[\lambda} \, L_{\rho]}{}^\mu\} + C^\mu{}_{\nu\lambda\rho}, \tag{2.7}$$

of the curvature tensor of ∇ in terms of the trace free conformal Weyl tensor $C^\mu{}_{\nu\lambda\rho}$ and the Schouten tensor

$$L_{\mu\nu} = \frac{1}{2} \, R_{\mu\nu} - \frac{1}{12} \, R \, g_{\mu\nu}, \tag{2.8}$$

which carries the information about the Ricci tensor $R_{\mu\nu} = R^\rho{}_{\mu\rho\nu}$, has an analogue for $\hat{\nabla}$ which takes the form

$$\hat{R}^\mu{}_{\nu\lambda\rho} = 2 \, \{g^\mu{}_{[\lambda} \, \hat{L}_{\rho]\nu} - g^\mu{}_\nu \, \hat{L}_{[\lambda\rho]} - g_{\nu[\lambda} \, \hat{L}_{\rho]}{}^\mu\} + C^\mu{}_{\nu\lambda\rho}, \tag{2.9}$$

where

$$\hat{L}_{\mu\nu} = \frac{1}{2}\hat{R}_{(\mu\nu)} - \frac{1}{4}\hat{R}_{[\mu\nu]} - \frac{1}{12}\hat{R}\,g_{\mu\nu}, \tag{2.10}$$

contains the information about the Ricci tensor $\hat{R}_{\mu\nu} = \hat{R}^\rho{}_{\mu\rho\nu}$ and the Ricci scalar $\hat{R} = g^{\mu\nu}\hat{R}_{\mu\nu}$. The tensors (2.8) and (2.10) are related by

$$\nabla_\mu f_\nu - f_\mu f_\nu + \frac{1}{2} g_{\mu\nu} f_\lambda f^\lambda = L_{\mu\nu} - \hat{L}_{\mu\nu}. \tag{2.11}$$

To take care of the specific direction dependence of the various fields near space-like infinity it is convenient to express the conformal field equations in terms of a suitably chosen orthonormal frame field. Let $\{e_k\}_{k=0,1,2,3}$ be a frame field satisfying $g_{ik} \equiv g(e_i, e_k) = \eta_{ik}$, denote by ∇_k and $\hat{\nabla}_k$ the covariant derivative with respect to ∇ and $\hat{\nabla}$ in the direction of e_k, and define the connection coefficients $\Gamma_i{}^j{}_k$ and $\hat{\Gamma}_i{}^j{}_k$ of ∇ and $\hat{\nabla}$ in this frame field by $\nabla_i e_k = \Gamma_i{}^j{}_k e_k$ and $\hat{\nabla}_i e_k = \hat{\Gamma}_i{}^j{}_k e_k$ respectively. Then $\hat{\Gamma}_i{}^j{}_k = \Gamma_i{}^j{}_k + \delta^j{}_i f_k + \delta^j{}_k f_i - g_{ik}\,g^{jl}\,f_l$ with $f_k = f_\mu e^\mu{}_k$, where $e^\mu{}_k = <dx^\mu, e_k>$ denote the frame coefficients with respect to an as yet unspecified coordinate system x^μ, $\mu = 0, 1, 2, 3$. We note that $f_i = \frac{1}{4}\hat{\Gamma}_i{}^k{}_k$ because $\Gamma_i{}^j{}_k\,g_{jl} + \Gamma_i{}^j{}_l\,g_{jk} = 0$.

If all tensor fields (except the e_k) are expressed in terms of the frame field and the corresponding connection coefficients, the *conformal field equations* used in the following are written as equations for the unknown

$$u = (e^\mu{}_k, \quad \hat{\Gamma}_i{}^j{}_k, \quad \hat{L}_{jk}, \quad W^i{}_{jkl}), \tag{2.12}$$

and are given by

$$[e_p, e_q] = (\hat{\Gamma}_p{}^l{}_q - \hat{\Gamma}_q{}^l{}_p)\,e_l, \tag{2.13}$$

$$e_p(\hat{\Gamma}_q{}^i{}_j) - e_q(\hat{\Gamma}_p{}^i{}_j) - \hat{\Gamma}_k{}^i{}_j(\hat{\Gamma}_p{}^k{}_q - \hat{\Gamma}_q{}^k{}_p) + \hat{\Gamma}_p{}^i{}_k\hat{\Gamma}_q{}^k{}_j - \hat{\Gamma}_q{}^i{}_k\hat{\Gamma}_p{}^k{}_j \tag{2.14}$$

$$= 2\{g^i{}_{[p}\hat{L}_{q]j} - g^i{}_j\hat{L}_{[pq]} - g_{j[p}\hat{L}_{q]}{}^i\} + \Theta W^i{}_{jpq},$$

$$\hat{\nabla}_p\hat{L}_{qj} - \hat{\nabla}_q\hat{L}_{pj} = d_i W^i{}_{jpq}, \tag{2.15}$$

$$\nabla_i W^i{}_{jkl} = 0. \tag{2.16}$$

The square brackets in the first equation denote the commutator of vector fields. The connection ∇, which appears in the last equation, can be expressed by the relations given above in terms of $\hat{\nabla}$ and f_k. The last equation, referred to in the following as the *Bianchi equation*, is in a sense the core of the system. It is obtained from the contracted vacuum Bianchi identity $\tilde{\nabla}_\mu C^\mu{}_{\nu\lambda\rho} = 0$ by using the specific conformal identity $\Omega^{-1}\tilde{\nabla}_\mu C^\mu{}_{\nu\lambda\rho} = \nabla_\mu(\Omega^{-1}C^\mu{}_{\nu\lambda\rho})$. The first three equations are then essentially the *structural equations* of the theory of normal conformal Cartan connections.

No equations are given so far for the fields Θ and $d_k = \Theta f_k + \nabla_k\Theta$. They reflect the *conformal gauge freedom* artificially introduced here into Einstein's field equations. These fields cannot be prescribed quite arbitrarily. For solution for which the limit $\Theta \to 0$ is meaningful the latter should imply $d_k \to \nabla_k\Theta$.

The theory of normal conformal Cartan connections associates with each conformal structure a distinguished class of curves which provides a useful way of dealing with the gauge freedom. A *conformal geodesic* for $(\tilde{M}, \tilde{g})$ is a curve $x(\tau)$ in $\tilde{M}$ which solves, together with a 1-form $\tilde{f} = \tilde{f}(\tau)$ along it, the system of ODE's

$$(\tilde{\nabla}_{\dot{x}} \dot{x})^{\mu} + S(\tilde{f})_{\lambda}{}^{\mu}{}_{\rho} \dot{x}^{\lambda} \dot{x}^{\rho} = 0, \tag{2.17}$$

$$(\tilde{\nabla}_{\dot{x}} \tilde{f})_{\nu} - \frac{1}{2} \tilde{f}_{\mu} S(\tilde{f})_{\lambda}{}^{\mu}{}_{\nu} \dot{x}^{\lambda} = \tilde{L}_{\lambda \nu} \dot{x}^{\lambda}, \tag{2.18}$$

where $S(\tilde{f})$ and $\tilde{L}$ are given by (2.4) and (2.8) with g replaced by $\tilde{g}$. For any given metric in the conformal class there are more conformal geodesics than metric geodesics because for given initial data $x_* \in M$, $\dot{x}_* \in T_{x_*} M$, $\tilde{f}_* \in T_{x_*}^* M$ there exists a unique solution $x(\tau)$, $\tilde{f}(\tau)$ to (2.17), (2.18) near x_* satisfying for given $\tau_* \in \mathbb{R}$

$$x(\tau_*) = x_*, \quad \dot{x}(\tau_*) = \dot{x}_*, \quad \tilde{f}(\tau_*) = \tilde{f}_*. \tag{2.19}$$

The sign of $\tilde{g}(\dot{x}, \dot{x})$ is preserved near x_* but not its modulus.

Conformal geodesics admit, unlike metric geodesics, general fractional linear maps as parameter transformations. They are *conformal invariants*. Denote by b a smooth 1-form field. Then, if $x(\tau)$, $\tilde{f}(\tau)$ solve the conformal geodesics equations (2.17), (2.18), the pair $x(\tau)$, $\tilde{f}(\tau) - b|_{x(\tau)}$ solves the same equations with $\tilde{\nabla}$ replaced by the connection $\hat{\nabla} = \tilde{\nabla} + S(b)$ and L by $\hat{L}$, i.e., the curve $x(\tau)$, and in particular its parameter τ, are independent of the Weyl connection in the conformal class which is used to write the equations (cf. [40]).

Let there be given a smooth congruence of conformal geodesics which covers an open set U of $\tilde{M}$ such that the associated 1-forms $\tilde{f}$ define a smooth field on U. Denote by $\hat{\nabla}$ the torsion free connection on U which has with the connection $\tilde{\nabla}$ difference tensor $\hat{\nabla} - \tilde{\nabla} = S(\tilde{f})$ and denote by $\hat{L}$ the tensor (2.10) derived from $\hat{\nabla}$. Comparing with (2.11), we find that equations (2.17), (2.18) can be written

$$\hat{\nabla}_{\dot{x}} \dot{x} = 0, \quad \hat{L}_{\lambda \nu} \dot{x}^{\lambda} = 0. \tag{2.20}$$

Let e_k be a frame field satisfying along the congruence

$$\hat{\nabla}_{\dot{x}} e_k = \tilde{\nabla}_{\dot{x}} e_k + \langle \tilde{f}, e_k \rangle \dot{x} + \langle \tilde{f}, \dot{x} \rangle e_k - \tilde{g}(\dot{x}, e_k) \tilde{g}^{\sharp}(\tilde{f}, \,.\,) = 0. \tag{2.21}$$

Suppose that $\tilde{S}$ is a hypersurface which is transverse to the congruence, meets each of the curves exactly once, and on which $\tilde{g}(e_i, e_k) = \Theta_*^2 \eta_{ik}$ with some function $\Theta_* > 0$. It follows that $\tilde{g}(e_i, e_k) = \Theta^2 \eta_{ik}$ on U with a function Θ which satisfies

$$\hat{\nabla}_{\dot{x}} \Theta = \Theta \langle \dot{x}, \tilde{f} \rangle, \quad \Theta|_{\tilde{S}} = \Theta_*. \tag{2.22}$$

The observations above allow us to construct a special gauge for the conformal equations. Let $\tilde{S}$ be a space-like hypersurface in the given vacuum solution $(\tilde{M}, \tilde{g})$. We choose on $\tilde{S}$ a positive 'conformal factor' Θ_*, a frame field e_{k*}, and a 1-form $\tilde{f}_*$ such that $\Theta_*^2 \tilde{g}(e_{i*}, e_{k*}) = \eta_{ik}$ and e_{0*} is orthogonal to $\tilde{S}$. Then there exists through each point $x_* \in \tilde{S}$ a unique conformal geodesic $(x(\tau), \tilde{f}(\tau))$ with $\tau = 0$ on $\tilde{S}$ which satisfies there the initial conditions $\dot{x} = e_{0*}$, $\tilde{f} = \tilde{f}_*$.

If all data are smooth these curves define in some neighborhood U of $\tilde{S}$ a smooth caustic free congruence which covers U. Furthermore, $\tilde{f}$ defines a smooth 1-form on U which supplies a Weyl connection $\hat{\nabla}$ as described above. A smooth frame field e_k and the related conformal factor Θ are then obtained on U by solving (2.21), (2.22) for given initial data $e_k = e_{k*}$, $\Theta = \Theta_*$ on $\tilde{S}$. The frame field is orthonormal for the metric $g = \Theta^2 \, \tilde{g}$. Dragging along local coordinates x^α, $\alpha = 1, 2, 3$, on $\tilde{S}$ with the congruence and setting $x^0 = \tau$ we obtain a coordinate system. This gauge is characterized on U by the explicit gauge conditions

$$\dot{x} = e_0 = \partial_\tau, \quad \hat{\Gamma}_0{}^j{}_k = 0, \quad \hat{L}_{0k} = 0. \tag{2.23}$$

Coordinates, a frame field, and a conformal factor as above are said to define a *conformal Gauss gauge*. Since metric Gauss systems are well known to quickly develop caustics, it may be mentioned that conformal Gauss systems can cover large space-time domains in a regular fashion (cf. [40]).

To obtain a closed system for all the fields entering equations (2.13)–(2.16), we could now supplement the latter by equations which are implied for the fields Θ and d_k in a conformal Gauss gauge. It turns out that such a gauge implies quite simple ordinary differential equations along the conformal geodesics defining the gauge, it holds in fact

$$\ddot{d}_0 = 0, \quad \dot{\Theta} = d_0, \quad \dot{d}_a = 0, \quad a = 1, 2, 3,$$

where the dot denotes differentiation with respect to the parameter τ.

Thus, the fields Θ and d_k given by a conformal Gauss can be determined in our situation explicitly ([36]): *If $\tilde{g}$ is a solution to Einstein's vacuum equations (2.1), the fields Θ and d_k are given by the explicit expressions*

$$\Theta = \Theta_* \left(1 + \tau \, \langle \tilde{f}, \dot{x} \rangle_* + \frac{\tau^2}{4} \, \Theta_*^2 \left((\tilde{g}(\dot{x}, \dot{x}))^2 \, \tilde{g}^\sharp(\tilde{f}, \tilde{f}) \right)_* \right) \tag{2.24}$$

$$= \Theta_* \left(1 + \tau \, \langle \tilde{f}, \dot{x} \rangle_* + \frac{\tau^2}{4} \left(g^\sharp(\tilde{f}, \tilde{f}) \right)_* \right),$$

$$d_0 = \dot{\Theta}, \quad d_a = \Theta_* \langle \tilde{f}_*, e_{a*} \rangle, \quad a = 1, 2, 3, \tag{2.25}$$

where $g^\sharp$ denotes the contravariant version of g and the quantities with a subscript star are considered as constant along the conformal geodesics and given by their values on $\tilde{S}$.

Assuming for Θ and d_k the expressions (2.24) and (2.25), equations (2.13)–(2.16) provide a complete system for u. In spite of the fact that we use a special gauge, we refer to this system as the *general conformal field equations* to indicate that they employ the full gauge freedom preserving conformal structures.

Equally important for us are the facts that the expression (2.24) offers the possibility to control in a conformal Gauss gauge the location of the set where $\Theta \to 0$ and that (2.24), (2.25) imply with the relation $d_k = \Theta \, f_k + \nabla_k \Theta$ in sufficiently regular situations that

$$\nabla_k \Theta \nabla^k \Theta \to 0 \quad \text{as} \quad \Theta \to 0. \tag{2.26}$$

2.2. Spinor version of the general conformal field equations

Writing the conformal equations in the spin frame formalism leads to various algebraic simplifications. We introduce here only the basic notions of this formalism and refer the reader to [61] for a comprehensive introduction. It should be noted, however, that our notation does not completely agree with that of [61]. In particular, if a specific frame field is used it will always be pointed out in the text but not be indicated by gothic indices.

Starting with the orthonormal frame introduced above we define null frame vector fields $e_{AA'} = \sigma^k{}_{AA'} e_k$ with constant van der Waerden symbols $\sigma^k{}_{AA'}$ (here and in the following indices $A, B, \ldots, A', B', \ldots$ take values 0 and 1 and the summation rule is assumed) such that

$$e_{00'} = \frac{1}{\sqrt{2}}(e_0 + e_3), \quad e_{11'} = \frac{1}{\sqrt{2}}(e_0 - e_3),$$

$$e_{01'} = \frac{1}{\sqrt{2}}(e_1 - i\,e_2), \quad e_{10'} = \frac{1}{\sqrt{2}}(e_1 + i\,e_2).$$

Then $e_{00'}, e_{11'}$ are real and $e_{01'}, e_{10'}$ are complex (conjugate) null vector fields and their scalar products are given by $g(e_{AA'}, e_{CC'}) = \epsilon_{AC}\,\epsilon_{A'C'}$ where $\epsilon_{AC}, \epsilon_{A'C'}, \epsilon^{AC}$, $\epsilon^{A'C'}$ denote the anti-symmetric spinor fields with $\epsilon_{01} = \epsilon_{0'1'} = \epsilon^{01} = \epsilon^{0'1'} = 1$. The latter are also used to raise and lower spinor indices according to the rules $X^A = \epsilon^{AB} X_B$ and $X_B = X^A \epsilon_{AB}$ so that $\epsilon_A{}^B = \epsilon_{AC}\,\epsilon^{BC}$ denotes the Kronecker spinor (similar rules hold for primed indices).

If connection coefficients $\Gamma_{AA'}{}^{BB'}{}_{CC'}$ are defined by writing $\nabla_{e_{AA'}} e_{CC'} = \Gamma_{AA'}{}^{BB'}{}_{CC'} e_{BB'}$, the spinor connection coefficients are given by $\Gamma_{AA'}{}^B{}_C = \frac{1}{2}\Gamma_{AA'}{}^{BE'}{}_{CE'}$ and one has

$$\Gamma_{AA'}{}^{BB'}{}_{CC'} = \Gamma_{AA'}{}^B{}_C \epsilon_{C'}{}^{B'} + \bar{\Gamma}_{AA'}{}^{B'}{}_{C'} \epsilon_C{}^B.$$

Here it is observed, as usual, that the relative order of primed and unprimed indices is irrelevant and that under complex conjugation primed indices are converted into unprimed indices and vice versa. Covariant derivatives of spinor fields are now given by similar rules as in the case of the standard frame formalism. Writing $\nabla_{AA'}$ for $\nabla_{e_{AA'}}$ we have, e.g., for a spinor field $X^A{}_B{}^{C'}$

$$\nabla_{DD'} X^A{}_B{}^{C'} = e_{DD'}(X^A{}_B{}^{C'}) + \Gamma_{DD'}{}^A{}_F X^F{}_B{}^{C'}$$

$$-\Gamma_{DD'}{}^F{}_B X^A{}_F{}^{C'} + \bar{\Gamma}_{DD'}{}^{C'}{}_{F'} X^A{}_B{}^{F'}.$$

We have similar rules for the connection $\hat{\nabla}$ and its associated connection coefficients $\hat{\Gamma}_{AA'BC}$ and it holds

$$\Gamma_{CC'AB} = \Gamma_{CC'BA}, \quad \hat{\Gamma}_{CC'AB} = \Gamma_{CC'AB} - \epsilon_{AC}\,f_{BC'}, \tag{2.27}$$

so that $\hat{\Gamma}_{CC'}{}^F{}_F = f_{CC'}$ gives the 1-form relating the connection $\hat{\nabla}$ to ∇.

The general conformal field equations are now written as equations for the unknowns

$$e_{AA'}, \quad \hat{\Gamma}_{AA'BC}, \quad \Theta_{AA'BB'}, \quad \phi_{ABCD}. \tag{2.28}$$

Here $\Theta_{AA'BB'}$ is the spinor representation of the tensor field $\hat{L}_{kj}$. It admits a decomposition of the form

$$\Theta_{AA'BB'} = \Phi_{AA'BB'} - \frac{1}{24} R \, \epsilon_{AB} \, \epsilon_{A'B'} + \Phi_{AB} \, \epsilon_{A'B'} + \bar{\Phi}_{A'B'} \, \epsilon_{AB},$$

where $\Phi_{AA'BB'} = \Phi_{BB'AA'} = \bar{\Phi}_{AA'BB'}$ represents the trace-free part of the tensor $\frac{1}{2} \hat{R}_{(jk)}$ provided by $\hat{\nabla}$ while $R = g^{jk} \hat{R}_{jk}$ is the Ricci scalar and the last two terms, with $\Phi_{ab} = \Phi_{(ab)}$, represent the anti-symmetric tensor $\frac{1}{4} \hat{R}_{[jk]}$. The symmetric spinor field $\phi_{ABCD} = \phi_{(ABCD)}$ represents the rescaled conformal Weyl tensor and is related to the latter by

$$W_{AA'BB'CC'DD'} = -\phi_{ABCD} \, \epsilon_{A'B'} \, \epsilon_{C'D'} - \bar{\phi}_{A'B'C'D'} \, \epsilon_{AB} \, \epsilon_{CD}.$$

The general conformal field equations in the order (2.13), (2.14), (2.15), (2.16) now take the form

$$[e_{BB'}, e_{CC'}] = (\Gamma_{BB'}{}^{AA'}{}_{CC'} - \Gamma_{CC'}{}^{AA'}{}_{BB'}) e_{AA'}, \tag{2.29}$$

$$e_{CC'}(\hat{\Gamma}_{DD'}{}^{A}{}_{B}) - e_{DD'}(\hat{\Gamma}_{CC'}{}^{A}{}_{B}) \tag{2.30}$$

$$-\hat{\Gamma}_{CC'}{}^{F}{}_{D}\,\hat{\Gamma}_{FD'}{}^{A}{}_{B} + \hat{\Gamma}_{DD'}{}^{F}{}_{C}\,\hat{\Gamma}_{FC'}{}^{A}{}_{B} - \bar{\hat{\Gamma}}_{CC'}{}^{F'}{}_{D'}\,\hat{\Gamma}_{DF'}{}^{A}{}_{B}$$

$$+\bar{\hat{\Gamma}}_{DD'}{}^{F'}{}_{C'}\,\hat{\Gamma}_{CF'}{}^{A}{}_{B} + \hat{\Gamma}_{CC'}{}^{A}{}_{F}\,\hat{\Gamma}_{DD'}{}^{F}{}_{B} - \hat{\Gamma}_{DD'}{}^{A}{}_{F}\,\hat{\Gamma}_{CC'}{}^{F}{}_{B}$$

$$= -\Theta_{BD'CC'}\,\epsilon_{D}{}^{A} + \Theta_{BC'DD'}\,\epsilon_{C}{}^{A} + \Theta\,\phi^{A}{}_{BCD}\,\epsilon_{C'D'},$$

$$\hat{\nabla}_{BB'}\Theta_{CC'AA'} - \hat{\nabla}_{AA'}\Theta_{CC'BB'} = d^{EE'}(\phi_{EABC}\epsilon_{E'C'}\epsilon_{A'B'} + \bar{\phi}_{E'A'B'C'}\epsilon_{EC}\epsilon_{AB}), \tag{2.31}$$

$$\nabla^{F}{}_{A'}\phi_{ABCF} = 0, \tag{2.32}$$

with the fields Θ, $d_{AA'}$ as given above. The simple form (2.32) of the spinor version of the Bianchi equation will be useful for us.

2.3. The reduced conformal field equations

The conformal Gauss gauge is not only distinguished by the fact that it is provided by the conformal structure itself and supplies explicit information on Θ and d_k, but also by a remarkable simplicity of the resulting evolution equations. Setting $p = 0$ in (2.13)–(2.16) and observing the gauge conditions (2.23) we obtain

$$\partial_\tau e^\mu{}_q = -\hat{\Gamma}_q{}^l{}_0 \, e^\mu{}_l, \tag{2.33}$$

$$\partial_\tau \hat{\Gamma}_q{}^i{}_j = -\hat{\Gamma}_k{}^i{}_j \hat{\Gamma}_q{}^k{}_0 + g^i{}_0 \hat{L}_{qj} + g^i{}_j \hat{L}_{q0} - g_{j0} \hat{L}_q{}^i + \Theta \, W^i{}_{j0q}, \tag{2.34}$$

$$\partial_\tau \hat{L}_{qj} + \hat{\Gamma}_q{}^k{}_0 \hat{L}_{kj} = d_i \, W^i{}_{j0q}, \tag{2.35}$$

$$\nabla_i W^i{}_{jkl} = 0.$$

While the first three equations are then ordinary differential equations along the conformal geodesics, we still have to deduce a suitable evolution system from the last equation. The Bianchi equation represents an overdetermined system of 16 equations for the 10 independent components of the rescaled conformal Weyl tensor. It implies a system of wave equations for $W^i{}_{jkl}$ which could be used as the evolution system. For the application studied in this article it turns out important, however, to use the first order system.

There are various ways of extracting from the Bianchi equations symmetric hyperbolic evolution systems but these are most easily found in the spin frame formalism. With the spinor field $\tau^{AA'} = \epsilon_0{}^A \epsilon_{0'}{}^{A'} + \epsilon_1{}^A \epsilon_{1'}{}^{A'}$ the gauge conditions (2.23) can be written

$$\tau^{AA'} e_{AA'} = \sqrt{2} \partial_\tau, \quad \tau^{AA'} \hat{\Gamma}_{AA'}{}^B{}_C = 0, \quad \tau^{BB'} \Theta_{AA'BB'} = 0. \tag{2.36}$$

Observing $\tau_{AA'} \tau^{BA'} = \epsilon_A{}^B$ and its complex conjugate version, one obtains from (2.27) and (2.36) the relation $\tau^{CC'} \Gamma_{CC'AB} = -\tau_A{}^{C'} f_{BC'}$ and thus

$$\hat{\Gamma}_{CC'AB} = \Gamma_{CC'AB} - \epsilon_{AC} \tau^{DD'} \Gamma_{DD'EB} \tau^E{}_{C'}, \tag{2.37}$$

which shows with (2.27) that $\hat{\Gamma}_{CC'AB}$ can be expressed in our gauge in terms of $\Gamma_{CC'AB}$ and vice versa.

Transvecting equations (2.29), (2.30), (2.31) suitably with $\tau^{EE'}$ thus gives the system of ODE's

$$\sqrt{2} \partial_\tau e^\mu{}_{CC'} = -\Gamma_{CC'}{}^{AA'}{}_{BB'} \tau^{BB'} e^\mu{}_{AA'}, \tag{2.38}$$

$$\sqrt{2} \partial_\tau \hat{\Gamma}_{DD'}{}^A{}_B + (\hat{\Gamma}_{DD'}{}^F{}_C \hat{\Gamma}_{FC'}{}^A{}_B + \hat{\bar{\Gamma}}_{DD'}{}^{F'}{}_{C'} \hat{\Gamma}_{CF'}{}^A{}_B) \tau^{CC'} \tag{2.39}$$

$$= \Theta_{BC'DD'} \tau^{AC'} + \Theta \phi^A{}_{BCD} \tau^C{}_{D'},$$

$$\sqrt{2} \partial_\tau \Theta_{CC'AA'} + (\hat{\Gamma}_{AA'}{}^F{}_B \Theta_{CC'FB'} + \hat{\bar{\Gamma}}_{AA'}{}^{F'}{}_{B'} \Theta_{CC'BF'}) \tau^{BB'} \tag{2.40}$$

$$= -d^{EE'}(\phi_{EABC} \epsilon_{E'C'} \tau^B{}_{A'} + \bar{\phi}_{E'A'B'C'} \epsilon_{EC} \tau_A{}^{B'}).$$

We set now $\Lambda_{ABCA'} \equiv \nabla^F{}_{A'} \phi_{ABCF}$. Equation (2.32) is then equivalent to $0 = \Lambda_{ABCD} \equiv \Lambda_{ABCA'} \tau_D{}^{A'}$. On the other hand we have the decomposition $\Lambda_{ABCD} = \Lambda_{(ABCD)} - \frac{3}{4} \epsilon_{D(C} \Lambda_{AB)F}{}^F$ with irreducible parts

$$\Lambda_{(ABCD)} = -\frac{1}{2} \left(P \phi_{ABCD} - 2 \mathcal{D}_{(D}{}^F \phi_{ABC)F} \right), \quad \Lambda_{ABF}{}^F = \mathcal{D}^{EF} \phi_{ABEF}, \tag{2.41}$$

where $P = \tau^{AA'} \nabla_{AA'} = \sqrt{2} \nabla_{e_0}$ and $\mathcal{D}_{AB} = \tau_{(A}{}^{A'} \nabla_{B)A'}$ denote covariant directional derivative operators such that $\mathcal{D}_{00} = -\nabla_{01'}$, $\mathcal{D}_{11} = \nabla_{10'}$, and $\mathcal{D}_{01} = \mathcal{D}_{10} = \frac{1}{\sqrt{2}} \nabla_{e_3}$, (cf. [35], [36] for more details of the underlying space-spinor formalism).

In a Cauchy problem one will in general assume e_0 to be the future directed normal to the initial hypersurface $\tilde{S}$. The operators $\mathcal{D}_{AB}$ then involve only differentiation in directions tangent to $\tilde{S}$ and the equations $\Lambda_{ABF}{}^F = 0$ are interior equations on $\tilde{S}$. They represent the six real constraint equations implied on $\tilde{S}$ by the Bianchi equation.

For a symmetric spinor field $\psi_{A_1 \ldots A_k}$ we define its (independent) *essential components* by $\psi_j = \psi_{(A_1 \ldots A_k)_j}$, where $0 \le j \le k$ and the brackets with subscript j indicate that j of the indices in the brackets are set equal to 1 while the others are set equal to 0.

The five equations $\Lambda_{(ABCD)} = 0$ for the components of ϕ_{ABCD} contain the operator P. Multiplying by suitable binomial coefficients (and considering the frame and connection coefficients as given), we find that the system

$$-\left(A + B + C + D\right) \Lambda_{(ABCD)} = 0, \qquad (2.42)$$

has the following properties. If ϕ denotes the transpose of the $\mathbb{C}^5$-valued 'vector' $(\phi_0, \phi_1, \phi_2, \phi_3, \phi_4)$, it takes the form

$$A^\mu \, \partial_\mu \, \phi = H(x, \phi),$$

with a $\mathbb{C}^5$-valued function $H(x, \phi)$ and 5×5-matrix-valued functions A^μ which are hermitian, i.e., $^T \bar{A}^\mu = A^\mu$, and for which there exists at each point a covector ξ_μ such that $A^\mu \xi_\mu$ is positive definite. The system (2.42) is thus *symmetric hyperbolic* ([44], see also [43] and the references given there).

While the constraints implied on a given space-like hypersurface are determined uniquely, there is a large freedom to select useful evolution systems. In fact, any system of the form

$$0 = 2 \, a \, \Lambda_{0001'},$$

$$0 = (c - d) \, \Lambda_{0011'} - 2 \, a \, \Lambda_{0000'},$$

$$0 = (c + d) \, \Lambda_{0111'} - (c - d) \, \Lambda_{0010'}, \qquad (2.43)$$

$$0 = 2 \, e \, \Lambda_{1111'} - (c + d) \, \Lambda_{0110'},$$

$$0 = -2 \, e \, \Lambda_{1110'},$$

with $a, c, e > 0$ and $-(2 \, e + c) < d < 2 \, a + c$, is symmetric hyperbolic (the system (2.42) occurs here as a special case). We note that only the characteristics of these systems which are null hypersurfaces are of physical significance.

Equations (2.33), (2.34), (2.35), respectively equations (2.38), (2.39), (2.40), combined with a choice of (2.43), will be referred to in the following as the (general) *reduced conformal field equations*. Solutions to these equations solve in fact the complete system (2.12), (2.13), (2.15), (2.16) if the solution admits a Cauchy hypersurface on which the latter equations hold ([36]). In other words, propagation by the reduced field equations preserves the constraints.

2.4. The conformal constraints

To analyze solutions to the conformal field equations in the context of a Cauchy problem one needs to study the conformal constraints implied on a space-like initial hypersurface $\tilde{S}$. It will be convenient to discuss the evolution equations in terms of a conformal factor Θ which differs on $\tilde{S}$ from the one used to analyze the constraints. We thus assume Einstein's equations (2.1), a conformal rescaling

$$g = \Omega^2 \tilde{g}, \qquad (2.44)$$

with a positive conformal factor Ω, and denote again the Levi-Civita connection of g by ∇. It is also convenient to derive the conformal constraints from the *metric conformal field equations*. The latter are written in terms of the unknown fields

g, Ω, $S \equiv \frac{1}{4}\nabla_\mu\nabla^\mu\Omega + \frac{1}{24}R\,\Omega$, $L_{\mu\nu}$ as in (2.8), and $W^\mu{}_{\nu\lambda\rho} = \Omega^{-1}C^\mu{}_{\nu\lambda\rho}$ and are given by equation (2.7), with $C^\mu{}_{\nu\lambda\rho} = \Omega\,W^\mu{}_{\nu\lambda\rho}$, and the equations

$$2\,\Omega\,S - \nabla_\mu\Omega\,\nabla^\mu\Omega = 0, \tag{2.45}$$

$$\nabla_\mu\nabla_\nu\Omega = -\Omega\,L_{\mu\nu} + S\,g_{\mu\nu}, \tag{2.46}$$

which are obtained by rewriting the trace and the trace free part of (2.2),

$$\nabla_\mu S = -L_{\mu\nu}\nabla^\nu\Omega, \tag{2.47}$$

$$\nabla_\lambda L_{\rho\nu} - \nabla_\rho L_{\lambda\nu} = \nabla_\mu\Omega\,W^\mu{}_{\nu\lambda\rho}, \tag{2.48}$$

which both can be obtained as integrability conditions of (2.46), and

$$\nabla_\mu W^\mu{}_{\nu\lambda\rho} = 0. \tag{2.49}$$

In these equations the Ricci scalar R is considered as the *conformal gauge source function*. Its choice, which is completely arbitrary in local studies, controls together with the initial data Ω and $d\Omega$ on $\tilde S$ the evolution of the conformal scaling.

To derive the constraints induced by these equations on $\tilde S$ we choose a g-orthonormal frame field $\{e_k\}_{k=0,1,2,3}$ near $\tilde S$ such that $n \equiv e_0$ is g-normal to $\tilde S$, write $\nabla_k \equiv \nabla_{e_k}$, $\nabla_k e_j = \Gamma_k{}^l{}_j e_l$, and express all fields (except the e_k) and equations in terms of this frame. We assume that indices $a, b, c, \dots$ from the beginning of the alphabet take values $1, 2, 3$ and that the summation convention also holds for these indices. The inner metric h induced by g on $\tilde S$ is then given by $h_{ab} = g(e_a, e_b) = -\delta_{ab}$ and the second fundamental form by

$$\chi_{ab} = g(\nabla_{e_a} n, e_b) = \Gamma_a{}^j{}_0\,g_{jb} = -\Gamma_a{}^0{}_b.$$

We set $\Sigma = \nabla_0\,\Omega$ and $W^*_{\mu\nu\lambda\rho} = \frac{1}{2}W_{\mu\nu\alpha\beta}\epsilon^{\alpha\beta}{}_{\lambda\rho}$. If the tensor fields

$$L_{\mu\nu}, \quad L_{\mu\nu}n^\nu, \quad W_{\mu\nu\lambda\rho}, \quad W_{\mu\nu\lambda\rho}n^\nu n^\rho, \quad W^*_{\mu\nu\lambda\rho}n^\nu n^\rho, \quad W_{\mu\nu\lambda\rho}n^\nu,$$

are projected orthogonally into $\tilde S$ and expressed in terms of the frame $\{e_a\}_{a=1,2,3}$ on $\tilde S$, they are given by (the left-hand sides of)

$$L_{ab}, \quad L_a \equiv L_{a0},$$

$$w_{abcd} \equiv W_{abcd}, \quad w_{ab} \equiv W_{a0b0}, \quad w^*_{ab} \equiv W^*_{a0b0}, \quad w_{abc} \equiv W_{a0bc},$$

respectively and satisfy the relations

$$R = 6\,L_\mu{}^\mu = 6\,(L_{00} + L_a{}^a), \tag{2.50}$$

$$w_{abcd} = -2\{h_{a[c}w_{d]b} + h_{b[d}w_{c]a}\}, \quad w^*_{ad}\,\epsilon^d{}_{bc} = w_{abc}, \quad w^*_{ad} = -\frac{1}{2}w_{abc}\,\epsilon_d{}^{bc},$$

$$w_{ab} = w_{ba}, \quad w_a{}^a = 0, \quad w^*_{ab} = w^*_{ba}, \quad w^*_a{}^a = 0,$$

$$w_{abc} = -w_{acb}, \quad w^a{}_{ac} = 0, \quad w_{[abc]} = 0,$$

where indices are moved with h_{ab} and ϵ_{abc} is totally antisymmetric with $\epsilon_{123} = 1$. The tensors w_{ab} and w^*_{ab} represent the n-electric and the n-magnetic part of $W^i{}_{jkl}$ on $\tilde S$ respectively.

Equation (2.7) implies Gauss' and Codazzi's equation on $\tilde{S}$

$$r_{ab} = -\Omega\, w_{ab} + L_{ab} + L_c{}^c\, h_{ab} + \chi_c{}^c\, \chi_{ab} - \chi_{ac}\, \chi_b{}^c, \tag{2.51}$$

$$D_b\, \chi_{ca} - D_c\, \chi_{ba} = \Omega\, w_{abc} + h_{ab}\, L_c - h_{ac}\, L_b, \tag{2.52}$$

while equations (2.45)–(2.49) imply the following interior equations which only involve derivatives in the directions of e_a, $a = 1, 2, 3$, tangent to $\tilde{S}$

$$2\,\Omega\, S - \Sigma^2 - D_a\Omega\, D^a\Omega = 0, \tag{2.53}$$

$$D_a\, D_b\, \Omega = -\Sigma\, \chi_{ab} - \Omega\, L_{ab} + S\, h_{ab}, \tag{2.54}$$

$$D_a\, \Sigma = \chi_a{}^c D_c\Omega - \Omega\, L_a, \tag{2.55}$$

$$D_a\, S = -D^b\,\Omega\, L_{ba} - \Sigma\, L_a, \tag{2.56}$$

$$D_a\, L_{bc} - D_b\, L_{ac} = D^e\Omega\, w_{ecab} - \Sigma\, w_{cab} - \chi_{ac}\, L_b + \chi_{bc}\, L_a, \tag{2.57}$$

$$D_a\, L_b - D_b\, L_a = D^e\Omega\, w_{eab} + \chi_a{}^c L_{bc} - \chi_b{}^c L_{ac}, \tag{2.58}$$

$$D^c\, w_{cab} = \chi^c{}_a\, w_{bc} - \chi^c{}_b\, w_{ac}, \tag{2.59}$$

$$D^a\, w_{ab} = \chi^{ac}\, w_{abc}, \tag{2.60}$$

where r_{ab} denotes the Ricci tensor of h_{ab}. These equations can be read as *conformal constraints* for the fields

$$\Omega,\ \Sigma,\ S,\ h_{ab},\ \chi_{ab},\ L_a,\ L_{ab},\ w_{ab},\ w_{ab}^*.$$

Alternatively, if a 'physical' solution $\tilde{h}_{ab}$, $\tilde{\chi}_{ab}$ to the vacuum constraints is given and a conformal factor Ω and functions Σ, R have been chosen, which are gauge dependent functions at our disposal, the equations above can be used to calculate

$$S,\ L_{\mu\nu},\ W^\mu{}_{\nu\lambda\rho},$$

from the conformal first and second fundamental forms h_{ab}, χ_{ab} of $\tilde{S}$, which are related to the physical data by

$$h_{ab} = \Omega^2\, \tilde{h}_{ab}, \quad \chi_{ab} = \Omega\, (\tilde{\chi}_{ab} + \Sigma\, \tilde{h}_{ab}). \tag{2.61}$$

The equations above will be discussed in more detail in Section 4.

3. Asymptotic simplicity

To characterize the fall-off behavior of asymptotically flat solutions at null infinity in terms of geometric concepts Penrose introduced the notion of *asymptotic simplicity* ([59], [60], cf. also [61] for further discussions and references).

Definition 3.1. *A smooth space-time $(\tilde{M}, \tilde{g})$ is called asymptotically simple if there exists a smooth, oriented, time-oriented, causal space-time (M, g) and a smooth function Ω on M such that:*

(i) *M is a manifold with boundary J,*

(ii) *$\Omega > 0$ on $M \setminus J$ and $\Omega = 0$, $d\Omega \neq 0$ on J,*

(iii) *there exists an embedding Φ of $\tilde{M}$ onto $\Phi(\tilde{M}) = M \setminus J$ which is conformal such that $\Omega^2\, \Phi^{-1*}\tilde{g} = g$,*

(iv) *each null geodesic of $(\tilde{M}, \tilde{g})$ acquires two distinct endpoints on J.*

We note that only the conformal class of $(\tilde{\mathcal{M}}, \tilde{g})$ enters the definition and it is only the conformal structure of $(\mathcal{M}, g)$ which is determined here. The set $\mathcal{J}$ is referred to as the *conformal boundary of* $(\tilde{\mathcal{M}}, \tilde{g})$ *at null infinity*. This definition is the mathematical basis for the

Penrose proposal: *Far fields of isolated gravitating systems behave like that of asymptotically simple space-times in the sense that they can be smoothly extended to null infinity, as indicated above, after suitable conformal rescalings.*

Since gravitational fields are governed by Einstein's equations, the proposal suggests a sharp characterization of the fall-off behavior implied by the field equations in terms of the purely geometrical definition (3.1).

We will be interested in the following in solutions to Einstein's vacuum field equations (2.1) which satisfy the conditions of definition (3.1) (or suitable generalizations). The two assumptions have important consequences for the structure of $(\mathcal{M}, g)$. We shall only quote those which will be used in the following discussion and refer the reader for further information to the references given above.

If the vacuum field equations hold near $\mathcal{J}$, the latter defines a smooth null hypersurface of $\mathcal{M}$ (cf. equation (2.45)). It splits into two components, $\mathcal{J}^+$ and $\mathcal{J}^-$, which are generated by the past and future endpoints of null geodesics in $\mathcal{M}$ and are thus called *future and past null infinity* (or *scri* $\pm$) respectively. Each of $\mathcal{J}^\pm$ is ruled by null generators, each set of null generators has topology S^2, and $\mathcal{J}^\pm$ have the topology of $\mathbb{R} \times S^2$. For the applications one will have to relax the conditions of definition (3.1). In particular condition (iv), which is important to obtain the result about the topology of $\mathcal{J}^\pm$, must be replaced by a different completeness condition if one wants to discuss space-times with black holes.

One of the main difficulties in developing a well-defined concept of *outgoing radiation* in the time-like cut model is related to the fact that there exists in general no distinguished null direction field along the time-like boundary $\mathcal{T}$. In contrast, the null generators of $\mathcal{J}^+$ define a unique causal direction field on $\mathcal{J}^+$, which is represented by $\nabla^{AA'}\Omega$. It turns out that the field $\phi_{ABCD}\, o^A\, o^B\, o^C\, o^D$ on $\mathcal{J}^+$, where o^A denotes a spinor field satisfying $o^A\, \bar{o}^{A'} = -\nabla^{AA'}\Omega$ on $\mathcal{J}^+$ and ϕ_{ABCD} the rescaled conformal Weyl spinor field, has a natural interpretation as the *outgoing radiation field*. Further important physical concepts can be associated with the hypersurface $\mathcal{J}^+$ or subsets of it and questions of interpretation have been extensively analyzed (cf. [4], [19], [46] and the references given there).

Critical however, and in fact a matter of controversy, have been the smoothness assumptions in the definition, which encode the fall-off behavior imposed on the physical fields. It is far from immediate that they are in harmony with the fall-off behavior imposed by the field equations. No problem arises if the proposal can be justified with C^∞ replaced by C^k with sufficiently large integer k. But there is a lower threshold for the differentiability, which is not easily specified, at which the definition will loose much of its elegance and simplicity.

In [60] it is assumed that

$$\mathcal{M} \text{ is of class } C^{k+1} \text{ and } g, \Omega \in C^k(\mathcal{M}), \quad k \geq 3. \tag{3.1}$$

The conformal Weyl spinor $\Psi_{ABCD} = \Omega\,\phi_{ABCD}$ is then in $C^{k-2}(\mathcal{M})$. Under the further assumption

$$\Omega\,\nabla_{EE'}\,\nabla^A{}_{A'}\,\Psi_{ABCD} \to 0 \quad \text{at} \quad \mathcal{J}^+, \tag{3.2}$$

which will certainly be satisfied if $k \geq 4$ in (3.1), and the natural assumption

$$\text{the set of null generators of } \mathcal{J}^+ \text{ has topology } S^2, \tag{3.3}$$

it is then shown that Ψ_{ABCD} vanishes on $\mathcal{J}^+$. The solution is thus asymptotically flat in the most immediate sense and the rescaled conformal Weyl spinor ϕ_{ABCD} extends in a continuous fashion to $\mathcal{J}^+$. As a consequence, it follows that the space-time satisfies the *Sachs peeling property* ([60], [63], [64]) which says that in a suitably chosen spin frame the components of the conformal Weyl spinor fall-off as $\Psi_{ABCD} = O(\tilde{r}^{A+B+C+D-5})$ (where A, B, C, D take values $0, 1$) along an outgoing null geodesic when its (physical) affine parameter $\tilde{r} \to \infty$ at $\mathcal{J}^+$.

Remarkable as it is that such a conclusion can be drawn for the spin-2 nature of the field Ψ_{ABCD} and its governing field equation $\tilde{\nabla}^F{}_{A'}\,\Psi_{ABCF} = 0$, there remains the question whether the long time evolution by the field equations is such that assumptions (3.1), (3.2) or the conclusion drawn from them can be justified.

As discussed in the introduction, we know by now that these conditions can be satisfied by non-trivial solutions to the vacuum field equations. What is not known, however, is how the solutions satisfying these conditions are to be characterized in terms of their Cauchy data, whether these conditions exclude solutions modelling important physical phenomena, and if they do, what exactly goes wrong. Obviously, these questions can only be answered by analyzing the Cauchy problem for Einstein' field equations with asymptotically flat Cauchy data in the large with the goal to derive sharp results on the behavior of the field at null infinity.

The results obtained so far on the existence of solutions which admit (partial) smooth boundaries at null infinity make it clear that the key problem here is the behavior of the fields near space-like infinity. We shall not consider any further the results which lead to this conclusion (cf. [38] for a discussion and the relevant references) but concentrate on this particular problem.

To begin with we have a look at the asymptotic region of interest here in the case of Minkowski space. If the latter is given in the form $(\tilde{\mathcal{M}} \simeq \mathbb{R}^4, \tilde{g} = \eta_{\mu\nu}\,dy^\mu\,dy^\nu)$, the coordinate transformation $\Phi : y^\mu \to x^\mu = (-y_\lambda y^\lambda)^{-1}\,y^\mu$ renders the metric in the domain $\mathcal{D} \equiv \{y_\lambda y^\lambda < 0\} = \{x_\lambda x^\lambda < 0\}$ in the form $\tilde{g} = \Omega^{-2}\,\eta_{\mu\nu}\,dx^\mu\,dx^\nu$ with $\Omega = -x_\lambda x^\lambda$. The metric

$$g = \Omega^2\,\tilde{g} = \eta_{\mu\nu}\,dx^\mu\,dx^\nu, \tag{3.4}$$

thus extends smoothly to the domain $\bar{\mathcal{D}}$ of points in $\{x_\lambda x^\lambda \leq 0\}$ which are obtained as limits of sequences in $\mathcal{D}$. The point $x^\mu = 0$ in this set then represents space-like infinity for Minkowski space. With this understanding it is denoted by i^0. The hypersurfaces $\mathcal{J}'^\pm = \{x_\lambda x^\lambda = 0, \pm x^0 > 0\} \subset \bar{\mathcal{D}}$ represent parts of future and past null infinity of Minkowski space close to space-like infinity and are generated by the future and past directed null geodesics of g through i^0.

Consider the Cauchy hypersurface $\tilde{S} = \{y^0 = 0\}$ of Minkowski space. The subset $\tilde{S} \cap \mathcal{D}$ is mapped by Φ onto $\{x^0 = 0, \, x^\mu \neq 0\}$. Extending the latter to include the point $x^\mu = 0$ amounts to a smooth compactification $\tilde{S} \to S = \tilde{S} \cup \{i\} \sim S^3$ such that the point i with coordinates $x^\mu = 0$ represents space-like infinity with respect the metric induced on $\tilde{S}$ by $\tilde{g}$. The distinction of space-like infinity i with respect to a Cauchy data set and space-like infinity i^0 with respect to the solution space-time will become important and much clearer later on.

Denote by $\tilde{h}_{\alpha\beta}$ and $\tilde{\chi}_{\alpha\beta}$ the metric and the extrinsic curvature induced by $\tilde{g}$ on $\tilde{S}$. A global representation of the conformal structure induced on S by $\tilde{h}_{\alpha\beta}$ is obtained by using a slightly different conformal rescaling than before. Set $h' = \Omega'^2 \tilde{h}$ with $\Omega' = 2(1 + |y|^2)^{-1}$ where $|y| = \sqrt{(y^1)^2 + (y^2)^2 + (y^3)^2}$. In terms of standard spherical coordinates θ, ϕ on $\tilde{S}$ and the coordinate χ defined by $\cot \frac{\chi}{2} = |y|$, $0 \leq \chi \leq \pi$, the conformal metric takes the form $h' = -(d\chi^2 + \sin^2 \chi \, d\sigma^2)$ of the standard metric on the 3-sphere and extends smoothly to the point i, which is given by the coordinate value $\chi = 0$ and distinguished by the property that $\Omega = 0$, $d\Omega = 0$, $Hess \, \Omega = c \, h'$, with $c \neq 0$ at i. Here $d\sigma^2 \equiv d\theta^2 + \sin^2 \theta \, d\phi^2$ denotes the standard line element on the 2-sphere S^2.

Since $\tilde{\chi}_{\alpha\beta} = 0$ and we are free to choose $\Sigma = 0$ in (2.61), we get $\chi'_{\alpha\beta} = 0$. By the formulas given in the previous section one can derive from the conformal Minkowski data (S, h', χ') and a suitable choice of initial data for the gauge defining fields (2.24), (2.25) a conformal initial data set for the reduced conformal field equations. These allow us then to recover the well-known conformal embedding of Minkowski space into the Einstein cosmos ([60]) as a smooth solution to the regular conformal field equations ([38], [40]).

We would like to control what happens if the conformal Minkowski data are subject to finite perturbations. Under which assumptions will the corresponding solutions preserve asymptotic simplicity? This or the apparently simpler question under which conditions the solutions will preserve near space-like infinity a reasonable amount of smoothness of the conformal boundary cannot be answered by immediate applications of the conformal field equations. The reason is that the conformal data will not be smooth at the point i. The structure of the conformal initial data as well as the initial value problem for the conformal field equations near space-like infinity thus require a careful and detailed analysis. This will be carried out to some extent in the next section.

4. Asymptotically flat Cauchy data

As indicated in the case of Minkowski space above we will assume that the data for the conformal field equations are given on a 3-dimensional manifold $S = \tilde{S} \cup \{i\}$ which is obtained from a 'physical' 3-manifold $\tilde{S}$ with an asymptotically flat end by adjoining a point i which represents space-like infinity. The data h_{ab}, χ_{ab} on S

are thought as being obtained from the physical data $\tilde{h}_{ab}$, $\tilde{\chi}_{ab}$ by equations (2.61) with suitable choices of Ω and Σ such that all fields extend with an appropriate behavior (to be specified more precisely below) to i and $\Sigma(i) = 0$, $\Omega > 0$ on $\tilde{S}$, $\Omega(i) = 0$, $D_a\,\Omega(i) = 0$, $D_a\,D_b\,\Omega(i) = -2\,h_{ab}$, where we assume the notation of Subsection 2.4.

The constraint equations (2.51)–(2.60) contain analogues of the vacuum constraints. The form of these equations suggests a solution procedure which does not require us to go back to the physical data. By taking the trace of equation (2.51), using (2.53) and the trace of (2.54), and writing $\Delta_h \equiv D_a\,D^a$, one gets

$$\Omega^2\,r = -4\,\Omega\,\Delta_h\,\Omega + 6\,D_a\Omega\,D^a\Omega - 4\,\Omega\,\Sigma\,\chi_c{}^c + \Omega^2((\chi_c{}^c)^2 - \chi_{ac}\,\chi^{ac}), \qquad (4.1)$$

where r denotes the Ricci scalar of h. With $\theta = \Omega^{-\frac{1}{2}}$ this equation takes the form of Lichnerowicz' equation

$$(\Delta_h - \frac{1}{8}\,r)\theta = -\frac{1}{8}\,\theta\,((\chi_c{}^c)^2 - \chi_{ab}\,\chi^{ab}) + \frac{1}{2}\,\theta^3\,\Sigma\,\chi_c{}^c. \qquad (4.2)$$

By taking the trace of (2.52) and using (2.55) one gets

$$D^b(\Omega^{-2}\,\chi_{bc}) = \Omega^{-2}\,D_c\,\chi_b{}^b - 2\,\Omega^{-3}\,D_c\Sigma. \qquad (4.3)$$

Equations (4.2) and (4.3) correspond to the Hamiltonian and the momentum constraint respectively. Assuming now

$$\chi_a{}^a = 0 \text{ and (the choice of gauge) } \Sigma = 0 \text{ on } S, \qquad (4.4)$$

which imply $\tilde{\chi}_a{}^a = 0$, equations (4.2) and (4.3) suggest to proceed as follows: (i) prescribe h on S and solve the equation $D^a\,\psi_{ab} = 0$ for a symmetric h-trace free tensor field ψ_{ab} on S, (ii) solve equation (4.2) with $\chi_{ab} = \theta^{-4}\psi_{ab}$ for a positive function θ, i.e., solve

$$(\Delta_h - \frac{1}{8}\,r)\theta = \frac{1}{8}\,\theta^{-7}\,\chi_{ab}\,\chi^{ab}, \qquad \theta > 0. \qquad (4.5)$$

The fields $\Omega = \theta^{-2}$, h_{ab}, and $\chi_{ab} = \Omega^2\psi_{ab}$ then solve (4.1) and (4.3). If it is required that

$$\rho\,\Theta \to 1, \quad \psi_{ab} = O(\frac{1}{\rho^4}) \quad \text{as } \rho \to 0, \qquad (4.6)$$

where $\rho(p)$ denotes near i the h-distance of a point p from i, the fields $\tilde{h}_{ab}$ and $\tilde{\chi}_{ab}$ related by (2.61) to h_{ab} and χ_{ab} satisfy the vacuum constraints and are asymptotically flat ([34]).

Using the conformal constraints to determine the remaining conformal fields one gets

$$S = \frac{1}{3}\,\Delta_h\,\Omega + \frac{1}{12}\,\Omega\,(r + \chi_{ab}\,\chi^{ab}) = \frac{1}{2\Omega}\,D_c\Omega\,D^c\Omega, \qquad (4.7)$$

$$L_a = \frac{1}{\Omega}\,D^c\Omega\,\chi_{ca}, \qquad (4.8)$$

$$L_{ab} - \frac{1}{3}\,L_c{}^c\,h_{ab} = -\frac{1}{\Omega}\left(D_a\,D_b\,\Omega - \frac{1}{3}\,\Delta_h\Omega\,h_{ab}\right), \qquad (4.9)$$

$$L_{00} = \frac{1}{6} R - L_c{}^c = \frac{1}{6} R - \frac{1}{4} \left(r + \chi_{ab} \chi^{ab} \right), \tag{4.10}$$

$$w_{ab} = -\frac{1}{\Omega^2} \left(D_a D_b \Omega - \frac{1}{3} \Delta_h \Omega\, h_{ab} \right) - \frac{1}{\Omega} \left(\chi_{ac} \chi_b{}^c - \frac{1}{3} \chi_{ce} \chi^{ce} h_{ab} + s_{ab} \right), \tag{4.11}$$

$$w_{ab}^* = -\frac{1}{\Omega} D_c \chi_{e(a} \epsilon_{b)}{}^{ce}, \tag{4.12}$$

where we set $s_{ab} = r_{ab} - \frac{1}{3} r\, h_{ab}$. The differential identities (2.56)–(2.60), which are not needed to get these expressions, will be then also be satisfied (cf. [31]).

In view of conditions (4.6) most of these fields will in general be singular at i. One will have $w_{ab} = O(r^{-3})$ near i whenever the ADM mass m of the initial data set $\tilde{h}_{ab}$, $\tilde{\chi}_{ab}$ does not vanish. Controlling the time evolution of these data requires a careful analysis of these singularities. As a simplifying hypothesis we assume, as in [37], that the data are time reflection symmetric and define a smooth conformal structure, i.e.,

$$h_{ab} \in C^\infty(\mathcal{S}), \qquad \chi_{ab} = 0. \tag{4.13}$$

We note that much of the following discussion can be extended to more general data such as those considered in [25] and the more general class of data discussed in [24], which includes the stationary data.

The Ricci scalar R is at our disposal. With $R = \frac{3}{2} r$ one gets on $\mathcal{S}$

$$L_{ab} = -\frac{1}{\Omega} \left(D_a D_b \Omega - \frac{1}{3} \Delta_h \Omega\, h_{ab} \right) + \frac{1}{12} r\, h_{ab}, \tag{4.14}$$

$$L_{00} = 0, \quad L_{0a} = 0, \quad w_{ab}^* = 0, \tag{4.15}$$

$$w_{ab} = -\frac{1}{\Omega^2} \left(D_a D_b \Omega - \frac{1}{3} \Delta_h \Omega\, h_{ab} + \Omega\, s_{ab} \right). \tag{4.16}$$

In spite of this simplification the crucial problem is still present; one finds that $w_{ab} = O(\rho^{-3})$ near i if $m \neq 0$ (cf. (5.28)).

4.1. Time reflection symmetric asymptotically flat Cauchy data

To allow for more flexibility in the following analysis, we also want to admit nontrivial cases with vanishing or negative mass. The positive mass theorem ([65]) then tells us that we must allow for non-compact $\mathcal{S}$. This will create no problems because we are interested only in the behavior of the fields near space-like infinity.

Let x^a, $a = 1, 2, 3$, denote h-normal coordinates defined on some convex open normal neighborhood $\mathcal{U}$ of i so that with $h = h_{ab}(x^c)\, dx^a\, dx^b$

$$x^a(i) = 0, \qquad x^a h_{ab}(x^c) = -x^a \delta_{ab} \quad \text{on} \quad \mathcal{U}. \tag{4.17}$$

All equations of this subsection will be written in these coordinates. We set $|x| = \sqrt{\delta_{ab} x^a x^b}$ and $\Upsilon = |x|^2 = \delta_{ab} x^a x^b$ so that

$$h^{ab} D_a \Upsilon D_b \Upsilon = -4\, \Upsilon, \tag{4.18}$$

and

$$\Upsilon(i) = 0, \quad D_a \Upsilon(i) = 0, \quad D_a D_c \Upsilon(i) = -2\, h_{ac}. \tag{4.19}$$

By taking derivatives of (4.18) and using (4.19) one obtains

$$D_a D_b D_c \Upsilon(i) = 0, \qquad D_a D_b D_c D_d \Upsilon(i) = -\frac{4}{3} r_{a(cd)b}[h](i), \qquad (4.20)$$

where the curvature tensor of h is given by

$$r_{abcd}[h] = 2\{h_{a[c}l_{d]b} + h_{b[d}l_{c]a}\}$$

with $l_{ab}[h] = r_{ab}[h] - \frac{1}{4} r[h] h_{ab}$ because $\dim(\mathcal{S}) = 3$. Proceeding further in this way on can determine an expansion of Υ in terms curvature terms at i. The relations above imply in particular

$$(\Delta_h \Upsilon + 6)(i) = 0, \qquad D_a(\Delta_h \Upsilon + 6)(i) = 0, \qquad D_a D_b(\Delta_h \Upsilon + 6)(i) = \frac{4}{3} r_{ab}(i). \qquad (4.21)$$

Equation (4.5) and the first of equations (4.6) can be combined under our assumptions into

$$(\Delta_h - \frac{1}{8} r) \theta = 4 \pi \delta_i,$$

where in the coordinates x^a the symbol δ_i denotes the Dirac-measure with weight 1 at $x^a = 0$. In a neighborhood of i there exists then a representation $\theta = \frac{U}{|x|} + W$ with functions U, W which satisfy

$$(\Delta_h - \frac{1}{8} r) \left(\frac{U}{|x|}\right) = 4 \pi \delta_i, \qquad (\Delta_h - \frac{1}{8} r) W = 0 \quad \text{near } i, \qquad (4.22)$$

and

$$U(i) = 1, \qquad W(i) = \frac{m}{2}, \qquad (4.23)$$

where m denotes the ADM-mass of the solution. The functions U, W are analytic on $\mathcal{U}$ if h is analytic ([45]) and smooth if h is C^∞ ([25]).

The function $\sigma \equiv \frac{\Upsilon}{U^2}$ is characterized uniquely by the conditions that it is smooth, satisfies the equation $(\Delta_h - \frac{1}{8} r) \sigma^{-1/2} = 4 \pi \delta_i$, and the relations

$$\sigma(i) = 0, \qquad D_a \sigma(i) = 0, \qquad D_a D_b \sigma(i) = -2 h_{ab}, \qquad (4.24)$$

hold, which follow from (4.19) and (4.23). If σ' is another function satisfying these conditions, then $\sigma' = \Upsilon U'^{-2}$ with $U' = 1 + O(|x|)$ by (4.24) and $U' \in C^\infty(\mathcal{U})$ by the results of [25]. The function $f = \sigma^{-1/2} - \sigma'^{-1/2}$ then solves $(\Delta_h - \frac{1}{8} r) f = 0$ and it follows that $f \in C^\infty(\mathcal{U})$ and $|x| f = U - U' \in C^\infty(\mathcal{U})$. Expanding f and $U - U'$ in terms of spherical harmonics it follows from the last equation that f vanishes at i at any order. Since f satisfies the conformal Laplace equation it follows that $f = 0$ on $\mathcal{U}$ by Theorem 17.2.6 of [51]. This implies that $\sigma' = \sigma$ on $\mathcal{U}$.

The first of equations (4.22) can be rewritten in the form

$$2 D^a \Upsilon D_a U + (\Delta_h \Upsilon + 6) U - 2 \Upsilon (\Delta_h - \frac{1}{8} r) U = 0. \qquad (4.25)$$

This allows us to determine from (4.24) recursively an asymptotic expansion of U, which is convergent if h is real analytic. The Hadamard parametrix construction is based on an ansatz

$$U = \sum_{p=0}^{\infty} U_p \, \Upsilon^p, \tag{4.26}$$

by which the calculation of U is reduced to an ODE problem. The functions U_p are obtained recursively by

$$U_0 = \exp\left\{ \frac{1}{4} \int_0^{\Upsilon^{\frac{1}{2}}} (\Delta_h \, \Upsilon + 6) \frac{d\rho}{\rho} \right\},$$

$$U_{p+1} = -\frac{U_0}{(4p-2)\,\Upsilon^{\frac{p+1}{2}}} \int_0^{\Upsilon^{\frac{1}{2}}} \frac{\Delta_h[U_p]\,\rho^p}{U_0} \, d\rho, \quad p = 0, 1, 2, \ldots,$$

where the integration is performed in terms of the affine parameter $\rho = \Upsilon^{\frac{1}{2}} = |x|$ along the geodesics emanating from i. It follows that

$$U(i) = 1, \quad D_a \, U(i) = 0, \quad D_a \, D_b \, U(i) = \frac{1}{6} \, l_{ab}[h], \tag{4.27}$$

which implies

$$D_a \, D_b \, D_c \sigma(i) = 0, \quad D_a \, D_b \, D_c \, D_d \sigma(i) = 2 \, (h_{cd} \, l_{ab} + h_{ab} \, l_{cd}). \tag{4.28}$$

Given h and the solution W of the conformal Laplace equation in (4.22), the considerations above show us how to determine an expansion of the function

$$\Omega = \theta^{-2} = \left(\frac{1}{\sqrt{\sigma}} + W \right)^{-2} = \frac{\Upsilon}{(U + \rho W)^2}, \tag{4.29}$$

in terms of ρ at all orders. Corresponding expansions can be obtained for the conformal data (4.7), (4.14), (4.15), (4.16).

While U is thus seen to be determined locally by the metric h, the function W carries non-local information. Cases where $\partial^{\alpha}_{x^a} W(i) = 0$ for all multiindices $\alpha = (\alpha^1, \alpha^2, \alpha^3) \in \mathbb{N}^3$ with $|\alpha| \equiv \alpha^1 + \alpha^2 + \alpha^3 \le N$ for some non-negative integer N or for $N = \infty$ will also be of interest in the following. In the latter case we have in fact ([51])

$$W = 0 \quad \text{near} \quad i. \tag{4.30}$$

For convenience this case will be referred to as the *massless case*.

A rescaling

$$h \to h' = \vartheta^4 \, h, \quad \Omega \to \Omega' = \vartheta^2 \, \Omega,$$

with a smooth positive factor ϑ satisfying $\vartheta(i) = 1$, leaves $\tilde{h} = \Omega^{-2} \, h$ unchanged but implies changes

$$\theta \to \theta' = \vartheta^{-1} \, \theta, \quad U \to U' = \frac{|x'|}{|x|} \, \vartheta^{-1} U, \quad W \to W' = \vartheta^{-1} \, W,$$

where $|x'|$ is defined in terms of h'-normal coordinates $x^{a'}$ as described above. Due to the conformal covariance of the operator on the left-hand sides of equations (4.22), relations (4.22), (4.23) will then also hold with all fields replaced by the primed fields.

To reduce this freedom it has been assumed in [37] that the metric h is given near i on S in the *cn-gauge*. By definition, this conformal gauge is satisfied by h if there exists a 1-form l_* at i such that the following holds. If $x(\tau)$, $l(\tau)$ solve the conformal geodesic equations (with respect to h) with $x(0) = i$, $l(0) = l_*$, and $h(\dot{x}, \dot{x}) = 1$ at i, then a frame e_a which is h-orthonormal at i and satisfies $\hat{D}_{\dot{x}} e_a = 0$ (with $\hat{D} - D = S(l)$), stays h-orthonormal near i. This gauge can be achieved without restrictions on the mass and fixes the scaling uniquely up to a positive real number and a 1-form given at i. It admits an easy discussion of limits where $m \to 0$.

If $m > 0$, it is convenient to set above $\vartheta = \frac{2}{m} W$. It follows then that $W' = \frac{m}{2}$, whence $0 = (\Delta_{h'} - \frac{1}{8} r[h']) W' = -\frac{m}{16} r[h']$. Thus, if $m > 0$, we can always assume h to be given such that

$$r[h] = 0, \quad \Omega = \frac{\sigma}{(1 + \sqrt{\mu \sigma})^2} \quad \text{with} \quad \sqrt{\mu} = \frac{m}{2}. \tag{4.31}$$

In this gauge the function σ satisfies near i the equation $\Delta_h (\sigma^{-1/2}) = 4\pi \delta_i$, which implies by (4.24)

$$2\sigma s = D_a \sigma D^a \sigma \quad \text{with} \quad s \equiv \frac{1}{3} \Delta_h \sigma, \tag{4.32}$$

(note that an analogous equation holds with σ replaced by Ω). Equation (4.32) implies in turn together with (4.24) the Poisson equation above.

For later reference we note the form of the conformal Schwarzschild data in this gauge. In isotropic coordinates the Schwarzschild line element is given by

$$d\tilde{s}^2 = \left(\frac{1 - m/2\tilde{r}}{1 + m/2\tilde{r}}\right)^2 dt^2 - (1 + m/2\tilde{r})^4 (d\tilde{r}^2 + \tilde{r}^2 d\sigma^2).$$

Expressing the initial data $\tilde{h}$, $\tilde{\chi}$ induced on $\{t = 0\}$ in terms of the coordinate $\rho = 1/\tilde{r}$, one finds that $\tilde{\chi} = 0$ and $\tilde{h} = \Omega^{-2} h$ with

$$h = -(d\rho^2 + \rho^2 d\sigma^2), \quad \Omega = \frac{\rho^2}{(1 + \frac{m}{2}\rho)^2}, \tag{4.33}$$

so that $\sigma = \Upsilon = \rho^2$ resp. $U = 1$. The metric h also satisfies a cn-gauge.

4.2. Static asymptotically flat Cauchy data

Static solutions to the vacuum field equations can be written in the form

$$\tilde{g} = v^2 dt^2 + \tilde{h},$$

with t-independent negative definite metric $\tilde{h}$ and t-independent norm

$$v = \sqrt{\tilde{g}(K, K)} > 0$$

of the time-like Killing field $K = \partial_t$. With the $\tilde{g}$-unit normal of a slice $\{t = \text{const.}\}$ being given by $\tilde{n} = v^{-1} K$ and the associated orthogonal projector by $\tilde{h}_\mu{}^\nu = \tilde{g}_\mu{}^\nu - \tilde{n}_\mu \tilde{n}^\nu$, one gets for the second fundamental form on this slice

$$\tilde{\chi}_{\mu\nu} = v^{-1} \tilde{h}_\mu{}^\rho \tilde{h}_\nu{}^\delta \tilde{\nabla}_\rho K_\delta = 0,$$

because it is symmetric by $\tilde{n}$ being hypersurface orthogonal while the second term is anti-symmetric by the Killing equation. The solutions are thus time reflection symmetric.

For these solutions the vacuum field equations are equivalent to the requirement that the *static vacuum field equations*

$$r_{ab}[\tilde{h}] = \frac{1}{v} \tilde{D}_a \tilde{D}_b v, \qquad \Delta_{\tilde{h}} v = 0, \tag{4.34}$$

hold on one and thus on any slice $\{t = \text{const.}\}$. In harmonic coordinates these equations become elliptic and $\tilde{h}$ and v thus real analytic.

We consider solutions $\tilde{h}$, v to equations (4.34) with non-vanishing ADM-mass which are given on a 3-manifold $\tilde{S}$ which is mapped by suitable coordinates $\tilde{x}^a$ diffeomorphically to $\mathbb{R}^3 \setminus \bar{B}$, where $\bar{B}$ is a closed ball in $\mathbb{R}^3$. We assume that $\tilde{h}$ satisfies in these coordinates the usual condition of asymptotic flatness and $v \to 1$ as $|\tilde{x}| \to \infty$. The work in [6] (cf. also [54] for a strengthening of this result) then implies that the conformal structure defined by $\tilde{h}$ can be extended analytically to space-like infinity. The physical 3-metric $\tilde{h}$ therefore belongs to the class of data considered above.

For such solutions it follows from the discussion in [7] that the gauge (4.31) is achieved if any of the equivalent equations

$$v = 1 - m\sqrt{\Omega} = \frac{1 - \sqrt{\mu\sigma}}{1 + \sqrt{\mu\sigma}}, \qquad \sigma = \left(\frac{2}{m}\frac{1-v}{1+v}\right)^2, \tag{4.35}$$

holds. The set $S = \tilde{S} \cup \{i\}$ can then be endowed with a differential structure such that the metric $h = \Omega^2 \tilde{h}$ extends as a real analytic metric to i. We shall consider in the following h-normal coordinates as in (4.17) such that the functions $\sigma(x^c)$, $h_{ab}(x^c)$ are then real analytic on $\mathcal{U}$. The first of the static vacuum field equations (4.34) then implies

$$0 = \Sigma_{ab} \equiv D_a D_b \sigma - s h_{ab} + \sigma(1 - \mu\sigma) r_{ab}[h], \tag{4.36}$$

where s is defined as in (4.32). The second of equations (4.34) implies $r[\tilde{h}] = 0$ and can thus be read as a conformally covariant Laplace equation for v. Using the transformation rule for this equation and observing (4.35), we find that it transforms into

$$0 = (\Delta_h - \frac{1}{8} r[h])(\theta v) = (\Delta_h - \frac{1}{8} r[h])(\theta - m) \quad \text{on} \ \tilde{S},$$

and is thus satisfied by our assumption $r[h] = 0$.

We shall repeat some of the considerations of [34] in the present conformal gauge. The fact that solutions to the conformal static field equations are real analytic and can be extended by analyticity into the complex domain allows us to use some very concise arguments. We note that the statements obtained here can also be obtained by recursive arguments. This will become important if some of the following considerations are to be transferred to C^∞ or C^k situations.

From (4.36) one gets

$$D_c \Sigma_{ab} = D_c D_a D_b \sigma - D_c s \, h_{ab} + \sigma \, (1 - \mu \sigma) \, D_c \, r_{ab} + (1 - 2\mu \sigma) \, D_c \sigma \, r_{ab}.$$

With the Bianchi identity, which takes in the present gauge the form $D^a r_{ab} = 0$, follow the integrability conditions

$$0 = \frac{1}{2} D^c \Sigma_{ca} = D_a s + (1 - \mu \sigma) \, r_{ab} \, D^b \sigma, \tag{4.37}$$

and

$$0 = D_{[c} \Sigma_{a]b} + \frac{1}{2} D^d \Sigma_{d[c} h_{a]b} \tag{4.38}$$

$$= \sigma \left\{ (1 - \mu \sigma) \, D_{[c} r_{a]b} - \mu \left(2 \, D_{[c} \sigma \, r_{a]b} + D^d \sigma \, r_{d[c} h_{a]b} \right) \right\}.$$

Equation (4.36) thus implies an expression for the Cotton tensor, which is given in the present gauge by $b_{bca} = D_{[c} r_{a]b}$, and for its dualized version, which is given by

$$b_{ab} = \frac{1}{2} b_{acd} \, \epsilon_b{}^{cd} = \frac{\mu}{1 - \mu \sigma} (D_c \sigma \, r_{da} \, \epsilon_b{}^{cd} - \frac{1}{2} r_{de} \, D^e \sigma \, \epsilon_{ba}{}^d). \tag{4.39}$$

It follows that

$$D_a (2\sigma \, s - D_c \sigma \, D^c \sigma) = \sigma \, D^c \Sigma_{ca} - 2 \, D^c \sigma \, \Sigma_{ca},$$

which shows that equation (4.32) is a consequence of equations (4.24) and (4.36) and that the latter contain the complete information of the conformal static field equations.

Let $e_a = e^c{}_a \partial_{x^c}$, $a = 1, 2, 3$, now denote the h-orthonormal frame field on $\mathcal{U}$ which is parallely transported along the h-geodesics through i and satisfies $e^c{}_a = \delta^c{}_a$ at i. In the following we assume all tensor fields, except the frame field e_a and the coframe field σ^c dual to it, to be expressed in term of this frame field and set $D_a \equiv D_{e_a}$. The coefficients of h are then given by $h_{ab} = -\delta_{ab}$. Any analytic tensor field $T^{b_1 \ldots b_q}_{a_1 \ldots a_p}$ on V has an expansion of the form (cf. [37])

$$T^{b_1 \ldots b_q}_{a_1 \ldots a_p}(x) = \sum_{k \geq 0} \frac{1}{k!} x^{c_k} \ldots x^{c_1} (D_{c_k} \ldots D_{c_1} T^{b_1 \ldots b_q}_{a_1 \ldots a_p})(i),$$

(where the summation rule ignores whether indices are bold face or not).

We want to discuss how expansions of this form can be obtained for the fields

$$\sigma, \quad s, \quad r_{ab},$$

which are provided by the solutions to the conformal static field equations. Once these fields are known, the coefficients of the 1-forms $\sigma^a = \sigma^a{}_b \, dx^b$, which provide the coordinate expression of the metric by the relation $h = -\delta_{ac} \sigma^a{}_b \sigma^c{}_d \, dx^b \, dx^d$, and the connection coefficients $\Gamma_{\mathbf{a}}{}^{\mathbf{b}}{}_{\mathbf{c}}$ with respect to $e_{\mathbf{a}}$ can be obtained from the structural equations in polar coordinates (cf. [50])

$$\frac{d}{d\rho}\left(\rho\,\sigma^{\mathbf{a}}{}_b(\rho\,x)\right) = \delta^{\mathbf{a}}{}_b + \rho\,\Gamma_{\mathbf{c}}{}^{\mathbf{a}}{}_{\mathbf{d}}(\rho\,x)\,x^d\,\sigma^{\mathbf{c}}{}_b(\rho\,x),$$

$$\frac{d}{d\rho}\left(\rho\,\Gamma_{\mathbf{a}}{}^{\mathbf{c}}{}_{\mathbf{e}}(\rho\,x)\,\sigma^{\mathbf{a}}{}_b(\rho\,x)\right) = \rho\,r^{\mathbf{c}}{}_{\mathbf{eda}}(\rho\,x)\,x^d\,\sigma^{\mathbf{a}}{}_b(\rho\,x).$$

For this purpose we consider the data

$$c_{\mathbf{a}_p\ldots\mathbf{a}_1\mathbf{bc}} = \mathcal{R}(D_{\mathbf{a}_p}\ldots D_{\mathbf{a}_1} r_{\mathbf{bc}})(i), \tag{4.40}$$

where $\mathcal{R}$ means 'trace free symmetric part of'. These data have the following interpretation. Since solutions to the conformal static field equations are real analytic in the given coordinates x^a, all the fields considered above can be extended into a complex domain $\mathcal{U}' \subset \mathbb{C}^3$ which comprises $\mathcal{U}$ as the subset of real points. The subset $\mathcal{N} = \{\Upsilon = 0\}$ of $\mathcal{U}'$, where we denote by Υ again the analytic extension of the real function denoted before by the same symbol, then defines the cone which is generated by the complex null geodesics $\mathbb{C} \supset \mathcal{O} \ni \zeta \to x^a(\zeta) = \zeta\,x^a_* \in \mathcal{U}'$ through i, where $x^a_* \neq 0$ is constant with $h_{ab}x^a_* x^b_* = 0$ at i. On $\mathcal{N}$ the field $D^a\Upsilon\,\partial_{x^a} = -2\,x^a\,\partial_{x^a}$ is tangent to the null generators of $\mathcal{N}$. The derivatives of $r_{ab}\,\dot{x}^a\,\dot{x}^b$ with respect to ζ at i are given by the complex numbers

$$x^{a_p}_*\ldots x^{a_1}_* x^b_* x^c_* D_{\mathbf{a}_p}\ldots D_{\mathbf{a}_1} r_{\mathbf{bc}}(i)$$

$$= \iota^{A_p}\iota^{B_p}\ldots\iota^{A_1}\iota^{B_1}\iota^C\ldots\iota^F D_{A_p B_p}\ldots D_{A_1 B_1} r_{CDEF}(i)$$

where the term on the left-hand side is rewritten on the right-hand side in space spinor notation and it is used that $x^{AB}_* \equiv \sigma^{AB}{}_a x^a_* = \iota^A \iota^B$ with some spinor ι^A because x^a_* is a null vector. Allowing x^a_* to vary over the null cone at i, i.e., allowing ι^A to vary over $P^1(\mathbb{C})$, we can extract from the numbers above the real quantities

$$c_{A_p B_p\ldots A_1 B_1 CDEF} = D_{(A_p B_p}\ldots D_{A_1 B_1} r_{CDEF)}(i), \tag{4.41}$$

which are equivalent to (4.40). Giving the data (4.40) is thus equivalent to giving $r_{ab}(\zeta\,x^a_*)\,\dot{x}^a\,\dot{x}^b$ where x^a_* varies over a cut of the complex null cone at i or to giving, up to a scaling, the restriction of $r_{ab}\,D^a\Upsilon\,D^b\Upsilon$ to $\mathcal{N}$. The data (4.40) are in one-to-one correspondence to the multipole moments considered in [6].

We consider now the Bianchi identity $D^a r_{ab} = 0$ and equation (4.38). In space spinor notation they combine into the concise form

$$(1 - \mu\,\sigma)\,D_A{}^E r_{BCDE} = 2\,\mu\,r_{E(BCD}\,D_{A)}{}^E\sigma. \tag{4.42}$$

Note that the contraction and symmetrization on the right-hand side project out precisely the information contained in $r_{ab}\,D^a\Gamma\,D^b\Gamma$ while the contraction which occurs on the left-hand side prevents us from using the equation to calculate any of the information in (4.41). We use equations (4.32), (4.37) in frame notation. By taking formal derivatives of these equations one can determine from (4.24) and the data (4.40) all derivatives of σ, s, and r_{ab} at i. The complete set of data (4.40) resp. (4.41) is required for this and these data determine the expansion uniquely.

This procedure has been formalized in the theory of 'exact sets of fields' discussed in [61], where equations of the type (4.42) are considered.

The formulation given above suggests proving a Cauchy-Kowalevska type results for equations (4.32), (4.37), (4.42) with data prescribed on $\mathcal{N}$. Although the existence of the vertex at i may create some difficulties in the present case, this problem has much in common with the characteristic initial value problem for Einstein's field equations for which the existence of analytic solutions has been shown ([30]). At present, no decay estimates for the $c_{\mathbf{a}_p\ldots\mathbf{a}_1\mathbf{bc}}$ as $p \to \infty$ are available which would ensure the convergence of these series. To simplify the following discussion we shall assume that the series considered above do converge.

We return to the coordinate formalism and show that this procedure provides a solution to the original equation (4.36), i.e., the quantity Σ_{ab} defined from the fields σ, s, and r_{ab} by the procedure above does vanish. We show first that $\Sigma_{ab} = 0$ on $\mathcal{N}$. Since $\sigma = 0$ on $\mathcal{N}$, this amounts to showing that $m_{ab} \equiv D_a D_b \sigma - s h_{ab}$ vanishes on $\mathcal{N}$. Differentiating twice the equation $D_a \sigma D^a \sigma - 2 \sigma s = 0$, which has been solved as part of the procedure above, observing that $D^c \Sigma_{cd} = 0$ and restricting the resulting equation to $\mathcal{N}$ gives the linear ODE

$$D^c \sigma D_c m_{ab} = -D_a D^c \sigma m_{cb},$$

along the null generators of $\mathcal{N}$. Observing that $D_a D_b \sigma = -2 h_{ab} + O(\Upsilon)$, this ODE can be written along the null geodesics $x^a(\zeta) = \zeta x_*^a$ considered above in the form

$$\frac{d}{d\zeta}(\zeta m_{ab}) = A_a^c \zeta m_{cb},$$

with a smooth function $A_a^c = A_a^c(\zeta)$. This implies the desired result. In view of (4.37), (4.38) it shows that we solved the problem

$$D^c \Sigma_{ca} = 0, \quad D_{[c} \Sigma_{a]b} = 0 \quad \text{near } i, \quad \Sigma_{ab} = 0 \quad \text{on } \mathcal{N}.$$

The first two equations combine in space spinor notation into $D_A{}^E \Sigma_{BCDE} = 0$ with symmetric spinor field Σ_{ABCD}. Following again the arguments of [61], we conclude that $\Sigma_{ab} = 0$.

Equation (4.39) implies

$$D^a \sigma D^b \sigma b_{ab} = 0 \quad \text{on } V. \tag{4.43}$$

A rescaling $h \to h' = \vartheta^4 h$ with a positive (analytic) conformal factor gives $\sigma \to \sigma' = \vartheta^2 \sigma$ and $b_{ab} \to b'_{ab} = \vartheta^{-2} b_{ab}$, whence

$$D^a \sigma D^b \sigma b_{ab} \to (D^a \sigma D^b \sigma b_{ab})' =$$

$$\vartheta^{-6} D^a \sigma D^b \sigma b_{ab} + 4 \sigma \vartheta^{-7} D^a \sigma D^b \vartheta b_{ab} + 4 \sigma^2 \vartheta^{-8} D^a \vartheta D^b \vartheta b_{ab}.$$

This shows that (4.43) is not conformally invariant, but it also shows that the relation

$$D^a \Upsilon D^b \Upsilon b_{ab}|_{\mathcal{N}} = 0, \tag{4.44}$$

implied by (4.43), is conformally invariant. Using again the argument which allowed us to get the quantities (4.41), we can translate this onto the equivalent relations

$$\mathcal{R}(D_{a_p} \cdots D_{a_1} b_{bc}(i)) = 0, \quad p = 0, 1, 2, \ldots, \tag{4.45}$$

which take in space spinor notation the form (cf. (5.89))

$$D_{(A_p B_p} \cdots D_{A_1 B_1} b_{CDEF)}(i) = 0, \quad p = 0, 1, 2, \ldots. \tag{4.46}$$

We note that for given integer $p_* > 0$ the string of such conditions with $0 \le p \le p_*$ is conformally invariant.

Since these condition have a particular bearing on the smoothness of gravitational fields at null infinity ([34], [37]) and it is not clear whether static equations are of a greater significance in this context than expected so far, we take a closer look at (4.43). If we apply the operators $D_a D_b$ and $D_a D_b D_c$ to (4.43) and restrict the resulting equation to i, we get the relations $b_{ab}(i) = 0$ and $D_{(a} b_{bc)}(i) = 0$ respectively, which agree with (4.45) at the corresponding orders because $D^a b_{ab} = 0$. However, if we proceed similarly with $D_a D_b D_c D_d$, we get

$$D_{(a} D_b \, b_{cd)}(i) = 0. \tag{4.47}$$

Since (4.45) with $p = 2$ can be written in the form

$$D_{(a} D_b \, b_{cd)}(i) = \frac{1}{7} h_{(ab} \Delta_h \, b_{cd)}(i),$$

the relation (4.47) implies in particular that $\Delta_h \, b_{cd}(i) = 0$. It appears that in general this equation cannot be deduced in the present gauge from known general identities and (4.44) alone. There will be similar such conditions at higher orders. While the particular form of them may depend on the conformal gauge, the existence of properties which go beyond (4.45) does not. In any case, these observations show that there is a gap between h satisfying the regularity conditions (4.45) and h being conformally static.

This situation is also illustrated by the following observation. If the data provided by h are conformally flat in a neighborhood of i they trivially satisfy conditions (4.45). Without further assumptions the solution θ to the Lichnerowicz equation which relates h to the induced vacuum data $\tilde{h} = \theta^4 h$ can still be quite general. However, if $\tilde{h}$ is static the function θ must be very special.

Lemma 4.1. *An asymptotically flat, static initial data set for the vacuum field equations with conformal metric h and positive ADM mass m is locally conformally flat if and only if it satisfies near i in the gauge (4.31) the equation $r_{ab}[h] = 0$ and thus in the normal coordinates (4.17)*

$$h = -\delta_{ab} \, d \, x^a \, d \, x^b, \quad U = 1, \quad \theta = \frac{1}{|x|} + \frac{m}{2}.$$

Remark: This tells us that *the only asymptotically flat, static vacuum data which are locally conformally flat near space-like infinity are the Schwarzschild data* (4.33). The result of ([71]), which suggests that conformal flatness of the data

h near i and the smoothness requirement on the functions u^p at $I^\pm$ imply that the solution be asymptotically Schwarzschild, can thus be reformulated as saying that for the given data the smoothness requirement implies the solution to be asymptotically static at space-like infinity.

Proof. By (4.38) the solution is locally conformally flat if and only if $2\,D_{[c}\sigma\,r_{a]b} = h_{b[c}\,r_{a]d}D^d\sigma$. Applying D_e to this equation and observing (4.36) and again $D_{[c}\,r_{a]b} = 0$, one gets after a contraction

$$D^c\sigma\,D_c\,r_{ab} = -3\,s\,r_{ab} + \sigma\,(1-\mu\,\sigma)\,(h_{ab}\,r_{cd}\,r^{cd} - 3\,r_{ac}\,r_b{}^c). \tag{4.48}$$

This equation can be read as an ODE along the integral curves of the vector field $D^c\,\sigma$. It follows from (4.32) that $u^a = (2\,\sigma\,|s|)^{-\frac{1}{2}}\,D^a\,\sigma$ is a unit vector field (with direction dependent limits at i). Because of

$$u^a\,D_a\,\Upsilon = -\left(\frac{2\,\Upsilon}{|s|}\right)^{1/2}\,(4\,U^{-1} + 2\,U^{-2}\,D^a\,U\,D_a\,\Upsilon) < 0,$$

its integral curves run into i and cover in fact a (possibly small) neighborhood $\mathcal{U}'$ of i. Equation (4.48) can be rewritten in the form

$$u^a\,D_a(\Upsilon^{3/2}\,r_{bc}) = A_{bc}^{de}\,(\Upsilon^{3/2}\,r_{de}),$$

with the matrix-valued function

$$A_{bc}^{de} = -\frac{3}{\sqrt{2\,\sigma\,|s|}}\,(s + 2\,U^{-1} + U^{-2}\,D^a U\,D_a\,\Upsilon)\,h^d{}_b\,h^e{}_c$$

$$+\frac{\sigma\,(1-\mu\,\sigma)}{\sqrt{2\,\sigma\,|s|}}\,(h_{bc}\,r^{de} - 3\,r^d{}_b\,h^e{}_c),$$

which is continuous on $\mathcal{U}'$. This implies that $r_{ab} = 0$ on $\mathcal{U}'$. The remaining statements follow immediately from (4.17) and (4.31).

Remark: We note that these data may be obtained in a different form if locally conformally flat data are given in the cn-gauge and one asks under which conditions they are conformally static. The data are then of the form

$$h_{ab} = -\delta_{ab}, \quad \Omega^{-\frac{1}{2}} = \theta = \frac{1}{|x|} + W, \quad \Delta_h\,W = 0, \quad W(i) = \frac{m}{2} > 0.$$

By a rescaling $h \to \vartheta^4\,h, \theta \to \vartheta^{-1}\theta = \frac{1}{\vartheta\,|x|} + \frac{m}{2}$ with $\vartheta = \frac{2}{m}\,W$ they are transformed into the present gauge. Assuming that these data satisfy the conformal static field equations and expressing the resulting equation again in terms of $h_{ab} = -\delta_{ab}$ one finds that the solution is static if and only if $2\,W\,D_a\,D_b\,W - 6\,D_a\,W\,D_b\,W + 2\,h_{ab}\,D_c\,W\,D^c\,W = 0$. Since $W > 0$ the equation can be rewritten in terms of $w = W^{-2}$, which gives

$$2\,w\,D_a\,D_b\,w = h_{ab}\,D_c\,w\,D^c\,w. \tag{4.49}$$

Applying D_c, multiplying with w, and using twice (4.49) again, we conclude that $D_a\,D_b\,D_c w = 0$, whence $w = k + k_a\,x^a + k_{ab}\,x^a\,x^b$ with some coefficients $k > 0$,

k_a, k_{ab}. This function satisfies (4.49) if $k_{ab} = h_{ab} \, k_c \, k^c/4 \, k$. With $j_a = k_a/2 \, k$ and $m = 2/\sqrt{k}$ this gives

$$W = \frac{m}{2} \frac{1}{\sqrt{1 + 2 \, j_a \, x^a + j_a \, j^a \, x_b \, x^b}},$$

with constant j^a.

That these data are equivalent to the ones considered above is seen by rescaling with $\vartheta = \frac{2}{m} \, W$. By this one achieves $W = \frac{m}{2}$. For the metric $\vartheta^4 \, h$ to acquire the flat standard form one needs to perform a coordinate transformation which is given by a special conformal transformation $x \to (I \circ T_c \circ I)(x)$ where I denotes the inversion $x^a \to x^a (\delta_{bc} \, x^b \, x^c)^{-1}$ and T_c a translation $x^a \to x^a + c^a$ with suitably chosen constant c^a.

5. A regular finite initial value problem at space-like infinity

In the conformal extension of Minkowski space described in Section 3 neighborhoods of space-like infinity, which are swept out by future complete outgoing and past complete incoming null geodesics, are squeezed into arbitrarily small neighborhoods of the point i^0. From the point of view of the causal structure it is natural to indicate space-like infinity by a point. The discussion in Section 4 shows, however, that in general i^0 cannot be a regular point of any smooth conformal extension. The condition for an extension to i^0 to be C^∞ (under our assumption (4.13)) is that the data are massless in the sense of (4.30) and that the free datum h satisfies the conditions (5.89) with $p_* = \infty$ ([34], [37]). Thus, smoothness at i^0 excludes the physically interesting cases.

A direct discussion of the initial value problem for the conformal field equations with initial data on an initial hypersurface $\mathcal{S} = \tilde{\mathcal{S}} \cup \{i\}$ such that $W^i{}_{jkl} = O(\rho^{-3})$ at i as discussed in Section 4 faces considerable technical problems. Not only the functional analytical treatment of a corresponding PDE problem poses enormous difficulties but already the choice of gauge becomes very subtle.

The setting described below has been arrived at by attempts to describe the structure of the singularity as clearly as possible and to deduce from the conformal field equations a formulation of the PDE problem which still preserves 'some sort of hyperbolicity' at space-like infinity. It is based on conformally invariant concepts so that possible singularities should be identifiable as defects of the conformal structure.

In a conformal Gauss gauge based on a Cauchy hypersurface $\tilde{\mathcal{S}}$ it turns out that after blowing up the point i into a sphere $\mathcal{I}^0$ and choosing the gauge suitably, one arrives at a formulation of the initial value problem near space-like infinity in which the data can be smoothly extended to and across $\mathcal{I}^0$. In that gauge also the evolution equations admit a smooth extension to space-like infinity. The evolution and extension process then generates from the set $\mathcal{I}^0$ a cylindrical piece of space-time boundary diffeomorphic to $]-1, 1[\times \mathcal{I}^0$, which is denoted by $\mathcal{I}$. It represents space-like infinity and can be considered as a blow-up of the point i^0. This boundary is neither postulated nor attached 'by hand'.

In this gauge the hypersurfaces $\mathcal{J}^\pm$ representing null infinity near space-like infinity are given by finite values of the coordinates which are explicitly known (it has to be shown, of course, that the evolution extends far enough). These hypersurfaces touch the cylinder $\mathcal{I}$ at sets $\mathcal{I}^\pm$ diffeomorphic to $\mathcal{I}^0$, which can be thought of as boundaries of $\mathcal{I}$ and of $\mathcal{J}^\pm$ respectively. The structure of the conformal field equations near the *critical sets* $\mathcal{I}^\pm$ appears to be the key to the question of asymptotic smoothness.

It may appear odd to squeeze space-time regions of infinite extend into arbitrarily small neighborhoods of a point i^0 and then perform a complicated blow-up to resolve the singularity on the initial hypersurface which has been created by the first step. The point of the construction is that the finiteness of the sets $\mathcal{I}^\pm$ allow us to disclose, to an extent that we can put our hands on it, a subtle feature of the field equations which otherwise would be hidden at infinity (in the standard vacuum representation) or in the singularity at i^0 (in the standard conformal rescaling).

In the following the setting indicated above and its various implications will be discussed in detail. While we shall add more recent results we shall follow to a large extent the original article [37]. For derivations and details we refer the reader to this or the articles quoted below.

5.1. The gauge on the initial slice and the blow-up at i

The non-smoothness of the conformal data (4.14), (4.7), (4.15), (4.16), (4.29) at i arises from the presence of various factors ρ in the explicit expressions. To properly take care of the specific radial and angular behavior of the various fields it is natural to choose the frame field in the general conformal field equations such that the spatial vector fields e_a, $a = 1, 2, 3$, are tangent to the initial hypersurface $\mathcal{S}$ and one of them, e_3 say, is radial. Since there is no preferred direction at i, this only makes sense if the frame is chosen on $\tilde{\mathcal{S}}$ such that it has direction dependent limits at i. This singular situation finds a well-organized description in terms of a smooth submanifold of the bundle of frames. To discuss the field equations in the spin frame formalism, we will consider in fact a submanifold $\mathcal{C}_e$ of the bundle of normalized spin frames over $\mathcal{S}$ near i. While the use of spinors leads to various simplifications, it should be mentioned that the construction could be carried out similarly in the standard frame formalism (cf. [38]).

5.1.1. The construction of $\mathcal{C}_e$.
Consider now $\mathcal{S}$ as a space-like Cauchy hypersurface of a 4-dimensional solution space-time $(\mathcal{M}, g)$ with induced metric h on $\mathcal{S}$. Denote by $SL(\mathcal{S})$ the set of spin frames $\delta = \{\delta_A\}_{A=0,1}$ on $\mathcal{S}$ which are normalized with respect to the alternating form ϵ, such that

$$\epsilon(\delta_A, \delta_B) = \epsilon_{AB}, \quad \epsilon_{01} = 1. \tag{5.1}$$

The group
$$SL(2,\mathbb{C}) = \{t^A{}_B \in GL(2,\mathbb{C}) \mid \epsilon_{AC} t^A{}_B t^C{}_D = \epsilon_{BD}\},$$
acts on $SL(\mathcal{S})$ by $\delta \to \delta \cdot t = \{\delta_A t^A{}_B\}_{B=0,1}$. The vector field $\tau = \sqrt{2}\,e_0$, with e_0 the future directed unit normal of $\mathcal{S}$, defines a subbundle $SU(\mathcal{S})$ of $SL(\mathcal{S})$ which is given by the spin frames in $SL(\mathcal{S})$ with
$$g(\tau, \delta_A \bar\delta_{A'}) = \epsilon_A{}^0 \epsilon_{A'}{}^{0'} + \epsilon_A{}^1 \epsilon_{A'}{}^{1'} \equiv \tau_{AA'}. \tag{5.2}$$

It has structure group
$$SU(2) = \{t^A{}_B \in SL(2,C) \mid \tau_{AA'} t^A{}_B \bar t^{A'}{}_{B'} = \tau_{BB'}\}.$$

In any frame in $SU(\mathcal{S})$ the vector τ is given by $\tau^{AA'}$. In the following we use the space spinor formalism in the notation of [36]. Using the van der Waerden symbols for space spinors
$$\sigma_a{}^{AB} = \sigma_a{}^{(A}{}_{A'} \tau^{B)A'}, \quad \sigma^c{}_{AB} = \tau_{(B}{}^{A'} \sigma^c{}_{A)A'}, \quad c = 1,2,3,$$
which satisfy
$$h_{ab} = \sigma_a{}_{AB} \sigma_b{}^{AB}, \quad \epsilon_A{}^B \epsilon_{A'}{}^{B'} = \frac{1}{2} \tau_{AA'} \tau^{BB'} + \sigma^a{}_{AF} \tau^F{}_{A'} \tau^{EB'} \sigma_a{}_E{}^B,$$
where
$$h_{ab} \sigma^a{}_{AB} \sigma^b{}_{CD} = -\epsilon_{A(C} \epsilon_{D)B} \equiv h_{ABCD} \quad \text{with} \quad h_{ab} = -\delta_{ab},$$
the covering map onto the connected component $SO(3)$ of the rotation group is given by
$$SU(2) \ni t^A{}_B \overset{\Psi}{\to} t^a{}_b = \sigma^a{}_{AB} t^A{}_C t^B{}_D \sigma_b{}^{CD} \in SO(3).$$
The induced isomorphism of Lie algebras will be denoted by Ψ_*.

The covering morphism of $SU(\mathcal{S})$ onto the bundle $O_+(\mathcal{S})$ of positively oriented orthonormal frames on $\mathcal{S}$ maps the frame $\delta \in SU(\mathcal{S})$ onto the frame with vectors $e_a = e_a(\delta) = \sigma_a{}^{AB} \delta_A \tau_B{}^{B'} \bar\delta_{B'}$ such that $h(e_a, e_b) = h_{ab}$. We use this map to pull back to $SU(\mathcal{S})$ the h-Levi-Civita connection form on $O_+(\mathcal{S})$. Combining this with the map Ψ_*^{-1}, the connection is represented by an $su(2)$-valued connection form $\check\omega^A{}_B$ on $SU(\mathcal{S})$. Similarly, pulling back the $\mathbb{R}^3$-valued solder form on $O_+(\mathcal{S})$ and contracting with the van der Waerden symbols results in a 1-form σ^{AB} on $SU(\mathcal{S})$ which is referred to as solder form on $SU(\mathcal{S})$.

Let $\check H$ denote the real horizontal vector field on $SU(\mathcal{S})$ satisfying $\langle \sigma^{AB}, \check H \rangle = \epsilon_0{}^{(A} \epsilon_1{}^{B)}$ or, equivalently,
$$T_\delta(\pi)\, \check H(\delta) = \delta_{(0} \tau_{1)}{}^{B'} \bar\delta_{B'} = \frac{1}{2}(\delta_0 \bar\delta_{0'} - \delta_1 \bar\delta_{1'}), \quad \delta \in SU(\mathcal{S}). \tag{5.3}$$

It follows that $T_{\delta t}(\pi)\, \check H(\delta t) = T_\delta(\pi)\, \check H(\delta)$ if and only if
$$t \in U(1) \equiv \left\{ t \in SU(2) \mid t = \begin{pmatrix} e^{i\phi} & 0 \\ 0 & e^{-i\phi} \end{pmatrix}, \phi \in \mathbb{R} \right\}.$$

The field $\check H$ will essentially correspond to the 'radial' vector field mentioned above.

We consider again the normal coordinates satisfying (4.17) near i, set $\mathcal{B}_e = \{p \in \mathcal{U} \mid |x(p)| < e\}$ with $e > 0$ chosen such that the closure of $\mathcal{B}_e$ in $\mathcal{S}$ is contained in $\mathcal{U}$, and denote by $(SU(\mathcal{B}_e), \pi)$ the restriction of $(SU(\mathcal{S}), \pi)$ to $\mathcal{B}_e$. Let δ^* be in the fiber $\pi^{-1}(i) \subset SU(\mathcal{B}_e)$ over i. The map $SU(2) \ni t \to \delta(t) \equiv \delta^* \cdot t \in \pi^{-1}(i)$ defines a smooth parametrization of $\pi^{-1}(i)$. We denote by $]-e, e[\ni \rho \to \delta(\rho, t) \in SU(\mathcal{B}_e)$ the integral curve of the vector field $\sqrt{2}\,\check{H}$ satisfying $\delta(0, t) = \delta(t)$ and set $\mathcal{C}_e = \{\delta(\rho, t) \in SU(\mathcal{B}_e) \mid |\rho| < e, \, t \in SU(2)\}$. This set defines a smooth submanifold of $SU(\mathcal{B}_e)$ which is diffeomorphic to $]-e, e[\times SU(2)$. The restriction of π to this set will be denoted by π'.

The symbol ρ, which has been introduced already in Section 4.1, is used here for the following reason. The integral curves of $\sqrt{2}\,\check{H}$ through $\pi^{-1}(i)$ project onto geodesics through i with h-unit tangent vector. Thus, the projection π' maps $\mathcal{C}_e$ onto $\mathcal{B}_e$. The action of $U(1)$ on $SU(\mathcal{B}_e)$ induces an action on $\mathcal{C}_e$. While $\mathcal{I}^0 \equiv \pi^{-1}(i) = \{\rho = 0\}$ is diffeomorphic to $SU(2)$, the fiber $\pi'^{-1}(p) \subset \mathcal{C}_e$ over a point p in the punctured disk $\tilde{\mathcal{B}}_e \equiv \mathcal{B}_e \setminus \{i\}$ coincides with an orbit of $U(1)$ in $SU(\mathcal{B}_e)$ on which $\rho = |x(p)|$ and another one on which $\rho = -|x(p)|$.

The map π' factorizes as $\mathcal{C}_e \xrightarrow{\pi_1} \mathcal{C}'_e \xrightarrow{\pi_2} \mathcal{B}_e$ with $\mathcal{C}'_e \equiv \mathcal{C}_e / U(1)$ diffeomorphic to $]-e, e[\times S^2$. For ρ_* with $0 < |\rho_*| < e$ the subsets $\{\rho = \rho_*\}$ of $\mathcal{C}_e$ are diffeomorphic to $SU(2)$ and the restrictions of the map π_1 to these sets define Hopf fibrations of the form

$$SU(2) \ni t \to \sqrt{2}\sigma^a{}_{AB}t^A{}_0 t^B{}_1 \in S^2 \subset \mathbb{R}^3. \tag{5.4}$$

The set $\pi_2^{-1}(\tilde{\mathcal{B}}_e)$ (resp. $\pi'^{-1}(\tilde{\mathcal{B}}_e)$) consists of two components $\mathcal{C}'^{\pm}_e$ (resp. $\mathcal{C}^{\pm}_e$) on which $\pm\rho > 0$ respectively. Each of the sets $\mathcal{C}'^{\pm}_e$ is mapped by π_2 diffeomorphically onto the punctured disk. If $\tilde{\mathcal{B}}_e$ is now identified via π_2 with $\mathcal{C}'^+_e$ the manifold $\tilde{\mathcal{B}}_e$ is embedded into $\mathcal{C}'_e$ such that it acquires the set $\pi_1(\mathcal{I}^0) = \pi_2^{-1}(i)$ as a boundary. The set $\bar{\mathcal{B}}_e \equiv \tilde{\mathcal{B}}_e \cup \pi_2^{-1}(i) \simeq [0, e[\times S^2$ is a smooth manifold with boundary. Viewing $\tilde{\mathcal{B}}_e$ again as the subset of $\tilde{\mathcal{S}} = \mathcal{S} \setminus \{i\}$, we get an extension $\bar{\mathcal{S}}$ of $\tilde{\mathcal{S}}$ which can be thought of as being obtained from $\mathcal{S}$ by blowing up the point i into a sphere. This is our desired extension of the physical initial manifold and the following discussion could be carried out in terms of the 3-dimensional manifold $\bar{\mathcal{B}}_e$.

It turns out more convenient, however, to use the 4-dimensional $U(1)$ bundle $\bar{\mathcal{C}}^+_e = \mathcal{C}^+_e \cup \mathcal{I}^0 = \{\delta \in \mathcal{C}_e \mid \rho(\delta) \geq 0\} \simeq [0, e[\times SU(2)$. It is a manifold with boundary smoothly embedded into $SU(\mathcal{B}_e)$, from which it inherits various structures. The set $\mathcal{C}_e$ is conveniently parametrized by ρ and the parallelizable group $SU(2)$. The solder and the connection form on $SU(\mathcal{B}_e)$ pull back to smooth 1-forms on $\mathcal{C}_e$. We denote the latter again by σ^{ab} and $\check{\omega}^a{}_b$ respectively. Any smooth spinor field ξ on $\mathcal{B}_e$ defines on $\mathcal{C}_e$ a smooth 'spinor-valued function' which is given at $\delta \in \mathcal{C}_e$ by the components of ξ in the frame defined by δ and denoted (in the case of a covariant field) by $\xi_{A_1 \ldots A_k, A'_1 \ldots A'_j}$. We shall refer to this function as to the 'lift' of ξ.

The structure equations induce on $\mathcal{C}_e$ the equations

$$d\sigma^{AB} = -\check{\omega}^A{}_E \wedge \sigma^{EB} - \check{\omega}^B{}_E \wedge \sigma^{AE}, \tag{5.5}$$

$$d\check{\omega}^A{}_B = -\check{\omega}^A{}_E \wedge \check{\omega}^E{}_B + \check{\Omega}^A{}_B, \tag{5.6}$$

where

$$\check{\Omega}^A{}_B = \frac{1}{2} r^A{}_{BCDF} \sigma^{CD} \wedge \sigma^{EF}$$

denotes the curvature form determined by the curvature spinor r_{ABCDEF}. It holds

$$r_{ABCDEF} = \left(\frac{1}{2} s_{ABCE} - \frac{r}{12} h_{ABCE}\right) \epsilon_{DF} + \left(\frac{1}{2} s_{ABDF} - \frac{r}{12} h_{ABDF}\right) \epsilon_{CE} \tag{5.7}$$

where $s_{ABCD} = s_{(ABCD)}$ is the trace free part of the Ricci tensor of h and r its Ricci scalar. The curvature tensor of h is given by

$$r_{AGBHCDEF} = -r_{ABCDEF}\epsilon_{GH} - r_{GHCDEF}\epsilon_{AB}.$$

and the Bianchi identity reads $6 D^{AB} s_{ABCD} = D_{CD} r$.

We use $t \in SU(2,C)$ and $x^1 \equiv \rho$ as 'coordinates' on $\mathcal{C}_e$. The vector field $\check{H}$ tangent to $\mathcal{C}_e$ then takes the form $\sqrt{2}\,\check{H} = \partial_\rho$. Consider now the basis

$$u_1 = \frac{1}{2}\begin{pmatrix} 0 & i \\ i & 0 \end{pmatrix}, \quad u_2 = \frac{1}{2}\begin{pmatrix} 0 & -1 \\ 1 & 0 \end{pmatrix}, \quad u_3 = \frac{1}{2}\begin{pmatrix} i & 0 \\ 0 & -i \end{pmatrix}, \tag{5.8}$$

of the Lie algebra $su(2)$. Here u_3 is the generator of the group $U(1)$. We denote by Z_{u_i}, $i = 0, 1, 2$, the Killing vector fields generated on $SU(\mathcal{B}_e)$ by u_i and the action of $SU(2)$. These fields are tangent to $\mathcal{I}^0$. We set there

$$X_+ \equiv -(Z_{u_2} + iZ_{u_1}), \quad X_- \equiv -(Z_{u_2} - iZ_{u_1}), \quad X \equiv -2i\,Z_{u_3},$$

and extend these fields smoothly to $\mathcal{C}_e$ by requiring

$$[\check{H}, X] = 0, \quad [\check{H}, X_\pm] = 0. \tag{5.9}$$

The vector fields $\check{H}$, X, X_+, X_- constitute a frame field on $\mathcal{C}_e$ which satisfies besides (5.9) the commutation relations

$$[X, X_+] = 2X_+, \quad [X, X_-] = -2X_-, \quad [X_+, X_-] = -X. \tag{5.10}$$

The vector field iX is tangent to the fibers defined by π_1. The complex vector fields X_+, X_- are complex conjugates of each other such that $\overline{X_- f} = X_+ f$ for any real-valued function f.

These vector fields are related to the 1-forms above by

$$\langle \sigma^{AB}, \check{H} \rangle = \epsilon_0{}^{(A} \epsilon_1{}^{B)}, \quad \langle \sigma^{AB}, X \rangle = 0 \quad \text{on} \quad \mathcal{C}_e, \tag{5.11}$$

$$\langle \sigma^{AB}, X_+ \rangle = \rho\,\epsilon_0{}^A \epsilon_0{}^B + O(\rho^2), \quad \langle \sigma^{AB}, X_- \rangle = -\rho\,\epsilon_1{}^A \epsilon_1{}^B + O(\rho^2), \tag{5.12}$$

$$\langle \check{\omega}^A{}_B, \check{H} \rangle = 0, \quad \langle \check{\omega}^A{}_B, X \rangle = \epsilon_0{}^A \epsilon_B{}^0 - \epsilon_1{}^A \epsilon_B{}^1 \quad \text{on} \quad \mathcal{C}_e, \tag{5.13}$$

$$\langle \check{\omega}^A{}_B, X_+ \rangle = \epsilon_0{}^A \epsilon_B{}^1 + O(\rho^2), \quad \langle \check{\omega}^A{}_B, X_- \rangle = -\epsilon_1{}^A \epsilon_B{}^0 + O(\rho^2), \tag{5.14}$$

as $\rho \to 0$.

To transfer the tensor calculus on $\mathcal{B}_e$ to $\mathcal{C}_e$ we define vector fields $c_{AB} = c_{(AB)}$ on $\mathcal{C}_e \setminus \mathcal{I}^0$ by requiring

$$\langle \sigma^{AB}, c_{CD} \rangle = \epsilon_{(C}{}^A \epsilon_{D)}{}^B, \quad c_{CD} = c^1{}_{CD}\partial_\rho + c^+{}_{CD} X_+ + c^-{}_{CD} X_-. \tag{5.15}$$

The first condition implies that $T_\delta(\pi') \, c_{AB} = \delta_{(A}\tau_{B)}{}^{B'}\bar\delta_{B'}$ for $\delta \in \mathcal{C}_e \setminus \mathcal{I}^0$, while the second removes the freedom for the vector fields to pick up an arbitrary component in the direction of X. It follows that

$$c^1{}_{AB} = x_{AB}, \quad c^+{}_{AB} = \frac{1}{\rho} z_{AB} + \check{c}^+{}_{AB}, \quad c^-{}_{AB} = \frac{1}{\rho} y_{AB} + \check{c}^-{}_{AB}, \qquad (5.16)$$

with smooth functions which satisfy

$$\check{c}^\alpha{}_{AB} = O(\rho), \quad \check{c}^\alpha{}_{01} = 0, \quad \alpha = 1, +, -, \qquad (5.17)$$

and

$$x_{AB} \equiv \sqrt{2}\epsilon_{(A}{}^0\epsilon_{B)}{}^1, \quad y_{AB} \equiv -\frac{1}{\sqrt{2}}\epsilon_A{}^1\epsilon_B{}^1, \quad z_{AB} \equiv \frac{1}{\sqrt{2}}\epsilon_A{}^0\epsilon_B{}^0. \qquad (5.18)$$

The connection coefficients with respect to c_{AB} satisfy

$$\gamma_{CD}{}^A{}_B \equiv \langle \check{\omega}^A{}_B, c_{CD} \rangle = \frac{1}{\rho} \overset{*}{\gamma}_{CD}{}^A{}_B + \check{\gamma}_{CD}{}^A{}_B \qquad (5.19)$$

with

$$\gamma^*{}_{ABCD} = \frac{1}{2}(\epsilon_{AC} x_{BD} + \epsilon_{BD} x_{AC}), \quad \check{\gamma}_{01CD} = 0, \quad \check{\gamma}_{ABCD} = O(\rho).$$

The smoothness of the 1-forms and the vector fields $\check{H}$, X_+, X_- implies that the vector fields $\rho\, c_{CD}$ and the functions

$$c^1{}_{CD}, \quad \rho c^+{}_{CD}, \quad \rho c^-{}_{CD}, \quad \rho \gamma_{CDAB},$$

extend smoothly to all of $\mathcal{C}_e$.

A smooth function F on an open subset of $\mathcal{C}_e$ is said to have spin weight s if

$$X(F) = 2s\, F \qquad (5.20)$$

on this set with $2s$ an integer. Any spinor-valued function induced by a spinor field on $\mathcal{B}_e$ has a well-defined spin weight, it holds, e.g.,

$$X \phi_{ABCD} = 2\,(2 - A - B - C - D)\,\phi_{ABCD}. \qquad (5.21)$$

It follows from the construction of $\mathcal{C}_e$ that this is also true for the functions considered above, it turns out that

$$X c^1{}_{AB} = 2\,(1 - A - B)\,c^1{}_{AB}, \quad X c^\pm{}_{AB} = 2\,(1 - (\pm 1) - A - B)\,c^\pm{}_{AB},$$

$$X \gamma_{ABCD} = 2\,(2 - A - B - C - D)\,\gamma_{ABCD} \quad \text{for} \quad A, B, C, D = 0, 1.$$

By our construction, equation (5.3), and the formula for $e_a(\delta)$ given above the vectors $T_\delta(\pi')\,(\sqrt{2}\,\check{H}(\delta)) = e_3(\delta)$ are tangent to and the frame $e_a(\delta(\rho, t))$ is parallely propagated along the geodesics $[-e, e[\ni \rho \to \pi'(\delta(\rho, t))$ through i. Thus we have constructed the type of frame field asked for in the beginning. Working on $\mathcal{C}_e$ has the advantage that ρ and $\sqrt{2}\,\check{H}$ define smooth fields and the smoothness of the various fields considered above can easily be discussed.

The transition from $\mathcal{B}_e$ to $\bar{\mathcal{C}}'_e$ respectively to $\bar{\mathcal{C}}^+_e$ amounts to a new choice of differential structure at space-like infinity. This change is reflected in the drop of rank of the map π' at the set $\mathcal{I}^0$. It follows from (5.11), (5.12), that at points over i the vectors X, $X_\pm$ project onto the zero vector while at points in $\pi'^{-1}(p)$ the real and imaginary parts of $\check{H}$, X_+, X_- have non-vanishing projections which span the tangent space $T_p\tilde{\mathcal{B}}_e$ if $p \in \tilde{\mathcal{B}}_e$. The relations (5.11), (5.12), (5.13), (5.14) show that the behavior of the map π' near $\mathcal{I}^0$ is encoded in the behavior of the solder and the connection form.

With the structures given above we can perform tensor calculations defined on $\tilde{\mathcal{B}}_e$ now also on $\mathcal{C}_e\setminus\mathcal{I}^0$ and they follow the 'usual' rules of the spin frame formalism. If F denotes the lift of a smooth function f on $\mathcal{B}_e$, the covariant differential Df is represented on $\mathcal{C}_e\setminus\mathcal{I}^0$ by the invariant function $D_{AB}f \equiv c_{AB}(F)$. In the following we shall use the same symbol for a function and its lift. If μ_{AB} is the invariant function induced by a spatial spinor field μ on $\tilde{\mathcal{B}}_e$ its covariant differential is given on $\mathcal{C}_e\setminus\mathcal{I}^0$ by the expression

$$D_{AB}\mu_{AD} = c_{AB}(\mu_{AD}) - \gamma_{AB}{}^E{}_C\,\mu_{ED} - \gamma_{AB}{}^E{}_D\,\mu_{CE}.$$

Analogous formulas hold for covariant differentials of spinor fields of higher valence.

In terms of ρ and $t = (t^A{}_B)$ on $\bar{\mathcal{C}}_e$ and the normal coordinates x^a satisfying (4.17) on $\mathcal{B}_e$, the projection π' has the local expression

$$\pi' : (\rho, t) \rightarrow x^a(\rho, t) = \rho\sqrt{2}\,\sigma^a{}_{CD}\,t^C{}_0\,t^D{}_1. \qquad (5.22)$$

This can be used to pull back the functions Ω, U, and W, which are related by (4.29), to functions of spin weight zero on $\bar{\mathcal{C}}_e$. The metric in (4.13) is built into our formalism and the second fundamental form lifts to a symmetric spinor-valued function χ_{ABCD} which vanishes everywhere. Using the fields

$$\check{c}^\pm{}_{AB}, \quad \check{\gamma}_{CDAB}, \quad s_{ABCD}, \quad r, \qquad (5.23)$$

given by (5.16), (5.19), and (5.7), one can determine

$$D_{AB}\,D_{CD}\,\Omega, \qquad (5.24)$$

on $\mathcal{C}_e^+$ and thus also the derived data (4.14), (4.15), (4.16).

In particular, a detailed expression for the rescaled conformal Weyl spinor ϕ_{ABCD} is obtained on $\mathcal{C}_e^+$ by using (4.16) and (4.29). It takes the form

$$\phi_{ABCD} = \phi'_{ABCD} + \phi^W_{ABCD}, \qquad (5.25)$$

where

$$\phi'_{ABCD} = \sigma^{-2}\left\{D_{(AB}\,D_{CD)}\,\sigma + \sigma\,s_{ABCD}\right\} \qquad (5.26)$$

$$= \frac{1}{\rho^4}\left\{U^2\,D_{(AB}\,D_{CD)}\,(\rho^2) - 8\,\rho\,U\,D_{(AB}\rho\,D_{CD)}U\right\}$$

$$-\frac{1}{\rho^2}\left\{2\,U\,D_{(AB}\,D_{CD)}\,U - 6\,D_{(AB}U\,D_{CD)}U - U^2\,s_{ABCD}\right\}$$

is derived from $\sigma = \rho^2\,U^{-2}$ and thus from the local geometry near i, while

$$\phi^W_{ABCD} = \frac{1}{\rho^3}\left\{-6\,U\,W\,D_{(AB}\,\rho\,D_{CD)}\,\rho + U\,W\,D_{(AB}\,D_{CD)}\,(\rho^2)\right\} \qquad (5.27)$$

$$+\frac{4}{\rho^2}\left(W\,D_{(AB}{}^\rho\,D_{CD)}U - 3\,U\,D_{(AB}{}^\rho\,D_{CD)}W\right)$$

$$-\frac{2}{\rho}\left(U\,D_{(AB}\,D_{CD)}\,W + W\,D_{(AB}\,D_{CD)}\,U - 6\,D_{(AB}U\,D_{CD)}\,W - U\,W\,s_{ABCD}\right)$$

$$-2\,W\,D_{(AB}\,D_{CD)}\,W + 6\,D_{(AB}W\,D_{CD)}\,W + W^2\,s_{ABCD},$$

is the part of the rescaled conformal Weyl spinor which depends on the non-local information in W and which vanishes in the massless case. Observing that

$$D_{AB}\,\rho = x_{AB},\quad D_{AB}\,D_{CD}\,(\rho^2) = -4\,\rho\,\mathring{\gamma}_{(AB}{}^E{}_C\,x_{D)E} = O(\rho^2),\quad D_{AB}\,U = O(\rho),$$

one finds that

$$\phi'_{ABCD} = O(\frac{1}{\rho^2}),\quad \phi^W_{ABCD} = -\frac{6\,m}{\rho^3}\,\epsilon^2_{ABCD} + O(\frac{1}{\rho^2}), \tag{5.28}$$

where we set $\epsilon^j_{ABCD} \equiv \epsilon_{(A}{}^{(E}\,\epsilon_B{}^F\,\epsilon_C{}^G\,\epsilon_{D)}{}^{H)j}$ for $j = 0,\dots,4$.

5.1.2. Normal expansions at $\mathcal{I}^0$ and the functions $T_m{}^j{}_k$. To analyze in detail the behavior of the various fields near space-like infinity it is convenient to study a particular type of expansion. It will be discussed here for an unprimed spinor field, similar expansions hold for other fields. In terms of the normal coordinates x^a on $\mathcal{B}_e$ define the radial vector field $V = x^a\,\partial_{x^a}$. Let $\delta^* = \delta^*(x^a)$ be the smooth spin frame field on $\mathcal{B}_e$ which satisfies $D_V\,\delta^* = 0$ on $\mathcal{B}_e$ and coincides with the spin frame at i chosen as the starting point for our construction of $\mathcal{C}_e$. Denote by e^*_{AB} the orthonormal frame associated with δ^* and write $V = V^{AB}\,e^*_{AB}$.

Suppose ξ is a smooth spinor field on $\mathcal{B}_e$ which is given in terms of the spin frame field δ^* by $\xi^*_{A_1\dots A_l} = \xi^*_{A_1\dots A_l}(x^a)$. Then its Taylor expansion at i is of the form

$$\xi^*_{A_1\dots A_l}(x^a) = \sum_{p=0}^{p=\infty}\frac{1}{p!}\,V^{B_pC_p}(x^a)\dots V^{B_1C_1}(x^a)\,D_{B_pC_p}\dots D_{B_1C_1}\xi^*_{A_1\dots A_l}(i). \tag{5.29}$$

To determine the lift $\xi_{A_1\dots A_l}$ of this field to $\mathcal{C}_e^+$ one has to observe its transformation behaviour $\xi^*_{A_1\dots A_l} \to \xi^*_{A_1\dots A_l}\,t^{B_1}{}_{A_1}\dots t^{B_l}{}_{A_l}$ under changes of the frame and the fact that the pull-back of the functions V^{AB} are given in view of (5.22) by

$$V^{AB}(x^a(\rho,t)) = \sqrt{2}\,\rho\,t^{(A}{}_0\,t^{B)}{}_1. \tag{5.30}$$

If the expansion coefficients $D_{B_pC_p}\dots D_{B_1C_1}\xi^*_{A_1\dots A_l}(i)$ are then decomposed into products of ϵ_{ab}'s and symmetric spinors at i, the essential components $\xi_j = \xi_{(a_1\dots a_l)j}$, $0 \le j \le l$, $0 \le j \le l$, which are of spin weight $s = \frac{l}{2} - j$, are obtained as expansion of the form

$$\xi_j = \sum_{p=0}^{\infty}\xi_{j,p}\,\rho^p \tag{5.31}$$

where

$$\xi_{j,p} = \sum_{m=\max\{|l-2j|,l-2p\}}^{2p+l} \sum_{k=0}^{m} \xi_{j,p;m,k} \, T_m{}^k{}_{\frac{m-l}{2}+j} \tag{5.32}$$

with complex coefficients $\xi_{j,p;m,k}$ and functions $T_m{}^j{}_k$ of t as discussed below.

We refer to this type of expansion as to the *normal expansion of ξ at I^0*. In the case considered above the lift of ξ to $\mathcal{C}_e^+$ has smooth limits at $\mathcal{I}^0$. Corresponding expansions in terms of ρ^k, $k \in \mathbb{Z}$, can also be obtained for fields such as ϕ_{ABCD} on $\mathcal{C}_e^+$ which are given by algebraic expressions of regular fields but which become singular at $\mathcal{I}^0$.

The functions $T_m{}^j{}_k$, arise (apart from some normalizing factors) naturally by the procedure indicated above. They are matrix elements of unitary representations

$$SU(2) \ni t \to T_m(t) = (T_m{}^j{}_k(t)) \in SU(m+1),$$

which are given by

$$T_0{}^0{}_0(t) = 1, \qquad T_m{}^j{}_k(t) = \binom{m}{j}^{\frac{1}{2}} \binom{m}{k}^{\frac{1}{2}} t^{(b_1}{}_{(a_1} \dots t^{b_m)_j}{}_{a_m)_k},$$

$$j, k = 0, \dots, m, \qquad m = 1, 2, 3, \dots.$$

The brackets with lower index now indicate symmetrization and taking 'essential components'. The expansions obtained above make sense under quite general assumptions; the functions $\sqrt{m+1}\, T_m{}^j{}_k(t)$ form a complete orthonormal set in the Hilbert space $L^2(\mu, SU(2))$ where μ denotes the normalized Haar measure on $SU(2)$.

Using the identification of $\mathcal{I}^0$ with $SU(2)$ built into our construction, we consider the $T_m{}^j{}_k$ as functions on $\mathcal{I}^0$ and extend them as ρ-independent functions to $\bar{\mathcal{C}}_e$. The vector fields $X_\pm$, X then act as left invariant vector fields and it holds

$$X \, T_m{}^k{}_j = (m - 2j) \, T_m{}^k{}_j, \tag{5.33}$$

$$X_+ \, T_m{}^k{}_j = \beta_{m,j} \, T_m{}^k{}_{j-1}, \qquad X_- \, T_m{}^k{}_j = -\beta_{m,j+1} \, T_m{}^k{}_{j+1} \tag{5.34}$$

for $0 \le k, j \le m$, $m = 0, 1, 2, \dots$, with $\beta_{m,j} = \sqrt{j\,(m-j+1)}$. It follows that functions f with spin weight s have expansions of the form

$$f = \sum_{m \ge |2s|} \sum_{k=0}^{m} f_{m,k} \, T_m{}^k{}_{\frac{m}{2}-s}, \tag{5.35}$$

where the m's are even if s is an integer and odd if s is a half-integer. All functions considered in the following have integer spin weight.

5.2. The regularizing gauge for the evolution equations

To obtain definite expressions for the expansions of the data at i and because
the terms of lower order are then simplified, it has been assumed in [37] that the
metric h is given in a cn-gauge near i. This will be assumed also here, though
the discussion of the static case given below will show that this is not necessary
for our construction. The coordinates ρ, t and the frame field constructed above
depend on the choice of scaling of the metric h on S. Most important is the fact
that $\Omega = O(\rho^2)$ near $\mathcal{I}^0$, it affects the definition of ρ in an essential way.

In analyzing the evolution of our data in time it turns out convenient to use
a different conformal factor Θ which is related to the conformal factor Ω by

$$\Theta = \kappa^{-1} \Omega \quad \text{on} \quad \bar{\mathcal{C}}_e, \tag{5.36}$$

with a function

$$\kappa = \rho \, \kappa' \quad \text{with} \quad \kappa' \in C^\infty(\bar{\mathcal{C}}_e), \quad \kappa' > 0, \quad X \kappa' = 0, \quad \kappa'|_{\mathcal{I}^0} = 1. \tag{5.37}$$

The value of κ' on $\mathcal{I}^0$ is chosen for convenience here, nothing is gained in the
following by requiring a different (positive) boundary value for it.

The change of the conformal factor implies a map $\Xi : \delta \rightarrow \kappa^{\frac{1}{2}} \delta$ which maps
the set $\mathcal{C}_e^+$ bijectively onto a smooth submanifold $\mathcal{C}^*$ of the bundle of conformal
spin frames over $\tilde{\mathcal{B}}_e$. We use the diffeomorphism Ξ to carry the coordinates ρ and
t and the vector fields $\partial_\rho, X, X_+, X_-$ to $\mathcal{C}^*$. The projection of $\mathcal{C}^*$ onto $\tilde{\mathcal{B}}_e$ will be
denoted again by π'.

Assuming a conformal Gauss system for the evolution in time as described
in Section 2.1, the evolution of the spin frames constituting $\mathcal{C}^*$ defines in the the
bundle of conformal frames over the space-time manifold $\tilde{\mathcal{M}}$ a smoothly embedded
5-dimensional manifold $\tilde{\mathcal{N}}$ which is again a $U(1)$ bundle over the space-time and
whose projection onto $\tilde{\mathcal{M}}$ we denote again by π'. The manifold $\mathcal{C}^*$ represents a
smooth hypersurface of $\tilde{\mathcal{N}}$.

By pushing forward the coordinates ρ, t and the vector fields ∂_ρ, X, $X_\pm$ with
the flow of the conformal geodesics ruling $\tilde{\mathcal{N}}$, these structures can be extended to
$\tilde{\mathcal{N}}$ such that $i X$ generates the kernel of π'. The parameter $x^0 \equiv \tau$ of the conformal
geodesics defines a further independent coordinate with $x^0 = \tau = 0$ on $\mathcal{C}^*$, so that
the tangent vector field of this congruence can be denoted by ∂_τ.

The reduced field equations (2.38), (2.39), (2.40), (2.42) (the latter specializa-
tion of (2.43) is chosen here for only definiteness) are now interpreted as equations
on $\tilde{\mathcal{N}}$ by assuming that the $e_{AA'}$ are vector fields on $\tilde{\mathcal{N}}$ which are defined at a
spin frame $\delta \in \tilde{\mathcal{N}}$ by the requirement that they project onto the frame defined by
δ on $\tilde{\mathcal{M}}$, i.e., $T_\delta \, \pi'(e_{AA'}) = \delta_A \bar{\delta}_{A'}$, and whose X-component is fixed by requiring
an expansion of the form

$$e_{AA'} = \frac{1}{\sqrt{2}} \tau_{AA'} \, \partial_\tau - \tau^B{}_{A'} \, e_{AB}, \tag{5.38}$$

with 'spatial vectors'

$$e_{AB} = e^0{}_{AB} \, \partial_\tau + e^1{}_{AB} \, \partial_\rho + e^+{}_{AB} \, X_+ + e^-{}_{AB} \, X_-. \tag{5.39}$$

The unknowns in the reduced field equations are then interpreted as spinor-valued functions on $\tilde{\mathcal{N}}$. It can be shown that spin weights are preserved under the evolution by the reduced system.

We have to express the initial data for the conformal field equations in terms of the new scaling. With κ, the fields (5.23), (5.24), and the associated covariant derivatives (carried over to $\mathcal{C}^*$, observing that the local expression of Ξ in the given coordinates is the identity) one gets for the curvature fields

$$\phi_{ABCD} = \frac{\kappa^3}{\Omega^2} \left(D_{(AB} D_{CD)}\Omega + \Omega\, s_{ABCD} \right), \tag{5.40}$$

$$\Theta_{AA'CC'} = -\kappa^2 \left(\frac{1}{\Omega} D_{(AB} D_{CD)}\Omega + \frac{1}{12} r\, h_{ABCD} \right) \tau^B{}_{A'} \tau^B{}_{C'}. \tag{5.41}$$

For the frame (5.38), one gets by (5.16)

$$e^0{}_{AB} = 0, \quad e^1{}_{AB} = \rho\,\kappa'\, x_{AB}, \tag{5.42}$$

$$e^+{}_{AB} = \kappa'\, z_{AB} + \kappa\,\check{c}^+{}_{AB}, \quad e^-{}_{AB} = \kappa'\, y_{AB} + \kappa\,\check{c}^-{}_{AB}. \tag{5.43}$$

For the conformal factor Θ we get

$$\Theta = \Theta_* \equiv \kappa^{-1}\Omega = \frac{\rho}{\kappa'\,(U + \rho W)^2} \quad \text{on} \quad \mathcal{C}^*. \tag{5.44}$$

We assume that initial data for the 1-form f, which will be related in the end to $\tilde{f}$ by the relation $f = \tilde{f} - \Theta^{-1}\, d\Theta$, satisfy

$$\langle f, \partial_\tau \rangle = 0, \qquad \text{pull-back of } f \text{ to } \mathcal{C}^* = \kappa^{-1}\, d\kappa. \tag{5.45}$$

It follows then that from (2.24) that Θ takes the form

$$\Theta = \Theta_* \left(1 - \tau^2\, \frac{\kappa_*^2}{\omega_*^2} \right) \quad \text{on} \quad \tilde{\mathcal{N}}, \tag{5.46}$$

with a function ω which is given by

$$\omega = \frac{2\Omega}{\sqrt{|D_a\Omega D^a\Omega|}} = \rho(U + \rho W)\left\{ U^2 + 2\rho U x^{AB} D_{AB} U - \rho^2 D^{AB} U D_{AB} U \right.$$
$$\left. + 2\rho^2 U x^{AB} D_{AB} W - 2\rho^3 D^{AB} U D_{AB} W - \rho^4 D^{AB} W D_{AB} W \right\}^{-\frac{1}{2}} \text{ on } \mathcal{C}^*. \tag{5.47}$$

Here the second member is given in the notation of Section 4.1 while the term on the right-hand side is given in the notation of Section 5.1.1. In (5.46) and in the following formulas the subscripts $*$ are saying that the corresponding functions are constant along the conformal geodesics.

For $d_{AA'}$ we get by (2.25) the explicit expression

$$d_{AA'} = \frac{1}{\sqrt{2}}\,\tau_{AA'}\,\dot{\Theta} - \tau^B{}_{A'}\, d_{AB} \quad \text{on} \quad \tilde{\mathcal{N}}, \tag{5.48}$$

where the dot denotes the derivative with respect to τ and

$$d_{AB} = 2\rho \left(\frac{U \,x_{AB} - \rho \, D_{AB}U - \rho^2 \, D_{AB}W}{(U + \rho W)^3} \right)_*, \tag{5.49}$$

where the notation of Section 5.1.1 is used on the right-hand side.

If one uses (5.42) and (5.43) to write for a given smooth function μ on C^*

$$\mu_{AB} \equiv \kappa^{-1} \left(e^1{}_{AB} \, \partial_\rho + e^+{}_{AB} \, X_+ + e^-{}_{AB} \, X_- \right) \mu,$$

one gets with the 1-form (5.45) and the spatial connection coefficients (5.19) the space-time connection coefficients in the form

$$\Gamma_{AA'CD} = \left(\frac{1}{2} \rho \left(\epsilon_{AC} \, \kappa'_{BD} + \epsilon'_{BD} \, \kappa'_{AC} \right) - \rho \kappa' \, \check{\gamma}_{ABCD} + \frac{1}{2} \epsilon_{AB} \, \kappa_{CD} \right) \tau^B{}_{A'}. \tag{5.50}$$

Note that the $\hat{\Gamma}_{AA'BC}$ in the reduced equations can be expressed by (2.37) in terms of the $\Gamma_{AA'BC}$.

Most important for us is the observation that *the functions given by* (5.40), (5.41), (5.42), (5.43), (5.46), (5.48), (5.50) *have smooth limits as $\rho \to 0$ and can in fact be smoothly extended into the coordinate range $\rho \le 0$*. For the unknowns in the new scaling we thus obtain normal expansion in terms of non-negative powers of ρ. In particular, one has

$$\phi_{ABCD} = \kappa^3 (\phi'_{ABCD} + \phi^W_{ABCD}), \tag{5.51}$$

with (5.26), (5.27) on the right-hand side. Pushing the expansion (5.28) a bit further and using (5.36) one gets in the cn-gauge (in which $h_{ab} = -\delta_{ab} + O(\rho^3)$)

$$\phi_{ABCD} = -\kappa'^3 \, 6 \, m \, \epsilon^2_{ABCD} \tag{5.52}$$

$$-\rho \kappa'^3 \, 12 \left(X_+ \, W_1 \, \epsilon^1{}_{ABCD} + 3 \, W_1 \, \epsilon^2{}_{ABCD} - X_- \, W_1 \, \epsilon^3{}_{ABCD} \right)$$

$$-\frac{\rho^2 \, \kappa'^3}{2} \sum_{k,j=0}^{4} \binom{4}{j} \left(4 \sqrt{6 \binom{4}{j}} \, W_{2;4,k} - \frac{2-j}{3} \sqrt{2 \binom{4}{k}} \, b^*_k(i) \right) T_4{}^k{}_j \, \epsilon^j_{ABCD},$$

$$+ O(\rho^3).$$

It is assumed here that W is an arbitrary solution to $(\Delta_h - \frac{1}{8} r) \, W = 0$ on $\mathcal{B}_e$. Its normal expansion takes in the cn-gauge the form

$$W = \sum_{p=0}^{2} \rho^p \, W_p + O(\rho^3) = \sum_{p=0}^{2} \rho^p \left(\sum_{k=0}^{2p} W_{p;2p,k} \, T_{2p}{}^k{}_p \right) + O(\rho^3)$$

with

$$W_{0;0,0} = W(i) = \frac{m}{2}, \qquad W_{1;2,k} = \binom{2}{k}^{\frac{1}{2}} D_{(ab)_k} W^*(i),$$

$$W_{2;4,k} = \binom{4}{2}^{-\frac{1}{2}} \binom{4}{k}^{\frac{1}{2}} D_{(ab} D_{cd)_k} W^*(i).$$

In the case where κ' is constant the right-hand side of (5.52) provides the terms of a normal expansion up to the quadrupole term. If κ depends on ρ and t the terms given above need to be expanded further to obtain the normal expansion.

The transition (5.36) to the conformal factor Θ corresponds to a transition $h \to h' = \kappa^{-2} h$ of the metric on $\mathcal{B}_e$ (assuming that κ' arise as a lift of a smooth positive function on $\mathcal{B}_e$ with $\kappa'(i) = 1$) in the sense that then $\Omega^{-2} h = \tilde{h} = \Theta^{-2} h'$. The coordinate ρ is then not adapted to the geometry defined by the metric h'. To illustrate the situation assume that h is flat. Then

$$h' = -\kappa'^{-2} \rho^{-2} \left(d\rho^2 + \rho^2 \, d\sigma^2\right) = -\kappa'^{-2} \left(dr^2 + d\sigma^2\right), \qquad (5.53)$$

with $r = -\log \rho$ near i. With respect to the new coordinate r, which is adapted to the geometry of h', the point i is shifted to infinity but the surface measure of any sphere around i remains finite and positively bounded from below. This behavior is reflected by the fact that the frame coefficient $e^1{}_{AB}$ in (5.42) vanishes while the frame coefficients $e^{\pm}{}_{AB}$ in (5.43) have finite and non-vanishing limits at $\mathcal{I}^0$. We shall keep the coordinate ρ because it ensures the finite coordinate representation of the boundary $\mathcal{I}^0$ as well as the smoothness of the data near $\mathcal{I}^0$.

With the gauge defined above the functions Θ and $d_{AA'}$ in equations (2.39), (2.40) are given by (5.46) and (5.48) and the finite regular initial value problem near space-like infinity for the reduced field equations (2.38), (2.39), (2.40), (2.42) is completely determined. We write this system schematically as system of equations for the unknown $u = (w, \phi)$ with $\phi = (\phi_{ABCD})$ and $w = (e_{AA'}, \hat{\Gamma}_{AA'BC}, \Theta_{AA'BB'})$ or, alternatively, $w = (e_{AA'}, \Gamma_{AA'BC}, \Theta_{AA'BB'})$. It takes the form

$$\partial_\tau w = F(x, w, \phi), \quad A^\mu \partial_{x^\mu} \phi = H(w) \phi, \qquad (5.54)$$

where the x-dependence in the first equation comes in here via the functions Θ and $d_{AA'}$.

Important for the following is that *with any choice of κ satisfying (5.37) the functions Θ and $d_{AA'}$ take smooth limits as $\rho \to 0$ and can be extended smoothly into a range where $\rho \leq 0$.* With the smooth extensibility of the initial data observed before, we find that *the initial value problem for the reduced field equations with the data prescribed above can be extended smoothly into a range where $\rho \leq 0$ so that the reduced equations form still a symmetric hyperbolic system.* It may be noted finally that the congruence of conformal geodesics (considered as point sets) underlying our gauge does not depend on the choice of κ, whereas the parameter τ depends on it in an essential way.

5.3. Specific properties of the regular finite initial value problem at space-like infinity

The nature of the initial value problem formulated above is conveniently discussed by considering certain subsets of $\mathbb{R} \times \mathbb{R} \times SU(2)$ which are defined by the range admitted for the coordinates (τ, ρ, t). We define 5-dimensional subsets

$$\tilde{\mathcal{N}} \equiv \{|\tau| < \frac{\omega}{\kappa}, \ 0 < \rho < e, \ t \in SU(2)\},$$

$$\bar{\mathcal{N}} \equiv \{|\tau| \leq \frac{\omega}{\kappa}, \ 0 \leq \rho < e, \ t \in SU(2)\},$$

where $\frac{\omega}{\kappa}$ is a function of ρ and t. It holds then

$$\bar{\mathcal{N}} = \tilde{\mathcal{N}} \cup \mathcal{J}^- \cup \mathcal{J}^+ \cup \mathcal{I} \cup \mathcal{I}^- \cup \mathcal{I}^+,$$

with 4-dimensional submanifolds

$$\mathcal{J}^\pm \equiv \{\tau = \pm \frac{\omega}{\kappa}, \ 0 < \rho < e, \ t \in SU(2)\}, \quad \mathcal{I} \equiv \{|\tau| < 1, \ \rho = 0, \ t \in SU(2)\},$$

and 3-dimensional submanifolds

$$\mathcal{I}^\pm \equiv \{|\tau| \pm 1, \ \rho = 0, \ t \in SU(2)\}, \quad \mathcal{I}^0 \equiv \{\tau = 0, \ \rho = 0, \ t \in SU(2)\},$$

where it has been observed that $\frac{\omega}{\kappa} \to 1$ as $\rho \to 0$. We note that

$$\Theta > 0 \text{ on } \tilde{\mathcal{N}}, \quad \Theta = 0, \ d\Theta \neq 0 \text{ on } \mathcal{J}^- \cup \mathcal{J}^+ \cup \mathcal{I}, \quad \Theta = 0, \ d\Theta = 0 \text{ on } \mathcal{I}^\pm.$$

The set $C^* = \{\tau = 0, \ 0 < \rho < e, \ t \in SU(2)\}$ defines a hypersurface of $\tilde{\mathcal{N}}$. Its closure in $\bar{\mathcal{N}}$ is given by

$$\bar{C} \equiv \{\tau = 0, \ 0 \leq \rho < e, \ t \in SU(2)\} = C^* \cup \mathcal{I}^0.$$

Factoring out the group $U(1)$ implies projections (denoted again by π') onto subsets $\mathbb{R} \times \mathbb{R} \times S^2$ which are of one dimension lower than the sets above. In particular, $\tilde{\mathcal{N}}$ projects onto a set $\tilde{M}$ which represents the 'physical space-time'. For convenience we will usually work with the manifolds above and use for them the same words as for the projections, so that $\tilde{\mathcal{N}}$ will be referred to as the 'physical space-time' etc.

For suitable $\epsilon > 0$ consider a smooth extension of the data given on C^* to the set $\bar{C}_{ext} = \{\tau = 0, \ -\epsilon < \rho < e, \ t \in SU(2)\}$ and an extension of the functions Θ, $d_{AA'}$ to the domain $\bar{\mathcal{N}}_{ext} = \{|\tau| < \frac{\omega}{\kappa}, \ -\epsilon < \rho < e, \ t \in SU(2)\}$, so that the reduced conformal field equations (2.38), (2.39), (2.40), (2.42) still represent a symmetric hyperbolic system of the form (5.54). Then there exists a neighborhood $\mathcal{V}$ of $\bar{C}_{ext}$ in $\bar{\mathcal{N}}_{ext}$ on which there exists a unique smooth solution $e_{AA'}$, $\hat{\Gamma}_{AA'BC}$ (resp. $\Gamma_{AA'BC}$), $\Theta_{AA'BB'}$, ϕ_{ABCD} to our extended initial value problem which satisfies the gauge conditions (2.36).

It turns out, that *the restriction of this solution to the set* $\mathcal{V} \cap \bar{\mathcal{N}}$ *is uniquely determined by the data on* C^*. The data on C^* have a unique smooth extension to $\bar{C}$ and it follows from (5.38), (5.39), (5.42), and (5.43) that $e^1{}_{CC'} \to 0$ as $\rho \to 0$. Equations (2.38) imply in particular

$$\sqrt{2} \, \partial_\tau \, e^1{}_{CC'} = -\Gamma_{CC'}{}^{AA'}{}_{BB'} \tau^{BB'} e^1{}_{AA'}. \tag{5.55}$$

It follows that $e^1{}_{CC'} = 0$ on $\mathcal{V} \cap \mathcal{I}$ and as a consequence that the matrices A^μ in (5.54) are such that

$$A^1 = 0 \text{ on } \mathcal{I}, \tag{5.56}$$

if the solution extends far enough. One can apply to the system (5.54) on subsets of $\mathcal{V} \cap \bar{\mathcal{N}}$ the standard method of deriving energy estimates. Without further information on the system the partial integration would yield contributions from boundary integrals over parts of $\mathcal{V} \cap \mathcal{I}$. Because of (5.56) these boundary integrals vanish and one obtains energy estimates which allow one to show the asserted uniqueness property. The extension above has been considered to simplify the argument. Alternatively, the space-time $\tilde{\mathcal{N}}$ can be thought of as a solution of a very specific 'maximally dissipative' initial boundary value problem where initial data are prescribed on $\bar{\mathcal{C}}$ and no data are prescribed on $\mathcal{I}$ because of (5.56) (cf. [42] and the existence theory in [48], [67]).

In the present gauge the set $\mathcal{I}$, which is generated from $\mathcal{I}^0$ by the extension and evolution process, can be considered as being obtained by performing limits of conformal geodesics. It represents a boundary of the space-time $\tilde{\mathcal{N}}$ which may be understood as a blow-up of the point i^0. We refer to it as the *cylinder at space-like infinity*.

Suppose that there exists on $\tilde{\mathcal{N}}$ a smooth solution $e_{AA'}$, $\hat{\Gamma}_{AA'BC}$ (resp. $\Gamma_{AA'BC}$), $\Theta_{AA'BB'}$, ϕ_{ABCD} of the reduced conformal field equations (2.38), (2.39), (2.40), (2.42) which satisfies the gauge conditions (2.36) on $\tilde{\mathcal{N}}$ and coincides on the initial hypersurface $C^* = \{\tau = 0\} \subset \tilde{\mathcal{N}}$ with the data given above. The projections $T\pi'(e_{AA'})$ then define a frame field on $\tilde{\mathcal{M}}$ for which exists a unique smooth metric g on $\tilde{\mathcal{M}}$ such that $g(T\pi'(e_{AA'}), T\pi'(e_{AA'})) = \epsilon_{AB}\,\epsilon_{A'B'}$. Denote by D' the domain of dependence in $\tilde{\mathcal{M}}$ with respect to g of the set $\pi'(C^*)$ and set $D = \pi'^{-1}(D')$. By the discussion above we can assume that the closure of D in $\tilde{\mathcal{N}}$ contains the set $\mathcal{I}$ and the solution extends smoothly to $\mathcal{I}$. It follows from the structure of the characteristics of the reduced equations, that the solution is determined on D uniquely by the data on $\mathcal{C}^*$ and it follows from the discussion in [36] and the fact that the data satisfy the constraints that the complete set (2.29), (2.30), (2.31), (2.32) of conformal field equations is satisfied on D. Since Θ has spin weight zero it descends to a function on $\tilde{\mathcal{M}}$ and $\tilde{g} = \Theta^{-2}\,g$ satisfies the vacuum field equations.

The restriction to D arises here because we only considered the data on $\mathcal{C}^*$. Observing that the latter were obtained by restricting the data given on the initial hypersurface $\mathcal{S}$ to $\mathcal{B}_e$ it is reasonable to assume that the conformal field equations hold everywhere on $\tilde{\mathcal{N}} \cup \mathcal{I}$ and $\tilde{g}$ defines a solution to the vacuum field equations on $\tilde{\mathcal{M}}$.

Assume u is a solution of a (possibly non-linear) hyperbolic system of partial differential equations of first order on some manifold. A hypersurface of this manifold is then called a *characteristic* of that system (with respect to u), if the system implies for some components of u non-trivial interior differential equations on the hypersurface. These interior equations are called *transport equations* (cf. [22]).

Because of (5.56) the set $\mathcal{I}$ is then a characteristic of the extended field equations. It is in fact of a very special type (i.e., a *total characteristic*), because the system (5.54) reduces on $\mathcal{I}$ to an interior symmetric hyperbolic system of transport equations for the *complete* system of unknowns. Together with the data on $\mathcal{I}^0$ it allows us to determine $u = (v, \phi)$ on $\mathcal{I}$.

Suppose that the solution extends in a C^1 fashion to the sets $\mathcal{J}^\pm$. Since $\Theta = 0$, $d\Theta \neq 0$ on $\mathcal{J}^\pm$ the sets $\pi'(\mathcal{J}^\pm)$ form (part of) the conformal boundary at null infinity for the vacuum solution $\tilde{g}$. Since ϕ_{ABCD} is C^1 one finds Sachs peeling. Of course, *it will be one of our main tasks to control under which assumptions the solutions will extend with a certain smoothness to the sets $\mathcal{J}^\pm$.*

As remarked before, we can expect the decision about the smoothness of the solution at null infinity to be made in the area where the latter 'touches' space-like infinity. This location has a precise meaning in the present setting. It is given by the *critical sets* $\mathcal{I}^\pm$, which can be considered either as boundaries of $\mathcal{J}^\pm$ or as the boundary components of $\mathcal{I}$. The nature of these sets is elucidated by studying *conformal Minkowski space* in the present setting.

We start with the line element given by (5.53) and choose $\kappa' = 1$. Since $\omega = \rho$ by (7.1) it follows that $\mathcal{J}^\pm = \{|\tau| = \pm 1, \, 0 < \rho < e, \, t \in SU(2)\}$ and $\mathcal{M} = \{|\tau| \leq 1, \, 0 \leq \rho < e\} \times S^2$. It will be useful to express the frames considered in the following in terms of the specific frame

$$v_0 = \partial_{\bar{\tau}}, \quad v_1 = \rho \, \partial_\rho, \quad v_\pm = X_\pm. \tag{5.57}$$

The complete solution to the conformal field equations then is given by

$$e^\star_{AA'} = \frac{1}{\sqrt{2}} \left\{ \left((1 - \tau) \epsilon_A{}^0 \epsilon_{A'}{}^{0'} + (1 + \tau) \epsilon_A{}^1 \epsilon_{A'}{}^{1'} \right) v_0 \right. \tag{5.58}$$

$$\left. + \left(\epsilon_A{}^0 \epsilon_{A'}{}^{0'} - \epsilon_A{}^1 \epsilon_{A'}{}^{1'} \right) v_1 - \epsilon_A{}^0 \epsilon_{A'}{}^{1'} v_+ - \epsilon_A{}^1 \epsilon_{A'}{}^{0'} v_- \right\}$$

$$\Gamma^\star_{AA'BC} = -\frac{1}{2} \tau_{AA'} x_{BC}, \tag{5.59}$$

$$\Theta^\star_{AA'BB'} = 0, \tag{5.60}$$

$$\phi^\star_{ABCD} = 0. \tag{5.61}$$

The conformal factor and the metric g implied by $e^\star_{AA'}$ are given by

$$\Theta^\star = \rho(1 - \tau^2), \quad g^\star = d\tau^2 + 2\frac{\tau}{\rho} \, d\tau \, d\rho - \frac{1 - \tau^2}{\rho^2} \, d\rho^2 - d\sigma^2. \tag{5.62}$$

With the coordinate transformation

$$r = \frac{1}{\rho(1 - \tau^2)}, \quad t = \frac{\tau}{\rho(1 - \tau^2)}, \tag{5.63}$$

one gets in fact the standard Minkowski metric $\tilde{g} = \Theta^{-2} g^\star = dt^2 - dr^2 - r^2 \, d\sigma^2$ in spherical coordinates. The flat metric corresponding to (3.4) is given by $\Omega^{\star 2}\tilde{g} = \rho^2 g^\star = d(\tau \rho)^2 - d\rho^2 - \rho^2 \, d\sigma^2$ with $\Omega^\star = \rho\Theta^\star = \rho^2 - (\tau \rho)^2$. For this metric the curves with constant coordinates ρ, θ, and ϕ are obviously conformal geodesics and because of their conformal invariance it follows that the corresponding curves for $g^\star$ are conformal geodesics with parameter τ. Equations (5.63) can be read as their parametrized version in Minkowski space.

The metric $g^\star$ given by (5.62) extends smoothly across null infinity but it has no reasonable limit at $\mathcal{I}$. Its contravariant version

$$g^{\star\sharp} = (1 - \tau^2)\,\partial_\tau^2 + 2\,\tau\,\partial_\tau\,(\rho\,\partial_\rho) - (\rho\,\partial_\rho)^2 - (d\,\sigma^2)^\sharp,$$

does extend smoothly to $\mathcal{I}$. While it drops rank in the limit, it does imply a smooth contravariant metric on $\mathcal{I}$ whose covariant version $l^\star = (1-\tau^2)^{-1}\,d\tau^2 - d\,\sigma^2$ defines a smooth conformally flat Lorentz metric on $\mathcal{I}$. The coordinate transformation $\tau = \sin\xi$ shows that this metric is not complete. The Killing fields of Minkowski space, which are conformal Killing fields for $g^\star$, extend smoothly to $\mathcal{I}$ such that they become tangent to $\mathcal{I}$, vanish there in the case of the translational Killing fields, and act as non-trivial conformal Killing fields for the metric $l^\star$ in the case of infinitesimal Lorentz transformations.

The fields (5.58), (5.59), (5.60) extend smoothly to all of $\bar{\mathcal{M}}$. The property (5.56) results from the fact that the fields $e^\star_{00'}$, $e^\star_{11'}$ become linear dependent on $\mathcal{I}$. Since they do not vanish there, this degeneracy does not cause any difficulties in the field equations. On $\mathcal{I}^+$ and $\mathcal{I}^-$ however, the field $e^\star_{00'}$ and $e^\star_{11'}$ respectively vanishes. This strong degeneracy has important consequences for the (extended) conformal field equations. To see this, we solve the transport equations on $\mathcal{I}$ to determine the matrices A^μ on $\mathcal{I}$ in the general case. Extending the data (5.40), (5.41), (5.42), (5.43), (5.50), one finds that they agree on $\mathcal{I}^0$, irrespective of the choice of κ' satisfying conditions of (5.37), with the implied Minkowski data. Since the extensions of the functions Θ and $d_{AA'}$ vanish on $\mathcal{I}$, the transport equations for the frame, connection, and Ricci tensor coefficients are independent of the choice of initial data. It follows that the restrictions of these coefficients to $\mathcal{I}$ agree with those of the Minkowski data given above. It follows in particular that $e^1{}_{AA'} = 0$ on $\mathcal{I}$. Applying formally the operator ∂_ρ to equation (5.55) (which is part of the reduced field equations), restricting to $\mathcal{I}$, and observing the data $\partial_\rho e^1{}_{AA'}|_{\mathcal{I}^0}$, one finds that $\partial_\rho e^1{}_{AA'} = \epsilon_A{}^0 \epsilon_{A'}{}^{0'} - \epsilon_A{}^1 \epsilon_{A'}{}^{1'}$ on $\mathcal{I}$. Writing

$$e_{AA'} = e^i{}_{AA'}\,v_i \quad \text{with} \quad i = 0, 1, +, -,$$

and assuming the summation rule, we find that *irrespective of the free datum h given on S and the choice of κ' the fields $e_{AA'}$, $\Gamma_{AA'BC}$, $\Theta_{AA'BB'}$ coincide at lowest order with the Minkowski fields above in the sense that $\Theta_{AA'BB'} = O(\rho)$ and*

$$e^i{}_{AA'} = e^{\star i}{}_{AA'} + \breve{e}^i{}_{AA'}, \quad \Gamma_{AA'BC} = \Gamma^\star_{AA'BC} + \breve{\Gamma}_{AA'BC}, \tag{5.64}$$

with

$$\breve{e}^i{}_{AA'} = O(\rho), \quad \breve{\Gamma}_{AA'BC} = O(\rho) \quad \text{as} \quad \rho \to 0. \tag{5.65}$$

Assuming $\kappa = \omega$ in the general case, which by (5.47) is consistent with (5.37) if e is chosen small enough, the similarity with the Minkowski case becomes even closer. Then $\Theta = f\,\Theta^\star$ with proportionality factor $f \equiv \frac{\Omega}{\rho\,\omega}$ which extends smoothly to $\bar{\mathcal{N}}$ such that $f \to 1$ on $\mathcal{I}$. The set $\mathcal{J}^\pm$ is given as in the Minkowski case above. The discussion below shows, however, that this particular choice of κ may not always be the most useful one.

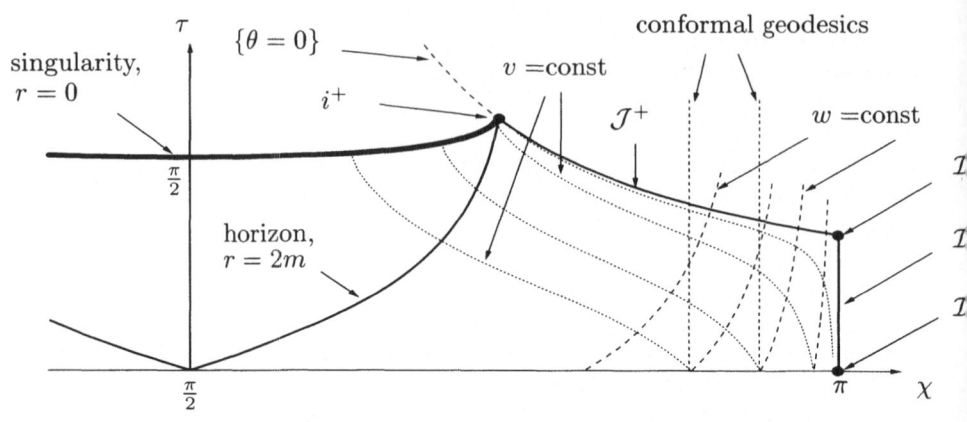

FIGURE 1. A Schwarzschild-Kruskal space-time in a conformal Gauss gauge. This is not a schematic picture but quantitatively correct. Each point in the figure corresponds to a 2-sphere. In Schwarzschild coordinates t and r the lower horizontal line, the initial hypersurface $\mathcal{S} \sim S^3$, corresponds for $\pi/2 < \chi < \pi$ to the hypersurface $\{t = 0\}$ of a Schwarzschild space-time. On $\mathcal{S}$ the coordinate χ satisfies $r = \tan(\chi/2)$, takes the value $\pi/2$ at the throat and the value π at one of the asymptotically flat ends. The parameter τ on the conformal geodesics vanishes on $\mathcal{S}$. With $\Omega = \sin^2 \chi / 2\,(1 + \sin \chi)$ the physical metric induced on $\tilde{\mathcal{S}}$ is $\tilde{h} = \Omega^{-2}\,d\omega^2$ with $d\omega^2$ the standard line element on S^3. The initial conditions of Section 5.2 are satisfied with $\kappa = \sin \chi$ so that $\Theta = \kappa^{-1}\Omega\{1 - \tau^2[\cos\chi(2 + \sin\chi)/2\,(1 + \sin\chi)]^2\}$. In the given gauge the maps $\chi \to \pi - \chi$ and $\tau \to -\tau$ are space-time symmetries. The rescaled space-time and the conformal Gauss gauge extend smoothly through null infinity, where $\Theta = 0$. The expression for Θ stops being meaningful when the conformal geodesics hit the singularity. The behavior of the hypersurfaces of constant retarded and advanced time w and v shows that along curves which approach the cylinder $\mathcal{I}$ the null cones collapse. Along curves on $\mathcal{J}^+$ which approach the critical set $\mathcal{I}^+$ this behavior does not occur. This indicates a degeneracy at $\mathcal{I}^+$ of the set of characteristics.

In the general case the first deviation from the Minkowski case is found in the value of the rescaled conformal Weyl spinor on $\mathcal{I}$. On $\mathcal{I}^0$ it is given by (5.52). Restricting the Bianchi equation to $\mathcal{I}$ and using the coefficients determined above one finds that

$$\phi_{ABCD} = -6\,m\,\epsilon^2_{ABCD} \quad \text{on } \mathcal{I}. \tag{5.66}$$

The discussion above shows that the matrices A^μ are determined on $\mathcal{I}$ by the Minkowski data and the structure of the characteristics of the evolution equations for the rescaled conformal Weyl tensor agrees on $\mathcal{I}$ with that of the equations which are obtained by linearizing the Bianchi equation on Minkowski space. These (overdetermined) spin-2 equations take in the gauge above the form

$$(1 + \tau)\,\partial_\tau \psi_k - \rho\,\partial_\rho \psi_k + X_+\,\psi_{k+1} + (2 - k)\,\psi_k = 0, \tag{5.67}$$

$$(1 - \tau)\,\partial_\tau \psi_{k+1} + \rho\,\partial_\rho \psi_{k+1} + X_-\,\psi_k + (1 - k)\,\psi_{k+1} = 0, \tag{5.68}$$

where $k = 0, 1, 2, 3$ and the ψ_j denote the essential components of the linearized conformal Weyl spinor.

The most conspicuous feature of these equations is the factor $(1+\tau)$ in (5.67), which vanishes on $\mathcal{J}^- \cup \mathcal{I}^-$, and the factor $(1 - \tau)$ in (5.68), which vanishes on $\mathcal{J}^+ \cup \mathcal{I}^+$. On $\mathcal{J}^\pm$ these factors arise because the coordinate τ is constant on $\mathcal{J}^\pm$ and these sets are characteristics for the equations. By choosing κ' differently, this degeneracy can be removed on $\mathcal{J}^\pm$ (cf. [39]). At $\mathcal{I}^\pm$, however, this degeneracy cannot be removed in the present setting. Any symmetric hyperbolic system extracted from these equations, like, e.g.,

$$(1 + \tau)\,\partial_\tau\,\psi_0 - \rho\,\partial_\rho\,\psi_0 + X_+\psi_1 = -2\,\psi_0,$$

$$(4 + 2\,\tau)\,\partial_\tau\,\psi_1 - 2\,\rho\,\partial_\rho\,\psi_1 + X_-\psi_0 + 3\,X_+\psi_2 = -4\,\psi_1,$$

$$6\,\partial_\tau\,\psi_2 + 3\,X_-\psi_1 + 3\,X_+\psi_3 = 0,$$

$$(4 - 2\,\tau)\,\partial_\tau\,\psi_3 + 2\,\rho\,\partial_\rho\,\psi_3 + 3\,X_-\psi_2 + X_+\psi_4 = 4\,\psi_3,$$

$$(1 - \tau)\,\partial_\tau\,\psi_4 + \rho\,\partial_\rho\,\psi_0 + X_+\psi_1 = 2\,\psi_4.$$

must contain such factors at least in the equations for ψ_0 and ψ_4. Writing this in the form $A^\mu\,\partial_\mu\,\psi = H\,\psi$, and writing $\xi_\tau = \langle \partial_\tau, \xi \rangle$, $\xi_\rho = \langle \partial_\rho, \xi \rangle$, $\xi_\pm = \langle \xi, X_\pm \rangle$ we find

$$\det(A^\mu\,\xi_\mu) = 24\,\xi_\tau\,(g^{\mu\nu}\,\xi_\mu\,\xi_\nu)\,(3\,\xi_\tau^2 + g^{\mu\nu}\,\xi_\mu\,\xi_\nu)$$

with

$$g^{\mu\nu}\,\xi_\mu\,\xi_\nu = (1 - \tau^2)\,\xi_\tau^2 + 2\,\tau\,\rho\,\xi_\tau\,\xi_\rho - \rho^2\,\xi_\rho^2 - \frac{1}{2}\,(\xi_+\,\xi_- + \xi_-\,\xi_+).$$

It follows that characteristics pertaining to the quadratic terms which start on $\mathcal{I}$, stay on $\mathcal{I}$ and that those starting in the physical space-time never end on $\mathcal{I} \cup \mathcal{I}^- \cup \mathcal{I}^+$ but always run out to $\mathcal{J}^\pm$. Most importantly however, and this also holds true for the general system (5.54), the quadratic form $g^{\mu\nu}\,\xi_\mu\,\xi_\nu$ degenerates at $\mathcal{I}^\pm$ and there is a loss of real characteristics (cf. Figure 1)[1]. This follows also directly from

$$\det(A^\tau) = 0 \quad \text{on } \mathcal{I}^\pm. \tag{5.69}$$

It appears that this *loss of hyperbolicity at the critical sets* $I^\pm$, is the key to the smoothness problem for the conformal structure at null infinity.

---

[1] I am grateful to Anil Zenginoglu for doing the calculations and preparing the figure for me.

5.4. The s-jet at space-like infinity

The relations (5.56) and (5.69) are the dominant features of the regular finite initial value problem at space-like infinity. The consequences of (5.69) are not deduced by the standard textbook analysis, we have to rely on the specific properties of our problem. It turns out that a considerable amount of information on the behavior of the solution near the critical sets can be obtained by exploiting (5.56). We know already that the solution is smooth in some neighborhood of $\bar{C}$ in $\bar{N}$ and that u can be calculated on $\mathcal{I}$ by solving intrinsic equations on $\mathcal{I}$. It will be shown in the following that a full formal expansion of u in terms of ρ can be calculated on $\mathcal{I}$ by solving certain transport equations.

The following notation will be convenient in the following. For $p = 0, 1, 2, \ldots$ and any sufficiently smooth (possibly vector-valued) function f defined on $\mathcal{N} \cup \mathcal{I}$ we write f^p for the restriction to $\mathcal{I}$ of the p-th radial derivative $\partial_\rho^p f$. The set of functions $f^0, f^1, \ldots, f^p$ on $\mathcal{I}$ will be denoted by $J_\mathcal{I}^p(f)$ and referred to as the *jet of order p of f on $\mathcal{I}$* (and similarly with $\mathcal{I}$ replaced by $\mathcal{I}^0$.) If $u = (w, \phi)$ is a *solution* of equations (5.54) we refer to $J_\mathcal{I}^p(u)$ (respectively $J_\mathcal{I}^p(w)$, $J_\mathcal{I}^p(\phi)$) as to *the s-jet of u (resp. w, ϕ) of order p* and to the *data* $J_{\mathcal{I}^0}^p(u)$ (respectively $J_{\mathcal{I}^0}^p(w)$, $J_{\mathcal{I}^0}^p(\phi)$) on $\mathcal{I}^0$ as to *the d-jet of u (resp. w, ϕ) of order p*. A s-jet $J_\mathcal{I}^p(u)$ (respectively $J_\mathcal{I}^p(w)$, $J_\mathcal{I}^p(\phi)$) of order p will be called *regular on*

$$\bar{\mathcal{I}} \equiv \mathcal{I} \cup \mathcal{I}^- \cup \mathcal{I}^+,$$

(or simply *regular*) if the corresponding functions on $\mathcal{I}$ extend smoothly to the critical sets $\mathcal{I}^\pm$.

An initial data set on $\mathcal{S}$ will be called *asymptotically static of order p*, where $p \in \mathbb{N} \cup \{\infty\}$, if its d-jet $J_{\mathcal{I}^0}^p(u)$ coincides with the d-jet of order p of some static asymptotically flat data set defined on some neighborhood of i in $\mathcal{S}$. It will be seen later that *asymptotic staticity* (of order p) is an important feature of initial data sets.

Applying the operator ∂_ρ^p formally to the first of equations (5.54) and restricting to $\mathcal{I}$, one obtains for w^p an equation of the form

$$\partial_\tau w^p = G(\tau, t, w^0, \ldots w^{p-1}, w^p, \phi^0, \ldots, \phi^{p-1}), \quad p = 1, 2, \ldots, \tag{5.70}$$

where the right-hand side is an affine function of w^p. The functions ϕ^p do not appear here, because the rescaled conformal Weyl spinor occurs in the equations for the frame, connection, and Ricci coefficients with the factors Θ and d_{AB}, which vanish on $\mathcal{I}$. It follows that the s-jet $J_\mathcal{I}^p(w)$ can be determined by the integration of an (easily solvable) linear system of ODE's, if the s-jet $J_\mathcal{I}^{p-1}(u)$ and the d-jet $J_{\mathcal{I}^0}^p(w)$ are known.

With the notation (5.64) the Bianchi equation can be written

$$\nabla^{\star F}{}_{A'} \phi_{BCDF} = -\phi_{A'BCD}, \tag{5.71}$$

where

$$\phi_{A'BCD} = \phi_{A'(BCD)} \equiv \check{e}^{iF}{}_{A'} v_i(\phi_{BCDF}) - 4 \check{\Gamma}^F{}_{A'}{}^E{}_{(B} \phi_{CDF)E}. \tag{5.72}$$

Then

$$(\sqrt{2}\,\nabla^{\star F}{}_{A'}\,\phi_{BCDF})^p = -\sqrt{2}\,\phi^p_{A'BCD}, \tag{5.73}$$

provides equations with left-hand sides given by

$$(1+\tau)\,\partial_\tau\,\phi^p_j + X_+\,\phi^p_{j+1} + (2-j-p)\,\phi^p_j = \dots, \tag{5.74}$$

$$(1-\tau)\,\partial_\tau\,\phi^p_{j+1} + X_-\,\phi^p_j + (1-j+p)\,\phi^p_{j+1} = \dots, \tag{5.75}$$

where $j = 0,\dots,3$, and right-hand sides given by (cf. (5.65))

$$\phi^p_{A'BCD} = \sum_{i=0,+,-}\ \sum_{j=1}^{p}\binom{p}{j}(\check{c}^{iF}{}_{A'})^j\,v_i\,(\phi^{p-j}_{BCDF}) \tag{5.76}$$

$$+\sum_{j=1}^{p}(p-j)\binom{p}{j}(\check{c}^{1F}{}_{A'})^j\,\phi^{p-j}_{BCDF} - 4\sum_{j=1}^{p}\binom{p}{j}(\check{\Gamma}^{F}{}_{A'}{}^{E}{}_{(B})^j\,\phi^{p-j}_{CDF)E}.$$

We note that these expressions depend on $J^p_{\mathcal{I}}(w)$ but only on $J^{p-1}_{\mathcal{I}}(\phi)$. Thus, given these s-jets, the s-jet $J^p_{\mathcal{I}}(\phi)$ can be obtained by solving a linear system of ODE's, if $J^p_{\mathcal{I}^0}(\phi)$ is given. *Because the system is singular at the critical sets it is not clear a priori that $J^p_{\mathcal{I}}(\phi)$ is regular, even if $J^p_{\mathcal{I}}(w)$ and $J^{p-1}_{\mathcal{I}}(\phi)$ are regular.*

To obtain more detailed information on the solutions, it is useful to consider a system system of second order. From (5.71) follows

$$\nabla^\star_{EE'}\,\nabla^{\star EE'}\,\phi_{ABCD} = 2\,\nabla^M_A{}^{E'}\,\nabla^{\star E}{}_{E'}\,\phi_{BCDE} = 2\,\nabla^M_A{}^{E'}\,\phi_{E'BCD},$$

which is equivalent to

$$\nabla^\star_{EE'}\,\nabla^{\star EE'}\,\phi_{ABCD} = f_{ABCD} \equiv -2\,\nabla^\star_{E'(A}\,\phi^{E'}{}_{BCD)}, \tag{5.77}$$

$$0 = g_{BC} \equiv \nabla^{\star A'A}\,\phi_{A'ABC}. \tag{5.78}$$

While the right-hand side of

$$(\nabla^\star_{EE'}\,\nabla^{\star EE'}\,\phi_{ABCD})^p = f^p_{ABCD}, \tag{5.79}$$

depends again, similar to (5.76), on $J^p_{\mathcal{I}}(w)$ and $J^{p-1}_{\mathcal{I}}(\phi)$, the left-hand side takes the decoupled form

$$(1-\tau^2)\,\partial^2_\tau\,\phi^p_j + 2\{(p-1)\,\tau - j + 2\}\,\partial_\tau\,\phi^p_j + C\,\phi^p_j - p\,(p-1)\,\phi^p_j = \dots \tag{5.80}$$

where the spin weight relations $X\,\phi_j = 2\,(2-j)\,\phi_j$ and the Casimir operator $C = -\frac{1}{2}\,(X_+\,X_- + X_-\,X_+) + \frac{1}{4}\,X^2$ on $SU(2)$ have been used to arrive at this expression.

The fields ϕ^p_j have expansions

$$\phi^p_j = \sum_{q=|2-j|}^{p}\phi^p_{j,q} \quad \text{where} \quad \phi^p_{j,q} = \sum_{k=0}^{2q}\phi^p_{j,q,k}\,T_{2q}{}^{k}{}_{q-2+j},$$

with coefficients $\phi^p_{j,q,k} = \phi^p_{j,q,k}(\tau)$. Since the Casimir operator satisfies

$$C\,(T_{2q}{}^{k}{}_{q-2+j}) = q\,(q+1)\,T_{2q}{}^{k}{}_{q-2+j},$$

equation (5.80) implies for $\phi_{j,q}^p$ ODE's of the form

$$D_{(n,\alpha,\beta)}\,\phi_{j,q}^p \equiv (1-\tau^2)\,\partial_\tau^2\,\phi_{j,q}^p + \{\beta - \alpha - (\alpha+\beta+2)\,\tau\}\,\partial_\tau\,\phi_{j,q}^p \qquad (5.81)$$
$$+n\,(n+\alpha+\beta+1)\,\phi_{j,q}^p = \ldots$$

with

$$\alpha = j - p - 2, \quad \beta = -j - p + 2, \quad n = n_1 \equiv p + q \ \text{ or } \ n = n_2 \equiv p - q - 1.$$

The equations above allow us to calculate recursively a formal expansion of the solution $u = (w, \phi)$ to (5.54) in a series of the form

$$u = \sum_{n=0}^{\infty} \frac{1}{n!}\,u^p\,\rho^p, \qquad (5.82)$$

on $\mathcal{I}$ (note the different meanings of the superscripts p) with coefficients $u^p = u^p(\tau, t) \in C^\infty(\mathcal{I})$. In some neighborhood of $\mathcal{I}^0$ in $\bar{\mathcal{N}}$ this series represents in fact the Taylor series of smooth functions and it converges near $\mathcal{I}^0$ if the datum h is real analytic. We shall try to deduce from it information on the behavior of u near the critical sets.

5.5. Behavior of the s-jets near the critical sets

Because $J_{\mathcal{I}}^0(u)$ is regular, the integration gives a regular s-jet $J_{\mathcal{I}}^1(w)$. The calculation of $J_{\mathcal{I}}^1(\phi)$ gives (in the cn-gauge and with $\kappa' = 1$) the regular solution

$$\phi_{ABCD}^1 = -\{W_1\,36\,(1-\tau^2) + m^2\,(18\,\tau^2 - 3\,\tau^4)\}\,\epsilon^2{}_{ABCD} \qquad (5.83)$$
$$-12\,(1-\tau)^2\,X_+\,W_1\,\epsilon^1{}_{ABCD} + 12\,(1+\tau)^2\,X_-\,W_1\,\epsilon^3{}_{ABCD}.$$

Thus $J_{\mathcal{I}}^2(w)$ will again be regular. It turns out that $J_{\mathcal{I}}^2(\phi)$ will not necessarily be regular. The integration (cn-gauge, $\kappa' = 1$) gives

$$\phi_{ABCD}^2 = \phi_{ABCD}^{ih\,2} + \phi_{ABCD}^{\breve{W}\,2} + \phi_{ABCD}^{\breve{}'\,2},$$

with

$$\phi_{(ABCD)_0}^{2\,ih} = 0, \quad \phi_{(ABCD)_2}^{2\,ih} = c_2(\tau)\,m\,W_1 + c_3(\tau)\,m^3, \quad \phi_{(ABCD)_4}^{2\,ih} = 0,$$

$$\phi_{(ABCD)_1}^{2\,ih} = c_1(\tau)\,m\,X_+W_1, \quad \phi_{(ABCD)_3}^{2\,ih} = -c_1(-\tau)\,m\,X_-W_1,$$

where the $c_i(\tau)$ are polynomials in τ of order ≤ 8,

$$\phi_{(ABCD)_j}^{\breve{W}\,2} = -4\,\sqrt{6\binom{4}{j}}\,(1+\tau)^j\,(1-\tau)^{4-j}\,\sum_{k=0}^{4} W_{2;4,k}\,T_4{}^k{}_j, \qquad (5.84)$$

and

$$\phi_{(ABCD)_j}^{\breve{}'\,2} = a_j(\tau)\,\frac{1}{3}\,\sum_{k=0}^{4}\,\sqrt{2\binom{4}{k}}\,b_{(EFGH)_k}^*\,T_4{}^k{}_j \qquad (5.85)$$

with

$$a_0(\tau) = 2\,(1-\tau)^4\,K(-\tau) = -a_4(-\tau),$$

$$a_1(\tau) = 4\,(1-\tau)^3\,(1+\tau)\,K(-\tau) - \frac{3}{1-\tau} = -a_3(-\tau),$$

$$a_2(\tau) = \sqrt{6}\,\{\frac{2-\tau}{(1+\tau)^2} - 2\,(1-\tau)^2\,(1+\tau)^2\,K(\tau)\} = -a_2(-\tau),$$

$$K(\tau) = 1 - 3\int_0^\tau \frac{ds}{(1-s)\,(1+s)^5}.$$

While the first two terms extend smoothly to $\mathcal{I}^\pm$, the third term has logarithmic singularities at the critical sets unless the *regularity condition* $b_{ABCD}(i) = 0$ is satisfied (the quadrupole term W_2, which looks so innocent here, reappears in obstructions to smoothness at higher order [71], [72]).

It is thus clearly important to control the behavior of the s-jets at $\mathcal{I}^\pm$ at all orders. Equations $D_{(n,\alpha,\beta)}u = 0$ are well known from the theory of Jacobi polynomials and they have been used in [37] to derive a certain representation of the solutions in terms of polynomials built from the generalized Jacobi polynomials $P_n^{(\alpha,\beta)}(\tau)$ ([68]). By the overdeterminedness of the system (5.74), (5.75) the problem can be reduced to the integration of the functions $\phi_{0,q}^p$, $\phi_{4,q}^p$. The functions $\phi_{1,q}^p$, $\phi_{2,q}^p$, $\phi_{3,q}^p$ can be calculated from them algebraically.

One finds for $p \geq 3$ and $q = p$ the representation

$$\phi_{0,p}^p = (1-\tau)^{p+2}\,(1+\tau)^{p-2}\,\big(\phi_{0,p\,*}^p + \tag{5.86}$$

$$\frac{(p+1)\,(p+2)}{4\,p}\,(\phi_{0,p\,*}^p - \phi_{4,p\,*}^p)\int_0^\tau \frac{d\tau'}{(1+\tau')^{p-1}\,(1-\tau')^{p+3}}\Big),$$

$$\phi_{4,p}^p = (1+\tau)^{p+2}\,(1-\tau)^{p-2}\,\big(\phi_{4,p\,*}^p - \tag{5.87}$$

$$\frac{(p+1)\,(p+2)}{4\,p}\,(\phi_{0,p\,*}^p - \phi_{4,p\,*}^p)\int_0^{-\tau} \frac{d\tau'}{(1+\tau')^{p-1}\,(1-\tau')^{p+3}}\Big),$$

where the subscript $*$ indicates initial data on $\mathcal{I}^0$.

Denoting by $y_{p,q}$ the column vector formed from $\phi_{0,q}^p$, $\phi_{4,q}^p$, one obtains for $p \geq 3$ and $0 \leq q \leq p-1$

$$y_{p,q}(\tau) = X_{p,q}(\tau)\,\left(X_{p,q\,*}^{-1}\,y_{p,q\,*} + \int_0^\tau X_{p,q}(\tau')^{-1}\,B_{p,q}(\tau')\,d\tau'\right). \tag{5.88}$$

The functions $B_{p,q}$ are derived from the right-hand sides of (5.73) and (5.79) and can thus be calculated from $J_{\mathcal{I}}^p(w)$ and $J_{\mathcal{I}}^{p-1}(\phi)$. The matrix-valued functions $X_{p,q}$ are given by

$$X_{p,0} = \left(\begin{array}{cc} (1+\tau)^{p-2}\,(p+\tau) & 0 \\ 0 & (1-\tau)^{p-2}\,(p-\tau) \end{array}\right),$$

$$X_{p,1} = \left(\begin{array}{cc} (1+\tau)^{p-2} & 0 \\ 0 & (1-\tau)^{p-2} \end{array}\right),$$

$$X_{p,q} = \left(\begin{array}{cc} Q_{1;p,q}(\tau) & (-1)^q\,Q_{3;p,q}(\tau) \\ (-1)^q\,Q_{3;p,q}(-\tau) & Q_{1;p,q}(-\tau) \end{array}\right), \quad 2 \leq q \leq p-1,$$

with polynomials

$$Q_{1;p,q}(\tau) = (\frac{1-\tau}{2})^{p+2} \, P_{q-2}^{(p+2,-p+2)}(\tau),$$

$$Q_{3;p,q}(\tau) = (\frac{1+\tau}{2})^{p-2} \, P_{q+2}^{(-p-2,p-2)}(\tau).$$

of degree $n_1 = p + q$.

The solutions to the transport equations can be calculated, order by order, explicitly. The only difficulty is the calculation of the functions

$$B_{p,q} = B_{p,q} \left[J_{\mathcal{I}}^{p}(w), \ J_{\mathcal{I}}^{p-1}(\phi) \right]$$

which become more and more complicated at each step.

The most conspicuous feature of these expressions is the occurrence of logarithmic singularities at $\mathcal{I}^{\pm}$. The latter can arise, as a consequence of the evolution process and the structure of the data, even under the strongest smoothness assumptions on the conformal datum h. We will have to discuss to what extent the occurrence of such singularities can be related to the structure of the initial data and whether it can be avoided by a judicious choice of the latter.

5.6. Regularity conditions

Expanding the integrals in (5.86), (5.87) one finds

$$\phi_{0,p}^{p} \approx (1-\tau)^{p+2} \, (1+\tau)^{p-2} \, \log(1-\tau) + \textit{analytic in } \tau \ \textit{ as } \tau \to 1$$

and a similar behavior for $\phi_{4,p}^{p}$ as $\tau \to -1$, unless the initial data on $\mathcal{I}^{0}$ satisfy the condition

$$\phi_{0,p\,*}^{p} = \phi_{4,p\,*}^{p}.$$

(Note that the singularities get less severe with increasing p.) This raises the question whether data can be given which satisfy these conditions. By a lengthy recursion argument it can be shown ([37]) that *for given integer $p_* \geq 0$ the fields $\phi_{j,p}^{p}$ resulting from (5.86), (5.87) extend smoothly to $\mathcal{I}^{\pm}$ for $2 \leq p \leq p_* + 2$ if and only if the free datum h satisfies the regularity condition*

$$D_{(A_q B_q} \ldots D_{A_1 B_1} b_{ABCD)}(i) = 0, \quad q = 0, 1, 2, \ldots, p_*. \tag{5.89}$$

By (4.46) these conditions are satisfied for static data with $p_* = \infty$. This allows one to construct a large class of data satisfying (5.89) by gluing with a partition of unity an asymptotically flat static end to a given time reflection symmetric data set and solving the Lichnerowicz equation.

Condition (5.89) has been observed as a regularity condition before. In [34] has been derived under the strong assumption that the solution be massless (cf. (4.30)) a *necessary and sufficient* condition on h that space-like infinity can be represented by a regular point i^0 in a smooth conformal space-time extension (so that $\mathcal{J}^{\pm}$ will be smooth near space-like infinity). This condition, referred to as *radiativity condition*, implies (5.89). It has been shown in [37] that these two conditions are in fact equivalent.

The first term in (5.88) is polynomial and thus regular. The second term is not so easy to handle. If (5.89) is not assumed the corresponding log-terms will enter the integral in a non-linear way and the solution will have at $\mathcal{I}^\pm$ polyhomogeneous expansions in terms of expressions $(1 \mp \tau)^k \log^j(1 \mp \tau)$ with $k, j \in \mathbb{N}_0$. We shall assume therefore that (5.89) holds with $p_* = \infty$.

From the expressions above it follows that the Wronskian $\det(X_{p,q})$ has a factor $(1 - \tau^2)^{p-2}$. The regularity of the integrals in (5.88) thus depends on the precise structure of the functions $B_{p,q}(\tau)$, which get quite complicate with increasing p. It has been shown in [37] and [41] that $J_\mathcal{I}^p(u)$ is regular for $p \leq 3$ if (5.89) is satisfied with $p_* \leq 1$.

Because the functions $B_{p,q}$ are getting increasingly complicated with p, J.A. Valiente-Kroon studied the case where h is conformallly flat on $\mathcal{B}_e$ with the help of an algebraic computer program ([71]). In that case condition (5.89) is trivially satisfied but there still exists a large class of non-trivial data for which h is not conformally flat outside $\mathcal{B}_e$. In the conformal factor (5.44) one has $U = 1$ on $\mathcal{B}_e$ but W will be a non-trivial solution to the conformally covariant Laplace equation with $m = 2W(i) \neq 0$. It turns out that $J_\mathcal{I}^4(u)$ is again regular. For $J_\mathcal{I}^5(u)$ however, logarithmic terms are observed. They come with certain coefficients which depend on the data. Choosing the data such that these coefficients vanish, still new logarithmic terms are observed for $J_\mathcal{I}^6(u)$. Restricting to the axially symmetric case to keep the expressions manageable, new logarithmic terms crop up for $p = 7$ and $p = 8$.

The form of the conditions obtained at these orders suggests a general formula which needs to be satisfied to excluded logarithmic terms at any given order p ([71]). If this formula is correct, all derivatives of W must vanish at i if the logarithmic terms are required to vanish at all orders. As a consequence the solution must become asymptotically Schwarzschild at i (cf. Lemma 4.1). Since W is governed on $\mathcal{B}_e$ by an elliptic equation with analytic coefficients it would follow that the solution is precisely Schwarzschild near i.

How seriously do we need to take the singularities at $\mathcal{I}^\pm$? To answer this question one needs to control the evolution of the field in a full neighborhood of $\tilde{\mathcal{I}}$ in $\tilde{\mathcal{M}}$. This has not been achieved yet. However, the analysis of the linearized setting, which is given by the spin-2 equations (5.67), (5.68) on Minkowski space in the gauge (5.58), gives some insight ([39]).

While the functions $B_{p,q}$ vanish in that case, the singularities arising from (5.86), (5.87) do in general survive the linearization process. The analysis then shows that for prescribed integer j the function

$$\psi_k - \sum_{p'=0}^{p-1} \frac{1}{p'!} \psi_k^{p'} \rho^{p'} \quad \text{on } \tilde{\mathcal{M}}$$

extends to a function of class C^j on $\bar{\mathcal{M}}$, if one chooses $p \geq j + 6$ in the expansion above. Here $\psi_k^{p'}$, $p' = 0, 1, \ldots, p - 1$, are understood as ρ-independent functions on $\tilde{\mathcal{M}} \cup \mathcal{I}$, which agree on $\mathcal{I}$ with the s-jet $J_\mathcal{I}^{p-1}(\psi)$ (defined by equations (5.67),

(5.68)). Note that the sum above provides the first terms of an asymptotic expansion of the solution at $\mathcal{J}^{\pm}$.

It follows that the solution will extend smoothly to all of $\bar{\mathcal{M}}$ if the linearized version of (5.89) is satisfied with $p_* = \infty$. If the condition is satisfied only with some finite $p_* \geq 2$ but violated at $p = p_* + 1$, the solution will develop a logarithmic singularity at $\mathcal{I}^{\pm}$ which will be transported along the null generators of $\mathcal{J}^{\pm}$ so that the solution will be only in $C^{p_* - 2}(\bar{\mathcal{M}})$. While it remains to be seen whether the solutions to the non-linear equations admit similar asymptotic expansions at $\mathcal{J}^{\pm}$, the discussion shows clearly that the regularity of the s-jets $J^p_{\mathcal{I}}(u)$ is a prerequisite for the smooth extensibility of the solutions to $\mathcal{J}^{\pm}$.

If the solutions to the non-linear equations show a singular behavior on $\mathcal{J}^{\pm}$ as indicated above, does it refer to something 'real' or to a failure of the gauge? If the underlying conformal structure where smooth at null infinity, the conformal geodesics should pass through $\mathcal{J}^{\pm}$ where $\Theta \to 0$ and the 1-form, the $\hat{\nabla}$-parallely transported frame, and therefore also the rescaled conformal Weyl spinor in that frame should be represented by smooth functions of τ along the conformal geodesics because these as well as their natural parameter τ depend only on the conformal structure. Singularities as indicated above therefore refer to intrinsic features of the underlying conformal structure.

The results of ([71]) show first of all that the regularity condition (5.89) with $p_* = \infty$ are *not sufficient* for the regularity of $J^p_{\mathcal{I}}(u)$, $p = 0, 1, 2, \ldots$. It appears that the Lichnerowicz equation, which breaks the conformal invariance by fixing the scaling of the physical metric $\tilde{h} = \Omega^{-2} h$, does play a role in the smoothness of the conformal structure at null infinity. This is remarkable because it shows that besides the local condition (5.89) there are other conditions to be observed which are 'not so local'. However, the Lichnerowicz equation is introduced only as a device to reduce the problem of solving the *underdetermined elliptic system* of constraints to an *elliptic* problem. The results of [16], [20], [21] exploit the underdeterminedness of the constraints in quite a different way. They teach us to be careful with the words 'local' and 'global' in the present context.

The main purpose of calculating $J^p_{\mathcal{I}}(u)$ for the first few p is to get an insight into (5.88) which would allow us to control the behavior of $J^p_{\mathcal{I}}(u)$ near $\mathcal{I}^{\pm}$ in dependence of the data given on $\mathcal{S}$. One may speculate that the results above are telling us that asymptotic staticity, or more generally asymptotic stationarity, at space-like infinity is of more importance in the present context than expected so far. Recent generalizations of the calculations in ([71]) to non-conformally flat data seem to support this view ([72]).

This raises the question whether the setting proposed in [37] is for static solutions as smooth as one would expect. This is far from obvious because of the loss of hyperbolicity at the critical sets. Giving an answer to this question for general static solutions will be the purpose of the following chapters.

6. Conformal extensions of static vacuum space-times

For static asymptotically flat vacuum solutions with positive ADM mass we shall construct in the following a conformal extension which will include null infinity and will also allow us to discuss the cylinder at space-like infinity. The extension will be defined in terms of explicitly given coordinates and conformal rescaling. In Section 7 will it be shown that it coincides with the extension (not the coordinates etc.) as defined in Section 5.3.

Because one expects usually 'not much to happen at space-like infinity' for static asymptotically flat solutions, one may wonder why the detailed discussion of the fields near space-like infinity should be so complicated. An obvious reason is that a gauge which is chosen to discuss space-like and null infinity must introduce a 'time dependence', it cannot be adapted to a Killing field whose flow lines run out to time-like infinity. However, the main reason is that the static field equations play an important role in discussing the regularity of the field near the critical sets; we will have to make extensive use of them.

The static vacuum solution is assumed in the form

$$\tilde{g} = v^2 \, dt^2 + \Omega^{-2} \, h,$$

with $v = v(x^c)$, $h = h_{ab}(x^c) \, dx^a \, dx^b$ and a conformal factor $\Omega = \Omega(x^c)$, where we assume h-normal coordinates x^a which satisfy (4.17) and the conformal gauge which achieves (4.31) on the set $\mathbb{R} \times \mathcal{U}$, where $\mathcal{U} = \{|x| < \bar{\rho}_*\}$ with a sufficiently small $\bar{\rho}_* > 0$. We set

$$\Upsilon = |x|^2, \quad e^a = \frac{x^a}{|x|} = -\frac{1}{2}\Upsilon^{-1/2} D^a \Upsilon \quad \text{for} \quad |x| > 0, \quad \bar{\rho} = \sqrt{\sum_{a=0}^{3}(x^a)^2}.$$

Coordinates ψ^A, $A = 2, 3$, on the sphere $S^2 = \{|x| = 1\}$ can be used to parametrize e^a and we write then $e^a = e^a(\psi^A)$ and $d\,e^a = e^a{}_{,\psi^A} \, d\psi^A$. For convenience the coordinates ψ^A will be assumed in the following to be real analytic. If $x^a = \bar{\rho}\, e^a(\psi^A)$, the metric h takes the form

$$h = -d\,\bar{\rho}^2 + \bar{\rho}^2 \, k,$$

with ($\bar{\rho}$-dependent) 2-metrics

$$k = k_{AC} \, d\psi^A \, d\psi^C \equiv h_{ac}(\bar{\rho}\, e^c) \, d e^a \, d e^c,$$

on the spheres $\bar{\rho} = \text{const.} > 0$. For $\bar{\rho} \to 0$ the metric k approaches the standard line element $d\sigma^2 = -k(0, \psi^A)$ on the 2-dimensional unit sphere in the coordinates ψ^A.

We write now $x^0 = t$ and $x^{0'} = \bar{\tau}$, $x^{1'} = \bar{\rho}$, $x^{A'} = \psi^A$ and consider the map $\Phi : x^{\mu'} \to x^\mu(x^{\mu'})$ defined by

$$t(x^{\mu'}) = \int\limits_{\bar{\rho}\,(1-\bar{\tau})}^{\bar{\rho}} \frac{d\,s}{(v\,\Omega)(s\,e^a(\psi^A))}, \qquad x^a(x^{\mu'}) = \bar{\rho}\,(1 - \bar{\tau})\, e^a(\psi^A). \qquad (6.1)$$

It follows that the four differentials

$$dx^a = ((1 - \bar{\tau}) \, d\bar{\rho} - \bar{\rho} \, d\bar{\tau}) \, e^a + \bar{\rho} \, (1 - \bar{\tau}) \, de^a, \tag{6.2}$$

$$dt = \left(\frac{1}{(v\,\Omega)(\bar{\rho}\,e^a)} - \frac{1 - \bar{\tau}}{(v\,\Omega)(\bar{\rho}\,(1 - \bar{\tau})\,e^a)} \right) d\bar{\rho} + \frac{\bar{\rho}}{(v\,\Omega)(\bar{\rho}\,(1 - \bar{\tau})\,e^a)} \, d\bar{\tau} + l, \tag{6.3}$$

with

$$l = l_A \, d\psi^A, \qquad l_A = \int_{\bar{\rho}\,(1-\bar{\tau})}^{\bar{\rho}} \left(\frac{1}{(v\,\Omega)(s\,e^a)} \right)_{,\psi^A} ds, \tag{6.4}$$

are independent for $0 \le \bar{\tau} < 1$ and $0 < \bar{\rho} < \bar{\rho}_*$ and we can consider the $x^{\mu'}$ as smooth coordinates on an open neighborhood of space-like infinity in $\{t \ge 0\}$. For $s > 0$ we set

$$h(s\,e^a) \equiv \frac{(v\,\Omega)(s\,e^a)}{s^2} = \frac{U(s\,e^a) - s\,\frac{m}{2}}{(U(s\,e^a) + s\,\frac{m}{2})^3}. \tag{6.5}$$

To indicate the different arguments replacing s in this and other functions of $s\,e^a$ or of s and ψ^A, we write out the argument replacing s explicitly but suppress the dependence on e^a or ψ^A. Thus $h(s)$ will be written for $h(s\,e^a)$ and $k(\bar{\rho})$ for $k(\bar{\rho}, \psi^A)$, etc.

With this notation and the conformal factor

$$\Lambda = \Omega \, \Upsilon^{-1/2},$$

a conformal representation of $\tilde{g}$ is defined by

$$\bar{g} \equiv \Phi^* \left(\Lambda^2 \, \tilde{g} \right) = 2 \left(\frac{h(\bar{\rho}\,(1 - \bar{\tau}))}{h(\bar{\rho})} \frac{d\bar{\rho}}{\bar{\rho}} + \bar{\rho}\,h(\bar{\rho}\,(1 - \bar{\tau}))\,l \right) d\bar{\tau} \tag{6.6}$$

$$-2\,(1 - \bar{\tau}) \left(\frac{h(\bar{\rho}\,(1 - \bar{\tau}))}{h(\bar{\rho})} \frac{d\bar{\rho}^2}{\bar{\rho}^2} + \bar{\rho}\,h(\bar{\rho}\,(1 - \bar{\tau}))\,l\,\frac{d\bar{\rho}}{\bar{\rho}} \right)$$

$$+ (1 - \bar{\tau})^2 \left(\frac{h(\bar{\rho}\,(1 - \bar{\tau}))}{h(\bar{\rho})} \frac{d\bar{\rho}}{\bar{\rho}} + \bar{\rho}\,h(\bar{\rho}\,(1 - \bar{\tau}))\,l \right)^2 + k(\bar{\rho}\,(1 - \bar{\tau})).$$

The new coordinates do not reflect the symmetries of the underlying space-time, but they are sufficient to discuss the part of the space-time in the future of the initial hypersurface $\{t = 0\}$. We replace $\mathcal{S}$ by the manifold with boundary $\bar{\mathcal{S}}$ introduced in Section 5.1.1. The points of $\partial\bar{\mathcal{S}}$ are thought of as ideal end points attached to the curves $\bar{\rho} \to x^a(\bar{\rho}) = \bar{\rho}\,e^a(\psi^A)$ in $\tilde{\mathcal{S}}$ as $\bar{\rho} \to 0$ for fixed value of ψ^A. The coordinates $\bar{\rho}$ and ψ^A extend (by definition) to analytic coordinates on $\bar{\mathcal{S}}$ with $\bar{\rho} = 0$ on $\partial\bar{\mathcal{S}}$. We set

$$\tilde{\mathcal{M}}' = \{0 \le \bar{\tau} < 1,\ 0 < \bar{\rho}\}, \quad \bar{\mathcal{M}}' = \tilde{\mathcal{M}}' \cup \mathcal{J}^{+'} \cup \mathcal{I}' \cup \mathcal{I}^{+'},$$

where it is understood that the unspecified coordinate systems ψ^A 'cover' the sphere S^2, and

$$\mathcal{J}^{+'} = \{\hat{\tau} = 1, \bar{\rho} > 0\}, \quad \mathcal{I}^{0'} = \partial\bar{\mathcal{S}} = \{\bar{\tau} = 0, \bar{\rho} = 0\},$$

$$\mathcal{I}' = \{0 \le \bar{\tau} < 1, \bar{\rho} = 0\}, \quad \mathcal{I}^{+'} = \{\bar{\tau} = 1, \bar{\rho} = 0\}, \quad \bar{\mathcal{I}}' = \mathcal{I}' \cup \mathcal{I}^{+'}.$$

While the notation alludes to related sets introduced in Section 5.3, the prime should warn the reader that the sets defined above differ in various aspects from those considered in 5.3. The range of $\bar{\rho}$ should be also bounded from above in these definitions. We leave this bound unspecified because its specific value is unimportant here, we will be concerned only with the behavior of the metric in a neighborhood of $\bar{\mathcal{I}}'$ in $\bar{\mathcal{M}}'$.

Important for the following are the observations:
(i) the function $h(s\,e^a(\psi^A))$ as given by the right-hand side of (6.5) and considered as function of s and ψ^A extends as a real analytic function into a domain where $s < 0$. This follows immediately from the values taken by U and its analyticity.
(ii) similarly, the 1-form l given by (6.4) extends as a real analytic function into a domain where $\bar{\rho} \leq 0$ and $\bar{\tau} \geq 1$. This follows from

$$\left(\frac{1}{v\,\Omega}\right)_{,\psi^A}(s) = \frac{1}{s^2}\,\frac{2\,(U(s)-s\,m)(U(s)+s\,\frac{m}{2})^2}{(U(s)-s\,\frac{m}{2})^2}\,U_{,\psi^A},$$

and (4.27) with $s^2 = \Upsilon$.

For the following it is convenient to slightly modify the frame (5.57) and set

$$v_0 = \partial_{\bar{\tau}},\ v_1 = \bar{\rho}\,\partial_{\bar{\rho}},\ v_A = \partial_{\psi^A}, \tag{6.7}$$

$$\alpha^0 = d\,\bar{\tau},\ \alpha^1 = \frac{1}{\bar{\rho}}\,d\,\bar{\rho},\ \alpha^B = d\,\psi^B,\quad A,B = 2,3.$$

One then gets $\bar{g} = \bar{g}_{ik}\,\alpha^i\,\alpha^k$ with metric coefficients

$$\bar{g}_{00} = 0,\quad \bar{g}_{01} = \frac{h(\bar{\rho}\,(1-\bar{\tau}))}{h(\bar{\rho})},\quad \bar{g}_{0A} = \bar{\rho}\,h(\bar{\rho}\,(1-\bar{\tau}))\,l_A,$$

$$\bar{g}_{11} = -(1-\bar{\tau})\,\frac{h(\bar{\rho}\,(1-\bar{\tau}))}{h(\bar{\rho})}\left(2-(1-\bar{\tau})\,\frac{h(\bar{\rho}\,(1-\bar{\tau}))}{h(\bar{\rho})}\right),$$

$$\bar{g}_{1A} = -(1-\bar{\tau})\,\bar{\rho}\,h(\bar{\rho}\,(1-\bar{\tau}))\left(1-(1-\bar{\tau})\,\frac{h(\bar{\rho}\,(1-\bar{\tau}))}{h(\bar{\rho})}\right)l_A,$$

$$\bar{g}_{AB} = \{\bar{\rho}\,(1-\bar{\tau})\,h(\bar{\rho}\,(1-\bar{\tau}))\}^2\,l_A\,l_B + k_{AB}(\bar{\rho}\,(1-\bar{\tau})).$$

In terms of the new coordinates the metric given by (6.6) extends analytically through the set $\mathcal{J}^{+'}$. The latter is a null hypersurface for the extended metric and represents future null infinity for the space-time defined by $\tilde{g}$. By contrast, the right-hand side of (6.6) does not extend smoothly to $\bar{\mathcal{I}}'$. However, the frame coefficients $\bar{g}_{ik}$ and their contravariant versions $\bar{g}^{ik}$ do extend analytically to all of $\bar{\mathcal{M}}'$. It will be shown later how $\mathcal{I}'$ relates to (part of) the cylinder at space-like infinity denoted in 5.3 by $\mathcal{I}$.

One has $\bar{g}_{ik} = g^*_{ik} + O(\bar{\rho}^2)$ with

$$g^*_{ik} = \begin{bmatrix} 0 & 1+2\,m\,\bar{\rho}\,\bar{\tau} & 0 & 0 \\ 1+2\,m\,\bar{\rho}\,\bar{\tau} & -(1-\bar{\tau})\,(1+\bar{\tau}+4\,m\,\bar{\rho}\,\bar{\tau}^2) & 0 & 0 \\ 0 & 0 & k_{22}(0) & k_{23}(0) \\ 0 & 0 & k_{32}(0) & k_{33}(0) \end{bmatrix}, \tag{6.8}$$

so that $\det(g^*_{ik}) < 0$ for $\bar{\rho} \geq 0$, $0 \leq \bar{\tau} \leq 1$ and $g^{ik} = g^{*ik} + O(\bar{\rho}^2)$ with

$$
g^{*ik} = \begin{bmatrix}
\frac{(1-\bar{\tau})(1+\bar{\tau}+4\,m\,\bar{\rho}\,\bar{\tau}^2)}{(1+2\,m\,\bar{\rho}\,\bar{\tau})^2} & \frac{1}{1+2\,m\,\bar{\rho}\,\bar{\tau}} & 0 & 0 \\
\frac{1}{1+2\,m\,\bar{\rho}\,\bar{\tau}} & 0 & 0 & 0 \\
0 & 0 & k^{22}(0) & k^{23}(0) \\
0 & 0 & k^{32}(0) & k^{33}(0)
\end{bmatrix}. \tag{6.9}
$$

Since the conformal factor Λ does not depend on t, the static Killing vector field represents a Killing field also for the metric $\bar{g}$. In the new coordinates it takes the form

$$
K = \frac{(v\,\Omega)(\bar{\rho})}{\bar{\rho}}\left\{(1-\bar{\tau})\,\partial_{\bar{\tau}} + \bar{\rho}\,\partial_{\bar{\rho}}\right\} = \bar{\rho}\,h(\bar{\rho})\left\{(1-\bar{\tau})\,v_0 + v_1\right\}, \tag{6.10}
$$

and extends smoothly to all of $\bar{\mathcal{M}}'$.

Denote by $\bar{\nabla}$ the Levi-Civita connection of $\bar{g}$. Since the commutators of the frame fields v_k vanish, the connection coefficients defined by $\bar{\nabla}_i v_j \equiv \bar{\nabla}_{v_i} v_j = \gamma_i{}^k{}_j\, v_k$ are given by the formula

$$
\gamma_i{}^k{}_j = \frac{1}{2}\,\bar{g}^{kl}\left(v_j(\bar{g}_{il}) + v_i(\bar{g}_{lj}) - v_l(\bar{g}_{ij})\right).
$$

Again, *the connection coefficients $\gamma_i{}^k{}_j$ in the frame v_k extend analytically through* $\{\bar{\rho} = 0\}$ *and* $\{\bar{\tau} = 1\}$. One finds

$$
\gamma_i{}^k{}_j = \frac{1}{2}\,g^{*kl}\left(v_j(g^*_{il}) + v_i(g^*_{lj}) - v_l(g^*_{ij})\right) + O(\bar{\rho}^2),
$$

which implies

$$
\gamma_i{}^k{}_j = \bar{\tau}\,\delta^k{}_0\left\{2\,\delta^0{}_{(i}\,\delta^1{}_{j)} - (1-\bar{\tau}^2)\,\delta^1{}_i\,\delta^1{}_j\right\} - \bar{\tau}\,\delta^k{}_1\,\delta^1{}_i\,\delta^1{}_j \quad \text{on} \quad \{\bar{\rho} = 0\}. \tag{6.11}
$$

As a consequence of the behavior of $\bar{g}_{ij}$ and $\gamma_i{}^k{}_j$ the components of all tensor fields in the frame v_k which are derived by standard formulas from the metric and the connection coefficients, such as those of the Ricci tensor and the conformal Weyl tensor of $\bar{g}$, extend analytically through $\mathcal{J}^{+'}$ and $\bar{\mathcal{I}}'$, i.e., *the metric $\bar{g}$ and its connection $\bar{\nabla}$ imply in the frame v_i a smooth frame formalism on $\bar{\mathcal{M}}'$.*

It follows that the coordinate expressions of these tensor fields, such as $R_{\mu'\nu'}[\bar{g}] = R_{ik}\,\alpha^i{}_{\mu'}\,\alpha^k{}_{\nu'}$, and, by the argument given in [60] (cf. also [38]), the rescaled conformal Weyl tensor $W^{\mu'}{}_{\nu'\lambda'\rho'}[\bar{g}] = \Lambda^{-1}\,C^{\mu'}{}_{\nu'\lambda'\rho'}[\bar{g}]$ extend smoothly to $\mathcal{J}^{+'}$. Unfortunately, this does not give us the needed details about the components R_{jk} and it does not tell us anything about the behavior of the frame components $W^i{}_{jkl}[\bar{g}]$ of the rescaled conformal Weyl tensor on $\mathcal{I}'$ and the critical set $\mathcal{I}^{+'}$. This requires detailed calculations. Only the analyticity of h near i is required to control the smoothness of the fields near $\mathcal{J}^{+'}$. This follows from the *ellipticity* of the conformal static field equations near i. To deduce the desired behavior near $\bar{\mathcal{I}}'$, however, one will have to invoke at least, as discussed in Section 5.6, the regularity condition (5.89) with $p_* = \infty$. The detailed form of the conformal static field equations will thus become much more important.

6.1. The Ricci tensor of $\bar{g}$ near $\bar{\mathcal{I}}'$

The tensor

$$L[\bar{g}]_{\rho'\nu'} = \frac{1}{2}\left(R[\bar{g}]_{\rho'\nu'} - \frac{1}{6}R[\bar{g}]\,\bar{g}_{\rho'\nu'}\right),$$

is needed to integrate the conformal geodesic equations which define the setting introduced in Section 5. The purpose of this section is to demonstrate

Lemma 6.1. *The frame components $L_{jk} = L[\bar{g}]_{\rho'\nu'}\,v^{\rho'}{}_j\,v^{\nu'}{}_k$ extend as real analytic functions to $\bar{\mathcal{I}}'$ with*

$$L_{0k} \to \frac{1}{2}\delta^1{}_k, \quad L_{11} \to -\frac{1-\bar{\tau}^2}{2}, \quad L_{1A} \to 0, \quad L_{AB} \to -\frac{1}{2}k_{AB}(0) \quad as \quad \bar{\rho} \to 0.$$

Proof. Under the rescaling $\tilde{g} \to \bar{g} = \Lambda^2\,\tilde{g}$ the tensor

$$L[\tilde{g}]_{\rho\nu} = \frac{1}{2}\left(R[\tilde{g}]_{\rho\nu} - \frac{1}{6}R[\tilde{g}]\,\tilde{g}_{\rho\nu}\right),$$

transforms into

$$L[\bar{g}]_{\rho\nu} = L[\tilde{g}]_{\rho\nu} - \frac{1}{\Lambda}\bar{\nabla}_\rho\bar{\nabla}_\nu\Lambda + \frac{1}{2\Lambda^2}\bar{\nabla}_\mu\Lambda\bar{\nabla}^\mu\Lambda\,\bar{g}_{\rho\nu}.$$

Suppose $\tilde{g} = v^2\,dt^2 + \tilde{h}$ is a static vacuum solution and $\bar{g} = \Lambda^2\,\tilde{g} = \Lambda^2\,(v^2\,dt^2 + \tilde{h}) = N^2\,dt^2 + h^*$ with

$$N = \Lambda\,v, \quad h^* \equiv \Lambda^2\,\tilde{h} = \Lambda^2\Omega^{-2}h = h^*_{ab}(x^c)\,dx^a\,dx^b,$$

$$\mu = \mu(x^a), \quad v = v(x^a), \quad \Omega = \Omega(x^a), \quad \Lambda = \Lambda(x^a).$$

In the following the gauge (4.31) and coordinates satisfying (4.17) will be assumed. The connection coefficients of the metric $\bar{g}$ in the coordinates $t,\ x^a$ are given by

$$\Gamma_a{}^b{}_c[\bar{g}] = \Gamma_a{}^b{}_c[h^*] \quad \text{(the Levi-Civita connection of } h^*\text{)},$$

$$\Gamma_t{}^a{}_t[\bar{g}] = 0, \quad \Gamma_t{}^a{}_t[\bar{g}] = -N\,h^{*ab}D_b\,N, \quad \Gamma_b{}^a{}_t[\bar{g}] = \Gamma_t{}^a{}_b[\bar{g}] = 0,$$

$$\Gamma_b{}^t{}_c[\bar{g}] = 0, \quad \Gamma_b{}^t{}_t[\bar{g}] = \Gamma_t{}^t{}_b[\bar{g}] = \frac{1}{N}\,D_b\,N.$$

and $L[\bar{g}]_{\rho\nu}$ is given by

$$L[\bar{g}]_{tt} = -\frac{v\,\Omega^2}{\Lambda^2}\,D_a\,N\,D^a\Lambda + \frac{v^2\,\Omega^2}{2\Lambda^2}\,D_a\Lambda\,D^a\Lambda, \tag{6.12}$$

$$L[\bar{g}]_{ta} = L[\bar{g}]_{at} = 0,$$

$$L[\bar{g}]_{ab} = -\frac{1}{\Lambda}\,D^*_a\,D^*_b\,\Lambda + \frac{1}{2\Lambda^2}\,D_c\Lambda\,D^c\Lambda\,h_{ab}, \tag{6.13}$$

where D and D^* denote the h- and h^*-Levi-Civita connections respectively. With $\Lambda = \Omega\,\Upsilon^{-1/2}$ and the map Φ defined by (6.1) one can determine from these formulas the frame coefficients

$$L_{ik} = \langle\Phi^*(L[\Lambda^2\,\tilde{g}]);\, v_i,\, v_k\rangle =$$

$$\langle(L[\bar{g}]_{tt}\circ\Phi)\,dt\,dt + (L[\bar{g}]_{ab}\circ\Phi)\,dx^a\,dx^b;\, v_i,\, v_k\rangle.$$

With equations (4.36) and the relations

$$D_a \Omega = \Omega^{\frac{3}{2}} \sigma^{-\frac{3}{2}} D_a \sigma = (1 + \sqrt{\mu\sigma})^{-3} D_a \sigma, \tag{6.14}$$

$$D_a D_b \Omega = \frac{1}{(1 + \sqrt{\mu\sigma})^3} D_a D_b \sigma - \frac{3}{2}\sqrt{\frac{\mu}{\sigma}} \frac{1}{(1 + \sqrt{\mu\sigma})^4} D_a \sigma D_b \sigma, \tag{6.15}$$

which are implied by (4.35), one gets

$$\Upsilon D^a N D_a \Lambda$$

$$= v\Omega \left\{ \frac{2s(1 - 2\sqrt{\mu\sigma})}{(1 - \sqrt{\mu\sigma})(1 + \sqrt{\mu\sigma})^4} - \frac{(2 - 3\sqrt{\mu\sigma})\Upsilon^{-1} D^c \Upsilon D_c \sigma}{2(1 - \sqrt{\mu\sigma})(1 + \sqrt{\mu\sigma})^3} - \frac{1}{U^2(1 + \sqrt{\mu\sigma})^2} \right\},$$

$$\Upsilon D^a \Lambda D_a \Lambda = \Omega \left\{ \frac{2s}{(1 + \sqrt{\mu\sigma})^4} - \frac{\Upsilon^{-1} D^c \Upsilon D_c \sigma}{(1 + \sqrt{\mu\sigma})^3} - \frac{1}{U^2(1 + \sqrt{\mu\sigma})^2} \right\},$$

$$\Upsilon^{1/2} D_a^* D_b^* \Lambda = \frac{s\, h_{ab} - \sigma(1 - \mu\sigma) R_{ab}}{(1 + \sqrt{\mu\sigma})^3} - \frac{3}{2} \frac{\sqrt{\mu\sigma}\, \sigma^{-1} D_a\sigma D_b\sigma}{(1 + \sqrt{\mu\sigma})^4}$$

$$- \frac{D_a D_b \Upsilon}{2\, U^2 (1 + \sqrt{\mu\sigma})^2} + \frac{e_a\, e_b}{U^2 (1 + \sqrt{\mu\sigma})^2}$$

$$- h_{ab} \left(\frac{1}{U^2 (1 + \sqrt{\mu\sigma})^2} + \frac{\Upsilon^{-1} D^c \Upsilon D_c \sigma}{2(1 + \sqrt{\mu\sigma})^3} \right),$$

which allow us to obtain the following expressions for the L_{jk}.

In the case of L_{00} there occurs a cancellation of the second terms in (6.12), (6.13) respectively, so that (with the understanding that $e^a \circ \Phi = e^a(\psi^A)$)

$$L_{00} = \bar{\rho}^2 \left(\frac{L[\bar{g}]_{tt}}{(v\,\Omega)^2} + L[\bar{g}]_{ab}\, e^a\, e^b \right) \circ \Phi \tag{6.16}$$

$$= -\bar{\rho}^2 \left(\frac{1}{v\,\Omega^2} \left\{ \Upsilon D_a N D^a \Lambda + v\,\Omega\, \Upsilon^{1/2} D_a^* D_b^* \Lambda\, e^a\, e^b \right\} \right) \circ \Phi,$$

with

$$\Upsilon D_a N D^a \Lambda + v\,\Omega\, \Upsilon^{1/2} D_a^* D_b^* \Lambda\, e^a\, e^b$$

$$= v\,\Omega \left\{ \frac{(1 - 4\sqrt{\mu\sigma} + \mu\sigma)(s + 2)}{(1 - \sqrt{\mu\sigma})(1 + \sqrt{\mu\sigma})^4} - \frac{\sigma(1 - \mu\sigma) R_{ab}\, e^a\, e^b}{(1 + \sqrt{\mu\sigma})^3} \right.$$

$$- \frac{6\sqrt{\mu\sigma}}{(1 + \sqrt{\mu\sigma})^4} \left[\left(\frac{1}{U} + \frac{D^a \Upsilon D_a U}{2\, U^2}\right)^2 - 1 \right]$$

$$\left. + \frac{(1 - 2\sqrt{\mu\sigma})}{(1 - \sqrt{\mu\sigma})(1 + \sqrt{\mu\sigma})^3} \left[\frac{2 - 2U^2}{U^2} + \frac{D^a \Upsilon D_a U}{U^3} \right] \right\}.$$

Since the term in curly brackets is of the order $O(\Upsilon)$, the function L_{00} extends smoothly to $\{\bar{\rho} = 0\}$ with $L_{00} \to 0$ as $\bar{\rho} \to 0$.

It holds

$$L_{01} = \frac{\bar{\rho}^2}{(v\,\Omega)(\bar{\rho})} \left(\frac{L[\bar{g}]_{tt}}{v\,\Omega} \right) \circ \Phi - (1 - \bar{\tau}) L_{00},$$

with

$$\frac{L[\bar{g}]_{tt}}{v\,\Omega} = \Omega^{-1}\,\Upsilon\,(\frac{v}{2}\,D_a\Lambda\,D^a\Lambda - D_a\,N\,D^a\,\Lambda)$$

$$= v\left\{\frac{1}{2\,U^2\,(1+\sqrt{\mu\,\sigma})^2} - \frac{(1-3\,\sqrt{\mu\,\sigma})\,s}{(1-\sqrt{\mu\,\sigma})\,(1+\sqrt{\mu\,\sigma})^4}\right.$$

$$\left. -\frac{1-2\,\sqrt{\mu\,\sigma}}{(1-\sqrt{\mu\,\sigma})\,(1+\sqrt{\mu\,\sigma})^3}\left[\frac{2}{U^2} + \frac{D^a\,\Upsilon\,D_a\,U}{U^3}\right]\right\},$$

so that L_{01} extends smoothly to $\{\bar{\rho}=0\}$ with $L_{01} \to \frac{1}{2}$ as $\bar{\rho}\to 0$.

$$L_{0A} = \bar{\rho}\left(\frac{L[\bar{g}]_{tt}}{v\,\Omega}\right)\circ\Phi\,l_A - \bar{\rho}^2\,(1-\bar{\tau})\,(L[\bar{g}]_{ab}\circ\Phi)\,e^a\,e^b_{,\psi A},$$

with

$$L[\bar{g}]_{ab}\,e^a\,e^b_{,\psi A}$$

$$= (1-\sqrt{\mu\,\sigma})\,R_{ab}\,e^a\,e^b_{,\psi A} + \frac{\mu\,U_{,\psi A}}{\sqrt{\mu\,\Upsilon\,\sigma}}\,\frac{1}{(1+\sqrt{\mu\,\sigma})^2}\left[\frac{1}{U^4} + 2\,\frac{\Upsilon\,D^a\,\Upsilon\,D_a\,U}{U^6}\right]$$

so that L_{0A} extends smoothly to $\{\bar{\rho}=0\}$ with $L_{0A}\to 0$ as $\bar{\rho}\to 0$.

$$L_{11} = \left(\frac{\bar{\rho}^2\,(v\,\Omega)(\bar{\rho}(1-\bar{\tau}))}{(v\,\Omega)^2(\bar{\rho})} - 2\,\frac{\bar{\rho}^2\,(1-\bar{\tau})}{(v\,\Omega)(\bar{\rho})}\right)\left(\frac{L[\bar{g}]_{tt}}{v\,\Omega}\right)\circ\Phi + (1-\bar{\tau})^2\,L_{00},$$

extends smoothly to $\{\bar{\rho}=0\}$ with $L_{11}\to -\frac{1-\bar{\tau}^2}{2}$ as $\bar{\rho}\to 0$.

$$L_{1A}$$

$$= \left(\frac{\bar{\rho}}{(v\,\Omega)(\bar{\rho})} - \frac{\bar{\rho}\,(1-\bar{\tau})}{(v\,\Omega)(\bar{\rho}(1-\bar{\tau}))}\right)L[\bar{g}]_{tt}\circ\Phi\,l_A - \bar{\rho}^2\,(1-\bar{\tau})^2\,(L[\bar{g}]_{ab}\circ\Phi)\,e^a\,e^b_{,\psi A},$$

extends smoothly to $\{\bar{\rho}=0\}$ with $L_{1A}\to 0$ as $\bar{\rho}\to 0$.

$$L_{AB} = L[\bar{g}]_{tt}\circ\Phi\,l_A\,l_B - (\Gamma\,L[\bar{g}]_{ab})\circ\Phi\,e^a_{,\psi A}\,e^b_{,\psi B}.$$

extends smoothly to $\{\bar{\rho}=0\}$ with $L_{AB}\to -\frac{1}{2}\,k_{AB}(0)$ as $\bar{\rho}\to 0$.

6.2. The rescaled conformal Weyl tensor of $\bar{g}$ near $\bar{\mathcal{I}}'$

In this section we shall make a few general observations concerning the rescaled conformal Weyl tensor and then specialize to the conformal static case. After a remark about the radiation field on $\mathcal{J}^{+'}$ we will analyze the smoothness of the rescaled conformal Weyl tensor near the set $\bar{\mathcal{I}}'$.

Let $\tilde{g}$ be a Lorentz metric and $\tilde{S}$ a space-like hypersurface with unit normal $\tilde{n}$ and induced metric $\tilde{h}_{\mu\nu} = \tilde{g}_{\mu\nu} - \tilde{n}_\mu\,\tilde{n}_\nu$. We set $\tilde{p}_{\mu\nu} = \tilde{h}_{\mu\nu} - \tilde{n}_\mu\,\tilde{n}_\nu$, $\tilde{\epsilon}_{\nu\lambda\rho} = \tilde{n}^\mu\,\tilde{\epsilon}_{\mu\nu\lambda\rho}$, and denote by $\tilde{c}_{\nu\rho} = C_{\mu\nu\lambda\rho}[\tilde{g}]\,\tilde{n}^\mu\tilde{n}^\lambda$ and $\tilde{c}^*_{\nu\rho} = C^*_{\mu\nu\lambda\rho}[\tilde{g}]\tilde{n}^\mu\,\tilde{n}^\lambda$ (the star on the right-hand side indicating the dual) the $\tilde{n}$-electric and the $\tilde{n}$-magnetic part of the conformal Weyl tensor respectively. The latter are symmetric, trace-free, and spatial, i.e., $\tilde{n}^\nu\,\tilde{c}_{\nu\rho} = 0$, $\tilde{n}^\nu\,\tilde{c}^*_{\nu\rho} = 0$. The conformal Weyl tensor of $\tilde{g}$ is then given in terms of its electric and the magnetic part by (cf. [42])

$$C_{\mu\nu\lambda\rho}[\tilde{g}] = 2\left(\tilde{p}_{\nu[\lambda}\,\tilde{c}_{\rho]\mu} - \tilde{p}_{\mu[\lambda}\,\tilde{c}_{\rho]\nu} - \tilde{n}_{[\lambda}\,\tilde{c}^*_{\rho]\delta}\,\epsilon^\delta\,_{\mu\nu} - \tilde{n}_{[\mu}\,\tilde{c}^*_{\nu]\delta}\,\epsilon^\delta\,_{\lambda\rho}\right). \tag{6.17}$$

Suppose that $\tilde{g}$ is a solution to the vacuum field equations. Then the first and second fundamental form $\tilde{h}_{ab}$ and $\tilde{\chi}_{ab}$ induced by $\tilde{g}$ on $\tilde{S}$ satisfy the Gauss and the Codazzi equation (expressing the pull-back of spatial tensors to $\tilde{S}$ in terms of spatial coordinates x^a)

$$r_{ab}[\tilde{h}] = -\tilde{c}_{ab} + \tilde{\chi}_c{}^c \tilde{\chi}_{ab} - \tilde{\chi}_{ca} \tilde{\chi}_b{}^c, \tag{6.18}$$

$$\tilde{D}_b \tilde{\chi}_{d(a} \tilde{\epsilon}_{c)}{}^{bd} = -\tilde{c}_{ac}^*. \tag{6.19}$$

This allows us to express the conformal Weyl tensor in terms of $\tilde{h}_{ab}$ and $\tilde{\chi}_{ab}$. If $\tilde{S}$ is a hypersurface of time reflection symmetry, so that $\tilde{\chi}_{ab} = 0$, these equations imply

$$r_{ab}[\tilde{h}] = -\tilde{c}_{ab}, \qquad \tilde{c}_{ac}^* = 0, \tag{6.20}$$

and the Weyl tensor assumes the form

$$C_{\mu\nu\lambda\rho}[\tilde{g}] = 2 \left(\tilde{p}_{\nu[\lambda} \tilde{c}_{\rho]\mu} - \tilde{p}_{\mu[\lambda} \tilde{c}_{\rho]\nu} \right) \equiv -(\tilde{p} \oslash \tilde{c})_{\mu\nu\lambda\rho}, \tag{6.21}$$

where $\oslash$ denotes the bi-linear Kulkarni-Nomizu product of two symmetric 2-tensors (cf. [8]).

If Λ is an arbitrary conformal factor, the rescaled conformal Weyl tensor of $\bar{g} = \Lambda^2 \tilde{g}$ is given by $W^\mu{}_{\nu\lambda\rho}[\bar{g}] = \Lambda^{-1} C^\mu{}_{\nu\lambda\rho}[\tilde{g}]$. In view of the behavior of the conformal Weyl tensor under conformal rescalings, one gets (observe the index positions)

$$W_{\mu\nu\lambda\rho}[\bar{g}] = \Lambda\, C_{\mu\nu\lambda\rho}[\tilde{g}]. \tag{6.22}$$

Its electric part with respect to the $\bar{g}$-unit vector $\Lambda^{-1}\tilde{n}$ is then given by

$$w_{\mu\nu}[\bar{g}] = \Lambda^{-1}\tilde{c}_{\mu\nu}[\tilde{g}] \tag{6.23}$$

With $h = \Omega^2\, \tilde{h}$, the gauge (4.31), the general transformation law

$$r_{ab}[\tilde{h}] = r_{ab}[h] + \Omega^{-1} D_a D_b \Omega + h_{ab} \left(\Omega^{-1} D_c D^c \Omega - 2\,\Omega^{-2} D_c \Omega D^c \Omega \right),$$

and the equation $2\,\Omega \Delta_h\,\Omega = 3\,D_a\Omega\,D^a\Omega$, one gets from (6.20) and (6.23) in the general time reflection symmetric case

$$w_{ab}[\bar{g}] = -(\Lambda\,\Omega)^{-1}(D_a\,D_b\,\Omega - \frac{1}{3}\,h_{ab}\,D_c\,D^c\,\Omega + \Omega\,r_{ab}[h]) \quad \text{on} \quad \mathcal{S} = \tilde{S} \cup \{i\}. \tag{6.24}$$

A conformal scaling which represents space-like infinity (with respect to the initial hypersurface $\tilde{S}$ and with respect to the solution space-time) by a point is achieved by choosing $\Lambda = \Omega$ on $\mathcal{S}$. With this particular choice one has

$$w_{ab}[\bar{g}] = -\Omega^{-2}(D_a\,D_b\,\Omega - \frac{1}{3}\,h_{ab}\,D_c\,D^c\,\Omega + \Omega\,r_{ab}[h]) \tag{6.25}$$

$$= O(\Upsilon^{-3/2}) \quad \text{as} \quad \Upsilon \to 0 \quad \text{unless} \quad m = 0.$$

We note that in the massless case the precise behavior depends on the freely prescribed metric h on $\mathcal{S}$ near i. In the massless case one has $\Omega = \sigma$ and the comparison of the expression for $w_{ab}[\bar{g}]$ with (4.36) shows that in the case where

h represents conformally static vacuum data one has

$$w_{ab}[\bar{g}] = -\mu\, r_{ab}[h], \tag{6.26}$$

i.e., the rescaled conformal Weyl tensor is smooth.

We return to the case where $\Lambda = \Omega\, \Upsilon^{-1/2}$. With (6.14), (6.15) we get then

$$w_{ab}[\bar{g}] = -\frac{\sqrt{\Gamma}}{\sigma^2}\left\{(1 + \sqrt{\mu\,\sigma})\,(D_a\, D_b\,\sigma - \frac{1}{3}\,\Delta_h\,\sigma\, h_{ab})\right. \tag{6.27}$$

$$-\frac{1}{2}\sqrt{\frac{\mu}{\sigma}}\,(3\,D_a\,\sigma\, D_b\,\sigma - D_c\,\sigma\, D^c\,\sigma\, h_{ab}) + \sigma\,(1 + \sqrt{\mu\,\sigma})^2\, r_{ab}\Big\}$$

$$= \frac{m}{4}\frac{U}{\sigma^2}\,(3\,D_a\,\sigma\, D_b\,\sigma - D_c\,\sigma\, D^c\,\sigma\, h_{ab}) - \frac{m}{2}\,U\,(1 + \sqrt{\mu\,\sigma})^2\, r_{ab}[h]$$

$$-\frac{\sqrt{\Upsilon}}{\sigma^2}\,(1 + \sqrt{\mu\,\sigma})\,\Sigma_{ab},$$

where we use Σ_{ab} as defined by the right-hand side of (4.36) without assuming h to be conformally static. If h is conformally static the electric part of the rescaled conformal Weyl tensor on $\mathcal{S}$ is given by the right-hand side of (6.27) with $\Sigma_{ab} = 0$. In the present conformal gauge, defined by (4.31), one has

$$\Sigma_{ab} = O(\Upsilon^{3/2}) \quad \text{near} \quad i,$$

for any time reflection symmetric initial data h.

If the solution is static and written again in the form $\tilde{g} = v^2\, dt^2 + \tilde{h}$, then equations (6.17), (6.20) hold with $\tilde{n} = \frac{1}{v}\,\partial_t$ and t-independent fields for each slice $\tilde{\mathcal{S}} = \{t = t_*\}$ with $t_* = const$. The relations above then imply for all (t, x^a)

$$W_{\mu\nu\lambda\rho}[\bar{g}] = -\Upsilon^{-1}(p \oslash w)_{\mu\nu\lambda\rho} \tag{6.28}$$

with

$$p_{\mu\nu} = h_{\mu\nu} - n_\mu\, n_\nu, \quad n_\mu = \Omega\,\tilde{n}_\mu.$$

With $\otimes$ denoting the tensor product, we write for arbitrary 1-forms a, c

$$a \otimes_s c = a \otimes c + c \otimes a, \quad a^2 = a \otimes a,$$

and note that the Kulkarni-Nomizu product is symmetric, i.e.,

$$m \oslash n = n \oslash m, \tag{6.29}$$

for symmetric 2-tensors m, n, and satisfies for arbitrary 1-forms a, c, e

$$(a \otimes a) \oslash (a \otimes_s c) = 0, \quad (a \otimes_s e) \oslash (a \otimes_s c) = -(a \otimes a) \oslash (c \otimes_s e). \tag{6.30}$$

We show how it follows in the present setting that the *radiation field* vanishes on $\mathcal{J}^{+'}$. Since the extended Killing vector field K is tangent to the null generators of $\mathcal{J}^{+'}$ without vanishing there, the complete information on the radiation field is contained in the field

$$K^\nu \, K^\rho \, W_{\mu\nu\lambda\rho}[\bar{g}] \, d\,x^\mu \, d\,x^\lambda = -\Upsilon^{-1} \, K^\nu \, K^\rho \, (p \oslash w)_{\mu\nu\lambda\rho} \, d\,x^\mu \, d\,x^\lambda$$

$$= -\Upsilon^{-1} \, K^\nu \, K^\rho \, p_{\nu\rho} \, w_{ab} \, d\,x^a \, d\,x^b = -\Upsilon^{-1} \, (v\,\Omega)^2 \, w_{ab} \, d\,x^a \, d\,x^b$$

$$= \frac{m}{4} \frac{(1-\sqrt{\mu\,\sigma})^2}{(1+\sqrt{\mu\,\sigma})^6} \left\{ \frac{1}{U} \, (2\,s\,h_{ab} - 3\,\sigma^{-1} \, D_a\,\sigma \, D_b\,\sigma) \, d\,x^a \, d\,x^b \right.$$

$$\left. + \frac{2\,\sigma}{U} (1+\sqrt{\mu\,\sigma})^2 \, r_{ab}[h] \, d\,x^a \, d\,x^b \right\}$$

with s as given in (4.32). Because of the relation

$$\sigma^{-1} \, D_a\,\sigma \, D_b\,\sigma = 4\,U^{-4} \, (U^2 \, e_a\,e_b - U\,\Upsilon^{1/2} \, (e_a\,D_b\,U + e_b\,D_a\,U) + \Upsilon\,D_a\,U\,D_b\,U),$$
$$(6.31)$$

and the factor σ in the second term it follows that

$$K^\nu \, K^\rho \, W_{\mu\nu\lambda\rho}[\bar{g}] \, d\,x^\mu \, d\,x^\lambda \to -2\,m\,\bar{\rho}^2 \, d\,\bar{\tau}^2 \quad \text{as } \bar{\tau} \to 1, \;\; \bar{\rho} > 0. \qquad (6.32)$$

Thus, the pull-back of $K^\nu \, K^\rho \, W_{\mu\nu\lambda\rho}[\bar{g}] \, d\,x^\mu \, d\,x^\lambda$ to $\mathcal{J}^+$, which provides the radiation field up to a scaling, vanishes everywhere on $\mathcal{J}^+$.

Lemma 6.2. *The components* $W_{ijkl}[\bar{g}] = \Lambda^{-1} \, C_{\mu'\nu'\lambda'\rho'}[\bar{g}] \, v^{\mu'}{}_i \, v^{\nu'}{}_j \, v^{\lambda'}{}_k \, v^{\rho'}{}_l$ *of the rescaled conformal Weyl tensor of* $\bar{g}$ *in the frame* v_k *extend as analytic functions to* $\bar{\mathcal{I}}'$.

Proof. In the coordinates $x^{\mu'}$ given by (6.1) the rescaled conformal Weyl tensor is obtained as the product of

$$-(\Upsilon \circ \Phi)^{-1} = -(\bar{\rho}\,(1-\bar{\tau}))^{-2},$$

with the Nomizu-Kulkarni product of

$$p' = (h_{ab} \circ \Phi) \, d\,x^a \, d\,x^b - ((v\,\Omega) \circ \Phi)^2 \, d\,t^2 = p'_1 + p'_2 + p'_3 + p'_4,$$

and $w' = w'_1 + w'_2 + w'_3 + w'_4 + w'_5$, where

$$p'_1 = -2\,((1-\hat{\tau})\,d\,\bar{\rho} - \bar{\rho}\,d\,\bar{\tau})^2,$$

$$p'_2 = ((1-\bar{\tau})\,d\,\bar{\rho} - \bar{\rho}\,d\,\bar{\tau}) \otimes_s \left(\frac{(v\,\Omega)(\bar{\rho}\,(1-\bar{\tau}))}{(v\,\Omega)(\bar{\rho})} \, d\,\bar{\rho} + (v\,\Omega)(\bar{\rho}\,(1-\bar{\tau}))\,l \right)$$

$$p'_3 = -\left(\frac{(v\,\Omega)(\bar{\rho}\,(1-\bar{\tau}))}{(v\,\Omega)(\bar{\rho})} \, d\,\bar{\rho} + (v\,\Omega)(\bar{\rho}\,(1-\bar{\tau}))\,l \right)^2,$$

$$p'_4 = \bar{\rho}^2 \, (1-\bar{\tau})^2 \, k,$$

$$w'_5 = -\left(\left\{ \frac{m}{2} \, U \, (1+\sqrt{\mu\,\sigma})^2 \, r_{ab}[h] \right\} \circ \Phi \right) d\,x^a \, d\,x^b,$$

and

$$w'_1 + w'_2 + w'_3 + w'_4 = -\left(\left\{ \frac{m}{4} \frac{U}{\sigma} \, (2\,s\,h_{ab} - 3\,\sigma^{-1} \, D_a\,\sigma \, D_b\,\sigma) \right\} \circ \Phi \right) d\,x^a \, d\,x^b,$$

with

$$w_1' = -\frac{m}{4} \left(\left\{ \frac{U}{\sigma} (s + 6\, U^{-2}) \right\} \circ \Phi \right) p_1',$$

$$w_2' = - \left(\left\{ \frac{m}{2}\, U^3\, s \right\} \circ \Phi \right) k,$$

$$w_3' = - \left(\left\{ \frac{3\, m}{U\, \sigma}\, \Upsilon^{1/2} \right\} \circ \Phi \right) \left((1 - \bar\tau)\, d\bar\rho - \bar\rho\, d\bar\tau \right) \otimes_s j,$$

$$w_4' = - \left(\left\{ \frac{m}{U} \right\} \circ \Phi \right) j^2.$$

We used above the relation (6.31) and set

$$j = (D_a\, U \circ \Phi)\, dx^a.$$

The desired result on the behavior of the rescaled conformal Weyl tensor near $\bar{\mathcal{I}}'$ is obtained now by showing that for arbitrary frame vector fields v_n one has

$$\langle p' \oslash w';\, v_i,\, v_j,\, v_k,\, v_l, \rangle = O(\bar\rho^2\, (1 - \bar\tau)^2).$$

From (6.29) it follows that

$$p_1' \oslash w_1' = p_1' \oslash w_3' = p_2' \oslash w_1' = 0.$$

Observing (4.27) one finds by inspection

$$\langle p_1';\, v_i,\, v_j \rangle = O(\bar\rho^2), \quad \langle p_M';\, v_i,\, v_j \rangle = O(\bar\rho^2\, (1 - \bar\tau)^2), \quad \text{for } M = 2, 3, 4,$$

$$\langle w_4';\, v_i,\, v_j \rangle = O(\bar\rho^2\, (1 - \bar\tau)^2), \quad \langle w_N';\, v_i,\, v_j \rangle = O(1), \quad \text{for } N = 2, 3, 5,$$

and thus

$$\langle p_M' \oslash w_N';\, v_i,\, v_j,\, v_k,\, v_l \rangle = O(\bar\rho^2\, (1 - \bar\tau)^2) \quad \text{for } M = 2, 3, 4, \quad N = 2, 3, 4, 5,$$

$$\langle p_1' \oslash w_4';\, v_i,\, v_j,\, v_k,\, v_l \rangle = O(\bar\rho^2\, (1 - \bar\tau)^2).$$

The remaining term is given by

$$p_1' \oslash (w_2' + w_5') + (p_3' + p_4') \oslash w_1' = p_1' \oslash m$$

with

$$m = w_2' + w_5' - \frac{m}{4} \left(\left\{ \frac{U}{\sigma} (s + 6\, U^{-2}) \right\} \circ \Phi \right) (p_3' + p_4')$$

$$= -\frac{m}{4} \left(\{ U\, (3\, s\, U^2 + 6) \} \circ \Phi \right) k - \left(\left\{ \frac{m}{2}\, U\, (1 + \sqrt{\mu\, \sigma})^2\, r_{ab}[h] \right\} \circ \Phi \right) dx^a\, dx^b$$

$$+ \frac{m}{4} \left(\{ s\, U^3 + 6\, U \} \circ \Phi \right) \left(\frac{(v\, \Omega)(\bar\rho\, (1 - \bar\tau))}{\bar\rho\, (1 - \bar\tau)\, (v\, \Omega)(\bar\rho)}\, d\bar\rho + \frac{(v\, \Omega)(\bar\rho\, (1 - \bar\tau))}{\bar\rho\, (1 - \bar\tau)}\, l \right)^2$$

For the three summands to be considered here we get the following. From $3\, s\, U^2 + 6 = O(\Upsilon)$ it follows that

$$\langle p_1' \oslash (\{ U\, (3\, s\, U^2 + 6) \} \circ \Phi)\, k;\, v_i,\, v_j,\, v_k,\, v_l \rangle = O(\bar\rho^2\, (1 - \bar\tau)^2).$$

Because of

$$dx^{(a}\, dx^{b)} = \bar\rho^2\, (1 - \bar\tau)^2\, d\, e^{(a} \otimes d\, e^{b)}$$

$$-\frac{1}{2}\, p_1'\, e^a\, e^b + \bar\rho\, (1 - \bar\tau)\, e^{(a}\, d\, e^{b)} \otimes_s ((1 - \hat\tau)\, d\rho - \rho\, d\hat\tau),$$

it follows by (6.30) that

$$\langle p_1' \oslash \left(\left\{ \frac{m}{2} U \left(1 + \sqrt{\mu \sigma} \right)^2 r_{ab}[h] \right\} \circ \Phi \right) d x^a \, d x^b; \, v_i, \, v_j, \, v_k, \, v_l \rangle$$

$$= O(\bar{\rho}^2 \, (1 - \bar{\tau})^2).$$

It holds that $s \, U^3 + 6 \, U = O(1)$ and by inspection it follows that

$$\langle p_1' \oslash \left(\frac{(v \, \Omega)(\bar{\rho} \, (1 - \bar{\tau}))}{\bar{\rho} \, (1 - \bar{\tau}) \, (v \, \Omega)(\bar{\rho})} \, d\bar{\rho} + \frac{(v \, \Omega)(\bar{\rho} \, (1 - \bar{\tau}))}{\bar{\rho} \, (1 - \bar{\tau})} \, l \right)^2; \, v_i, \, v_j, \, v_k, \, v_l \rangle$$

$$= O(\bar{\rho}^2 \, (1 - \bar{\tau})^2).$$

7. Static vacuum solutions near the cylinder at space-like infinity

The conformal extension considered in the previous section relies on specific features of static fields. We use it to show that the construction of the cylinder at space-like infinity in Section 5, which is based on general concepts and applies to general solutions, is for static vacuum solutions as smooth as can be expected.

Theorem 7.1. *For static vacuum solutions which are asymptotically flat the construction of Section 5 is analytic in the sense that in the frame (5.57) all conformal fields, including the rescaled conformal Weyl tensor, extend to analytic fields on some neighborhood $\mathcal{O}$ of $\bar{\mathcal{I}}$ in $\bar{\mathcal{N}}$. This statement does not depend on a particular choice of (analytic) scaling of the (analytic) free datum h on $\mathcal{S}$.*

This result will be obtained as a consequence of Lemmas 7.2, 7.3, and 7.4 below.

The construction of Section 5 will be discussed here for static solutions in terms of the initial data h and Ω in the gauge given by (4.31), and the field $\bar{g}$ given on $\bar{\mathcal{M}}'$ in the coordinates defined by (6.1). The effect of a rescaling of h will be discussed separately because it is of interest in itself.

The conformal factor Θ is assumed in the form (5.44), (5.46) with

$$\kappa = \omega = 2 \Omega \, |D_a \Omega D^a \Omega|^{-\frac{1}{2}} = 2 \Omega \, (1 + \sqrt{\mu \sigma})^{-3} \sqrt{2 \, |s| \, \sigma}. \tag{7.1}$$

It follows that $\omega = \Upsilon^{1/2} + O(\Upsilon)$ and $\Theta = 1/2 \, |D_a \Omega D^a \Omega|^{\frac{1}{2}} = \Upsilon^{1/2} + O(\Upsilon)$ so that

$$\text{on } \bar{\mathcal{S}}: \quad \lim_{\bar{\rho} \to 0} \Upsilon^{1/2} \, \omega^{-1} = \lim_{\bar{\rho} \to 0} \omega \, \Lambda^{-1} = \lim_{\bar{\rho} \to 0} \Theta \, \Lambda^{-1} = 1. \tag{7.2}$$

The metric $\bar{h}$ induced by $g = \Theta^2 \, \tilde{g}$ on $\bar{\mathcal{S}}$ is given by $\omega^{-2} \, h$.

The main ingredient of the gauge for the evolution equations used in Section 5.2 are the conformal geodesics generating the conformal Gauss system described in Section 2.1. We shall try to control their evolution on $\bar{\mathcal{M}}'$ near $\bar{\mathcal{I}}'$. Following the prescription in Section 2.1, we assume that the tangent vectors $\dot{x} = d x / d \tau$ of the conformal geodesics with parameter τ satisfy

$$\dot{x} \perp \tilde{\mathcal{S}}, \quad \Theta^2 \, \tilde{g}(\dot{x}, \dot{x}) = 1 \quad \text{on} \quad \tilde{\mathcal{S}}. \tag{7.3}$$

With the frame (6.7) and the coordinates (6.1) this translates into the initial condition

$$\dot{x} = \frac{\omega(\bar{\rho})}{\bar{\rho}} \, \partial_{\bar{\tau}} + \omega(\bar{\rho}) \, \partial_{\bar{\rho}} = \frac{\omega(\bar{\rho})}{\bar{\rho}} \, (v_0 + v_1) \equiv X^i \, v_i \quad \text{at} \quad \bar{\tau} = 0, \qquad (7.4)$$

with

$$X^i = \delta^i{}_0 + \delta^i{}_1 + O(\bar{\rho}) \quad \text{as } \bar{\rho} \to 0.$$

For the following we need to observe besides $\bar{g} = \Lambda^2 \, \tilde{g}$ the relations

$$g = \Pi^2 \, \bar{g} = \Theta^2 \, \tilde{g}, \quad \text{with} \quad \Pi \equiv \Lambda^{-1} \, \Theta.$$

For the connections $\hat{\nabla}$, ∇, and $\bar{\nabla}$ of $\tilde{g}$, g, and $\bar{g}$ respectively we have relations

$$\hat{\nabla} = \tilde{\nabla} + S(\tilde{f}), \qquad \hat{\nabla} = \nabla + S(f), \qquad \hat{\nabla} = \bar{\nabla} + S(\bar{f}),$$

$$\nabla = \tilde{\nabla} + S(\Theta^{-1} d\,\Theta),, \qquad \bar{\nabla} = \tilde{\nabla} + S(\Lambda^{-1} d\,\Lambda).$$

The comparison gives $f = \tilde{f} - \Theta^{-1} d\,\Theta$ and $\bar{f} = \tilde{f} - \Lambda^{-1} d\,\Lambda$ which imply the relation

$$\bar{f} = f + \Theta^{-1} d\,\Theta - \Lambda^{-1} d\,\Lambda = f + \Pi^{-1} d\,\Pi, \qquad (7.5)$$

between the 1-form $\bar{f}$ which is obtained if the conformal geodesic equations are written in terms of the metric $\bar{g}$, the 1-form f which is supplied by the conformal geodesic equations written in terms of the metric g, and the conformal factor which relates $\bar{g}$ to g.

By the choices of Section 5.2 we have $\langle f, \dot{x} \rangle = 0$ everywhere on the space-time and $\langle d\,\Theta, \dot{x} \rangle = 0$ on $\tilde{S}$. Since Λ has been chosen to be independent of t and ∂_t is orthogonal $\tilde{S}$, it follows that $\langle d\,\Lambda, \dot{x} \rangle = 0$ and thus $\langle \bar{f}, \dot{x} \rangle = 0$ on $\tilde{S}$. Observing the pull-back of f to $\tilde{S}$ given by (5.45) and the relation

$$\Pi = \Upsilon^{1/2} \, \omega^{-1} \quad \text{on} \quad \tilde{S}, \qquad (7.6)$$

we find that the pull-back of $\bar{f}$ to $\tilde{S}$ is given by $1/2 \, \Upsilon^{-1} d\,\Upsilon$. From this one gets in the frame (6.7) and the coordinates (6.1) with $\bar{\tau} = 0$

$$\bar{f} = (1/2 \, \Upsilon^{-1} \, D_a \, \Upsilon) \circ \Phi \, dx^a = \bar{f}_i \, \alpha^i \quad \text{with} \quad \bar{f}_i = -\delta^0{}_i + \delta^1{}_i \quad \text{on} \quad \tilde{S}. \qquad (7.7)$$

The relation $\langle f, \dot{x} \rangle = 0$ and equation (7.5) imply the ODE

$$\dot{\Pi} = \Pi \langle \bar{f}, \dot{x} \rangle, \qquad (7.8)$$

along the conformal geodesics, which, together with (7.6), will allow one to determine Π once $\langle \bar{f}, \dot{x} \rangle$ is known.

7.1. The extended conformal geodesic equation on I

With respect to the metric (6.6) a solution to the conformal geodesic equations is given by a space-time curve $x^\mu(\tau) = (\bar\tau(\tau), \bar\rho(\tau), \psi^A(\tau))$ and along that curve a vector field $X(\tau)$ and a 1-form $\bar f(\tau)$ such that

$$\dot x = X,$$

$$\bar\nabla_X X = -2\langle \bar f, X\rangle X + \bar g(X, X)\, \bar f^\sharp,$$

$$\bar\nabla_X \bar f = \langle \bar f, X\rangle \bar f - \frac{1}{2}\, \bar g(\bar f, \bar f)\, X^\flat + L(X, \cdot).$$

With the expansions $X = X^i\, v_i$, $\bar f = \bar f_i\, \alpha^i$, $\bar g = \bar g_{jk}\, \alpha^j\, \alpha^k$, $\bar g^\sharp = \bar g^{jk}\, v_j\, v_k$, $L = L_{jk}\, \alpha^j\, \alpha^k$, $e_k = e^i{}_k\, v_i$, the equations above take in the domain where $\bar\rho > 0$ the form

$$\frac{d}{d\tau}\bar\tau = X^0, \quad \frac{d}{d\tau}\bar\rho = \bar\rho\, X^1, \quad \frac{d}{d\tau}\psi^A = X^A,$$

which is the equation $\dot x^\mu = X^i\, v^\mu{}_i$, relating the coordinate to the frame expressions,

$$\frac{d}{d\tau} X^i + \gamma_j{}^i{}_k\, X^j\, X^k = -2\, \bar f_k\, X^k\, X^i + \bar g_{jk}\, X^j\, X^k\, \bar g^{il}\, \bar f_l,$$

$$\frac{d}{d\tau}\bar f_k - \gamma_j{}^i{}_k\, \xi^j\, \bar f_i = \bar f_l\, X^l\, \bar f_k - \frac{1}{2}\, \bar g^{lj}\, \bar f_l\, \bar f_j\, \bar g_{lk}\, X^l + L_{jk}\, X^j.$$

Note that the functions $\bar g_{jk}$, $\bar g^{il}$, $\gamma_j{}^i{}_l$, L_{jk} entering these equations extend by analyticity through $\bar{\mathcal{I}}'$ into a domain where $\bar\rho < 0$. Assuming such an extension, we get the *extended conformal geodesic equations*. Since also the data are analytic on $\bar{\mathcal{S}}$, it makes sense to consider these equations in a neighborhood of $\bar{\mathcal{I}}'$.

Lemma 7.2. *With the values of L_{jk} on $\bar{\mathcal{I}}'$ found in Lemma 6.1, the initial data $x = (0, 0, \psi^{A'})$ and (cf. (7.4), (7.7)) $X^i = \delta^i{}_0 + \delta^i{}_1$, $\bar f_i = -\delta^0{}_i + \delta^1{}_i$ on $\mathcal{I}^{0'}$ determine a solution $x(\tau)$, $X(\tau)$, $\bar f(\tau)$ of the extended conformal geodesic equations with $\tau = 0$ on $\mathcal{I}^{0'}$ and*

$$x(\tau) = (\bar\tau(\tau), \bar\rho(\tau), \psi^A(\tau)) = (\tau, 0, \psi^{A'}).$$

By analyticity it extends as a solution into a domain $0 \le \tau \le 1 + 2\epsilon$ for some $\epsilon > 0$. The extension to $\bar{\mathcal{I}}'$ of the conformal factor Π which is determined by (7.6) and (7.8) takes the value $\Pi = 1$ on $\bar{\mathcal{I}}'$.

Proof. With the ansatz $x(\tau) = (\bar\tau(\tau), 0, \psi^{A'})$, $X(\tau) = X^0(\tau)\, v_0 + X^1(\tau)\, v_1$, $\bar f = \bar f_0(\tau)\, \alpha^0 + \bar f_1(\tau)\, \alpha^1$ those of the extended conformal geodesic equations which are not identically satisfied because of (6.8), (6.9), (6.11) are given by

$$\frac{d}{d\tau}\bar\tau = X^0,$$

$$\frac{d}{d\tau} X^0 + 2\bar\tau\, X^0\, X^1 - \bar\tau\, (1 - \bar\tau^2)\, X^1\, X^1$$
$$= -2\, (\bar f_0\, X^0 + \bar f_1\, X^1)\, X^0 + (2\, X^0\, X^1 - (1 - \bar\tau^2)X^1\, X^1)\, ((1 - \bar\tau^2)\, \bar f_0 + \bar f_1),$$

$$\frac{d}{d\tau} X^1 - \bar\tau\, X^1\, X^1 = -2\, (\bar f_0\, X^0 + \bar f_1\, X^1)\, X^1 + (2\, X^0\, X^1 - (1 - \bar\tau^2)X^1\, X^1)\, \bar f_0,$$

$$\frac{d}{d\tau}\,\bar{f}_0 - \bar{\tau}\,\bar{f}_0\,X^1 = (\bar{f}_0\,X^0 + \bar{f}_1\,X^1)\,\bar{f}_0 - \frac{1}{2}\,((1-\bar{\tau}^2)\,\bar{f}_0\,\bar{f}_0 + 2\,\bar{f}_0\,\bar{f}_1)\,X^1 + \frac{1}{2}\,X^1,$$

$$\frac{d}{d\tau}\,\bar{f}_1 - \bar{\tau}\,\bar{f}_0\,(X^0 - (1-\bar{\tau}^2)\,X^1) + \bar{\tau}\,\bar{f}_1\,X^1 = (\bar{f}_0\,X^0 + \bar{f}_1\,X^1)\,\bar{f}_1$$

$$-\frac{1}{2}\,((1-\bar{\tau}^2)\,\bar{f}_0\,\bar{f}_0 + 2\,\bar{f}_0\,\bar{f}_1)\,(X^0 - (1-\bar{\tau}^2)\,X^1) + \frac{1}{2}\,X^0 - \frac{1-\bar{\tau}^2}{2}\,X^1.$$

A calculation shows that the solution of this system for the prescribed initial is given by

$$\bar{\tau} = \tau, \quad X^0 = 1, \quad X^1 = \frac{1}{1+\bar{\tau}}, \quad \bar{f}_0 = -\frac{1}{1+\bar{\tau}}, \quad \bar{f}_1 = 1. \tag{7.9}$$

This proves the first assertion. With the solution above equation (7.8) reads $\dot{\Pi} = 0$ and we have $\Pi = 1$ on I^0 by (7.2). This proves the second assertion.

Remark: The ODE above is sufficiently complicated so that giving the solution explicitly deserves an explanation. In (5.62) is given the conformal factor and the conformal representation of Minkowski space which result from the general procedure of Section 5. In (5.63) is given the coordinate transformation which, together with the conformal factor, relates the conformal metric to the standard representation of Minkowski space in coordinates t and r.

If the Minkowski values $m = 0$, $U = 1$, $h_{ab} = -\delta_{ab}$ are assumed in Section 6 the metric $\bar{g}$ reduces by (6.8) to the metric $g_{ik}^*\,\alpha^i\,\alpha^k$ with $m = 0$. One can consider this as the lowest order (in $\bar{\rho}$) approximation of the general version of $\bar{g}$. Tracing back how the functions $\bar{\tau}$, $\bar{\rho}$, Λ in Section 6 are related in the flat case to t and r, one finds

$$r = \frac{1}{\bar{\rho}\,(1-\bar{\tau})}, \quad t = \frac{\bar{\tau}}{\bar{\rho}\,(1-\bar{\tau})}, \quad \Lambda = \frac{1}{r}.$$

The conformal factors in the conformal representations thus agree but the coordinates are related by the transformation

$$\bar{\tau} = \tau, \quad \bar{\rho} = \rho\,(1+\tau). \tag{7.10}$$

This implies

$$2\,\frac{d\bar{\rho}}{\bar{\rho}}\,d\bar{\tau} - (1-\bar{\tau}^2)\,\left(\frac{d\bar{\rho}}{\bar{\rho}}\right)^2 - d\sigma^2 = d\tau^2 + 2\,\tau\,\frac{d\rho}{\rho}\,d\tau - (1-\tau^2)\,\left(\frac{d\rho}{\rho}\right)^2 - d\sigma^2. \tag{7.11}$$

The left-hand side is the conformal Minkowski metric (6.8) with $m = 0$ while the right-hand side is the metric g^* given by (5.62). The conformal geodesics underlying (5.62) have tangent vector $X = \partial_\tau$ and 1-form $f = \frac{d\rho}{\rho}$. With (7.10) these transform into

$$X = \partial_{\bar{\tau}} + \frac{1}{1+\bar{\tau}}\,\bar{\rho}\,\partial_{\bar{\rho}} = v_0 + \frac{1}{1+\bar{\tau}}\,v_1,$$

$$f = \frac{d\bar{\rho}}{\bar{\rho}} - \frac{1}{1+\bar{\tau}}\,d\bar{\tau} = -\frac{1}{1+\bar{\tau}}\,\alpha_0 + \alpha_1,$$

from which one can read off (7.9).

The hypersurfaces $\{\bar{\rho} = \bar{\rho}_{\#} = \text{const.} > 0\}$ are in general time-like for the metric (6.6). The form of $\bar{g}^{\sharp}$ suggests that these hypersurfaces approximate null hypersurfaces in the limit as $\bar{\rho}_{\#} \to 0$, but the conclusion is delicate because of the degeneracy of $\bar{g}^{\sharp}$ on $\bar{\mathcal{I}}'$. The discussion above shows that they do become null asymptotically in the sense that for the metric on the left-hand side of (7.11) the hypersurfaces $\{\bar{\rho} = \text{const.}\}$ are in fact null. To some extent this explains why the coordinates given by (6.1) had a chance to extend smoothly to $\mathcal{J}^+$ and to provide a description of the cylinder at space-like infinity.

7.2. The smoothness of the gauge of Section 5 for static asymptotically flat vacuum solution near $\mathcal{I}$

Let $\bar{\mathcal{S}}_{\text{ext}}$ denote an analytic extension of $\tilde{\mathcal{S}}$ into a range where $\bar{\rho} < 0$ so that $\bar{\rho}$, ψ^A extend to analytic coordinates. If the set $\bar{\mathcal{S}}_{\text{ext}} \setminus \bar{\mathcal{S}}$ is sufficiently small, the following statements make sense. The initial conditions (7.4), (7.7) extend analytically to $\bar{\mathcal{S}}_{\text{ext}}$ and determine near $\bar{\mathcal{S}}_{\text{ext}}$ an analytic congruence of solutions to the extended conformal geodesic equations. It therefore follows from Lemma 7.2 and well-known results on ODE's that, with the ϵ of Lemma 7.2, there exists a $\rho_{\#} > 0$ such that for initial data $\bar{\tau}(0) = 0$, $\bar{\rho}(0) = \rho'$ with $|\rho'| < \rho_{\#}$, $\psi^A(0) = \psi^{A'}$ and those implied at these points by (7.4), (7.7) the solution

$$\bar{\tau} = \bar{\tau}(\tau, \rho', \psi^{A'}), \quad \bar{\rho} = \bar{\rho}(\tau, \rho', \psi^{A'}), \quad \psi^A = \psi^A(\tau, \rho', \psi^{A'}),$$

$$X^i = X^i(\tau, \rho', \psi^{A'}), \quad \bar{f}_k = \bar{f}_k(\tau, \rho', \psi^{A'}),$$

of the extended conformal geodesic equations exists for the values $0 \leq \tau \leq 1 + \epsilon$ of their natural parameter and the function Π is positive in the given range of ρ' and τ.

Taking a derivative of the equation satisfied by $\bar{\rho}$ and observing (7.9) gives

$$\frac{d}{d\tau}\left(\frac{\partial \bar{\rho}}{\partial \rho'}|_{\rho'=0}\right) = \left(\frac{\partial \bar{\rho}}{\partial \rho'}|_{\rho'=0}\right)\frac{1}{1+\bar{\tau}},$$

which implies by (7.9)

$$\left(\frac{\partial \bar{\rho}}{\partial \rho'}|_{\rho'=0}\right) = 1 + \tau \geq 1.$$

It follows that the Jacobian of the analytic map

$$(\tau, \rho', \psi^{A'}) \to x^{\mu}(\tau, \rho', \psi^{A'}),$$

takes the value $1 + \bar{\tau}$ on $\bar{\mathcal{I}}'$ and for sufficiently small $\rho_{\#} > 0$ the Jacobian does not vanish in the range $0 \leq \tau \leq 1 + \epsilon$, $|\bar{\rho}| \leq \rho_{\#}$. The relations $\Lambda = \Theta \Pi^{-1}$, $\Pi > 0$, and $\Theta = (\omega^{-1}\Omega)_* (1 - \tau^2)$ imply that the curves with $\rho' > 0$ cross $\mathcal{J}^{+'}$ for $\tau = 1$. It follows that τ, ρ', and $\psi^{A'}$ define an analytic coordinate system in a certain neighborhood $\mathcal{O}'$ of $\bar{\mathcal{I}}'$ in $\bar{\mathcal{M}}'$, such that (suppressing again the upper bounds for ρ') $\mathcal{O}' \cap \mathcal{J}^{+'} = \{\tau = 1, \rho' > 0\}$, $\mathcal{I}' = \{0 \leq \tau < 1, \rho' = 0\}$, $\mathcal{I}^{+'} = \{\tau = 1, \rho' = 0\}$, and $\mathcal{O}'$ is ruled by conformal geodesics.

The metric $g = \Pi^2 \, \bar{g}$, the connection coefficients of the connection $\hat{\nabla}$ and the tensor fields (cf. (2.11))

$$\hat{L}_{\mu\nu} = L_{\mu\nu}[\bar{g}] - \nabla_\mu \bar{f}_\nu + \bar{f}_\mu \bar{f}_\nu - \frac{1}{2} \bar{g}_{\mu\nu} \bar{f}_\lambda \bar{f}^\lambda,$$

$$f = \bar{f} - \Pi^{-1} d\Pi, \quad W_{\mu\nu\rho\lambda}[g] = \Pi \, W_{\mu\nu\rho\lambda}[\bar{g}].$$

in the frame (6.7) extend in the new coordinates as analytic fields to $\mathcal{O}'$.

Given these structures and the conformal geodesics on $\mathcal{O}'$, the construction of the manifold $\bar{\mathcal{N}}$ as described in Section 5 poses no problems. With the given analytic initial data on $\bar{S}$ it only involves solving linear ODE's corresponding to (2.21), such as

$$\frac{d}{d\tau} e^i{}_k + \gamma_j{}^i{}_l X^j e^l{}_k = -\bar{f}_l X^l e^i{}_k - \bar{f}_l e^l{}_k X^i + \bar{g}_{jl} X^j e^l{}_k \bar{g}^{im} \bar{f}_m, \qquad (7.12)$$

or its spinor analogue, along the conformal geodesics. This allows us to conclude

Lemma 7.3. *Starting with static asymptotically flat initial data in the gauge (4.31), the construction of Section 5 leads to a conformal representation of the static vacuum space-time which is real analytic in a neighborhood $\mathcal{O}$ of the set $\bar{I}$ in $\bar{\mathcal{N}}$.*

7.3. Changing the conformal gauge on the initial slice

It will be shown now how the construction described in Section 5 depends for static vacuum solutions on rescalings

$$\omega^{-2} h \to h' = \vartheta^2 \, \omega^{-2} \, \bar{h}, \quad \Omega \to \Omega' = \vartheta \, \Omega \quad \text{on} \quad \mathcal{S},$$

with analytic, positive conformal factors ϑ.

There are harmless consequences such as the change of the normal coordinates $x^a \to x'^a = x'^a(x^c)$ with $x'^a(0) = 0$ and a related change $e_a \to e'_a = \vartheta^{-1} s^c{}_a e_c$ of the frame vector fields tangent to $\bar{S}$. Here $s^c{}_a$ denotes an analytic function on $\bar{S}$ with values in $SO(3)$ such that $s^c{}_a \to \delta^c{}_a$ as $\bar{\rho} \to 0$. These changes will simply be propagated along the new conformal geodesics.

Critical is the transition from the congruence of conformal geodesics related to Ω (the Ω-*congruence*) to the new one related to Ω' (the Ω'-*congruence*). If the curves are considered as point sets, the two families of curves will be different if $\Omega'^{-1} d\Omega' - \Omega^{-1} d\Omega = \vartheta^{-1} d\vartheta \neq 0$ (cf. [40]).

The rescaling above implies on $\bar{S}$ the transitions

$$\|d\Omega\|_h \to \|d\Omega'\|_{h'} = \xi \, \|d\Omega\|_h,$$

$$\omega = \frac{2\Omega}{\|d\Omega\|_h} \to \omega' = \frac{2\Omega'}{\|d\Omega'\|'_h} = \frac{\omega \, \delta}{\xi},$$

$$\Theta|_{\bar{S}} = \omega^{-1} \Omega \to \Theta'|_{\bar{S}} = \omega'^{-1} \Omega' = \xi \, \Theta|_{\bar{S}},$$

with the function

$$\xi = \left| 1 - 3 \vartheta^{-1} \frac{D_a\Omega \, D^a\vartheta}{\Delta_h \Omega} - \frac{3}{2} \vartheta^{-2} \Omega \frac{D_a\vartheta \, D^a\vartheta}{\Delta_h \Omega} \right|^{\frac{1}{2}},$$

which extends to $\bar{S}$ as an analytic function of $\bar{\rho}$ and ψ^A.

It follows from the initial conditions for the Ω-congruence that

$$\xi^{-1}\dot{x} \perp \tilde{S}, \qquad \Theta'^2\,\tilde{g}(\xi^{-1}\dot{x},\xi^{-1}\dot{x}) = 1, \tag{7.13}$$

and for the transformed 1-form that

$$\langle f',\dot{x}\rangle = 0, \qquad f_{\tilde{s}} \to f'_{\tilde{s}} = \omega'^{-1}\,d\omega' = f_{\tilde{s}} + \vartheta^{-1}\,d\vartheta - \xi^{-1}\,d\xi,$$

where the subscripts indicate the pull-back to $\tilde{S}$. These two lines give the initial data for the Ω'-congruence if the conformal geodesic equations are expressed with respect to the rescaled metric g' and its connection ∇'.

To compare the Ω'-congruence with the Ω-congruence we observe the conformal invariance of conformal geodesics (cf. [38]) and express the equations for the Ω'-congruence in terms of g and its connection ∇. The space-time curves, including their parameter τ', then remain unchanged. The 1-form is transformed because of $g = (\Theta\,\Theta'^{-1})^2\,g'$ according to $f' \to f^* = f' - (\Theta\,\Theta'^{-1})^{-1}\,d\,(\Theta\,\Theta'^{-1})$, which implies $\langle f^*,\dot{x}\rangle = 0$, $f^*_{\tilde{s}} = \bar{f}_{\tilde{s}} + \vartheta^{-1}\,d\vartheta$ on $\tilde{S}$. If this 1-form is expressed in terms of the g-orthonormal frame e_k with $e_0 \perp \tilde{S}$, one finds

$$f^*_0 \equiv \langle f^*, e_0\rangle = 0, \qquad f^*_a \equiv \langle f^*, e_a\rangle = f_a + \vartheta^{-1}\langle d\vartheta, e_a\rangle \ \ a = 1,2,3. \tag{7.14}$$

The fields $\xi^{-1}\dot{x}$, f^*_k are the initial data for the Ω'-congruence in terms of g, e_k, and ∇. Since $\xi \to 1$ and $\langle d\vartheta, e_a\rangle = O(\bar{\rho})$ as $\bar{\rho} \to 0$, it follows that

$$\frac{\Theta'}{\Theta} \to 1, \qquad \dot{x} - \xi^{-1}\dot{x} \to 0, \qquad f^*_k - f'_k \to 0 \quad \text{as } \bar{\rho} \to 0.$$

As a consequence, the initial data for the Ω'- and the Ω-congruence have coinciding limits on $\mathcal{I}^{0'}$ and the corresponding curves are identical on $\bar{\mathcal{I}}'$.

Assuming now the conditions of Section 7.2 and using arguments similar to the ones used there, we conclude that in a certain neighborhood $\mathcal{O}'$ of $\bar{\mathcal{I}}'$ in $\mathcal{M}'$ the gauge related to the Ω'-congruence is as smooth and regular as the one related to the Ω-congruence. Thus we have

Lemma 7.4. *In the case of static asymptotically flat space-times the construction of the set $\bar{\mathcal{I}}'$ is independent of the choice of Ω and the set $\mathcal{I}'$ introduced in Section 6 coincides with the projection $\pi'(\mathcal{I})$ of the cylinder at space-like infinity as defined in Section 5.*

We note that the comparison of the Ω'- with the Ω-congruence leads in the case where the solution is not static and thus not necessarily analytic still to similar results if the solution acquires a certain smoothness near $\mathcal{J}^{\pm} \cup \mathcal{I}^{\pm}$. In the case of low smoothness, however, the detailed behavior of the different congruences needs to be analyzed in the context of an existence theorem.

8. Concluding remarks

Concerning the regularity conditions we have now the following situation. For *static* asymptotically flat solutions with $m \neq 0$ the conformal extensions to $\bar{\mathcal{I}}$ are smooth (in the sense discussed above) and their data satisfy the regularity condition (5.89) with $p_* = \infty$. In the *massless case* condition (5.89) with $p_* = \infty$ is necessary and sufficient for space-like infinity to be represented by a regular point in a smooth conformal extension. In the *general time reflection case with* $m \neq 0$ conditions (5.89) are necessary but not sufficient for the s-jets $J_{\mathcal{I}}^p(u)$, $p \in \mathbb{N}$, to be regular at the critical sets $\mathcal{I}^\pm$. Thus, the mass $m = 2W(i)$ and also the derivatives of $\partial_{x^a}^\alpha W(i)$, $\alpha \in \mathbb{N}^3$, play a crucial role for the behavior of the $J_{\mathcal{I}}^p(u)$ at $\mathcal{I}^\pm$. The mechanism which decides on the smoothness remains to be understood.

Only the d-jet $J_{\mathcal{I}^0}^p(u)$ and the s-jets $J_{\mathcal{I}}^{p-1}(u)$ are needed to obtain $J_{\mathcal{I}}^p(u)$ by integrating the transport equations on $\mathcal{I}$. Since the left-hand sides of the transport equations are universal in the sense that they do not depend on the data, it follows that $J_{\mathcal{I}}^p(u)$ is uniquely determined by $J_{\mathcal{I}^0}^p(u)$ for $p \in \mathbb{N}$. In the static case the s-jets $J_{\mathcal{I}}^p(u)$ are regular. In [16] has been exhibited a class of data which are asymptotically static of order p for given $p \in \mathbb{N} \cup \{\infty\}$ and which are essentially arbitrary on given compact sets. It follows that for prescribed differentiability order p there exists a large class of data for which the s-jet $J_{\mathcal{I}}^p(u)$ is regular on $\mathcal{I}$.

We expect there to be a threshold in p beyond which the regularity of $J_{\mathcal{I}}^p(u)$ ensures peeling resp. asymptotic smoothness of a given order of differentiability and below which the singularity of $J_{\mathcal{I}}^p(u)$ implies a failure of peeling. This order is likely to be low enough such that the behavior of $J_{\mathcal{I}}^q(u)$ with $q \leq p$ can be controlled by a direct, though tedious, calculation. However, if asymptotic staticity does play a role here, one should try to understand the underlying mechanism. It would be quite a remarkable feature of Einstein's equations if asymptotic staticity could be *deduced* from asymptotic regularity at null infinity.

References

[1] L. Andersson, P.T. Chruściel. On hyperboloidal Cauchy data for the vacuum Einstein equations and obstructions to the smoothness of scri. Comm. Math. Phys. 161 (1994) 533–568.

[2] L. Andersson, P.T. Chruściel. Solutions of the constraint equations in general relativity satisfying hyperboloidal boundary conditions. Dissertationes Mathematicae Polska Akademia Nauk, Inst. Matem., Warszawa, 1996.

[3] L. Andersson, P.T. Chruściel, H. Friedrich. On the regularity of solutions to the Yamabe equation and the existence of smooth hyperboloidal initial data for Einstein's field equations. Comm. Math. Phys. 149 (1992) 587–612.

[4] A. Ashtekar. Asymptotic properties of isolated systems: Recent developments. In: B. Bertotti et al. (eds.) *General Relativity and Gravitation* Dordrecht, Reidel, 1984.

[5] R.A. Bartnik, A.H. Norton. Numerical experiments at null infinity. In: J. Frauendie-ner, H. Friedrich (eds.), *The Conformal Structure of Spacetime: Geometry, Analysis, Numerics*. Springer, Berlin, 2002.

[6] R. Beig, W. Simon. Proof of a multipole conjecture due to Geroch. Comm. Math. Phys. 79 (1981) 581–589.

[7] R. Beig, B. Schmidt. Time-independent gravitational fields. In: B.G. Schmidt (ed.): Einstein's field equations and their physical implications. Springer, Berlin, 2000.

[8] A.L. Besse. Einstein Manifolds. Springer, Berlin, 1987.

[9] L. Blanchet. Post-Newtonian Gravitational Radiation. In: B. Schmidt (ed.), *Einstein's Field Equations and Their Physical Implications*. Berlin, Springer, 2000.

[10] H. Bondi, M.G.J. van der Burg, A.W.K. Metzner. Gravitational waves in general relativity VII. Waves from axi-symmetric isolated systems. Proc. Roy. Soc A 269 (1962) 21–52.

[11] Y. Choquet-Bruhat. Théorèmes d'existence pour certains systèmes d'équations aux dérivées partielles non linéaires. *Acta. Math.* 88 (1952) 141–225.

[12] M.W. Choptuik. Universality and scaling in gravitational collapse of a massless scalar field. *Phys. Rev. Lett.* 70 (1993) 9.

[13] D. Christodoulou. The Global Initial Value Problem in General Relativity. In: V.G. Gurzadyan et al. (eds.) Proceedings of the 9th Marcel Grossmann Meeting World Scientific, New Jersey, 2002.

[14] D. Christodoulou, S. Klainerman. The Global Nonlinear Stability of the Minkowski Space. Princeton University Press, Princeton, 1993.

[15] P.T. Chruściel, E. Delay. Existence of non-trivial, vacuum, asymptotically simple spacetimes. *Class. Quantum Grav.*, 19 (2002) L 71–L 79. Erratum *Class. Quantum Grav.*, 19 (2002) 3389.

[16] P.T. Chruściel, E. Delay. On mapping properties of the general relativistic constraints operator in weighted function spaces, with application. *Mém. Soc. Math. France* submitted. http://xxx.lanl.gov/abs/gr-qc/0301073

[17] P.T. Chruściel, M.A.H. MacCallum, D. B. Singleton. Gravitational Waves in General Relativity. XIV: Bondi Expansions and the "Polyhomogeneity" of *Scri. Phil. Trans. Royal Soc., London* A 350 (1995) 113–141.

[18] P.T. Chruściel, O. Lengard. Solutions of wave equations in the radiation regime. Preprint (2002). http://xxx.lanl.gov/archive/math.AP/0202015

[19] P.T. Chruściel, J. Jezierski, J. Kijowski. Hamiltonian Field Theory in the Radiating Regime. Springer, Berlin, 2002.

[20] J. Corvino. Scalar curvature deformation and a gluing construction for the Einstein constraint equations. Comm. Math. Phys. 214 (2000) 137–189.

[21] J. Corvino, R. Schoen On the Asymptotics for the Vacuum Einstein Constraint Equations. http://xxx.lanl.gov/abs/gr-qc/0301071

[22] R. Courant, D. Hilbert. Methods of mathematical physics, Vol II. J. Wiley, New York, 1962.

[23] S. Dain. Initial data for stationary space-times near space-like infinity. *Class. Quantum Grav.*, 18 (2001) 4329–4338.

[24] S. Dain. Asymptotically Flat Initial Data with Prescribed Regularity II. In preparation

[25] S. Dain, H. Friedrich. Asymptotically Flat Initial Data with Prescribed Regularity. Comm. Math. Phys. 222 (2001) 569–609.

[26] Ellis, G.F.R. (1984) Relativistic Cosmology: Its Nature, Aims, and Problems, *General Relativity and Gravitation*, B. Bertotti et al. (eds.), Reidel, Dordrecht.

[27] J. Frauendiener. Some aspects of the numerical treatment of the conformal field equations. In: J. Frauendiener, H. Friedrich (eds.), *The Conformal Structure of Spacetime: Geometry, Analysis, Numerics*. Springer, Berlin, 2002.

[28] H. Friedrich. On the regular and the asymptotic characteristic initial value problem for Einstein's vacuum field equations. Proceedings of the 3rd Gregynog Relativity Workshop on Gravitational Radiation Theory MPI-PEA/Astro 204 (1979) 137–160 and Proc. Roy. Soc., 375 (1981) 169–184.

[29] H. Friedrich. The asymptotic characteristic initial value problem for Einstein's vacuum field equations as an initial value problem for a first-order quasilinear symmetric hyperbolic system. Proc. Roy. Soc., A 378 (1981) 401–421.

[30] H. Friedrich. On the Existence of Analytic Null Asymptotically Flat Solutions of Einstein's Vacuum Field Equations. Proc. Roy. Soc. Lond. A 381 (1982) 361–371.

[31] H. Friedrich. Cauchy Problems for the Conformal Vacuum Field Equations in General Relativity. Comm. Math. Phys. 91 (1983) 445–472.

[32] H. Friedrich. On the hyperbolicity of Einstein's and other gauge field equations. Comm. Math. Phys. 100 (1985) 525–543.

[33] H. Friedrich. On the existence of n-geodesically complete or future complete solutions of Einstein's equations with smooth asymptotic structure. Comm. Math. Phys. 107 (1986) 587–609.

[34] H. Friedrich. On static and radiative space-times. Comm. Math. Phys., 119 (1988) 51–73.

[35] H. Friedrich. On the global existence and the asymptotic behaviour of solutions to the Einstein-Maxwell-Yang-Mills equations. *J. Diff. Geom.*, 34 (1991) 275–345.

[36] H. Friedrich. Einstein equations and conformal structure: existence of anti-de Sitter-type space-times. J. Geom. Phys., 17 (1995) 125–184.

[37] H. Friedrich. Gravitational fields near space-like and null infinity. J. Geom. Phys. 24 (1998) 83–163.

[38] H. Friedrich. Conformal Einstein Evolution In: J. Frauendiener, H. Friedrich (eds.), *The Conformal Structure of Spacetime: Geometry, Analysis, Numerics*. Springer, Berlin, 2002.

[39] H. Friedrich. Spin-2 fields on Minkowski space near spacelike and null infinity. *Class. Quantum. Grav.* 20 (2003) 101–117.

[40] H. Friedrich. Conformal geodesics on vacuum space-times. *Commun. Math. Phys.* 235 (2003) 513–543.

[41] H. Friedrich, J. Kánnár Bondi systems near space-like infinity and the calculation of the NP-constants. J. Math. Phys. 41, (2000), 2195–2232.

[42] H. Friedrich, G. Nagy. The initial boundary value problem for Einstein's vacuum field equations. Comm. Math. Phys. 201 (1999) 619–655.

[43] H. Friedrich, A. Rendall. The Cauchy Problem for the Einstein Equations. In: B. Schmidt (ed.), *Einstein's Field Equations and Their Physical Implications*. Berlin, Springer, 2000.

[44] K.O. Friedrichs. Symmetric hyperbolic linear differential equations. Comm. Pure Appl. Math. 7 (1954) 345–392.

[45] P. Garabedian. *Partial Differential Equations*. J. Wiley, New York, 1964.

[46] R. Geroch. Asymptotic structure of space-time. In: F.P. Esposito, L. Witten (eds.) *Asymptotic Structure of Space-Time*. New York, Plenum, 1977.

[47] R.P. Geroch, G. Horowitz. Asymptotically simple does not imply asymptotically Minkowskian. Phys. Rev. Lett. 40 (1978) 203–206.

[48] O. Guès. Problème mixte hyperbolique quasi-linéaire charactéristique. *Commun. Part. Diff. Equ.*, 15 (1990) 595–645.

[49] C. Gundlach. Critical Phenomena in gravitational collapse. *Physica Reports* 367 (2003) 339–405.

[50] S. Helgason. *Differential Geometry and Symmetric Spaces*. Academic Press, New York, 1962.

[51] L. Hörmander. The Analysis of Linear Partial Differential Operators III. Springer, Berlin, 1985.

[52] P. Hübner. From now to timelike infinity on a finite grid. Class. Quantum Grav. 18 (2001) 1871–1884.

[53] S. Husa. Problems and successes in the numerical approach to the conformal field equations. In: J. Frauendiener, H. Friedrich (eds.), *The Conformal Structure of Space-time: Geometry, Analysis, Numerics*. Springer, Berlin, 2002.

[54] D. Kennefick, N. O'Murchadha. Weakly decaying asymptotically flat static and stationary solutions to the Einstein equations. Class. Quantum. Grav. 12 (1995) 149–158.

[55] S. Klainerman, F. Nicolò. The Evolution Problem in General Relativity. Birkhäuser, Basel, 2003.

[56] S. Klainerman, F. Nicolò. Peeling properties of asymptotically flat solutions to the Einstein vacuum equations. *Class. Quantum Grav.* 20 (2003) 3215–3257.

[57] O. Lengard. Solution of the Einstein equation, wave maps, and semilinear waves in the radiation regime. Ph. D. thesis, Université de Tours, 2001. http://www/phys.uni-tours.fr/ piotr/papers/batz

[58] E.T. Newman, R. Penrose. *An approach to gravitational radiation by a methods of spin coefficients*. J. Math. Phys. 3 (1962) 566–578.

[59] R. Penrose. Asymptotic properties of fields and space-time. Phys. Rev. Lett., 10 (1963) 66–68.

[60] R. Penrose. Zero rest-mass fields including gravitation: asymptotic behavior. Proc. Roy. Soc. Lond., A 284 (1965) 159–203.

[61] R. Penrose, W. Rindler. *Spinors and space-time*, Vol. 1 and 2. Cambridge University Press, 1984.

[62] F.A.E. Pirani. Invariant Formulation of Gravitational Radiation Theory. *Phys. Rev.* 105 (1957) 1089–1099.

[63] R.K. Sachs. Gravitational waves in general relativity VI. The outgoing radiation condition. Proc. Roy. Soc A 264 (1961) 309–338.

[64] R.K. Sachs. Gravitational waves in general relativity VIII. Waves in asymptotically flat space-time. Proc. Roy. Soc A 270 (1962) 103–126.

[65] R. Schoen, S.-T. Yau. Proof of the positive mass theorem II. *Commun. Math. Phys.*, 79 (1981) 231–260.

[66] B.F. Schutz. Mathematical and Physical Perspectives on Gravitational Radiation. Talk given at the summer school "50 years of the Cauchy problem in general relativity", Cargèse July 29–August 10, 2002. http://fanfreluche.math.univ-tours.fr

[67] P. Secchi. Well-Posedness of Characteristic Symmetric Hyperbolic Systems. *Arch. Rational Mech. Anal.*, 134 (1996) 155–197.

[68] G. Szegö. Orthogonal Polynomials. 4th Edition. A.M.S. Colloqu. Publ. Vol. 23, Providence, 1978.

[69] A. Trautman. Radiation and boundary conditions in the theory of gravitation. *Bull. Acad. Pol. Sci., Série sci. math., astr. et phys.* VI (1958) 407–412.

[70] J.A. Valiente Kroon. Can one detect a non-smooth null infinity? *Class. Quantum Grav.* 18 (2001) 4311–4316.

[71] J.A. Valiente Kroon. A new class of obstructions to the smoothness of null infinity. *Commun. Math. Phys.* 244 (2004) 133–156.

[72] J.A. Valiente Kroon. Does asymptotic simplicity allow for radiation near spatial infinity? *Commun. Math. Phys.* (2004) to appear.

[73] J. Winicour. Logarithmic Asymptotic Flatness. *Foundations of Physics* 15 (1985) 605–616.

Helmut Friedrich
Max-Planck-Institut für Gravitationsphysik
Am Mühlenberg 1
D-14476 Golm, Germany
e-mail: hef@aei-potsdam.mpg.de

Status Quo and Open Problems in the Numerical Construction of Spacetimes

Luis Lehner and Oscar Reula

Abstract. The possibility of explicitly constructing solutions to the Einstein equations via computer simulations represents a relatively new, powerful approach to investigate the theory. Numerical relativity is the branch of General Relativity whose goal is to obtain such solutions. Unfortunately, despite several remarkable successes, the field has not yet reached the mature level required to successfully tackle a number of interesting problems. Several open issues slow down the progress in the field and their resolution will be key in realizing its full possibilities. This article presents a brief introduction to the discipline, placing special emphasis on some of the main open conceptual problems.

1. Introduction

Computer simulations are revolutionizing science in profound terms as they, in principle, allow one to treat problems which are too difficult to solve analytically. Einstein's theory is described by arguably one of the most complicated partial differential equations systems, and this has prevented researchers from unravelling the full implications of this beautiful theory. Undoubtedly, considerable advances have been obtained in several regimes based on purely analytical considerations. However, these require different non-generic assumptions and it is safe to say that there are a number of issues we can not yet, at the analytical level, begin to understand; numerical simulations can open the door to these issues. Furthermore, there is no telling what surprises might be hidden in the intricacies of non linear behavior of the solutions. Perhaps the prime example of this is the discovery of *Critical Phenomena in General Relativity* by Choptuik[16] where simulations uncovered a rich and unexpected phenomena at the threshold of black hole formation. Not only does the mass of the (arbitrarily small) black hole respect a universal scaling relation, the spacetime also exhibits a remarkable self-similar behavior. Furthermore, the numerical results provided enough insight in the solution structure to enable analytical analysis to be carried out, which explained this behavior in terms of dynamical systems.

The promise of unravelling what the theory has so far kept hidden is certainly more than enough incentive to pursue the numerical construction of spacetimes. Unfortunately, this is far from a straightforward enterprise as there are a number of obstacles in our path. The construction of a robust numerical implementation can be paralleled to building a 'three-legged table', as there are three crucial ingredients that need to be addressed for a successful simulation. First, there is an "analytical" leg where the properties of the system must be understood from an analytical point of view; in particular, the needed ingredients to present the problem in a well-posed way. Second, a "numerical" leg where the search for a convenient *discrete* approximation to the problem must be addressed, to produce a *stable* simulation which can reproduce fairly the systems of interest. Finally, the third leg refers to having sufficient computational resources (and their efficient usage) for a successful implementation, which might condition the number of available options. Unfortunately all three legs presently suffer from shortcomings.

In the present article we review the status of mainly the first two, presenting in each case the latest advances which address some previously unresolved issues while commenting on those which are, for the most part, open.

1.1. Setting the stage

The present article intends to be a contribution for readers interested in learning about the issues involved in the numerical construction of spacetimes. Special emphasis is placed in highlighting several new techniques which have not yet appeared in a review of the subject (see for instance [36, 4]) and so should be taken as complementary to these. Additionally, it hopes to reach out to the more mathematically oriented community and point out a number of conceptually open issues whose solutions will mean a significant contribution to the goal of achieving robust implementation of Einstein equations.

In an oversimplified picture, one could state that most $3D$ simulations of Einstein equations display instabilities when pushed hard enough (i.e., in large curvature regions, in long simulations, etc.). These instabilities cause the implementation to crash after some finite time; this sometimes being too short to extract sensible physical information from the phenomena under consideration. Unfortunately, it is hard to find the culprit of these instabilities as several factors are at play and interact with each other. Among these factors one finds: a bad choice of evolution system, use of unstable boundary conditions, an inconvenient foliation of the spacetime, constraint violations which are not kept under control, the presence of artificial shocks, an unfortunate choice of numerical algorithm, etc.

The discrimination of what causes the observed instabilities is central for a healthier development of the subject. The initial step, of course, is to start with a system of equations with a well-understood initial boundary value problem. To date we have many systems with a well-known initial value problem but only one with fully understood initial boundary value problem (some others are beginning to be grasped). The issues involved in assessing the well-posedness of an initial boundary value problem are involved enough that, to date, no 3D simulations (in the Cauchy approach) has been performed with the boundary problem well understood. While this problem is being studied one can address other possible sources of instabilities like numerical ones. For instance, the use of numerical methods which are proven stable for the linearized equations (off arbitrary backgrounds) sets the stage for diminishing the sources of numerical instabilities. Subsequently one is faced with a choice of gauge or coordinate conditions. This inescapable issue can by itself make the difference between having a chance to evolve the system under consideration or not. Ideally one would like to choose coordinates so as to simplify the dynamics of the variables which are compatible with well-posedness. Last one might need to worry about constraint violations. It these are grossly violated, one is far away from a physical solution and thus no physical argument can be used to estimate its behavior.

These different but related issues are discussed throughout the article paying particular attention on present strategies to address them and discuss their main open issues. To set the stage we mention here the main goals,

- *To have sharp growth estimates for Einstein equations in the Cauchy approach.* As discussed later, this can be exploited in the numerical implementation and be useful to discern between solutions.
- *To understand the constraint behavior off the mass shell and whether it can be controlled with a different reformulation of the equations, or by interchanging (some of) the evolution equations by constraints.* In generic simulations the constraints are seen to diverge exponentially. Since initial data is never in the mass shell (as they contain round-off errors at least), a bad behavior off the constraint surface is a bad sign for a numerical simulations.
- *To deal with the initial boundary value problem for Einstein's equations in a well-posed way. Furthermore, to be able to prescribe physically motivated boundary conditions; for example no incoming radiation through the boundaries.* In most cases, to deal with a finite computational domain timelike boundaries are introduced. These are certainly artificial and its spurious influence in the numerical solution should be minimized.
- *To find a viable way to handle the initial data problem for the conformal equations (beyond them being satisfied in the physical spacetime).*
- *To assess the well-posedness of the characteristic initial value problem of General Relativity.*

In what follows we comment on reasons for considering these goals, some partial answers and their role in the numerical efforts.

2. Einstein's equations in the computer. "Formal considerations"

Einstein's equations in their basic form, $G_{ab} = 8\pi T_{ab}$, are not well adapted for a numerical treatment, for it requires setting up an initial value problem (IVP). The work of Choquet-Bruhat half a century ago provided a Cauchy-type formulation. This became the launching platform for a large number of reformulations of Einstein equations, providing different expressions for the IVP. At the basic level, an IVP requires that a notion of 'time' evolution be defined which is done through a particular foliation of the spacetime $(\mathcal{M}, g_{ab})$. The character of this foliation defines the type of IVP under consideration. For instance, with spacelike foliations a *Cauchy* IVP is obtained, while null foliations give a *characteristic* IVP. Naturally one can define more generic foliations, but the prevalent options in the numerical implementations are the two mentioned.

There is another possibility to be considered when attempting to numerically construct a spacetime. Namely, one can opt to solve for a conformally related spacetime $(\tilde{\mathcal{M}}, \Omega^2 g_{ab})$ first, where $\mathcal{M} \subset \tilde{\mathcal{M}}$. Then, the spacetime of interest is recovered a posteriori.

(Note: throughout this work we will restrict to vacuum scenarios.)

1. Cauchy approach

In this approach, a foliation of the spacetime is introduced whose leaves (parametrized by t) are spacelike. Each hypersurface is endowed with an Euclidean metric γ_{ij} and its embedding in $\mathcal{M}$ is described by the second fundamental form K_{ij} ($i = 1 \ldots 3$). The normal to each hypersurface can be used to relate the time difference between the freely specifiable spacelike hypersurfaces. This freedom is encoded in the *lapse function*, α, defined so as to give the proper time along the normal direction. Additionally, points on each hypersurface are given labels x^i. Since points' labels need not be equally assigned on each hypersurface, three more functions, encoded in the so-called *shift vector* β^i are used to 'measure' the shift coordinates at a given point on the hypersurface Σ_t, have with respect to those in Σ_{t+dt}. With this convention, the metric of the spacetime can be written as:

$$ds^2 = -\alpha^2 dt^2 + \gamma_{ij}(dx^i + \beta^i dt)(dx^j + \beta^j dt). \tag{1}$$

Before commenting on the equations themselves, let us mention an important point. A convenient choice of $\{\alpha, \beta\}$ can greatly simplify the description of the spacetime, and exploiting this freedom has proved extremely valuable in the analytical arena. In the numerical one, this choice is crucial to simplify the numerical treatment (and even make it possible!). Unfortunately, little is known on how to exploit this freedom for numerical purposes (aside from some cases [1, 27, 7, 28, 60]). The prescription of appropriate coordinates for generic simulations is quite an open problem.

A common way of writing Einstein equations is obtained from the different projections of the equations $R_{ab} - 8\pi(T_{ab} - 1/2 g_{ab}T) = 0$ (R_{ab} being the Ricci tensor of g_{ab}). These projections yield equations that are quite different in nature. Namely, the full projections onto Σ_t provide "essentially" hyperbolic equations for K_{ij}, while the definition of K_{ij} in terms of derivatives normal to the foliation gives an evolution equation for γ_{ij}. These equations are called, in the numerical relativity community, the ADM evolution equations (in a slight abuse of notation as K_{ij} is not the canonical conjugate to γ_{ij}):

$$
\begin{aligned}
(\partial_t - \mathcal{L}_\beta)\gamma_{ij} &= 2\alpha K_{ij}\,; \\
(\partial_t - \mathcal{L}_\beta)K_{ij} &= \nabla_i\nabla_j\alpha + \alpha(R_{ij} + KK_{ij} - 2K_{im}K_j^m)\,.
\end{aligned} \tag{2}
$$

(with $\mathcal{L}_\beta$ the Lie derivative along β^i). The equations resulting from the projection orthogonal to the hypersurface (either one or both indices) contain neither derivatives with respect to t nor do they involve α or β^i. These latter equations are indeed the Gauss-Codacci equations describing the embedding of the three-dimensional hypersurface onto the higher-dimensional one:

$$
\begin{aligned}
{}^3R + K^2 - K_{ij}K^{ij} &= 0\,, \\
\nabla_j K^{ij} - \gamma^{ij}\nabla_j K &= 0
\end{aligned} \tag{3}
$$

(with ${}^3R$ the Ricci scalar of γ_{ij}). Note that this system of equations, called the 'constraint system' is unique (up to linear combinations), while the evolution system is not, as one can arbitrarily add the constraints to them to obtain different looking evolution equations [25, 49]. A solution to the first system would also be a solution to the second system, as long as the constraint equations are also satisfied. Thus all these different systems of evolution equations are equivalent "on the mass shell", i.e., on the constraint submanifold of the manifold of solutions to the evolution equations[1]. Outside it, the evolution equation systems can be very different; some might have too many solutions and some too few. Additionally, these solutions may or may not depend continuously on the initial data given. As we will discuss later, these differences can have a strong impact on a numerical implementation.

This fact has been recognized in the last decade and considerable effort has been made in finding **"good"** evolution systems. By good evolution systems we refer to those whose (unique) solutions exhibit continuous dependence on initial data. Note that this is a necessary condition to have robust numerical implementations; however, as we shall discuss later, it is not sufficient to guarantee it. Systems with these properties are called strongly hyperbolic, and they are characterized by certain **algebraic properties** of the equations. To be more explicit, consider a system of equations in a space-time $(\mathcal{M}, g_{ab})$ of the form:

$$
M^a_{\alpha\beta}\nabla_a\phi^\beta = j_\alpha.
$$

where the Greek index of ϕ^β is used to denote a set of tensor fields in the space-time[2], and $M^a_{\alpha\beta}$, j_α are tensor fields in space-time. The latter might also depend on ϕ^β (but not on their derivatives) in which case we say that the system is quasi-linear. If the tensor field $M^a_{\alpha\beta}$ is symmetric in the Greek indices, and if for some n_a, $M^a_{\alpha\beta}n_a$ is positive definite we say that the system is symmetric hyperbolic. Causal

---

[1] In fact this bigger subspace is not unique, it depends on the form of the evolution equations one is choosing.

[2] For a more general setting see the article of Geroch in this volume.

systems, in the sense normally used in general relativity, requires furthermore that positivity holds for all n_a that are time-like.

For the IVP it is well known that symmetric hyperbolicity implies the existence of a norm in ϕ^α space such that given any solution of the above system, ϕ^α, with initial data $ID := \{\phi_0^\alpha\}$, and any time T for which ϕ^α exists, then there is a neighborhood of ID such that any solution with data in that neighborhood is bounded by its initial data during the whole time interval; i.e.,

$$\|\phi_T^\alpha\| \leq C(T)\|\phi_0^\alpha\|.$$

It is clear then that the solution depends continuously on the initial data under this norm. Many evolution systems in general relativity are symmetric hyperbolic, or more generally, strongly hyperbolic (a generalization of symmetric hyperbolicity, see for instance [49] for the convention commonly followed in Numerical Relativity). However many are not. It should be noted that it is in general not necessary to turn the evolution system into an explicit first order system to study its well-posedness properties, more general tools to explore this have been developed and applied to study directly well-posedness of modifications of the ADM system, and those called BSSN, see for instance [35, 43].

Considerable efforts have been made to classify the possible reformulations of Einstein equations into a single class motivated by the following questions: *Among the class of all evolution systems, can one find all those with good properties?* or in algebraic terms, *Among the class of all evolution systems, which ones are strongly, or better yet, symmetric hyperbolic?*. To date we know of a large number of different systems with good evolution properties (even after considering two evolution systems equivalent if they differ by a field transformation). Yet, these families are likely not exhaustive and thus we still do not have a complete answer to these questions. In particular, we do not yet have full control over the freedom to construct possible systems and how to exploit the gauge freedom to our advantage (see, for instance, the review article [25]).

II. Characteristic Approach

In this approach [6, 50], the foliation is composed of null hypersurfaces, labeled in this case say by u. These hypersurfaces emanate from a central timelike or null worldtube (or geodesic). By choosing x^A to label the null rays on each hypersurface, the (degenerate) metric on the initial null hypersurface $\mathcal{N}_{u=u_o}$ is given by h_{AB}. To complete the metric for the full spacetime, we assign a coordinate r that gives some notion of distance on each ray (here we choose it to be areal). The metric can then be written as:

$$ds^2 = -e^{2\beta}(V/r)du^2 + r^2 h_{AB}(dx^A - U^A du)(dx^B - U^B du) - 2e^{2\beta}dudr. \quad (4)$$

Immediate inspection of the above line element at an $r = const$ surface provides an analogy of combinations of β, V and U^A with the lapse and shift functions previously discussed (see [5]). However, in this case the equations themselves restrict these functions. Namely, components of the projection of R_{ab} give two 'evolution equations' for h_{AB}, while the four remaining ones are essentially ODE's equations for V, β and U^A. This reflects the fact that the coordinate freedom in this approach is more restricted to remain consistent with a null foliation. Einstein's equations, in this formulation, have a radically different nature from that found in the Cauchy case. Basically they reduce to

$$
\begin{aligned}
(h_{AB})_{,ur} &= F_{AB}(h_{CD}, V, \beta, U^E), \\
\beta_{,r} &= F_\beta(h_{CD}), \\
U^A_{,rr} &= F^A_U(\beta, h_{CD}), \\
V_{,r} &= F_V(U^C, \beta, h_{CD});
\end{aligned}
\tag{5}
$$

where the $F's$ denote functions of the specified variables and some of their derivatives in the spatial directions (for the full expressions see [5]). These equations are obtained from the completely orthogonal projection to the direction $n_a = \nabla_a u$. The remaining ones, having at least one projected component onto n_a, are preserved by the above equations if they are satisfied at a given $r = const$ surface. Note that contrary to the Cauchy case, the equations for the intrinsic metric to the hypersurface are explicitly first order in time. Another important difference is that data on a single null hypersurface does not suffice to determine the solution at a point p to the future of it, irrespective of how close p is to the surface. The reason for this is that the domain of dependence of p will be finite only if data is also given at another surface: timelike, null or spacelike intersecting the initial one. In the first two cases one is forced to consider the proper setting of initial and boundary data to asses well-posedness of a given problem. Unfortunately, the theory for such a task is considerably less developed than in the Cauchy case, especially when the other surface is timelike. Indeed, only recently has the well-posedness of the timelike-null version of Maxwell's [3] equations been established. Despite the lack of stronger rigorous mathematical background, numerical implementations within the characteristic framework have shown remarkable robustness. This behavior can be explained in part by the fact that at least one of the boundaries (the one placed at future null infinity) is a causal boundary, and signals reaching it abandon the computational domain. An unfortunate shortcoming of this approach is that caustics render the coordinates singular, as the coordinates are tied to null rays. Hence, a straightforward application of this approach can only be performed in caustic/crossover free scenarios. More generic cases require considerable further care in the application.

Numerical simulations are, for the most part, restricted to the no caustic case though a few proposals have been presented to go around this difficulty[26, 37].

Note. The equations corresponding to the **conformal approach to Einstein's equations** are discussed elsewhere in this volume by Friedrich. In this article we will make some brief remarks of this approach, especially when further developments in either the standard Cauchy or characteristic approaches will benefit the conformal one as well.

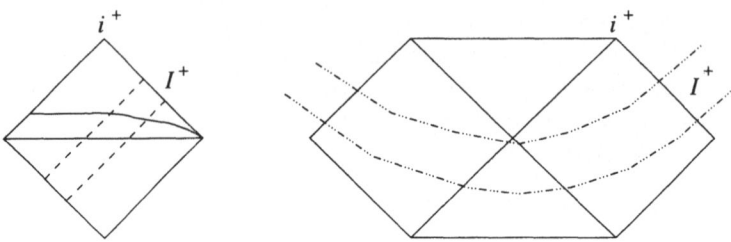

FIGURE 1. Schematic representation of the three approaches being considered in numerical relativity. In the left figure, the "physical" sector of Schwarzschild's spacetime is presented showing two leaves of possible foliations corresponding to the standard Cauchy approach (solid line) and characteristic approaches (dashed lines). In the right figure we show two leaves of a possible foliation in the conformal (Cauchy-type) approach to Einstein equations.

2.1. Initial data

In the **standard Cauchy approach** to GR, consistent initial data must satisfy the constraint equations[3]. These equations are 'essentially' elliptic in nature and provide a way to solve for four variables if the remaining eight are somehow defined. A problem that arises here as in what variables to solve for (as there certainly are no preferred ones!) One way to approach this problem, which at the same time renders the system of equations explicitly elliptic, was presented by Lichnerowicz[38] and later extended by York[59]. Basically one gives two symmetric tensors on the initial surface, $\{\tilde{\gamma}_{ij} \; \tilde{A}_{ij}\}$ and then express $\gamma_{ij} = \Psi^4 \tilde{\gamma}_{ij}$ and $K_{ij} = F(\Psi)(\tilde{A}_{ij}) + G(\Psi)(\tilde{\gamma}_{ij} K) + 2\tilde{\nabla}_{(i} X_{j)} - 2/3 \tilde{\gamma}_{ij} \tilde{\nabla}^i X^j$ (with $\tilde{\nabla}_a$ the covariant derivative compatible $\tilde{\gamma}_{ij}$ and $K = \gamma^{ij} K_{ij}$). Thus, replacing γ_{ij} and K_{ij} in the constraints with these expressions, one obtains (after appropriate cancellations) elliptic equations for Ψ and the vector X^i. The well-posedness of this problem has been the subject of intense research for more than half a century (see the article of Bartnik and Isenberg in this volume). Under many circumstances this problem can be solved if appropriate boundary conditions are provided[4].

Among the interesting scenarios one would like to consider are black hole spacetimes. This poses the interesting challenge of dealing with the singularity present inside each black hole. There are, roughly, two alternatives to deal with this issue. The first one relies on knowing the analytical behavior of the variables at the singularity, and 'extracting' it via a 'regularization' procedure[9]. The second one, which is more generic, reduces to defining an inner boundary either at or inside the black hole (whose location is estimated by the apparent horizon or other trapped

---

[3]This is not the case in the characteristic approach as the initial hypersurface is itself characteristic.

[4]There is a rather complete theory about the solvability of this system, but the discussion of this exceeds the scope of this work.

surfaces). Here, appropriate boundary data is obtained by identifications with other isometric space-times, or by a suitable ansatz read-off from single black hole analytical solutions[40] (Note, in the latter case no rigorous result exists supporting this strategy, rather the justification of this method has so far relied on being able to solve the equations numerically.) Boundary data at large distances (outer boundary) are traditionally defined by certain boundary conditions prescribing specific fall-off conditions for the field based on asymptotically flat assumptions[20, 18, 42, 21]. An alternate (but complementary) way has been presented recently where the construction of initial data can be done such that outside some large radius the solution can be made to agree exactly with that of a Kerr black hole[19]. This approach has not yet been implemented numerically.

The generation of initial data in the Cauchy approach to the **conformal equations** is considerably less developed. There are necessarily more equations that consistent initial data must satisfy (as there are more variables). Unfortunately, beyond one-dimensional scenarios, it is not known how to present these equations in a way where standard numerical techniques can directly be applied to solve them. An alternative, indirect, way to obtain consistent initial data was presented in[2], where the constraint equations are solved restricted to the physical spacetime and then extended to the unphysical one (in that case solutions for some of the fields automatically give solutions for the rest of the system). An unfortunate feature of this construction, particularly from the numerical point of view, is that the equations exhibit 0/0 behavior at future null infinity which is delicate to deal with[23]. Aside from this technical difficulty, the resulting equations are quite similar to those solved by well-known numerical techniques in the standard Cauchy approach.

Consistent initial data for the **characteristic formulation** is straightforward to provide. The initial hypersurface is characteristic and initial data corresponds to the (freely specifiable) intrinsic metric on the whole hypersurface; the rest of the metric functions are obtained via radial integrations (if their values are given at either the origin or the intersection of the initial hypersurfaces).

Irrespective of the above considerations, the main open question when posing initial data is how it conforms to the physical situation in mind. Here, the problem lies in the difficulty of controlling spurious amounts of gravitational radiation (which can be present even in vacuum scenarios). There is a fundamental obstacle in trying to do this since at a given point, interior to the spacetime, it is generically impossible to distinguish between incoming and outgoing radiation. This can only be done for spacetimes which are asymptotically flat at future null infinity. Such setting is used by some authors as the definition of a situation with purely outgoing radiation. Naturally, this is impossible to asses in a Cauchy approach (in the standard picture) where hypersurfaces never reach $\mathcal{I}^+$. In the characteristic and conformal approaches, although it is at least possible to calculate the amount of outgoing gravitational radiation on the initial hypersurface, one cannot estimate the incoming amount. Therefore the task is left to the simulations to recognize this via comparison of the evolution of different initial data.

2.2. Particulars: Outer Boundaries

The simplest boundary treatment can be found in the **characteristic formulation of GR** as the outer boundary is usually set to correspond to $\mathcal{I}^+$. This boundary is indeed the causal boundary of the initial data, therefore one is restricting the simulation to the domain of dependence of the initial hypersurface and so it should be a well-posed problem. Furthermore, the interpretation of the solutions obtained is simplified by the fact that at $\mathcal{I}^+$ gravitational radiation can be given a rigorous physical meaning by constructing the Bondi news function.

In the case of the **conformal equations** the physically relevant region of space time is inside the domain of dependence of the initial surface; hence, in principle, no boundary condition is needed to evolve the fields in just the physically relevant region. Unfortunately, as seen from the unphysical spacetime, the problem has boundaries; therefore one also deals with an initial boundary value problem. However, the outer boundary conditions in this case need not be physically motivated as they cannot influence the solution in the physical spacetime (which is 'shielded' by $\mathcal{I}^+$).

The picture in the **standard Cauchy formulation**, i.e., that restricted to the physical spacetime, is considerably more complicated. In the standard approach, all hypersurfaces meet at i^o where the equations become singular. This reason, coupled to the (always) limited computational resources, generically prompts the introduction of an outer boundary which is usually chosen to be timelike so as not to loose resolution and be able to evolve for long times. Therefore, one is forced to deal with an initial-boundary value problem.[5]

Hence, implementations of the standard Cauchy approach must handle these boundaries appropriately (i.e., in a well-posed way) and for the standard (non-conformal) approach boundary data must be physically relevant. Unfortunately, the well-posed handling of the boundaries in hyperbolic systems is still under development, both analytic and numerically, and for the case of general relativity is very much in its infancy. Even in a linear theory specification of boundary values is a complicated task. No analytical theory exists which covers the general relativistic case, except for the case of the Friedrich-Nagy frame system, where such a specification has been found and proven to be well posed[24]. For the type of systems most used in numerical computations the problem is very much open.

For definiteness consider a boundary value problem for linear symmetric hyperbolic systems with smooth coefficients.

$$M^a_{\alpha\beta}\nabla_a\phi^\beta = j_\alpha.$$

Let T be a time-like hyper-surface, with unit normal (outer) n_a, and let t^a be a time like unit vector tangent to T. Defining $A^\alpha{}_\beta := A^{a\alpha}{}_\beta n_a = -(M^b t_b{}^{-1})^{\alpha\gamma} M^a_{\gamma\beta} n_a$ we can compute the eigenvectors and eigenvalues of this linear map; since $M^a_{\alpha\beta}$ is

---

[5]In the conformal approach it is possible, in principle, to extend the integration region beyond the physically relevant region and modify the data (in a way which would no longer satisfy the constraints outside that domain) in such a way as to consider a periodic problem, for which the mathematical theory is much better known.

symmetric and $M^a_{\alpha\beta}t_a$ positive definite it has a complete set of eigenvectors and its eigenvalues are all real. Eigenvectors corresponding to positive eigenvalues are called incoming modes, since small perturbations with values in these modes propagate from the boundary towards the inside of the integration region. Eigenvectors corresponding to negative eigenvalues are called outgoing modes for their linear perturbation travel away from the integration region. Eigenvectors corresponding to zero eigenvalues are called zero modes, and they propagate tangent to the boundary. To prescribe values to the outgoing or zero modes at boundary points is clearly physically incorrect, and therefore also mathematically. For we know the values those fields take there are already determined from the interior (or boundary) propagation. On the other hand, incoming modes can be specified arbitrarily as if they were initial data. They can even be given as functions depending on the outgoing fields, for example to have a mirror type boundary condition. In this generality the existence theory for solutions to the boundary value problem is only for weak solutions, but it is simple. At each point on the boundary one can choose any subspace H^+ with the following two properties: i) For any vector $\psi^\alpha \in H^+$, $M^a_{\gamma\beta}n_a\psi^\alpha\psi^\beta \geq 0$, and ii) The dimension of H^+ is maximal; that is, there is no larger subspace with the same property.

Thus we have a distribution of subspaces at each point of T. We shall assume that such a distribution is piece-wise smooth. The choice of the subspaces H^+ corresponds, in the usual language to a choice of matrix S in a boundary condition of the form $V^+ - SV^- = 0$ where V^+ (V^-) represent the vector of coefficients of the positive (negative) eigenvalues of the eigenvector decomposition of $\phi|_T$. (Note that there is no-condition on the zero eigenvalue eigenvector coefficients.) One can show that [10] given any smooth field ϕ^α_0 at the boundary there exists a unique solution $\phi^\alpha \in L^2$ such that $\phi^\alpha|_T - \phi^\alpha_0 \in H^+$. (The proof of this statement is a generalization of the one given in [33] using Riesz's representation theorem.)

This theorem can be extended to the case of an initial boundary value problem. In this case, as n_a points downwards at the initial surface, $M^a_{\gamma\beta}n_a$ is negative definite and so H^+ contains just the zero element, thus $\phi^\alpha|_T - \phi^\alpha_0 = 0$ there and the solution at the initial surface coincides with the data given. At the final time, on the other hand, H^+ is just the whole space and so the above condition is empty; thus, no data can be prescribed there.

If no zero modes are present (in the literature boundaries allowing zero modes are called "characteristic") then one can also estimate the L^2 norms of all derivatives and so, using Sobolev embedding Lemma obtain classical regularity estimates. Furthermore this allows generalize some results to the non-linear case, where the above assertion on the linear problem is used in an iterative procedure. The presence of zero modes prevents estimating normal derivatives at boundary points and the theory beyond L^2 becomes cumbersome, in particular for nonlinear equations [29].

It should be mentioned that most physical systems do have zero modes and so their boundary value theory is not simple. Fortunately those theories also have constraints, and it is the information contained in their constraints which allows one to complete the normal derivative estimates. Unfortunately, a general enough theory about these cases is not at hand.

Note that constraints also make, in many cases, the problem considerably more difficult; this is particularly true in the case of Einstein equations. In the *initial value formulation*, the constraints equations are differential relations between the fields at each given time foliation. If these relations are satisfied at the initial hypersurface, they are then satisfied at later times. This is so because it can be shown (case-by-case) that the constraint equations, thought of not as equations but as the values these relations take, also satisfy evolution equations, and their character is the same as that of the original evolution system. That is, if the evolution system is strongly hyperbolic then the constraint system is also strongly hyperbolic[6]. Furthermore, in all known cases the characteristic cone of the constraint propagation equations is a subset of the characteristic cone of the evolution equations. Uniqueness of solutions to symmetric hyperbolic equations imply that if initially satisfied, the constraint equations are satisfied on the whole domain of dependence of the evolution system (note that it might not coincide with the domain of dependence of the metric solution of the system, for there are evolution systems whose propagation velocities are larger that the speed of light).

When dealing with a boundary value problem this is no longer the case, for in general the domain of dependence of the IBVP does not coincide with the domain of dependence of the initial surface. Thus, unless the constraint propagation equations have appropriate boundary data so as to guarantee that the only possible solution to the constraint equation is the trivial one, the constraints would not remain satisfied. This uniqueness requirement is necessary; however, in many cases it *is in general not a differential relation among the incoming modes of the evolution system* since it also contains outgoing and zero modes. Thus, it is not even clear how to *solve* for the constraints there. The equations that result from requiring such uniqueness do not result in a well-posed boundary value problem for most Einstein's evolution systems. Among all of the ones examined so far there is a single one where we can correctly specify non-incoming boundary conditions for the physical modes and still preserve constraint propagation [24]. For all other systems examined [54, 55, 13], the known freedom to specify boundary data for Einstein's equations is not enough for imposing non-incoming radiation conditions. More analytical work is urgently needed in this area in order to fully understand the problem.

2.3. Particulars: Inner boundaries

Among the most interesting problems in G.R. are those where regions of strong gravity and extreme dynamics are considered. These systems usually have black holes, or will likely form them, during the evolution. One is then faced with the

---

[6]There is no general proof to this fact, but it is observed in all known cases where it has been computed. Furthermore, it is observed that the characteristic cone of the constraint system is a subset of the characteristic cone of the evolution system, so constraint quantities propagate at some of the same speeds as the evolution fields.

need to handle the singularities contained in these black holes or the simulation will break. The prevalent way to treat this problem relies on the assumption that cosmic censorship holds, and consequently the dynamics inside the black hole is hidden from the outside world. This property can, in principle, be exploited by placing an inner boundary inside the event horizon which effectively *excises* the singularities from the computational domain. However, by placing an artificial inner boundary in the spacetime under consideration, special care must be taken to maintain both the well-posedness of the problem and the stability of the implementation. Fortunately, the causal structure inside the event horizon can be exploited to satisfy the former condition. Namely, the inner boundary can be chosen carefully so as to ensure it is of 'inflow' type; i.e., all characteristic propagate towards this boundary. Hence, no boundary data is required there as these are determined by the equations themselves. Satisfying the 'inflow-boundary' requirement restricts the type and size of boundary one must consider. To give an idea of the nature of the problem, if one wishes to use a cubical excision region in a Schwarzschild spacetime (in Painleve-Gulstrand or Kerr-Schild coordinates), that cube can not have sides larger than about $0.7M$[7]. The reason for this, is that surfaces inside the black hole are not necessarily spacelike. The spacelike condition ensures the boundary is of 'inflow' type when employing a formulation of Einstein equations with physical characteristic speeds. Note that even when this requirement is satisfied, numerical diffusion might still lead to the interior of the black hole actually influencing the outside of it. (Note that in practice the dynamics in that region is so strong, that numerical modes propagating at very large speeds are generated and the influence of the interior calculation is felt outside. It is expected that with better numerical resolution such problems would be minimized, but at present, and for several years of expected computer power increase, this will not be the case, and extra care must be given to this problem.)

Here we must make an observation, that the previous statement is applicable when considering 'free evolution schemes' systems where only hyperbolic equations are involved in updating the field variables. On the other hand, if a 'constrained evolution' is considered, that is, where some variables are obtained after solving (perhaps some of) the constraint equations, boundary conditions must be provided for these. It might seem that we have encountered a stumbling block; however, one could use a 'free scheme' to update the boundary values and once this is performed use them in the constraint equations. This procedure would need to be iterated. Little is known on the formal side of this strategy, but preliminary numerical investigations are certainly encouraging [17].

---

[7]One could consider larger cubes and give some (stable) boundary data at the regions of the faces which have outgoing modes, unfortunately there is no theory for boundary regions where the number of incoming modes changes along it. So we fall in trial and error without any guiding principle. Alternatively, one can consider excision boundaries which are more complicated than cubes, for instances hexagons or even spheres (with a superimposed spherical grid) this way the code becomes considerably more complicated, but at least for those cases a bit more of theoretical understanding of the numerics is known.

3. Einstein equations in the computer. "Practical considerations"

In numerically solving the evolution equations one would ideally like to apply standard integration schemes for which, at least when applied to simpler related systems, a well-understood convergence and stability theory exists. If that is the case one can with some confidence concentrate on the analytical aspects of the problem and discard schemes with possible numerical instabilities. Furthermore, one not only worries about the stability of the obtained solution, but also on obtaining (small) reliable error bounds. This is not straightforward and requires further considerations at the implementation level. It is important to point out that the numerical treatment of hyperbolic systems of equations has only recently reached a certain maturity level. Indeed, even today there is a considerable body of works dealing with rather basic problems on areas such as boundary conditions, phase accuracy, constraint propagation, etc. In what follows we shall discuss some topics which are currently under scrutiny.

3.1. Reduction/formulation: Is there a preferred one?

As mentioned, Einstein's equations have constraints. This implies that even after imposing gauge conditions the evolution equations are not uniquely defined: given a set of evolution equations one can find another by just adding constraint equations to them. Thus, when referring to Einstein evolution equations one is really implying a large family of equations. Some of these have nice properties but, surprisingly, some have undesirable properties, such as admitting solutions that do not depend continuously on the initial data. Naturally, the latter class of equations can not be solved for numerically. One must, therefore, restrict attention to systems with nice properties. Unfortunately, as stated earlier, a complete classification of these systems is still incomplete.

Even when restricting to 'nice' systems, having an evolution system with continuous dependence on the initial data is not good enough in the numerical arena. Here one further needs systems to be stable in a very "practical" numerical sense, namely systems without exponential growth (if none is expected) on time scales comparable to those of the physical phenomena under study. Clearly, although well-posedness is not sufficient for numerical purposes, it is certainly the necessary starting point in the search for good systems[8].

---

[8]If a system is not well posed then no consistent numerical method can converge, for if it were, any numerical solution would converge to the analytical one in the same norm as the initial data of one would to the other. Hence via the numerical solution we would conclude that the analytical one depends continuously on the initial data, which is not the case. An interesting case is that of *weakly hyperbolic* systems. Some of these systems have continuous dependence on the initial data, but on different norms. Typically one needs to control a larger number derivatives of the initial data than the ones included in the norm for the solution. The problem is that these systems are not structurally stable, and a change on the lower order terms of the system can render these systems unstable. In general relativity, where the number of lower order terms is very large, weakly hyperbolic systems should be avoided. For instance, it has been recently shown that some numerical simulations of the ADM equations, a weakly hyperbolic system, exhibit weak numerical instabilities [11].

After a choice of evolution equations has been made one can ask whether the rest of Einstein's equations, namely the constraints equations, would be satisfied. Via the Bianchi identities and the evolution equations, it is usually the case that the constraints will be satisfied at later times in the domain of dependence of the initial data when dealing with 'nice' systems. However, no result is known when dealing with weakly hyperbolic evolution equations. Furthermore, in the presence of boundaries this is in fact an open problem and very limited boundary conditions are known which preserve the constraints.

Suppose now one deals with a system that, at the analytical level, preserves the constraints. Here again, in translating to the numerical arena, this will generically not be the case. In fact, there is always the risk that the system will depart exponentially fast from the constraint surface, rendering the numerical solution of little use soon after[9]. As mentioned, the search for a reformulation of the equations with nice properties that exhibits good solution and constraint behavior at the numerical level is a very active field of research. Different efforts have made use of the freedom in the choice of evolutions equations to search for better behaved simulations. For instance, in[34], a code implementing Einstein equations with the addition of constraints premultiplied by constants was run for different values of these constants. It was observed that depending on the choice of parameters, accurate evolutions were obtained lasting long times while for other values only lasted considerable shorter times. Note that the system of equations used was symmetric hyperbolic, with physical characteristic speeds, in all cases considered. In these tests, the principal part of the system remained unchanged (via a redefinition of variables) and the change in parameters only affected lower order terms. The message from these and other efforts is that there is an urgent need for a way to analyze the system at hand which takes into account the influence of the lower order terms and yield sharp growth estimates. To reiterate, in a numerical implementation, knowledge of the well-posedness of a problem is not enough. Preliminary attempts to obtain a priori estimates of more convenient choice of parameters are under investigation. Formally, a bound for the growth-rate of allowed solutions is provided by the standard energy estimates used when proving well-posedness. This provides an estimate for the growth of the *energy* of the system. Unfortunately, as the evaluation of this requires knowledge of the solution (since it depends on the boundaries) it cannot be evaluated directly. In [39], a method to produce some rough estimate was presented where a bound for the growth rate is obtained by projecting the whole term inside the energy estimate to those satisfying no incoming modes. Thus, an expression is obtained which can be integrated as it only involves the initial data. Unfortunately, it is difficult to assess whether this is a sharp estimate since in order to evaluate the integral further structure needs to be

---

[9]Note that if this happens in a region of space time where there is no physical reason to have large curvatures, then this growth must be due to gauge. Constraint quantities are not scalars, they are just pieces of a four-dimensional tensor, and thus subject to gauge transformations.

introduced, like an ansatz for the expected gradients of the variables under consideration. It should be noted however, that in cases tried numerically, the bound obtained by this method seems to agree reasonably well with the behavior of the solution under the few cases examined. It is thus interesting to investigate whether this method is indeed robust or the agreement is coincidence.

Another option being developed relies on analyzing the growth behavior of useful quantities (like constraints and/or suitably defined geometric scalars) during the evolution and choosing parameters which minimize their growth. The rationale behind this strategy is that it is generically impossible to determine whether any growth observed in the solutions is itself the true behavior or not. For instance, in the case of where constraints are monitored, as they are supposed to remain zero they can be used to determine the quality of the evolution and adjust 'on the fly' the free parameters accordingly[57].

Clearly, the proposed methods have their shortcomings and there is an urgent need for more robust ones. The ideal method would be to study the system at the analytical level, and predict sharp growth estimates which can then be exploited at the numerical level. However this is quite a complicated task which is not only initial data but also gauge dependent.

3.2. Boundary values specification

Let us briefly summarize what was said in Section II.1. Specification of boundary values in a non linear theory is always a complicated task. No analytical theory exists which would cover the general relativistic case, except for the case of the Friedrich-Nagy[24] frame system, where such a specification has been found and proven to be well posed. As mentioned, for the type of systems most used in numerical computations the problem is very much open. If the system is symmetric hyperbolic then necessary and sufficient conditions for well-posedness are known from the linear theory: incoming modes can be freely prescribed while nothing should be done with the outgoing ones. In General Relativity we face the additional problem of dealing with a constrained system, which prevents one from *just* providing data to all incoming modes. The constraint equations imply the incoming modes are not all independent; some need to be chosen so that the constraints will be preserved.

The complications found in this problem prompted the use of *approximate boundary conditions* which, at the very core, were inspired by wave propagation analysis. However, these conditions are applied to all (or most) variables, which is generically inconsistent. Problems observed at the boundaries are dealt with numerical dissipation and placing the boundaries as far as possible from the region of interest to minimize their influence. Assuming these inconsistencies are not responsible for crashing the code, this strategy would be a way to go around a series of conceptual problems. However, in most applications this is hardly the case as computational resources quickly become scarce and boundaries are too close.

A road towards achieving well-behaved implementations in the presence of boundaries can be designed starting at the continuum limit and constructing the implementation so as to reproduce desired analytical properties [12, 44]. To fix ideas, we will restrict to first order symmetric hyperbolic systems (as higher order ones can be reduced to these), which can be schematically expressed as:

$$\partial_t \phi^\alpha = -A^{i\alpha}_\beta \partial_i \phi^\beta + j^\alpha, \tag{6}$$

As we asserted before, these systems have an "energy norm" on the solution which is bounded in time by some function of time and the energy norm of the initial data. From there we get the analytical well-posedness of the Cauchy problem, and in some cases the well-posedness of the IBVP. The strategy now is to try to reproduce (as close as possible) the basic ingredients required for assessing the well-posedness of the system at the analytical level and translate them into the numerical arena. To that goal one can adopt algorithms containing the following ingredients:

- A discrete analogue of integration by parts
- A consistent boundary treatment
- A combination of variables yielding at *the numerical level* a growth estimate which approaches the one expected at the continuum.
- A conveniently chosen time integration operator.

The above-mentioned ingredients are naturally motivated by the goal of reproducing, at the discrete level, what is expected from the analysis at the continuum level. The choice of algorithms is therefore delicate. One way to assessing whether the algorithms adopted do satisfy the desired goal is to consider them in a global sense. That is, consider the full discretization problem (which combines the boundary treatment and the discretization of the equations in the interior) and analyze the obtained system. This strategy is what is entailed in the Gustaffson-Kreiss-Sundrum method [30]. Another way relies on considering separately the semidiscrete problem, including the boundary treatment while regarding time as continuous) and finally adopting a convenient time integrator. The former is more involved and requires a case by case analysis. The latter is naturally related to the use of the method of lines, which consists first in discretizing space so as to regard the systems as a collection of coupled ordinary differential equations. One then shows how to produce discrete energy estimates, and when discretizing in time, a stable numerical method is obtained via the use of an appropriate numerical time integrator [56].

In what follows, we briefly describe each of these properties and their relevance in achieving a robust implementation of hyperbolic system of equations. We here concentrate on the case of finite difference techniques.

Finite difference analogue of integration by parts: Summation by parts. A way of constructing difference operators that yield a stable semidiscrete problem is via the use of difference operators satisfying *summation by parts* (SBP). This is the discrete version of the integration by parts (a main ingredient when getting the energy estimate at the continuum). Employing difference operators satisfying SBP and having a well-posed initial-boundary value problem at the continuum provides a semidiscrete energy estimate and numerical stability follows.

In practice, the idea behind schemes respecting SBP is to reproduce, at the discrete level, the equality

$$\int_b^a (uv_{,x} + vu_{,x})\, dx = u(a)v(a) - u(b)v(b) \tag{7}$$

When using finite difference schemes (with grid functions centered at points $x^i = a + (i)(b-a)/N$, $i = 0 \ldots N$), satisfying SBP at the discrete levels implies

$$\int_b^a (uv_{,x} + vu_{,x})\, dx \rightarrow \Sigma_{i=0}^N \sigma_i (u_i Dv|_i + v_i Du|_i)\delta x = u(N)v(N) - u(0)v(0), \tag{8}$$

where σ_i is a weighting factor, for instance, for a scalar grid product that is second order accurate one has $\sigma_0 = \sigma_N = 1/2$, $\sigma_i = 1$ $i = 1 \ldots N-1$. Preserving this condition forces the discrete derivative operator D to be defined in a specific way (depending, naturally, on the order of accuracy of the scheme). A straightforward path to achieving this condition is to employ standard centered discrete derivative operators at interior points but modifying them near boundaries appropriately so that the condition holds. More accurate discrete derivative operators than the one above mentioned have been found, see for instance [32, 52, 30]. Many of them can be directly generalized to higher-dimensional rectangular domains. In more complicated computational domains, especially those with inner boundaries the operators need to be modified properly. Derivative operators satisfying this property were constructed by Engquist [22] for some 2D domains and recently extended to 3D domains with inner cubic boundaries [12, 44].

Consistent boundary treatment. Writing the continuum equations as $\partial_t \phi^\alpha = A^{i\alpha}{}_\beta \partial_i \phi^\beta + j^\alpha$ and solving the semidiscrete problem $\partial_t \phi^\alpha = A^{i\alpha}{}_\beta D_i \phi^\beta + j^\alpha$ with schemes satisfying SBP does not suffice to correctly pose the problem, because boundary conditions must still be consistently implemented. Namely, given any boundary (which could be physical or artificially placed when using domain decomposition, mesh refinement etc.), a hyperbolic system will exhibit characteristic modes propagating from, towards, or parallel to the boundary. At the continuum level, a consistent treatment requires data be given *only* to modes propagating from the boundary inwards, while modes propagating parallel or outwards should be left alone for their values are defined via the evolution equations. Naturally, this construction must be respected at the numerical level or inconsistencies will be introduced which, at the very least, will lead to spurious solutions. It might seem surprising that most simulations do not make this distinction and instead boundary data is provided to all variables. The reason for the widespread use of such simplistic boundary treatments lies not only in the complications found in some systems to find out which ones are the outgoing modes, but also, when modes are identified, in the difficulties of implementing this analytically motivated strategy at the numerical level. Fortunately, recent developments in applied mathematics are paving the way to finally handling the boundary data specification consistently. Of course, one would like to impose boundary conditions without spoiling the SBP property, crucial for the discrete energy estimate. Consider a boundary condition of the form, $L_\alpha \phi^\alpha = 0$ (equivalent to $\phi \in H^+$). This is just a way of writing $V^+ = SV^-$, where V^+ and V^- are eigenmodes of the boundary matrix $A^i n_i$ (with n^i the normal to the boundary). The treatment for inhomogeneous boundary conditions follows along similar lines once the fields are redefined to absorb the inhomogeneous term.

Following Olsson [46, 47], one can consistently implement the boundary conditions at the discrete level as orthogonal projections. That is, *at boundary points* one introduces an operator P defined as,

$$P = I - L^T (LL^T)^{-1} L \tag{9}$$

and, for homogeneous boundary conditions, solves the system,

$$\partial_t \phi = P(A^i D_i \phi + j), \tag{10}$$

This method provides a natural, and efficient, way to deal with such boundary conditions and have yielded excellent results in many applications. This is not surprising, for its main property is that together with SBP gives the required discrete energy estimates. A related approach to impose the boundary conditions has been proposed by Carpenter et al. [15], known as "simultaneous approximation term" (SAT). Its applicability and efficiency has not yet been tried out in numerical relativity, but its applications in simpler settings yielded excellent results [53, 31, 14]).

These techniques are certainly a powerful way to handle boundaries in a stable way. However, the implementation of boundary conditions in numerical relativity is complicated by two facts. First, the constraint conditions on the boundary data imply that further evolution equations along the boundary be satisfied for some of the fields [24, 55, 13]. Second, the projection operators depend on the metric, which itself is being evolved.

3.3. Sharp growth estimates

Usually one wants not only the scheme to be stable (i.e., bounded by the initial data multiplied by an exponential in time), but also that the numerical solution, at fixed resolution, be as close as possible to the continuum one. This is a crucial ingredient for a numerical implementation to reproduce what the analytical expectation is and avoid weak solutions which might diverge grossly from the sought-after long evolutions. In 3D, even when there were no spurious solutions, the very low resolution practically available requires a careful choice of discretization. A way to achieve this is to re-express the system of equations in terms of another one (obtained by a suitable combination of variables) which behaves better at the numerical level. A possible way to approach this problem is to follow a 'pseudo-analytical' way to obtain better behaved systems: it relies on the use of schemes satisfying SBP and reproducing at the discrete level the continuous energy estimates. The basic strategy behind this method is to study the discrete energy estimates given by the same analytical system but under different combination of variables or discretizations. It can be seen then that the expected growth of the associated 'discrete energy' can be very different. The underlying reason for this is the fact that the Leibniz rule is only satisfied approximately in a numerical implementation (i.e., $D(u^2) = 2uD(u) + \epsilon(u_{,x}, u_{,xx})\Delta^n)$, where Δ is a measure of the grid spacing and n is given by the accuracy of the scheme employed). This in turn can give rise to discrete energy estimates which do not imply conservation (in the case the analytical system had such conservation) or whose bound is far too large (at a fixed Δ) compared to the discrete one. The analysis of which option is preferable can be done a priori. To illustrate this, consider the following example:

$$u_{,t} = 2vu_{,x} + uv_{,x} \tag{11}$$
$$= (uv)_{,x} + vu_{,x}, \tag{12}$$

where $v = v(x)$ is a given smooth function. At the continuum level both ways of writing the equations are of course equivalent. But the straightforward discretizations of them (namely the direct substitution $\partial_i \rightarrow D_i$) are not. We shall see that the discretized version of (12) is more convenient [44]. Consider the (conserved) energy of the system ($E \equiv \int_a^b u^2 dx$) and take its time derivative. Then substitute $u_{,t}$ by each right-hand side, and replace ∂_x by the discrete approximation D_x to obtain an expression for what is expected from the numerical evolution (thought still as continuum in the time variable). We shall analyze each case separately. First, for (11) we have

$$E_{,t} = \int_a^b u\,(2vDu + uDv)\,dx \tag{13}$$
$$= -2\int_a^b uD(uv)dx + 2B - \int_a^b D(u^2)v + B \tag{14}$$
$$= B + \int_a^b \epsilon\Delta^n v dx\,; \tag{15}$$

with $B = u^2 v|_1^N$. Thus, the estimate contains a non-negligible contribution, at a fixed resolution, given by the last term of equation (15). This term can be written as the integral of $(Dv)(\Delta^n u^2)$. Therefore, the energy estimate obtained for system (11) contains an additional term which although can be controlled by the energy, at a given resolution, the energy of the system is not necessarily conserved at the discrete level. On the other hand, for system (12)

$$E_{,t} = \int_a^b u\,(D(uv) + vDu)\,dx \tag{16}$$
$$= -\int_a^b D(u)uvdx + B + \int uvDudx = B\,. \tag{17}$$

Hence its growth is only governed, just as at the continuum level, by boundary conditions. We thus see that system (12) reproduces the continuous estimate
and its energy decreases or is conserved according to the boundary conditions
imposed on the system. (This is confirmed by numerical experiments.) A semidiscrete system whose energy growth estimate reproduces that of the continuum
equation, in a suitable norm, is said to be *strictly stable*. Note the technique presented here not only applies to cases where the energy is conserved. Indeed, it
can be seen that given a system defined by $u_{,t} = 2fu_{,x}$, re-expressing it in the
form $u_{,t} = (fu)_{,x} + fu_{,x} - uf_{,x}$ gives rise to a discrete energy reproducing the
continuous one (when SBP is respected). As mentioned, an attractive feature of
this technique is that it can be used to first study the system at the pre-coding
level and obtain sharp growth estimates which could then be used as a measure
for which reformulation of equations to use. At first, it might seem that for complicated systems, like general relativity, this procedure is hopeless, but experience
with simpler systems indicate that indeed this procedure imposes a better, perhaps
more physical, choice of variables and grouping of them [12, 44]. It is important to
remind the reader here that in General Relativity, unfortunately, we do not have
a useful physical notion of energy which could be exploited at the numerical level
as explained above. An energy can indeed be defined for the system by choosing
an appropriate norm for the solution but this will not be a physical one and offers
only partial guidance for the numerical implementation.

3.4. Initial data

Although consistent initial data exists, in most cases, it is unclear how physically
realistic these are. A main reason behind this is that in a *single* hypersurface one
can not elucidate the amount of spurious radiation present in the initial data. In
the standard Cauchy approach, the hypersurfaces do not even reach future null
infinity, and therefore it is much more difficult to make sense of radiation. In the
case of the conformal Einstein equations, the hypersurfaces indeed cross future null
infinity making the problem clearer. In the characteristic approach, an alternative
often pursued is to provide data that satisfies (in the NP notation[45]) $\Psi^0 = 0$.
This choice is motivated by the fact that at large distances, Ψ^0 gives an indication
of the amount of incoming radiation crossing the initial null hypersurface.

Although the two latter cases give somehow better control on the radiation
content at large distances, it is safe to say that no approach can gauge the physical
relevance of the data used based solely on the initial value problem setting (unless
data is given at $\mathcal{I}^-$). The ultimate resolution of this problem requires obtaining
at least a partial history of the spacetime and analyzing the outcome. It should be
noted here that this can either be done with the evolution equations, i.e., evolving
the equations under different initial data sets, or dealing with an alternative way
of describing the system that (i) covers the spacetime at times earlier than the
initial hypersurface chosen for the numerical evolution and (ii) describes correctly
the physics of the system. An example of this in the case of binary neutron stars
is the use of Post Newtonian orbits when the stars are far enough apart. This

strategy would extract information from the 'pre-numerical epoch' and use it to seed the initial data setting appropriately. There are still a number of unresolved issues in this strategy, like how to match the typical 'orbital information' found in these approaches to the seed metric variables for the full relativistic problem. Work in different fronts is underway to study these issues.

4. Final words

For most of this work, we have restricted our attention to open problems in numerical relativity and to highlight useful and robust techniques which can play a major role in future developments in the field. These novel techniques provide a systematic way to treat Einstein's evolution systems based on recently obtained rigorous results which can aid tremendously in achieving a stable implementation, *or at the very least* rule out basic possible sources of instability. In the past, the search for robust schemes relied on experience and trial and error. Such methods for searching for appropriate schemes is a viable strategy in lower-dimensional settings as the turn-around for each test is quite acceptable. Prime examples of achievements obtained this way are the discovery of critical phenomena [16]; the resolution of the fate of the Cauchy horizon in charged black hole spacetimes [48, 8]; the head-on collision of black holes [41]; etc.

Unfortunately, in higher-dimensional scenarios the 'experimental' search for convenient schemes requires even more 'heroic' efforts where only a limited search can be performed. In some particular 3D systems (with little dynamics and based on stationary spacetimes, or slight departures off these) techniques have been recently presented [58, 1, 51] which improve the treatment of these systems. However, it is unclear that these schemes will perform as well when generic settings are considered and long enough evolutions are sought after. The techniques mentioned here should prove very useful in this enterprise. Coupled to advances in our understanding of boundary treatments; coordinate conditions and improved computational power (including a more efficient use of it via, for instance, adaptive mesh refinement and multigrid techniques) the future construction of numerical spacetimes promises to be an ideal tool to investigate the theory and to be fundamental in practical applications in astrophysics and gravitational wave phenomena.

Acknowledgments
We wish to thank the Relativity group at LSU for many useful discussions. In particular, the application of techniques guaranteeing strict stability and boundary treatment for the Cauchy problem is an outgrowth of research efforts by this group presented in [12, 44]. We thank Frans Pretorius for a careful reading of the manuscript. This work was supported in part by the Horace Hearne Jr. Institute for Theoretical Physics and Fundación Antorchas. O.R. is a Conicet Fellow. L.L is an Alfred P. Sloan Fellow. This research was supported in part by the NSF under grant numbers NSF-PHY0244699 and NSF-PHY0326311 to Louisiana State University, PHY99-07949 to the University of California at Santa Barbara and by the Horace Hearne Institute for Theoretical Physics.

References

[1] Alcubierre, M., *et al.*, *Gauge conditions for long-term numerical black hole evolutions without excision*, 2003, Phys. Rev. **D67**, 084023.

[2] Anderson, L., P. Chrusciel, and H. Friedrich, 1992, Comm. Math. Phys. **149**, 587.

[3] Balean, R.M., 1996, Ph. D. Dissertation, University of New England.

[4] T.W. Baumgarte and S.L. Shapiro, *Numerical relativity and compact binaries*, Phys. Rept. **376** (2003) 41–131.

[5] Bishop, N.T., R. Gomez, L. Lehner, M. Maharaj, and J. Winicour, *High powered gravitational news*, 1997, Phys. Rev. **D56**, 6298–6309.

[6] Bondi, H., van der Burg, M., and Metzner, A., *Gravitational waves in general relativity VII. Waves from axi-symmetric isolated systems*, 1962, Proc. Roy. Soc. London Ser. A **270**, 21–52.

[7] Brady, P.R., J.D.E. Creighton, and K.S. Thorne, *Computing the merger of black-hole binaries: the IBBH problem*, 1998, Phys. Rev. **D58**, 061501.

[8] Brady, P.R., and J.D. Smith, *Black hole singularities: a numerical approach*, 1995, Phys. Rev. Lett. **75**, 1256–1259.

[9] Brandt, S., and B. Bruegman, *A simple construction of initial data for multiple black holes*, 1997, Phys. Rev. Lett. **78**, 3606–3609.

[10] Cagliero, L., 1993, Master Thesis; FaMAF-UNC, *Operadores simetricos hiperbolicos y una aplicacion a las Ecuaciones de Maxwell*

[11] Calabrese, G., J. Pullin, O. Sarbach, and M. Tiglio, *Convergence and stability in numerical relativity*, 2002, Phys. Rev. **D66**, 041501.

[12] G. Calabrese, L. Lehner, D. Neilsen, J. Pullin, O. Reula, O. Sarbach and M. Tiglio, *Novel finite-differencing techniques for numerical relativity: application to black hole excision*, Class. Quant. Grav. **20** (2003) L245-L251.

[13] Calabrese G., J. Pullin, O. Sarbach, M. Tiglio and O. Reula, *Well posed constraint-preserving boundary conditions for the linearized Einstein equations*, 2003, Comm. Math. Phys. **240**, 377–395.

[14] Carpenter, M., J. Nordstrom, and D. Gottlieb, 1998, *A stable and conservative interface treatment of arbitrary spatial accuracy*, 1999, J. Comp. Phys. **148**, 341–365.

[15] Carpenter, M., Gottlieb D., and Abarbanel S., *Time-stable boundary conditions for finite difference schemes solving hyperbolic systems: Methodology and application to higher order compact schemes*, 1994, J. Comp. Phys. **111**, 220–236.

[16] Choptuik, M.W., 1993, Phys. Rev. Lett., *Universality and scaling in gravitational collapse of a massless scalar field*, **70**, 9–12.

[17] Choptuik, M.W., E.W. Hirschmann, S.L. Liebling, and F. Pretorius, *Critical collapse of the massless scalar field in axisymmetry*, 2003, Phys. Rev. D **68**, 044007.

[18] Cook, G., *Initial data for numerical relativity*, 2000, Living Rev. Rel. **5**, 1.

[19] Corvino, J., and R.M. Schoen, , 2003, `gr-qc/0301071`.

[20] Dain, S., 2002, , Lect. Notes Phys. **604**, 161–182.

[21] Dain, S., and Nagy, G., arXiv:gr-qc/0308009.

[22] Engquist, B., 1978, *A difference method for initial boundary value problems in general domains in two space dimensions*, DCG Progress Report, Dept. of Computer Sciences, Uppsala University.

[23] Frauendiener, J., *Calculating initial data for the conformal Einstein equations by pseudo-spectral methods*, 1998, Preprint gr-qc/9806103.

[24] Friedrich, H., and G., Nagy, *The initial boundary value problem for Einstein's vacuum field equations*, 1999, Communications in Mathematical Physic **201, Issue 3**, 619.

[25] Friedrich, H., and A.D. Rendall, *The Cauchy problem for the Einstein equations*, 2000, Lect. Notes Phys. **540**, 127–224.

[26] Friedrich, H., and J.M. Stewart, *Characteristic initial data and wavefront singularities in general relativity*, 1983, Proc. R. Soc. A **385**, 345.

[27] Garfinkle, D., and C. Gundlach, *Symmetry-seeking spacetime coordinates*, 1999, Class. Quant. Grav. **16**, 4111–4123.

[28] Gourgoulhon, E., P. Grandclement, K. Taniguchi, J.-A. Marck, and S. Bonazzola, *Quasiequilibrium sequences of synchronized and irrotational binary neutron stars in general relativity. I. Method and tests*, 2001, Phys. Rev. **D63**, 064029.

[29] Gues, O., 1990, Commun. Part. Diff. Eqs. **15**, 595.

[30] Gustaffson, B., H.-O. Kreiss, and J. Oliger, 1995, *Time-Dependent Problems and Difference Methods* (Wiley, New York, USA).

[31] Gustafsson, B., *On the implementation of boundary conditions for the method of lines*, 1998, BIT **38**(2), 293.

[32] H.-O., Kreiss, and G. Scherer, 1977, in *Mathematical aspects of finite elements in partial differential equations*.

[33] John, F., 1982, *Partial Differential equations* (Springer Verlag, New York, USA).

[34] Kidder, L. E., M.A. Scheel, and S.A. Teukolsky, *Extending the lifetime of 3D black hole computations with a new hyperbolic system of evolution equations* 2001, Phys. Rev. **D64**, 064017.

[35] Kreiss, H.O., and O.E. Ortiz, *Some mathematical and numerical questions connected with first and second order time dependent systems of partial differential equations*, 2002, Lect. Notes Phys. **604**, 359.

[36] Lehner, L. *Numerical Relativity: A review*, Class. Quant. Grav. **18**, R25-R86 (2001) [arXiv:gr-qc/0106072].

[37] Lehner, L., *Matching characteristic codes: exploiting two directions*, 2000, Int. J. Mod. Phys. **D9**, 459.

[38] Lichnerowicz, A., 1944, J. Math. Pures et Appl. **23**, 37.

[39] Lindblom, L., and M.A. Scheel, *Energy Norms and the Stability of the Einstein Evolution Equations*, 2002, Phys. Rev. **D66**, 084014.

[40] Marronetti, P., and R.A. Matzner, *Solving the Initial Value Problem of two Black Holes*, 2000, Phys. Rev. Lett. **85**, 5500–5503.

[41] Matzner, R.A., 1995, Science **270**, 941.

[42] D. Maxwell, *Solutions of the Einstein Constraint Equations with Apparent Horizon Boundary*, arXiv:gr-qc/0307117.

[43] Nagy, G., O. Ortiz, and O. Reula, 2003, *On the hyperbolicity of the BSSN equations, a second order approach.*, in preparation.

[44] Neilsen, G., *et al.*, 2003a.

[45] Newman, E.T., and R. Penrose, *An approach to gravitational radiation by a method of spin coefficients*, 1962, J. Math. Phys. **3**, 566–578.

[46] Olsson, P., Math. Comp., *Summation by parts, projection and stability I*, **64**, 1035–1065 (1995).

[47] Olsson, P., Math. Comp., *Summation by parts, projection and stability II*, **64**, 1473–1493 (1995).

[48] Poisson, E., and W. Israel, 1989, Phys. Rev. Lett. **63**, 1796–1799.

[49] Reula, O. A., *Hyperbolic methods for Einstein equations*, 1998, Living Rev. Rel. **1**, 3.

[50] Sachs, R., *Gravitational waves in general relativity VIII. Waves in asymptotically flat space-times.*, 1962, Proc. Roy. Soc. A **270**, 103.

[51] D. Shoemaker, K. Smith, U. Sperhake, P. Laguna, E. Schnetter and D. Fiske, *Moving black holes via singularity excision*, Class. Quant. Grav. **20**, 3729–3744 (2003).

[52] Strand, B., *Summation by parts for finite difference approximations for d/dx*, 1994, J. Comp. Phys. **110**, 47–67.

[53] Strand, B., 1998, Applied Numerical Mathematics **26**, 497–521.

[54] Szilagyi, B., B. Schmidt, and J. Winicour, *Boundary conditions in linearized harmonic gravity*, 2002, Phys. Rev. **D65**, 064015.

[55] Szilagyi, B., and J. Winicour, *Well-posed initial boundary evolution in general relativity*, 2002, Phys. Rev. **D68**, 041501.

[56] Tadmor, E., 2002, Proceedings in Appl. Math. **109**, 25.

[57] Tiglio, M., *Dynamical control of the constraints growth in free evolutions of Einstein's equations*, 2003, gr-qc/0304062.

[58] Yo, H.-J., T.W. Baumgarte, and S.L. Shapiro, *Improved numerical stability of stationary black hole evolution calculations*, 2002, Phys. Rev. **D66**, 084026.

[59] York, J., 1973, *Conformally invariant orthogonal decompositions of symmetric tensors of Riemannian manifolds and the initial value problem*, J. Math. Phys. **14**, 456–464.

[60] York, J., 1978, in *Sources of Gravitational Radiation*, edited by L. Smarr (Cambridge University Press., Seattle), 83–126

Luis Lehner
Department of Physics and Astronomy
Louisiana State University
Baton Rouge, LA 70810, USA
e-mail: `lehner@lsu.edu`

Oscar Reula
FAMAF
Universidad Nacional de Cordoba
Cordoba 5000, Argentina
e-mail: `reula@fis.uncor.edu`

The Einstein-Vlasov System

Alan D. Rendall

Abstract. Rigorous results on solutions of the Einstein-Vlasov system are surveyed. After an introduction to this system of equations and the reasons for studying it, a general discussion of various classes of solutions is given. The emphasis is on presenting important conceptual ideas, while avoiding entering into technical details. Topics covered include spatially homogeneous models, static solutions, spherically symmetric collapse and isotropic singularities.

1. Introduction

The basic equations of general relativity are the Einstein equations coupled to some other partial differential equations describing the matter content of spacetime. There are many choices of matter model which are of physical interest and Yvonne Choquet-Bruhat has published fundamental results on the Cauchy problem for the Einstein equations coupled to a wide variety of matter models. One of these is collisionless matter described by the Vlasov equation. It is the subject of these lectures.

The Vlasov equation arises in kinetic theory. It gives a statistical description of a collection of particles. It is distinguished from other equations of kinetic theory by the fact that there is no direct interaction between particles. In particular, no collisions are included in the model. Each particle is acted on only by fields which are generated collectively by all particles together. The fields which are taken into account depend on the physical situation being modelled. In plasma physics, where this equation is very important, the interaction is electromagnetic and the fields are described either by the Maxwell equations or, in a quasi-static approximation, by the Poisson equation [32]. In gravitational physics, which is the subject of the following, the fields are described by the Einstein equations or, in the Newtonian approximation, by the Poisson equation. (There is a sign difference in the Poisson equation in comparison with the electromagnetic case due to the replacement of a repulsive by an attractive force.) The best-known applications of the Vlasov equation to self-gravitating systems are to stellar dynamics [3]. It can also be applied to cosmology. In the first case the systems considered are galaxies or parts of galaxies where there is not too much dust or gas which would require a hydrodynamical treatment. Possible applications are to globular clusters, elliptical galaxies and the central bulge of spiral galaxies. The 'particles' in all these cases are stars. In the cosmological case they might be galaxies or even clusters of galaxies. The fact that they are modelled as particles reflects the fact that their internal structure is believed to be irrelevant for the dynamics of the system as a whole. The Vlasov equation is also used in cosmology to model non-baryonic dark matter ([4], p. 323). In that case the 'particles' are elementary particles.

These lectures are concerned not with the above physical applications but with some basic mathematical aspects of the Einstein-Vlasov system. First the definition and general mathematical properties of this system of partial differential equations are discussed and then the Cauchy problem for the system is formulated. The central theme in what follows is the global Cauchy problem, where 'global' means global in time. The known results on this and related problems are surveyed and important methods used are highlighted. Further information on kinetic theory in general relativity may be found in [7].

Let $(M, g_{\alpha\beta})$ be a spacetime, i.e., M is a four-dimensional manifold and $g_{\alpha\beta}$ is a metric of Lorentz signature $(-, +, +, +)$. Note that $g_{\alpha\beta}$ denotes a geometric object here and not the components of the geometric object in a particular coordinate system. In other words the indices are abstract indices. (See [34], Section 2.4 for a discussion of this notation.) It is always assumed that the metric is time-orientable, i.e., that the two halves of the light cone at each point of M can be labelled past and future in a way which varies continuously from point to point. With this global direction of time, it is possible to distinguish between future-pointing and past-pointing timelike vectors. The worldline of a particle of non-zero rest mass m is a timelike curve in spacetime. The unit future-pointing tangent vector to this curve is the 4-velocity v^{α} of the particle. Its 4-momentum p^{α} is given by mv^{α}. There are different variants of the Vlasov equation depending on the assumptions made. Here it is assumed that all particles have the same mass m but it would also be possible to allow a continuous range of masses. When all the masses are equal, units can be chosen so that $m = 1$ and no distinction need be made between 4-velocity and 4-momentum. There is also the possibility of considering massless particles, whose wordlines are null curves. In the case $m = 1$ the possible values of the four-momentum are precisely all future-pointing unit timelike vectors. These form a hypersurface P in the tangent bundle TM called the mass shell. The distribution function f, which represents the density of particles with given spacetime position and four-momentum, is a non-negative real-valued function on P. A basic postulate in general relativity is that a free particle travels along a geodesic. Consider a future-directed timelike geodesic parametrized by proper time. Then its tangent vector at any time is future-pointing unit timelike. Thus this geodesic has a natural lift to a curve on P, by taking its position and tangent vector together. This defines a flow on P. Denote the vector field which generates this flow by X. (This vector field is what is sometimes called the geodesic spray in the mathematics literature.) The condition that f represents the distribution of a collection of particles moving freely in the given spacetime is that it should be constant along the flow, i.e., that $Xf = 0$. This equation is the Vlasov equation, sometimes also known as the Liouville or collisionless Boltzmann equation.

To get an explicit expression for the Vlasov equation, it is necessary to introduce local coordinates on the mass shell. In the following local coordinates x^α on spacetime are always chosen such that the hypersurfaces x^0=const. are spacelike. (Greek and Latin indices take the values $0, 1, 2, 3$ and $1, 2, 3$ respectively.) Intuitively this means that x^0, which may also be denoted by t, is a time coordinate and that the x^a are spatial coordinates. A timelike vector is future-pointing if and only if its zero component in a coordinate system of this type is positive. It is not assumed that the vector $\partial/\partial x^0$ is timelike. One way of defining local coordinates on P is to take the spacetime coordinates x^α together with the spatial components p^a of the four-momentum in these coordinates. Then the explicit form of the Vlasov equation is:

$$\partial f/\partial t + (p^a/p^0)\partial f/\partial x^a - (\Gamma^a_{\beta\gamma}p^\beta p^\gamma/p^0)\partial f/\partial p^a = 0 \qquad (1)$$

where $\Gamma^\alpha_{\beta\gamma}$ are the Christoffel symbols associated to the metric $g_{\alpha\beta}$. Here it is understood that p^0 is to be expressed in terms of p^a and the metric using the relation $g_{\alpha\beta}p^\alpha p^\beta = -1$. An alternative form of the Vlasov equation which is often useful is obtained by coordinatizing the mass shell using the components of the four-momentum in an orthonormal frame instead of the coordinate components.

The Vlasov equation can be coupled to the Einstein equations as follows, giving rise to the Einstein-Vlasov system. The unknowns are a 4-manifold M, a (time orientable) Lorentz metric $g_{\alpha\beta}$ on M and a non-negative real-valued function f on the mass shell defined by $g_{\alpha\beta}$. The field equations consist of the Vlasov equation defined by the metric $g_{\alpha\beta}$ for f and the Einstein equation $G_{\alpha\beta} = 8\pi T_{\alpha\beta}$. (Units are chosen here so that the speed of light and the gravitational constant both have the numerical value unity.) To obtain a complete system of equations it remains to define $T_{\alpha\beta}$ in terms of f and $g_{\alpha\beta}$. It is defined as an integral over the part of the mass shell over a given spacetime point with respect to a measure which will now be defined. The metric at a given point of spacetime defines in a tautological way a metric on the tangent space at that point. The part of the mass shell over that point is a submanifold of the tangent space and as such has an induced metric, which is Riemannian. The associated measure is the one which we are seeking. It is evidently invariant under Lorentz transformations of the tangent space, a fact which may be used to simplify computations in concrete situations. In the coordinates (x^α, p^a) on P the explicit form of the energy-momentum tensor is:

$$T_{\alpha\beta} = -\int f p_\alpha p_\beta |g|^{1/2}/p_0 dp^1 dp^2 dp^3 \qquad (2)$$

A simple computation in normal coordinates based at a given point shows that $T_{\alpha\beta}$ defined by (2) is divergence-free, independently of the Einstein equations being satisfied. This is of course a necessary compatibility condition in order for the Einstein-Vlasov system to be a reasonable set of equations. Another important quantity is the particle current density, defined by:

$$N^\alpha = -\int f p^\alpha |g|^{1/2}/p_0 dp^1 dp^2 dp^3 \tag{3}$$

A computation in normal coordinates shows that $\nabla_\alpha N^\alpha = 0$. This equation is an expression of the conservation of the number of particles. There are some inequalities which follow immediately from the definitions (2) and (3). Firstly $N_\alpha V^\alpha \leq 0$ for any future-pointing timelike or null vector V^α, with equality only if $f = 0$ at the given point. Hence unless there are no particles at some point, the vector N^α is future-pointing timelike. Next, if V^α and W^α are any two future-pointing timelike vectors then $T_{\alpha\beta}V^\alpha W^\beta \geq 0$. This is the dominant energy condition ([9], p. 91). Finally, if X^α is a spacelike vector then $T_{\alpha\beta}X^\alpha X^\beta \geq 0$. This is the non-negative pressures condition. This condition, the dominant energy condition and the Einstein equations together imply that the Ricci tensor satisfies the inequality $R_{\alpha\beta}V^\alpha V^\beta \geq 0$ for any timelike vector V^α. The last inequality is called the strong energy condition. These inequalities constitute one of the reasons which mean that the Vlasov equation defines a well-behaved matter model in general relativity. However this is not the only reason. A perfect fluid with a reasonable equation of state or matter described by the Boltzmann equation also have energy-momentum tensors which satisfy these inequalities.

The Vlasov equation in a fixed spacetime is a linear hyperbolic equation for a scalar function and hence solving it is equivalent to solving the equations for its characteristics. In coordinate components these are:

$$\begin{aligned} dX^a/ds &= P^a, \\ dP^a/ds &= -\Gamma^a_{\beta\gamma}P^\beta P^\gamma \end{aligned} \tag{4}$$

Let $X^a(s, x^\alpha, p^a)$, $P^a(s, x^\alpha, p^a)$ be the unique solution of (4) with initial conditions $X^a(t, x^\alpha, p^a) = x^a$ and $P^a(t, x^\alpha, p^a) = p^a$. Then the solution of the Vlasov equation can be written as:

$$f(x^\alpha, p^a) = f_0(X^a(0, x^\alpha, p^a), P^a(0, x^\alpha, p^a)) \tag{5}$$

where f_0 is the restriction of f to the hypersurface $t = 0$. This function f_0 serves as initial datum for the Vlasov equation. It follows immediately from this that if f_0 is bounded by some constant C, the same is true of f. This obvious but important property of the solutions of the Vlasov equation is used frequently without comment in the study of this equation.

The above calculations involving $T_{\alpha\beta}$ and N^α were only formal. In order that they have a precise meaning it is necessary to impose some fall-off in the momentum variables on f so that the integrals occurring exist. The simplest condition to impose is that f has compact support for each fixed t. This property holds if the initial datum f_0 has compact support and if each hypersurface $t = t_0$ is a Cauchy hypersurface. For by the definition of a Cauchy hypersurface, each timelike curve which starts at $t = 0$ hits the hypersurface $t = t_0$ at a unique point. Hence the geodesic flow defines a continuous mapping from the part of the mass shell over the initial hypersurface $t = 0$ to the part over the hypersurface $t = t_0$. The support of $f(t_0)$, the restriction of f to the hypersurface $t = t_0$ is the image of the support of f_0 under this continuous mapping and so is compact. Let $P(t)$ be the supremum of the values of $|p^a|$ attained on the support of $f(t)$. It turns out that in many cases controlling the solution of the Vlasov equation coupled to some field equation in the case of compactly supported initial data for the distribution function can be reduced to obtaining a bound for $P(t)$. An example of this is given below.

The data in the Cauchy problem for the Einstein equations coupled to any matter source consist of the induced metric g_{ab} on the initial hypersurface, the second fundamental form k_{ab} of this hypersurface and some matter data. In fact these objects should be thought of as objects on an abstract 3-dimensional manifold S. Thus the data consist of a Riemannian metric g_{ab}, a symmetric tensor k_{ab} and appropriate matter data, all defined intrinsically on S. The nature of the initial data for the matter will now be examined in the case of the Einstein-Vlasov system. It is not quite obvious what to do, since the distribution function f is defined on the mass shell and so the obvious choice of initial data, namely the restriction of f to the initial hypersurface, is not appropriate. For it is defined on the part of the mass shell over the initial hypersurface and this is not intrinsic to S. This difficulty can be overcome as follows. Let ϕ be the mapping which sends a point of the mass shell over the initial hypersurface to its orthogonal projection onto the tangent space to the initial hypersurface. The map ϕ is a diffeomorphism. The abstract initial datum f_0 for f is taken to be a function on the tangent bundle of S. The initial condition imposed is that the restriction of f to the part of the mass shell over the initial hypersurface should be equal to f_0 composed with ϕ. An initial data set for the Einstein equations must satisfy the constraints and in order that the definition of an abstract initial data set for the Einstein equations be adequate it is necessary that the constraints be expressible purely in terms of the abstract initial data. The constraint equations are:

$$R - k_{ab}k^{ab} + (\mathrm{tr}k)^2 \;=\; 16\pi\rho \tag{6}$$

$$\nabla_a k_b^a - \nabla_b(\mathrm{tr}k) \;=\; 8\pi j_b \tag{7}$$

Here R denotes the scalar curvature of the metric g_{ab}. If n^α denotes the future-pointing unit normal vector to the initial hypersurface and $h^{\alpha\beta} = g^{\alpha\beta} + n^\alpha n^\beta$ is the orthogonal projection onto the tangent space to the initial hypersurface then $\rho = T_{\alpha\beta}n^\alpha n^\beta$ and $j^\alpha = -h^{\alpha\beta}T_{\beta\gamma}n^\gamma$. The vector j^α satisfies $j^\alpha n_\alpha = 0$ and so can be naturally identified with a vector intrinsic to the initial hypersurface, denoted here by j^a. What needs to be done is to express ρ and j_a in terms of the intrinsic initial data. They are given by the following expressions:

$$\rho \;=\; \int \mathring{f}_0(p^a)p^a p_a/(1 + p^a p_a)^{1/2}(^{(3)}g)^{1/2}dp^1 dp^2 dp^3 \tag{8}$$

$$j_a \;=\; \int f_0(p^a)p_a(^{(3)}g)^{1/2}dp^1 dp^2 dp^3 \tag{9}$$

If a three-dimensional manifold on which an initial data set for the Einstein-Vlasov system is defined is mapped into a spacetime by an embedding ψ then the embedding is said to induce the given initial data on S if the induced metric and second fundamental form of $\psi(S)$ coincide with the results of transporting g_{ab} and k_{ab} with ψ and the relation $f = f_0 \circ \phi$ holds, as above. A form of the local existence and uniqueness theorem can now be stated. This will only be done for the case of smooth (i.e., infinitely differentiable) initial data although versions of the theorem exist for data of finite differentiability.

Theorem 1.1 *Let S be a 3-dimensional manifold, g_{ab} a smooth Riemannian metric on S, k_{ab} a smooth symmetric tensor on S and f_0 a smooth non-negative function of compact support on the tangent bundle TS of S. Suppose further that these objects satisfy the constraint equations. Then there exists a smooth spacetime $(M, g_{\alpha\beta})$, a smooth distribution function f on the mass shell of this spacetime and a smooth embedding ψ of S into M which induces the given initial data on S such that $g_{\alpha\beta}$ and f satisfy the Einstein-Vlasov system and $\psi(S)$ is a Cauchy hypersurface. Moreover, given any other spacetime $(M', g'_{\alpha\beta})$, distribution function f' and embedding ψ' satisfying these conditions, there exists a diffeomorphism χ from an open neighborhood of $\psi(S)$ in M to an open neighborhood of $\psi'(S)$ in M' which satisfies $\chi \circ \psi = \psi'$ and carries $g_{\alpha\beta}$ and f to $g'_{\alpha\beta}$ and f' respectively.*

The formal statement of this theorem is rather complicated, but its essential meaning is as follows. Given an initial data set (satisfying the constraints) there exists a corresponding solution of the Einstein-Vlasov system and this solution is locally unique up to diffeomorphism. There also exists a global uniqueness statement which uses the notion of the maximal Cauchy development of an initial data set, but this is not required in the following. The first proof of a theorem of the above kind for the Einstein-Vlasov system was given by Yvonne Choquet-Bruhat in [5].

The problem of extending this local theorem to one which is in some sense global is a very difficult one. In fact with presently available mathematical techniques it is too difficult. One way of making some progress in understanding the general problem is to study the simplified cases obtained by imposing various symmetries on the solutions. Note that if a symmetry is imposed on the initial data for the Cauchy problem this is inherited by the corresponding solutions. (See [8], Section 5.6 for a discussion of this.) This ensures the consistency of restricting the problem to a particular symmetry class.

In the following different symmetry classes will be considered in turn, proceeding from the strongest to the weakest assumptions. First spatially homogeneous solutions are discussed. These are simple enough that it is possible to make statements about the Einstein equations coupled to a general class of matter models. After this has been done, further results which can be obtained in the particular case where the matter is described by the Vlasov equation are presented. The first inhomogeneous solutions to be discussed are those which are static and spherically symmetric. Apart from their intrinsic interest these are of relevance for the study of spherically symmetric collapse which is discussed next. Brief comments are made on dynamical cosmological solutions before concentrating on one question where there are results on solutions of the Einstein-Vlasov system without any symmetry assumptions being required. This concerns the construction of solutions with an isotropic singularity.

2. Spatially homogeneous solutions I: general matter models

A solution of the Einstein equations coupled to some matter equations is said to be symmetric under the action of a Lie group G if G acts by isometries of the metric which also leave the matter fields invariant. A solution of the Einstein-matter equations is called spatially homogeneous if it is symmetric under the action of a Lie group whose orbits are spacelike hypersurfaces. If we think of these as hypersurfaces of constant time then the metric only depends on time and the Einstein equations reduce to ordinary differential equations, an enormous simplification. The equations of motion of matter fields which are defined on spacetime also reduce to ODE's. Since the Vlasov equation is defined on the mass shell it in general still contains derivatives with respect to the momenta and thus remains a partial differential equation in the spatially homogeneous case.

The spatially homogeneous spacetimes can be classified into various types according to the Lie group involved. The conventional terminology is that there are nine Bianchi types I-IX and one additional type, the Kantowski-Sachs models. The latter will not be discussed further here. We might like to use spatially homogeneous spacetimes as cosmological models. Since there is in a sense no spatial dependence we do need to worry about spatial boundary conditions. Eventually, however, we would like to consider more realistic cosmological models which are inhomogeneous perturbations of the Bianchi spacetimes, and then boundary conditions become important. One simple condition to pose is that the spacetimes involved should contain a compact Cauchy surface. Then there is no danger of extra information coming in from infinity. Supposing that it is desired to impose this condition of spatial compactness the question arises if all Bianchi types are compatible with it. Unfortunately this is not the case. In fact the only ones which are are types I and IX.

A larger class of spatially compact spacetimes where the Einstein equations reduce to ODE's are the locally spatially homogeneous ones, as defined in [23]. The idea is to require that the spacetime itself be spatially compact while its universal cover is spatially homogeneous. For details see [23]. This allows a much bigger class of Bianchi types to be included.

The idea now is to consider solutions of the Einstein-matter equations which are spatially compact and locally spatially homogeneous only assuming some general conditions on the matter model. These conditions are satisfied in the case of the Vlasov equation but also, for example, in the case of the Euler equation describing a perfect fluid with a physically reasonable equation of state. This generality is a luxury we can only afford due to the assumption of (local) spatial homogeneity. In general solutions of the Euler equations can be expected to form shocks which leads to a breakdown of the solution of the evolution equations. In the homogeneous case this possibility does not arise. It is the absence of shock formation which makes the Vlasov equation particularly convenient to work with when studying inhomogeneous spacetimes.

The matter models to be considered will be defined in terms of some general properties. As usual $T^{\alpha\beta}$ denotes the energy-momentum tensor. When a specific matter model has been chosen $T^{\alpha\beta}$ will be a functional of some matter variables, denoted collectively by F, and the spacetime metric $g_{\alpha\beta}$. In the following another quantity N^{α} (called the particle current density) will be required. It is also assumed to be a functional of F and $g_{\alpha\beta}$. Now various properties which will be assumed at appropriate points will be listed.

(1) $T^{\alpha\beta}V_{\alpha}W_{\beta} \geq 0$ for all future-pointing timelike vectors V^{α} and W^{α} (dominant energy condition)

(2) $T^{\alpha\beta}(g_{\alpha\beta} + V_{\alpha}V_{\beta}) \geq 0$ for all unit timelike vectors V^{α} (non-negative sum pressures condition)

(3) for any F and $g_{\alpha\beta}$ the conditions $\nabla_{\alpha}N^{\alpha} = 0$ and $\nabla_{\alpha}T^{\alpha\beta} = 0$ are satisfied (conservation conditions)

(4) for any F and $g_{\alpha\beta}$ the vector N^{α} is future-pointing timelike or zero

(5) for any constant $C_1 > 0$ there exists a positive constant C_2 such that for any F and $g_{\alpha\beta}$ with $-N_{\alpha}N^{\alpha} \leq C_1$ and any timelike vector V^{α} the following inequality holds:
$$T^{\alpha\beta}V_{\alpha}V_{\beta} \geq C_2(N^{\alpha}V_{\alpha})^2$$

(6) for any constant $C_1 > 0$ there exists a positive constant $C_2 < 1$ such that for any F and $g_{\alpha\beta}$ with $-N^{\alpha}N_{\alpha} \leq C_1$ and any unit timelike vector V^{α}
$$(g_{\alpha\beta} + V_{\alpha}V_{\beta})T^{\alpha\beta} \leq 3C_2T^{\alpha\beta}V_{\alpha}V_{\beta}$$

(7) if a solution with the given symmetry of the Einstein equations coupled to the given matter model is such that the time coordinate defined above takes all values in the interval (t_1, t_2), if it is not possible to extend the spacetime so as to make this interval longer and if t_1 or t_2 is finite then $\operatorname{tr}k(t)$ is unbounded in a neighborhood of t_1 or t_2 respectively.

(8) for any constant $C_1 > 0$ there exists a constant $C_2 > 0$ such that $T_{\alpha\beta}T^{\alpha\beta} \leq C_1$ implies $-N_{\alpha}N^{\alpha} \leq C_2$

Some comments will now be made concerning the physical motivation of some of these conditions. If a given type of matter can be considered as being made up of particles then a particle current density N^{α} is defined. If the particles have positive rest mass then this vector is future pointing timelike or zero as required by condition (4). If the particles are massless then this condition is still satisfied except for very special types of matter where N^{α} might be null. If particles cannot be created or destroyed then N^α is divergenceless as required in condition (3). It is not easy to give an intuitive interpretation of conditions (5) and (6). The meaning of (5) is roughly as follows. If matter is observed from a boosted frame then the particle density is multiplied by a γ-factor arising from the effect of Lorentz contraction on the volume element. The observed energy density is also affected in this way but picks up an additional γ-factor. Hence when a given matter distribution is considered from a boosted frame the multiplicative factor in the observed energy density behaves like the square of that in the observed particle density. As for condition (6), for a perfect fluid it is related to the condition that the speed of sound should be bounded away from the speed of light. The given symmetry referred to in condition (7) will be the Bianchi symmetry being considered.

In a spatially compact spacetime an interesting quantity is the volume of the hypersurfaces of constant time. The Friedman-Robertson-Walker (FRW) models generally used in cosmology have the property that this volume either increases at all times (open models) or increases to maximum after which it decreases again (closed models). The possibility that it always decreases, a case which is allowed mathematically, is usually ignored since we know that our universe is expanding at the present time. In any case these models can be obtained from those where the volume is always increasing by reversing the direction of time. Thus they present no essentially new mathematical phenomena.

In local spatially homogeneous spatially compact spacetimes with reasonable matter the same pattern is found. It can be proved that they also share some other significant physical properties with the FRW models. These spacetimes can be parametrized by a Gaussian time coordinate based on a (locally) homogeneous hypersurface. Suppose that the solution is maximal in the sense that it cannot be extended to a larger interval of Gaussian time. Then we would like to know two things. Firstly, if the volume is always increasing then the time of existence in the future is infinite and the spacetime is future geodesically complete. Secondly, if the volume is increasing at some time then the time of existence in the past is finite and as the time of breakdown is approached some geometrical invariant of the spacetime geometry diverges. This rules out the possibility of extending the spacetime in some way which is not globally hyperbolic. This is desirable from the point of view of the strong cosmic censorship hypothesis.

In [23] theorems of the desired type were proved. The first says that if conditions (1), (2) and (7) above are satisfied by some matter model then inextendible locally spatially homogeneous spatially compact solutions of the Einstein-matter equations where the volume is always increasing satisfy the first conclusion mentioned above. In particular, they are future geodesically complete. The second says that if (1)–(8) are satisfied and the spacetime is not vacuum then the curvature invariant $G^{\alpha\beta}G_{\alpha\beta}$ diverges as the finite past limit of the domain of existence is approached. The fundamental intuitive reason for this is that a finite amount of matter is being squeezed to zero volume as the singularity is approached.

These theorems apply to the Einstein-Vlasov system since in that case, as will now be discussed, conditions (1)–(8) are satisfied. In that case $T^{\alpha\beta}$ and N^{α} have been defined above. It has also been stated that (1) and (3) hold. Condition (2) is a consequence of the non-negative pressures condition mentioned above. To check condition (4) it is merely necessary to observe that if V^{α} is a future-pointing timelike vector then $N_{\alpha}V^{\alpha} < 0$ unless $N^{\alpha} = 0$ at a given spacetime point. Condition (7) is an existence theorem which says, roughly speaking, that as long as the geometry does not break down too badly, the solution of the matter equations (in this case the Vlasov equation) cannot break down in finite time. Here the mean curvature $\mathrm{tr}k$ plays the role of a controlling quantity whose boundedness ensures the continued existence of the solution. It remains to check conditions (5), (6) and (8) and to do this we may choose a frame whose timelike member is V^{α} in order to do the calculation. Then the inequalities of (5) and (6) become $\hat{T}^{00} \geq C_2(\hat{N}^0)^2$ and $\delta^{ij}\hat{T}_{ij} \leq 3C_2\hat{T}^{00}$. (The hats here indicate the use of indices associated to an orthonormal frame.)

$$
\begin{aligned}
-N_{\alpha}N^{\alpha} &= -\left(\int f(p^a)p_{\alpha}/p^0 dp^1 dp^2 dp^3\right)\left(\int f(q^a)q^{\alpha}/q^0 dq^1 dq^2 dq^3\right) \\
&= -\int\int f(p^a)f(q^a)p^{\alpha}q_{\alpha}/(p^0 q^0)dp^1 dp^2 dp^3 dq^1 dq^2 dq^3 \\
&\geq \int\int f(p^a)f(q^a)/(p^0 q^0)dp^1 dp^2 dp^3 dq^1 dq^2 dq^3 \\
&= \left(\int f(p^a)/p^0 dp^1 dp^2 dp^3\right)^2
\end{aligned}
$$

Hence

$$
\begin{aligned}
\hat{N}^0 &= \int f(p^a)dp^1 dp^2 dp^3 \\
&\leq \left(\int f(p^a)p^0 dp^1 dp^2 dp^3\right)^{1/2}\left(\int f(p^a)/p^0 dp^1 dp^2 dp^3\right)^{1/2} \\
&\leq (\hat{T}^{00})^{1/2}(-N_{\alpha}N^{\alpha})^{1/4}
\end{aligned}
$$

This shows that (5) holds. It follows directly from the definitions that the inequality of (6) holds with $C_2 = 1/3$ even without restricting $N_{\alpha}N^{\alpha}$ to be bounded. Finally

$$
\begin{aligned}
T_{\alpha\beta}T^{\alpha\beta} &= \left(\int f(p^a)p_{\alpha}p_{\beta}/p^0 dp^1 dp^2 dp^3\right)\left(\int f(q^a)q^{\alpha}q^{\beta}/q^0 dq^1 dq^2 dq^3\right) \\
&= \int\int f(p^a)f(q^a)(p^{\alpha}q_{\alpha})^2/(p^0 q^0)dp^1 dp^2 dp^3 dq^1 dq^2 dq^3 \\
&\geq -\int\int f(p^a)f(q^a)(p^{\alpha}q_{\alpha})/(p^0 q^0)dp^1 dp^2 dp^3 dq^1 dq^2 dq^3 \\
&= -N_{\alpha}N^{\alpha}
\end{aligned}
$$

Thus the general theorems apply to give results on the dynamics of locally spatially homogeneous solutions of the Einstein-Vlasov system. It is known that Bianchi type IX solutions cannot expand forever, while models of the other types force the volume to be monotone. Thus we can say the following about inextendible non-vacuum spatially compact solutions of the Einstein-Vlasov system with Bianchi symmetry. If the Bianchi type is IX they have curvature singularities after finite proper time both in the past and in the future. For all other Bianchi types models which are expanding at some time have a curvature singularity at a finite time in the past and are future geodesically complete. The statement about geodesic completeness also holds in the vacuum case. The statement about curvature singularities, however, does not. In some cases there is a Cauchy horizon. This has been discussed in detail in [6]. See also [30].

3. Spatially homogeneous solutions II: application of dynamical systems

We have now obtained a crude picture of the dynamics of spatially homogeneous cosmological models of the Einstein-matter equations for a variety of matter models. It is reasonable to hope that in the case of the Vlasov equation this can be considerably refined, in order to get a detailed picture of the asymptotics of the models near an initial singularity or in a phase of unlimited expansion. This has not yet been achieved in general but for certain cases results were obtained in [28]. These extended theorems concerning the simpler case of massless particles obtained in [27]. The models considered were of the simplest Bianchi types I, II and III. It was assumed that a further symmetry is present so that the spacetimes have a total of four Killing vectors. These are the so-called LRS models (locally rotationally symmetric). The reason for making this assumption is that then the Vlasov equation can be solved explicitly and only the Einstein equations remain to be handled. The equations can be reduced to a system of ODE's in contrast to the general case, although the coefficients of the system involve one function which is not known explicitly. Fortunately it suffices to know certain qualitative features of this function in order to determine the asymptotic behavior of the solutions of the ODE's at early and late times. In fact to do this analysis it was also necessary to assume invariance under certain reflections, a piece of information which will be suppressed in the following for simplicity.

To describe the results it is useful to introduce the generalized Kasner exponents p_i. Each of the homogeneous hypersurfaces has an induced metric g_{ab} and a second fundamental form k_{ab}. Let λ_i denote the eigenvalues of the second fundamental form with respect to the metric. By definition this means that they are the solutions of the eigenvalue equation $\det(k_{ab} - \lambda g_{ab}) = 0$. Suppose now that $\mathrm{tr} k$ is non-zero. Then we can define $p_i = \lambda_i / (\sum_j \lambda_j)$. The quantities p_i are functions of t and their sum is equal to one. The Kasner solution of the vacuum Einstein equations is given by

$$ds^2 = -dt^2 + t^{2p_1} dx^2 + t^{2p_2} dy^2 + t^{2p_3} dz^2 \qquad (10)$$

for the Kasner exponents p_i, which are constants satisfying $\sum_j p_j = 1$ and $\sum_j p_j^2 = 1$. These equations are called the first and second Kasner relations. In this case the generalized Kasner exponents are equal to the quantities p_i in this metric form and this explains their name.

Consider a solution of the Einstein-Vlasov system of Bianchi type I which is LRS. Choose the time of the initial singularity to be $t = 0$. The following statements hold. For each i the quantity $p_i(t)$ converges to $1/3$ as $t \to \infty$. This is the value of the generalized Kasner exponents in a spatially flat FRW model. This means that the spacetime isotropizes at late times. At early times there are three possibilities. The first is that the p_i are identically equal to $1/3$ at all times. This is the FRW case. The second is that they have limits $(1/2, 1/2, 0)$. The third, which is the generic case, is that they have the limits $(2/3, 2/3, -1/3)$. In this last case the second Kasner relation is satisfied asymptotically and so, in a certain sense, the solution with matter is approximated near the singularity by a vacuum solution.

In type III the initial singularity is similar to that in type I but in type II it is oscillatory. As $t \to 0$ the generalized Kasner exponents approach both the values $(2/3, 2/3, -1/3)$ and $(0, 0, 1)$ as closely as desired in any neighborhood of $t = 0$. In the expanding direction the type II solution approaches an explicitly known self-similar dust spacetime whose generalized Kasner exponents have the values $(3/8, 3/8, 1/4)$. In type III the generalized Kasner exponents approach $(0, 0, 1)$ in the expanding direction. It is possible to get more detailed information about the asymptotics of these models in the limits $t \to 0$ and $t \to \infty$. The proofs of all these statements make use of the fact that for ODE's the highly developed theory of dynamical systems is available. Getting out the finer features of the expanding phase of Bianchi III models is more delicate than the other cases just discussed and was carried out using centre manifold theory in [25].

A key aspect of the work on the asymptotics of types I, II and III was writing the dynamical system in cleverly chosen variables. As we go to a singularity or to infinity in a phase of unlimited expansion the most obvious variables go to zero or infinity. If dimensionless variables can be found which, at least for some of the solutions, converge to finite non-zero limits in these regimes then this is a great help in analyzing the asymptotics. Often the original dynamical system can be extended to a smooth dynamical system on a compact region. This avoids the problem, found with many choices of variables, that the solutions expressed in the given variables run off to infinity in a way which is hard to control. The strategy of dimensionless variables and compactification has been carried much further in the case of spatially homogeneous solutions of the Einstein-Euler system. For an account of this see [33].

4. Static solutions

This section is concerned with static spherically symmetric solutions of the Einstein-Vlasov system. These may play a role in describing the long-time behavior of solutions of the full dynamical equations and so they have a natural place in these lectures. Before coming to the Einstein-Vlasov system it is worth spending a little time thinking about the corresponding non-relativistic problem, where a lot more is known. In that case the gravitational potential U satisfies the Poisson equation $\Delta U = 4\pi\rho$ where

$$\rho(t,x) = \int f(t,x,v)dv \tag{11}$$

and the Vlasov equation is

$$\partial f/\partial t + v^a \partial f/\partial x^a - \nabla^a U \partial f/\partial v^a = 0 \tag{12}$$

These equations together constitute the Vlasov-Poisson system. There are results on static solutions of the Vlasov-Poisson system which are not spherically symmetric and stationary solutions which are not static [16]. Nothing comparable has yet been done in the case of the Einstein-Vlasov system. This is a gap which should be filled. From now on we restrict consideration to the static spherically symmetric case.

There are two methods which have been used to construct static spherically symmetric solutions of the Vlasov-Poisson system. The first may be called the ODE method. In a spherically symmetric static spacetime there are two constants of motion of the particles, namely the energy $E = (1/2)|v|^2 + U$ and the modulus L of the angular momentum defined by $L^2 = |x|^2|v|^2 - (x \cdot v)^2$, which are useful in constructing solutions of the Vlasov equation. In fact E and L, like any quantity conserved along geodesics, are solutions of the Vlasov equation. The same is true of any function $\Phi(E, L)$ of these quantities. In other words a function f of the form

$$f(x,v) = \Phi(E(x,v), L(x,v)) \tag{13}$$

is a time-independent solution of the Vlasov equation with the potential U. Jeans' theorem says that conversely in a spherically symmetric static solution of the Vlasov-Poisson system the distribution function is a function of E and L. Thus a natural procedure is to make an ansatz for the distribution function by choosing a particular function Φ. The Einstein equations then reduce to a system of integro-differential equations for the metric coefficients as functions of a radial coordinate. What remains to be done is to analyze the global properties of solutions of this system.

The second method is a variational one. The Vlasov-Poisson system can be expressed as an infinite-dimensional Hamiltonian system. (Cf. [10].) It is degenerate in the sense that instead of a symplectic structure there is only a Poisson structure. This leads to a large class of conserved quantities known as Casimir invariants. In a Hamiltonian system a minimum of the Hamiltonian is a time-independent solution of the equations of motion. This suggests a variational route to finding static solutions. When Casimir invariants are present there are more general possibilities. If C is a Casimir invariant then a minimum of $H + C$ is a time-independent solution of the equations of motion. An advantage of this method is that apart from giving results on the existence of static solutions it can also provide information on the stability of the solutions obtained. This method has been applied extensively to the Vlasov-Poisson system by Guo and Rein (see [17] and references therein).

Returning to the Einstein-Vlasov system, there is a paper [35] where the energy-Casimir method has been applied but it seems to be much harder than the Vlasov-Poisson case and the results are much more limited. More straightforward is the ODE method. One cautionary note is in order. The direct analogue of Jeans' theorem is not true in the case of the Einstein-Vlasov system. Counterexamples were constructed by Schaeffer [31]. Nevertheless we can still assume a distribution function of the form $\Phi(E, L)$ and proceed from there. A theorem on global existence in the radial coordinate for a rather general choice of Φ was obtained in [14]. There is a difficulty concerning the physical relevance of these solutions. If we would like to use them to model globular clusters, for instance, then we would like to obtain configurations of finite total mass. The easiest way to prove this is if the spatial density has compact support. The general existence theorem does not give any information on this. In fact whether it is true or not depends on the choice of the function Φ in quite a delicate way. A criterion for the finiteness of the mass in a large class of functions Φ was given in [19].

All known static solutions of the Einstein-Euler system with a physically reasonable equation of state are spherically symmetric and the density is a monotone decreasing function of the radius. In the case of the Einstein-Vlasov system another kind of configuration is possible where the support of the density is a thick shell, i.e., the region between two concentric spheres. In order to achieve this the function Φ must depend on the angular momentum. If it only depends on the energy then the system is equivalent to a solution of the Einstein-Euler system with an equation of state which is in general not explicitly known. The existence of shell solutions was proved in [15].

5. Spherically symmetric collapse

An interesting situation to consider is that of an isolated system consisting of matter undergoing gravitational collapse. The traditional model for this, following Oppenheimer and Snyder, is the collapse of a homogeneous spherical cloud of dust. Unfortunately when inhomogeneities are introduced into the Oppenheimer-Snyder model they often lead to pathologies such as shell-crossing singularities. The advantage of using collisionless matter is that it avoids some of (and perhaps all) the problems associated with dust.

A natural first step towards understanding spherical collapse is to fully understand the case where there is no collapse. If we have only a small amount of matter then it is to be expected that its self-gravitation will not suffice to keep it together and that it will spread out and disperse to infinity. For collisionless matter this has been proved, as described in more detail later. For dust it is not true since even small amounts of matter can develop shell-crossing singularities. Even dust without gravitation can do so and so this effect has nothing to do with gravity at all.

Consider initial data for the Einstein-Vlasov system which are spherically symmetric and asymptotically flat. In a suitable coordinate system only data for the distribution function need be given since the metric can then be determined by solving equations on the hypersurfaces of constant time. This is a reflection of the familiar statement that there is no gravitational radiation in spherical symmetry. Now let us make the initial data small in the following sense. In the presence of fixed bounds on the extent of the support of the initial data for f in position and velocity space we require the maximum of f to be small. For small data it can be shown that the solution exists globally in a suitable time coordinate and, more importantly, that it is geodesically complete. Moreover, various quantities such as the energy density of the matter decay to zero as $t \to \infty$ [18]. Thus it can be seen that for small data the solution disperses and the situation is completely under control.

What happens for large data? It is known that for large data a trapped surface (and presumably a black hole) can form [22]. We might nevertheless get global existence in a singularity-avoiding time slicing like maximal or polar slicing. The latter is also sometimes called a Schwarzschild time coordinate. There is a theorem [20] which says that if a singularity forms in a solution of the spherically symmetric Einstein-Vlasov system then the first singularity (as measured in Schwarzschild time) occurs at the centre of symmetry. Note that the shell-crossing singularities of dust occur away from the centre. A corresponding result for maximal slicing has been proved in [24].

A general mathematical result on the behavior of spherically symmetric asymptotically flat solutions of the Einstein-Vlasov system has not yet been obtained. In the absence of further analytical progress, attempts have been made to study the problem numerically. One theme which plays an important role is that of critical collapse. Suppose that we have a family of data depending on a parameter λ for the spherically symmetric Einstein-Vlasov system which interpolates between weak and strong data, with $\lambda = 0$ corresponding to data for flat space. For λ sufficiently small the theorem already mentioned tells us that the matter disperses. For λ sufficiently large we might expect collapse to a black hole and this is indeed seen numerically. More precisely, it is seen that for λ smaller than a certain value λ_* the matter disperses while for $\lambda > \lambda_*$ it collapses to a black hole. In general some of the matter falls into the black hole while some escapes. Let $M(\lambda)$ be the mass of the black hole formed when the initial data corresponds to the parameter value λ. If no black hole is formed $M(\lambda)$ is defined to be zero. One of the questions which comes up in the study of critical collapse is whether the function $M(\lambda)$ is continuous at λ_* or not.

In [21] numerical evidence was presented that $M(\lambda)$ is not continuous. In other words, the limit of $M(\lambda)$ as $\lambda \to \lambda_*$ from above is strictly positive. This is different from what is found for some other matter models, such as the massless scalar field. Olabarrieta and Choptuik [13] confirmed this finding and were able to present a more detailed picture of what happens. It is convenient for the numerical calculations to take initial data where there are no particles at the centre and no particles on purely radial orbits. In that case as long as the solution remains regular no particle can reach the centre due to conservation of angular momentum. Thus we have a dynamical configuration of matter with a hole in the middle. This allows difficulties with the singularity of polar coordinates at $r = 0$ to be avoided. It is found in [13] that the solution evolves towards an unstable static shell solution before turning away again and dispersing or collapsing. The mass of the shell solution sets the mass gap in the graph of $M(\lambda)$. The connection between the shell solutions observed numerically in collapse calculations and those whose existence has been shown rigorously is not clear.

6. Isotropic singularities and Fuchsian methods

In the last section results for certain asymptotically flat solutions of the Einstein-Vlasov system were described. These are spherically symmetric and hence have three Killing vectors. There seem to be no other symmetry assumptions on asymptotically flat spacetimes which can be usefully studied at the present time. The obvious symmetry class which comes to mind is axial symmetry. In that case, however, there is only one Killing vector, which is very little, and even that has fixed points, which leads to singularities of the equations obtained when the symmetry is factored out. At the moment it seems to offer no advantage over the general case. In the case of spacetimes evolving from data on a compact Cauchy surface (cosmological spacetimes) there is a variety of interesting symmetry types with two or three Killing vectors and a number of papers on solutions of the Einstein-Vlasov system with these symmetries. They will not be reviewed here since some choices had to be made in order to limit the volume of the lectures. A good point of entry into the literature is [1].

There is one mathematical result on the Einstein-Vlasov system which does not require any symmetry assumptions and it will be the subject of the remainder of this section. It concerns solutions of the Einstein-Vlasov system with massless particles and it would be interesting to know if analogous results hold for massive particles. The idea is to construct large classes of solutions of the equations whose singularities have a particular structure, the isotropic singularities.

Given a spacetime with a foliation by spacelike hypersurfaces we can define the generalized Kasner exponents as in Section 3 in terms of the eigenvalues of the second fundamental form. In the inhomogeneous case these are functions p_i on spacetime which in general depend on both the time and space coordinates.

The condition for an isotropic singularity (at least intuitively) is that all generalized Kasner exponents should tend to 1/3 as the singularity is approached. Thus the solution looks like an isotropic FRW model near the singularity. The actual definition used in the theorem is a different one. A spatially flat FRW model is conformally flat. In the definition of an isotropic singularity it is assumed that the given metric is conformal to a metric which is smooth at the singularity.

The requirement of an isotropic singularity is a restriction on the spacetimes considered but it has been proved by Anguige [2] that there is a very large class of solutions of the Einstein-Vlasov system with isotropic singularities. In particular, he does not have to make any symmetry assumptions. The solutions can be parametrized by certain data on the singularity which can be given freely. The method of proving this is to use Fuchsian methods (although Anguige does not use this terminology). This technique is of wider importance in studying singularities of solutions of the Einstein-matter equations and will now be discussed in a more general context.

Let a system of partial differential equations with smooth coefficients be given and suppose we would like to investigate the existence of solutions which become singular on a certain hypersurface. For simplicity assume that this is the coordinate hypersurface $t = 0$. Suppose that in some way it was possible to guess the asymptotic behavior of the solutions in the approach to the singularity. This might be done by studying explicit solutions or by using trial and error to get a formally consistent asymptotic expansion. Then express the solution u to be constructed in terms of an explicit function u_0 having the expected asymptotics near the singularity and a remainder v which is expected to be regular and vanish at $t = 0$. Now rewrite the original equation for the unknown u as an equation for v whose coefficients depend on u_0. Since u_0 is singular it is to be expected that the equation for v is singular at $t = 0$. Thus the problem of finding a singular solution of a regular equation has been replaced by that of finding a regular solution of a singular equation.

In favorable cases the singular equation obtained by this method is a Fuchsian equation of the form

$$t\partial_t v + N(x)v = tf(t, x, v, v_x) \tag{14}$$

where the matrix-valued function N has some positivity property. There are theorems which guarantee that an equation of this kind has a unique solution v which is regular and vanishes at $t = 0$. One theorem of this kind was proved in [12], where it was applied to study singularities in Gowdy spacetimes. Since then there have been a number of other applications. (See [26], Section 6.2, for more information on this.) Unfortunately the Einstein-Vlasov system does not fit into the framework of this theorem due to the fact that it is an integrodifferential equation rather than a differential equation. For this reason Anguige had to prove his theorem by doing a direct iteration.

In general the reduction of a given system to Fuchsian form requires a considerable amount of algebraic manipulation. In particular, if the original system includes second-order equations they must be reduced to first-order by introducing additional variables. A detailed exposition of the procedure is given in [11]. Examples related to the Einstein equations are discussed in [29]

7. Outlook

In these lectures a selection of work on the Einstein-Vlasov system has been surveyed. Although the results were only discussed on a very general level, without getting into details, it was still necessary to leave out a lot of interesting topics. Some of them were mentioned briefly, some not at all. It should be clear that this is an area of research where there are many open problems and many promising directions to be explored. The references given here should provide a good starting point for those wanting to follow this road.

References

[1] Andréasson, H., Rein, G. and Rendall, A.D. 2003 On the Einstein-Vlasov system with hyperbolic symmetry. Math. Proc. Camb. Phil. Soc. 134, 529–549.

[2] Anguige, K. 2000 Isotropic cosmological singularities 3: The Cauchy problem for the inhomogeneous conformal Einstein-Vlasov equations. Ann. Phys. (NY) 282, 395–419.

[3] Binney, J., Tremaine, S. 1987 Galactic dynamics. Princeton University Press, Princeton.

[4] Börner, G. 1993 The early universe. Facts and fiction. Springer, Berlin.

[5] Choquet-Bruhat, Y. 1971 Problème de Cauchy pour le système intégro différentiel d'Einstein-Liouville. Ann. Inst. Fourier 21, 181–201.

[6] Chruściel, P.T. and Rendall, A.D. 1995 Strong cosmic censorship in vacuum spacetimes with compact locally homogeneous Cauchy surfaces. Ann. Phys. (NY) 242, 349–385.

[7] Ehlers, J. 1973 Survey of general relativity theory. In: Israel, W. (ed.) Relativity, Astrophysics and Cosmology. Reidel, Dordrecht.

[8] Friedrich, H., Rendall, A.D. 2000 The Cauchy problem for the Einstein equations. In B.G. Schmidt (ed) Einstein's Field Equations and Their Physical Implications. Lecture Notes in Physics 540. Springer, Berlin.

[9] Hawking, S.W., Ellis, G.F.R. 1973 The large-scale structure of space-time. Cambridge University Press, Cambridge.

[10] Holm, D.D., Marsden, J.E., Ratiu, T. and Weinstein, A. 1985 Nonlinear stability of fluid and plasma equilibria. Phys. Rep. 123, 1–116.

[11] Kichenassamy, S. 1996 Nonlinear wave equations. Marcel Dekker, New York.

[12] Kichenassamy, S., Rendall, A.D. 1998 Analytic description of singularities in Gowdy spacetimes. Class. Quantum Grav. 15, 1339–1355.

[13] Olabarrieta, I., Choptuik, M.W. 2002 Critical phenomena at the threshold of black hole formation for collisionless matter in spherical symmetry. Phys. Rev. D65, 024007.

[14] Rein, G. 1994 Static solutions of the spherically symmetric Vlasov-Einstein system. Math. Proc. Camb. Phil. Soc. 115, 559–570.

[15] Rein, G. 1999 Static shells for the Vlasov-Poisson and Vlasov-Einstein systems. Indiana University Math. J. 48, 335–346.

[16] Rein, G. 2000 Stationary and static stellar dynamic models with axial symmetry. Nonlinear Analysis; Theory, Methods and Applications 41, 313–344.

[17] Rein, G. 2002 Stability of spherically symmetric steady states in galactic dynamics against general perturbations. Arch. Rat. Mech. Anal. 161, 27–42.

[18] Rein, G., Rendall, A. D. 1992 Global existence of solutions of the spherically symmetric Vlasov-Einstein system with small initial data. Commun. Math. Phys. 150:561–583

[19] Rein, G., Rendall, A.D. 2000 Compact support of spherically symmetric equilibria in non-relativistic and relativistic galactic dynamics. Math. Proc. Camb. Phil. Soc. 128, 363–380.

[20] Rein, G., Rendall, A.D. and Schaeffer, J. 1995 A regularity theorem for solutions of the spherically symmetric Vlasov-Einstein system. Commun. Math. Phys. 168, 467–478.

[21] Rein, G., Rendall, A.D. and Schaeffer, J. 1998 Critical collapse of collisionless matter – a numerical investigation. Phys. Rev. D58, 044007

[22] Rendall, A.D. 1992 Cosmic censorship and the Vlasov equation. Class. Quantum Grav. 9, L99–L104.

[23] Rendall, A.D. 1995 Global properties of locally spatially homogeneous cosmological models with matter. Math. Proc. Camb. Phil. Soc. 118, 511–526.

[24] Rendall, A.D. 1997 An introduction to the Einstein-Vlasov system. Banach Centre Publications 41, 35–68.

[25] Rendall, A.D. 2002 Cosmological models and centre manifold theory. Gen. Rel. Grav. 34, 1277–1294.

[26] Rendall, A.D. 2002 Theorems on existence and global dynamics for the Einstein equations. Living Reviews in Relativity 5, 6.

[27] Rendall, A.D., Tod, K.P. 1999 Dynamics of spatially homogeneous solutions of the Einstein-Vlasov equations which are locally rotationally symmetric. Class. Quantum Grav. 16, 1705–1726.

[28] Rendall, A.D., Uggla, C. 2000 Dynamics of spatially homogeneous locally rotationally symmetric solutions of the Einstein-Vlasov equations. Class. Quantum Grav. 17, 4697–4714.

[29] Rendall, A.D., Blow-up for solutions of hyperbolic PDE and spacetime singularities. In Depauw, N., Robert, D. and Saint Raymond, X. (eds.) Journées Equations aux Dérivées Partielles, Nantes, 2000. CNRS, Nantes.

[30] Ringström, H. 2000 Curvature blow up in Bianchi VIII and IX vacuum spacetimes. Class. Quantum Grav. 17, 713–731.

[31] Schaeffer, J. 1999 A class of counterexamples to Jeans' theorem for the Vlasov-Einstein system. Commun. Math. Phys. 204, 313–327.

[32] Swanson, D.G., 1989 Plasma Waves. Academic Press, Boston.

[33] Wainwright, J., Ellis, G.F.R. 1997 Dynamical systems in cosmology. Cambridge University Press, Cambridge.

[34] Wald, R.M. 1984 General Relativity. Chicago University Press, Chicago.

[35] Wolansky, G. 2001 Static solutions of the Vlasov-Einstein system. Arch. Rat. Mech. Anal. 156, 205–230.

Alan D. Rendall
Max-Planck-Institut für Gravitationsphysik
Am Mühlenberg 1
D-14476 Golm, Germany
e-mail: `rendall@aei.mpg.de`

Future Complete U(1) Symmetric Einsteinian Spacetimes, the Unpolarized Case

Yvonne Choquet-Bruhat

1. Introduction

In this paper I generalize the non linear stability theorem obtained in collaboration with V. Moncrief (CB-M1, CB-M2) for vacuum Einsteinian 4-manifolds $(V,^{(4)}g)$ where $V = M \times R$ with M a circle bundle over a compact, orientable surface Σ of genus greater than 1. The Lorentzian metric $^{(4)}g$ admits a Killing symmetry along the (spacelike) circular fibers . I remove the so-called polarization condition, i.e., the orthogonality of the fibers to quotient 3-manifolds. The reduced field equations take now the form of a wave map equation, instead of a linear wave equation in the polarized case, coupled to $2 + 1$ gravity. I use results on wave maps from curved manifolds obtained in CB1, CB2. Like in CB-M2 we do not restrict the conformal geometry of Σ to avoid those regions of Teichmüller space for which the lowest positive eigenvalues of the scalar Laplacian lie, in our normalization, in the gap $(0, \frac{1}{8}]$. A consequence is that the asymptotic behavior of the wave map field does not exhibit a universal rate of decay but instead develops a decay rate which depends upon the asymptotic values of the lowest eigenvalue.

Under the Kaluza-Klein reduction which one carries out in the presence of the assumed spacelike Killing field one is first led to field equations of the type of an Einstein-Maxwell-Jordan system on the 3-manifold $\Sigma \times R$. To transform this to a more convenient Einstein-wave map system one needs a further topological restriction on the fields allowed. The need for this arises from considering the constraint equation for the effective $2 + 1$-dimensional electric type field density $\tilde{e} = e^a \frac{\partial}{\partial x^a}$ which reads $e^a_{,a} = 0$. On a higher genus surface Σ the general solution of this equation (which results from a Hodge decomposition of one-forms on Σ) takes the form $e^a = \varepsilon^{ab}(\omega_{,b} + h_b)$ where $h_b dx^b$ is a harmonic one-form on Σ. A consistent simplification results from setting this harmonic contribution to zero so that $e^a \frac{\partial}{\partial x^a}$ becomes expressible purely in terms of the so-called twist potential ω. Taken together with the norm of the ($U(1)$ generating) Killing field $\tilde{Y}$, conveniently expressed via $\tilde{Y}.\tilde{Y} = e^{2\gamma}$, the twist potential ω and the function γ provide a map from $\Sigma \times R$ to R^2. When expressed in terms of the pair (γ, ω) Einstein's equations take the form of a wave map from a $2 + 1$ Lorentzian manifold $(\Sigma \times R, {}^{(3)}g)$ into the Poincaré plane with its standard metric $2d\gamma^2 + \frac{1}{2}e^{-4\gamma}d\omega^2$; the metric ${}^{(3)}g$ satisfies the $2 + 1$ Einstein equations on $\Sigma \times R$, with source the wave map. These $2 + 1$ Einstein equations, supplemented by suitable coordinate conditions to fix the gauge, reduce to an elliptic system on each slice Σ_t of $\Sigma \times R$ for the lapse, the shift, and the conformal factor of a 2-dimensional metric, together with an ordinary differential system for the Teichmüller parameters which determine the conformal geometry. The wave map field and the Teichmüller parameters represent therefore the true propagating gravitational degrees of freedom of the original problem.

The basic methods we use to prove existence for an infinite proper time involve the construction of higher-order energies to control the Sobolev norms of the wave map and the solution of the differential system[1] satisfied by the Teichmüller parameters degrees of freedom. A subtlety is that the most obvious definition of wave map energies does not lead to a well-defined rate of decay so that suitable corrected energies must be developed which exploit information about the lowest eigenvalues of the spatial Laplacian which appears in the relevant wave operators. The eigenvalues vary with position in Teichmüller space and thus evolve along with Teichmüller parameters. If the lowest (non-trivial) eigenvalue asymptotically avoids a well-known gap in the spectrum (the gap $(0, \frac{1}{8}]$ in our normalization which has the more familiar form $(0, \frac{1}{4}]$ if one instead normalizes the Gauss curvature on the higher genus surface) then we obtain a universal rate of decay for the energies asymptotically. If the lowest eigenvalue however drifts into this gap and remains there asymptotically then the rate of decay of the energies will depend upon the asymptotic value of this lowest eigenvalue and will no longer be universal. We need slightly different forms for the corrected energies to handle these different eventualities (universal versus non universal rates of decay). In all cases the conformal geometry of our circle bundles undergoes a kind of Cheeger-Gromov collapse in which the circular fibers (after a conformal rescaling needed to take out the overall expansion) collapse to zero length asymptotically while only the conformal 2 - geometry remains well behaved. In our set up the Sobolev constants depend only on the conformal 2-geometry (i.e., upon the Teichmüller parameters) and, so long as the evolution remains in a compact subspace of Teichmüller space, these constants remain under control.

The sense in which our solutions are global in the expanding direction is that they exhaust the maximal range allowed for the mean curvature function on a manifold of negative Yamabe type, for which a zero mean curvature cannot be achieved but only asymptotically approached. In addition however our estimates prove that the normal trajectories to our spatial slices all have infinite future proper time length, and allow us to establish, using [CB-C], causal geodesic completeness in the expanding direction.

---

[1] We use here directly this system, instead of introducing the Dirichlet energy as in CB-M 1 and CB-M 2.

If the harmonic one-form H discussed above were allowed to be non zero it would disturb the pure wave map character of the reduced field equations. On the other hand it seems plausible that energy arguments could still be made to work in the presence of H. Alternatively one might simply refrain from trying to force the reduced field equations into a wave map framework and instead develop energy arguments for the Einstein-Maxwell-Jordan type system itself which require no splitting of $\tilde{e}$ into twist potential and harmonic contributions. I shall not however pursue either of these possibilities here but leave them for further study.

I need as in CB-M1 and CB-M2 a smallness condition on suitably defined energies which control the norms of the evolving (here wave map) field and for this reason I continue to restrict my attention to trivial S^1 bundles over Σ (i.e., to those for which $M = S^1 \times \Sigma$). The reason for this is that the curvature of the $U(1)$ connection and its assumed ($U(1)$-generating) Killing field has a quantized integral over Σ and, in the case of a non trivial bundle when this integral is not zero, cannot be adjusted to satisfy the smallness condition needed for the energy argument. It seems plausible that one could probably substract off this unavoidable topological contribution to curvature and work with suitable energies defined for the substracted fields to handle the case of non-trivial bundles but I shall not attempt to do so here.

Another approach (suggested by V. Moncrief) to treating solutions on non-trivial S^1 bundles involves applying a well-known action of $SL(2,R)$ (the isometry group of the Poincaré plane which plays the role of target for our wave map fields) to the fields defined on the base manifold $\Sigma \times R$. In certain cases this group action can be used to transform solutions which lift to the trivial S^1 bundle over $\Sigma \times R$ to other solutions which lift instead to another, non-trivial bundle. There is an obstruction to obtaining such solutions in this way since a certain Casimir invariant (which is of course preserved under the group action) is necessarily positive for solutions which lift to the trivial bundle (it can be negative for a subset of solutions which lift to non-trivial bundles). This formulation has so far only been developed for the case of circle bundles over $S^2 \times R$ but can most likely be generalized to the cases of bundles over $\Sigma \times R$ where Σ is either a torus or a higher genus surface. That possibility is left for further study.

The small data future global existence theorem for solutions of Einstein's equations of Andersson and Moncrief, this volume, makes no symmetry assumption whatsoever, but treats a different class of spatial 3-manifolds which are taken to be compact hyperbolic. The results of their analysis show that the standard hyperbolic (i.e., constant negative curvature) metric on such a manifold serves as an attractor for the conformal geometry under the (future) Einstein flow. In other words the evolving conformal geometry has a well-behaved limit in that problem. This fact plays a crucial role in their analysis since various Sobolev "constants" (which are in fact functionals of the geometry) which are needed in the associated energy estimates are asymptotically under control since they are tending toward their (regular) limiting values for the hyperbolic metric. Thus the difficulty of degenerating Sobolev constants, avoided by the introduction of a conformal 2 metric in the case of our $U(1)$ symmetry assumption, never arises in the Andersson-Moncrief work.

Besides the fact that the $U(1)$ symmetric case is not included in the no symmetry case treated by Andersson and Moncrief, an interest of the $U(1)$ case is that in our problem the number of effective spatial dimensions is two, and also that there is no known "physical" reason why large data solutions should develop singularities in the direction of cosmological expansion. Black hole formation seems to be suppressed by the topological character of the assumed Killing symmetry (which is of translational rather than rotational type and excludes the appearance of an axis of symmetry) and the big bang singularity is avoided by considering the future evolution from an initially expanding Cauchy hypersurface. Any possible big crunch is excluded by our requirement that the spatial manifold M is of negative Yamabe type (which is true of all circle bundles over higher genus manifolds). Such manifolds are incompatible (in the vacuum and electrovacuum cases for example) with the development of a maximal hypersurface which would be a necessary prelude to the "recollapse" of an expanding universe towards a hypothetical big crunch singularity. At a maximal hypersurface the scalar curvature of M would have to be everywhere positive – an impossibility on any manifold of negative Yamabe type. Thus it is conceivable that for large data future global existence holds for our problem. Up to now the only large data global results require simplifying assumptions so stringent that they effectively reduce the number of spatial dimensions to one (e.g., Gowdy models and their generalizations, plane symmetric gravitational waves, spherically symmetric matter coupled with gravity) or zero (e.g., Bianchi models, $2+1$ gravity). Unfortunately we have at present no way of proving this global existence, even in the polarized case for which the wave map equation reduces to a wave equation for a scalar function, because the reduced field equations are non local in character. The "background" spacetime on which the scalar field evolves is not given a priori but is instead a certain functional (obtained by solution of elliptic equations) of the evolving field (and the Teichmüller parameters) itself. In the unpolarized case, the problem of global existence of strong solutions for wave maps on a fixed background in $2+1$ dimensions is still unsolved. However there is a proof [M-S] of global existence of a weak solution (with no uniqueness) for wave maps from Minkowski spacetime. Any progress on the large data global existence, even of weak solutions, for the $U(1)$-symmetric problem would represent a "quantum jump" forward in our understanding of long time existence problems for Einstein's equations.

It is worth mentioning here that, again with suitable topological restrictions, using the reduction obtained by V. Moncrief [M2], an analogous Einstein-wave map form of the reduced field equations can be obtained even when one begins with the full Einstein-Maxwell system in $3+1$ dimensions.

Some steps of the proof given here have been obtained independently, using other notations, by V. Moncrief. I thank him for communicating his manuscript to me, and for numerous conversations on the subject.

2. S^1 invariant Einsteinian universes

2.1. Definition.

The spacetime manifold V is a principal fiber bundle with Lie group S^1 and base $\Sigma \times R$, with Σ a smooth orientable 2-dimensional manifold which we suppose here to be compact and of genus greater than one.

The spacetime metric $^{(4)}g$ is invariant under the action of S^1, the orbits are the fibers of V and are supposed to be space like. We write it in the form adapted to the bundle structure[2]:

$$^{(4)}g = e^{-2\gamma}\,^{(3)}g + e^{2\gamma}(\theta)^2,$$

where γ and $^{(3)}g$ can be identified respectively with a scalar function and a Lorentzian metric on the base manifold $\Sigma \times R$. In coordinates (x^3, x^α) adapted to a local trivialization of the bundle, with x^3 a coordinate on S^1 (i.e., with $x^3 = 0$ and $x^3 = 2\pi$ identified) and $(x^\alpha) \equiv (x^a, t)$, $a = 1, 2$ coordinates on $\Sigma \times R$, it holds that:

$$^{(3)}g = -N^2 dt^2 + g_{ab}(dx^a + \nu^a dt)(dx^b + \nu^b dt)$$

equivalently, in terms of a moving frame

$$^{(3)}g = -N^2(\theta^0)^2 + g_{ab}\theta^a\theta^b, \quad \theta^0 \equiv dt, \quad \theta^a \equiv dx^a + \nu^a dt$$

where N and ν are respectively the lapse and shift of $^{(3)}g$, and

$$g = g_{ab}dx^a dx^b$$

is a Riemannian metric on Σ, depending on t. The 1-form θ, S^1 connection on the fiber bundle V, is represented by

$$\theta = dx^3 + A_\alpha dx^\alpha.$$

The 1-form $A \equiv A_\alpha dx^\alpha$ depends on the trivialization of V, it is only a locally defined 1-form on $\Sigma \times R$ if the bundle is not trivial.

2.2. Twist potential

2.2.1. Definition. The curvature of the connection locally represented by A is a 2-form F on $\Sigma \times R$, given by, if the equations $^{(4)}R_{\alpha 3} = 0$ are satisfied,

$$F_{\alpha\beta} = (1/2)e^{-4\gamma}\eta_{\alpha\beta\lambda}E^\lambda$$

with η the volume form of the metric $^{(3)}g$, and E an arbitrary closed 1-form. Hence if Σ is compact

$$E = d\omega + H$$

where ω is a scalar function on V, called the twist potential, and H a representative of the 1-cohomology class of $\Sigma \times R$, for instance defined by a 1-form on Σ, harmonic for some given Riemannian metric m. For simplicity we take $H = 0$.

---

[2]See for instance CB-DM Kaluza Klein theories p. 286. The Lorentzian metric on the base manifold $\Sigma \times R$ is weighted by $e^{-2\gamma}$ in order to obtain equations which split in a nice hyperbolic-elliptic coupled form (see M1, CB-M3).

2.2.2. Construction of A. The connection 1-form θ can be constructed, when F is known if the following integrability condition is satisfied (see [M1], [CB-M3])

$$\int_{\Sigma_t} F = \frac{1}{2} \int_{\Sigma_t} F_{ab} dx^a \wedge dx^b = -\int_{\Sigma_t} e^{-4\gamma} N^{-1} \partial_0 \omega \mu_g = 2\pi n, \qquad (2.1)$$

where n is the Chern number of the bundle over $\Sigma \times R$.

We will suppose here that this bundle is trivial, i.e., $n = 0$; this value of n is the only one compatible with the smallness assumptions on the energy that we will make. The 1-form $A \equiv A_\alpha dx^\alpha$ is then defined globally on $\Sigma \times R$. It satisfies the equation

$$dA = F. \qquad (2.2)$$

We denote by $\tilde{A}$ and $\tilde{F}$ the t dependent 1-form and 2-form on Σ given by

$$\tilde{A} \equiv A_a dx^a, \qquad \tilde{F} \equiv \frac{1}{2} F_{ab} dx^a \wedge dx^b. \qquad (2.3)$$

Equation 2.2 splits into

$$d\tilde{A} = \tilde{F}, \qquad \text{i.e.,} \quad \partial_a A_b - \partial_b A_a = F_{ab} \qquad (2.4)$$

and, denoting by $F_{(t)}$ the 1-form $F_{ta} dx^a$:

$$\partial_t A_a - \partial_a A_t = F_{ta}, \qquad \text{i.e.,} \quad \partial_t \tilde{A} - dA_t = F_{(t)}. \qquad (2.5)$$

We solve 2.4 by introducing a smooth metric m on Σ. We denote by δ_m and $\tilde{\Delta}_m \equiv \delta_m d + d\delta_m$ the codifferential and the de Rham-Laplace operator in this metric. If we suppose that $\tilde{A}$ satisfies the Coulomb gauge condition:

$$\delta_m \tilde{A} = 0. \qquad (2.6)$$

Equations 2.4 and 2.6 imply that

$$\tilde{\Delta}_m \tilde{A} = \delta_m \tilde{F}. \qquad (2.7)$$

The general solution of this equation is the sum of the unique solution $\hat{A}$ which is $L^2(m)$-orthogonal to the elements $H_{(i)}$ of a basis of harmonic 1-forms, and an arbitrary harmonic 1-form, that is:

$$\tilde{A} = \hat{A} + \sum_i c_i H_{(i)}, \qquad (\hat{A}, H_{(i)})_{L^2(m)} = 0 \qquad (2.8)$$

where the c_i are t-dependent numbers. The solution $\hat{A}$ satisfies a Sobolev inequality

$$\|\hat{A}\|_{H_2(m)} \leq C_m \|\delta_m \tilde{F}\|_{L^2(m)}, \qquad \|\delta_m \tilde{F}\|_{L^2(m)} \leq C_m \|\tilde{F}\|_{H_1(m)}. \qquad (2.9)$$

A solution $\tilde{A}$ of 2.7 satisfies 2.4 and 2.6 because 2.7 implies

$$d\tilde{\Delta}_m \tilde{A} \equiv \tilde{\Delta}_m d\tilde{A} = d\delta_m \tilde{F} \equiv \tilde{\Delta}_m \tilde{F} \qquad (2.10)$$

since $\tilde{F}$ is closed, and

$$\delta_m \tilde{\Delta}_m \tilde{A} \equiv \tilde{\Delta}_m \delta_m \tilde{A} = 0. \qquad (2.11)$$

2.10 implies that $d\tilde{A} - \tilde{F}$ is a harmonic 2-form on Σ, it is zero because its period, i.e., its integral on the unique 2-cycle Σ, is zero (equation 2.1). 2.11 implies that the scalar function $\delta_m \tilde{A}$ is harmonic, but its integral on Σ is zero, since it is a divergence, therefore $\delta_m \tilde{A} = 0$.

Equation 2.5 can be satisfied by choice of A_t, a t dependent function on Σ, if the 1-form $\partial_t A - F_{(t)}$ is an exact differential. The commutation of partial derivatives and the closure of F show that a solution $\tilde{A}$ of 2.4 satisfies the equation

$$\partial_a \partial_t A_b - \partial_b \partial_t A_a = \partial_t F_{ab} = \partial_a F_{tb} - \partial_b F_{ta}, \quad \text{i.e.,} \quad d(\partial_t \tilde{A} - F_{(t)}) = 0. \quad (2.12)$$

Since the form $\partial_t \tilde{A} - F_{(t)}$ is closed it will be the differential of a function A_t if and only if it is $L^2(m)$ orthogonal to the harmonic 1-forms, that is, because $\partial_t \hat{A}$ is like $\hat{A}$ $L^2(m)$ orthogonal to the $H_{(i)}$ which do not depend on t :

$$\left(\sum_j \frac{dc_j}{dt} H_{(j)} - F_{(t)}, H_{(i)} \right)_{L^2(m)} = 0. \quad (2.13)$$

Choosing the $H'_{(i)}s$ to be $L^2(m)$ orthonormal, this equation reduces to:

$$\frac{dc_i}{dt} = (F_{(t)}, H_{(i)})_{L^2(m)}. \quad (2.14)$$

These equations determine c_i by integration on t through its initial value $c_i(t_0)$.

We complete the determination of the scalar function A_t by remarking that for such a function Equation 2.5 implies[3], using 2.6,

$$\Delta_m A_t = -\delta_m F_{(t)}, \quad (2.15)$$

an equation which determines uniquely A_t if we impose that its integral on Σ_t is zero. It satisfies then the inequality

$$\|A_t\|_{H_2(m)} \leq C_m \|\delta_m F_{(t)}\|_{L^2(m)}, \quad \|\delta_m F_{(t)}\|_{L^2(m)} \leq C_m \|F_t\|_{H_1(m)}. \quad (2.16)$$

Remark 2.1. *We can, instead of the Coulomb gauge, determine $\tilde{A}$ in temporal gauge, i.e., impose $A_t = 0$. We determine the 1-form $\tilde{A}_0$, the value of $\tilde{A}$ for $t = t_0$ by the relation 2.4 through the value of $\tilde{F}_0$ as above. Equation 2.5 is, when A_t and $F_{(t)}$ are known, an ordinary differential equation for $\tilde{A}$ which can be solved by integration on t:*

$$\tilde{A} = \tilde{A}_0 + \int_{t_0}^t F_{(t)}. \quad (2.17)$$

When 2.5 is satisfied it implies, whatever A_t may be using the commutation of partial derivatives and the closure of F the equation

$$\partial_t(\partial_a A_b - \partial_b A_a) = \partial_a F_{tb} - \partial_b F_{ta} = \partial_t F_{ab}, \quad (2.18)$$

i.e.,

$$\partial_t(d\tilde{A} - \tilde{F}) = 0 \quad (2.19)$$

hence $d\tilde{A} - \tilde{F} = 0$ for all t if it is so for $t = t_0$.

The disadvantage of the temporal gauge is that it gives only H_1 estimates for $\tilde{A}$.

---

[3]For a scalar function it holds that $\delta_m f \equiv 0$ and $\tilde{\Delta}_m f \equiv \Delta_m f$.

2.3. Wave map equation

The fact that F is a closed form together with the equation $^{(4)}R_{33} = 0$ imply (with the choice $H = 0$ in the definition of ω) that the pair $u \equiv (\gamma, \omega)$ satisfies a wave map equation from $(\Sigma \times R, ^{(3)}g)$ into the Poincaré plane (R^2, G),

$$G = 2(d\gamma)^2 + (1/2)e^{-4\gamma}(d\omega)^2.$$

It is a system of hyperbolic type when $^{(3)}g$ is a known Lorentzian metric which reads, denoting by $^{(3)}\nabla_\alpha$ the components of covariant derivatives of tensors on $\Sigma \times R$ in the metric $^{(3)}g$ in the moving frame (θ^a, dt):

$$^{(3)}\nabla^\alpha \partial_\alpha \gamma + \frac{1}{2}e^{-4\gamma} g^{\alpha\beta} \partial_\alpha \omega \partial_\beta \omega = 0$$

$$^{(3)}\nabla^\alpha \partial_\alpha \omega - 4g^{\alpha\beta} \partial_\alpha \omega \partial_\beta \gamma = 0.$$

The integral 2.1 is independent of t if F is closed, hence if the wave map equation is satisfied.

Remark 2.2. *The non-zero Christoffel symbols of the metric G are*

$$G_{22}^1 \equiv G_{\omega\omega}^\gamma = \frac{1}{2}e^{-4\gamma}, \qquad G_{12}^2 = G_{21}^2 = G_{\gamma\omega}^\omega = -2.$$

The scalar and Riemann curvature are:

$$R_{12,12} = -2e^{-4\gamma}, \qquad R = -4.$$

2.4. 3-dimensional Einstein equations

When $^{(4)}R_{3\alpha} = 0$ and $^{(4)}R_{33} = 0$ the Einstein equations $^{(4)}R_{\alpha\beta} = 0$ are equivalent to Einstein equations on the 3-manifold $\Sigma \times R$ for the metric $^{(3)}g$ with source the stress energy tensor of the wave map:

$$^{(3)}R_{\alpha\beta} = \rho_{\alpha\beta} \equiv \partial_\alpha u.\partial_\beta u$$

where a dot denotes a scalar product in the metric of the Poincaré plane:

$$\partial_\alpha u.\partial_\beta u \equiv 2\partial_\alpha \gamma \partial_\beta \gamma + \frac{1}{2}e^{-4\gamma} \partial_\alpha \omega \partial_\beta \omega.$$

In dimension 3 the Einstein equations are non dynamical, except for the conformal class of g determined by Teichmüller parameters. They decompose into:

a. Constraints.

b. Equations for lapse and shift to be satisfied on each Σ_t. These equations, as well as the constraints, are of elliptic type.

c. Evolution equations for the Teichmüller parameters, ordinary differential equations.

2.4.1. Constraints on Σ_t. One denotes by k the extrinsic curvature of Σ_t as submanifold of $(\Sigma \times R, {}^{(3)}g)$. Then, with ∇ the covariant derivative in the metric g,

$$k_{ab} \equiv (2N)^{-1}(-\partial_t g_{ab} + \nabla_a \nu_b + \nabla_b \nu_a),$$

the equations (momentum constraint)

$${}^{(3)}R_{0a} \equiv N(-\nabla_b k_a^b + \partial_a \tau) = \partial_0 u.\partial_a u \tag{2.20}$$

and (Hamiltonian constraint, ${}^{(3)}S_{00} \equiv {}^{(3)}R_{00} + \frac{1}{2}N^2 {}^{(3)}R$)

$$2N^{-2(3)}S_{00} \equiv R(g) - k_b^a k_a^b + \tau^2 = N^{-2}\partial_0 u.\partial_0 u + g^{ab}\partial_a u.\partial_b u \tag{2.21}$$

do not contain second derivatives transversal to Σ_t of g or u. They are the constraints. To transform the constraints into an elliptic system one uses the conformal method. We set

$$g_{ab} = e^{2\lambda}\sigma_{ab},$$

where σ is a Riemannian metric on Σ, depending on t, on which we will comment later, and

$$k_{ab} = h_{ab} + \frac{1}{2}g_{ab}\tau$$

where τ is the g-trace of k, hence h is traceless.

We denote by D a covariant derivation in the metric σ. We set

$$u' = N^{-1}\partial_0 u$$

with ∂_0 the Pfaff derivative of u, namely

$$\partial_0 = \frac{\partial}{\partial t} - \nu^a \partial_a \text{ with } \partial_a = \frac{\partial}{\partial x^a}$$

and

$$\dot{u} = e^{2\lambda}u'.$$

The momentum constraint on Σ_t reads if τ is constant in space, a choice which we will make

$$D_b h_a^b = L_a, L_a \equiv -D_a u.\dot{u}. \tag{2.22}$$

This is a linear equation for h, with left-hand side independent of λ. The general solution is the sum of a transverse traceless tensor $h_{TT} \equiv q$ (see 2.28 below) and a conformal Lie derivative r. Such tensors are L^2-orthogonal on (Σ, σ).

The Hamiltonian constraint reads as the semilinear elliptic equation in λ:

$$\Delta\lambda = f(x, \lambda) \equiv p_1 e^{2\lambda} - p_2 e^{-2\lambda} + p_3, \tag{2.23}$$

with $\Delta \equiv \Delta_\sigma$ the Laplacian in the metric σ and:

$$p_1 \equiv \frac{1}{4}\tau^2, \quad p_2 \equiv \frac{1}{2}(|\dot{u}|^2 + |h|^2), \quad p_3 \equiv \frac{1}{2}(R(\sigma) - |Du|^2).$$

2.4.2. Equations for lapse and shift. Lapse and shift are gauge parameters for which we obtain elliptic equations on each Σ_t as follows.

We impose that the $\Sigma_t's$ have constant (in space) mean curvature, namely that τ is a given increasing function of t. The lapse N satisfies then the linear elliptic equation

$$\Delta N - \alpha N = -e^{2\lambda}\frac{\partial \tau}{\partial t} \tag{2.24}$$

with ($|.|$ pointwise norm in the metric σ)

$$\alpha \equiv e^{-2\lambda}(|h|^2 + |\dot{u}|^2) + \frac{1}{2}e^{2\lambda}\tau^2. \tag{2.25}$$

The equation to be satisfied by the shift ν results from the expression for h deduced from the definition of k

$$h_{ab} \equiv (2N)^{-1}[-(\partial_t g_{ab} - \frac{1}{2}g_{ab}g^{cd}\partial_t g_{cd}) + \nabla_a\nu_b + \nabla_b\nu_a - g_{ab}\nabla_c\nu^c]$$

which implies, if $g_{ab} \equiv e^{2\lambda}\sigma_{ab}$ and if $n_a \equiv e^{-2\lambda}\nu_a$ denotes the covariant components of the shift vector ν in the metric σ (thus $n^a = \nu^a$)

$$h_{ab} \equiv (2N)^{-1}e^{2\lambda}[-(\partial_t\sigma_{ab} - \frac{1}{2}\sigma_{ab}\sigma^{cd}\partial_t\sigma_{cd}) + D_a n_b + D_b n_a - \sigma_{ab}D_c n^c].$$

The equation is therefore:

$$(L_\sigma n)_{ab} \equiv D_a n_b + D_b n_a - \sigma_{ab}D_c n^c = f_{ab} \tag{2.26}$$

$$f_{ab} \equiv 2Ne^{-2\lambda}h_{ab} + \partial_t\sigma_{ab} - \frac{1}{2}\sigma_{ab}\sigma^{cd}\partial_t\sigma_{cd}. \tag{2.27}$$

The homogeneous associated operator, the conformal Killing operator L_σ, has injective symbol, and it has a kernel zero, since manifolds of genus greater than 1 admit no conformal Killing fields.

The kernel of the dual of L_σ is the space of transverse traceless symmetric 2-tensors, i.e., symmetric 2-tensors T such that

$$\sigma^{ab}T_{ab} = 0, \quad D^a T_{ab} = 0. \tag{2.28}$$

These tensors are usually called TT tensors. The spaces of TT tensors are the same for two conformal metrics.

2.5. Teichmüller parameters

On a compact 2-dimensional manifold of genus G ≥ 2 the space T_{eich} of conformally inequivalent Riemannian metrics, called Teichmüller space, can be identified (Fisher and Tromba, see [F-T] or [CB-DM]) with M_{-1}/D_0, the quotient of the space of metrics with scalar curvature -1 by the group of diffeomorphisms homotopic to the identity. $M_{-1} \to T_{eich}$ is a trivial fiber bundle whose base can be endowed with the structure of the manifold R^n, with $n = 6G - 6$.

We require the metric σ_t to be in some chosen cross section $Q \to \psi(Q)$ of the above fiber bundle. Let $Q^I, I = 1, \ldots, n$ be coordinates in T_{eich}, then $\partial\psi/\partial Q^I$ is a known tangent vector to M_{-1} at $\psi(Q)$, that is a symmetric 2-tensor field on Σ, the sum of a transverse traceless tensor field $X_I(Q)$ and of the Lie derivative of a vector field on the manifold $(\Sigma, \psi(Q))$. The tensor fields $X_I(Q), I = 1, \ldots, n$ span the space of transverse traceless tensor fields on $(\Sigma, \psi(Q))$. The matrix with elements

$$\int_\Sigma X_I^{ab} X_{Jab} \mu_{\psi(Q)}$$

is invertible.

We have found in [CB-M1] an ordinary differential system satisfied by $t \mapsto Q(t)$ by using on the one hand the solvability condition for the shift equation which determines dQ^I/dt in terms of h_t which reads

$$\int_{\Sigma_t} f_{ab} X_J^{ab} \mu_{\sigma_t} = 0, \quad J = 1, \ldots, 6G - 6, \tag{2.29}$$

and on the other hand the necessary and sufficient conditions for the previous equations to imply also the remaining equations $^{(3)}R_{ab} - \rho_{ab} = 0$, that is:

$$\int_{\Sigma_t} N(^{(3)}R_{ab} - \rho_{ab}) X_J^{ab} \mu_{\sigma_t} = 0, \text{ for } J = 1, 2, \ldots, 6G - 6. \tag{2.30}$$

We have used the expression

$$\partial_t \sigma_{ab} = \frac{dQ^I}{dt} X_{I,ab} + C_{ab}$$

where C_{ab} is a Lie derivative, L^2 orthogonal to TT tensors, together with the decomposition $h = q + r$, with r a tensor in the range of the conformal Killing operator and q a TT tensor. This last tensor can be written with the use of the basis X_I of such tensors, the coefficients P^I depending only on t:

$$q_{ab} = P^I(t) X_{I,ab}.$$

The orthogonality condition 2.29 reads, using the expression 2.27 of f_{ab} and the fact that the transverse tensors X_I are orthogonal to Lie derivatives and are traceless:

$$\int_{\Sigma_t} [2Ne^{-2\lambda}(r_{ab} + P^I X_{I,ab}) + (dQ^I/dt) X_{I,ab}] X_J^{ab} \mu_\sigma = 0.$$

The tangent vector dQ^I/dt to the curve $t \to Q(t)$ and the tangent vector $P^I(t)$ to T_{eich} are therefore linked by the linear system

$$X_{IJ} \frac{dQ^I}{dt} + Y_{IJ} P^I + Z_J = 0$$

with

$$X_{IJ} \equiv \int_{\Sigma_t} X_I^{ab} X_{J,ab} \mu_\sigma, \tag{2.31}$$

$$Y_{IJ} \equiv \int_{\Sigma_t} 2Ne^{-2\lambda} X_I^{ab} X_{J,ab} \mu_\sigma, \quad Z_J \equiv \int_{\Sigma_t} 2Ne^{-2\lambda} r_{ab} X_J^{ab} \mu_\sigma. \tag{2.32}$$

While, using

$$^{(3)}R_{ab} \equiv R_{ab} - N^{-1}\bar{\partial}_0 k_{ab} - 2k_{ac} k_b^c + \tau k_{ab} - N^{-1} \nabla_a \partial_b N \tag{2.33}$$

where

$$R_{ab} \equiv \frac{1}{2}Rg_{ab}, \quad \rho_{ab} \equiv \partial_a u.\partial_b u, \quad k_{ab} \equiv P^I X_{I,ab} + r_{ab} + \frac{1}{2}g_{ab}\tau$$

and $\bar{\partial}_0$ is an operator on time dependent space tensors defined by, with $\mathcal{L}_\nu$ the Lie derivative in the direction of ν,

$$\bar{\partial}_0 \equiv \partial_t - \mathcal{L}_\nu$$

gives[4] for 2.30 the expression:

$$\int_{\Sigma_t} (-\bar{\partial}_0 k_{ab} - 2Ne^{2\lambda}h_{ac}h_b^c + \tau N h_{ab} - \nabla_a\partial_b N - \partial_a u.\partial_b u)X_J^{ab}\mu_{\sigma_t} = 0. \quad (2.34)$$

We have thus obtained an ordinary differential system of the form

$$X_{IJ}\frac{dP^I}{dt} + \Phi_J(P, \frac{dQ}{dt}) = 0$$

where Φ is a polynomial of degree 2 in P and dQ/dt with coefficients depending smoothly on Q and directly but continuously on t through the other unknowns, namely:

$$\Phi_J \equiv A_{JIK}P^I P^K + B_{JIK}P^I\frac{dQ^K}{dt} + C_{JI}P^I + D_J$$

with

$$A_{JIK} \equiv \int_{\Sigma_t} 2Ne^{2\lambda}X_{I,a}^c X_{K,bc}X_J^{ab}\mu_{\sigma_t}$$

$$B_{JIK} \equiv \int_{\Sigma_t} \frac{\partial X_{I,ab}}{\partial Q^K}X_J^{ab}\mu_{\sigma_t}$$

$$C_{JI} \equiv \int_{\Sigma_t} [(-\mathcal{L}_\nu X_I)_{ab} + 4Ne^{-2\lambda}r_b^c X_{I,ac} - \tau N X_{I,ab}]X_J^{ab}\mu_{\sigma_t}$$

and, using integration by parts and the transverse property of the X_I to eliminate second derivatives of N (recall that $\nabla_a\partial_b N \equiv D_a\partial_b N - 2\partial_a\lambda\partial_b N$)

$$D_J \equiv \int_{\Sigma_t} (-\bar{\partial}_0 r_{ab} - 2Ne^{-2\lambda}r_{ac}r_b^c + \tau N r_{ab} + 2\partial_a\lambda\partial_b N - \partial_a u.\partial_b u)X_J^{ab}\mu_{\sigma_t}.$$

3. Cauchy problem

3.1. Cauchy data

The Cauchy data on Σ_{t_0} are:

1. A C^∞ Riemannian metric σ_0 which projects onto a point $Q(t_0)$ of T_{eich} and a C^∞ tensor q_0 which is TT in the metric σ_0.
2. Cauchy data for u and $\dot{u}$ on Σ_{t_0}, i.e.,

$$u(t_0,.) = u_0, \quad \dot{u}(t_0,.) = \dot{u}_0.$$

---

[4]In the formula 2.33 indices are raised with g, in 2.34 they are raised with σ.

We say that a pair of scalar functions, $u \equiv (\gamma, \omega)$ or $\dot{u} \equiv (\dot{\gamma}, \dot{\omega})$ belongs to W_s^p if it is so of each of the scalars; W_s^p and $H_s \equiv W_s^2$ are the usual Sobolev spaces of scalar functions on the Riemannian manifold (Σ, σ_0). We suppose that

$$u_0 \in H_2, \qquad \dot{u}_0 \in H_1.$$

From these data one determines the values on Σ_0 of the auxiliary unknown, $h_0 \in W_2^p$, $1 < p < 2$, the conformal factor, lapse and shift $\lambda_0, N_0, \nu_0 \in W_3^p$.

One deduces then the usual Cauchy data for the wave map by

$$(\partial_t u)_0 = e^{-2\lambda_0} N_0 \dot{u}_0 + \nu_0^a \partial_a u_0. \tag{3.1}$$

It holds that

$$(\partial_t u)_0 \in H_1. \tag{3.2}$$

We suppose that the initial data satisfy the integrability condition 2.1 and we deduce from them an admissible $\tilde{A}_0$.

3.2. Local in time existence theorem

The following theorem is a consequence of previous results (see CB-M2, CB1, CB-M1).

Theorem 3.1. *The Cauchy problem with the above data for the Einstein equations with S^1 isometry group has, if $T - t_0$ is small enough, a solution with $u \in C^0([t_0, T), H_2)$ $\dot{u} \in C^1([t_0, T), H_1)$; $\lambda, N, \nu \in C^0([t_0, T), W_3^p) \cap C^1([t_0, T), W_2^p)$, $1 < p < 2$ and $N > 0$ while $\sigma \in C^1([t_0, T), C^\infty)$ with σ_t uniformly equivalent to σ_0. This solution is unique up to t parametrization of τ, choice of A_t, and choice of a cross section of M_{-1} over T_{eich}.*

3.3. Scheme for global existence

If the universe is expanding the mean curvature τ starts negative and increases, the universe attains a moment of maximum expansion if it exists up to $\tau = 0$. We choose the time parameter t by requiring that

$$t = -\frac{1}{\tau}. \tag{3.3}$$

Then t increases from $t_0 > 0$ to infinity when τ increases from $\tau_0 < 0$ to zero.

It results from the local existence theorem and a standard argument that the solution of the Einstein equations exists on $[t_0, \infty)$ if the curve $t \mapsto Q(t)$ remains in a compact subset of T_{eich} and the norms $||\gamma(t, .), \omega(t, .)||_{H_2}$, $||\partial_t \gamma(t, .), \partial_t \omega(t, .)||_{H_1}$ do not blow up for any finite t.

It will result from the following sections that these norms do not blow up if it is so of the energies that we will now define. However this non-blow up will be proved only for small initial data and the proof of the boundedness of Q will require the consideration of corrected energies, analogous to the corrected energies introduced in [CB-M1], but linked with the wave map structure and more complicated to estimate.

In Section 4 the first and second energies are defined. A proof is given that the first energy is non-increasing. Some preliminary properties of gauge covariant derivatives are given, but the estimate of the second energy is postponed after the estimates of the coefficients of the wave map equation through the elliptic equations they satisfy.

In Sections 5 and 6 we obtain these elliptic estimates for the difference of various quantities with what will be their asymptotic value, in terms of the energies previously defined. We choose σ_t such that $R(\sigma_t) = -1$, we suppose that the energies are bounded by some number H_E and that the projection of σ_t on the Teichmüller space remains in a compact subset. We first obtain bounds in H_1 for h, and λ in H_2, then bounds for h in W_2^p, $2 - N$ and ν in W_3^p, $1 < p < 2$. We also bound[5] $\partial_t \sigma_t$.

In Section 8 we use the estimates found on the coefficients of the wave map equation to obtain a non-linear differential inequality for the second energy. We could deduce from it, by a continuity argument, an a priori bound for this energy also (the first energy has been shown to be non-decreasing) if we knew that the metrics σ_t are all uniformly equivalent. To obtain such a result we must prove the decay of the energies. This decay is proved in Section 9 and 10 through the introduction of modified energies. The proof is more involved than in the polarized case, but follows essentially the same lines. All the obtained estimates lead to a global existence theorem by a continuity argument.

4. Energies

4.1. First energy

4.1.1. Definition. We denote below by $|.|$ a norm in the metric G and $|.|_g$ a norm in the metrics g and G, in particular:

$$|u'|^2 \equiv 2(\gamma')^2 + \frac{1}{2}e^{-4\gamma}(\omega')^2 \ , \quad |Du|_g^2 \equiv g^{ab}(2D_a\gamma D_b\gamma + \frac{1}{2}e^{-4\gamma}D_a\omega D_b\omega). \quad (4.1)$$

The $2 + 1$-dimensional Einstein equations with source the stress energy tensor of the wave map u contain the following equation (Hamiltonian constraint)

$$2N^{-2}(T_{00} - {}^{(3)}S_{00}) \equiv |u'|^2 + |Du|_g^2 + |k|_g^2 - R(g) - \tau^2 = 0. \quad (4.2)$$

The splitting of the covariant 2-tensor k into a trace and a traceless part:

$$k_{ab} = h_{ab} + \frac{1}{2}g_{ab}\tau \quad (4.3)$$

gives that:

$$|k|_g^2 = g^{ac}g^{bd}k_{ab}k_{cd} = |h|_g^2 + \frac{1}{2}\tau^2 \quad (4.4)$$

and the Hamiltonian constraint equation reads

$$|u'|^2 + |Du|_g^2 + |h|_g^2 = R(g) + \frac{1}{2}\tau^2. \quad (4.5)$$

---

[5] We did not need this bound in CB-M1 due to the consideration of a special class of initial data, whose property (equation 47) was conserved in time: this conservation does not hold in the unpolarized case.

Inspired by this equation, we define the first energy by the following formula

$$E(t) \equiv \int_{\Sigma_t} (I_0 + I_1 + \frac{1}{2}|h|_g^2)\mu_g$$

with

$$I_0 \equiv \frac{1}{2} \mid u' \mid^2 \equiv (\gamma')^2 + \frac{1}{4}e^{-4\gamma}(\omega')^2, \qquad I_1 \equiv \frac{1}{2} \mid Du \mid_g^2 \equiv \mid D\gamma \mid_g^2 + \frac{1}{4}e^{-4\gamma} \mid D\omega \mid_g^2$$

that is:

$$E(t) \equiv \frac{1}{2}\{\parallel u' \parallel_g^2 + \parallel Du \parallel_g^2 + \parallel h \parallel_g^2\} \tag{4.6}$$

with $\parallel . \parallel_g^2$ the square of the integral in the metric g of $|.|_g^2$. This energy is the first energy of the wave map u completed by the square of the $L^2(g)$-norm of h.

4.2. Bound of the first energy

The integration of the Hamiltonian constraint on (Σ_t, g) using the constancy of τ and the Gauss-Bonnet theorem which reads, with χ the Euler characteristic of Σ

$$\int_{\Sigma_t} R(g)\mu_g = 4\pi\chi \tag{4.7}$$

shows that

$$E(t) = \frac{\tau^2}{4}Vol_g(\Sigma_t) + 2\pi\chi, \quad Vol_g(\Sigma_t) = \int_{\Sigma_t} \mu_g.$$

Recall that on a compact manifold

$$\frac{dVol_g\Sigma_t}{dt} = \frac{1}{2}\int_{\Sigma_t} g^{ab}\frac{\partial g_{ab}}{\partial t}\mu_g = -\tau\int_{\Sigma_t} N\mu_g.$$

We use the equation

$$N^{-1(3)}R_{00} \equiv \Delta_g N - N|k|_g^2 + \partial_t\tau = N|u'|^2$$

together with the splitting of k to write after integration, since τ is constant in space,

$$\frac{1}{2}\tau^2\int_{\Sigma_t} N\mu_g = \frac{d\tau}{dt}Vol_g(\Sigma_t) - \int_{\Sigma_t} N(|h|_g^2 + |u'|^2)\mu_g.$$

We then find as in [CB-M1] that it simplifies to:

$$\frac{dE(t)}{dt} = \frac{1}{2}\tau\int_{\Sigma_t} (|h|_g^2 + |u'|^2)N\mu_g. \tag{4.8}$$

We see that $E(t)$ is a non-increasing[6] function of t if τ is negative. We remark that, due to the use of the constraints DN does not appear in 4.8, as it would have if we had used only the wave map energy.

We set

$$\varepsilon \equiv \{E(t)\}^{\frac{1}{2}}, \quad \varepsilon_0 \equiv \{E(t_0)\}^{\frac{1}{2}}, \tag{4.9}$$

we have proved that if $\tau \le 0$ then

$$\varepsilon \le \varepsilon_0. \tag{4.10}$$

---

[6] The absence of the term $|Du|^2$ prevents the use of this equality to obtain a decay estimate.

4.3. Second energy

4.3.1. Notation. We denote by $\hat{\nabla}$ a covariant derivative in the metrics g and G, for t dependent sections of the fiber bundle $E^{p,q}$ with base Σ and fiber $\otimes^p T_x^* \Sigma \otimes^q T_{u(x)} P$, with P the Poincaré plane. That is we set, for $\partial_c u^A$, a section of $E^{1,1}$,

$$\hat{\nabla}_b \partial_c u^A \equiv \partial_b \partial_c u^A - \Gamma_{bc}^a \partial_a^A + G_{BC}^A \partial_b u^B \partial_c u^C \tag{4.11}$$

where Γ_{bc}^a and G_{BC}^A denote respectively the connection coefficients of the metric g, and of G given in Remark 2.2. For u'^A, section of $E^{0,1}$, we have:

$$\hat{\nabla}_a u'^A = \partial_a u'^A + G_{BC}^A \partial_a u^B u'^C, \tag{4.12}$$

while for G_{AB}, section of $E^{0,2}$ it holds that

$$\hat{\nabla}_a G_{AB} = 0. \tag{4.13}$$

On the other hand we define by $\hat{\partial}_0$ a differential operator mapping a t-dependent section of a bundle $E^{p,q}$ into another such section by the formula:

$$\hat{\partial}_0 \hat{\nabla}^p u^A = \bar{\partial}_0 \hat{\nabla}^p u^A + G_{BC}^A \partial_0 u^B \hat{\nabla}^p u^C \tag{4.14}$$

with[7]

$$\bar{\partial}_0 \equiv \partial_t - \mathcal{L}_\nu \tag{4.15}$$

where $\mathcal{L}_\nu$ denotes the Lie derivative with respect to the shift ν. In particular:

$$\hat{\partial}_0 u'^A = \partial_0 u'^A + G_{BC}^A \partial_0 u^B u'^C \tag{4.16}$$

and

$$\hat{\partial}_0 G^{AB} = 0. \tag{4.17}$$

With these notations the wave map equation reads:

$$-\hat{\partial}_0 u'^A + \hat{\nabla}^a (N \partial_a u^A) + N \tau u' = 0. \tag{4.18}$$

We will use the following lemma, which can be foreseen, and also checked[8] by direct computation.

Lemma 4.1. *The following commutation relations are satisfied:*

$$\hat{\partial}_0 \partial_a u^A = \hat{\nabla}_a \partial_0 u^A, \tag{4.19}$$

$$\hat{\partial}_0 \hat{\nabla}_a \partial_0 u^A - \hat{\nabla}_a \hat{\partial}_0 \partial_0 u^A = R_{CB}{}^A{}_D \partial_0 u^C \partial_a u^B \partial_0 u^D, \tag{4.20}$$

$$\hat{\partial}_0 \hat{\nabla}_a \partial_b u^A - \hat{\nabla}_a \hat{\partial}_0 \partial_b u^A = R_{CB}{}^A{}_D \partial_0 u^C \partial_a u^B \partial_b u^D - \partial_c u^A \hat{\partial}_0 \Gamma_{ab}^c. \tag{4.21}$$

We recall the identities

$$\hat{\partial}_0 g^{ab} = \bar{\partial}_0 g^{ab} = 2N k^{ab}, \tag{4.22}$$

$$\hat{\partial}_0 \Gamma_{ab}^c = \bar{\partial}_0 \Gamma_{ab}^c = \nabla^c (N k_{ab}) - \nabla_a (N k_b^c) - \nabla_b (N k_a^c). \tag{4.23}$$

---

[7]Operator on tensors denoted $\hat{\partial}_0$ in [CB-Yo]. Note that only $\partial_0 u^A$ is defined, since u is a mapping, not a tensor, and that $\bar{\partial}_0 u'^A \equiv \partial_0 u'^A$.
[8]See CB1.

4.3.2. Definition. We define the second energy by the following formula

$$E^{(1)}(t) \equiv \int_{\Sigma_t} (J_0 + J_1)\mu_g \tag{4.24}$$

with

$$J_1 = \frac{1}{2} \mid \hat{\Delta}_g u \mid^2 \equiv \frac{1}{2}\{(2(\hat{\Delta}_g\gamma)^2 + \frac{1}{2}e^{-4\gamma}(\hat{\Delta}_g\omega)^2\} \tag{4.25}$$

$$J_0 = \frac{1}{2} \mid \hat{\nabla} u' \mid_g^2 \equiv \frac{1}{2}\{2|\hat{\nabla}\gamma'|_g^2 + \frac{1}{2}e^{-4\gamma}|\hat{\nabla}\omega'|_g^2\}. \tag{4.26}$$

4.3.3. Estimate. We postpone the computation and estimate of the derivative of $E^{(1)}(t)$ until after the estimates of h, λ, N and ν. We set:

$$E(t) \equiv \varepsilon^2, \quad E^{(1)}(t) \equiv \tau^2\varepsilon_1^2. \tag{4.27}$$

4.4. Norms

We suppose chosen a smooth cross section $Q \to \psi(Q)$ of M_{-1} over the Teichmüller space T_{eich}, together with a C^1 curve $t \to Q(t)$. We are then given by lift to M_{-1} a regular metric σ_t for $t \in [t_0, T]$, with scalar curvature -1.

Definition 4.1. *Hypothesis H_σ: the curve is contained in a compact subset of T_{eich}.*

Under the hypothesis H_σ the metric σ_t is uniformly equivalent to the metric $\sigma_0 \equiv \sigma_{t_0}$. A t-dependent Sobolev constant on (Σ, σ_t) is uniformly equivalent to a number. We denote by C_σ any such number, which depends only on the considered compact subset of T_{eich}.

The spaces $W_s^p(\sigma_t)$ are the usual Sobolev spaces of tensor fields on the Riemannian manifold (Σ, σ_t). By the hypothesis on σ_t their norms are uniformly equivalent for $t \in [t_0, T]$ to the norm in $W_s^p(\sigma_0)$ denoted simply W_s^p. We set $W_s^2 = H_s$.

We denote now by $|.|$ a pointwise norm in the σ and G metrics; $\| . \|$ and $\| . \|_p$ denote L^2 and L^p-norms in the σ metric.

We denote by $\hat{D}$ a covariant derivative relative to the metrics σ and G.

A lower case index m or M denotes respectively the lower or upper bound of a scalar function on Σ_t. It may depend on t.

Lemma 4.2. *It holds that:*

1. $$\|Du\|^2 \equiv \|Du\|_g^2 \leq 2\varepsilon^2, \quad \|u'\|^2 \leq e^{-2\lambda_m}\|u'\|_g^2 \leq 2e^{-2\lambda_m}\varepsilon^2. \tag{4.28}$$

2. $$\|\hat{D}Du\|^2 \leq 2e^{2\lambda_M}\tau^2\varepsilon_1^2 + \varepsilon^2. \tag{4.29}$$

Proof. 1. Results directly from the definitions.
2. By definition

$$\|\hat{D}Du\|^2 = \int_\Sigma \hat{D}^a D^b u.\hat{D}_a D_b u \mu_\sigma \tag{4.30}$$

$$= \int_\Sigma \{\hat{D}^a(D^b u.\hat{D}_a D_b u) - D^b u.\hat{D}^a \hat{D}_a D_b u\}\mu_\sigma = - \int_\Sigma D^b u.\hat{D}^a \hat{D}_a D_b u \mu_\sigma \tag{4.31}$$

(since $D^b u.\hat{\nabla}_a D_b u$ is an ordinary covariant vector on Σ its divergence integrates to zero).

The Ricci commutation formula gives that, with $\rho_{ab} = -\frac{1}{2}\sigma_{ab}$ the Ricci curvature of the metric σ:

$$\hat{D}^a \hat{D}_a D_b u^C = \hat{D}^a \hat{D}_b D_a u^C = \hat{D}_b \hat{\Delta} u^C + \rho_b{}^c D_c u^C + D^a u^A D_b u^B R_{AB,}{}^C{}_D D_a u^D. \tag{4.32}$$

By another integration by parts 4.31 gives then

$$\int_\Sigma -D^b u.\hat{D}^a \hat{D}_a D_b u \mu_\sigma = \int_\Sigma \{|\hat{\Delta} u|^2 - D^b u.(\rho_b^c D_c u^C + D^a u^A D_b u^B R_{AB,\,D} D_a u^D)\}\mu_\sigma.$$

On a 2-dimensional manifold the Riemann curvature is given by:

$$R_{AB,}{}^C{}_D = \frac{1}{2}R(G)\{\delta_A^C G_{BD} - \delta_B^C G_{AD}\}. \tag{4.33}$$

A straightforward computation gives therefore

$$D^b u.D^a u^A D_b u^B R_{AB,\,D} D_a u^D \tag{4.34}$$

$$= \frac{1}{2}R(G)D^b u^E D^a u^A D_b u^B D_a u^D \{G_{AE}G_{BD} - G_{BE}G_{AD}\} \tag{4.35}$$

$$= \frac{1}{2}R(G)\{(D^b u.D^a u)(D_b u.D_a u) - (D^b u.D_b u)(D^a u.D_a u)\} \tag{4.36}$$

$$= \frac{1}{2}R(G)\{|Du.Du|^2 - |\,|Du|^2|^2\}. \tag{4.37}$$

In the case of the Poincaré plane $R(G) = -4$, hence

$$-D^b u.D^a u^A D_b u^B R_{AB,\,D} D_a u^D = 2\{|Du.Du|^2 - |\,|Du|^2|^2\} \le 0, \tag{4.38}$$

because

$$|Du.Du| \le |Du|^2. \tag{4.39}$$

$\square$

5. First elliptic estimates

The equations for h, λ, N, and ν are elliptic equations on (Σ_t, σ_t), identical with those written in [CB-M1], except that in the coefficients $Du.\dot{u}$, $|Du|^2$, $|\dot{u}|^2$ which appear in these equations u is now a wave map and not a scalar function. The estimates obtained in [CB-M1] in terms of ε and ε_1 will be valid if the new coefficients satisfy the same estimates in terms of our new ε and ε_1.

5.1. Basic bounds on N and λ

The generalized maximum principle[9] applied to Equations 2.24 and 2.23 satisfied respectively by N and λ shows that, with our choice of the time parameter

$$t = -\tau^{-1}, \tag{5.1}$$

it holds that

$$0 \leq N_m \leq N \leq N_M \leq 2, \tag{5.2}$$

$$e^{-2\lambda_M} \leq e^{-2\lambda} \leq e^{-2\lambda_m} \leq \frac{1}{2}\tau^2. \tag{5.3}$$

Definition 5.1. *We say that the hypothesis[10] H_λ is satisfied if there exists a number $c_\lambda > 1$, independent of t, such that*

$$\frac{1}{\sqrt{2}}e^{\lambda_M}|\tau| \leq c_\lambda. \tag{5.4}$$

and we denote by C_λ any positive continuous function of $c_\lambda \in R^+$.

5.2. L^2 estimates of $\||Du|^2\|$ and $\||\dot{u}|^2\|$

Under the hypothesis H_σ and H_λ there exist numbers C_σ, C_λ such that u satisfies the same inequalities as in the polarized case, that is:

Lemma 5.1.

$$\||Du|^2\| \leq C_\sigma C_\lambda \{\varepsilon^2 + \varepsilon\varepsilon_1\}. \tag{5.5}$$

$$\||u'|^2\| \leq C_\sigma C_\lambda \tau^2 \varepsilon(\varepsilon + \varepsilon_1). \tag{5.6}$$

$$\||\dot{u}|^2\| \leq C_\sigma C_\lambda \tau^{-2} \varepsilon(\varepsilon + \varepsilon_1). \tag{5.7}$$

Proof. A Sobolev embedding theorem applied to the scalar function $|u'|^2$ gives that:[11]

$$\||u'|^2\| \leq C_\sigma(\||u'|^2\|_1 + \|D|u'|^2\|_1).$$

It holds that

$$D|u'|^2 \equiv 2u'.\hat{D}u'$$

therefore, since $\hat{D}u' \equiv \hat{\nabla}u'$,

$$\|D|u'|^2\|_1 \leq 2\|u'\| \, \|\hat{D}u'\| \leq 2e^{-\lambda_m}\|u'\|_g \, \|\hat{\nabla}u'\|_g.$$

Hence, since $\||u'|^2\|_1 \equiv \|u'\|^2$, and using the bound 5.3 of λ

$$\||u'|^2\| \leq C_\sigma e^{-\lambda_m}\|u'\|_g(e^{-\lambda_m}\|u'\|_g + \|\hat{D}u'\|_g) \leq C_\sigma|\tau|^2\varepsilon(\varepsilon + \varepsilon_1).$$

By the definition of $\dot{u}$ it holds that

$$\||\dot{u}|^2\| \leq e^{4\lambda_M} \, \||u'|^2\|,$$

hence

$$\||\dot{u}|^2\| \leq C_\sigma e^{4\lambda_M}\tau^2\varepsilon(\varepsilon + e^{\lambda_M}|\tau|\varepsilon_1) \leq C_\sigma C_\lambda \tau^{-2}\varepsilon(\varepsilon + \varepsilon_1).$$

---

[9]The coefficients in these equations belong to the same functional spaces as in [CB-M1], as will be proved in the next subsection which will also estimate them.

[10]This hypothesis replaces the hypothesis H_c made on v in [CB-M1].

[11]See C.B2, here case $n = 2$.

On the other hand, since
$$D|Du|^2 \equiv 2Du.\hat{D}Du,\tag{5.8}$$
it holds that, using again the Sobolev embedding theorem
$$\| \, | \, Du \, |^2 \, \| \leq C_\sigma \|Du\|(\|Du\| + \|\hat{D}Du\|)$$
which gives using Lemma 4.2
$$\| \, | \, Du \, |^2 \, \| \leq C_\sigma \varepsilon (\varepsilon + e^{\lambda_M}|\tau|\varepsilon_1)$$
hence under the hypothesis H_λ :
$$\| \, | \, Du \, |^2 \, \| \leq C_\sigma C_\lambda \varepsilon (\varepsilon + \varepsilon_1).\qquad\square$$

Lemma 5.2.
$$\| \, |Du|^2 \|_g \leq C_\sigma C_\lambda |\tau|\varepsilon \{\varepsilon + \varepsilon_1\}.\tag{5.9}$$
$$\| \, | \, u' \, |^2 \, \|_g \leq C_\sigma C_\lambda |\tau|\varepsilon (\varepsilon + \varepsilon_1).\tag{5.10}$$

Proof. The inequalities 5.5 and 5.6 imply that
$$\| \, | \, u' \, |^2 \, \|_g \equiv \left(\int_{\Sigma_t} |u'|^4 \mu_g \right)^{\frac{1}{2}} \leq e^{\lambda_M} \| \, | \, u' \, |^2 \, \| \leq C_\lambda C_\sigma |\tau|\varepsilon (\varepsilon + \varepsilon_1)$$
and, using the lower bound of λ,
$$\| \, | \, Du \, |^2 \, \|_g \equiv \left(\int_{\Sigma_t} |Du|_g^4 \mu_g \right)^{\frac{1}{2}} \leq e^{-\lambda_m} \| \, | \, Du \, |^2 \, \| \leq C_\sigma C_\lambda |\tau|\varepsilon (\varepsilon + \varepsilon_1).\qquad\square$$

5.3. Estimate of h in H_1
We have defined the auxiliary unknown h by
$$h_{ab} \equiv k_{ab} - \frac{1}{2}g_{ab}\tau.$$

5.3.1. Estimate of $\|h\|$.
The L^2-norm of h on (Σ, σ) is bounded in terms of the first energy and an upper bound λ_M of the conformal factor since we have
$$\| \, h \, \|^2 = \int_{\Sigma_t} \sigma^{ac}\sigma^{bd}h_{ab}h_{cd}\mu_\sigma \leq e^{2\lambda_M} \| \, h \, \|_{L^2(g)}^2 \leq 2e^{2\lambda_M}\varepsilon^2.$$

5.3.2. Estimate of $\| \, Dh \, \|$.
The tensor h satisfies the equations
$$D_a h_b^a = L_b \equiv -\partial_a u.\dot{u}.$$
It is the sum of a TT (transverse, traceless) tensor $h_{TT} \equiv q$ and a conformal Lie derivative r:
$$h \equiv q + r.$$
It results from elliptic theory that on each Σ_t the tensor r satisfies the estimate
$$\| \, r \, \|_{H_1} \leq C_\sigma \| \, Du.\dot{u} \, \| \leq C_\sigma \| \, |Du|^2 \, \|^{\frac{1}{2}} \| \, |\dot{u}|^2 \, \|^{\frac{1}{2}} .$$
It results from the inequalities of Lemma 5.2 that
$$\| \, r \, \|_{H_1} \leq C_\sigma C_\lambda e^{2\lambda_M} |\tau|\varepsilon \{(\varepsilon + \varepsilon_1)(\varepsilon + \varepsilon_1 e^{\lambda_M}|\tau|)\}^{\frac{1}{2}}.$$

We recall that for the transverse part $h_{TT} = q$ it holds that

$$\| \, Dq \, \| = \| \, q \, \| \leq \| \, h \, \| + \| \, r \, \|$$

therefore

$$\| \, Dh \, \| \leq e^{\lambda_M} \varepsilon \{ \sqrt{2} + C_\sigma e^{\lambda_M} |\tau| (\varepsilon + \varepsilon_1 e^{\lambda_M} |\tau|)^{\frac{1}{2}} (\varepsilon + \varepsilon_1)^{\frac{1}{2}} \}.$$

Definition 5.2. *We say that the hypothesis H_E is satisfied if there exists a positive number c_E such that $\varepsilon + \varepsilon_1 \leq c_E$. We denote by C_E any continuous and positive function of $c_E \in R^+$.*

If the hypothesis H_σ, H_λ and H_E are satisfied, h verifies the inequality:

$$\| \, Dh \, \| \leq C_\lambda |\tau|^{-1} \varepsilon (1 + C_\sigma C_\lambda C_E).$$

5.4. Estimates for the conformal factor λ

The conformal factor λ satisfies on each Σ_t the equation

$$\Delta \lambda = f(\lambda) \equiv p_1 e^{2\lambda} - p_2 e^{-2\lambda} + p_3$$

where the coefficients p_i are given by

$$p_1 = \frac{1}{4}\tau^2, p_2 = \frac{1}{2}(| \, h \, |^2 + | \, \dot{u} \, |^2), p_3 = -\frac{1}{2}(1 + | \, Du \, |^2).$$

The equation admits the subsolution λ_- given by

$$e^{-2\lambda_-} = \frac{1}{2}\tau^2$$

and it holds that

$$\lambda_- \leq \lambda_m \leq \lambda \leq \lambda_M \leq \lambda_+$$

with λ_+ a supersolution, for example:

$$\lambda_+ = \theta + v - \min v$$

where v is the solution with mean value zero on Σ_t of the linear equation

$$\Delta v = f(\theta) \equiv p_1 e^{2\theta} - p_2 e^{-2\theta} + p_3$$

with[12] $e^{2\theta}$ a t-dependent number, positive solution of the equation

$$\bar{p}_1 e^{4\theta} + \bar{p}_3 e^{2\theta} - \bar{p}_2 = 0,$$

where $\bar{f}$ denotes the mean value on (Σ, σ) of a function f:

$$\bar{f} \equiv \frac{1}{V_\sigma} \int_\Sigma f \mu_\sigma, \quad V_\sigma \equiv \int_\Sigma \mu_\sigma = -4\pi\chi.$$

Using the expressions of $\bar{p}_2, \bar{p}_3$ and $\bar{p}_1 = \frac{1}{4}\tau^2$, together with

$$\| \, \dot{u} \, \|^2 \leq e^{2\lambda_M} \| \, u' \, \|_g^2, \text{ and } \| \, h \, \|^2 \leq e^{2\lambda_M} \| \, h \, \|_g^2$$

---

[12] We have renamed θ the function called ω in [CB-M1].

and the expression of $\varepsilon^2 \equiv E(t)$ we have found in [CB-M1], Section 8.1 (recall that $V_\sigma = -4\pi\chi$, a constant) that (we have renamed θ the t-dependent number ω of CB-M1):

$$0 \leq \frac{1}{2}\tau^2 e^{2\theta} - 1 \leq V_\sigma^{-1}(1 + \frac{\tau^2}{2}e^{2\lambda_M})\varepsilon^2 \leq C_\lambda\varepsilon^2. \tag{5.11}$$

Lemma 5.3. *The following inequalities hold*

1.
$$0 \leq \lambda_M - \theta \leq 2 \| v \|_{L^\infty}. \tag{5.12}$$

2.
$$1 \leq \frac{1}{2}\tau^2 e^{2\lambda_M} \leq 1 + C_\lambda\varepsilon^2 + C_\lambda C_E \| v \|_{L^\infty} e^{4\|v\|_{L^\infty}}. \tag{5.13}$$

Proof. 1. It holds that

$$\lambda_M \leq \sup \lambda_+ = \theta + \max v - \min v, \tag{5.14}$$

from which results 5.12.

2. The inequality 5.12 implies by elementary calculus

$$e^{2(\lambda_M - \theta)} \leq 1 + 4\|v\|_{L^\infty} e^{4\|v\|_{L^\infty}}. \tag{5.15}$$

The inequalities 5.11 and 5.15 imply that

$$\frac{1}{2}\tau^2 e^{2\lambda_M} \leq (1 + C_\lambda\varepsilon^2)(1 + 4 \| v \|_{L^\infty} e^{4\|v\|_{L^\infty}}), \tag{5.16}$$

from which the inequality 5.13 follows. □

5.4.1. Estimate of v. The equation satisfied by v implies

$$\int_\Sigma |Dv|^2 \mu_\sigma = -\int_\Sigma f(\theta) v \mu_\sigma$$

hence

$$\| Dv \|^2 \leq \| f(\theta) \| \| v \|$$

but the Poincaré inequality applied to the function v which has mean value 0 on Σ gives

$$\| v \|^2 \leq [\Lambda_\sigma]^{-1} \| Dv \|^2$$

where Λ_σ is the first (positive) eigenvalue of $-\Delta_\sigma$ for functions on Σ_t with mean value zero. Therefore on each Σ_t

$$\| Dv \| \leq [\Lambda_\sigma]^{-1/2} \| f(\theta) \| .$$

We use Ricci identity and $R(\sigma) = -1$ to obtain that

$$\| \Delta_\sigma v \|^2 = \| D^2 v \|^2 - \frac{1}{2} \| Dv \|^2 .$$

The equation satisfied by v implies then, as in [1],

$$\|D^2 v\|^2 = \|f(\theta)\|^2 + \frac{1}{2}\|Dv\|^2.$$

Assembling these various inequalities gives that:

$$\| v \|_{H_2} \leq [1 + 3/(2\Lambda_\sigma) + 1/\Lambda_\sigma^2]^{1/2} \| f(\theta) \| .$$

The Sobolev inequality

$$\| v \|_{L^\infty} \leq C_\sigma \| v \|_{H_2}$$

gives then a bound on the L^∞-norm of v on Σ_t in terms of the L^2-norm of $f(\theta)$, a Sobolev constant C_σ and the lowest eigenvalue Λ_σ of $-\Delta_\sigma$, which is itself a number C_σ.

We now estimate the L^2-norm of $f(\theta)$.

$$f(\theta) \equiv f = p_1 e^{2\theta} - p_2 e^{-2\theta} + p_3.$$

By the isoperimetric inequality, and since $\bar{f} = 0$, there exists a constant C_σ such that:

$$\|f\| \leq C_\sigma \|Df\|_1.$$

We want to bound the right-hand side in terms of the first and second energies of the wave map. We have by the definition of f and the expression of the $p's$ that:

$$\|Df\|_1 \leq \frac{1}{2}\{\|D|Du|^2\|_1 + e^{-2\theta}(\|D|h|^2\|_1 + \|D|\dot{u}|^2\|_1)\}.$$

Lemma 5.4. *The following estimate holds under the hypothesis H_λ:*

$$\frac{1}{2} \| D|Du|^2 \|_1 \leq C_\lambda(\varepsilon^2 + \varepsilon\varepsilon_1).$$

Proof. We have:

$$D|Du|^2 = 2Du.\hat{D}^2 u$$

hence

$$\|D|Du|^2\|_1 \leq 2\|Du\|\|\hat{D}^2 u\|.$$

We have seen that

$$\|Du\| = \| Du \|_g, \quad \|\hat{D}^2 u\| \leq e^{\lambda_M}\|\hat{\Delta}_g u\|_g + (1/\sqrt{2})\|Du\|_g \qquad (5.17)$$

which implies the given result under the hypothesis H_λ and the definitions of ε and ε_1. $\square$

Lemma 5.5. *The following estimates hold under the hypothesis H_σ, H_E and H_λ:*

1.
$$\frac{1}{2}e^{-2\theta} \| D|h|^2 \|_1 \leq C_\sigma C_E C_\lambda \varepsilon^2. \qquad (5.18)$$

2.
$$\frac{1}{2}e^{-2\theta}\|D|\dot{u}|^2\|_1 \leq C_E C_\lambda C_\sigma(\varepsilon^2 + \varepsilon\varepsilon_1). \qquad (5.19)$$

Proof. 1. We have:

$$\| D|h^2| \|_1 \leq 2 \| h \| \| Dh \|.$$

Using the inequalities of Sections 5.3.1 and 5.3.2 we find that

$$\| D|h^2| \|_1 \leq C_E C_\lambda \tau^{-2}\varepsilon^2.$$

The given result follows from the bound 5.11 of $e^{-2\theta}$.

2. The estimate given in [CB-M1], Lemma 21, when u is a scalar function holds when u is a wave map, with the same proof which, as far as u is concerned, contains only norms. It gives the announced inequality. $\square$

We denote by $C_{E,\lambda,\sigma}$ a number depending only on c_E, c_λ and the considered compact domain of T_{eich}.

Lemma 5.6. *There exists a number $C_{E,\lambda,\sigma}$ such that the L^∞-norm of v is bounded by the following inequality*

$$\| v \|_\infty \leq C_{E,\lambda,\sigma}(\varepsilon^2 + \varepsilon\varepsilon_1). \tag{5.20}$$

Proof. Recall that there exists a Sobolev constant C_σ such that

$$\|v\|_\infty \leq C_\sigma\{\||D|Du|^2\|_1 + e^{-2\omega}(\||D|h|^2\|_1 + \||D|\dot{u}|^2\|_1)\}.$$

The three terms in the sum have been evaluated in the Lemma 5.4. □

Theorem 5.7.

1. *It holds that:*

$$1 \leq \frac{1}{2}\tau^2 e^{2\lambda_M} \leq 1 + C_{E,\lambda,\sigma}(\varepsilon + \varepsilon_1)^2. \tag{5.21}$$

2. *There exists a number $\eta_1 > 0$ such that the hypothesis H_λ is satisfied, i.e.:*

$$1 \leq \frac{1}{\sqrt{2}}|\tau|e^{\lambda_M} \leq c_\lambda, \quad c_\lambda > 1, \tag{5.22}$$

as soon as

$$\varepsilon + \varepsilon_1 \leq \eta_1. \tag{5.23}$$

Proof. 1. Lemmas 5.3 and 5.6.

2. By 5.23 it holds that, with $C_{E,\lambda,\sigma}$ the number of that inequality,

$$1 \leq \frac{1}{\sqrt{2}}|\tau|e^{\lambda_M} < c_\lambda \quad \text{if} \quad (\varepsilon + \varepsilon_1)^2 < \frac{c_\lambda^2 - 1}{C_{E,\lambda,\sigma}}. \tag{5.24}$$

The result follows from a continuity argument. □

5.4.2. Bound of λ in H_1. The following theorem holds, with the same proof as Theorem 23 of [CB-M1].

Theorem 5.8. *The following estimate holds*

$$\|D\lambda\|_{H_1} \leq C_{E,\lambda,\sigma}(\varepsilon^2 + \varepsilon\varepsilon_1). \tag{5.25}$$

6. Estimates in W_s^p

6.1. Estimates for h in W_2^p

The estimates of h in W_2^p, with $1 < p < 2$ (for definiteness we will choose $p = \frac{4}{3}$) will be obtained using estimates for the conformal factor λ which have been obtained by using the H_1-norm of h.

Theorem 6.1. *Under the H hypotheses there exists a positive number $C_{E,\lambda,\sigma}$ such that the W_2^p-norm of h, choosing to be specific $p = \frac{4}{3}$, is bounded by*

$$\| h \|_{W_2^p} \leq C_{E,\lambda,\sigma}|\tau|^{-1}(\varepsilon + \varepsilon_1).$$

Corollary 6.2. *It holds that*

$$|\tau|\,\|\,h\,\|_\infty \le C_{E,\lambda,\sigma}(\varepsilon + \varepsilon_1) \qquad \text{and that} \qquad \|\,h\,\|_{L^\infty(g)} \le C_{E,\lambda,\sigma}|\tau|(\varepsilon + \varepsilon_1).$$

Proof. The inequalities satisfied by $||h||_{W_2^p}$ in [CB-M1], with $p = \frac{4}{3}$, are still valid when u is a wave map, with the same proof, in particular because $\|\,Du.\dot{u}\,\|_{\frac{4}{3}}$ and $\|\,D(Du.\dot{u})\,\|_{\frac{4}{3}}$ satisfy estimates of the same type as in [CB-M1]; indeed, using Section 5.2:

$$\|\,Du.\dot{u}\,\|_{\frac{4}{3}} \le \|\,Du\,\|\|\,\dot{u}\,\|_4 \le C_{E,\lambda,\sigma}|\tau|^{-1}\varepsilon^{\frac{3}{2}}(\varepsilon + \varepsilon_1)^{\frac{1}{2}}.$$

On the other hand a straightforward calculation gives

$$D(Du.\dot{u}) \equiv \hat{D}(Du).\dot{u} + Du.\hat{D}\dot{u} \tag{6.1}$$

hence

$$\|\,D(Du.\dot{u})\,\|_{\frac{4}{3}} \le \|\,\hat{D}^2 u\,\|\|\,\dot{u}\,\|_4 + \|\,|Du|\,\|_4\|\,\hat{D}\dot{u}\,\|$$

which gives, using previous estimates

$$\|\,D(Du.\dot{u})\,\|_{\frac{4}{3}} \le C_\lambda C_\sigma e^{\lambda_M}\{\varepsilon^{\frac{1}{2}}(\varepsilon + \varepsilon_1)^{\frac{3}{2}} + \varepsilon^{\frac{3}{2}}(\varepsilon + \varepsilon_1)^{\frac{1}{2}}\}.$$

The result of the theorem follows from the bound of ε by $\varepsilon + \varepsilon_1$.

The corollary is a consequence of the Sobolev embedding theorem,

$$\|\,h\,\|_\infty \le C_\sigma\,\|\,h\,\|_{W_2^p} \qquad \text{if} \qquad p > 1,$$

and the estimate

$$\|\,h\,\|_{L^\infty(g)} = Sup_\Sigma\{g^{ac}g^{bd}h_{ab}h_{cd}\}^{\frac{1}{2}} \le e^{-2\lambda_m}\,\|\,h\,\|_\infty \le \frac{1}{2}\tau^2\,\|\,h\,\|_\infty. \qquad \square$$

6.2. W_3^p estimates for N

6.2.1. H_2 estimates of N.

Theorem 6.3. *There exists a number $C_{E,\lambda,\sigma}$ such that the H_2-norm of N satisfies the inequality*

$$\|\,2 - N\,\|_{H_2} \le C_{E,\lambda,\sigma}(\varepsilon^2 + \varepsilon\varepsilon_1).$$

Corollary 6.4.

a. *It holds that:*

$$\|\,2 - N\,\|_{L^\infty} \le C_{E,\lambda,\sigma}(\varepsilon^2 + \varepsilon\varepsilon_1). \tag{6.2}$$

b. *There exists $\eta_2 > 0$ such that*

$$\varepsilon + \varepsilon_1 \le \eta_2 \tag{6.3}$$

implies the existence of a positive number N_m (independent of t) such that

$$N \ge N_m > 0.$$

Proof. We write, as in [CB-M1] the equation satisfied by N in the form

$$\Delta(2 - N) - (2 - N) = \beta$$

with, having chosen the parameter t such that $\partial_t \tau = \tau^2$,

$$\beta \equiv (2 - N)(e^{2\lambda}\frac{1}{2}\tau^2 - 1) - N(e^{2\lambda} \mid u' \mid^2 + e^{-2\lambda} \mid h \mid^2).$$

The standard elliptic estimate applied to the form given to the lapse equation gives

$$\parallel 2 - N \parallel_{H_2} \leq C_\sigma \parallel \beta \parallel .$$

Since $0 < N \leq 2$ and $e^{-2\lambda} \leq \frac{1}{2}\tau^2$ it holds that

$$\parallel \beta \parallel \leq 2(\frac{1}{2}e^{2\lambda_M}\tau^2 - 1)V_\sigma^{1/2} + 2(e^{2\lambda_M} \parallel |u'|^2 \parallel + \frac{1}{2}\tau^2 \parallel |h|^2 \parallel). \qquad (6.4)$$

The L^4-norms of h and u' as well as $\frac{1}{2}e^{2\lambda_M}\tau^2 - 1$ have been estimated in the section on the conformal factor estimate. We deduce from these estimates the bound

$$\parallel \beta \parallel \leq C_{E,\lambda,\sigma}(\varepsilon^2 + \varepsilon\varepsilon_1)$$

which gives the result of the theorem.

The corollary a. is a consequence of the Sobolev embedding theorem, b. is a consequence of a. □

6.2.2. L^∞ estimate of DN.

Theorem 6.5. *Under the hypotheses H there exist numbers C_E, C_λ and C_σ such that if $1 < p < 2$, for instance $p = \frac{4}{3}$*

$$\parallel 2 - N \parallel_{W_3^p} \leq C_\lambda C_\sigma C_E(\varepsilon^2 + \varepsilon\varepsilon_1). \qquad (6.5)$$

Corollary 6.6. *The gradient of N satisfies the inequality:*

$$\parallel DN \parallel_{L^\infty(g)} \mid \leq C_\lambda C_\sigma C_E |\tau|(\varepsilon^2 + \varepsilon\varepsilon_1). \qquad (6.6)$$

Proof. The proof, essentially the same as in [CB-M1] rests on the W_1^p estimate of β, since applying the standard elliptic estimate gives

$$\parallel 2 - N \parallel_{W_3^p} \leq C_\sigma \parallel \beta \parallel_{W_1^p} . \qquad (6.7)$$

The estimate of

$$\parallel \beta \parallel_p \leq V_\sigma^{\frac{1}{p} - \frac{1}{2}} \parallel \beta \parallel$$

is the same as in [CB-M1] Theorem 28. In the estimate $\parallel D\beta \parallel_p$ the difference could be only in the estimate of the term $\parallel D|u'|^2 \parallel_p$. We have here

$$D|u'|^2 \equiv 2u'.\hat{D}u' \qquad (6.8)$$

hence, with $p = \frac{4}{3}$

$$\parallel D|u'|^2 \parallel_p \leq 2 \parallel u' \parallel_4 \parallel Du' \parallel + 2e^{4(\gamma_M - \gamma_m)}||D\gamma||_4||e^{-2\gamma}\omega'||_4^2$$

which leads to the same estimate as in [CB-M1]:

$$||\beta||_{W_1^p} \leq C_{E,\lambda,\sigma}(\varepsilon^2 + \varepsilon\varepsilon_1).$$

The corollary is a consequence of the Sobolev embedding theorem and the relation between σ and g-norms:

$$\parallel DN \parallel_{L^\infty(g)} \leq e^{-\lambda_m} \parallel DN \parallel_\infty \leq e^{-\lambda_m} C_\sigma \parallel DN \parallel_{W_2^p} \leq C_{E,\lambda,\sigma}|\tau|(\varepsilon^2 + \varepsilon\varepsilon_1). \quad □$$

7. $\partial_t \sigma$ and shift estimates

7.1. $\partial_t \sigma$ estimate

We have chosen a section ψ of M_{-1} over Teichmüller space, denoted $\sigma \equiv \psi(Q)$ and supposed (Hypothesis H_σ) that Q remains in a compact subset of T_{eich}.

We then have:

$$\partial_t \sigma_{ab} = \frac{\partial \psi_{ab}}{\partial Q^I} \frac{dQ^I}{dt} \tag{7.1}$$

where $\frac{\partial \psi}{\partial Q^I}$ is uniformly bounded. Hence it holds that

$$|\partial_t \sigma| \leq C_\sigma |\frac{dQ}{dt}|. \tag{7.2}$$

We recall that Q satisfies the differential equation

$$X_{IJ} \frac{dQ^I}{dt} + Y_{IJ} P^I + Z_J = 0$$

where $X_{IJ} \equiv \int_{\Sigma_t} X_I^{ab} X_{J,ab} \mu_{\sigma_t}$ is a matrix X with uniformly bounded inverse while Y_{IJ} and Z_J admit the following bounds, deduced from the basic estimates of N, λ and the L^2 bound of r:

$$|Y_{IJ}| \equiv |\int_{\Sigma_t} 2Ne^{-2\lambda} X_I^{ab} X_{J,ab} \mu_{\sigma_t}| \leq C_\sigma \tau^2$$

$$|Z_J| \equiv |\int_{\Sigma_t} 2Ne^{-2\lambda} r_{ab} X_J^{ab} \mu_{\sigma_t}| \leq C_\sigma \tau^2 ||r|| \leq C_\sigma C_E |\tau| \varepsilon (\varepsilon + \varepsilon_1).$$

On the other hand we recall that

$$q_{ab} \equiv h_{ab}^{TT} \equiv X_{I,ab} P^I \tag{7.3}$$

hence

$$P^I \equiv (X^{-1})^{IJ} \int_{\Sigma_t} X_J^{ab} q_{ab} \mu_\sigma. \tag{7.4}$$

Therefore:

$$|P^I| \leq C_\sigma ||q|| \leq C_\sigma (||h|| + ||r||) \leq C_\sigma C_E |\tau|^{-1} (\varepsilon + \varepsilon_1) \tag{7.5}$$

hence

$$|Y_{IJ} P^J| \leq C_\sigma C_E |\tau| (\varepsilon + \varepsilon_1). \tag{7.6}$$

We have obtained inequalities of the following type:

$$|\frac{dQ}{dt}| \leq C_\sigma C_E |\tau| (\varepsilon + \varepsilon_1), \quad |\partial_t \sigma| \leq C_\sigma C_E |\tau| (\varepsilon + \varepsilon_1). \tag{7.7}$$

The derivatives $D^k \partial_t \sigma$ satisfy inequalities of the same type.

7.2. Shift estimate

The equation to be satisfied by the shift ν reads, with n_a the covariant components of the vector ν in the metric σ, i.e., $n_a \equiv \sigma_{ab}\nu^b \equiv e^{-2\lambda}g_{ab}\nu^b$:

$$(L_\sigma n)_{ab} \equiv D_a n_b + D_b n_a - \sigma_{ab} D_c n^c = f_{ab} \tag{7.8}$$

$$f_{ab} \equiv 2Ne^{-2\lambda}h_{ab} + \partial_t \sigma_{ab} - \frac{1}{2}\sigma_{ab}\sigma^{cd}\partial_t\sigma_{cd}.$$

The elliptic theory for this first-order system gives the estimate

$$||n||_{W_3^p} \equiv ||\nu||_{W_3^p} \leq C_\sigma ||f||_{W_2^p} \tag{7.9}$$

with, if $p > 1$, using the bound 5.3 of $e^{-2\lambda}$,

$$||f||_{W_2^p} \leq C_\sigma C_E \{\tau^2 ||N||_{W_2^p}||\lambda||_{W_2^p}||h||_{W_2^p} + ||\partial_t\sigma||_{W_2^p}\}. \tag{7.10}$$

Hence, using previous estimates

$$||f||_{W_2^p} \leq C_\sigma C_E |\tau|(\varepsilon + \varepsilon_1). \tag{7.11}$$

8. Second energy estimate

We have defined the energy $E^{(1)}(t)$ of gradient u by the formula

$$\tau^2 \varepsilon_1^2 \equiv E^{(1)}(t) \equiv \int_{\Sigma_t} (J_0 + J_1)\mu_g \tag{8.1}$$

with

$$J_1 = \frac{1}{2} \mid \hat{\Delta}_g u \mid^2 \equiv \frac{1}{2}\{2(\hat{\Delta}_g\gamma)^2 + \frac{1}{2}e^{-4\gamma}(\hat{\Delta}_g\omega)^2\} \tag{8.2}$$

$$J_0 = \frac{1}{2} \mid \hat{D}u' \mid_g^2 \equiv \frac{1}{2}\{2|\hat{D}\gamma'|_g^2 + \frac{1}{2}e^{-4\gamma}|\hat{D}\omega'|_g^2\}. \tag{8.3}$$

8.1. Second energy equality

We have:

$$\frac{d}{dt}\int_{\Sigma_t}(J_1 + J_0)\mu_g = \int_{\Sigma_t}\{\partial_t(J_1 + J_0) - (N\tau - \nabla_a\nu^a)(J_1 + J_0)\}\mu_g. \tag{8.4}$$

On a compact manifold Σ, divergences integrate to zero, which leads to the following formula where the shift does not appear explicitly:

$$\frac{d}{dt}\int_{\Sigma_t}(J_1 + J_0)\mu_g = \int_{\Sigma_t}\{\partial_0(J_1 + J_0) - N\tau(J_1 + J_0)\}\mu_g \tag{8.5}$$

with, since $\hat{\partial}_0 G_{AB} = 0$,

$$\partial_0 J_1 = \hat{\partial}_0\hat{\Delta}_g u.\hat{\Delta}_g u. \tag{8.6}$$

We deduce from the commutation relation of Lemma 4.1 that

$$\hat{\partial}_0\hat{\Delta}_g u^A = g^{ab}(\hat{\nabla}_a\hat{\partial}_0\partial_b u^A - \partial_c u^A\hat{\partial}_0\hat{\Gamma}_{ab}^c) + \hat{\partial}_0 g^{ab}\hat{\nabla}_a\partial_b u^A + \hat{F}_1^A \tag{8.7}$$

with

$$\hat{F}_1^A \equiv g^{ab}R_{CB}{}^A{}_D{}_0\partial_0 u^C\partial_a u^B\partial_b u^D. \tag{8.8}$$

Hence, using the identities 4.23:

$$\hat{\partial}_0 \hat{\Delta}_g u^A = g^{ab} \hat{\nabla}_a \hat{\partial}_0 \partial_b u^A + N\tau \hat{\Delta}_g u^A + F_1^A + \hat{F}_1^A, \qquad (8.9)$$

with

$$F_1^A \equiv 2\partial_c u^A (h_g^{ac} \partial_a N + N\nabla_a k^{ac}) + 2N h_g^{ab} \hat{\nabla}_a \partial_b u^A. \qquad (8.10)$$

We have therefore, using Stokes formula, and $\hat{\partial}_0 \partial_b u \equiv \hat{\nabla}_b (Nu')$

$$\int_{\Sigma_t} \partial_0 J_1 \mu_g = \int_{\Sigma_t} \{-N\hat{\nabla}_a u'.\hat{\nabla}^a \hat{\Delta}_g u + 2N\tau J_1 - \partial_a Nu'.\hat{\nabla}^a \hat{\Delta}_g u + (F_1 + \hat{F}_1).\hat{\Delta}_g u\}\mu_g.$$

$$(8.11)$$

On the other hand

$$\partial_0 J_0 = g^{ab} \hat{\partial}_0 \hat{\nabla}_a u'.\hat{\nabla}_b u' + N(h_g^{ab} + \frac{1}{2}g^{ab}\tau)\partial_a u'.\partial_b u'$$

where we have used the identity, h_g^{ab} denoting the contravariant components of h_{ab} computed with the metric g,

$$\hat{\partial}_0 g^{ab} = 2N k^{ab} \equiv 2N h_g^{ab} + N g^{ab}\tau.$$

The commutation relation 4.20 gives that:

$$g^{ab} \hat{\partial}_0 \hat{\nabla}_a u'.\hat{\nabla}_b u' \equiv \hat{\partial}_0 \hat{\nabla}_a u'.\hat{\nabla}^a u' = \hat{\nabla}_a \hat{\partial}_0 u'.\hat{\nabla}^a u' + \hat{F}_0, \qquad (8.12)$$

with

$$\hat{F}_0 \equiv R_{AB,CD} u'^D \partial_0 u^A \partial_a u^B \hat{\nabla}^a u'^C. \qquad (8.13)$$

Therefore, using the wave map equation

$$-\hat{\partial}_0 u' + N\hat{\Delta}_g u + \partial^a N \partial_a u + N\tau u' = 0. \qquad (8.14)$$

we find that

$$\hat{\partial}_0 \hat{\nabla}_a u'.\hat{\nabla}^a u' = \hat{\nabla}_a [N\hat{\Delta}_g u + \partial^c N \partial_c u + N\tau u'].\hat{\nabla}^a u' + \hat{F}_0. \qquad (8.15)$$

Therefore

$$\int_{\Sigma_t} \partial_0 J_0 \mu_g = \int_{\Sigma_t} \{N\hat{\nabla}_a \hat{\Delta}_g u.\hat{\nabla}^a u' + 3N\tau J_0 + F_0 + \hat{F}_0\}\mu_g \qquad (8.16)$$

with

$$F_0 \equiv [\partial^a N \hat{\Delta}_g u + \hat{\nabla}^a (\partial^c N \partial_c u)].\hat{\nabla}_a u' + \tau \partial^a N u'.\hat{\nabla}_a u' + N h_g^{ab} \hat{\nabla}_a u'.\hat{\nabla}_b u'. \qquad (8.17)$$

We see that the third-order terms in u disappear from the integral of $\partial_0 (J_0 + J_1)$ which reduces to

$$\int_{\Sigma_t} \partial_0 (J_0 + J_1)\mu_g = \int_{\Sigma_t} \{3N\tau J_0 + 2N\tau J_1\}\mu_g + Z_1 \qquad (8.18)$$

with

$$Z_1 \equiv \int_{\Sigma_t} \{(F_1 + \hat{F}_1).\hat{\Delta}_g u + F_0 + \hat{F}_0\}\mu_g. \qquad (8.19)$$

We have obtained

$$\frac{dE^{(1)}}{dt} = \int_{\Sigma_t} N\tau (2J_0 + J_1)\mu_g + Z_1 \qquad (8.20)$$

which we write

$$\frac{dE^{(1)}}{dt} - 2\tau E^{(1)} = \tau \int_{\Sigma_t} N J_0 \mu_g + Z_2 + Z_1 \qquad (8.21)$$

with

$$Z_2 \equiv \tau \int_{\Sigma_t} (N - 2)(2J_0 + J_1)\mu_g. \qquad (8.22)$$

8.2. Second energy inequality

Since τ is negative (and N positive) the equality 8.21 implies the inequality

$$\frac{dE^{(1)}}{dt} - 2\tau E^{(1)}(t) \leq Z_1 + Z_2. \qquad (8.23)$$

We now estimate the various terms of Z_1, Z_2, called non-linear terms because they are all homogeneous and cubic in h_g^{ab}, $N - 2$, and the derivatives of N and u. These estimates are essentially the same as the ones given in [CB-M1], due to the estimates of the previous section. We first write, using the estimate of $||N-2||_{L^\infty(g)}$ and the definition of ε_1 :

$$|Z_2| \equiv |\tau \int_{\Sigma_t} (N - 2)(2J_0 + J_1)\mu_g| \leq C_\lambda C_\sigma C_E |\tau|^3 (\varepsilon + \varepsilon_1)^4. \qquad (8.24)$$

We now estimate the different terms of Z_1, beginning with the terms X_1 and X_2 coming from F_0:

$$|X_1| \equiv |\int_{\Sigma_t} \{(\partial^a N \hat{\Delta}_g u + \partial^c N \hat{\nabla}^a \partial_c u).\hat{\nabla}_a u'\}\mu_g|.$$

A proof analogous to the proof of Lemma 4.2 gives

$$\int_{\Sigma_t} |\hat{\nabla} Du|^2_g \mu_g \leq \int_{\Sigma_t} \{|\hat{\Delta}_g u|^2_g - \frac{1}{2} R(g)|Du|^2_g\}\mu_g. \qquad (8.25)$$

The Hamiltonian constraint 2.21 implies that

$$R(g) = |u'|^2 + |Du|^2_g + |h|^2_g - \frac{1}{2}\tau^2 \geq -\frac{1}{2}\tau^2 \qquad (8.26)$$

therefore

$$||\hat{\nabla} Du||^2_g \leq ||\hat{\Delta}_g u||^2_g + \frac{1}{4}\tau^2 ||Du||^2_g. \qquad (8.27)$$

Using the estimate 6.6 of DN gives then

$$|X_1| \leq C_{E,\lambda,\sigma} |\tau|^3 (\varepsilon + \varepsilon_1)^4. \qquad (8.28)$$

The remaining, X_2, of the integral of F_0 is estimated as follows

$$|X_2| = |\int_{\Sigma_t} (\hat{\nabla}_a \partial^c N)\partial_c u)].\hat{\nabla}_b u'\mu_g| \leq (||\nabla^2 N||_{L^4(g)}|| |Du| ||_{L^4(g)}||\hat{\nabla}u'||_g. \qquad (8.29)$$

The estimates of $||\nabla^2 N||_{L^4(g)}$ and $|| |Du| ||_{L^4(g)}$ given in [CB-M1], Section 10.2.2, for the estimate of Y_2 applies here, due to Lemma 5.1, and give

$$|X_2| \leq C_{E,\lambda,\sigma} |\tau|^3 (\varepsilon + \varepsilon_1)^4. \qquad (8.30)$$

The terms in

$$X_3 \equiv \int_{\Sigma_t} F_1.\hat{\Delta}_g u \mu_g \equiv \int_{\Sigma_t} [2\partial_c u(h_g^{ac}\partial_a N + N\nabla_a k^{ac}) + 2N h_g^{ab}\hat{\nabla}_a \partial_b u].\hat{\Delta}_g u \mu_g$$

(8.31)

are analogous to terms found in [CB-M1] and can be estimated similarly, giving an inequality of the form

$$|X_3| \leq C_{E,\lambda,\sigma}|\tau|^3(\varepsilon + \varepsilon_1)^3.$$

(8.32)

The new terms, in our unpolarized case, are

$$X_4 \equiv \int_{\Sigma_t} \hat{F}_1.\hat{\Delta}_g u \mu_g \equiv \int_{\Sigma_t} g^{ab} R_{CB,AD}\partial_0 u^C \partial_a u^B \partial_b u^D \hat{\Delta}_g u^A \mu_g$$

(8.33)

and

$$X_5 \equiv \int_{\Sigma_t} \hat{F}_0 \mu_g \equiv \int_{\Sigma_t} R_{AB,CD} u'^D \partial_0 u^A \partial^a u^B \hat{\nabla}_a u'^C \mu_g.$$

(8.34)

We have here $|Riemann(G)| = 4$, therefore, using $\partial_0 u \equiv N u'$ and $0 < N \leq 2$, and the Hölder inequality:

$$|X_4| \leq 8\int_{\Sigma_t} |u'| \, |Du|_g^2|\hat{\Delta}_g u|_g \mu_g \leq 8|\tau|\varepsilon_1 ||u'||_{L^6(g)}|| \, |Du|^2 \, ||_{L^3(g)}$$

(8.35)

and

$$|X_5| \leq 8||u'||_{L^6(g)}^2|| \, |Du|_g \, ||_{L^6(g)}|\tau|\varepsilon_1.$$

(8.36)

The $L^6(g)$-norms can be estimated as follows. It results from the definitions that:

$$|| \, |Du|_g \, ||_{L^6(g)} = \{\int_{\Sigma_t} e^{-4\lambda}|Du|^6\mu_\sigma\}^{\frac{1}{6}} \leq e^{-\frac{2}{3}\lambda_m}|| \, |Du| \, ||_6$$

(8.37)

while, by the Sobolev embedding theorem

$$|| \, |Du| \, ||_6 \leq C_\sigma(||Du|| + || \, D|Du| \, ||).$$

(8.38)

It holds that

$$D|Du| = \frac{D|Du|^2}{2|Du|} = \frac{\hat{D}Du.Du}{|Du|}$$

(8.39)

hence

$$|D|Du| \, | \leq |\hat{D}Du|$$

(8.40)

and

$$|| \, |Du| \, ||_6 \leq C_\sigma(||Du|| + || \, \hat{D}Du \, ||).$$

(8.41)

The inequalities of Lemma 5.1 and the lower bound of λ_m give then

$$|| \, |Du| \, ||_{L^6(g)} \leq C_\sigma C_\lambda|\tau|^{\frac{2}{3}}(\varepsilon + \varepsilon_1)).$$

(8.42)

An analogous proof gives that

$$||u'||_{L^6(g)} \leq e^{\frac{1}{3}\lambda_M}||u'||_6$$

(8.43)

and

$$||u'||_6 \leq C_\sigma(||u'|| + ||D|u'| \, ||),$$

(8.44)

with

$$|D|u'| | \leq |\hat{D}u'|, \quad ||\hat{D}u'|| = ||\hat{D}u'||_g \tag{8.45}$$

therefore, using again Lemma 5.1

$$||u'||_{L^6(g)} \leq C_\sigma e^{\frac{1}{3}\lambda_M} |\tau|(\varepsilon + \varepsilon_1) \leq C_\sigma C_\lambda |\tau|^{\frac{2}{3}} (\varepsilon + \varepsilon_1). \tag{8.46}$$

These estimates imply that

$$|X_4| \leq C_\sigma C_\lambda |\tau|^3 (\varepsilon + \varepsilon_1)^3 \varepsilon_1 \tag{8.47}$$

and the same inequality for X_5. We have proved the following theorem

Theorem 8.1. *The second energy satisfies an equality of the form*

$$\frac{dE^{(1)}}{dt} - 2\tau E^{(1)}(t) \leq |\tau|^3 B_1 \tag{8.48}$$

with $|B_1| \leq C_\sigma C_\lambda C_E (\varepsilon + \varepsilon_1)^3$.

9. Corrected first energy

9.1. Definition and lower bound

One defines as follows a corrected first energy where α is a constant, which we will choose positive:

$$E_\alpha(t) = E(t) - \alpha\tau E_c(t), \quad E_c(t) \equiv \int_{\Sigma_t} (u - \bar{u}).u' \mu_g \tag{9.1}$$

where we have set:

$$(u - \bar{u}).u' \equiv 2(\gamma - \bar{\gamma})\gamma' + \frac{1}{2}e^{-4\gamma}(\omega - \bar{\omega})\omega' \tag{9.2}$$

and denoted by $\bar{f}$ the mean value on (Σ_t, σ) of a scalar function f:

$$\bar{f} = \frac{1}{Vol_\sigma(\Sigma_t)} \int_{\Sigma_t} f\mu_\sigma \equiv -\frac{1}{4\pi\chi} \int_{\Sigma_t} f\mu_\sigma.$$

An estimate of E_α will involve second derivatives of u, it cannot alone give a bound of the first energy E.

The Cauchy-Schwarz inequality on (Σ, g) and the relation between g and σ imply that:

$$\left| \int_{\Sigma_t} (\gamma - \bar{\gamma})\gamma' \mu_g \right| \leq ||\gamma - \bar{\gamma}||_g \, || \gamma' \,||_g \leq e^{\lambda_M} ||\gamma - \bar{\gamma}|| \, ||\gamma'||_g \, . \tag{9.3}$$

Using the Poincaré inequality and recalling that Λ_σ denotes the first positive eigenvalue of the Laplacian Δ_σ on functions with mean value zero gives the majoration (recall that $|| Df ||=|| Df ||_g$ if f is a scalar function):

$$\left| \int_{\Sigma_t} (\gamma - \bar{\gamma})\gamma' \mu_g \right| \leq e^{\lambda_M} \Lambda_\sigma^{-1/2} \, || D\gamma \,|||| \, \gamma' \,||_g \tag{9.4}$$

an analogous procedure gives, with γ_m and γ_M the lower and upper bounds of γ:

$$|\int_{\Sigma_t} e^{-4\gamma}(\omega - \tilde{\omega})\omega'\mu_g| \le e^{2(\gamma_M - \gamma_m)}e^{\lambda_M}\Lambda_\sigma^{-1/2} \parallel e^{-2\gamma}D\omega \parallel \parallel e^{-2\gamma}\omega' \parallel_g \qquad (9.5)$$

since

$$\parallel D\omega \parallel \le e^{2\gamma_M} \parallel e^{-2\gamma}D\omega \parallel .$$

Using the definition of the G-norm we see that the inequalities 9.4, 9.5 imply:

$$|\int_{\Sigma_t} (u - \tilde{u}).u'\mu_g| \le e^{\lambda_M}\Lambda_\sigma^{-\frac{1}{2}}e^{2(\gamma_M - \gamma_m)} \parallel Du \parallel \parallel u' \parallel_g \qquad (9.6)$$

with (Theorem 5.7)

$$e^{\lambda_M} \le |\tau|^{-1}\{\sqrt{2} + C_\sigma C_E C_\lambda(\varepsilon + \varepsilon_1)\}. \qquad (9.7)$$

Lemma 9.1. *It holds that:*

$$\gamma_M - \gamma_m \le C_\sigma C_E C_\lambda\{\varepsilon + \varepsilon_1\}.$$

Proof. We have:

$$0 \le \gamma_M - \gamma_m \le 2 \parallel \gamma - \bar{\gamma} \parallel_{L^\infty} .$$

The Sobolev embedding theorem gives therefore that:

$$\gamma_M - \gamma_m \le 2C_\sigma \parallel \gamma - \bar{\gamma} \parallel_{H_2}$$

hence, using again the Poincaré inequality to estimate $||\gamma - \bar{\gamma}||$,

$$\gamma_M - \gamma_m \le 2C_\sigma\{(\Lambda_\sigma^{-1} + 1) \parallel D\gamma \parallel + \parallel D^2\gamma \parallel\}.$$

It results from the definitions of the G-norm and of ε that

$$||D\gamma||^2 \le \frac{1}{2}||Du||^2 \le \varepsilon^2. \qquad (9.8)$$

On the other hand since γ is a scalar function on a 2-manifold with constant scalar curvature -1, it holds that:

$$||D^2\gamma||^2 = ||\Delta\gamma||^2 + \frac{1}{2}||D\gamma||^2. \qquad (9.9)$$

We have $\Delta\gamma = \hat{\Delta}\gamma - \frac{1}{2}e^{-4\gamma}|D\omega|^2$, hence

$$||\Delta\gamma|| \le ||\hat{\Delta}\gamma|| + \frac{1}{2}|| \, |e^{-2\gamma}D\omega|^2|| \qquad (9.10)$$

with

$$||\hat{\Delta}\gamma||^2 \le \frac{1}{2}||\hat{\Delta}u||^2 \le \frac{1}{2}e^{2\lambda_M}||\hat{\Delta}_g u||^2 \le C_\lambda \varepsilon_1^2, \qquad (9.11)$$

and, using Lemma 5.1

$$|| \, |e^{-2\gamma}D\omega|^2|| \le || \, |Du|^2|| \le C_\sigma C_\lambda\{\varepsilon^2 + \varepsilon\varepsilon_1\}. \qquad (9.12)$$

Hence:

$$||D^2\gamma||^2 \le C_E C_\lambda C_\sigma(\varepsilon_1^2 + \varepsilon^2) \qquad (9.13)$$

Using the hypothesis H_E we deduce from all these inequalities the announced result.   □

We deduce from this lemma and the elementary calculus formula:

$$e^{2(\gamma_M - \gamma_m)} \leq 1 + 2(\gamma_M - \gamma_m)e^{2(\gamma_M - \gamma_m)}$$

that there exists an inequality of the form

$$e^{2(\gamma_M - \gamma_m)} \leq 1 + C_{E,\lambda,\sigma}(\varepsilon + \varepsilon_1). \tag{9.14}$$

We have proved that

$$\left| \tau \int_{\Sigma_t} (u - \tilde{u}).u'\mu_g \right| \leq \sqrt{2}\Lambda_\sigma^{-\frac{1}{2}} \parallel Du \parallel \parallel u' \parallel_g + A_1 \tag{9.15}$$

with

$$|A_1| \leq C_{E,\lambda,\sigma}\varepsilon^2(\varepsilon + \varepsilon_1). \tag{9.16}$$

Therefore:

$$E_\alpha(t) \geq \frac{1}{2}\|h\|_g^2 + Q_{\alpha,\Lambda}(x_0, x_1) - C_{E,\lambda,\sigma}\varepsilon^2(\varepsilon + \varepsilon_1), \tag{9.17}$$

where $\parallel Du \parallel = x_1$, $\parallel u' \parallel_g = x_0$, $x = (x_0, x_1)$ and $Q_{\alpha,\Lambda}$ is the quadratic form

$$Q_{\alpha,\Lambda}(x) \equiv \frac{1}{2}(x_0^2 + x_1^2) - \alpha\sqrt{2}\Lambda_\sigma^{-\frac{1}{2}}x_0 x_1. \tag{9.18}$$

The right-hand side of the inequality 9.17 can be positive only if this quadratic form is positive, that is if:

$$\alpha < \frac{\Lambda_\sigma^{\frac{1}{2}}}{\sqrt{2}}. \tag{9.19}$$

There exists a number $K > 0$ such that:

$$Q_{\alpha,\Lambda}(x) \geq \frac{1}{2}K(x_0^2 + x_1^2) \tag{9.20}$$

if the following quadratic form Q_K is positive definite

$$Q_K(x) \equiv (1 - K)(x_0^2 + x_1^2) - 2\alpha\sqrt{2}\Lambda_\sigma^{-\frac{1}{2}}x_0 x_1) \tag{9.21}$$

that is, under the condition 9.19 on α,

$$0 < K < 1 - \alpha\sqrt{2}\Lambda_\sigma^{-\frac{1}{2}}. \tag{9.22}$$

There will then exist a number $0 < \ell < K$ such that

$$E_\alpha(t) \geq \ell E(t) \tag{9.23}$$

as soon as

$$\varepsilon + \varepsilon_1 \leq \eta_3 \tag{9.24}$$

with ($C_{\sigma,E,\lambda}$ denotes the coefficient of this type in 9.17)

$$\eta_3 = \frac{K - \ell}{C_{\sigma,E,\lambda}} < \frac{1 - \alpha\sqrt{2}\Lambda_\sigma^{-\frac{1}{2}}}{C_{\sigma,E,\lambda}}. \tag{9.25}$$

9.2. Time derivative of the corrected first energy

We have (recall that terms involving the shift are exact divergences which integrate to zero, and we have set $\frac{d\tau}{dt} = \tau^2$):

$$\frac{dE_\alpha}{dt} = \frac{dE}{dt} - \alpha\tau\mathcal{R} \quad \text{with} \quad \mathcal{R} \equiv \frac{dE_c}{dt} + \tau E_c,$$

that is:

$$\mathcal{R} \equiv \int_{\Sigma_t} \{\hat{\partial}_0 u'.(u - \bar{u}) + u'.\hat{\partial}_0(u - \bar{u}) - N\tau u'.(u - \bar{u}) + \tau u'.(u - \bar{u})\}\mu_g. \quad (9.26)$$

The mapping u satisfies the wave map equation:

$$-\hat{\partial}_0 u' + \hat{\nabla}^a(N\partial_a u) + \tau N u' = 0, \quad (9.27)$$

therefore, performing an integration by parts where we derivate $u - \bar{u}$ as if it were a section of $E^{0,1}$ we obtain that:

$$\int_{\Sigma_t} (\hat{\partial}_0 u' - N\tau u').(u - \bar{u})\mu_g = \int_{\Sigma_t} \hat{\nabla}^a(N\partial_a u).(u - \bar{u})\mu_g \quad (9.28)$$

$$= \int_{\Sigma_t} -Ng^{ab}\partial_a u.\hat{\partial}_b(u - \bar{u})\mu_g \quad (9.29)$$

where

$$\hat{\partial}_b(u^B - \bar{u}^B) \equiv \partial_b(u^B - \bar{u}^B) + G^B_{CD}\partial_b u^C(u^D - \bar{u}^D) \quad (9.30)$$

that is, due to the values of the coefficients G^A_{BC} (Remark 2.2):

$$\hat{\partial}_b(\gamma - \bar{\gamma}) \equiv \partial_b(\gamma - \bar{\gamma}) + \frac{1}{2}e^{-4\gamma}\partial_b\omega(\omega - \bar{\omega}), \quad (9.31)$$

$$\hat{\partial}_b(\omega - \bar{\omega}) \equiv \partial_b(\omega - \bar{\omega}) - 2e^{-4\gamma}[\partial_b\omega(\gamma - \bar{\gamma}) + \partial_b\gamma(\omega - \bar{\omega})] \quad (9.32)$$

hence, using the expression of the metric G:

$$\int_{\Sigma_t} -Ng^{ab}\partial_a u.\hat{\partial}_b(u - \bar{u})\mu_g = -\int_{\Sigma_t} N\{|Du|^2_g - e^{-4\gamma}g^{ab}\partial_a\omega\partial_b\omega(\gamma - \bar{\gamma})\}\mu_g. \quad (9.33)$$

The non-linear term can be estimated using previous results, namely (recall $g^{ab}\mu_g = \sigma^{ab}\mu_\sigma$):

$$|\int_{\Sigma_t} Ne^{-4\gamma}g^{ab}\partial_a\omega\partial_b\omega(\gamma - \bar{\gamma})\mu_g| \leq 2||\gamma - \bar{\gamma}|| \; ||\; |e^{-2\gamma}D\omega|^2||. \quad (9.34)$$

It holds that

$$||\gamma - \bar{\gamma}|| \leq \Lambda_\sigma^{-\frac{1}{2}}||D\gamma||, \quad ||D\gamma||^2 \leq \varepsilon^2 \quad (9.35)$$

and, due to the definition of the norm in G and Lemma 5.1,

$$||\; |e^{-2\gamma}D\omega|^2|| \leq 2||\; |Du|^2|| \leq C_\sigma C_\lambda(\varepsilon + \varepsilon_1)^2. \quad (9.36)$$

On the other hand:

$$\hat{\partial}_0(u - \bar{u})^A = \partial_0(u - \bar{u})^A + G^A_{CD}\partial_0 u^C(u - \bar{u})^D. \quad (9.37)$$

A straightforward computation using the values of the coefficients G_{CD}^A gives that

$$\int_{\Sigma_t} u'.\hat{\partial}_0(u - \bar{u})\mu_g = \int_{\Sigma_t} \{N|u'|^2 - Ne^{-4\gamma}\omega'^2(\gamma - \bar{\gamma})\}\mu_g$$

$$-\partial_t\bar{\gamma}\int_{\Sigma_t}\frac{1}{2}\gamma'\mu_g - \partial_t\bar{\omega}\int_{\Sigma_t}2e^{-4\gamma}\omega'\mu_g. \tag{9.38}$$

The non-linear term can be estimated as before:

$$\int_{\Sigma_t} Ne^{-4\gamma}\omega'^2(\gamma - \bar{\gamma})\mu_g \le 2e^{2\lambda_M}\Lambda_\sigma^{-\frac{1}{2}}\varepsilon|| |u'|^2|| \le C_\sigma C_\lambda \varepsilon^2(\varepsilon + \varepsilon_1). \tag{9.39}$$

To bound the remaining terms we observe that for a scalar function f, since $V_\sigma = -4\pi\chi$ is a constant, it holds that

$$\partial_t\bar{f} = \frac{1}{V_\sigma}\partial_t\int_{\Sigma_t} f\mu_\sigma = \frac{1}{V_\sigma}\int_{\Sigma_t}\{\partial_0 f + v^a\partial_a f + \frac{1}{2}f\sigma^{ab}\partial_t\sigma_{ab}\}\mu_\sigma \tag{9.40}$$

with, for the considered metric σ

$$\int_{\Sigma_t}\sigma^{ab}\partial_t\sigma_{ab}\mu_\sigma = 0. \tag{9.41}$$

We write $\partial_t\bar{f}$ under the form (recall that $f' \equiv N^{-1}\partial_0 f$)

$$\partial_t\bar{f} = \frac{1}{V_\sigma}\int_{\Sigma_t}\{2f' + (N - 2)f' + v^a\partial_a f + \frac{1}{2}(f - \bar{f})\sigma^{ab}\partial_t\sigma_{ab}\}\mu_\sigma \tag{9.42}$$

with

$$\frac{1}{V_\sigma}\int_{\Sigma_t}2f'\mu_\sigma \equiv \frac{\tau^2}{V_\sigma}\int_{\Sigma_t}f'\mu_g + \frac{1}{V_\sigma}\int_{\Sigma_t}(2 - e^{2\lambda}\tau^2)f'\mu_\sigma. \tag{9.43}$$

We deduce from these equalities that

$$-\partial_t\bar{\gamma}\int_{\Sigma_t}\gamma'\mu_g = -\frac{2}{V_\sigma}(\int_{\Sigma_t}\gamma'\mu_g)^2 + \frac{1}{V_\sigma}X \tag{9.44}$$

with:

$$X \equiv \{\int_{\Sigma_t}[(2-e^{2\lambda}\tau^2+N-2)\gamma'+v^a\partial_a\gamma+\frac{1}{2}(\gamma-\bar{\gamma})\sigma^{ab}\partial_t\sigma_{ab}]\mu_\sigma\}\{\int_{\Sigma_t}\gamma'\mu_g\}. \tag{9.45}$$

Equality 9.44 implies the inequality:

$$-\partial_t\bar{\gamma}\int_{\Sigma_t}\gamma'\mu_g \le \frac{1}{V_\sigma}X. \tag{9.46}$$

All the terms in X are non-linear in the energies and can be estimated. Indeed:

$$|\int_{\Sigma_t}\gamma'\mu_g| \le V_g^{\frac{1}{2}}||\gamma'||_g \le e^{\lambda_M}V_\sigma||\gamma'||_g \tag{9.47}$$

while (Theorems 5.7 and Corollary 6.4)

$$|2 - e^{2\lambda}\tau^2 + N - 2| \le C_E C_\lambda C_\sigma(\varepsilon + \varepsilon_1)^2 \tag{9.48}$$

and

$$\int_{\Sigma_t}|\gamma'|\mu_\sigma \le V_\sigma^{\frac{1}{2}}||\gamma'|| \le V_\sigma^{\frac{1}{2}}e^{-\lambda_m}||\gamma'||_g. \tag{9.49}$$

Also

$$||\nu|| \leq C_E C_\sigma |\tau| (\varepsilon + \varepsilon_1) \tag{9.50}$$

and

$$||D\gamma|| = ||D\gamma||_g \leq \varepsilon. \tag{9.51}$$

Using Section 7.1 we find that:

$$|\sigma^{ab} \partial_t \sigma_{ab}| \leq C_\sigma C_E |\tau| (\varepsilon + \varepsilon_1). \tag{9.52}$$

The same type of inequalities applies to the scalar function $\partial_t \bar{\omega}$, but we must now use also the identities

$$e^{-2\bar{\gamma}} \int_{\Sigma_t} \omega' \mu_g \equiv \int_{\Sigma_t} e^{-2\gamma} \omega' e^{2(\gamma - \bar{\gamma})} \mu_g \equiv \int_{\Sigma_t} \{e^{-2\gamma} \omega' + (e^{2(\gamma - \bar{\gamma})} - 1) e^{-2\gamma} \omega'\} \mu_g$$

$$e^{2\bar{\gamma}} \int_{\Sigma_t} e^{-4\gamma} \omega' \mu_g \equiv \int_{\Sigma_t} e^{-2\gamma} \omega' e^{2(\bar{\gamma} - \gamma)} \mu_g \equiv \int_{\Sigma_t} \{e^{-2\gamma} \omega' + (e^{-2(\gamma - \bar{\gamma})} - 1) e^{-2\gamma} \omega'\} \mu_g$$

to obtain an inequality which bounds $-\partial_t \bar{\omega} \int_{\Sigma_t} e^{-4\gamma} \omega' \mu_g$ with higher-order terms in the energies, using the bound

$$|e^{\pm 2(\gamma - \bar{\gamma})} - 1| \leq 2|\gamma - \bar{\gamma}| e^{2|\gamma - \bar{\gamma}|} \leq C_{E,\lambda,\sigma} (\varepsilon + \varepsilon_1). \tag{9.53}$$

We have proved that

$$\mathcal{R} \leq \int_{\Sigma_t} \{-N|Du|^2 + N|u'|^2 + \tau u'.(u - \bar{u})\} \mu_g + A_2 \tag{9.54}$$

with:

$$|A_2| \leq C_{E,\lambda,\sigma} (\varepsilon + \varepsilon_1)^3. \tag{9.55}$$

Theorem 9.2. *There exist numbers $\alpha > 0$ and $k > 0$ such that*

$$\frac{dE_\alpha}{dt} - k\tau E_\alpha \leq |\tau A| \tag{9.56}$$

with

$$|\tau A| \leq |\tau| C_{E,\lambda,\sigma} (\varepsilon + \varepsilon_1)^3. \tag{9.57}$$

1. *If $\Lambda_\sigma > \frac{1}{8}$ the best choice is*

$$\alpha = \frac{1}{4}, \quad k = 1. \tag{9.58}$$

2. *$\Lambda_\sigma \leq \frac{1}{8}$. Then α and k are such that:*

$$0 < \alpha < \frac{4}{8 + \Lambda_\sigma^{-1}} \leq \frac{1}{4}, \quad 0 < k < 1 - \frac{1 - 4\alpha}{(1 - 2\Lambda_\sigma^{-1}\alpha^2)^{\frac{1}{2}}}. \tag{9.59}$$

A number α satisfying the conditions of the above theorem is also such that
$$0 < \alpha < \frac{\Lambda_\sigma^{\frac{1}{2}}}{\sqrt{2}}.$$

Proof. Using 10.54 and the expression 4.8 of $\frac{dE}{dt}$ we find that:

$$\frac{dE_\alpha}{dt} \leq \tau \int_{\Sigma_t} \{|h|^2 + (1-2\alpha)|u'|^2 + 2\alpha|Du|_g^2] - \alpha\tau u'.(u-\bar{u})\}\mu_g + |\tau A| \quad (9.60)$$

where

$$A \equiv \alpha A_1 + A_2 \quad (9.61)$$

with

$$A_2 \equiv \int_{\Sigma_t} \frac{1}{2}(N-2)[|h|^2 + (1-2\alpha)|u'|^2 + 2\alpha|Du|_g^2]\mu_g. \quad (9.62)$$

We deduce from Corollary 6.4 (L^∞ estimate of $N-2$) that A_2 satisfies the same type of estimate than A_1, hence:

$$|\tau A| \leq |\tau|C_E C_\lambda C_\sigma(\varepsilon + \varepsilon_1)^3. \quad (9.63)$$

We look for a positive number k such that the difference $\frac{dE_\alpha}{dt} - k\tau E_\alpha$ can be estimated with higher-order terms in the energies. We deduce from 9.60 that:

$$\frac{dE_\alpha}{dt} - k\tau E_\alpha \leq \tau\{||h||_g{}^2 + (1-2\alpha-\frac{k}{2})||u'||_g{}^2 + (2\alpha-\frac{k}{2})||Du||_g^2 \quad (9.64)$$

$$+\alpha\int_{\Sigma_t} |\tau|(1-k)u'.(u-\bar{u})\mu_g\} + |\tau A|. \quad (9.65)$$

We have seen that:

$$|\tau\int_{\Sigma_t} u'.(u-\bar{u})\mu_g| \leq \sqrt{2}\Lambda_\sigma^{-1/2}||u'||_g||Du||_g + A_1. \quad (9.66)$$

Since $\tau < 0$, it will hold that

$$\frac{dE_\alpha}{dt} - k\tau E_\alpha \leq |\tau A|, \quad A \equiv A_1 + A_2 + A_3. \quad (9.67)$$

if the quadratic form

$$Q_{\alpha,k}(x) \equiv (1-2\alpha-\frac{k}{2})x_0^2 + (2\alpha-\frac{k}{2})x_1^2 - \alpha(1-k)\sqrt{2}\Lambda_\sigma^{-1/2}x_0 x_1 \quad (9.68)$$

is non-negative.

The quadratic form $Q_{\alpha,k}$ is non negative if:

$$k \leq 4\alpha, \quad \text{and} \quad k \leq 2(1-2\alpha) \quad (9.69)$$

and k is such that its discriminant is negative, that is:

$$2\alpha^2\Lambda_\sigma^{-1}(1-k)^2 - 4(2\alpha-\frac{k}{2})(1-2\alpha-\frac{k}{2}) \leq 0. \quad (9.70)$$

The inequalities 9.69 imply

$$k \leq 1. \quad (9.71)$$

The inequality 9.70 reads

$$(1-2\Lambda_\sigma^{-1}\alpha^2)k^2 - (1-2\Lambda_\sigma^{-1}\alpha^2)2k - 2\Lambda_\sigma^{-1}\alpha^2 + 8\alpha(1-2\alpha) > 0. \quad (9.72)$$

We have already supposed that $1 - 2\Lambda_\sigma^{-1}\alpha^2 > 0$, the inequality above is therefore equivalent to:

$$k^2 - 2k + 1 - \frac{(1 - 4\alpha)^2}{(1 - 2\Lambda_\sigma^{-1}\alpha^2)} > 0 \tag{9.73}$$

that is

$$k < 1 - \frac{1 - 4\alpha}{(1 - 2\Lambda_\sigma^{-1}\alpha^2)^{\frac{1}{2}}} . \tag{9.74}$$

There will exist such a $k > 0$ if

$$\frac{1 - 4\alpha}{(1 - 2\Lambda_\sigma^{-1}\alpha^2)^{\frac{1}{2}}} < 1. \tag{9.75}$$

Since $\alpha > 0$ this inequality reduces to:

$$-2\Lambda_\sigma^{-1}\alpha - 16\alpha + 8 > 0, \tag{9.76}$$

i.e.,

$$\alpha < \frac{4}{8 + \Lambda_\sigma^{-1}} . \tag{9.77}$$

We remark that this inequality imposes the hypothesis first made on α, since elementary calculus shows that, for any Λ, it holds that:

$$\frac{\Lambda^{\frac{1}{2}}}{\sqrt{2}} \le \frac{4}{8 + \Lambda^{-1}}, \tag{9.78}$$

the equality being attained only for $\Lambda = \frac{1}{8}$.

We distinguish two cases

1. $\Lambda_\sigma > \frac{1}{8}$. In this case it is possible to take $\alpha = \frac{1}{4}$, $k = 1$ and obtain immediately

$$\frac{dE_{\frac{1}{4}}}{dt} - \tau E_{\frac{1}{4}} \le |\tau A|. \tag{9.79}$$

2. $\Lambda_\sigma \le \frac{1}{8}$. We have then:

$$\frac{4}{8 + \Lambda_\sigma^{-1}} \le \frac{1}{4} . \tag{9.80}$$

We choose α such that it satisfies the inequality 9.77, which implies in this case $\alpha < \frac{1}{4}$, and then $k > 0$ such that it satisfies 9.74. $\square$

10. Corrected second energy

We define a **corrected second energy** $E_\alpha^{(1)}$ by the formula

$$E_\alpha^{(1)}(t) = E^{(1)}(t) + \alpha\tau E_c^{(1)}(t), \quad E_c^{(1)}(t) \equiv \int_{\Sigma_t} \hat{\Delta}_g u.u' \mu_g. \tag{10.1}$$

10.1. Lower bound

We have, according to previous notations,

$$\hat{\Delta}_g u.u' \equiv 2\Delta_g \gamma\gamma' + \frac{1}{4}e^{-4\gamma}\Delta_g \omega\omega' + b_1 \tag{10.2}$$

with

$$b_1 \equiv e^{-4\gamma}g^{ab}(\partial_a\omega\partial_b\omega\gamma' - \partial_a\gamma\partial_b\omega\omega'). \tag{10.3}$$

Hence, using Lemma 5.2:

$$B_1 \equiv |\int_{\Sigma_t} b_1\mu_g| \leq |\tau|C_{E,\lambda,\sigma}(\varepsilon + \varepsilon_1)^3. \tag{10.4}$$

The Cauchy-Schwarz inequality and the Poincaré inequality ($\bar{\gamma}'$ is a constant on Σ_t and on a compact manifold $\int_{\Sigma_t} \Delta_g\gamma\mu_g = 0$) give that:

$$|\int_{\Sigma_t} \Delta_g\gamma\gamma'\mu_g| = |\int_{\Sigma_t} \Delta_g\gamma(\gamma' - \bar{\gamma}')\mu_g| \leq e^{\lambda_M}\Lambda_\sigma^{-1/2}||\Delta_g\gamma||_g||D\gamma'|| \tag{10.5}$$

while

$$|\int_{\Sigma_t} (\Delta_g\omega)e^{-4\gamma}\omega'\mu_g| = |\int_{\Sigma_t} \Delta_g\omega(e^{-4\gamma}\omega' - \overline{e^{-4\gamma}\omega'})\mu_g|$$

$$\leq e^{\lambda_M}\Lambda_\sigma^{-1/2}||\Delta_g\omega||_g||D(e^{-4\gamma}\omega')||.$$

It holds that

$$||D(e^{-4\gamma}\omega')|| = ||e^{-4\gamma}(D\omega' - 4D\gamma\omega')|| \tag{10.6}$$

$$\leq e^{-2\gamma_m}(||e^{-2\gamma}D\omega'|| + 4||e^{-2\gamma}\omega'||_4||D\gamma||_4 \tag{10.7}$$

while

$$||\Delta_g\omega||_g \leq e^{2\gamma_M}||e^{-2\gamma}\Delta_g\omega||_g. \tag{10.8}$$

Using the bound (Lemma 9.1) of $\gamma_M - \gamma_m$ and the inequalities on the L^4-norms $||.||_4$ (Lemma 5.1), we find an inequality of the form:

$$|\int_{\Sigma_t} (\Delta_g\omega)e^{-4\gamma}\omega'\mu_g|| \leq e^{\lambda_M}\Lambda_\sigma^{-1/2}||e^{-2\gamma}\Delta_g\omega||_g||e^{-2\gamma}D\omega')|| + B_2$$

where B_2 satisfies an inequality of the same type as B_1. We have shown that

$$|\int_{\Sigma_t} \Delta_g u.u'\mu_g| \leq e^{\lambda_M}\Lambda_\sigma^{-1/2}||\Delta_g u||_g||Du'|| + B_1 + B_2.$$

The estimates of Lemma 5.1 and the inequalities

$$||\Delta_g u||_g \leq ||\hat{\Delta}_g u||_g + || |Du|^2||_g, \quad ||Du'|| \leq ||\hat{\nabla}u'|| + || |Du|^2||_g^{\frac{1}{2}}|| |u'|^2||_g^{\frac{1}{2}} \tag{10.9}$$

give

$$e^{\lambda_M}\Lambda_\sigma^{-1/2}||\Delta_g u||_g||Du'|| \leq e^{\lambda_M}\Lambda_\sigma^{-1/2}||\hat{\Delta}_g u||_g||\hat{\nabla}u'|| + B_3$$

with

$$B_3 \leq |\tau|C_{E,\lambda,\sigma}(\varepsilon + \varepsilon_1)^3. \tag{10.10}$$

Using the estimate 5.21 of $e^{\lambda_M}|\tau| - \sqrt{2}$ we have:

$$\left| \int_{\Sigma_t} \Delta_g u. u' \mu_g \right| \leq \sqrt{2} |\tau|^{-1} \Lambda_\sigma^{-1/2} ||\hat{\Delta}_g u||_g ||\hat{\nabla} u'|| + B_1 + B_2 + B_3 + B_4 \quad (10.11)$$

with

$$B_4 = |e^{\lambda_M} - \sqrt{2}|\tau|^{-1}|\Lambda_\sigma^{-1/2}||\hat{\Delta}_g u||_g||\hat{\nabla} u'|| \leq |\tau| C_\sigma (\varepsilon + \varepsilon_1)^3.$$

We deduce from these estimates, with $Q_{\alpha,\Lambda}$ the same quadratic form as in 9.18 but with $y_1 \equiv |\tau|^{-1}||\hat{\Delta}_g u||_g$, $y_0 \equiv |\tau|^{-1}||\hat{\nabla} u'||$, that:

$$\tau^{-2} E_\alpha^{(1)}(t) \geq Q_{\alpha,\Lambda}(y_0, y_1) - C_{E,\sigma,\lambda}(\varepsilon + \varepsilon_1)^3. \quad (10.12)$$

Theorem 10.1. *If α is chosen satisfying 9.19 there exists a number $\eta_4 > 0$ and $L > 0$ such that*

$$E_\alpha + \tau^{-2} E_\alpha^{(1)} \geq L(\varepsilon^2 + \varepsilon_1^2) \quad (10.13)$$

as soon as

$$\varepsilon + \varepsilon_1 \leq \eta_4. \quad (10.14)$$

Proof. We have found that

$$\psi(t) \equiv E_\alpha(t) + \tau^{-2} E_\alpha^{(1)}(t) \geq Q_{\alpha,\Lambda}(y, x) - (A + B) \quad (10.15)$$

where $Q_{\alpha,\Lambda}(x, y)$ is the quadratic form

$$Q_{\alpha,\Lambda}(x, y) \equiv Q_{\alpha,\Lambda}(x) + Q_{\alpha,\Lambda}(y). \quad (10.16)$$

and $A + B$ admits a bound of the form

$$|A + B| \leq C_{E,\sigma,\lambda}(\varepsilon^2 + \varepsilon_1^2)^{\frac{3}{2}}. \quad (10.17)$$

We have

$$Q_{\alpha,\Lambda}(x, y) > K(\varepsilon^2 + \varepsilon_1^2) \equiv \frac{1}{2} K(x_0^2 + x_1^2 + y_0^2 + y_1^2) \quad (10.18)$$

if the quadratic form Q_K defined in Section 9.1 is positive definite. The conditions on α and the corresponding limitation on K are the same as in Section 9.1, and the proof continues along the same line. $\square$

10.2. Decay of the second corrected energy

We have (recall $\frac{d\tau}{dt} = \tau^2$) :

$$\frac{dE_\alpha^{(1)}}{dt} \equiv \frac{dE^{(1)}}{dt} + \alpha\tau\mathcal{R}^{(1)} \quad \mathcal{R}^{(1)} \equiv \frac{d}{dt} E_c^{(1)} + \tau E_c^{(1)}$$

that is:

$$\mathcal{R}^{(1)} = \int_{\Sigma_t} \{\hat{\partial}_0 \hat{\Delta}_g u. u' + \hat{\Delta}_g u.(\hat{\partial}_0 u' - N\tau.u' + \tau u')\}\mu_g. \quad (10.19)$$

We have found in Lemma 4.1 that

$$\hat{\partial}_0 \hat{\Delta}_g u^A \equiv g^{ab} \hat{\nabla}_a \hat{\partial}_0 \partial_b u^A + N\tau \hat{\Delta}_g u^A + F_1^A + \hat{F}_1^A, \quad (10.20)$$

with

$$\hat{F}_1^A \equiv g^{ab} R_{CB}{}^A{}_D \partial_0 u^C \partial_a u^B \partial_b u^D$$

and

$$F_1^A \equiv 2\partial_c u^A (h_g^{ac}\partial_a N + N\nabla_a k^{ac}) + 2Nh_g^{ab}\hat{\nabla}_a\partial_b u^A \qquad (10.21)$$

that is, using the equation

$$^{(3)}R_0^c \equiv -N\nabla_a k^{ac} = \partial_0 u.\partial^c u,$$

$$F_1^A \equiv 2\partial_c u^A (h_g^{ac}\partial_a N - \partial_0 u.\partial^c u) + 2Nh_g^{ab}\hat{\nabla}_a\partial_b u^A. \qquad (10.22)$$

Partial integration gives, using also the identity $\hat{\partial}_0\partial_b u \equiv \hat{\nabla}_b\partial_0 u \equiv \hat{\nabla}_b(Nu')$,

$$\int_{\Sigma_t} (\hat{\partial}_0\hat{\Delta}_g u).u'\mu_g = \int_{\Sigma_t} \{-N|\hat{\nabla}u'|_g^2 - \partial^a N\hat{\nabla}_a u'.u' + N\tau\hat{\Delta}_g u.u' + (F_1 + \hat{F}_1).u'\}\mu_g.$$
$$\qquad (10.23)$$

On the other hand, if u satisfies the equation

$$-\hat{\partial}_0 u' + \hat{\nabla}^a(N\partial_a u) + \tau Nu' = 0 \qquad (10.24)$$

it holds that:

$$\int_{\Sigma_t} \hat{\Delta}_g u.(\hat{\partial}_0 u' - \tau Nu')\mu_g = \int_{\Sigma_t} \{N|\hat{\Delta}_g u|^2 + \partial^a N\partial_a u.\hat{\Delta}_g u\}\mu_g.$$

We have found that:

$$\mathcal{R}^{(1)} = \int_{\Sigma_t} \{-N|\hat{D}u'|^2 + N|\hat{\Delta}_g u|^2 + \tau(N+1)\hat{\Delta}_g u.u'\}\mu_g + \tilde{\mathcal{R}}^{(1)} \qquad (10.25)$$

with

$$\tilde{\mathcal{R}}^{(1)} \equiv \int_{\Sigma_t} \{-\partial^a N\hat{\nabla}_a u'.u' + (F_1 + \hat{F}_1).u' + \partial^a N\partial_a u.\hat{\Delta}_g u\}\mu_g. \qquad (10.26)$$

Using the expression of $\frac{dE^{(1)}}{dt}$ we find that:

$$\frac{dE_\alpha^{(1)}}{dt} = \tau\int_{\Sigma_t} \{N(2-2\alpha)J_0 + N(2\alpha+1)J_1 + (N+1)\alpha\tau\hat{\Delta}_g u.u']\}\mu_g + Z_1 + \alpha\tau\tilde{\mathcal{R}}^{(1)}$$

which implies:

$$\frac{dE_\alpha^{(1)}}{dt} - (2+k)\tau E_\alpha^{(1)} = \tau\{\int_{\Sigma_t} \{(2N-2-k-2\alpha N)J_0 + (2\alpha N + N - 2 - k)J_1$$
$$+\alpha\tau(N+1-2-k)\hat{\Delta}_g u.u'\}\mu_g + Z_1 + \alpha\tau\tilde{\mathcal{R}}^{(1)} \qquad (10.27)$$

which we write:

$$\frac{dE_\alpha^{(1)}}{dt} - (2+k)\tau E_\alpha^{(1)} \leq \tau\{\int_{\Sigma_t} \{(2-k-4\alpha)J_0 + (4\alpha - k)J_1$$
$$+\alpha\tau(1-k)\hat{\Delta}_g u.u'\}\mu_g + Z_1 + \alpha\tau\tilde{\mathcal{R}}^{(1)} + Z_2 \qquad (10.28)$$

with

$$Z_2 \equiv \tau\int_{\Sigma_t} \{(N-2)(2-2\alpha)J_0 + (N-2)(2\alpha+1)J_1 + (N-2)\alpha\tau\hat{\Delta}_g u.u']\}\mu_g. \qquad (10.29)$$

We have found in Section 8.2 that

$$|Z_1| \leq |\tau|^3 C_{E,\lambda,\sigma}(\varepsilon + \varepsilon_1)^3. \tag{10.30}$$

It results immediately from the estimate of $N - 2$, Section 6.2, that Z_2 satisfies an inequality of the same type.

Some terms of $\tilde{\mathcal{R}}^{(1)}$ are bounded using the $L^\infty(g)$ estimate 6.6 of DN, which gives that

$$|\int_{\Sigma_t} \{-\partial^a N \hat{\nabla}_a u'.u' + \partial^a N \partial_a u.\hat{\Delta}_g u\}\mu_g| \leq \tau^2 C_{E,\sigma,\lambda}\varepsilon^2 \varepsilon_1(\varepsilon + \varepsilon_1). \tag{10.31}$$

The estimate of the remaining ones uses similar techniques as those of Section 9 and lead to an inequality of the form

$$|\tilde{\mathcal{R}}^{(1)}| \leq \tau^2 C_{E,\sigma,\lambda}(\varepsilon + \varepsilon_1)^4. \tag{10.32}$$

Theorem 10.2. *Under the conditions on α and k given in Theorem 9.2 the following inequality holds:*

$$\frac{dE_\alpha^{(1)}}{dt} - (2 + k)\tau E_\alpha^{(1)} \leq |\tau|^3 B \tag{10.33}$$

with

$$|B| \leq C_{E,\lambda,\sigma}(\varepsilon + \varepsilon_1)^3.$$

Proof. We have seen that (10.11)

$$|\int_{\Sigma_t} \tau \hat{\Delta}_g u.u' \mu_g| \leq \sqrt{2}\Lambda_\sigma^{-\frac{1}{2}} ||\hat{\Delta}_g u||_g \, ||\hat{\nabla}u'||_g + Z_3 \tag{10.34}$$

with

$$|Z_3| \leq \tau^2 C_{E,\sigma,\lambda}(\varepsilon + \varepsilon_1)^3. \tag{10.35}$$

Therefore we deduce from 10.28 and the definition of y_0, y_1 that

$$\frac{dE_\alpha^{(1)}}{dt} - (2 + k)\tau E_\alpha^{(1)} \leq \tau^3 Q_{\alpha,k}^{(1)}(y) + |\tau|^3 B$$

with

$$B \equiv Z_1 + \alpha\tau\tilde{\mathcal{R}}^{(1)} + Z_2 + |\tau|\alpha Z_3, \quad |B| \leq C_{E,\sigma,\lambda}(\varepsilon^2 + \varepsilon_1^2)^{\frac{3}{2}} \tag{10.36}$$

and

$$Q_{\alpha,k}^{(1)}(y) \equiv (1 - \frac{k}{2} - 2\alpha)y_0^2 + (2\alpha - \frac{k}{2})y_1^2 + \sqrt{2}\Lambda_\sigma^{-\frac{1}{2}}\alpha\tau(1 - k)y_0 y_1. \tag{10.37}$$

This quadratic form in y is non-negative under the same conditions as the form $Q_{\alpha,k}(x)$. The conclusion follows, since $\tau < 0$. $\qquad \square$

11. Decay of the total energy

We make the following a priori hypothesis, for all $t \geq t_0$ for which the considered quantities exist

- **Hypothesis H_σ:** 1. The t-dependent numbers C_σ are uniformly bounded by a constant M.

  2. There exist $\Lambda > 0$ such that $\Lambda_\sigma \geq \Lambda$.

- We choose α such that

$$\alpha = \frac{1}{4} \quad \text{if} \quad \Lambda > \frac{1}{8}, \quad \alpha < \frac{4}{8 + \Lambda^{-1}} \leq \frac{1}{4} \quad \text{if} \quad \Lambda \leq \frac{1}{8}. \tag{11.1}$$

- **Hypotheses H_E^η:** The t-dependent energies ε^2 and ε_1^2 having been supposed bounded by a number c_E we suppose, moreover that they satisfy the inequalities 5.23, 6.3, 9.25, 10.14.

  We have seen (Theorem 5.7) that the hypothesis H_λ is then satisfied.

We denote by M_i any given positive number dependent on the bounds of these H's hypothesis but independent of t.

We have defined $\psi(t)$ to be the **total corrected energy** namely:

$$\psi(t) \equiv E_\alpha(t) + \tau^{-2} E_\alpha^{(1)}.$$

We have seen (10.13) that $\psi(t)$ bounds the total energy $\phi(t) \equiv \varepsilon^2 + \varepsilon_1^2$ by an inequality of the form:

$$\phi(t) \equiv \varepsilon^2 + \varepsilon_1^2 \leq M_0 \psi(t), \quad M_0 = L^{-1}. \tag{11.2}$$

Lemma 11.1. *Under the hypotheses H_σ and H_E^η the function ψ satisfies a differential inequality of the form*

$$\frac{d\psi}{dt} \leq -\frac{k}{t}(\psi - M_1 \psi^{3/2}). \tag{11.3}$$

Proof. The inequalities 9.56 and 10.33 together with the choice $\tau = -\frac{1}{t}$, and the bound 10.13. $\qquad \square$

Theorem 11.2. *Under the hypotheses H_σ and H_E^η there exists a number $k > 0$ such that the total energy $E_{tot}(t) \equiv \phi(t) \equiv \varepsilon^2 + \varepsilon_1^2$ satisfies an estimate of the form*

$$t^k \phi(t) \leq M_2 \phi(t_0) \tag{11.4}$$

if it is small enough initially.

Proof. We suppose that $\psi_0 \equiv \psi(t_0)$ satisfies

$$\psi_0^{1/2} < \frac{1}{M_1}. \tag{11.5}$$

Then ψ starts decreasing, continues to decrease as long as it exists, therefore $(\psi - M_1 \psi^{3/2}) > 0$ and the inequality 11.3 is equivalent to

$$\frac{dz}{z - M_1 z^2} + \frac{k}{2}\frac{dt}{t} \leq 0, \quad \text{with} \quad \psi = z^2.$$

This inequality gives by integration:

$$\log\{\frac{z}{(1-M_1 z)z_0}\frac{(1-M_1 z_0)}{z_0}\} + \frac{1}{2}k\log\frac{t}{t_0} \leq 0$$

equivalently

$$\{\frac{z}{(1-M_1 z)}\frac{(1-M_1 z_0)}{z_0}\}\{\frac{t}{t_0}\}^{\frac{1}{2}k} \leq 1 \tag{11.6}$$

and, a fortiori[13],

$$t^k\psi \leq \frac{t_0^k\psi_0}{(1-M_1 z_0)^2}. \tag{11.7}$$

Hence, using 11.2 and an analogous converse bound between ϕ and ψ, immediate consequence of their definitions, the announced **decay estimate** follows.   □

12. Teichmüller parameters

Instead of considering as in [CB-M1], [CB-M2] the Dirichlet energy of the metric σ we use directly the estimate 7.7 of dQ/dt which we now write, using 11.8:

$$|\frac{dQ}{dt}| \leq C_{\sigma,E}t^{-(1+\frac{k}{2})}\phi(t_0)^{1/2}. \tag{12.1}$$

Therefore:

Theorem 12.1. *There exists M_3 such that*

$$|Q(t) - Q(t_0)| \leq M_3\phi(t_0)^{1/2}. \tag{12.2}$$

13. Global existence

Theorem 13.1.   *Let $(\sigma_0, q_0) \in C^\infty(\Sigma_0)$ and $(u_0, \dot{u}_0) \in H_2(\Sigma_0, \sigma_0) \times H_1(\Sigma_0, \sigma_0)$ be initial data for the Einstein equations with U(1) isometry group on the initial manifold $M_0 \equiv \Sigma_0 \times U(1)$, with Σ_0 compact, orientable and of genus greater than one, σ_0 chosen such that $R(\sigma_0) = -1$. Suppose the initial integral condition 2.1 (with $n = 0$) satisfied. Then there exists a number $\eta_0 > 0$ such that if*

$$\phi(t_0) \equiv E_{tot}(t_0) < \eta_0 \tag{13.1}$$

these Einstein equations have a solution on $M \times [t_0, \infty)$, with initial values determined by $\sigma_0, q_0, u_0, \dot{u}_0$. The parameter t is $t = -\tau^{-1}$, with τ the mean extrinsic curvature of $\Sigma \times \{t\}$ in the Lorentzian metric $^{(3)}g$ on $\Sigma \times [t_0, \infty)$.

   This solution is unique[14] up to the choice of a section of Teichmüller space and a gauge choice for A.

---

[13]If for instance we choose $z_0 \leq M_1/2$ it holds that $t^k\psi \leq 4t_0^k\psi_0$.
[14]The global uniqueness theorem of CB-Geroch says that it is geometrically unique in the class of globally hyperbolic spacetimes.

Proof. We first prove that $E_{tot}(t)$ is uniformly bounded, and decays to zero (without a priori hypothesis). We have obtained in the previous sections, under the hypotheses H_E^η and H_σ, the following result: there are numbers M_i depending only on c_E and c_σ such that

$$t^k E_{tot}(t) \le M_2 E_{tot}(t_0) \tag{13.2}$$

and

$$|Q(t) - Q(t_0)| \le M_3 \phi(t_0)^{1/2}. \tag{13.3}$$

Now consider the pair of t dependent numbers

$$(\phi(t), \zeta(t)), \quad \zeta(t) \equiv |Q(t) - Q(t_0)|$$

The inequalities 13.2, 13.3 show that the hypothesis (where c_E satisfies H_E^η)

$$\phi(t) \le c_E, \ \zeta(t) \le c_\sigma$$

imply that there exists $\eta_0 > 0$ such that $\phi(t_0) \le \eta_0$ implies that the pair belongs to the subset $U_1 \subset R^2$ defined by the inequalities:

$$U_1 \equiv \{ \phi(t) < c_E, \ \zeta(t) < c_\sigma \}.$$

Therefore for such an η_0 the pair belongs either to U_1 or to the subset U_2 defined by

$$U_2 \equiv \{ \phi(t) > c_E \quad \text{or} \quad \zeta(t) > c_\sigma \}$$

These subsets are disjoint. We have supposed that for $t = t_0$ it holds that $(\phi(t_0), \zeta(t_0)) \in U_1$ hence, by continuity in t, $(\phi(t), \zeta(t)) \in U_1$ for all t.

We have now proved that the total energy is uniformly bounded, and σ_t uniformly equivalent to σ_0.

To complete the proof of existence of the spacetime for $t \in [t_0, \infty)$ we need the following lemma.

Lemma. The H_2-norm of the pair of scalar functions (γ, ω) is uniformly bounded, as well of the H_1-norm of $(\partial_t \gamma, \partial_t \omega)$.

Proof of the lemma. We have already proven in Section 9 the uniform bound of $||D\gamma||$ and $||D^2\gamma||$ in terms of the total energy. On the other hand it holds that

$$\gamma - \gamma_0 = \int_{t_0}^t \partial_t \gamma dt \tag{13.4}$$

hence

$$||\gamma - \gamma_0|| \le \int_{t_0}^t ||\partial_t \gamma|| dt. \tag{13.5}$$

Using previous estimates, the fall off of the energy and the property

$$||\partial_t \gamma|| \le e^{-\lambda_m} ||\partial_t \gamma||_g \tag{13.6}$$

we find that there exists a number M such that

$$||\gamma - \gamma_0|| \le M \int_{t_0}^t t^{-(1+k)} dt, \tag{13.7}$$

which completes the proof of the uniform bound of $||\gamma||_{H_2}$, hence also of γ in C^0.

When γ_M is uniformly bounded one can bound $||D\omega|| \leq e^{2\gamma_M}||e^{-2\gamma}D\omega||$ with the first energy and $||D^2\omega||$ with the total energy, in a manner analogous as the one used for $||D^2\gamma||$. We just recalled the estimate of $||\partial_t\gamma||$, the estimate of $||\partial_t\omega||$ is similar, when γ has been bounded. It is also easy to bound $||D\partial_t\omega||$ and $||D\partial_t\gamma||$. $\qquad\square$

Corollary 13.2.

1. *This solution is globally hyperbolic, future timelike and null complete.*
2. *It is asymptotic to a flat solution:*

$$^{(4)}g = -4dt^2 + 2t^2\sigma_\infty + \theta_\infty^2 \tag{13.8}$$

with σ_∞ a metric on Σ independent of t and of scalar curvature -1, and θ_∞ a 1-form on $\Sigma \times S^1$ of the type

$$\theta_\infty = C(dx^3 + H), \tag{13.9}$$

where C is a constant and H is a harmonic 1-form on (Σ, σ_0).

Proof. 1. The orthogonal trajectories to the space sections $M \times \{t\}$ have an infinite proper length since the lapse N is bounded below by a strictly positive number. It can be checked that the conditions given in C.B and Cotsakis for global hyperbolicity, and for future and null completeness are satisfied by $(\Sigma \times R, ^{(3)}g)$.

2. Theorem 5.7 and the decay of $\varepsilon + \varepsilon_1$ show that λ tends to $2t^2$ in C^0-norm when t tends to infinity.

The decay estimate of $\frac{dQ}{dt}$ show that Q tends to a point Q_∞ in T_{eich} when t tends to infinity, σ tends to $\sigma(Q_\infty)$.

The lapse and shift estimates 6.2 and 7.9, 7.11 show that N tends to 2 and ν tends to zero in C^0-norm when t tends to infinity.

The integral formula for γ shows that $\gamma(t,.) - \gamma_0(.)$ tends to a function on Σ, $\hat{\gamma}_\infty(.)$, in L^2-norm when t tends to infinity, hence γ tends to $\gamma_\infty = \gamma_0 + \hat{\gamma}_\infty$ in this norm, therefore a fortiori $\gamma(t,.)$ tends to $\gamma_\infty(.)$ in the sense of distributions on Σ. We know on the other hand that $||D\gamma||$ tends to zero, hence $D\gamma$ tends to zero in the sense of distributions. Since derivation in this sense is a continuous operator it holds that $D\gamma_\infty = 0$, therefore γ_∞ is a constant.

An analogous reasoning holds for ω. The value of ω_∞ does not appear in the expression of F.

The estimates of Section 2.2 of $\hat{A}$ and A_t (in Coulomb gauge) show that they both tend to zero in C^0-norm on Σ. The differential formula giving the $c_i(t)$ shows then that the 1-form $\tilde{A}$ tends in C^0-norm to the harmonic form on Σ, $H_\infty = c_{i,\infty}H_{(i)}$. The spacetime metric is asymptotic to the metric

$$^{(4)}g = e^{-2\gamma_\infty}(-4dt^2 + 2t^2\sigma_\infty) + e^{2\gamma_\infty}(dx^3 + H_\infty)^2 \tag{13.10}$$

which takes the indicated form by rescaling of t. $\qquad\square$

Acknowledgements

We thank the University of the Aegean in Samos, the Schrödinger Institute in Vienna and the Institut des Hautes Etudes Scientifiques in Bures-sur-Yvette which made fruitful discussions with V. Moncrief possible.

References

[A-M-T] L. Andersson, V. Moncrief and A., Tromba On the global evolution problem in 2 + 1 gravity, J. Geom. Phys., 23 1997, n°3–4, 1991–205.

[CB1] Y. Choquet-Bruhat, Global wave maps on curved spacetimes, in "Mathematical and Quantum Aspects of Relativity and Cosmology", Cotsakis and Gibbons, eds., LNP 535, Springer 1998, 1–30.

[CB2] Y. Choquet-Bruhat, Wave Maps in General Relativity in "On Einstein path", A. Harvey, ed., Springer 1998, 161–185.

[CB-C] Y. Choquet-Bruhat and S. Cotsakis, Global hyperbolicity and completeness, J. Geom. Phys. 43 $n°4$, 2002, 345–350.

[CB-DM] Y. Choquet-Bruhat and C. DeWitt-Morette, Analysis Manifolds and Physics, II, enlarged edition (2000).

[CB-M 1] Y. Choquet-Bruhat and V. Moncrief, Future global in time Einsteinian spacetimes with U(1) isometry group, Ann. Henri Poincaré 2 (2001), 1007–1064.

[CB-M 2] Y. Choquet-Bruhat and V. Moncrief, Non-linear stability of Einsteinian spacetimes with U(1) isometry group, in Partial differential equations and mathematical physics, in honor of J. Leray, Kajitani and Vaillant, eds., to appear 2003, Birkhäuser.

[CB-M 3] Y. Choquet-Bruhat and V. Moncrief, Existence theorem for solutions of Einstein equations with 1 parameter spacelike isometry group, Proc. Symposia in Pure Math, 59, 1996, H. Brezis and I.E. Segal, eds., 67–80.

[CB-Y] Y. Choquet-Bruhat and J.W. York, Geometrical well-posed system for the Einstein equations, C.R. Acad. Sciences Paris 321, 1995, 1089–1095.

[F-T] A. Fisher and A. Tromba, Teichmüller spaces, Math. Ann. 267, 1984, 311–345.

[M1] V. Moncrief, Reduction of Einstein equations for vacuum spacetimes with U(1) spacelike isometry group, Annals of Physics 167 (1986), 118–142.

[M2] V. Moncrief, Reduction of the Einstein-Maxwell and the Einstein-Maxwell-Higgs equations for cosmological spacetimes with U(1) spacelike isometry group. Class. Quantum Grav. 7 (1990) 329–352.

[M-S] S. Müller and M. Struwe, Global existence of wave maps in 2 + 1 dimensions with finite energy data, Top. Met. in non lin. An. 7, n°2 1996, 245–261.

Yvonne Choquet-Bruhat
YCB Université Paris 6
4 Place Jussieu
F-75232 Paris, France
e-mail: YCB@CCR.jussieu.fr

Future Complete Vacuum Spacetimes

Lars Andersson[1] and Vincent Moncrief[2]

Abstract. In this paper we prove a global existence theorem, in the direction of cosmological expansion, for sufficiently small perturbations of a family of spatially compact variants of the $k = -1$ Friedmann–Robertson–Walker vacuum spacetime. We use a special gauge defined by constant mean curvature slicing and a spatial harmonic coordinate condition, and develop energy estimates through the use of the Bel-Robinson energy and its higher-order generalizations. In addition to the smallness condition on the data, we need a topological constraint on the spatial manifold to exclude the possibility of a non-trivial moduli space of flat spacetime perturbations, since the latter could not be controlled by curvature-based energies such as those of Bel-Robinson type. Our results also demonstrate causal geodesic completeness of the perturbed spacetimes (in the expanding direction) and establish precise rates of decay towards the background solution which serves as an attractor asymptotically.

1. Introduction

In this paper we establish global existence and asymptotic behavior, in the cosmologically expanding direction, for a family of spatially compact, vacuum solutions to the $3 + 1$-dimensional Einstein equations for sufficiently small perturbations of certain known "background" solutions. The backgrounds we consider are the spatially compactified variants of the familiar vacuum $k = -1$ Friedmann–Robertson–Walker (FRW) solution, which exist on any 4-manifold $\bar{M}$ of the form $(0, \infty) \times M$, where M is a compact hyperbolic 3-manifold (i.e., a manifold admitting a Riemannian metric with constant negative sectional curvature).

Let γ be the standard hyperbolic metric with sectional curvature -1 on M. Then $(\bar{M}, \bar{\gamma})$ given by

$$\bar{M} = (0, \infty) \times M, \qquad \bar{\gamma} = -d\rho \otimes d\rho + \rho^2 \gamma$$

1) Supported in part by the Swedish Natural Sciences Research Council (SNSRC), contract no. R-RA 4873-307 and NSF, contract no. DMS 0104402.

2) Supported in part by the NSF, with grants PHY-9732629 and PHY-0098084 to Yale University.

is a flat spacetime, locally isometric to the $k = -1$ vacuum FRW model, which we shall call a hyperbolic cone spacetime. It has a big bang singularity as $\rho \searrow 0$ but expands to infinite volume as $\rho \nearrow \infty$. The vector field $\rho \frac{\partial}{\partial \rho}$ is a timelike homothetic Killing field on $(\bar{M}, \bar{\gamma})$ so that these backgrounds are continuously self-similar. We shall be considering sufficiently small perturbations of such hyperbolic cone spacetimes to the future of an arbitrary ρ =constant Cauchy surface under the additional topological restriction that $(\bar{M}, \bar{\gamma})$ be "rigid" in a sense that we shall define more fully below. The rigidity assumption serves to eliminate the possibility of making non-trivial but still flat perturbations of the chosen backgrounds.

Our main result treats the vacuum Einstein equations on $\bar{M}$ and proves global existence in the expanding direction for initial data sufficiently close to data for $(\bar{M}, \bar{\gamma})$. More precisely, we show that the maximal globally hyperbolic future vacuum development $(\bar{M}, \bar{g})$ of such data is causally geodesically complete and globally foliated by constant mean curvature (CMC) hypersurfaces in the expanding direction. We further show that the metric $\bar{g}$ decays asymptotically to $\bar{\gamma}$ at a well-defined rate (that correctly predicted by linearized theory) and give the sharp rate of decay. In this sense our result may be viewed as a nonlinear stability result for the future evolution. We could also view it as implying nonlinear instability for the past evolution but, since our arguments are insufficient to treat global evolution in the past direction, we shall concentrate here on the expanding direction. Since the formation of black holes would be expected to violate geodesic completeness towards the future, we can also interpret our smallness condition in the data as sufficient to exclude the formation of black holes.

We work in a specific gauge defined by constant mean curvature slicing and a spatial harmonic coordinate condition which serves to kill off certain second-order terms in the spatial Ricci tensor, reducing it to a nonlinear elliptic operator on the metric. This in turn effectively reduces the evolution equations for the spatial metric to nonlinear wave equations wherein, however, the lapse function and shift vector field are determined by an associated set of (linear) elliptic equations. Local existence and well-posedness for the Einstein equations in this gauge was established in [1] along with a continuation principle which provides the needed criterion for proving global existence.

The main tool we employ for our global existence proof is an energy argument based on the Bel–Robinson energy and its higher-order generalization, which we define. The Bel–Robinson energy for a vacuum spacetime is basically an L^2-norm of spacetime curvature on a given Cauchy hypersurface, and its higher-order generalization incorporates the L^2-norm of the spatial gradient of this same curvature. One of the key steps in our proof will be to show that, in our chosen gauge, this generalized Bel–Robinson energy bounds an $H^3 \times H^2$-norm of the perturbed first and second fundamental forms of a CMC slice in the spacetime $(\bar{M}, \bar{g})$.

Nontrivial spacetime perturbations which preserve flatness are invisible to such purely curvature based energies, and this is the reason we have been forced to impose an additional rigidity condition upon the hyperbolic manifolds that we consider. Already by Mostow rigidity one cannot perturb the flat metric $\bar{\gamma} = -d\rho \otimes d\rho + \rho^2\gamma$ to another flat one by simply deforming the hyperbolic metric γ on M, but there can be more subtle ways of deforming $\bar{\gamma}$ on $\bar{M}$ that preserve flatness. These arise whenever (M, γ) admits so-called nontrivial traceless Codazzi tensors. Our rigidity requirement is that (M, γ) be such as to exclude such tensors – a condition which is known to be satisfied for a non-empty set of hyperbolic manifolds.

The Bel–Robinson energy is of course not a conserved quantity but, together with its higher-order generalization, can actually be shown to decay in the expanding direction for sufficiently small perturbations of a hyperbolic cone spacetime. The main source of this decay is the overall expansion of the universe which leads to an omnipresent term of good sign, proportional to the energy itself, in the time derivative of this energy. A corresponding result holds for the generalized energy. The remaining terms in the time derivative in general have no clear sign but fortunately can be bounded by a power greater than unity of the generalized energy itself. When the initial value of the generalized energy is sufficiently small this implies decay to the future at an asymptotically well-defined rate and leads to our main result.

While we shall not pursue this issue here, there seems to be a straightforward way to remove the rigidity constraint and thereby deal with arbitrary hyperbolic M. This involves supplementing the Bel–Robinson energies considered here by another non-curvature-based energy called the reduced Hamiltonian. As discussed in [8] this quantity is always monotonically decaying towards the future (even for large data) but bounds at most the rather weak $H^1 \times L^2$-norm of the CMC Cauchy data. However this should more than suffice to control the finite-dimensional space of moduli parameters which arises in the case of non-rigid M but is invisible to the Bel–Robinson energies.

Apart from general Lorentzian geometry results such as singularity theorems and conclusions drawn from the study of explicit solutions, very little is known about the global properties of generic $3 + 1$-dimensional Einstein spacetimes, with or without matter, and present PDE technology is far from being applicable to the study of such global questions, except in the case of small data.

In [6] Christodoulou and Klainerman proved the nonlinear stability of $3 + 1$-dimensional Minkowski space, i.e., a small data global existence result together with precise statements about the asymptotic decay of the metric to the Minkowski metric. This proof was based on a bootstrap argument using decay estimates for suitably defined Bel–Robinson energies. A central element in the proof was the construction of approximate Killing and conformal Killing fields, which were then used in a way which is analogous to the way in which true Killing and conformal Killing fields of Minkowski space are used in the proof of the Klainerman Sobolev inequalities for solutions of the wave equation on Minkowski space.

In still earlier work [10] Friedrich had proven global existence to the future of a Cauchy surface for the development of data sufficiently close to that of a hyperboloid in Minkowski space, with asymptotic behavior compatible with a regular conformal compactification in the sense of Penrose. This result used the fact that the conformal compactification of such spacetimes has a regular null boundary (Scri) and exploited Cauchy stability for a conformally regular first-order symmetric hyperbolic system of field equations deduced from the Einstein equations. Roughly speaking, local existence for the conformally regular system can correspond, for sufficiently small data, to global existence for the conformally related, physical spacetime.

Our argument is close in spirit to that of Christodoulou and Klainerman but is much simpler than theirs by virtue of the universal energy decay described above. The source of this decay can easily be seen in linear perturbation theory by exploiting the fact that $\rho\frac{\partial}{\partial\rho}$ is a timelike homothetic Killing field in the background. One readily constructs from this an exactly conserved quantity for the linearized equations which differs from the (linearized analogue of the) Bel–Robinson energy we consider by a multiplicative factor in the time variable ρ. This gives immediately the specific decay rate predicted by linearized theory and our arguments ultimately show that this is the precise decay rate asymptotically realized by solutions to the (small data) nonlinear problem.

Our arguments are also similar in spirit to those used in the papers [4] by Choquet-Bruhat and Moncrief and [3] by Choquet-Bruhat, which treat perturbations of certain $\mathcal{U}(1)$-symmetric vacuum spacetimes on $\mathbb{R} \times \Sigma \times S^1$, where Σ is a surface of genus greater than one, and in which the $\mathcal{U}(1)$ (Killing) symmetry is imposed along the fibers of a trivial circle bundle $\mathbb{R} \times \Sigma \times S^1 \to \mathbb{R} \times \Sigma$. The paper [4] deals with the polarized case, where the bundle is a product, and [3] deals with the general case. Their results use energy arguments which exploit the universal expansion to obtain decay for small data.

For the case of linearized perturbations, Fischer and Moncrief [9] have analyzed the stability of higher-dimensional analogues of the hyperbolic cone spacetimes described above wherein the hyperbolic metric γ is replaced by an arbitrary Einstein metric with negative Einstein constant. These of course include the higher-dimensional hyperbolic metrics but in fact comprise a much larger set. It now seems likely that the nonlinear stability problem for these spacetimes can be handled by a combination of the methods employed herein and in the article by Choquet–Bruhat.

We now give a more precise description of our main results. Let g be a Riemannian metric on M and let k be a symmetric covariant 2-tensor on M. We call (M, g, k) a vacuum data set for the Einstein equations if (g, k) satisfy the vacuum constraint equations, reviewed in Section 2.1 below. Given such a vacuum data set there is a unique maximal Cauchy development $(\bar{M}, \bar{g})$ of (M, g, k) which contains the latter as an embedded Cauchy hypersurface. Our results concern the structure of $(\bar{M}, \bar{g})$, especially to the future of the Cauchy hypersurface, for (g, k) sufficiently close to the data corresponding to a rigid hyperbolic cone spacetime $(\bar{M}, \bar{\gamma})$. We show in this case that, in the expanding direction, $(\bar{M}, \bar{g})$ is globally foliated by hypersurfaces of constant mean curvature and that $(\bar{M}, \bar{g})$ is causally geodesically complete in this (future) direction. In particular, $(\bar{M}, \bar{g})$ is inextendible in the expanding direction and thus our results support the strong cosmic censorship hypothesis.

Our main result is summarized as follows.

Theorem 1.1. *Let* (M, γ) *be a compact hyperbolic 3-manifold and assume that* (M, γ) *is rigid (i.e., admits no nontrivial traceless Codazzi tensors). Assume that* (M, g^0, k^0) *is a CMC vacuum data set with* $(g^0, k^0) \in H^s \times H^{s-1}$, $s \geq 3$, *having* $t_0 = \mathrm{tr}_{g^0} k^0 = constant < 0$. *Then there is an* $\epsilon > 0$ *so that if*

$$\left\| \frac{t_0^2}{9} g^0 - \gamma \right\|_{H^3} + \left\| \frac{t_0}{3} k^0 - \gamma \right\|_{H^2} < \epsilon$$

then

1. *The maximal Cauchy development* $(\bar{M}, \bar{g})$ *of the vacuum data set* (M, g^0, k^0) *has a global CMC foliation in the expanding direction (to the future of* M_{t_0} *in CMC time* $t = \mathrm{tr}_g k$).
2. $(\bar{M}, \bar{g})$ *is future causally geodesically complete.*

Remark 1.1.

1. *Under our conventions, cf. Section 2, the standard hyperboloid* $\{\langle x, x \rangle = -1\}$ *in* $I^+(\{0\}) \subset \mathbb{R}^{3,1}$ *has mean curvature* -3 *and* $\mathrm{Vol}(M, g)$ *increases as* $t \nearrow 0$.
2. $(\frac{t^2}{9} g, \frac{|t|}{3} k)$ *are rescaled Cauchy data that reduce to* $(\gamma, -\gamma)$ *for the background solution. Our energy arguments show that the rescaled data approach their background values at a well-defined asymptotic rate as* $t = \mathrm{tr}_g k \nearrow 0$.
3. *By exploiting the scaling with respect to* t *at* t_0 *one can satisfy the smallness condition for initial data* (g^0, k^0) *corresponding to arbitrarily large initial spacetime curvature. In this sense one can choose the initial hypersurface to be "close to the singularity".*

In outline our paper proceeds as follows. Some preliminaries and a discussion of the Einstein equations in our chosen gauge including a review of the local existence theorem proven in [1], are given in Sections 2 and 2.1. Sections 2.2–2.4 discuss the background spacetimes, the constraint set for the perturbed spacetimes and the rigidity condition needed to exclude the occurrence of a moduli space of flat perturbations. Section 3 introduces Weyl fields in the spirit of Christodoulou and Klainerman and presents the field equations they satisfy when Einstein's equations are imposed. Section 4 discusses the Bel–Robinson energy and its higher-order generalization and computes the time derivative of these quantities in the chosen gauge. Section 4.1 describes the scale-free variants of these energies that are used in our estimates and Section 4.2 gives the calculation which shows how these energies actually bound Sobolev norms of the perturbed data in the rigid case. Sections 5 and 5.1 discuss estimates and the differential inequalities satisfied by our rescaled Bel–Robinson energies. The global existence proof is completed in Section 6 and Section 6.1 establishes causal geodesic completeness. A number of useful definitions and identities are collected in the appendix.

2. Preliminaries

Let $\bar{M}$ be a spacetime, i.e., an $n+1$-dimensional manifold with Lorentz metric $\bar{g}$ of signature $-+\cdots+$ and covariant derivative $\bar{\nabla}$. We denote by $\langle\cdot,\cdot\rangle$ the scalar product defined by $\bar{g}$ on $T\bar{M}$. Let $M \subset \bar{M}$ be a spacelike hypersurface with timelike normal T, $\langle T,T\rangle = -1$, and let t be a time function on a neighborhood of M. Then we can introduce local coordinates $(t, x^i, i = 1,\ldots,n)$ on $\bar{M}$ so that x^i are coordinates on the level sets M_t of t. We will often drop the subscript t on M_t and associated fields.

Let $\partial_t = \partial/\partial t$ be the coordinate vector field corresponding to t. The lapse function N and shift vector field X of the foliation $\{M_t\}$ are defined by $\partial_t = NT + X$. Assume T is future directed so that $N > 0$. The space-time metric $\bar{g}$ takes the form

$$\bar{g} = -N^2 dt \otimes dt + g_{ij}(dx^i + X^i dt) \otimes (dx^j + X^j dt). \tag{2.1}$$

Let $\{e_i\}_{i=1,\ldots,n}$ be a Fermi-propagated orthonormal frame tangent to M_t, i.e., $\langle\bar{\nabla}_T e_i, e_j\rangle = 0$, $\forall i,j$, with dual frame $\{e^i\}_{i=1}^n$. If one drops the assumption that the frame is Fermi propagated, then in general $\bar{\nabla}_T e_i = \bar{\nabla}_T^{//} e_i + (N^{-1}\nabla_i N)T$, where $\bar{\nabla}_T^{//} e_i$ denotes the tangential part of $\bar{\nabla}_T e_i$. With $e_0 = T$, $\{e_\mu\}_{\mu=0}^n$ is an ON frame on $\bar{M}$, adapted to the foliation $\{M_t\}$. We will use the convention that lower case Latin indices run over over $1,\ldots,n$, while Greek indices run over $0,\ldots,n$. Our conventions for curvature as well as some useful identities are given in Appendix A.1. The index T in a tensor expression denotes contraction with T, for example $\bar{\nabla}_T A_\alpha = T^\beta \bar{\nabla}_\beta A_\alpha$.

The second fundamental form k_{ij} of M_t is given by $k_{ij} = -\frac{1}{2}(\mathcal{L}_T \bar{g})_{ij}$. In terms of the Fermi-propagated frame $\{e_i\}$ we have the following relations between N, T and k_{ij}.

$$\bar{\nabla}_i e_j = \nabla_i e_j - k_{ij} T, \qquad \bar{\nabla}_i T = -k_{ij} e_j, \tag{2.2a}$$

$$\bar{\nabla}_T e_i = (N^{-1}\nabla_i N)T, \qquad \bar{\nabla}_T T = (N^{-1}\nabla_i N)e_i. \tag{2.2b}$$

In computations we frequently make use of equations (2.2) to do an $n+1$ split, for example $\bar{\nabla}_i A_j = \nabla_i A_j + k_{ij} A_T$. The trace of a tensor h on M is denoted $\text{tr}_g h = g^{ij} h_{ij}$. When there is no room for confusion, we will drop reference to the metric and write for example $\text{tr}k$ for $\text{tr}_g k$.

2.1. The vacuum Einstein equations

The vacuum Einstein equations

$$\bar{R}_{\alpha\beta} = 0, \tag{2.3}$$

consist after an $n+1$ split of the constraint equations

$$R - |k|^2 + (\text{tr}k)^2 = 0, \tag{2.4a}$$

$$\nabla_i \text{tr}k - \nabla^j k_{ij} = 0, \tag{2.4b}$$

and the evolution equations[1]

$$\mathcal{L}_{\partial_t} g_{ij} = -2Nk_{ij} + \mathcal{L}_X g_{ij}, \tag{2.5a}$$

$$\mathcal{L}_{\partial_t} k_{ij} = -\nabla_i \nabla_j N + N(R_{ij} + \mathrm{tr} k k_{ij} - 2k_{im} k^m_j) + \mathcal{L}_X k_{ij}. \tag{2.5b}$$

We will call a solution (g, k) to the Einstein vacuum constraint equations on M, a **vacuum data set**. If in addition $\nabla \mathrm{tr} k = 0$, then (g, k) is a CMC vacuum data set. A curve $t \mapsto (g, k, N, X)$ solving the Einstein vacuum evolution and constraint equations corresponds to a vacuum space-time metric $\bar{g}$ via (2.1). A vacuum space-time $(\bar{M}, \bar{g})$ with an isometric embedding of a vacuum data set (g, k) on M is said to be a **vacuum extension** of (g, k).

Let $\hat{g}$ be a fixed C^∞ Riemann metric on M with Levi–Civita covariant derivative $\hat{\nabla}$ and Christoffel symbol $\hat{\Gamma}^k_{ij}$. Define the vector field V^k by

$$V^k = g^{ij} e^k (\nabla_i e_j - \hat{\nabla}_i e_j). \tag{2.6}$$

In terms of a coordinate frame, $V^k = g^{ij}(\Gamma^k_{ij} - \hat{\Gamma}^k_{ij})$. The identity map $\mathrm{Id} : (M, g) \to (M, \hat{g})$ is harmonic exactly when $V^k = 0$, see [1] for discussion.

A vacuum data set (g, k) is in **CMCSH gauge** with respect to $\hat{g}$ if

$$\mathrm{tr} k = t \qquad \text{(Constant Mean Curvature)}, \tag{2.7a}$$

$$V^k = 0 \qquad \text{(Spatial Harmonic coordinates)}. \tag{2.7b}$$

A foliation $\{M_t, \ t \in (T_-, T_+)\}$ in $(\bar{M}, \bar{g})$ is called a CMC foliation if $\nabla \mathrm{tr} k = 0$ for all $t \in (T_-, T_+)$. If the induced data (g, k) on M_t is in CMCSH gauge for all $t \in (T_-, T_+)$, then $\{M_t\}$ is called a CMCSH foliation. The CMCSH gauge conditions imply the following elliptic equations for the lapse and shift

$$-\Delta N + |k|^2 N = 1, \tag{2.8a}$$

$$\Delta X^i + R^i_{\ f} X^f - \mathcal{L}_X V^i = (-2Nk^{mn} + 2\nabla^m X^n) e^i (\nabla_m e_n - \hat{\nabla}_m e_n)$$
$$+ 2\nabla^m N k^i_m - \nabla^i N k^m_m, \tag{2.8b}$$

where $\Delta X^i = g^{mn} \nabla_m \nabla_n X^i$. The ellipticity constant $\Lambda[g]$ of g is defined as the least $\Lambda \geq 1$ so that

$$\Lambda^{-1} g(Y, Y) \leq \hat{g}(Y, Y) \leq \Lambda g(Y, Y), \quad \forall Y \in TM. \tag{2.9}$$

Let $\bar{g}$ defined in terms of g, N, X by (2.1). Define $\Lambda[\bar{g}]$ by

$$\Lambda[\bar{g}] = \Lambda[g] + ||N||_{L^\infty} + ||N^{-1}||_{L^\infty} + ||X||_{L^\infty}. \tag{2.10}$$

Then $\bar{g}$ is a non-degenerate Lorentz metric, as long as $\Lambda[\bar{g}]$ is bounded.

We refer to [1] for the background and proof of the following theorem and for the analysis concepts used in the present paper.

---

[1] For a tensor on M with frame components s_{ij}, $\mathcal{L}_{\partial_t} s_{ij} = \partial_t(s_{ij}) - s_{mj} e^m([\partial_t, e_i]) - s_{im} e^m([\partial_t, e_j])$. If we specialize to a time-independent frame, $\mathcal{L}_{\partial_t} s_{ij} = \partial_t s_{ij}$.

Theorem 2.1 ([1]). *Assume that M is of hyperbolic type with hyperbolic metric $\hat{g}$ of unit negative sectional curvature. Let $(g^0, k^0) \in H^s \times H^{s-1}$, $s > n/2+1$, s integer, be a vacuum data set on M in CMCSH gauge with respect to $\hat{g}$. Let $t_0 = \operatorname{tr}_{g^0} k^0$. The following holds.*

1. **Existence:** *There are $T_- < t_0 < T_+ \leq 0$ so that there is a vacuum extension $(\bar{M}, \bar{g})$ of (g^0, k^0), $\bar{M} = (T_-, T_+) \times M$, $\bar{g} \in H^s(\bar{M})$, and such that the foliation $\{M_t = \{t\} \times M, \ t \in (T_-, T_+)\}$, is CMCSH.*
2. **Continuation:** *Suppose that (T_-, T_+) is maximal among all intervals satisfying point 1. Then either $(T_-, T_+) = (-\infty, 0)$ or*
$$\limsup \left(\Lambda[\bar{g}] + \|D\bar{g}\|_{L^\infty} + \|k\|_{L^\infty} \right) = \infty$$
as $t \nearrow T_+$ or as $t \searrow T_-$.
3. **Cauchy stability:** *Let $\bar{g}$ be the space-time metric constructed from the solution (g, k, N, X) to the Einstein vacuum equations in CMCSH gauge. The map $(g^0, k^0) \to \bar{g}$ is continuous $H^s \times H^{s-1} \to H^s((t_-, t_+) \times M)$, for all t_-, t_+, satisfying $T_- < t_- < t_+ < T_+$.*

2.2. Hyperbolic cone space-times

Let (M, γ) be a compact manifold of hyperbolic type, of dimension $n \geq 2$, with hyperbolic metric γ of sectional curvature -1. The **hyperbolic cone space-time** $(\bar{M}, \bar{\gamma}_0)$ with spatial section M is the Lorentzian cone over (M, γ), i.e.,

$$\bar{M} = (0, \infty) \times M, \qquad \bar{\gamma} = -d\rho^2 + \rho^2 \gamma.$$

Let $(\bar{M}, \bar{\gamma})$ be a hyperbolic cone spacetime of dimension $n + 1$. The family of hyperboloids M_ρ given by $\rho =$constant has normal

$$T = \partial_\rho.$$

Here T is future directed w.r.t. the time function ρ. Construct an adapted ON frame T, e_i on $\bar{M}$. A calculation gives

$$k_{ij} = -\frac{1}{\rho} g_{ij},$$

and the mean curvature is given by $\operatorname{tr} k = -n/\rho$. The mean curvature time is defined by setting $t = \operatorname{tr} k$ and the t-foliation has lapse

$$N = -\langle \partial_t, T \rangle = \frac{n}{t^2}.$$

In terms of the mean curvature time we have

$$g(t) = \frac{n^2}{t^2} \gamma, \qquad k(t) = \frac{n}{t} \gamma. \tag{2.11}$$

In the rest of this section we will consider CMCSH foliations, with the reference metric $\hat{g}$ chosen as $\hat{g} = \gamma$.

2.3. The constraint set and the slice

Let M be a compact manifold of hyperbolic type, of dimension $n \geq 2$ with hyperbolic metric γ of sectional curvature -1.

For $s > n/2$, let $\mathcal{M}^s$ be the manifold of Riemann metrics of Sobolev class H^s on M. Then $\mathcal{M}^s$ is a smooth Hilbert manifold and the group $\mathcal{D}^{s+1}$ of H^{s+1} diffeomorphisms acts on $\mathcal{M}^s$. A symmetric 2-tensor h on (M, g), which satisfies $\mathrm{tr} h = 0$, $\mathrm{div} h = 0$, is called a TT-tensor.

Lemma 2.2. *Let $s > n/2 + 1$ and fix $\tau \in \mathbb{R}$, $\tau \neq 0$. There is an open neighborhood $\mathcal{U}_\tau^s \subset \mathcal{M}^s$ of $\frac{n^2}{\tau^2}\gamma$, so that for all $g \in \mathcal{U}_\tau^s$, there is a unique $\phi \in \mathcal{D}^{s+1}(M)$, so that $\phi : (M, g) \to (M, \gamma)$ is harmonic.*

Proof. M is compact and γ has negative sectional curvature. Then there is a unique harmonic map $\phi \in H^{s+1}(M; M)$ from (M, g) to (M, γ) [7]. For g close to γ, the implicit function theorem shows ϕ is close to the identity map Id and hence $\phi \in \mathcal{D}^{s+1}(M)$. $\square$

Let $\mathcal{U}_\tau^s$ be as in Lemma 2.2. Let $\mathcal{S}_\tau^s \subset \mathcal{M}^s$ be defined by

$$\mathcal{S}_\tau^s = \{g \in \mathcal{U}_\tau^s : \mathrm{Id} : (M, g) \to (M, \gamma) \text{ is harmonic}\}. \tag{2.12}$$

For $g \in \mathcal{U}_\tau^s$, if ϕ is the harmonic map provided by Lemma 2.2, $\phi_* g \in \mathcal{S}_\tau^s$. By uniqueness for harmonic maps with target (M, γ), it follows that $\mathcal{S}_\tau^s$ is a local slice for the action of $\mathcal{D}^{s+1}$ on $\mathcal{M}$. Let

$$\mathcal{C}_\tau^s = \{(g, k) \in H^s \times H^{s-1}, \quad \mathrm{tr} k = \tau,$$

$$(g, k) \text{ solves the constraint equations (2.4)}\}, \tag{2.13}$$

be the set of solutions to the vacuum Einstein constraint equations, with $\mathrm{tr} k = \tau$. As M is a manifold of hyperbolic type, $\mathcal{C}_\tau^s$ is a smooth Hilbert submanifold of $\mathcal{M}^s \times H^{s-1}$. To see this, recall that $\mathcal{M}_{-1}^s$, the space of H^s metrics with scalar curvature -1, is a smooth Hilbert manifold, and that the space of TT tensors of class H^{s-1} on (M, g) is a closed subspace of H^{s-1}, which depends smoothly on g. The standard Lichnerowicz-York construction of vacuum data from conformal data now shows that $\mathcal{C}_\tau^s$ is smooth. See [8] for a discussion of the C^∞ case. The action of $\mathcal{D}^{s+1}$ on $\mathcal{C}_\tau^s$ is the lift of the action on $\mathcal{M}$, and therefore the local slice $\mathcal{S}_\tau^s \subset \mathcal{M}^s$ lifts to a local slice Σ_τ^s, at $(\frac{n^2}{\tau^2}\gamma, k) \in \mathcal{C}_\tau^s$,

$$\Sigma_\tau^s = \{(g, k) : (g, k) \in \mathcal{C}_\tau^s \text{ and } g \in \mathcal{S}_\tau^s\}.$$

The slice Σ_τ^s is a smooth Hilbert submanifold of $\mathcal{C}_\tau^s$.

In the rest of this section, let $\mathbf{D}$ denote the Fréchet derivative in the direction $(h, p) \in T_{(\gamma, -\gamma)} \mathcal{C}_{-n}$. It is important to keep in mind that expressions involving $\mathbf{D}$ are evaluated at $(\gamma, -\gamma)$.

Lemma 2.3.

$$T_{(\frac{n^2}{\tau^2}\gamma, \frac{n}{\tau}\gamma)} \Sigma_\tau^s = \{(h, p) \in H^s \times H^{s-1}, \quad h, p \text{ TT-tensors w.r.t. } \gamma\}. \tag{2.14}$$

Proof. We give the proof assuming $\tau = -n$, the general case follows by scaling. First note

$$0 = \mathbf{D}\mathrm{tr}k = \mathrm{tr}_\gamma h + \mathrm{tr}_\gamma p. \tag{2.15}$$

Since $k\big|_{(\gamma,-\gamma)} = -\gamma$, (2.15) implies $\mathbf{D}|k|^2 = 0$. Let $H = R + (\mathrm{tr}k)^2 - |k|^2$, so that the Hamiltonian constraint (2.4a) is $0 = H$. By the above,

$$0 = \mathbf{D}H = \mathbf{D}R.$$

Any symmetric 2-tensor can be decomposed as

$$h = \psi\gamma + h_{\mathrm{TT}} + \mathcal{L}_Y\gamma,$$

where ψ is a function, h_{TT} is a TT-tensor and Y is a vector field. As γ is hyperbolic, $R[\gamma]$ is constant, and due to covariance of R, $\mathbf{D}R.\mathcal{L}_Y\gamma = YR[\gamma] = 0$. The Fréchet derivative of the scalar curvature is the operator

$$DR.u = -\nabla^k\nabla_k u_i{}^i + \nabla^i\nabla^j u_{ij} - R_{ij}u^{ij},$$

which by the above gives

$$\mathbf{D}R.h = \mathbf{D}R.(\psi\gamma)$$
$$= -(n-1)\Delta_\gamma\psi + n(n-1)\psi.$$

In view of the fact that Δ is negative semidefinite, $0 = \mathbf{D}R$ implies $\psi = 0$. Thus,

$$h = h_{\mathrm{TT}} + \mathcal{L}_Y\gamma. \tag{2.16}$$

Let V be given by (2.6). Then with $\hat{g} = \gamma$,

$$\mathbf{D}V^i = \nabla^j h_j{}^i - \frac{1}{2}\nabla^i\mathrm{tr}h,$$

where $\nabla, \Gamma, \mathrm{tr}$ are defined w.r.t. γ. By definition, $V = 0$ on $\mathcal{S}_\tau$, and therefore $(h, p) \in T_{(\gamma,-\gamma)}\Sigma_\tau$ implies using (2.16)

$$\mathbf{D}V = \mathbf{D}V.\mathcal{L}_Y\gamma,$$

which by the uniqueness of harmonic maps with target γ implies that $Y = 0$ and hence $h = h_{\mathrm{TT}}$. In particular $\mathrm{tr}h = 0$ and therefore by (2.15), $\mathrm{tr}p = 0$.

Let $C_i = \nabla_i\mathrm{tr}k - \nabla^j k_{ji}$ so that the momentum constraint (2.4b) is $0 = C_i$. By assumption, $\nabla_i\mathrm{tr}k = 0$, which using $h = h_{\mathrm{TT}}$ and the momentum constraint (2.4b) gives

$$0 = \nabla^j p_{ji},$$

where ∇ is the covariant derivative w.r.t. γ. By the above, $\mathrm{tr}p = 0$ so p is a TT-tensor w.r.t. γ. $\qquad\square$

2.4. Flat space-times

Let $(\bar{M}, \bar{\gamma})$ be a hyperbolic cone space-time of dimension $n+1$, $n \geq 2$, with spatial section (M, γ). Consider a vacuum metric $\bar{g}$ on $\bar{M}$. Then, $C_{\alpha\beta\gamma\delta} = \bar{R}_{\alpha\beta\gamma\delta}$ is the Weyl tensor and by the structure equations,

$$C_{iTjT} = R_{ij} - k_{im}k^m_j + k_{ij}\text{trk} \qquad (2.17)$$

$$C_{mTij} = d^\nabla k_{mij} \qquad (2.18)$$

where the covariant exterior derivative $d^\nabla u$ on symmetric 2-tensors is

$$(d^\nabla u)_{ijk} = \nabla_k u_{ij} - \nabla_j u_{ik}$$

Let $E_{ij} = C_{iTjT}$ and $F_{mij} = C_{mTij}$, considered as tensors on M, and define the second-order elliptic operator A on symmetric 2-tensors by

$$Au = \nabla^* \nabla u - nu, \qquad (2.19)$$

so that $(Au)_{ij} = -\nabla^k \nabla_k u_{ij} - nu_{ij}$. By [12, Lemma 4]

$$\ker A = \ker \text{tr} \cap \ker d^\nabla. \qquad (2.20)$$

An element of $\ker d^\nabla$ is called a **Codazzi tensor**, i.e., the kernel of A consists of the trace-free Codazzi tensors. Clearly, a trace-free Codazzi tensor is also a TT-tensor. Let **D** denote the Fréchet derivative in the direction

$$(h, p) \in T_{(\gamma, -\gamma)} \Sigma_\tau,$$

as in Section 2.3.

Lemma 2.4.

$$2\mathbf{D}E(\gamma, -\gamma)(h, p) = Ah - (n-2)h - 2(n-2)p, \qquad (2.21a)$$

$$\mathbf{D}F(\gamma, -\gamma)(h, p) = d^\nabla(p + h). \qquad (2.21b)$$

Proof. By Lemma 2.3, (h, p) are TT-tensors w.r.t. γ. If h is a TT-tensor, w.r.t. γ, then

$$\mathbf{D}R_{ij}.h = \frac{1}{2}\nabla^* \nabla h_{ij} - nh_{ij} = -\frac{1}{2}\nabla^k \nabla_k h_{ij} - nh_{ij}, \qquad (2.22)$$

which gives (2.21a) after simplification. The Fréchet derivative of Γ^i_{jk} is given by

$$D\Gamma^i_{jk}.h = \frac{1}{2}g^{im}(\nabla_j h_{km} + \nabla_k h_{jm} - \nabla_m h_{jk}). \qquad (2.23)$$

A computation using (2.23) and (2.18) yields

$$\mathbf{D}C_{mTij} = \nabla_j h_{im} - \nabla_i h_{jm} + \nabla_j p_{im} - \nabla_i p_{jm},$$

which gives (2.21b). $\qquad \square$

Consider a curve $\bar{g}_\lambda$ of vacuum metrics on $\bar{M}$, $\bar{g}_0 = \bar{\gamma}$, such that $\{M_t\}$ is CMCSH foliation with respect to $\bar{g}_\lambda$, and let g_λ, k_λ be the induced data on M_{-n}. Then

$$(h, p) = \frac{\partial}{\partial\lambda}(g_\lambda, k_\lambda)\Big|_{\lambda=0},$$

satisfy $(h, p) \in T_{(\gamma, -\gamma)} \Sigma_{-n}$. As above, let $\mathbf{D}$ denote the Fréchet derivative in the direction (h, p). If we further assume that $\bar{g}_\lambda$ is a family of flat metrics, then $\mathbf{D}E = 0$ and $\mathbf{D}F = 0$.

Decompose h, p using the L^2-orthogonal direct sum decomposition $\ker A \oplus \ker^T A$, and write $h = h^0 + h^1$, $p = p^0 + p^1$, with $h^0, p^0 \in \ker A$, $h^1, p^1 \in \ker^T A$. Then $\mathbf{D}E = 0$ is equivalent to the system of equations

$$(n - 2)[h^0 + 2p^0] = 0, \tag{2.24}$$

$$Ah^1 - (n - 2)h^1 - 2(n - 2)p^1 = 0. \tag{2.25}$$

By Lemma 2.3, h, p are TT-tensors on (M, γ). Therefore, by (2.21b), $h + p \in \ker \operatorname{tr} \cap \ker d^\nabla$, and hence, by (2.20), $h^1 + p^1 = 0$. This means that equation (2.25) is equivalent to

$$0 = Ah^1 + (n - 2)h^1.$$

The restriction of A to $\ker^T A$ is positive definite, so it follows that $h^1 = 0$. However, we know that $h^1 + p^1 = 0$, and hence $h^1 = p^1 = 0$. Thus we have shown that $(h, p) = (h^0, p^0)$. It remains to make use of (2.24). In case $n = 2$, this is trivial, while if $n \geq 3$, $h^0 + 2p^0 = 0$ follows.

By construction, $\ker \mathbf{D}E \cap \ker \mathbf{D}F$ is precisely the formal tangent space $T_{\bar{\gamma}} \mathbb{F}(\bar{M})$ at $\bar{\gamma}$, of the space of flat Lorentz metrics $\mathbb{F}(\bar{M})$ on $\bar{M}$. Recalling that in dimension 2, TT-tensors are precisely trace-free Codazzi tensors [2], we have proved

Lemma 2.5. If $n = 2$, $T_{\bar{\gamma}} \mathbb{F}(\bar{M})$ is isomorphic to the direct sum of the space of TT-tensors on M with itself, while for $n \geq 3$, $T_{\bar{\gamma}} \mathbb{F}(\bar{M})$ is isomorphic to the space of trace-free Codazzi tensors on M.   □

In case $n = 2$, M is a Riemann surface of genus ≥ 2, and in this case $T_{\bar{\gamma}} \mathbb{F}(\bar{M})$ has dimension $12\operatorname{genus}(M) - 12$, while for $n \geq 3$, $T_{\bar{\gamma}} \mathbb{F}(\bar{M})$ is trivial in case (M, γ) has no non-vanishing trace-free Codazzi tensors, a topological condition. This motivates the following definition.

Definition 2.6. A hyperbolic manifold (M, γ) of dimension 3, is **rigid** if it admits no non-zero Codazzi tensors with vanishing trace. A hyperbolic cone space-time $(\bar{M}, \bar{\gamma})$ is called rigid if (M, γ) is rigid.

A computation [12] shows that (M, γ) is rigid in the sense of Definition 2.6, if and only if the formal tangent space at γ, of the space of flat conformal structures on M is trivial. Kapovich [11, Theorem 2] proved the existence of compact hyperbolic 3-manifolds which are rigid w.r.t. infinitesimal deformations in the space of flat conformal structures. We formulate this as

Proposition 2.7. The class of rigid hyperbolic 3-manifolds (M, γ) (and rigid standard space-times $(\bar{M}, \bar{\gamma})$), in the sense of Definition 2.6, is non-empty.   □

3. Weyl fields

In this section, and in the rest of the paper, let $n = 3$. A tracefree 4-tensor W with the symmetries of the Riemann tensor is called a Weyl field. We define the left and right Hodge duals of W by

$$^*W_{\alpha\beta\gamma\delta} = \frac{1}{2}\epsilon_{\alpha\beta\mu\nu}W^{\mu\nu}{}_{\gamma\delta}, \tag{3.1}$$

$$W^*{}_{\alpha\beta\gamma\delta} = W_{\alpha\beta}{}^{\mu\nu}\frac{1}{2}\epsilon_{\mu\nu\gamma\delta}. \tag{3.2}$$

If W is a Weyl field, then $^*W = W^*$ and $W = -*(^*W)$. Define the tensors J and J^* by

$$\bar{\nabla}^\alpha W_{\alpha\beta\gamma\delta} = J_{\beta\gamma\delta}, \tag{3.3a}$$

$$\bar{\nabla}^{\alpha *}W_{\alpha\beta\gamma\delta} = J^*_{\beta\gamma\delta}. \tag{3.3b}$$

Then

$$J^*_{\beta\gamma\delta} = \frac{1}{2}J_\beta{}^{\mu\nu}\epsilon_{\mu\nu\gamma\delta},$$

and

$$\bar{\nabla}_{[\mu}W_{\gamma\delta]\alpha\beta} = \frac{1}{3}\epsilon_{\nu\mu\gamma\delta}J^{*\nu}{}_{\alpha\beta}, \tag{3.4a}$$

$$\bar{\nabla}_{[e}{}^*W_{\gamma\delta]\alpha\beta} = -\frac{1}{3}\epsilon_{\nu\mu\gamma\delta}J^\nu{}_{\alpha\beta}. \tag{3.4b}$$

The electric and magnetic parts $E(W)$, $B(W)$ of the Weyl field W, with respect to the foliation M_t are defined by

$$E(W)_{\alpha\beta} = W_{\alpha\mu\beta\nu}T^\mu T^\nu, \qquad B(W)_{\alpha\beta} = {}^*W_{\alpha\mu\beta\nu}T^\mu T^\nu. \tag{3.5}$$

The tensors E and B are t-tangent, i.e., $E_{\alpha\beta}T^\beta = B_{\alpha\beta}T^\beta = 0$ and tracefree, $\bar{g}^{\alpha\beta}E_{\alpha\beta} = \bar{g}^{\alpha\beta}B_{\alpha\beta} = 0$. It follows that $g^{ij}E_{ij} = g^{ij}B_{ij} = 0$.

In case $(\bar{M}, \bar{g})$ is vacuum, i.e., $\bar{R}_{\alpha\beta} = 0$, the Weyl tensor $C_{\alpha\beta\gamma\delta}$ of $(\bar{M}, \bar{g})$ satisfies $C_{\alpha\beta\gamma\delta} = \bar{R}_{\alpha\beta\gamma\delta}$ the Gauss and Codazzi equations can be written in terms of E and B to give

$$\nabla_i k_{jm} - \nabla_j k_{im} = \epsilon_{ij}{}^l B(C)_{lm}, \tag{3.6a}$$

$$R_{ij} - k_{im}k^m{}_j + k_{ij}\mathrm{tr}k = E(C)_{ij}. \tag{3.6b}$$

Note that from the definition (A.18) of d^∇, (3.6a) is equivalent to $d^\nabla k_{mij} = -\epsilon_{ij}{}^l B_{lm}$. Using the definition (A.13) of curl and (A.19) we get the alternate form of (3.6a), valid if (g, k) satisfies the vacuum constraint equations (2.4),

$$-(\mathrm{curl}k)_{ij} = B(C)_{ij}. \tag{3.7}$$

The following identities relate $W, {}^*W, E = E(W), B = B(W)$, cf. [6, eq. (7.2.1), p. 169]

$$\begin{aligned} W_{ijkT} &= -\epsilon_{ij}{}^m B_{mk}, & {}^*W_{ijkT} &= \epsilon_{ij}{}^m E_{mk}, \\ W_{ijk\ell} &= -\epsilon_{ijm}\epsilon_{k\ell n}E^{mn}, & {}^*W_{ijk\ell} &= -\epsilon_{ijm}\epsilon_{k\ell n}B^{mn}. \end{aligned} \tag{3.8}$$

The tensors $\bar{\nabla}_T E$, $\bar{\nabla}_T B$ have the property

$$g^{ij}\bar{\nabla}_T E_{ij} = 0, \qquad g^{ij}\bar{\nabla}_T B_{ij} = 0,$$

i.e., the pull-back of $\bar{\nabla}_T E$, $\bar{\nabla}_T B$ to M is trace-free, but $\bar{\nabla}_T E$, $\bar{\nabla}_T B$ are not t-tangent in general. The following result allows us to express derivatives of the Weyl field W in terms of $E(W), B(W), J(W), J^*(W)$. See Appendix A.2 for the definition of div and curl.

Proposition 3.1. *Let E, B be the electric and magnetic parts of a Weyl-field W and let J, J^* be defined from W by (3.4). Then*

$$\mathrm{div}E_i = +(k \wedge B)_i + J_{TiT}, \tag{3.9a}$$

$$\mathrm{div}B_i = -(k \wedge E)_i + J^*_{TiT}. \tag{3.9b}$$

$$\bar{\nabla}_T E_{ij} - \mathrm{curl}B_{ij} = -N^{-1}(\nabla N \wedge B)_{ij} - \frac{3}{2}(E \times k)_{ij} + \frac{1}{2}(\mathrm{tr}k)E_{ij} - J_{iTj}, \tag{3.10a}$$

$$\bar{\nabla}_T B_{ij} + \mathrm{curl}E_{ij} = +N^{-1}(\nabla N \wedge E)_{ij} - \frac{3}{2}(B \times k)_{ij} + \frac{1}{2}(\mathrm{tr}k)B_{ij} - J^*_{iTj}. \tag{3.10b}$$

Written in terms of $\mathcal{L}_{\partial_t}$, (3.10) becomes

$$N^{-1}\mathcal{L}_{\partial_t} E_{ij} = +\mathrm{curl}B_{ij} - N^{-1}(\nabla N \wedge B)_{ij}$$
$$- \frac{5}{2}(E \times k)_{ij} - \frac{2}{3}(E \cdot k)g_{ij} - \frac{1}{2}\mathrm{tr}k E_{ij} + N^{-1}\mathcal{L}_X E_{ij} - J_{iTj}, \tag{3.11a}$$

$$N^{-1}\mathcal{L}_{\partial_t} B_{ij} = -\mathrm{curl}E_{ij} + N^{-1}(\nabla N \wedge E)_{ij}$$
$$- \frac{5}{2}(B \times k)_{ij} - \frac{2}{3}(B \cdot k)g_{ij} - \frac{1}{2}(\mathrm{tr}k)B_{ij} + N^{-1}\mathcal{L}_X B_{ij} - J^*_{iTj}. \tag{3.11b}$$

Proof. We write $\bar{\nabla}^\alpha E_{\alpha i}$ in two ways. First,

$$\mathrm{div}E_i = \bar{\nabla}^\alpha E_{\alpha i} - N^{-1}\nabla^j N E_{ji}.$$

Secondly, by (3.8) and (A.10),

$$\bar{\nabla}^\alpha E_{\alpha i} = \bar{\nabla}^\alpha W_{\alpha TiT} + \bar{g}^{\alpha\beta}W_{\alpha\gamma i\delta}\bar{\nabla}_\beta T^\gamma T^\delta + \bar{g}^{\alpha\beta}W_{\alpha\gamma i\delta}T^\gamma\bar{\nabla}_\beta T^\delta$$
$$= J_{TiT} + (k \wedge B)_i + N^{-1}\nabla^j N E_{ji}.$$

This gives (3.9a) and the argument for (3.9b) is similar. To prove (3.10), first note the identities

$$\bar{\nabla}_k W_{iTjT} = \nabla_k E_{ij} - (\epsilon_{il}{}^m B_{mj} + \epsilon_{jl}{}^m B_{mi})k^l{}_k, \tag{3.12a}$$

$$\bar{\nabla}_k{}^* W_{iTjT} = \nabla_k B_{ij} + (\epsilon_{il}{}^m E_{mj} + \epsilon_{jl}{}^m E_{mi})k^l{}_k. \tag{3.12b}$$

From this we get, after expanding the covariant derivative, and rewriting using (3.8)

$$\epsilon_i{}^{mn}\bar\nabla_n W_{mTjT} + \epsilon_j{}^{mn}\bar\nabla_n W_{mTiT} = 2(\mathrm{curl}E)_{ij} + 3(B \times k)_{ij} - (\mathrm{tr}k)B_{ij}, \tag{3.13a}$$

$$\epsilon_i{}^{mn}\bar\nabla_n{}^* W_{mTjT} + \epsilon_j{}^{mn}\bar\nabla_n{}^* W_{mTiT} = 2(\mathrm{curl}B)_{ij} - 3(E \times k)_{ij} + (\mathrm{tr}k)E_{ij}. \tag{3.13b}$$

The Bianchi equations (3.4) imply

$$\epsilon_i{}^{mn}\bar\nabla_T W_{mnjT} = 2\epsilon_i{}^{mn}\bar\nabla_n W_{mTjT} - 2J^*_{ijT}, \tag{3.14a}$$

$$\epsilon_i{}^{mn}\bar\nabla_T{}^* W_{mnjT} = 2\epsilon_i{}^{mn}\bar\nabla_n{}^* W_{mTjT} + 2J_{ijT}. \tag{3.14b}$$

Using (3.8) and (2.2) we get

$$\epsilon_i{}^{mn}\bar\nabla_T W_{mnjT} = -2\bar\nabla_T B_{ij} + 2N^{-1}(\nabla N \wedge E)_{ij}, \tag{3.15a}$$

$$\epsilon_i{}^{mn}\bar\nabla_T{}^* W_{mnjT} = 2\bar\nabla_T E_{ij} + 2N^{-1}(\nabla N \wedge B)_{ij}. \tag{3.15b}$$

Using (3.15), multiplying by $\frac{1}{2}$, taking the symmetric parts of (3.14) and using (3.13) now gives the identities (3.10). It is straightforward to derive (3.11) from (3.10) using (A.17). □

Given a Weyl field W, the covariant derivative $\bar\nabla_T W$ is again a Weyl field. Proposition 3.1 gives the following expressions for $E(\bar\nabla_T W), B(\bar\nabla_T W)$.

Corollary 3.2.

$$E(\bar\nabla_T W)_{ij} = +\mathrm{curl}B_{ij} - \frac{3}{2}(E \times k)_{ij} + \frac{1}{2}(\mathrm{tr}k)E_{ij} - J_{iTj}, \tag{3.16a}$$

$$B(\bar\nabla_T W)_{ij} = -\mathrm{curl}E_{ij} - \frac{3}{2}(B \times k)_{ij} + \frac{1}{2}(\mathrm{tr}k)B_{ij} - J^*_{iTj}. \tag{3.16b}$$

where in the right-hand side, E, B, J, J^* are defined w.r.t. W.

Proof. From the definition and using (2.2) we have, taking into account the fact that E is t-tangent,

$$E(\bar\nabla_T W)_{ij} = T^\gamma T^\delta T^\nu \bar\nabla_\nu W_{i\gamma j\delta}$$
$$= T^\nu \bar\nabla_\nu E(W)_{ij} - N^{-1}\nabla^m N W_{imjT} - N^{-1}\nabla^n N W_{iTjn}$$

using (3.8) and (A.11)

$$= T^\nu \bar\nabla_\nu E(W)_{ij} + N^{-1}(\nabla N \wedge B(W))_{ij},$$

which gives (3.16a) using (3.10a). The proof of (3.16b) is similar. □

4. The Bel–Robinson Energy

Given a Weyl field W we can associate to it a fully symmetric and traceless tensor $Q(W)$, the Bel–Robinson tensor, given by

$$Q(W)_{\alpha\beta\gamma\delta} = W_{\alpha\mu\gamma\nu}W_{\beta}{}^{\mu}{}_{\delta}{}^{\nu} + {}^*W_{\alpha\mu\gamma\nu}{}^*W_{\beta}{}^{\mu}{}_{\delta}{}^{\nu}. \tag{4.1}$$

$Q(W)$ is positive definite in the sense that $Q(X, Y, X, Y) \geq 0$ whenever X, Y are timelike vectors, with equality only if W vanishes, cf. [5, Prop. 4.2]. Let $E = E(W)$, $B = B(W)$. The following identities relate $Q(W)$ to E and B.

$$Q(W)_{TTTT} = |E|^2 + |B|^2, \tag{4.2a}$$

$$Q(W)_{iTTT} = 2(E \wedge B)_i, \tag{4.2b}$$

$$Q(W)_{ijTT} = -(E \times E)_{ij} - (B \times B)_{ij} + \frac{1}{3}(|E|^2 + |B|^2)g_{ij}, \tag{4.2c}$$

where $|E|^2 = E^{ij}E_{ij} = |E|_g^2$, and similarly for $|B|^2$. From equations (3.8) and (4.2a) it follows that $Q(W)_{TTTT} = 0$ if and only if $W = 0$. The divergence of the Bel–Robinson tensor takes the form [6, Prop. 7.1.1]

$$\begin{aligned}\bar{\nabla}^{\alpha}Q(W)_{\alpha\beta\gamma\delta} = W_{\beta}{}^{\mu}{}_{\delta}{}^{\nu}J(W)_{\mu\gamma\nu} + W_{\beta}{}^{\mu}{}_{\gamma}{}^{\nu}J(W)_{\mu\delta\nu} \\ + {}^*W_{\beta}{}^{\mu}{}_{\delta}{}^{\nu}J^*(W)_{\mu\gamma\nu} + {}^*W_{\beta}{}^{\mu}{}_{\gamma}{}^{\nu}J^*(W)_{\mu\delta\nu},\end{aligned} \tag{4.3}$$

and the definition of $E(W)$ and $B(W)$ gives

$$\bar{\nabla}^{\alpha}Q(W)_{\alpha TTT} = 2E^{ij}(W)J(W)_{iTj} + 2B^{ij}(W)J^*(W)_{iTj}. \tag{4.4}$$

Let W be a Weyl field and let $Q(W)$ be the corresponding Bel–Robinson tensor. Then working in a foliation M_t, we define the Bel–Robinson energy $\mathcal{Q}(t, W)$ by

$$\mathcal{Q}(t, W) = \int_{M_t} Q(W)_{TTTT} d\mu_{M_t}.$$

By the Gauss law, this has the evolution equation

$$\begin{aligned}\partial_t \mathcal{Q}(t, W) = &-\int_{M_t} N\bar{\nabla}^{\alpha}Q(W)_{\alpha TTT} d\mu_{M_t} \\ &-3\int_{M_t} NQ(W)_{\alpha\beta TT}\pi^{\alpha\beta} d\mu_{M_t},\end{aligned} \tag{4.5}$$

where π denotes the "deformation tensor" of T,

$$\pi_{\alpha\beta} = \bar{\nabla}_{\alpha}T_{\beta}. \tag{4.6}$$

The components of π in terms of an adapted, Fermi-propagated frame are as follows:

$$\pi_{ij} = -k_{ij}, \qquad\qquad \pi_{iT} = 0, \tag{4.7a}$$

$$\pi_{Ti} = N^{-1}\nabla_i N, \qquad\qquad \pi_{TT} = 0. \tag{4.7b}$$

We will need control of $\bar{g}$ in H^3, and for this purpose we consider in addition to the Bel–Robinson energy of order zero, $\mathcal{Q}_0(t, W) = \mathcal{Q}(t, W)$, the first-order Bel–Robinson energy $\mathcal{Q}_1(t, W) = \mathcal{Q}(t, \bar{\nabla}_T W)$. In the vacuum case, $J(W) = J^*(W) =$

0, so we may view $\mathcal{Q}_1$ as a function on the set of solutions to the Einstein vacuum constraint equations, by using Corollary 3.2 to compute $E(\bar{\nabla}_T W), B(\bar{\nabla}_T W)$.

Expanding $\partial_t \mathcal{Q}(t, W)$ using (4.2), (4.4), (4.7), gives

$$\partial_t \mathcal{Q}(t, W) = -3 \int_{M_t} N[(E \times E) \cdot k + (B \times B) \cdot k - \frac{1}{3}(|E|^2 + |B|^2)\mathrm{tr}k$$
$$- 2N^{-1}\nabla^i N (E \wedge B)_i] d\mu_{M_t}$$
$$- 2 \int_{M_t} N(E^{ij} J_{iTj} + B^{ij} J^*_{iTj}) d\mu_{M_t}$$

(perform a partial integration and use (A.15))

$$= -3 \int_{M_t} N[(E \times E) \cdot k + (B \times B) \cdot k - \frac{1}{3}(|E|^2 + |B|^2)\mathrm{tr}k$$
$$- 2\mathrm{curl}E \cdot B + 2E \cdot \mathrm{curl}B] d\mu_{M_t}$$
$$- 2 \int_{M_t} N(E^{ij} J_{iTj} + B^{ij} J^*_{iTj}) d\mu_{M_t}. \tag{4.8}$$

It is straightforward to show that this expression agrees with that obtained after a direct computation of $\partial_t \mathcal{Q}(t, W)$ using (3.11).

Let $\tau = \mathrm{tr}k$ and specialize to a constant mean curvature foliation $\{M_\tau\}$ in the following. With the discussion in Subsection 2.2 as a guide we introduce the following quantities which vanish when evaluated in the standard foliation on a hyperbolic cone space-time, namely the "trace free" part $\hat{\pi}$ of π,

$$\hat{\pi}_{\alpha\beta} = \pi_{\alpha\beta} + \frac{\tau}{3}(\bar{g}_{\alpha\beta} + T_\alpha T_\beta) \tag{4.9}$$

and the "perturbed" part of the lapse,

$$\hat{N} = N - \frac{3}{\tau^2}. \tag{4.10}$$

In the following Lemma we record the form of $\partial_\tau \mathcal{Q}_i$, $i = 0, 1$ which will be used in the global existence proof.

Lemma 4.1. *In a vacuum space time, the Bel–Robinson energies $\mathcal{Q}_j(t, W)$, $j = 0, 1$ satisfy the following evolution equations.*

$$\partial_\tau \mathcal{Q}_0(\tau, W) = \frac{3}{\tau}\mathcal{Q}_0(t, W) - 3 \int_{M_\tau} N Q(W)_{\alpha\beta\gamma\delta} \hat{\pi}^{\alpha\beta} T^\gamma T^\delta d\mu_{M_\tau}$$
$$+ \tau \int_{M_\tau} \hat{N} Q(W)_{TTTT} d\mu_{M_\tau}, \tag{4.11}$$

$$\partial_\tau \mathcal{Q}_1(\tau, W) = \frac{5}{\tau} \mathcal{Q}_1(t, W) - 2 \int_{M_\tau} N\mathcal{G}_1(W) d\mu_{M_\tau}$$

$$- 3 \int_{M_t} NQ(\bar{\nabla}_T W)_{\alpha\beta\gamma\delta} \hat{\pi}^{\alpha\beta} T^\gamma T^\delta d\mu_{M_\tau} \qquad (4.12)$$

$$+ \frac{5\tau}{3} \int_{M_\tau} \hat{N} Q(\bar{\nabla}_T W)_{TTTT} d\mu_{M_\tau},$$

where

$$\mathcal{G}_1(W) = E(\bar{\nabla}_T W)^{ij} \left(J(\bar{\nabla}_T W)_{iTj} + \frac{\tau}{3} E(\bar{\nabla}_T W)_{ij} \right)$$

$$+ B(\bar{\nabla}_T W)^{ij} \left(J^*(\bar{\nabla}_T W)_{iTj} + \frac{\tau}{3} B(\bar{\nabla}_T W)_{ij} \right). \qquad (4.13)$$

In particular,

$$J(\bar{\nabla}_T W)_{iTj} + \frac{\tau}{3} E(\bar{\nabla}_T W)_{ij} = \hat{\pi}^{\alpha\mu} \bar{\nabla}_\mu W_{\alpha iTj} + \frac{3}{2}(E \times E)_{ij} - \frac{3}{2}(B \times B)_{ij}$$

$$(4.14a)$$

$$J^*(\bar{\nabla}_T W)_{iTj} + \frac{\tau}{3} B(\bar{\nabla}_T W)_{ij} = \hat{\pi}^{\alpha\mu} \bar{\nabla}_\mu {}^* W_{\alpha iTj} + 3(E \times B)_{ij} \qquad (4.14b)$$

where

$$\hat{\pi}^{\alpha\mu} \bar{\nabla}_\mu W_{\alpha iTj} = \hat{k}^{rs} \nabla_s (-\epsilon_{ri}{}^n B_{nj})$$

$$- \hat{k}^{rs} k_{rs} E_{ij} + \hat{k}^{rs} k_{si} E_{rj} + \hat{k}^{rs} k_s{}^m \epsilon_{ri}{}^n \epsilon_{mj}{}^p E_{np}$$

$$- N^{-1} \nabla^r N (\nabla_r E_{ij} + k_r{}^s \epsilon_{si}{}^n B_{nj} + k_r{}^s \epsilon_{sj}{}^n B_{ni}) \qquad (4.15a)$$

$$\hat{\pi}^{\alpha\mu} \bar{\nabla}_\mu {}^* W_{\alpha iTj} = \hat{k}^{rs} \nabla_s (\epsilon_{ri}{}^n E_{nj})$$

$$- \hat{k}^{rs} k_{rs} B_{ij} + \hat{k}^{rs} k_{si} B_{rj} + \hat{k}^{rs} k_s{}^m \epsilon_{ri}{}^n \epsilon_{mi}{}^p B_{np}$$

$$- N^{-1} \nabla^r N (\nabla_r B_{ij} - k_r{}^s \epsilon_{si}{}^n E_{nj} - k_r{}^s \epsilon_{sj}{}^n E_{ni}) \qquad (4.15b)$$

Remark 4.1. *In the proof of the main theorem, it is of central importance that the terms in* (4.14) *are quadratic in* $\hat{\pi}, \bar{\nabla} W, W$. *This has the consequence that the term given by* (4.13) *can be treated as a perturbation term in case of small data. In particular the terms* $\hat{\pi}^{\alpha\mu} \bar{\nabla}_\mu W_{\alpha iTj}$ *and* $\hat{\pi}^{\alpha\mu} \bar{\nabla}_\mu {}^* W_{\alpha iTj}$ *when expanded are seen to be of third-order in* $N^{-1} \nabla_i N, k_{ij}, E_{ij}, B_{ij}$ *and of second-order in* $\hat{k}_{ij}, \nabla_i E_{jk}, \nabla_i B_{jk}$. *We will not make use of the explicit expression for* $\partial_t \mathcal{Q}_1$, *but for completeness, it is given in equation* (4.19) *below.*

Proof. In order to evaluate $\text{Div} Q(\bar{\nabla}_T W)(T, T, T)$, we need

$$J(\bar{\nabla}_T W)_{iTj} \qquad \text{and} \qquad J^*(\bar{\nabla}_T W)_{iTj}.$$

A computation gives

$$J(\bar{\nabla}_T W)_{\beta\gamma\delta} = \bar{\nabla}^\alpha \bar{\nabla}_T W_{\alpha\beta\gamma\delta} = \pi^{\alpha\nu} \bar{\nabla}_\nu W_{\alpha\beta\gamma\delta} + T^\nu \bar{\nabla}_\nu J(W)_{\beta\gamma\delta}$$

$$+ T^\nu \bar{R}_\alpha{}^{\mu\alpha}{}_\nu W_{\mu\beta\gamma\delta} + T^\nu \bar{R}_\beta{}^{\mu\alpha}{}_\nu W_{\alpha\mu\gamma\delta} \qquad (4.16)$$

$$+ T^\nu \bar{R}_\gamma{}^{\mu\alpha}{}_\nu W_{\alpha\beta\mu\delta} + T^\nu \bar{R}_\delta{}^{\mu\alpha}{}_\nu W_{\alpha\beta\gamma\mu},$$

Note that in vacuum, $J(W) = 0$ and $\bar{R}_{\alpha\beta\gamma\delta} = W_{\alpha\beta\gamma\delta}$. Substituting $\bar{R}$ for W in (4.16) and using (3.8) to rewrite the terms quadratic in W gives (4.14a). A similar calculation for $J^*(\bar{\nabla}_T W)$, taking into account the fact that in this case, $\bar{R} = W$ and $^*W$ are distinct, gives (4.14b). It is now straightforward to check that (4.11) and (4.12) hold, given the definition of $\mathcal{G}_1$ in (4.13). $\qquad\square$

One can decompose equations (4.14) into symmetric and antisymmetric parts to obtain the analogues of equations (3.9) and (3.10). Setting $J(W) = J^*(W) = 0$ for the vacuum case, defining $\tilde{E}_{ij} = E(\bar{\nabla}_T W)_{ij}$ and $\tilde{B}_{ij} = B(\bar{\nabla}_T W)_{ij}$ and writing $E_{ij|k}$ and $B_{ij|k}$ for $\nabla_k E_{ij}$ and $\nabla_k B_{ij}$ respectively we get, first for the symmetric parts (the analogues of equations (3.11)),

$$
\begin{aligned}
N^{-1}\mathcal{L}_{\partial_t}\tilde{E}_{ij} &= k_{ij}(k^{lm}E_{lm}) + g_{ij}[E^{lm}E_{lm} + k^{ls}k_s{}^m E_{lm} + k^{lm}\tilde{E}_{lm} - B^{lm}B_{lm}] \\
&\quad + 2E_{ij}(k^{lm}k_{lm}) - 3E_{il}E^l{}_j - 3k_i{}^m\tilde{E}_{mj} \\
&\quad - 3k_j{}^m\tilde{E}_{mi} - k_i{}^m k_j{}^l E_{lm} + 3B_{il}B^l{}_j \\
&\quad + \frac{1}{2}g_{jk}\epsilon^{lkm}(\tilde{B}_{im|l} - k_{ms}B^s{}_{i|l}) + \frac{1}{2}g_{ik}\epsilon^{lkm}(\tilde{B}_{jm|l} - k_{ms}B^s{}_{j|l}) \\
&\quad + N^{-1}N^{|l}E_{ij|l} - \frac{5}{2}k_j{}^s k_s{}^m E_{im} - \frac{5}{2}k_i{}^s k_s{}^m E_{jm} \\
&\quad + (\mathrm{tr}k)[3\tilde{E}_{ij} + \frac{5}{2}k_i{}^m E_{jm} + \frac{5}{2}k_j{}^m E_{im} - 2(\mathrm{tr}k)E_{ij} - g_{ij}k^{lm}E_{lm}] \\
&\quad + N^{-1}N^{|l}[B^r{}_i\epsilon_{kjr}k_l{}^k + B^r{}_j\epsilon_{kir}k_l{}^k - \epsilon_{rjl}\tilde{B}^r{}_i - \epsilon_{ril}\tilde{B}^r{}_j] \\
&\quad + N^{-1}\mathcal{L}_X\tilde{E}_{ij} \qquad\qquad\qquad\qquad\qquad\qquad\qquad\qquad (4.17a)
\end{aligned}
$$

$$
\begin{aligned}
N^{-1}\mathcal{L}_{\partial_t}\tilde{B}_{ij} &= 2B_{ij}k^{lm}k_{lm} + g_{ij}[2B^{lm}E_{lm} + k^{ls}k_s{}^m B_{lm} + k^{lm}\tilde{B}_{lm}] \\
&\quad + N^{-1}N^{|l}B_{ij|l} - 3B_j^m E_{im} - 3B_i^m E_{jm} - 3k_i{}^l\tilde{B}_{lj} \\
&\quad - 3k_j{}^l\tilde{B}_{li} - k_i{}^l k_j{}^s B_{ls} + k_{ij}k^{lm}B_{lm} \\
&\quad - \frac{5}{2}(k_{jm}k^{ml}B_{li} + k_{im}k^{ml}B_{lj}) \\
&\quad - \frac{1}{2}g_{ik}\epsilon^{lkm}(\tilde{E}_{mj|l} - k_m{}^s E_{sj|l}) - \frac{1}{2}g_{jk}\epsilon^{lkm}(\tilde{E}_{mi|l} - k_m{}^s E_{si|l}) \\
&\quad + (\mathrm{tr}k)[3\tilde{B}_{ij} + \frac{5}{2}k_i{}^l B_{lj} + \frac{5}{2}k_j{}^l B_{li} - 2(\mathrm{tr}k)B_{ij} - g_{ij}k^{lm}B_{lm}] \\
&\quad - N^{-1}N^{|l}[E^r{}_i\epsilon_{kjr}k^k{}_l + E^r{}_j\epsilon_{kir}k^k{}_l - \epsilon_{rjl}\tilde{E}^r{}_i - \epsilon_{ril}\tilde{E}^r{}_j] \\
&\quad + N^{-1}\mathcal{L}_X\tilde{B}_{ij} \qquad\qquad\qquad\qquad\qquad\qquad\qquad\qquad (4.17b)
\end{aligned}
$$

and then for the antisymmetric part (the analogues for equations (3.9)),

$$
\tilde{E}^i{}_j{}^{|j} = k_j{}^m\tilde{B}_{mr}\epsilon^{ijr} - k_j{}^m E^i{}_m{}^{|j} + (\mathrm{tr}k)\epsilon^{imn}k_m{}^s B_{sn} - k^i{}_r k_l{}^s B_{sj}\epsilon^{rlj} \qquad (4.18a)
$$

$$
\tilde{B}^i{}_j{}^{|j} = -k_j{}^m\tilde{E}_{mr}\epsilon^{ijr} - k_j{}^m B^i{}_m{}^{|j} - (\mathrm{tr}k)\epsilon^{imn}k_m{}^s E_{sn} + k^i{}_r k_l{}^s E_{sj}\epsilon^{rlj} \qquad (4.18b)
$$

The formula for $\partial_t \mathcal{Q}_1$ that is analogous to that given above for $\partial_t \mathcal{Q}$ is given explicitly as follows,

$$
\begin{aligned}
\frac{\partial}{\partial t} \int_M & \mu_g (\tilde{E}_i^\ell \tilde{E}_\ell^i + \tilde{B}_i^\ell \tilde{B}_\ell^i) = \partial_t \mathcal{Q}_1 \\
= 2 \int_M & \left\{ N g_{jk} \epsilon^{\ell k m} \left[\hat{k}_m^s (\tilde{B}^{ij} E_{si|\ell} - \tilde{E}^{ij} B_{si|\ell}) \right] \mu_g \right. \\
& + N \mu_g (\operatorname{trk}) \left[\frac{5}{6} \tilde{E}^{ij} \tilde{E}_{ij} + \frac{5}{6} \tilde{B}^{ij} \tilde{B}_{ij} \right] \\
& + \mu_g N^{|\ell} \left(\tilde{E}^{ij} E_{ij|\ell} + \tilde{B}^{ij} B_{ij|\ell} \right) \\
& + 2 N \mu_g \left(\hat{k}_{mn} \hat{k}^{mn} \right) \left(\tilde{E}^{ij} E_{ij} + \tilde{B}^{ij} B_{ij} \right) \\
& - 4 N \mu_g \hat{k}_j^\ell \left[\tilde{E}_\ell^i \tilde{E}_i^j + \tilde{B}_\ell^i \tilde{B}_i^j \right] \\
& + N \mu_g \left[(\tilde{E}^{ij} \hat{k}_{ij})(E^{mn} \hat{k}_{mn}) + (\tilde{B}^{ij} \hat{k}_{ij})(B^{mn} \hat{k}_{mn}) \right] \\
& - 3 N \mu_g \left[\tilde{E}^{ij} E_{i\ell} E_j^\ell - \tilde{E}^{ij} B_{i\ell} B_j^\ell + 2 \tilde{B}^{ij} B_j^\ell E_{i\ell} \right] \\
& - 3 N_{|\ell} \epsilon^{rj\ell} \tilde{E}_j^i \tilde{B}_{ri} \mu_g + 2 N^{|\ell} \left[\tilde{E}_j^i B_{ri} \epsilon^{kjr} (\hat{k}_{\ell k} + \frac{1}{3} g_{lk}(\operatorname{trk})) \right. \\
& \left. - \tilde{B}_j^i E_{ri} \epsilon^{kjr} (\hat{k}_{k\ell} + \frac{1}{3} g_{kl}(\operatorname{trk})) \right] \mu_g \\
& - N \mu_g \hat{k}_i^m \hat{k}_j^\ell \left[\tilde{E}^{ij} E_{lm} + \tilde{B}^{ij} B_{\ell m} \right] \\
& \left. - 5 N \mu_g \hat{k}_j^s \hat{k}_s^m (\tilde{E}^{ij} E_{im} + \tilde{B}^{ij} B_{im}) \right\}
\end{aligned}
\tag{4.19}
$$

4.1. The scale-free Bel–Robinson energy

In the rest of Section 4, let $n = 3$, and assume $(\bar{M}, \bar{g})$ is a vacuum space-time with a CMC foliation $\{M_\tau\}$ with $\tau = \operatorname{trk} < 0$. Let $W_{\alpha\beta\gamma\delta} = \bar{R}_{\alpha\beta\gamma\delta}$ be the Weyl tensor of $(\bar{M}, \bar{g})$. Then W satisfies the homogeneous Bianchi identities, i.e., $J(W) = J^*(W) = 0$.

Since we will be estimating geometric quantities in terms of $\mathcal{Q}_0, \mathcal{Q}_1$ via Sobolev inequalities which depend on scale, we need scale-free versions of these energies. It follows from the definitions that the following variables are scale-free if λ has dimensions $(length)^{-1}$. Here indices refer to a coordinate frame.

$$
\begin{array}{ll}
\tilde{g}_{ab} = \lambda^2 g_{ab}, & \tilde{T}^\alpha = \lambda^{-1} T^\alpha, \\
\tilde{k}_{ab} = \lambda k_{ab}, & \widetilde{\operatorname{trk}} = \lambda^{-1} \operatorname{trk}, \\
\tilde{N} = \lambda^2 N, & \widetilde{\mu_g} = \lambda^3 \mu_g.
\end{array}
\tag{4.20a}\tag{4.20b}\tag{4.20c}
$$

Note that trk has dimensions of $(length)^{-1}$ while we treat spatial coordinates as dimensionless quantities. The Weyl tensor in the (3,1)-form is conformally invariant,

and hence the Bel–Robinson tensor is also conformally invariant, and in particu-
lar scale-free. From this can be seen that the Bel–Robinson energies $\mathcal{Q}_0, \mathcal{Q}_1$ have
dimensions *(length)$^{-1}$* and *(length)$^{-3}$* respectively, so that the expressions

$$\widetilde{\mathcal{Q}}_i = \lambda^{-1-2i}\mathcal{Q}_i, \qquad i = 0, 1,$$

are scale-free, i.e., $\widetilde{\mathcal{Q}}_i$ is precisely given by $\mathcal{Q}_i$ evaluated on the scale-free variables
$\tilde{g}, \tilde{k}$ etc. In the following we will use the scale factor λ, defined by

$$\lambda = \frac{|\mathrm{tr}k|}{3} = -\frac{\mathrm{tr}k}{3}. \tag{4.21}$$

The scale-free energy function which will be used in the proof of global existence is the sum of $\widetilde{\mathcal{Q}}_0$ and $\widetilde{\mathcal{Q}}_1$,

$$\mathcal{E} = \widetilde{\mathcal{Q}}_0 + \widetilde{\mathcal{Q}}_1. \tag{4.22}$$

It is convenient to introduce the logarithmic time $\sigma = -\ln(-\tau)$. Then $\sigma \nearrow \infty$ as
$\tau \nearrow 0$. The logarithmic time σ has the property that $\partial_\sigma = -\tau\partial_\tau$ is scale-free, so
that for example $\partial_\sigma\mathcal{E}$ is scale-free.

4.2. The Hessian of the Bel–Robinson energy

Let M be a compact 3-dimensional manifold of hyperbolic type and let γ be the
standard hyperbolic metric on M. Let $(\frac{9}{\tau^2}\gamma, \frac{3}{\tau}\gamma)$ be data with mean curvature τ
for the hyperbolic cone space-time $(\bar{M}, \bar{\gamma})$. For $s \geq 3$, let Σ_τ^s be the local slice
for the action of $\mathcal{D}^{s+1}$ on the constraint set C_τ, at the hyperbolic cone data with
mean curvature τ.

The energies $\widetilde{\mathcal{Q}}_0, \widetilde{\mathcal{Q}}_1, \mathcal{E}$ may be thought of as functions on the constraint set
C_τ^s by using equations (3.6) and (3.8). The following lemma is a straightforward
consequence of the Sobolev embedding theorems.

Lemma 4.2. *The scale-free energies $\widetilde{\mathcal{Q}}_0(\tau, W), \widetilde{\mathcal{Q}}_1(\tau, W), \mathcal{E}(\tau, W)$, are C^∞ functions on C_τ^3 and Σ_τ^3.*

It is clear that at data corresponding to a flat spacetime, $\widetilde{\mathcal{Q}}_0, \widetilde{\mathcal{Q}}_1, \mathcal{E}$ and
their first Fréchet derivatives are zero. Let $\mathrm{Hess}\widetilde{\mathcal{Q}}_0(\gamma, -\gamma)$ denote the Hessian of
the function $\widetilde{\mathcal{Q}}_0$, evaluated at $(\gamma, -\gamma)$. A computation shows that for $(h, p) \in$
$T_{(\gamma,-\gamma)}T\Sigma_{-n}$,

$$\mathrm{Hess}\widetilde{\mathcal{Q}}_0(\gamma, -\gamma)((h, p), (h, p)) = \frac{1}{2}\|Ah\|_{L^2}^2 + (Ap, p)_{L^2} + \frac{1}{2}\|h + 2p\|_{L^2}^2,$$

where A is given by (2.19), and $\|\cdot\|_{L^2}, (\cdot, \cdot)_{L^2}$ denote the L^2-norm and inner
product defined with respect to γ. Recall that $\ker A = \{0\}$ if and only if M is
rigid. It is now straightforward to prove the following Lemma.

Lemma 4.3. *Let M be a compact hyperbolic 3-manifold. The Hessian of the scale-
free Bel–Robinson energy $\widetilde{\mathcal{Q}}_0$, defined by equation (4.22), considered as a function*

on Σ_{-3}, evaluated at the standard data $(\gamma, -\gamma)$, satisfies the inequality

$$\text{Hess}\,\widetilde{Q}_0(\gamma, -\gamma)((h, p), (h, p)) \geq C(||h||^2_{H^2} + ||p||^2_{H^1}), \qquad (4.23)$$

if and only if M is rigid. The constant C depends only on the topology of M. $\square$

Consider a solution $\bar{h}$ of the linearized Einstein equations on the hyperbolic cone space-time $(\bar{M}, \bar{\gamma})$. The derivative of the Weyl tensor in the direction of h,

$$W' = DW[\bar{\gamma}].\bar{h},$$

is a Weyl field on $(\bar{M}, \bar{\gamma})$ which satisfies the homogeneous Bianchi equations,

$$J(W') = J^*(W') = 0, \qquad (4.24)$$

Let $E(W'), B(W')$ be the electric and magnetic parts of W' at $M_{-3} \subset \bar{M}$. Then

$$DE(\bar{\nabla}_T W)|_{\bar{\gamma}}\bar{h} = E(\bar{\nabla}_T W'), \qquad DB(\bar{\nabla}_T W)|_{\bar{\gamma}}\bar{h} = B(\bar{\nabla}_T W').$$

Recall that the second fundamental form of M_{-3} is $k = -\gamma$. This implies using (3.9), (4.24),

$$\text{div}E(\bar{\nabla}_T W') = \text{div}B(\bar{\nabla}_T W') = 0, \qquad (4.25)$$

which shows that $E(\bar{\nabla}_T W')$ and $B(\bar{\nabla}_T W')$ are TT-tensors. It follows from Corollary 3.2, using $k = -\gamma$ and (4.24),

$$E(\bar{\nabla}_T W') = +\text{curl}B(W') + \frac{1}{2}\text{trk}E(W'),$$

$$B(\bar{\nabla}_T W') = -\text{curl}E(W') + \frac{1}{2}\text{trk}B(W').$$

The scale-free Bel–Robinson energy $\mathcal{E}$ is a smooth function on Σ^3_τ. Using the above it is straightforward, using (A.23) to prove that the Hessian of $\mathcal{E}$ is positive definite on $H^3 \times H^2$ in case M is rigid. We state this as

Theorem 4.4. *The Hessian $\text{Hess}\mathcal{E}$ on Σ_τ, evaluated at the standard data $(\frac{\tau^2}{9}\gamma, \frac{\tau}{3}\gamma)$, satisfies the inequality*

$$\text{Hess}\mathcal{E}(\frac{9}{\tau^2}\gamma, \frac{3}{\tau}\gamma)((h, p), (h, p)) \geq C(||h||^2_{H^3} + ||p||^2_{H^2}), \qquad (4.26)$$

if and only if M is rigid. The constant C depend only on the topology of M.

Results analogous to Theorem 4.4 can easily be proved for even higher-order Bel–Robinson type energies. This will not be needed in this paper.

5. Estimates

In this section we will introduce a "smallness condition" on the vacuum data (g, k), under which we are able to control all relevant geometric quantities in terms of the energy function $\mathcal{E}$ defined in Section 4.1. Recall the definition of the slice Σ_τ in Section 2.3. In particular, vacuum data $(g, k) \in \Sigma_\tau$ satisfy the CMCSH gauge conditions.

Definition 5.1. *Let* (g, k) *be a vacuum data set on* M *with mean curvature* τ. *Let* λ *be given by* (4.21) *and let* $(\tilde{g}, \tilde{k})$ *be the rescaled metric and second fundamental form as defined in* (4.20). *Let* $\mathcal{B}(\alpha)$ *be the set of* $(g, k) \in \Sigma_\tau^3$ *so that*

$$||\tilde{g} - \gamma||_{H^3}^2 + ||\tilde{k} + \gamma||_{H^2}^2 < \alpha.$$

We will say that (g, k) *satisfies the smallness condition if* $(g, k) \in \mathcal{B}(\alpha)$. □

The smoothness of the scale-free Bel–Robinson energy $\mathcal{E}$, Lemma 4.2, together with the fact that the Hessian of $\mathcal{E}$, the scale-free Bel–Robinson energy restricted to the slice Σ_τ, is positive definite with respect to the Sobolev norm $H^3 \times H^2$, Theorem 4.4, implies, by Taylor's theorem, the following estimate.

Theorem 5.2. *Assume that* M *is rigid. There is an* $\alpha > 0$ *so that for* $(g, k) \in \mathcal{B}(\alpha)$, *there is a constant* $D(\alpha) < \infty$, *depending only on* α *and the topology of* M, *such that*

$$D(\alpha)^{-1}\mathcal{E}(g, k) \leq ||\tilde{g} - \gamma||_{H^3}^2 + ||\tilde{k} + \gamma||_{H^2}^2 \leq D(\alpha)\mathcal{E}(g, k). \tag{5.1}$$

In view of the analysis of the elliptic defining equations for N, X in [1], there is a neighborhood of $(\gamma, -\gamma)$ in Σ_{-3}^3, such that $\hat{N}, X$ are small in $W^{1,\infty}$-norm, defined by $||f||_{W^{1,\infty}} = ||f||_{L^\infty} + ||Df||_{L^\infty}$. It is straightforward to check

Corollary 5.3. *Let* $\alpha > 0$ *be such that the conclusion of Theorem 5.2 holds. There is a constant* $\delta > 0$ *so that for* $(g, k) \in \mathcal{B}(\alpha)$ *with* $\mathcal{E}(g, k) < \delta$, *it holds that*

$$\max(\Lambda, ||\hat{N}||_{W^{1,\infty}}, ||X||_{W^{1,\infty}}, ||g||_{W^{1,\infty}}, ||k||_{L^\infty}) < 1/\delta.$$

Lemma 5.4. *Let* $\alpha > 0$ *be small enough so that the conclusion of Theorem 5.2 holds. Let* $(g, k) \in \mathcal{B}(\alpha)$ *and let* N, X *be the corresponding solutions of the defining equations* (2.8). *Then there is a constant* C *such that*

$$||\tilde{k}||_{L^\infty} \leq C\mathcal{E}^{\frac{1}{2}}, \tag{5.2a}$$

$$||\tilde{N}||_{L^\infty} \leq C\mathcal{E}, \tag{5.2b}$$

$$||\widetilde{\nabla N}||_{L^\infty} \leq C\mathcal{E}, \tag{5.2c}$$

$$||\tilde{\pi}||_{L^\infty} \leq C\mathcal{E}^{\frac{1}{2}}. \tag{5.2d}$$

Proof. The inequality (5.2a) follows from the definition of $\mathcal{B}$ and Sobolev embedding. The Lapse equation (2.8a) implies by the maximum principle,

$$|\hat{N}| \leq \frac{3}{\tau^2} \frac{||\hat{k}||_{L^\infty}^2}{||k||_{L^\infty}^2},$$

which gives (5.2b) after rescaling. A standard elliptic estimate gives (5.2c). Finally, (4.7) together with the above estimates yield $||\tilde{\pi}||_{L^\infty} \leq C(\mathcal{E}^{\frac{1}{2}} + \mathcal{E})$, which after using the smallness assumption and redefining C gives (5.2d). □

5.1. Differential inequalities for the rescaled Bel–Robinson energies

In this section we will estimate the time derivatives $\partial_\sigma \widetilde{\mathcal{Q}}_i$ of the scale-free Bel–Robinson energies with respect to the logarithmic time σ defined in Section 4.1.

Lemma 5.5. *Assume* $(g, k) \in \mathcal{B}(\alpha)$ *for* α *sufficiently small so that the conclusion of Theorem 5.2 holds. Then*

$$\partial_\sigma \widetilde{\mathcal{Q}}_0(\sigma, W) \leq -(2 - 2C\mathcal{E}^{1/2})\widetilde{\mathcal{Q}}_0, \tag{5.3}$$

Proof. Replacing all the fields in the RHS of (4.11) by their scale-free versions, noting in particular that $\tilde{\tau} = -3$ and $t < 0$, we get

$$\partial_\sigma \widetilde{\mathcal{Q}}_0 = -2\widetilde{\mathcal{Q}}_0 + \tilde{F}_1$$

where $\tilde{F}_1$ is the scalefree version of

$$F_1 = -9 \int_{M_\tau} N Q_{abcd} \hat{\pi}^{ab} T^c T^d d\mu_{M_\tau} + 3\tau \int_{M_\tau} \hat{N} Q_{TTTT}$$

The maximum principle applied to the Lapse equation (2.8a) implies, after a rescaling, that $\tilde{N} \leq \frac{1}{3}$. This gives the estimate

$$\tilde{F}_1 \leq C(||\hat{\tilde{\pi}}||_{L^\infty} + ||\hat{\tilde{N}}||_{L^\infty})\widetilde{\mathcal{Q}}_0(\tau, W).$$

To finish the proof note that by Lemma 5.4,

$$||\hat{\tilde{\pi}}||_{L^\infty} + ||\hat{\tilde{N}}||_{L^\infty} \leq C\mathcal{E}^{1/2},$$

using the smallness assumption. We write the resulting inequality in the form (5.3) for convenience. □

Remark 5.1. *The proof of Lemma 5.5 gives the inequality*

$$\partial_\sigma \widetilde{\mathcal{Q}}_0 \leq -(2 - C(||\hat{\tilde{\pi}}||_{L^\infty} + ||\hat{\tilde{N}}||_{L^\infty}))\widetilde{\mathcal{Q}}_0,$$

which is valid without *the smallness assumption.* □

Lemma 5.6. *Assume* $(g, k) \in \mathcal{B}(\alpha)$ *for* α *sufficiently small so that the conclusion of Theorem 5.2 holds. Then*

$$\partial_\sigma \mathcal{E} \leq -(2 - 2C\mathcal{E}^{\frac{1}{2}})\mathcal{E}. \tag{5.4}$$

Proof. In view of the smallness condition, the inequality (5.2b) and Lemma 5.5, we only need to consider $\partial_\sigma \widetilde{\mathcal{Q}}_1$. We proceed as in the proof of Lemma 5.5, using the scale-free version of (4.12) taking into account $\tilde{\tau} = -3$, we get

$$\partial_\sigma \widetilde{\mathcal{Q}}_1 = -2\widetilde{\mathcal{Q}}_1 + \tilde{F}_2$$

where $\tilde{F}_2$ is the scale-free version of

$$F_2 = -6 \int_{M_\tau} N \mathcal{G}_1(W) d\mu_{M_\tau} - 9 \int_{M_\tau} N Q(\bar{\nabla}_T W)_{abcd} \hat{\pi}^{ab} T^c T^d d\mu_{M_\tau}$$

$$+ 5\tau \int_{M_\tau} \hat{N} Q(\bar{\nabla}_T W)_{TTTT}$$

We see that in order to estimate $\tilde{F}_2$, we need to estimate $\mathcal{G}_1(W)$, which is given in Lemma 4.1. Taking into account the detailed structure of $\mathcal{G}_1$, cf. Remark 4.1 we get the estimate

$$\int N\mathcal{G}_1(W) \leq C\Big[||\hat{\pi}||_\infty(1 + ||k||_\infty)(\mathcal{Q}_0 + \mathcal{Q}_1)$$
$$+ \int_{M_\tau} N|E(\bar{\nabla}_T W)|(|E(W)|^2 + |B(W)|^2)d\mu_{M_\tau}\Big].$$

By the Hölder inequality,

$$\int |E(\bar{\nabla}_T W)|(|E(W)|^2 + |B(W)|^2) \leq C\mathcal{Q}_1^{1/2}(||E(W)||_{L^4}^2 + ||B(W)||_{L^4}^2).$$

By the Sobolev inequality, we may bound the scale-free version of

$$||E(W)||_{L^4}^2 + ||B(W)||_{L^4}^2$$

by $C\mathcal{E}$. We now have the estimate

$$\int \tilde{N}\tilde{\mathcal{G}}_1 \leq C\left(||\tilde{\hat{\pi}}||_\infty(1 + ||\tilde{k}||_\infty)\mathcal{E} + \mathcal{E}^{3/2}\right) \tag{5.5}$$

$$\leq C\mathcal{E}^{3/2}, \tag{5.6}$$

for $(g, k) \in \mathcal{B}(\alpha)$. Proceeding similarly with the other terms in $\tilde{F}_2$ yields an inequality which we write in the form (5.4) for convenience. □

6. Global existence

Fix $\tau_0 < 0$ and let $(g(\tau_0), k(\tau_0))$ be data for Einstein's equations with mean curvature τ_0 and assume that $(g(\tau_0), k(\tau_0)) \in \mathcal{B}(\alpha)$ for an $\alpha > 0$ small enough so that the conclusion of Theorem 5.2 holds.

We have seen above that for small data, the second-order scale-free Bel–Robinson energy $\mathcal{E}$ satisfies the differential inequality (5.4). We will use this to prove

Theorem 6.1 (Global existence for small data). *Assume that M is rigid. Let $\alpha > 0$ be such that the conclusion of Theorem 5.2 holds. There is an $\epsilon \in (0, \alpha)$ small enough that if $(g^0, k^0) \in \mathcal{B}(\epsilon)$, then the maximal existence interval in mean curvature time τ, for the vacuum Einstein equations in CMCSH gauge, with data (g^0, k^0) is of the form $(T_-, 0)$. In particular, the CMCSH vacuum Einstein equations have global existence in the expanding direction for initial data in $\mathcal{B}(\epsilon)$.*

Here ϵ can be chosen as

$$\epsilon = D(\alpha)^{-1} \min(\delta, C^{-2}),$$

where $D(\alpha)$ is defined in Theorem 5.2, $\delta > 0$ is given by Corollary 5.3, and C is the constant in (5.4).

Proof. Under the assumptions of the theorem, by Theorem 5.2,

$$\mathcal{E}(g^0, k^0) < \min(\hat{\delta}, C^{-2}) \tag{6.1}$$

holds. Thus the conclusion of Corollary 5.3 holds. By the definition of $\mathcal{B}(\alpha)$ we may apply Theorem 2.1 to conclude that there is a nontrivial maximal existence interval (T_-, T_+) in mean curvature time τ, with $T_+ \leq 0$, for (g^0, k^0) in $H^3 \times H^2$. We will assume $T_+ < 0$ and prove that this leads to a contradiction, using energy estimates and the continuation principle.

Let $y(\sigma)$ be the solution to the initial value problem

$$\frac{dy}{d\sigma} = -2y + 2Cy^{3/2}, y(\sigma_0) = y_0. \tag{6.2}$$

Then if $y_0 = \mathcal{E}(g^0, k^0)$, is such that $y(\sigma) < \infty$ for $\sigma \in [\sigma_0, \sigma_+]$, we have

$$\mathcal{E}(\sigma) \leq y(\sigma), \quad \sigma \in [\sigma_0, \sigma_+).$$

The solution to (6.2) is

$$y^{-1/2} = C + e^{\sigma - \sigma_0}(y_0^{-1/2} - C),$$

if $y_0 < C^{-2}$, and in this case $y(\sigma) < y(\sigma_0)$ for $\sigma \in (\sigma_0, \infty)$. This means that if (6.1) holds at $\sigma = \sigma_0$, it holds for $\sigma \in [\sigma_0, \sigma_+)$. By Theorem 5.2, this implies that $\|\tilde{g} - \gamma\|_{H^3} + \|\tilde{k} + \gamma\|_{H^2}$ is uniformly bounded for $\sigma \in [\sigma_0, \sigma_+)$. By Corollary 5.3, this implies that the inequality

$$\sup_{\sigma \in [\sigma_0, \sigma_+)} (\Lambda[\bar{g}] + \|D\bar{g}\|_{L^\infty} + \|k\|_{L^\infty}) < \delta^{-1},$$

holds.

In view of the continuation principle, Point 2 of Theorem 2.1, this contradicts the assumption that (T_-, T_+) is the maximal existence interval in mean curvature time τ, with $T_+ < 0$. It follows that $T_+ = 0$ which completes the proof. □

6.1. Geodesic completeness

Theorem 6.2. *Let (M, g^0, k^0) and $(\bar{M}, \bar{g})$ be as in Theorem 6.1. Then $(\bar{M}, \bar{g})$ is causally geodesically complete in the expanding direction.*

Proof. By Theorem 6.1, $(\bar{M}, \bar{g})$ is globally foliated by CMC hypersurfaces to the future of (M, g^0, k^0), i.e., in the expanding direction, with $t = \text{trk} \nearrow t_* = 0$.

Let $c(\lambda)$ be a future directed causal geodesic, with affine parameter λ. Let

$$u = \frac{dc}{d\lambda}, \quad \langle u, u \rangle = \begin{cases} -1 \\ 0 \end{cases},$$

be the normalized velocity, where $\langle \cdot, \cdot \rangle = \bar{g}(\cdot, \cdot)$. The geodesic equation is

$$\bar{\nabla}_u u = 0. \tag{6.3}$$

As c is causal, we may use t as parameter. Let

$$u^0 = dt(u) = \frac{dt}{d\lambda}.$$

In order to prove geodesic completeness in the expanding direction, it is sufficient to prove that the solution to the geodesic equation exists for an infinite interval of the affine parameter, i.e., $\lim_{t \nearrow t_*} \lambda(t) = \infty$ or

$$\lim_{t \nearrow t_*} \int_{t_0}^t \frac{d\lambda}{dt} dt = \infty,$$

or using the definition of u^0,

$$\lim_{t \nearrow t_*} \int_{t_0}^t \frac{1}{u^0} dt = \infty. \qquad (6.4)$$

Suppose that we are able to prove that Nu^0 is bounded from above as $t \nearrow t_*$. Then (6.4) holds precisely when

$$\lim_{t \nearrow t_*} \int_{t_0}^t N dt = \infty.$$

For a function f on $\bar{M}$, we have

$$\frac{df(c(t))}{dt} = \frac{d\lambda}{dt} \frac{df(c(t))}{d\lambda} = \left(\frac{dt}{d\lambda} \right)^{-1} \frac{df(c(t))}{d\lambda} = \frac{1}{u^0} \nabla_u f. \qquad (6.5)$$

A calculation in local coordinates using the $3 + 1$-form of $\bar{g}$, gives $T_\mu = -N\delta_\mu^0$ where δ_μ^ν is the Kronecker delta, i.e., $\langle T, V \rangle = -Ndt(V)$ for any V. This shows that

$$-Nu^0 = \langle u, T \rangle,$$

or

$$u = Nu^0 T + Y, \qquad (6.6)$$

where Y is tangent to M. Let $\epsilon = 0, 1$. Then by assumption, $\langle u, u \rangle = -|\epsilon|$ which using (6.6) gives

$$|Y|_g^2 = N^2(u^0)^2 - |\epsilon|.$$

In particular, we get the inequality

$$|Y|_g \leq Nu^0. \qquad (6.7)$$

A computation in a Fermi propagated frame gives using (6.6),

$$\bar{\nabla}_u T = Nu^0 N^{-1} \nabla_i N e_i - k_{ij} Y^j e_i. \qquad (6.8)$$

By our choice of time orientation we have $N > 0$ and $u^0 > 0$ and $\mathrm{tr}k < 0$.

We now compute using $\bar{\nabla}_u u = 0$, (6.6) and (6.8)

$$\frac{d}{dt} \ln(Nu^0) = -\frac{1}{Nu^0} \frac{d}{dt} \langle u, T \rangle = -\frac{1}{N(u^0)^2} \langle u, \bar{\nabla}_u T \rangle$$

$$= -\frac{1}{N(u^0)^2} (u^0 \nabla_Y N - k_{ij} Y^i Y^j)$$

$$= -\frac{\nabla_Y N}{Nu^0} + N\hat{k}_{ij} \frac{Y^i}{Nu^0} \frac{Y^j}{Nu^0} + N \frac{\mathrm{tr}k}{3} \frac{|Y|_g^2}{N^2(u^0)^2}$$

use (6.7) and $\text{tr}k < 0$,

$$\leq ||\nabla N||_{L^\infty;g} + ||N\hat{k}||_{L^\infty;g}$$

use scaling properties of $N, \hat{k}, g$, cf. Section 4.1,

$$\leq ||\widetilde{\nabla N}||_{L^\infty;\tilde{g}} + \lambda^{-1}||\tilde{N}\tilde{\hat{k}}||_{L^\infty;\tilde{g}},$$

with $\lambda = \text{tr}k/3 = t/3$. By the proof of the global existence result Theorem 6.1, we have $\mathcal{E}(t) \leq Ct^2$ and $\tilde{g}$ is close to γ. Therefore by Sobolev embedding, we can relate the norms w.r.t. $\tilde{g}$ to the norms w.r.t. γ and we get

$$\frac{d}{dt}\ln(Nu^0) \leq C\left(||\widetilde{\nabla N}||_{L^\infty} + \lambda^{-1}||\tilde{N}\tilde{\hat{k}}||_{L^\infty}\right).$$

Now an application of the estimate (5.2) together with the decay of $\mathcal{E}$ gives

$$||\widetilde{\nabla N}||_{L^\infty} \leq Ct^2,$$

$$||\tilde{N}\tilde{\hat{k}}||_{L^\infty} \leq Ct,$$

which in view of $\lambda^{-1}t = 3$ gives

$$\ln(Nu^0)(t) - \ln(Nu^0)(t_0) \leq C,$$

and hence $\ln(Nu^0) \leq C$ for some constant C as $t \nearrow t_* = 0$.

We have now proved that Nu^0 is bounded from above and therefore it is sufficient to prove that

$$\lim_{t \nearrow t_*} \int_{t_0}^{t} N dt = \infty. \qquad (6.9)$$

Write $N = \hat{N} + \frac{3}{t^2}$ as in (4.10). Using (5.2) and the scaling rule (4.20) to estimate $\hat{N}$ gives $N \geq C/t^2$ as $t \nearrow t_* = 0$. This shows that (6.9) holds and completes the proof of Lemma 6.2. $\qquad\qquad\qquad\qquad\qquad\qquad\qquad\qquad$ □

Acknowledgements

The authors are grateful to the Erwin Schrödinger Institute in Vienna, l'Institute des Hautes Études Scientifiques in Bures-sur-Yvette, the Albert Einstein Institute in Golm, the Université Paris VI and the Institute of Theoretical Physics in Santa Barbara for hospitality and support while most of this work was being carried out.

Appendix A. Basic definitions and identities

A.1. Conventions

We begin by recalling some basic facts and definitions. We use the following conventions for curvature.

The Riemann tensor is defined by

$$R(X,Y)Z = \nabla_X \nabla_Y Z - \nabla_Y \nabla_X Z - \nabla_{[X,Y]}Z.$$

In a coordinate frame $\{e_a\}$ we have

$$R^d{}_{cab}Z^c = \nabla_a \nabla_b Z^d - \nabla_b \nabla_a Z^d.$$

This gives the conventions for index calculations

$$[\nabla_a, \nabla_b]t_c = R_{abc}{}^d t_d, \tag{A.1}$$

$$[\nabla_a, \nabla_b]t^c = R^c{}_{dab}t^d. \tag{A.2}$$

The Ricci curvature and the scalar curvature are defined (in an ON frame) by

$$\mathrm{Ric}(X,Y) = \sum_i \langle R(e_i, X)Y, e_i \rangle, \tag{A.3}$$

$$\mathrm{Scal} = \sum_i \mathrm{Ric}(e_i, e_i), \tag{A.4}$$

or in index notation

$$R_{ij} = g^{kl}R_{ikjl}, \qquad R = g^{ij}R_{ij}.$$

Note also

$$\mathrm{Ric}(X,Y) = \mathrm{tr}(Z \mapsto R(Z,X)Y).$$

The Riemann tensor satisfies the Bianchi identities

$$\nabla_{[e}R_{ab]cd} = \frac{1}{3}(\nabla_e R_{abcd} + \nabla_a R_{becd} + \nabla_b R_{eacd}) = 0.$$

The trace free part of the Riemann tensor in an n-dimensional manifold is

$$\begin{aligned}
C_{abcd} = R_{abcd} &- \frac{1}{n-2}(g_{ac}R_{bd} + g_{bd}R_{ac} - g_{bc}R_{ad} - g_{ad}R_{bc}) \\
&+ \frac{1}{(n-1)(n-2)}(g_{ac}g_{bd} - g_{ad}g_{bc})R.
\end{aligned} \tag{A.5}$$

The totally anti-symmetric tensor ϵ in dimension $3+1$ satisfies the identities

$$\begin{aligned}
\epsilon^{\alpha_1\alpha_2\alpha_3\alpha_4}\epsilon_{\beta_1\beta_2\beta_3\beta_4} &= -\det(\delta^{\alpha_i}_{\beta_j})_{i,j=1,\dots,4}, \\
\epsilon^{\alpha_1\alpha_2\alpha_3\alpha_4}\epsilon_{\alpha_1\beta_2\beta_3\beta_4} &= -\det(\delta^{\alpha_i}_{\beta_j})_{i,j=2,\dots,4}, \\
\epsilon^{\alpha_1\alpha_2\alpha_3\alpha_4}\epsilon_{\alpha_1\alpha_2\beta_3\beta_4} &= -2\det(\delta^{\alpha_i}_{\beta_j})_{i,j=3,\dots,4}, \\
\epsilon^{\alpha_1\alpha_2\alpha_3\alpha_4}\epsilon_{\alpha_1\alpha_2\alpha_3\beta_4} &= -6\delta^{\alpha_4}_{\beta_4}, \\
\epsilon^{\alpha_1\alpha_2\alpha_3\alpha_4}\epsilon_{\alpha_1\alpha_2\alpha_3\alpha_4} &= -24.
\end{aligned} \tag{A.6a}$$

In an ON frame adapted to a spacelike hypersurface M in a $3 + 1$-dimensional manifold we define (c.f. [6, p. 144])

$$\epsilon_{ijk} = \epsilon_{Tijk}. \tag{A.7}$$

$$\epsilon^{i_1 i_2 i_3} \epsilon_{j_1 j_2 j_3} = \det(\delta^{i_k}_{j_l})_{k,l=1,2,3} \qquad = 6\delta^{[i_1}_{j_1} \delta^{i_2}_{j_2} \delta^{i_3]}_{j_3}, \tag{A.8a}$$

$$\epsilon^{i_1 i_2 i_3} \epsilon_{i_1 j_2 j_3} = \det(\delta^{i_k}_{j_l})_{k,l=2,3}) \qquad = 2\delta^{[i_2}_{j_2} \delta^{i_3]}_{j_3}, \tag{A.8b}$$

$$\epsilon^{i_1 i_2 i_3} \epsilon_{i_1 i_2 j_3} = 2\delta^{i_3}_{j_3}, \tag{A.8c}$$

$$\epsilon^{i_1 i_2 i_3} \epsilon_{i_1 i_2 i_3} = 6. \tag{A.8d}$$

In dimension 3 we have the duality relations

$$\xi_{ab} = \epsilon_{ab}{}^{m} \eta_m,$$

for $\xi_{ab} = \xi_{[ab]}$, where

$$\eta_m = \frac{1}{2} \epsilon_m{}^{ab} \xi_{ab}.$$

A.2. Operations on symmetric 2-tensors

Define the following operations on symmetric 2-tensors on a 3-dimensional Riemann manifold:

$$A \cdot B = A_{ab} B^{ab}, \tag{A.9}$$

$$(A \wedge B)_a = \epsilon_a{}^{bc} A_b{}^d B_{dc}, \tag{A.10}$$

$$(v \wedge A)_{ab} = \epsilon_a{}^{cd} v_c A_{db} + \epsilon_b{}^{cd} v_c A_{ad}, \tag{A.11}$$

$$(A \times B)_{ab} = \epsilon_a{}^{cd} \epsilon_b{}^{ef} A_{ce} B_{df} + \frac{1}{3}(A \cdot B) g_{ab} - \frac{1}{3}(\text{tr}A)(\text{tr}B) g_{ab}, \tag{A.12}$$

$$\text{curl}A_{ab} = \frac{1}{2}(\epsilon_a{}^{cd} \nabla_d A_{cb} + \epsilon_b{}^{cd} \nabla_d A_{ca}), \tag{A.13}$$

$$\text{div}A_a = \nabla^b A_{ab}. \tag{A.14}$$

The operation $\wedge$ is skew symmetric, while $\times$ is symmetric, and the identities

$$A \cdot (v \wedge B) = -2v \cdot (A \wedge B)$$
$$A \cdot (B \times C) = (A \times B) \cdot C \qquad \text{(if } \text{tr}A = \text{tr}C = 0)$$

hold. The expression $A \times B$ can be expanded as

$$(A \times B)_{ab} = A_a{}^c B_{cb} + A_b{}^c B_{ca}$$
$$- \frac{2}{3}(A \cdot B) g_{ab} + \frac{2}{3}(\text{tr}A)(\text{tr}B) g_{ab} - (\text{tr}A) B_{ab} - (\text{tr}B) A_{ab}.$$

A computation shows

$$\text{div}(A \wedge B) = -(\text{curl}A) \cdot B + A \cdot (\text{curl}B). \tag{A.15}$$

Let A be a symmetric covariant 2-tensor on $\bar{M}$ and suppose A is t-tangent, i.e., $A_{\alpha\beta}T^\beta = 0$. Then in a Fermi propagated frame,

$$\bar{\nabla}_T A_{ij} = TA_{ij} \tag{A.16}$$

$$\mathcal{L}_{\partial_t} A_{ij} = N\bar{\nabla}_T A_{ij} - N\left((k \times A)_{ij} + \frac{2}{3}(k \cdot A)g_{ij}\right.$$
$$\left. - \frac{2}{3}(\text{tr}k)(\text{tr}A)g_{ij} + (\text{tr}A)k_{ij} + (\text{tr}k)A_{ij}\right). \tag{A.17}$$

Define the covariant exterior derivative $d^\nabla u$ on symmetric 2-tensors by

$$(d^\nabla u)_{ijk} = \nabla_k u_{ij} - \nabla_j u_{ik}. \tag{A.18}$$

The operators curl, div, d^∇ are related by [6, p. 103]

$$d^\nabla u_{kij} = \left(\text{curl}u_{kl} + \frac{1}{2}(\text{div}u_m - \nabla_m \text{tr}u)\epsilon^m_{kl}\right)\epsilon^l_{ij}. \tag{A.19}$$

Taking into account the symmetry of curl this implies

$$|d^\nabla u|^2 = 2(|\text{curl}u|^2 + \frac{1}{2}|\text{div}u - \nabla\text{tr}u|^2). \tag{A.20}$$

If u has compact support then in dimension n,

$$\int_M |d^\nabla u|^2 = 2\int_M |\nabla u|^2 - 2\int_M |\text{div}u|^2 + 2\int_M (u_k{}^l R^k{}_{ijl} + u_{ik}R^k{}_j)u^{ij}. \tag{A.21}$$

This leads to, if $\text{tr}u = 0$,

$$\int_M (|\nabla u|^2 + 3R_{ij}u^{ki}u_k{}^j - \frac{1}{2}R|u|^2) = \int_M (|\text{curl}u|^2 + \frac{3}{2}|\text{div}u|^2), \tag{A.22}$$

in case M is of dimension 3. If we further restrict to (M, γ) with γ hyperbolic, so that $R[\gamma] = -6$, we get

$$\int_M (|\text{curl}u|^2 + \frac{3}{2}|\text{div}u|^2) = \int_M (|\nabla u|^2 - 3|u|^2). \tag{A.23}$$

References

[1] Lars Andersson and Vincent Moncrief, *Elliptic-hyperbolic systems and the Einstein equations*, submitted to Ann. Henri Poincaré, 2001.

[2] Lars Andersson, Vincent Moncrief, and Anthony J. Tromba, *On the global evolution problem in 2 + 1 gravity*, J. Geom. Phys. **23** (1997), no. 3–4, 191–205.

[3] Yvonne Choquet-Bruhat, *Future complete Einsteinian space times with U(1) symmetry, the unpolarized case*, article in this volume.

[4] Yvonne Choquet-Bruhat and Vincent Moncrief, *Future global in time Einsteinian spacetimes with* U(1) *isometry group*, Ann. Henri Poincaré **2** (2001), no. 6, 1007–1064.

[5] Demetrios Christodoulou and Sergiu Klainerman, *Asymptotic properties of linear field equations in Minkowski space*, Comm. Pure Appl. Math. **43** (1990), no. 2, 137–199.

[6] Demetrios Christodoulou and Sergiu Klainerman, *The global nonlinear stability of the Minkowski space*, Princeton University Press, Princeton, NJ, 1993.

[7] James Eells and Luc Lemaire, *A report on harmonic maps*, Bull. London Math. Soc. **10** (1978), no. 1, 1–68.

[8] Arthur E. Fischer and Vincent Moncrief, *Hamiltonian reduction of Einstein's equations of general relativity*, Nuclear Phys. B Proc. Suppl. **57** (1997), 142–161, Constrained dynamics and quantum gravity 1996 (Santa Margherita Ligure).

[9] _____, *Hamiltonian reduction and perturbations of continuously self-similar (n + 1)-dimensional Einstein vacuum spacetimes*, Classical Quantum Gravity **19** (2002), no. 21, 5557–5589.

[10] Helmut Friedrich, *On the existence of n-geodesically complete or future complete solutions of Einstein's field equations with smooth asymptotic structure*, Comm. Math. Phys. **107** (1986), no. 4, 587–609.

[11] Michael Kapovich, *Deformations of representations of discrete subgroups of* SO(3, 1), Math. Ann. **299** (1994), no. 2, 341–354.

[12] Jacques Lafontaine, *Modules de structures conformes plates et cohomologie de groupes discrets*, C.R. Acad. Sci. Paris Sér. I Math. **297** (1983), no. 13, 655–658.

Lars Andersson
Department of Mathematics
University of Miami
Coral Gables, FL 33124, USA
e-mail: larsa@math.miami.edu

Vincent Moncrief
Department of Physics
Yale University
P.O. Box 208120
New Haven, CT 06520, USA
e-mail: vincent.moncrief@yale.edu

The Cauchy Problem on Spacetimes That Are Not Globally Hyperbolic

John L. Friedman

Abstract. The initial value problem is well defined on a class of spacetimes broader than the globally hyperbolic geometries for which existence and uniqueness theorems are traditionally proved. Simple examples are the time-nonorientable spacetimes whose orientable double cover is globally hyperbolic. These spacetimes have generalized Cauchy surfaces on which smooth initial data sets yield unique solutions. A more difficult problem is to characterize the class of spacetimes with closed timelike curves that admit a well-posed initial value problem. Examples of spacetimes with closed timelike curves are given for which smooth initial data at past null infinity has been recently shown to yield solutions. These solutions appear to be unique, and uniqueness has been proved in particular cases. Other examples, however, show that confining closed timelike curves to compact regions is not sufficient to guarantee uniqueness. An approach to the characterization problem is suggested by the behavior of congruences of null rays. Interacting fields have not yet been studied, but particle models suggest that uniqueness (and possibly existence) is likely to be lost as the strength of the interaction increases.

1. Introduction

Motivating the definition of a *globally hyperbolic* spacetime are two facts: On globally hyperbolic spacetimes, wave equations have a well-defined initial value formulation; and the ordinary causal structure of a globally hyperbolic spacetime mirrors the ordinary causal structure observed in the universe. The requirement that the initial value problem be well defined, however, picks out a broader class of spacetimes. And the causal structure of the physical universe on the largest and smallest scales may not conform to that of a globally hyperbolic spacetime.

In particular, a class of spacetimes that are not globally hyperbolic and nevertheless admit a well-defined initial value problem are Lorentzian 4-geometries with a single spacelike boundary – Lorentzian *universes-from-nothing* (Friedman and Higuchi [1, 2]). These are the metrics that arise in a Lorentzian path-integral construction of the Hartle-Hawking wave function of the Universe. There is large class of such geometries, spacetimes that are compact on one side of a spacelike boundary. A simple two-dimensional example, considered in more detail below, is a Möbius strip with a flat Lorentz metric, for which the direction orthogonal to the median circle is timelike. On each underlying manifold of such a spacetime, one can choose metrics that have no closed timelike curves and for which the boundary remains spacelike; time nonorientability is then its only causal pathology. With metrics so chosen, these spacetimes provide the only examples of topology change in which one has a smooth, nondegenerate Lorentzian metric without closed timelike curves. The initial value problem for these spacetimes is discussed in Section II.

The more difficult problem is to characterize the class of spacetimes that have closed timelike curves (CTCs) and that nevertheless allow a well-defined initial value problem for hyperbolic systems. This brief review outlines recent work that has been done in proving existence and uniqueness of solutions to the scalar wave equation on a class of spacetimes with closed timelike curves. Section III introduces the subject with several two-dimensional examples that are easily analyzed and illustrate the obstacles to the existence of a generalized initial value problem. These obstacles are less severe in four dimensions, and Section IV considers four-dimensional examples in which existence and uniqueness theorems have been proved. These 4-dimensional examples are so far restricted to stationary, asymptotically flat wormhole spacetimes, with data specified at past null infinity: Friedman and Morris [24] proved existence and a weak uniqueness theorem for a massless free field on a class of such spacetimes, in which the CTCs loop through a static wormhole. More recently, Bachelot[19] proved existence and uniqueness for another class of stationary, asymptotically flat spacetimes. Finally, Section V outlines the heuristic arguments that have been made for the existence of a well-defined initial value problem in a much broader class of spacetimes. The section concludes with two conjectures that would partly characterize such a class.

Let us first generalize the definition of a Cauchy surface to allow spacetimes that are not globally hyperbolic. Let M, g be a smooth spacetime, a manifold M together with a Lorentz-signature metric g. Recall that a set $S \subset M$ is *achronal* if no two points are timelike separated.[3])

Definition. A *generalized Cauchy surface* Σ is an achronal hypersurface of M for which the initial value problem for the scalar wave equation is well defined:
For any smooth data in $L_2(\Sigma)$ with finite energy, for a scalar field Φ, there is a unique solution Φ on M.

One can then ask what the class is of spacetimes for which the initial value problem is well defined. To see that this class is larger than the class of globally hyperbolic spacetimes, we begin with the time non-orientable spacetimes mentioned above.

2. Time-nonorientable spacetimes with a well-defined initial value problem

We begin with a two-dimensional example and then describe a general class of geometries on a countably infinite set of 4-manifolds. Consider the Möbius strip M with a metric for which its median circle is spacelike. One can construct such a 2-dimensional spacetime from a cylinder $\overline{M} = \mathbb{R} \times S^1$, with an obvious choice of Minkowski metric that makes copies of $\mathbb{R}$ into timelike lines orthogonal to copies of S^1. In terms of the natural chart t, ϕ, on $\overline{M}$, the metric is $-dt^2 + a^2 d\phi^2$, some length a. To construct the Möbius strip, identify each point (t, ϕ), with its antipodal point

$$A(t, \phi) = (-t, \phi + \pi). \tag{1}$$

The Möbius strip is the quotient space $M = \overline{M}/A$; because A is an isometry, M inherits a flat Lorentzian metric for which its median circle Σ (the image of the circle $t = 0$) is spacelike.

It is easy to see that the median circle Σ is a generalized Cauchy surface (as is any boosted image of it). Initial data is a pair $\Phi, \nabla\Phi$ (with ∇ the 2-dimensional gradient) on Σ. Data on Σ lifts to initial data on a Cauchy surface $\overline{\Sigma}$ of the cylinder, and the data is antipodally symmetric. Because the the cylinder is globally hyperbolic, there is a unique solution $\overline{\Phi}$ to the wave equation with this data, and that solution is itself antipodally symmetric. Thus there is a field Φ on the Möbius strip, whose lift to the cylinder is $\overline{\Phi}$; and Φ is the unique solution to the wave equation on M with the specified initial data.

In this flat example, the Möbius strip has closed timelike curves (CTCs) *and* is time-nonorientable. If one chooses a deSitter metric on the cylinder instead of the flat metric, the antipodal map remains an isometry, and the Möbius strip inherits a local deSitter metric, a metric for which it has no CTCs. (CTCs arise from timelike lines that emerge in opposite directions – forward and backwards with respect to a locally defined time direction; in the deSitter geometry circles far from the median circle are large, and they expand fast enough that the timelike lines never meet.) In both the flat and the deSitter case (i.e., with and without CTCs), because the orientable double-cover of M, g is globally hyperbolic, the spacetime has a generalized Cauchy surface.

More generally let $\overline{\Sigma}$ be any 3-manifold that admits a free involution, a diffeo I that has no fixed points. There are countably many spherical spaces and countably many hyperbolic spaces that admit such involutions. As in the above construction, one defines on a cylinder $\overline{M} = \mathbb{R} \times \overline{\Sigma}$ an antipodal map $A = T \times I$, where $T : \mathbb{R} \to \mathbb{R}$ is time reversal:

$$A(t, p) = (-t, I(p)). \tag{2}$$

A is again a free involution, and the manifold of the spacetime is again the quotient

$$M = \overline{M}/A. \tag{3}$$

We will choose a metric for which the 3-manifold $\Sigma = \{0\} \times \overline{\Sigma}/A$ is a generalized Cauchy surface. Any metric g on M, for which each of the surfaces corresponding to $\{t\} \times \Sigma$ are spacelike, will do. The pull-back of g to $\overline{M}$ is a metric for which M is foliated by the spacelike hypersurfaces $\{t\} \times \overline{\Sigma}$ and is therefore globally hyperbolic. (For example, let ^3g be any Riemannian metric on Σ, $^3\overline{g}$ its pull-back to $\overline{M}$. The metric $-dt^2 + {}^3\overline{g}$ on $\overline{M}$ is antipodally symmetric and induces a suitable metric on M.)

By our construction of $\overline{g}$, the antipodal map A is an isometry. Again, the lift to $\overline{\Sigma}$ of data $\Phi, \nabla\Phi$ on Σ is an initial data set $\overline{\Phi}, \nabla\overline{\Phi}$ that is antipodally symmetric (invariant under A). Because $\overline{M}, \overline{g}$ is globally hyperbolic, this data has a unique time evolution, $\overline{\Phi}$; because both the data and the spacetime $\overline{M}, \overline{g}$ are antipodally symmetric, $\overline{\Phi}$ is antipodally symmetric. The field $\overline{\Phi}$ is therefore the lift to $\overline{M}$ of a solution Φ to the scalar wave equation on M. Finally, Φ is unique because $\overline{\Phi}$ is unique.

Classically, spacetimes of the kind considered in this section (nonorientable spacetimes whose orientable double cover is globally hyperbolic), are locally indistinguishable from their covering spacetimes. Treatments of classical spinor fields and of quantum field theory on such spacetimes are given in Refs. [4, 5, 1]

3. Two-dimensional spacetimes with closed timelike curves

On spacetimes with CTCs, the initial value problem is more subtle. Simple examples show that some spacetimes with CTCs have a generalized Cauchy surface; but the Cauchy problem is not well defined in generic two-dimensional spacetimes, and other examples in this section illustrate several essentially different ways in which CTCs can block the existence of smooth solutions or allow more than one solution for the same initial data on a spacelike surface.

We again begin with two-dimensional spacetimes, built from Minkowski space. Obstacles are most severe here, and we will see that some can be overcome in higher dimensions. First a familiar case in which one's naive expectation of the way CTCs prevent solutions is fullfilled. Identify the edges of the strip of Minkowski space between two parallel, straight timelike lines, $t = 0$ and $t = T$:

$$(t = 0, x) \equiv (t = T, x). \tag{4}$$

Here the only candidates for an initial value surface are spacelike lines Σ extending to spatial infinity. Minkowski space $\overline{M}$ covers this spacetime, and data on any one of the spacelike lines of $\overline{M}$ that covers Σ can be uniquely evolved in the covering space, but the resulting solution corresponds to a solution on the original spacetime only if it is suitably periodic. It must have the same value on each covering line $\overline{\Sigma}$; and almost no data yields such a solution. This is essentially the grandfather paradox: Locally one can construct a unique solution on M; but when extended, the locally evolved solution returns to Σ with a value that is inconsistent with its initial data.

Flat cylinders with infinite extent in a timelike direction are obtained by identifying the left and right edges of the strip of Minkowski space between two parallel straight spacelike lines, $x = 0$ and $x = d$, after a time translation by τ:

$$(t, x = 0) \equiv (t + \tau, x = d). \tag{5}$$

For $\tau > d$, lines joining identified points are CTC's. These spacetimes are everywhere dischronal: A CTC passes through every point, and there is no candidate for an initial value surface – no complete spacelike surface transverse to all timelike curves. On spacetimes with no generalized Cauchy surface, one can ask a related question, more closely tied to our knowledge of the universe's causal structure: whether the spacetime is *benign* [6]. A spacetime is benign if, at each point x, there is a finite spacelike surface S containing x for which arbitrary smooth data on S can be extended to a solution on the spacetime. For the massive wave equation, the cylinder spacetimes seem not even to be benign: A single massive particle leaving any spacelike surface S can be aimed to return to the surface at a point different from the one from which it left.

Problem: Prove (or disprove) the conjecture that the flat cylinders, given by the identifications (5), are not benign for the massive scalar wave equation.

One spacetime M, g that avoids the problems so far encountered – the grandfather paradox and the lack of a candidate Cauchy surface – is akin to spaces discussed by Geroch and Horowitz [10] and by Politzer[8]. Heuristically, as illustrated in Fig. 1, one removes from Minkowski space two parallel, timelike slits that are related by translation along a different timelike direction. The inner edges of the two slits are then glued; and the outer edges of the two slits are similarly glued. The formal construction and details of the initial value problem outlined below are given in Friedman and Morris [24]. Analysis of the initial value problem for a related spacetime with spacelike slits is given by Goldwirth et al.[11]

Because corresponding points on the left and right slits are related by a timelike translation, closed timelike curves extend from the left to the right slit, e.g., from the point labelled Q on the left to the identified point labelled $\mathcal{T}(Q)$ on the right. The *dischronal* region A of a spacetime is the set of points through which there are closed timelike or null curves. Here it is a bounded region within the intersection of the past light cone of the top slit endpoint and the future light cone of the bottom slit endpoint. A hypersurface Σ that lies in the past of A and is a Cauchy surface of Minkowski space is an obvious candidate for a generalized Cauchy surface of M, g. In fact, it is easy to see that initial data in $L_2(\Sigma)$ leads to a solution in $L_2(M)$. In the past of A, solutions to the massless wave equation can be written as the sum $f(t - x) + g(t + x)$ of a right-moving and left-moving solution. To obtain a solution in the spacetime M, one simply propagates left moving data that encounters the slit in the obvious way. For example, if a left-moving wave enters the left slit at Q, it emerges unaltered from the right slit at $\mathcal{T}(Q)$. The solution is unique. But it is discontinuous along future-directed null rays that extend from the endpoints of the slits, because the result of the wave propagation is to piece together solutions from disjoint parts of the initial data

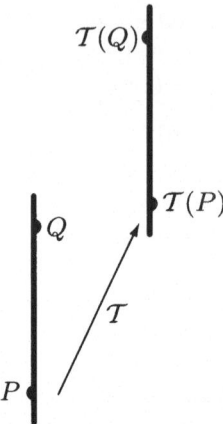

FIGURE 1. A simple spacetime with CTCs and a generalized Cauchy surface is shown in this figure. Two parallel segments of equal length are removed from Minkowski space, two disjoint edges are joined to the left and right sides of each slit, and edge points related by the timelike translation $\mathcal{T}$ are then identified.

surface. For example, the right-going solutions on adjacent sides of a right-directed null ray from any endpoint are Minkowski space solutions obtained from data that came from segments of Σ that are not adjacent.

The existence of solutions in L_2, however, is not a generic property of two-dimensional spacetimes with a Cauchy horizon. If, for example, the two timelike slits were not parallel, the resulting spacetime would have an unstable Cauchy horizon: If Σ is a Cauchy surface for the past of A, data on Σ leads to a solution that diverges on the boundary of the past of A. The paradigm for this generic case is *Misner space* [7, 3, 15].

Misner space can be constructed by identifying the edges of a strip of Minkowski space between two parallel *null* lines. As in the previous example, the CTCs of Misner space are confined to a spatially bounded region, and one can ask whether spacelike surfaces lying to the past or future of the dischronal region A are generalized Cauchy surfaces.

To construct the space, let $u = t - x$, $v = t + x$, and consider the null strip $u_0 < u < Bu_0$, where $u_0 > 0$, and a boost B of Minkowski space corresponding to velocity $V > 0$ is given by

$$u \to Bu, \qquad v \to B^{-1}v, \tag{6}$$

with

$$B = \sqrt{\frac{1 + V}{1 - V}}. \tag{7}$$

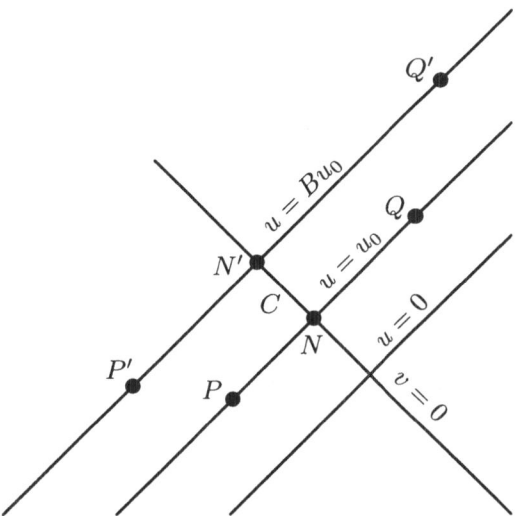

FIGURE 2.   Misner space is the region between the two null rays
$u = u_0$ and $u = Bu_0$, with points of the null boundaries identified
by the boost B. The curve $C = NN'$ is a chronology horizon, a
closed null geodesic that separates the dischronal region above it
from the globally hyperbolic spacetime to its past.

Points at the boundaries of the strip are identified after a boost:

$$(u_0, v) \equiv (Bu_0, B^{-1}v). \qquad (8)$$

Identified points are spacelike separated for $v < 0$ (e.g., P and P' in Fig. 2), null
separated at $v = 0$ (e.g., N and N'), and timelike separated for $v > 0$ (e.g., Q and
Q'). Closed timelike curves (e.g., the segment QQ') thus pass though each point
of the region $v > 0$. Misner space has a single closed null geodesic, $C = NN'$, and
the past $\mathcal{P}$ of C is globally hyperbolic. The future of C is dischronal, so C is a
chronology horizon, a Cauchy horizon that bounds the dischronal region. Initial
data for the scalar-wave equation can be posed on a Cauchy surface Σ of $\mathcal{P}$, but
solutions have divergent energy on the chronology horizon.

   This globally hyperbolic past part of Misner space can obtained from a 1-
dimensional room whose walls are moving toward each other – by identifying left
and right walls at the same proper time read by clocks on each wall (see, e.g.,
Thorne [1994]). This construction makes it obvious that light rays are boosted
each time they traverse the space, in the same way that a light ray is boosted
when reflected by a moving mirror.

   The reason solutions diverge is then clear in the geometrical optics limit. A
light ray γ, starting from Σ, loops about the space and is boosted each time it
loops. That is, trajectories of a (locally-defined) timelike Killing vector cross the
null geodesic at a sequence of points. The Killing vector can be used to compare the

affine parameter at successive crossing points by time-translating a segment of the geodesic to successively later segments. Compared in this way, the affine parameter of a given segment will will be less than that of the next segment by the blueshift factor $[(1 + V)/(1 - V)]^{1/2}$. Because γ loops an infinite number of times before reaching $\mathcal{C}$, its frequency and energy diverge as it approaches the horizon. The ray γ is an incomplete geodesic: It reaches the horizon in finite affine parameter length, because each boost decreases the affine parameter by the blueshift factor $[(1 + V)/(1 - V)]^{1/2}$.

This behavior is not unique to Misner space: A theorem due to Tipler [16] shows that geodesic incompleteness is generic in spacetimes like Misner space in which CTCs are "created" – spacetimes whose dischronal region lies to the future of a spacelike hypersurface. A similar argument by Hawking underpins the classical part of his Chronology Protection Conjecture. [17] (See also Chrusciel and Isenberg [18], who show that the generic, compactly generated horizon has generators whose structure is more complex than that considered by Hawking.)

When the horizon is not compactly generated, classical fields need not diverge, and a class of Gott spacetimes [9] serve as an example. Cutler [12] shows that a spacelike hypersurface Σ extends to spatial infinity and lies to the past of the dischronal region. Here CTCs run to spatial infinity. These characteristics hold for the particular Gott spacetime introduced here, but an additional key feature is that its covering space is three-dimensional Minkowski space (with images of the string singularities removed). Carinhas [13] has shown for the massless scalar wave equation that data on Σ satisfying suitable asymptotic conditions leads to solutions on a set of Gott spacetimes (see also Boulware [14]).

4. Existence and uniqueness theorems for some four-dimensional spacetimes with CTCs

In four spacetime dimensions, existence and uniqueness theorems have been proved for a class of stationary, asymptotically flat spacetimes.[24, 19] In these spacetimes, the dischronal region is bounded in space, but there is no Cauchy horizon, and CTCs are always present. Because the spacetimes are asymptotically flat, one can define future and past null infinity $\mathscr{I}^{\pm}$. [3] In Minkowski space $\mathscr{I}^{-}$ is a generalized Cauchy surface for massless wave equations, and the goal here is to show that $\mathscr{I}^{-}$ is a also generalized Cauchy surface for a class of spacetimes with CTCs.

We first review in some detail work by Friedman and Morris [24] on spacetimes with topology $\mathcal{N} = \mathcal{M} \times \mathbb{R}$, where $\mathcal{M}$ is a hyperplane with a handle (wormhole) attached: $\mathcal{M} = \mathbb{R}^3 \# (S^2 \times S^1)$. The metric $g_{\alpha\beta}$ on $\mathcal{N}$ is smooth (C^∞), and, for simplicity in treating the asymptotic behavior of the fields, we will assume that outside a compact region $\mathcal{R}$ the geometry is flat, with metric $\eta_{\alpha\beta}$.

One can construct the 3-manifold $\mathcal{M}$ from $\mathbb{R}^3$ by removing two balls and identifying their spherical boundaries, Σ_I and Σ_{II}, as shown in Fig. 3. The sphere obtained by the identification will be called the "throat" of the handle. (Its location

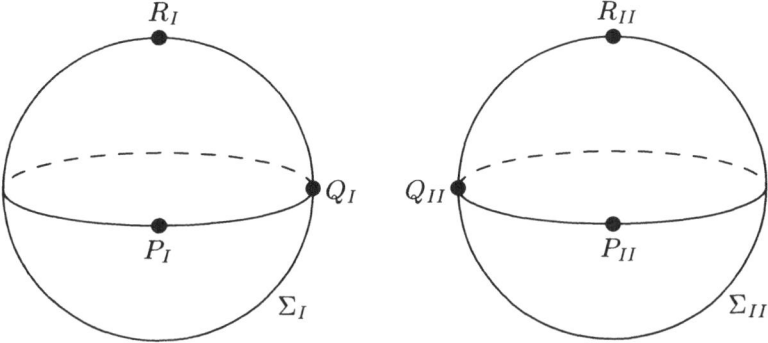

FIGURE 3.   An orientable 3-manifold M is constructed by identifying points of Σ_I and points of Σ_{II} that are labelled by the same letter, with subscripts I and II.

is arbitrary: After removing any sphere, Σ, from the handle of $\mathcal{M}$ one is left with a manifold homeomorphic to $\mathbb{R}^3\backslash(B^3\#B^3)$, whose boundary is the disjoint union of two spheres.) One can similarly construct the spacetime $\mathcal{N}$ from $\mathbb{R}^4$ by removing two solid cylinders and identifying their boundaries C_I and C_{II}. We will denote by $\mathcal{T}$ the map from C_I to C_{II} that relates identified points. For the spacetimes we will consider, the identified points will be timelike separated.

A static metric on $\mathcal{N}$ is given by

$$g_{\alpha\beta} = -e^{-2\nu}t_\alpha t_\beta + h_{\alpha\beta}, \tag{9}$$

where $h_{\alpha\beta}t^\beta = 0$.

If the Minkowski coordinate t is extended to $\mathcal{N}\backslash C$ by making $\mathcal{M}_t$ a $t = $ constant surface, then $t^\alpha\nabla_\alpha t = 1$, $\nabla^\alpha t = -e^{-2\nu}t^\alpha$, and the metric (9) can be written on $\mathcal{N}\backslash C$ in the form

$$g_{\alpha\beta} = -e^{2\nu}\partial_\alpha t\partial_\beta t + h_{\alpha\beta}. \tag{10}$$

It will be convenient to single out a representative hypersurface,

$$\mathcal{M} := \mathcal{M}_0. \tag{11}$$

We will denote by h_{ab} the corresponding spatial metric on $\mathcal{M}$; that is, h_{ab} is the pull-back of $h_{\alpha\beta}$ (or $g_{\alpha\beta}$) to $\mathcal{M}$.

We consider the wave equation

$$\Box\,\Phi \equiv \nabla^\alpha\nabla_\alpha\Phi = 0, \tag{12}$$

for a massless scalar field Φ.

Initial data in Minkowski space on a future null cone is simply a specification of the field Φ on that cone. On $\mathscr{I}^-$, the field itself vanishes, but the field rescaled by a radial coordinate is finite; initial data on $\mathscr{I}^-$ can then be written in terms of the standard ingoing null coordinate $v = t + r$, radial coordinate r, and unit radial vector $\hat{\mathbf{r}}$ as

$$f(v, \hat{\mathbf{r}}) = \lim_{r \to \infty} r\Phi(v, r\hat{\mathbf{r}}). \tag{13}$$

The same definition of data on $\mathscr{I}^-$ can be used for the spacetimes $\mathcal{N}, g$, and the data again determines a solution to the wave equation in these spacetimes.

For readers unfamiliar with $\mathscr{I}$, the following description suffices for this review. Denote by $u = t - r$ the outgoing null coordinate. One can adjoin a future causal boundary $\mathscr{I}^+$ to Minkowski space by adding a future endpoint $u = \infty, v, \theta, \phi$ to each outgoing null geodesic v, θ, ϕ constant. Similarly, one can adjoin a past causal boundary $\mathscr{I}^+$, adding a past endpoint $v = \infty, u, \theta, \phi$ to each ingoing null geodesic, $u, \theta, \phi =$ constant. The simple, standard, formal construction of $\mathscr{I}$ (see, e.g., [3]) exploits the fact that spacetimes with conformally related metrics have the same set of null geodesics; one introduces a metric that is conformally related to Minkowski space and for which Minkowski space is a compact region bounded by future and past null cones. We will not, however, use the formal construction of $\mathscr{I}$. All that is needed is a definition of initial data at past null infinity $\mathscr{I}^-$ for a scalar field Φ as a limit (13).

Proposition 1. [24] *For almost all spacetimes $\mathcal{N}, g$ of the kind just described (for almost all parameters τ), the following existence theorem holds. Let f be initial data on $\mathscr{I}^-$ for which f and all its derivatives are in $L_2(\mathscr{I}^-)$. Then there exists a solution Φ to the scalar wave equation that is smooth and asymptotically regular at null and spatial infinity and that has f as initial data.*

Because the geometry is static, we can express solutions as a superposition of functions with harmonic time dependence. The fact that there is no foliation by spacelike slices leads to a lack of orthogonality of the eigenfunctions, and the spectral theorem cannot be used. Instead we explicitly prove convergence of a superposition of the form

$$\Phi(t, x) = \int d\omega \ \phi(\omega, x) \ e^{-i\omega t}. \tag{14}$$

Here x is naturally a point of the manifold of trajectories of t^α, but we can identify it with a point of a simply connected spacelike hypersurface $\mathcal{M}$, with spherical boundaries Σ_I and Σ_{II}. Let (t, x_I) and $(t + \tau, x_{II})$ be points of $\mathcal{N} \backslash C$ that are identified in $\mathcal{N}$. Continuity of Φ and its normal derivative at the identified points is expressed by

$$\Phi(x_{II}) = \Phi(x_I) \tag{15}$$

$$\hat{n}_{II} \cdot \nabla\Phi(x_{II}) = -\hat{n}_I \cdot \nabla\Phi(x_I). \tag{16}$$

The harmonic components of Φ on $\mathcal{N}$ can be regarded as fields $\phi(\omega, x)$ on $\mathcal{M}$ satisfying elliptic equations of the form

$$(\omega^2 + \mathcal{L})\phi = 0, \tag{17}$$

where $\mathcal{L}$ can be defined by the action of $\nabla_\alpha \nabla^\alpha$ on time independent fields f on $\mathcal{N}$:

$$\mathcal{L} = e^\nu D^a e^\nu D_a, \tag{18}$$

and D_a is the covariant derivative of the 3-metric h_{ab} on $\mathcal{M}$. The major difficulty lies in the fact that, because the boundary conditions (15,16) involve a time-translation by Killing parameter τ, the corresponding boundary conditions on the harmonic components ϕ depend on the frequency ω, via a phase $\eta = \omega \tau$:

$$\phi(\omega, x_{II}) = e^{i\eta} \phi(\omega, x_I), \tag{19}$$

$$\hat{n}_{II} \cdot \nabla \phi(\omega, x_{II}) = -e^{i\eta} \hat{n}_I \cdot \nabla \phi(\omega, x_I). \tag{20}$$

As a result, eigenfunctions associated with different frequencies are eigenfunctions of different operators; they are not orthogonal, and their completeness is not guaranteed by the spectral theorem.

Instead, the following steps outline the construction of a solution. (The proof involves a standard relation between the norm of a function in a Sobolev space H_s and the norm of its fourier transform in a weighted L_2 space $L_{2,s}$. See the following footnote and references mentioned there.)[1]

1. For a fixed value η of the phase, *the operator $\mathcal{L}_\eta$ with boundary conditions (19, 20) is self-adjoint on the space $L_2(\mathcal{M})$ with domain $\mathbb{H}_2$.*

Here the boundary conditions enforce the symmetry of the operator

$$\langle f \mid \mathcal{L}_\eta g \rangle = \langle \mathcal{L}_\eta f \mid g \rangle = \langle h \mid g \rangle, \tag{21}$$

by requiring that the current entering Σ_I coincide with the current leaving Σ_{II}:

$$\int_{\Sigma_{II}} dS_a e^{-\nu} (\bar{f} D^a g - \bar{g} D^a f) + \int_{\Sigma_I} dS_a e^{-\nu} (\bar{f} D^a g - \bar{g} D^a f). \tag{22}$$

2. Eigenfunctions exist whose incoming part coincides with the incoming part of a plane wave for each wave vector k. These are solutions $F(\eta, k, x)$ to Eq. (17) that, for $r > R$, have the form

$$F = (2\pi)^{-3/2} e^{ik \cdot x} + \text{outgoing waves.}, \tag{23}$$

Existence is proved, following Wilcox [26], by the limiting absorption method: One adds an imaginary part to the frequency $\omega = |k|$. Because $\mathcal{L}_\eta$ is self-adjoint, $\mathcal{L}_\eta + i\epsilon$ is invertible in L_2. One can rewrite the homogeneous equation $\mathcal{L}_{\eta+i\epsilon} F = 0$, for F with asymptotic behavior (23), as an inhomogeneous equation $\mathcal{L}_{\eta+i\epsilon} F_{\text{out}} = \rho$, with F_{out} purely outgoing for $r > R$. The sign of the imaginary part of the frequency enforces an outgoing solution, and F is then found from the limit, as the imaginary part goes to zero, of a family of functions F_{out} in L_2.

---

[1] The Sobolev spaces $H_s(\mathbb{R}^3)$ are Hilbert spaces of functions with norm $\|f\|_s = [\int d\hat{k} |\hat{f}(k)|^2 (1 + k^2)^{s/2}]^{1/2}$, where $\hat{f}$ is the Fourier transform of f. The spaces $L_{2,r}$ are similarly Hilbert spaces with norm $\|f\|_{2,r} = [\int d^3 x \, |f(x)|^2 (1 + x^2)^r]^{1/2}$. The extension of these definitions to functions on manifolds is straightforward.[24, 25]

3. In flat space, the solution for data f on $\mathscr{I}^-$ can be written in terms of the Fourier transform $\tilde{f}$ of f in the form

$$\Phi(t,x) = \mathrm{Re}\, 2 \int d^3k\, a(k) e^{i(k\cdot x - \omega t)},$$

where

$$\tilde{f}(\omega, \hat{r}) = i\omega a(-\omega\hat{r}), \quad \omega \geq 0.$$

Here, $e^{ik\cdot x}$ is replaced by $F(\eta = \omega\tau, k, x)$, and one shows convergence of the corresponding superposition,

$$\Phi(t,x) = \mathrm{Re}\, 2 \int dk\, F(\eta = \omega\tau, k, x) a(k). \tag{24}$$

Although one cannot directly use the spectral theorem, convergence of the integral does rely on a related unitarity relation for the eigenfunctions F of the Hermitian operator $\mathcal{L}_\eta$ for fixed boundary phase η. That is, regarded as a map from $L_2(\mathcal{M})$ to $L_2(\mathbb{R}^3)$, F is norm-preserving:

$$\|g\|_{L_2(\mathcal{M})} = \left\| \int dV_x F(\eta, k, x) g(x) \right\|_{L_2(\mathbb{R}^3)}. \tag{25}$$

This allows us to bound the norm of a truncated Fourier transform of F: Let χ be a smooth step function, satisfying

$$\chi(x) = \begin{cases} 0, & r > R + \epsilon \\ 1, & r < R \end{cases}.$$

Then

$$\hat{F}(\eta, k, y) := \int dV_x\, e^{-\nu}\, F(\eta, k, x) \chi(x) \tag{26}$$

has uniformly bounded norm in k-space,

$$\left\| \hat{F}(\eta, \cdot, y) \right\|_{L_2(\mathbb{R}^3)} \leq CR^{3/2}, \quad \forall \eta, y. \tag{27}$$

From this uniform bound, one can show

$$\int d\tau dk dy\, \frac{\omega^{2n} |\hat{F}(\omega\tau, k, y)|^2}{(1+\omega^2)^n (1+y^2)^{3/2+\epsilon}} < \infty.$$

This inequality, in turn implies

$$\omega^n \hat{F}(\omega\tau, k, y) \in L_2(I) \otimes L_{2,-n}(\mathbb{R}^3) \otimes L_{-3/2-\epsilon}(\mathbb{R}^3)$$
$$\implies \quad \nabla^n F(\omega\tau, k, x) \in L_2(I) \otimes L_{2,-n}(\mathbb{R}^3) \otimes H_{-3/2-\epsilon}(\mathcal{M}_D)$$
$$\implies \quad F(\omega\tau, k, x) \in L_2(I) \otimes L_{2,-n}(\mathbb{R}^3) \otimes H_{n-3/2-\epsilon}(\mathcal{M}_D). \tag{28}$$

Thus, for almost all τ,

$$F(\omega\tau, k, x) \in L_{2,-n}(\mathbb{R}^3) \otimes H_{n-3/2-\epsilon}(\mathcal{M}_D)$$
$$\implies \quad \int dk\, a(k) F(\omega\tau, k, x) \in H_{n-3/2-\epsilon}(\mathcal{M}_D). \tag{29}$$

Finally, $f \in H_n(\mathscr{I}^-)$, all n, implies $\tilde{f} \in L_{2,-n}$, all n, whence Φ given by Eq. (24) is smooth.

4. Asymptotic regularity of Φ follows from its explicit form for $r > R$ in terms of the value of Φ and $\nabla\Phi$ at $r = R$. That is, one can use the flat-space Green function to write Φ outside $r = R$.

More recently, Bachelot [19] has proved a similar existence theorem and a strong uniqueness theorem for another family of stationary, four-dimensional spacetimes that are flat outside a spatially compact region. These spacetimes have Euclidean topology and their dischronal regions have topology (solid torus) $\times\mathbb{R}$. The metric is axisymmetric, with one free function a that describes the tipping of the light cones in the direction of the rotational Killing vector ∂_ϕ.

$$g = -(dt - a\,d\phi)^2 + dr^2 + r^2 d\theta^2 + r^2 \sin^2\theta\, d\phi^2. \tag{30}$$

Circles about the axis of symmetry are CTCs when ∂_ϕ is spacelike – that is, when $r\sin\theta < a$. By choosing $a = 0$ outside a torus, one can restrict CTCs to the interior of a smaller torus. Again data at $\mathscr{I}^-$ for Φ yields a smooth, asymptotically regular solution Φ, and Bachelot shows that Φ is unique. This is a significantly stronger result than the weak uniqueness obtained for the wormhole spacetimes described above, and it suggests that a strong uniqueness theorem should hold for those spacetimes as well.

5. Conjectures for more general four-dimensional spacetimes

As noted in Section 3 four-dimensional spacetimes that have Cauchy horizons and satisfy the null energy condition are geodesically incomplete. In two dimensions, an incomplete null geodesic that approaches a closed null geodesic as it approaches the chronology horizon leads to instability of that horizon. In four dimensions, however, an incomplete null geodesic γ does not always imply that the chronology horizon is unstable. This is because there may be only a set of measure zero of such geodesics, so that the energy may remain finite on the chronology horizon. For the time-tunnel spacetimes considered in references [20, 21, 22], a congruence of null rays initially parallel to γ spreads as the rays approach the chronology horizon. When the spreading of the rays overcomes the successive boosts (when the fractional decrease in flux is greater than the fractional increase in squared frequency), the horizon is stable in the geometrical optics approximation, and we will call it *optically stable*. (A precise, but long-winded definition of optical stability is given in Ref. [24]; a similar definition, applicable in a more restricted context, is given by Hawking[17].) Because the instability of the chronology horizon (or of the spacetime to its future) appears to be the obstacle to existence of solutions for data on candidate generalized Cauchy surfaces, we are led to a conjecture that relates optical stability to the existence of solutions.

Existence Conjecture. Let $\mathcal{N}, g$ be a smooth, asymptotically flat spacetime for which past and future regions $\mathcal{P} = \mathcal{N}\backslash J^+(A)$ and $\mathcal{F} = \mathcal{N}\backslash J^-(A)$ of a compact 4-dimensional submanifold A are globally hyperbolic. If $\mathcal{N}, g$ is optically stable, solutions to massless wave equations (for scalar, Maxwell, and Weyl fields) exist on $\mathcal{N}, g$ for smooth data on a Cauchy surface for $\mathcal{P}$.

A conjecture relating uniqueness for massless fields to uniqueness in a geo-metric-optics sense is easier to formulate.

Uniqueness Conjecture. Again let $\mathcal{N}, g$ be a smooth, asymptotically flat spacetime for which past and future regions $\mathcal{P} = \mathcal{N} \backslash J^+(A)$ and $\mathcal{F} = \mathcal{N} \backslash J^-(A)$ of a compact 4-dimensional submanifold A are globally hyperbolic. Let $S_\pm$ be Cauchy surfaces for $\mathcal{N} \backslash J^\pm(A)$. If all but a set of measure zero of null geodesics intersect S_+ and S_-, then solutions to massless wave equations on $\mathcal{N}$ are unique for initial data on S_- (and for initial data on S_+).

If one omits the restriction on null geodesics, uniqueness fails: It is not difficult to construct spacetimes satisfying the remaining conditions of the conjecture for which solutions to the massless scalar wave equation have support on a compact region.[24] One example begins with a 4-torus with flat Lorentz metric chosen to make two of the generators null and the other two spacelike. The metric allows a nonzero plane-wave solution whose support is not the entire torus. On can smoothly glue the torus to an asymptotically flat Lorentzian spacetime without altering the metric on the support of the scalar field.

For no asymptotically flat spacetime in 4 dimensions, in which CTCs are confined to a compact region, am I aware of a rigorous demonstration that finite-energy solutions to the scalar wave equation do exist for arbitrary initial data, or that solutions are unique.

Still less is known about interacting fields.

The well-known billiard-ball examples of Echeverria *et al.* [22, 27] are the basis for our present intuitive understanding. These examples exhibit a multiplicity of solutions for the same initial data, suggesting that uniqueness in spacetimes with CTCs is likely to hold only for free or weakly interacting fields. Because solutions seem always to exist for the billiard ball examples in the spacetimes they considered, it may be that classical interacting fields have solutions on spacetimes for which solutions to the free field equations exist.

Fewster, Higuchi and Wells [28] looked at a model of an interacting field theory in which space is discrete, and time is identified to obtain a discrete version of 2-dimensional Minkowski space with two horizontal slits removed and opposite edges of the slits identified. The field ψ satisfies an equation of the form

$$\partial_t \psi = L\psi + \lambda \psi^\dagger \psi \psi,$$

where L is a linear operator, and λ is real.

Fewster et al. find that solutions exist for arbitrary data and arbitrary λ and that they are unique for small λ. For large λ, however, uniqueness is lost.

An obvious question is whether generalized Cauchy surfaces for free fields similarly serve as generalized Cauchy surfaces for weakly interacting fields; and whether, as the toy models suggest, uniqueness is fails above some critical value of the interaction parameter.

References

[1] J.L. Friedman and A. Higuchi, Phys. Rev D **52**, 5687 (1995).

[2] J.L. Friedman, "Lorentzian universes-from-nothing," Classical and Quantum Gravity, **15**, 2639 (1998).

[3] S.W. Hawking and G.F.R. Ellis, *The Large Scale Structure of Space-time*, 1973 (Cambridge: Cambridge University Press).

[4] J.L. Friedman, "Two-component spinor fields on a class of time-nonorientable spacetime," Class. Quantum Grav. **12**, 2231 (1995).

[5] A. Chamblin and G.W. Gibbons, "A judgement on sinors," Class. Quantum Grav. **12**, 2243 (1995).

[6] U. Yurtsever, Class. Quantum Grav., **11**, 999 (1994).

[7] C.W. Misner, in *Relativity Theory and Astrophysics I. Relativity and Cosmology*, ed. J. Ehlers, 1967 (Providence: Amer. Math. Soc.).

[8] Politzer, H.D., "Simple quantum systems in spacetimes with closed timelike curves." Phys. Rev. D **46**, 4470–4476 (1992).

[9] Gott, J.R. , "Closed timelike curves produced by pairs of moving cosmic strings: Exact solutions." Phys. Rev. Lett., **66**, 1126–1129 (1991).

[10] R. Geroch, and G. Horowitz: Global structure of spacetimes. In: S.W. Hawking, W. Israel (eds.), General Relativity. Cambridge, 1979 pp. 212–193.

[11] D.S. Goldwirth, M.J. Perry, T. Piran, and K.S. Thorne, Phys. Rev. D**49**, 3951 (1994).

[12] C. Cutler, "Global structure of Gott's two-string spacetime," Phys. Rev. D, **45**, 487 (1992).

[13] P. Carinhas, Doctoral Dissertation, University of Wisconsin-Milwaukee (1993); "Cauchy Problem for Gott Spacetime," gr-qc/9507020 (1995).

[14] D. Boulware, Phys. Rev. D**46**, 4421 (1992).

[15] K.S. Thorne, in Directions in General Relativity, v. 1, eds. B.L. Hu, M.P. Ryan, and C.V. Vishveshwara, 1994 (Cambridge: Cambridge Univ. Press).

[16] Tipler, F.J.: Singularities and Causality Violation. Ann. Phys. **108**, 1–36 (1977)

[17] Hawking, S.W.: Chronology protection conjecture, Phys. Rev. Lett. **46**, 603 (1992)

[18] P.T. Chrusciel, J. Isenberg, "On the dynamics of generators of Cauchy horizons," *Proceedings of the Kananaskis conference on chaos in general relativity*, ed. D. Hobill, A. Burd and A. Coley, pp. 113–125, (Plenum) 1994; preprint, gr-qc/9401015 (1994).

[19] A. Bachelot, "Global properties of the wave equation on non-globally hyperbolic manifolds," J. Math. Pures Appl. **81**, 35–65 (2002).

[20] Morris, M.S., Thorne, K.S. Wormholes in spacetime and their use for interstellar travel: a tool for teaching general relativity. Am. J. Phys. **56**, 395–412 (1988).

[21] Morris, M.S., Thorne, K.S., Yurtsever, U.: Wormholes, time machines, and the weak energy condition. Phys. Rev. Lett. **61**, 1446–1449 (1988).

[22] Friedman, J.L., Morris, M.S., Novikov, I.D., Echeverria, F., Klinkhammer, G., Thorne, K.S., Yurtsever, U.: Cauchy problem on spacetimes with closed timelike curves. Phys. Rev. D, **42**, 1915–1930 (1990).

[23] Friedman, J.L., Morris, M.S.: The Cauchy problem for the scalar wave equation is well defined on a class of spacetimes with closed timelike curves. Phys. Rev. Lett.**66**, 401–404 (1991).

[24] Friedman, J.L. , Morris, M.S.: Existence and uniqueness theorems for massless fields on a class of spacetimes with closed timelike curves. Comm. Math. Phys. **186**, 495 (1997).

[25] Palais, R.S.: Seminar on the Atiyah-Singer index theorem. Princeton: Princeton University Press, 1965, Chap. X.

[26] Wilcox, C. H.:Scattering Theory for the D'Alembert Equation in Exterior Domains. New York: Springer-Verlag, 1975.

[27] Echeverria, F. Klinkhammer, G. & Thorne, K.S.: Billiard balls in wormhole spacetimes with closed timelike curves: Classical theory. Phys. Rev. D, **44**, 1077–1099 (1991).

[28] C.J. Fewster, A. Higuchi, and C.G. Wells, "Classical and quantum initial value problems for models of chronology violation," Phys. Rev. D**54**, 3806–3854 (1996).

[29] R. Geroch, private conversation.

[30] A. Chamblin, G.W. Gibbons, and A.R. Steif, Kinks and Time Machines, DAMPT preprint (and electronic preprint gr-qc/9405001)(1994).

John L. Friedman
Department of Physics
University of Wisconsin-Milwaukee
P.O. Box 413
Milwaukee, WI 53201, USA
e-mail: friedman@thales.phys.uwm.edu

Cheeger-Gromov Theory and Applications to General Relativity

Michael T. Anderson

Abstract. This paper surveys aspects of the convergence and degeneration of Riemannian metrics on a given manifold M, and some recent applications of this theory to general relativity. The basic point of view of convergence/degeneration described here originates in the work of Gromov, cf. [31]–[33], with important prior work of Cheeger [16], leading to the joint work of [18].

This Cheeger-Gromov theory assumes L^∞ bounds on the full curvature tensor. For reasons discussed below, we focus mainly on the generalizations of this theory to spaces with L^∞, (or L^p) bounds on the Ricci curvature. Although versions of the results described hold in any dimension, for the most part we restrict the discussion to 3 and 4 dimensions, where stronger results hold and the applications to general relativity are most direct. The first three sections survey the theory in Riemannian geometry, while the last three sections discuss applications to general relativity.

1. Background: Examples and definitions

The space $\mathbb{M}$ of Riemannian metrics on a given manifold M is an infinite-dimensional cone, (in the vector space of symmetric bilinear forms on M), and so is highly non-compact. Arbitrary sequences of Riemannian metrics can degenerate in very complicated ways.

On the other hand, there are two rather trivial but nevertheless important sources of non-compactness.

• *Diffeomorphisms.* The group $\mathcal{D}$ of diffeomorphisms of M is non-compact and acts properly on $\mathbb{M}$ by pull-back. Hence, if g is any metric in $\mathbb{M}$ and ϕ_i is any divergent sequence of diffeomorphisms, then $g_i = \phi_i^* g$ is a divergent sequence in $\mathbb{M}$, (at least if the manifold M is compact for instance). However, all the metrics g_i are isometric, and so are indistinguishable metrically. In terms of a local coordinate

Partially supported by NSF Grant DMS 0072591.

representation, the metrics g_i locally are just different representatives of the fixed metric g.

Thus, for most problems, one considers only equivalence classes of metrics $[g]$ in the moduli space

$$\mathcal{M} = \mathbb{M}/\mathcal{D}.$$

(A notable exception is the Yamabe problem, which is not well defined on $\mathcal{M}$, since it is not invariant under $\mathcal{D}$.)

• *Scaling.* For a given metric g and parameter $\lambda > 0$, let $g_\lambda = \lambda^2 g$ so that all distances are rescaled by a factor of λ. If $\lambda \to \infty$, or $\lambda \to 0$, the metrics g_λ diverge. In the former case, the manifold (M, g_λ), say compact, becomes arbitrarily large, in that global invariants such as diameter, volume, etc. diverge to infinity; there is obviously no limit metric. In the latter case, (M, g_λ) converges, as a family of metric spaces, to a single point. Again, there is no limiting Riemannian metric on M.

Although one has divergence in both cases described above, they can be combined in natural ways to obtain convergence. Thus, for g_λ as above, suppose $\lambda \to \infty$, and choose any fixed point $p \in M$. For any fixed $k > 0$, consider the geodesic ball $B_p = B_p(k/\lambda)$, so the g-radius of this ball is $k/\lambda \to 0$, as $\lambda \to \infty$. On the other hand, in the metric g_λ, the ball B_p is a geodesic ball of fixed radius k. Since k/λ is small, one may choose a local coordinate system $\mathcal{U} = \{u_i\}$ for B_p, with p mapped to the origin in $\mathbb{R}^n$. Let $u_i^\lambda = \lambda u_i = \phi_\lambda \circ u_i$, where $\phi_\lambda(x) = \lambda x$. Thus ϕ_λ is a divergent sequence of diffeomorphisms of $\mathbb{R}^n$, and $\mathcal{U}_\lambda = \{u_i^\lambda\}$ is a new collection of charts. One then easily sees that

$$g_\lambda(\partial/\partial u_i^\lambda, \partial/\partial u_j^\lambda) = g(\partial/\partial u_i, \partial/\partial u_j) = g_{ij}. \qquad (1.1)$$

As $\lambda \to \infty$, the ball B_p shrinks to the point p and the coefficients g_{ij} tend to the constants $g_{ij}(p)$. On the other hand, the metrics g_λ are defined on the intrinsic geodesic ball of radius k. Since k is arbitrary, the metrics $\phi_\lambda^* g_\lambda$ converge smoothly to the limit flat metric g_0 on the tangent space $T_p(M)$, induced by the inner product g_p on $T_p(M)$,

$$(M, \phi_\lambda^* g_\lambda) \to (T_p M, g_0). \qquad (1.2)$$

This process is called "blowing up", since one restricts attention to smaller and smaller balls, and blows them up to a definite size. Note that the part of M at any definite g-distance to p escapes to infinity, and is not detected in the limit g_0. Thus, it is important to attach base points to the blow-up construction; different base points may give rise to different limits, (although in this situation all pointed limits are isometric).

There is an analogous, although more subtle blowing up process for Lorentzian metrics due to Penrose, where the limits are non-flat plane gravitational waves, cf. [42].

If (M, g) is complete and non-compact, one can carry out a similar procedure with $\lambda \to 0$, called "blowing down", where geodesic balls, (about a given point), of large radius $B_p(k/\lambda)$ are rescaled down to unit size, i.e., size k. This is of importance in understanding the large scale or asymptotic behavior of the metric and will arise in later sections.

This discussion leads to the following definition for convergence of metrics. Let Ω be a domain in $\mathbb{R}^n$ and let $C^{k,\alpha}$ denote the usual Hölder space of C^k functions on Ω with α-Hölder continuous k^{th} partial derivatives. Similarly, let $L^{k,p}$ denote the Sobolev space of functions with k weak derivatives in L^p. Since one works only locally, we are only interested in the local spaces $C^{k,\alpha}_{loc}$ and $L^{k,p}_{loc}$ and corresponding local norms and topology.

Definition 1.1. A sequence of metrics g_i on n-manifolds M_i is said to **converge in the $L^{k,p}$ topology** to a limit metric g on the n-manifold M if there is a locally finite collection of charts $\{\phi_k\}$ covering M, and a sequence of diffeomorphisms $F_i : M \to M_i$, such that

$$(F_i^* g_i)_{\alpha\beta} \to g_{\alpha\beta}, \tag{1.3}$$

in the $L^{k,p}_{loc}$ topology. Here $(F_i^* g_i)_{\alpha\beta}$ and $g_{\alpha\beta}$ are the local component functions of the metrics $F_i^* g_i$ and g in the charts ϕ_k.

The same definition holds for convergence in the $C^{k,\alpha}$ topology, as well as the weak $L^{k,p}$ topology. (Recall that a sequence of functions $f_i \in L^p(\Omega)$ converges weakly in L^p to a limit $f \in L^p(\Omega)$ iff $\int f_i g \to \int fg$, for all $g \in L^q(\Omega)$, where $p^{-1} + q^{-1} = 1$.)

It is easily seen that this definition of convergence is independent of the choice of charts $\{\phi_k\}$ covering M. The manifolds M and M_i are not required to be compact. When M is non-compact, the convergence above is then uniform on compact subsets.

In order to obtain local control on a metric, or sequence of metrics, one assumes curvature bounds. The theory described by Cheeger-Gromov requires a bound on the full Riemann curvature tensor

$$|\operatorname{Riem}| \leq K, \tag{1.4}$$

for some $K < \infty$. Since the number of components of the Riemann curvature is much larger than that of the metric tensor itself, (in dimensions ≥ 4), this corresponds to an overdetermined set of constraints on the metric and so is overly restrictive. It is much more natural to impose bounds on the Ricci curvature

$$|\operatorname{Ric}| \leq k, \tag{1.5}$$

since the Ricci curvature is a symmetric bilinear form, just as the metric is. Of course, assuming bounds on Ricci is natural in general relativity, via the Einstein equations. Thus throughout the paper, we emphasize (1.5) over (1.4) whenever possible.

The Cheeger-Gromov theory may be viewed as a vast generalization of the basic features of Teichmüller theory to higher dimensions and variable curvature, (although it was not originally phrased in this way). Recall that Teichmüller theory describes the moduli space $\mathcal{M}_c$ of constant curvature metrics on surfaces, cf. [46] and references therein. On closed surfaces, one has a *basic trichotomy* for the behavior of sequences of such metrics, normalized to unit area:

• *Compactness/Convergence.* A sequence $g_i \in \mathcal{M}_c$ has a subsequence converging smoothly, (C^∞), to a limit metric $g \in \mathcal{M}_c$. As in the definition above, the convergence is understood to be modulo diffeomorphisms. For instance this is always the case on S^2, since the moduli space $\mathcal{M}_c$ is a single point for S^2.

• *Collapse.* The sequence $g_i \in \mathcal{M}_c$ collapses everywhere, in that

$$\mathrm{inj}_{g_i}(x) \to 0, \tag{1.6}$$

at every x, where inj_{g_i} is the injectivity radius w.r.t. g_i. This collapse occurs only on the torus T^2 and such metrics become very long and very thin, (if the area is normalized to 1). There is no limit metric on T^2. Instead, by choosing (arbitrary) base points x_i, one may consider based sequences (T^2, g_i, x_i), whose limits are then the "collapsed" space $(\mathbb{R}, g_\infty, x_\infty)$. Here $\mathbb{R}$ is the real line, and g_∞ is any Riemannian metric on $\mathbb{R}$; recall that all metrics on $\mathbb{R}$ are isometric. The convergence here is that of metric spaces, i.e., in the Gromov-Hausdorff topology, cf. [31], [43].

• *Cusp Formation.* This is a mixture of the two previous cases, and occurs only for hyperbolic metrics, i.e., on surfaces Σ_g of genus $g \geq 2$. In this case, there are based sequences (Σ_g, g_i, x_i) which converge to a limit $(\Sigma, g_\infty, x_\infty)$ which is a complete non-compact hyperbolic surface of finite volume, hence with a finite number of cusp ends $S^1 \times \mathbb{R}^+$. The convergence is smooth, and uniform on compact subsets. As one goes to infinity in any such cusp end $S^1 \times \mathbb{R}^+$, the limit metric collapses in the sense that $\mathrm{inj}_{g_\infty}(z_k) \to 0$, as $z_k \to \infty$. There are other based sequences (Σ, g_i, y_i) which collapse, i.e., (1.6) holds on domains of arbitrarily large but bounded diameter about y_i. As before, limits of such sequences are of the form $(\mathbb{R}, g_\infty, y_\infty)$.

2. Convergence/compactness

To prove the (pre)-compactness of a family of metrics, or the convergence of a sequence of metrics, the main point is to establish a lower bound on the radius of balls on which one has a priori control of the metric in a given topology, say $C^{k,\alpha}$ or $L^{k,p}$. Given such uniform local control, it is then usually straightforward to obtain global control, via suitable global assumptions on the volume or diameter. (Alternately, one may work instead on domains of bounded diameter.)

To obtain such local control, the first issue is to choose a good "gauge", i.e., representation of the metric in local coordinates. For this, it is natural to look at coordinates built from the geometry of the metric itself. In the early stages of development of the theory, geodesic normal coordinates were used. Later, Gromov [31] used suitable distance coordinates. However, both these coordinate systems entail loss of derivatives – two in the former case, one in the latter. It is now well known that Riemannian metrics have optimal regularity properties in harmonic coordinates, cf. [23]; this is due to the special form of the Ricci curvature in harmonic coordinates, known to relativists long ago.

Given the choice of harmonic gauge, it is natural to associate a harmonic radius $r_h : M \to \mathbb{R}^+$, which measures the size of balls on which one has harmonic coordinates in which the metric is well controlled. The precise definition, cf. [1], is as follows.

Definition 2.1. Fix a function topology, say $L^{k,p}$, and a constant $c_o > 1$. Given $x \in (M, g)$, define the $L^{k,p}$ harmonic radius to be the largest radius $r_h(x) = r_h^{k,p}(x)$ such that on the geodesic ball $B_x(r_h(x))$ one has a harmonic coordinate chart $U = \{u_\alpha\}$ in which the metric $g = g_{\alpha\beta}$ is controlled in $L^{k,p}$-norm: thus,

$$c_o^{-1}\delta_{\alpha\beta} \le g_{\alpha\beta} \le c_o\delta_{\alpha\beta}, \quad \text{(as bilinear forms)}, \tag{2.1}$$

$$[r_h(x)]^{kp-n} \int_{B_x(r_h(x))} |\partial^k g_{\alpha\beta}|^p dV \le c_o - 1. \tag{2.2}$$

Here, it always assumed that $kp > n = \dim M$, so that $L^{k,p}$ embeds in C^0, via Sobolev embedding. The precise value of c_o is usually unimportant, but is understood to be fixed once and for all. Both estimates in (2.1)–(2.2) are scale invariant, (when the harmonic coordinates are rescaled as in (1.1)), and hence the harmonic radius scales as a distance.

Note that if $r_h(x)$ is large, then the metric is close to the flat metric on large balls about x, while if $r_h(x)$ is small, then the derivatives of $g_{\alpha\beta}$ up to order k are large in L^p on small balls about x. Thus, the harmonic radius serves as a measure of the degree of concentration of $g_{\alpha\beta}$ in the $L^{k,p}$-norm.

It is important to observe that the harmonic radius is continuous with respect to the (strong) $L^{k,p}$ topology on the space of metrics, cf. [1], [3]. In general, it is not continuous in the weak $L^{k,p}$ topology.

One may define such harmonic radii w.r.t. other topologies, for instance $C^{k,\alpha}$ in a completely analogous way; these have the same properties.

Suppose g_k is a sequence of metrics on a manifold M, (possibly open), with a uniform lower bound on r_h. On each ball, one then has $L^{k,p}$ control of the metric components. The well-known Banach-Alaoglu theorem, (bounded sequences are weakly compact in reflexive Banach spaces), then implies that the metrics on the ball have a weakly convergent subsequence in $L^{k,p}$, so one obtains a limit metric on each ball. Using elliptic regularity associated with harmonic functions, it is straightforward to verify that the overlaps of these charts are in $L^{k+1,p}$, and so one has a limit $L^{k,p}$ metric on M. The convergence to limit is in the weak $L^{k,p}$ topology and uniform on compact subsets. Strictly speaking, one also has to prove that the harmonic coordinate charts for g_k also converge, or more precisely may be replaced by a fixed coordinate chart, but this also is not difficult, cf. [1], [3] for details.

The same type of arguments hold w.r.t. the $C^{k,\alpha}$ topology, via the Arzela-Ascoli theorem; here weak $L^{k,p}$ convergence is replaced by convergence in the $C^{k,\alpha'}$ topology, for $\alpha' < \alpha$.

Thus, the main issue in obtaining a convergence result is to obtain a lower bound on a suitable harmonic radius r_h under geometric bounds. The following result from [1] is one typical example.

Theorem 2.2. (Convergence I). *Let M be a closed n-manifold and let $\mathcal{M}(\lambda, i_o, D)$ be the space of Riemannian metrics such that*

$$|\operatorname{Ric}| \leq k, \ \operatorname{inj} \geq i_o, \ \operatorname{diam} \leq D. \tag{2.3}$$

Then $\mathcal{M}(\lambda, i_o, D)$ is precompact in the $C^{1,\alpha}$ and weak $L^{2,p}$ topologies, for any $\alpha < 1$ and $p < \infty$.

Thus, for any sequence, there is a subsequence which converges, in these topologies, to a limit $C^{1,\alpha} \cap L^{2,p}$ metric g_∞ on M.

Sketch of Proof: As discussed above, it suffices to prove a uniform lower bound on the $L^{2,p}$ harmonic radius $r_h = r_h^{2,p}$, i.e.,

$$r_h(x) \geq r_o = r_o(k, i_o, D), \tag{2.4}$$

under the bounds (2.3).

Overall, the proof of (2.4) is by contradiction. Thus, if (2.4) is false, there is a sequence of metrics g_i on M, satisfying the bounds (2.3), but for which $r_h(x_i) \to 0$, for some points $x_i \in M$. Without loss of generality, (since M is closed), assume that the base points x_i realize the minimal value of r_h on (M, g_i). Then rescale the metrics g_i by this minimal harmonic radius, i.e., set

$$\bar{g}_i = r_h(x_i)^{-2} \cdot g_i. \tag{2.5}$$

If $\bar{r}_h$ denotes the harmonic radius w.r.t. $\bar{g}$, by scaling properties one has

$$\bar{r}_h(x_i) = 1, \ \text{and} \ \bar{r}_h(y_i) \geq 1, \tag{2.6}$$

for all $y_i \in (M, \bar{g}_i)$. By the remarks preceding the proof, the pointed Riemannian manifolds $(M, \bar{g}_i, x_i)$ have a subsequence converging in the *weak $L^{2,p}$* topology to a limit $L^{2,p}$ Riemannian manifold $(N, \bar{g}_\infty, x_\infty)$. (Again, this convergence is understood to be modulo diffeomorphisms, as in Definition 1.1.) Of course $\operatorname{diam}_{\bar{g}_i} M \to \infty$, so that the complete open manifold N is distinct from the original compact manifold M. The convergence is uniform on compact subsets.

So far, nothing essential has been done – the construction above more or less amounts to just renormalizations. There are two basic ingredients in obtaining further control however, one geometric and one analytic.

We begin with the geometric argument. The limit space $(N, \bar{g}_\infty)$ is Ricci-flat, since the bound (2.3) on the Ricci curvature of g_i becomes in the scale $\bar{g}_i$,

$$|\operatorname{Ric}_{\bar{g}_i}| \leq k \cdot r_h(x_i) \to 0, \ \text{as} \ i \to \infty. \tag{2.7}$$

Actually, it is Ricci-flat in a weak sense, since the convergence is only in weak $L^{2,p}$. However, it is easy to see, (cf. also below), that weak $L^{2,p}$ solutions of the (Riemannian) Einstein equations are real-analytic, and so the limit is in fact a smooth Ricci-flat metric.

Next, by (2.3), the injectivity radius of $\bar{g}_i$ satisfies

$$\text{inj}_{\bar{g}_i} \geq i_o \cdot r_h(x_i)^{-1} \to \infty, \quad \text{as } i \to \infty, \tag{2.8}$$

so that, roughly speaking, the limit $(N, \bar{g}_\infty)$ has infinite injectivity radius at every point. More importantly, the bound (2.8) implies that $(M, \bar{g}_i)$ contains arbitrarily long, (depending on i), minimizing geodesics in any given direction through the center point x_i. It follows that the limit $(N, \bar{g}_\infty)$ has infinitely long minimizing geodesics in every direction through the base point x_∞. This means that $(N, \bar{g}_\infty)$ contains a line in every direction through x_∞.

Now the well-known Cheeger-Gromoll splitting theorem [17] states that a complete manifold with non-negative Ricci curvature splits isometrically along any line. It follows that $(N, \bar{g}_\infty)$ splits isometrically in every direction through x_∞, and hence $(N, \bar{g}_\infty) = (\mathbb{R}^n, g_0)$, where g_0 is the flat metric on $\mathbb{R}^n$.

Now of course $(\mathbb{R}^n, g_0)$ has infinite harmonic radius. If the convergence of $(N, \bar{g}_i)$ to the limit $(\mathbb{R}^n, g_0)$ can be shown to be in the **strong** $L^{2,p}$ topology, then the continuity of r_h in this topology immediately gives a contradiction, since by (2.6), the limit $(N, \bar{g}_\infty)$ has $r_h(x_\infty) = 1$.

The second or analytic part of the argument is to prove strong $L^{2,p}$ convergence to the limit. The idea here is to use elliptic regularity to bootstrap or improve the smoothness of the convergence.

In harmonic coordinates, the Ricci curvature of a metric g has the following especially simple form:

$$-\frac{1}{2}\Delta g_{\alpha\beta} + Q_{\alpha\beta}(g, \partial g) = \text{Ric}_{\alpha\beta}, \tag{2.9}$$

where $\Delta = g^{\alpha\beta}\partial_\alpha\partial_\beta$ is the Laplacian w.r.t. the metric g and Q is quadratic in g, its inverse, and ∂g. In particular, if $r_h(x) = 1$ and $r_h(y) \geq r_o > 0$, for all $y \in \partial B_x(1)$, then one has a uniform $L^{1,p}$ bound on Q and uniform $L^{2,p}$ bounds on the coefficients for the Laplacian within $B_x(1 + \frac{1}{2}r_o)$.

If now Ric is uniformly bounded in L^∞, then standard elliptic regularity applied to (2.9) implies that $g_{\alpha\beta}$ is uniformly controlled in $L^{2,q}$, for any $q < \infty$, (in particular for $q > p$). More importantly, if g_i is a sequence of metrics for which $(\text{Ric}_{g_i})_{\alpha\beta}$ converges strongly in L^p to a limit $(\text{Ric}_{g_\infty})_{\alpha\beta}$, then elliptic regularity again implies that the metrics $(g_i)_{\alpha\beta}$ converge strongly in $L^{2,p}$ to the limit $(g_\infty)_{\alpha\beta}$. For the metrics $\bar{g}_i$, (2.7) implies that Ric $\to 0$ in L^∞, and so Ric $\to 0$ strongly in L^q, for any $q < \infty$.

These remarks essentially prove that the $L^{2,p}$ harmonic radius is continuous w.r.t. the strong $L^{2,p}$ topology. Further, when applied to the sequence $\bar{g}_i$ and using (2.6), they imply that the metrics $\bar{g}_i$ converge strongly in $L^{2,p}$ to the limit $\bar{g}_\infty$. This completes the proof.

It is easy to see from the proof that the lower bound on the injectivity radius in (2.3) can be considerably weakened. For instance, define the 1-cross $\mathrm{Cro}_1(x)$ of (M, g) at x to be the length of the longest minimizing geodesic in (M, g) with center point x and set

$$\mathrm{Cro}_1(M, g) = \inf_x \mathrm{Cro}_1(x).$$

We introduce this notion partly because it has a natural analogue in Lorentzian geometry, when a minimizing geodesic is replaced by a maximizing time-like geodesic, cf. §5. Then one has the following result on 4-manifolds, cf. [4].

Theorem 2.3. (Convergence II). *Let M be a 4-manifold. Then the conclusions of Theorem 2.2 hold under the bounds*

$$|\mathrm{Ric}| \leq k, \mathrm{Cro}_1 \geq c_o, \mathrm{vol} \geq v_o, \mathrm{diam} \leq D. \tag{2.10}$$

The proof is the same as that of Theorem 2.2. The lower bound on Cro_1 implies that on the blow-up limit $(N, \bar{g}_\infty, x_\infty)$ above, one has a line. Hence, the splitting theorem implies that $N = N' \times \mathbb{R}$. It follows that N' is Ricci-flat and hence, since $\dim N' = 3$, N' is flat. Using the volume bound in (2.10), it follows that $(N, \bar{g}_\infty) = (\mathbb{R}^4, g_0)$, cf. (2.12)–(2.13) below. (The volume bound rules out the possibility that N' is a non-trivial flat manifold of the form $\mathbb{R}^3/\Gamma$.) This gives the same contradiction as before.

Of course, in dimension 3 any Ricci-flat manifold is necessarily flat, and so the same proof shows that one has $C^{1,\alpha}$ and $L^{2,p}$ precompactness within the class of metrics on 3-manifolds satisfying

$$|\mathrm{Ric}| \leq k, \mathrm{vol} \geq v_o, \mathrm{diam} \leq D. \tag{2.11}$$

Thus, no assumptions on inj or Cro_1 are needed in dimension 3.

Remark 2.4.

(i) Although (2.4) gives the existence of a lower bound on r_h in terms of the bounds k, i_o and D, currently there is no proof of an effective or computable bound. Equivalently, there is no direct proof of Theorem 2.2, which does not involve a passage to limits and invoking a contradiction. This is closely related to the fact there is currently no *quantitative* or *finite* version of the Cheeger-Gromoll splitting theorem, where one can deduce definite bounds on the metric in the presence of (a collection of) minimizing geodesics of a finite but definite length.

If however the bound on $|\mathrm{Ric}|$ in (2.3) is strengthened to a bound on $|\mathrm{Riem}|$, as in (1.4), then it is not difficult to obtain an effective or computable lower bound on r_h, cf. [36].

(ii) The proof above can be easily adapted to give a similar result if the L^∞ bound on Ric is replaced by an L^q bound, for some $q > n/2$; one then obtains convergence in weak $L^{2,q}$.

In the opposite direction, the convergence can be improved if one has bounds on the derivatives of the Ricci curvature. This will be the case if Ric satisfies an elliptic system of PDE, for instance the Einstein equations. In this case, one obtains C^∞ convergence to the limit.

(iii) The assumption that M is closed in Theorem 2.2 is merely for convenience, and an analogous result holds for open manifolds, away from the boundary.

The bounds on injectivity radius in (2.3), or even the 1-cross in (2.10), are rather strong and one would like to replace them with just a lower volume bound, as in (2.11).

An elementary but important result, the volume comparison theorem of Bishop-Gromov [31], [43], states that if Ric $\geq (n-1)k$, for some k, on (M,g), $n = \dim M$, then the ratio

$$\frac{\text{vol}\, B_x(r)}{\text{vol}\, B_k(r)} \tag{2.12}$$

is monotone non-increasing in r; here $\text{vol}\, B_k(r)$ is the volume of the geodesic r-ball in the n-dimensional space form of constant curvature k. In particular, if the bounds (2.11) hold, in dimension n, then (2.12) gives a lower bound on the volumes of balls on *all* scales:

$$\text{vol}\, B_x(r) \geq \frac{\text{vol}\, M}{\text{vol}\, B_k(D)} \cdot \text{vol}\, B_k(r). \tag{2.13}$$

Note that the estimate (2.13) also implies that, for any fixed $r > 0$, if $\text{vol}\, B_x(r) \geq v_0 > 0$, then $\text{vol}\, B_y(r) \geq v_1 > 0$, where v_1 depends only on v_0 and $\text{dist}_g(x,y)$. Thus, the ratio of the volumes of unit balls cannot become arbitrarily large or small on domains of bounded diameter.

Now a classical result of Cheeger [16] implies that if (2.11) is strengthened to

$$K_P \geq -K, \text{vol} \geq v_o, \text{diam} \leq D, \tag{2.14}$$

where K_P is the sectional curvature of any plane P in the tangent bundle TM, then one has a lower bound on the injectivity radius, $\text{inj}_g(M) \geq i_o(K, v_o, D)$. However, it was observed in [2] that this estimate fails under the bounds (2.11). It is worthwhile to exhibit a simple concrete example illustrating this.

Example 2.5. Let g_λ be the family of Eguchi-Hanson metrics on the tangent bundle TS^2 of S^2. The metrics g_λ are given explicitly by

$$g_\lambda = [1 - (\frac{\lambda}{r})^4]^{-1} dr^2 + r^2 [1 - (\frac{\lambda}{r})^4] \theta_1^2 + r^2 (\theta_2^2 + \theta_3^2). \tag{2.15}$$

Here $\theta_1, \theta_2, \theta_3$ are the standard left-invariant coframing of $SO(3) = \mathbb{RP}^3$, (the sphere bundles in TS^2) and $r \geq \lambda$. The locus $r = \lambda$ is the image of the 0-section and is a totally geodesic round $S^2(\lambda)$ of radius λ.

The metrics g_λ are Ricci-flat, and are all homothetic, i.e., are rescalings (via diffeomorphisms) of a fixed metric; in fact,

$$g_\lambda = \lambda^2 \cdot \psi_\lambda^*(g_1), \tag{2.16}$$

where $\psi_\lambda(r) = \lambda r$, and ψ_λ acts trivially on the $SO(3)$ factor. As $\lambda \to 0$, i.e., as one blows down the metrics, g_λ converges to the metric g_0, the flat metric on the cone

$C(\mathbb{RP}^3)$. The convergence is smooth in the region $r \geq r_o$, for any fixed $r_o > 0$, but is not smooth at $r = 0$. Since $S^2(\lambda)$ is totally geodesic, the injectivity radius at any point of $S^2(\lambda)$ is $2\pi\lambda$, which tends to 0. On the other hand, the volumes of unit balls, or balls of any definite radius, remain uniformly bounded below.

One sees here that the metrics (TS^2, g_λ) converge as $\lambda \to 0$ to a limit metric on a singular space $C(\mathbb{RP}^3)$. The limit is an orbifold $\mathbb{R}^4/\mathbb{Z}_2$, where $\mathbb{Z}_2$ acts by reflection in the origin.

The Eguchi-Hanson metric is the first and simplest example of a large class of Ricci-flat ALE (asymptotically locally Euclidean) spaces, whose metrics are asymptotic to cones $C(S^3/\Gamma)$, $\Gamma \subset SO(4)$, on spherical space forms. This is the family of ALE gravitational instantons, studied in detail by Gibbons and Hawking, cf. [30] and references therein, in connection with Hawking's Euclidean quantum gravity program.

It is straightforward to modify the construction in Example 2.5 to obtain orbifold degenerations on compact 4-manifolds satisfying the bounds (2.11). Thus, one does not have $C^{1,\alpha}$ or even C^0 (pre)-compactness of the space of metrics on M under the bounds (2.11). Singularities can form in passing to limits, although the singularities are of a relatively simple kind. The next result from [1] shows that this is the only kind of possible degeneration or singularity formation.

Theorem 2.6. (Convergence III). *Let $\{g_i\}$ be a sequence of metrics on a 4-manifold, satisfying the bounds*

$$|\operatorname{Ric}| \leq k, \operatorname{vol} \geq v_o, \operatorname{diam} \leq D. \qquad (2.17)$$

Then a subsequence converges, (in the Gromov-Hausdorff topology), to an orbifold (V, g), with a finite number of singular points $\{q_j\}$. Each singular point q has a neighborhood homeomorphic to a cone $C(S^3/\Gamma)$, for Γ a finite subgroup of $SO(4)$. The metric g is $C^{1,\alpha}$ or $L^{2,p}$ on the regular set

$$V_0 = V \setminus \cup\{q_j\},$$

and extends in a local uniformization of a singular point to a C^0 Riemannian metric. Further, there are embeddings

$$F_i : V_0 \to M$$

such that $F_i^*(g_i)$ converges in the $C^{1,\alpha}$ topology to the metric g.

Here, convergence in the Gromov-Hausorff topology means convergence as metric spaces, cf. [31], [43]. We mention only a few important issues in the proof of Theorem 2.6. First, the Chern-Gauss-Bonnet formula implies that for metrics with bounded Ricci curvature and volume on 4-manifolds, one has an a priori bound on the L^2-norm of the full curvature tensor:

$$\frac{1}{8\pi^2} \int_M |R|^2 dV \leq \chi(M) + C(k, V_o),$$

where $C(k, V_o)$ is a constant depending only on k from (2.17) and an upper bound V_o on $\operatorname{vol}_g M$: $\chi(M)$ is the Euler characteristic of M. Second, with each singular point $q \in V$, there is a associated a sequence of rescalings $\bar{g}_i = \lambda_i^2 g_i$, $\lambda_i \to \infty$, and base points $x_i \in M$, $x_i \to q$, such that a subsequence of $(M, \bar{g}_i, x_i)$ converges in $C^{1,\alpha} \cap L^{2,p}$ to a non-trivial Ricci-flat ALE space $(N, \bar{g}_\infty)$ as above. It is not difficult to see that any such ALE space has a definite amount of curvature in L^2. This implies basically that there are only a finite number of such singular points. Further, the ALE spaces N are embedded in M, in a topologically essential way.

3. Collapse/formation of cusps

In this section, we consider what happens when

$$\text{vol} \to 0 \quad \text{or} \quad \text{diam} \to \infty$$

in the bounds (2.11). This involves the notion of Cheeger-Gromov collapse, or collapse with bounded curvature.

For simplicity, we restrict the discussion to dimension 3. While there is a corresponding theory in higher dimensions, cf. [18], there are special and advantageous features that hold only in dimension 3 in general. Further, the relations with general relativity are most direct in dimension 3, in that the discussion can be applied to the behavior of space-like hypersurfaces in a given space-time.

The simplest non-trivial example of collapse is the Berger collapse of the 3-sphere along S^1 fibers of the Hopf fibration. Thus, consider the family of metrics on S^3 given by

$$g_\lambda = \lambda^2 \theta_1^2 + (\theta_2^2 + \theta_3^2), \tag{3.1}$$

where $\theta_1, \theta_2, \theta_3$ are the standard left-invariant coframing of S^3. The metrics g_λ have an isometric S^1 action, with Killing field K dual to θ_1, with length of the S^1 orbits given by $2\pi\lambda$. Thus, in letting $\lambda \to 0$, one is blowing down the metric in *one* direction. (This is exactly what occurs on approach to the horizon of the Taub-NUT metric, cf. [35].) A simple calculation shows that the curvature of g_λ remains uniformly bounded as $\lambda \to 0$. Clearly $\text{vol}_{g_\lambda} S^3 \sim \lambda \to 0$. The metrics g_λ collapse S^3 to a limit space, in this case S^2.

This same procedure may be carried out, with the same results, on any 3-manifold (or n-manifold) which has a free or locally free isometric S^1 action; locally free means that the isotropy group of any orbit is a finite subgroup of S^1, i.e., there are no fixed points of the action. Similarly, one may collapse along the orbits, as in (3.1), of a locally free T^k-action, where T^k is the k-torus. Remarkably, Gromov [32] showed that more generally one may collapse along the orbits of an isometric nilpotent group action, and furthermore, such groups are *only* groups which allow such a collapse with bounded curvature. Thus for instance collapsing along the orbits of an isometric G-action, where G is semi-simple and non-abelian, increases the curvature without bound.

A 3-manifold which admits a locally free S^1 action is called a **Seifert fibered space**. Such a space admits a fibration over a surface V, with S^1 fibers. Where the action is free, this fibration is a circle bundle. There may exist an isolated collection of non-free orbits, corresponding to isolated points in V. Topologically, a neighborhood of such an orbit is of the form $D^2 \times S^1$, where the S^1 acts by rotation on the S^1 factor and by rotation through a rational angle about $\{0\}$ in D^2.

The collection of Seifert fibered spaces falls naturally into 6 classes, according to the topology of the base surface V, i.e., $V = S^2$, T^2, or Σ_g, $g \geq 2$, and according to whether the S^1 bundle is trivial or not trivial. These account for 6 of the 8 possible geometries of 3-manifolds in the sense of Thurston [46]. These geometries are: $S^2 \times \mathbb{R}$, $\mathbb{R}^3$, $\mathbb{H}^2 \times \mathbb{R}$, S^3, Nil, and $\widetilde{SL(2,\mathbb{R})}$, respectively. The two remaining geometries are Sol, corresponding to non-trivial torus bundles over S^1, and the hyperbolic geometry $\mathbb{H}^3$.

Now suppose N is a compact Seifert fibered space with boundary. The boundary is a finite collection of tori, on which one has a free S^1 action. In a neighborhood of the boundary, this S^1 action then embeds in the standard free T^2 action on $T^2 \times I$. Given a collection of such spaces N_i, one may then glue the toral boundaries together by automorphisms of the torus, i.e., by elements of $SL(2,\mathbb{Z})$. For example, the gluing may interchange the fiber and base circles.

Definition 3.1. *A* graph manifold G *is a 3-manifold obtained by gluing Seifert fibered spaces by toral automorphisms of the boundary tori.*

Thus, a graph manifold has a decomposition into two types of regions,

$$G = S \cup L. \tag{3.2}$$

Each component of S is a Seifert fibered space, while each component of L is $T^2 \times I$, and glues together different boundary components of elements in S. The exceptional case of gluing two copies of $T^2 \times I$ by toral automorphisms of the boundary is also allowed; this defines the class of Sol manifolds, up to finite covers. The Seifert fibered components have a locally free S^1 action, the $T^2 \times I$ components have a free T^2 action; in general, these group actions do not extend to actions on topologically larger domains.

Graph manifolds are an especially simple class of 3-manifolds; they were introduced, and their structure was completely classified, by Waldhausen [48]. The terminology comes from the fact that one may associate a graph to G, by assigning a vertex to each component of S, and an edge to each component of L which connects a pair of components in S.

It is not difficult to generalize the construction above to show that any closed graph manifold G admits a sequence of metrics g_i which collapse with uniformly bounded curvature, i.e.,

$$|\operatorname{Ric}_{g_i}| \leq k, \quad \operatorname{vol}_{g_i} G \to 0. \tag{3.3}$$

The metrics g_i collapse the Seifert fibered pieces along the S^1 orbits, while collapsing the toral regions $T^2 \times I$ along the tori. Thus the collapse is rank 1 along S, while rank 2 along L. (Of course a bound on the full curvature is the same as a bound on the Ricci curvature in dimension 3.)

If the graph manifold is Seifert fibered, then the collapse (3.3) may be carried out with bounded diameter,

$$\text{diam}_{g_i} S \leq D, \text{ for some } D < \infty. \tag{3.4}$$

In fact, if S is a Nil-manifold, then the collapse may be carried out so that $\text{diam}_{g_i} S \to 0$, cf. [32].

On the other hand, suppose G is non-trivial in that it has both S and L components. If N denotes any S or L component, then it follows from work of Fukaya [27] that

$$\text{diam}_{g_i} N \to \infty \tag{3.5}$$

under the bounds (3.3). This phenomenon can be viewed as a refinement of the remark following (2.13), in that one has uniform control on the relative size of the injectivity radius on domains of bounded diameter, (cf. also [31]). In particular, the transition from Seifert fibered domains to toral domains takes longer and longer distance the more collapsed the metrics are. One obtains different collapsed "limits" depending on choice of base point. This "pure" behavior on regions of bounded diameter is special to dimension 3, cf. [5] for further details.

The discussion above shows that one may collapse graph manifolds with bounded curvature. The Cheeger-Gromov theory, [18], see also [33], implies that the converse also holds.

Theorem 3.2. (Collapse). *If M is a closed 3-manifold which collapses with bounded curvature, i.e., there is a sequence of metrics such that (3.3) holds, then M is a graph manifold.*

In fact, this result holds if M admits a sufficiently collapsed metric, i.e., $|\text{Ric}_g| \leq k$ and $\text{vol}_g M \leq \varepsilon_o$, for some $\varepsilon = \varepsilon_o(k)$ sufficiently small. Note of course that a collapsing sequence of metrics g_i is not *necessarily* invariant under the S^1 or T^2 actions associated with the graph manifold structure; these local group actions are smooth actions, but need not be isometric w.r.t. a highly collapsed metric.

In a certain sense, the vast majority of 3-manifolds are not graph manifolds, and so Theorem 3.2 gives strong topological restrictions on the existence of sufficiently collapsed metrics.

Idea of proof: First, it is easy to see from elementary comparison geometry, cf. [43], that

$$\text{vol}_{g_i} B_x(1) \to 0 \Rightarrow \text{inj}_{g_i}(x) \to 0.$$

At any x, rescale the metrics g_i to make $\text{inj}(x) = 1$, i.e., set

$$\bar{g}_i = [\text{inj}_{g_i}(x)]^{-2} \cdot g_i.$$

Now the bound (3.3) gives $|\text{Ric}_{\bar{g}_i}| \sim 0$. Thus, the metrics $\bar{g}_i$ are close to flat metrics on $\mathbb{R}^3/\Gamma$, where Γ is a non-trivial discrete group of Euclidean isometries, (by Theorem 2.2 for instance). Thus, essentially, $\mathbb{R}^3/\Gamma = \mathbb{R}^2 \times S^1$, or $\mathbb{R} \times S^1 \times S^1$. It follows that the local geometry, i.e., the geometry on the scale of the injectivity radius, is modeled by *non-trivial, flat* 3-manifolds. One then shows that these local structures for the geometry and topology can be glued together consistently to give a global graph manifold structure.

If S is a closed Seifert fibered space, the orbits of the S^1 action always inject in $\pi_1(S)$, i.e.,

$$\pi_1(S^1) \hookrightarrow \pi_1(S),$$

unless $S = S^3/\Gamma$. In case S has non-empty toral boundary components, the tori in ∂S always inject in $\pi_1(S)$ except in the single case of $S = D^2 \times S^1$, cf. [41]. Thus, if a graph manifold G is not a spherical space form, or does not have a solid torus component in its Seifert fibered decomposition (3.2), then the fibers of the decomposition, namely circles and tori, always inject in π_1:

$$\pi_1(\text{fiber}) \hookrightarrow \pi_1(G). \tag{3.6}$$

Hence, in this situation, one can pass to covering spaces to *unwrap* any collapse. If g_i is a collapsing sequence of metrics, by passing to larger and larger covering spaces, based sequences will always have convergent subsequences (in domains of arbitrary but bounded diameter). In addition, the isometric covering transformations on the covers have displacement functions converging uniformly to 0 on compact subsets. Hence, all such limits have a free isometric S^1 or T^2 action, depending on whether the collapse is rank 1 or 2 on the domains. This means that the limits have an **extra symmetry** not necessarily present on the initial collapsing sequence. Again, this feature of being able to unwrap collapse by passing to covering spaces is special to dimension 3, cf. [5] for further discussion and applications.

Finally, we discuss the third possibility, the formation of cusps. This case, although the most general, corresponds to a mixture of the two previous cases convergence/collapse, and so no essentially new phenomenon occurs. To start, given a complete Riemannian manifold (M, g), choose $\varepsilon > 0$ small, and let

$$M^\varepsilon = \{x \in M : \text{vol } B_x(1) \geq \varepsilon\}, \ M_\varepsilon = \{x \in M : \text{vol } B_x(1) \leq \varepsilon\}. \tag{3.7}$$

M^ε is called the ε-thick part of (M, g), while M_ε is the ε-thin part.

Now suppose g_i is a sequence of complete Riemannian metrics on the manifold M.

- If $x_i \in M^\varepsilon$, for some fixed $\varepsilon > 0$, then one has convergence, (in subsequences), in domains of arbitrary but bounded diameter about $\{x_i\}$, see the discussion concerning (2.13). Essentially, the bounds (2.11) hold on such domains in this case.
- If $y_i \in M_{\varepsilon_o}$, for ε_o sufficiently small, then domains of bounded, depending on ε_o, diameter about $\{y_i\}$ are graph manifolds, in fact Seifert fibered spaces.
- If $z_i \in M_{\varepsilon_i}$, $\varepsilon_i \to 0$, then domains of arbitrary but bounded diameter about $\{z_i\}$ are collapsing.

If $(M_\varepsilon, g_i) = \emptyset$, for some fixed $\varepsilon > 0$, then one is in the convergence situation. If $(M^\varepsilon, g_i) = \emptyset$, for all $\varepsilon > 0$ sufficiently small, depending on i, then one is in the collapsing situation. The only remaining possibility is that, for any fixed small $\varepsilon > 0$,

$$(M^\varepsilon, g_i) \neq \emptyset, \text{ and } (M_\varepsilon, g_i) \neq \emptyset. \tag{3.8}$$

This is equivalent to the existence of base points x_i, y_i, such that,

$$\text{vol } B_{x_i}(1) \geq \varepsilon_1, \quad \text{vol } B_{y_i}(1) \to 0, \tag{3.9}$$

for some $\varepsilon_1 > 0$. Observe that the volume comparison theorem (2.13) implies that $\text{dist}_{g_i}(x_i, y_i) \to \infty$ as $i \to \infty$, so that these different behaviors become further and further distant as $i \to \infty$.

This leads to the following result, cf. [5], [18] for further details.

Theorem 3.3. (Cusp Formation). *Let M be a 3-manifold and g_i a sequence of unit volume metrics on M with uniformly bounded curvature, and satisfying (3.8). Then pointed subsequences (M, g_i, p_i) converge to one of the following:*

• *complete cusps (N, g_∞, p_∞). These are complete, open Riemannian 3-manifolds, of finite volume and with graph manifold ends, which collapse at infinity. The convergence is in the $C^{1,\alpha}$ and weak $L^{2,p}$ topologies, uniform on compact subsets.*

• *Graph manifolds collapsed along local S^1 or T^2 actions to lower-dimensional metric spaces of infinite diameter. The convergence is in the Gromov-Hausdorff topology.*

In contrast to the topological implications of collapse in Theorem 3.2, (i.e., collapse implies M is a graph manifold), in general there are no a priori topological restrictions on M imposed by Theorem 3.3. To illustrate, let M be an arbitrary closed 3-manifold and let $\{C_k\}$ be a collection of disjoint solid tori $D^2 \times S^1$ embedded in M; for example $\{C_k\}$ may be a tubular neighborhood of a (possibly trivial) link in M. Then it is not difficult to construct a sequence of metrics of bounded curvature which converge to a collection of complete cusps on $M \setminus \cup C_k$ and collapse along the standard graph manifold structure on each C_k.

The ends of the cusp manifolds N in Theorem 3.3, i.e., the graph manifolds, necessarily have embedded tori. If such tori are essential in M, i.e., inject on the π_1 level, then Theorem 3.3. does imply strong topological constraints on the topology of M; cf. §6 for some further discussion.

Remark 3.4. We point out that there are versions of Theorems 3.2 and 3.3 also in dimension 4, as well as in higher dimensions. The concept of graph manifold is generalized to manifolds having an "F-structure", or an "N-structure" (F is for flat, N is for nilpotent), cf. [18], provided bounds are assumed on the full curvature, as in (1.4). In dimension 4, this can be relaxed to bounds on the Ricci curvature, as in (1.5), provided one allows for a finite number of singularities in F-structure, as in Theorem 2.6.

4. Applications to static and stationary space-times

In this section, we discuss applications of the results of §2–3 to static and stationary space-times, i.e., space-times $(\mathbf{M}, \mathbf{g})$ which admit a time-like Killing field K. These space-times are viewed as being the end or final state of evolution of a (time-dependent) gravitational field. Since they are time-independent in a natural sense, they may be analyzed by methods of Riemannian geometry, which are not available in general for Lorentzian manifolds.

Throughout this section, we assume that $(\mathbf{M}, \mathbf{g})$ is chronological, i.e., $(\mathbf{M}, \mathbf{g})$ has no closed time-like curves, and that K is a complete vector field.

Let Σ be the orbit space of the isometric $\mathbb{R}$-action generated by the Killing field K, and let $\pi : \mathbf{M} \to \Sigma$ be the projection to the orbit space. The 4-metric $\mathbf{g}$ has the form

$$\mathbf{g} = -u^2(dt + \theta)^2 + \pi^*(g), \qquad (4.1)$$

where $K = \partial/\partial t$, θ is a connection 1-form for the bundle π, $u^2 = -\mathbf{g}(K, K) > 0$ and $g = g_\Sigma$ is the metric induced on the orbit space.

The vacuum Einstein equations are equivalent to an elliptic system of P.D.E's in the data (Σ, g, u, θ). Let ω be the twist 1-form on Σ, given by $2\omega = *(\kappa \wedge d\kappa) = -u^4 * d\theta$, where $\kappa = -u^2(dt + \theta)$ is the 1-form dual to K. (The first $*$ operator is on $\mathbf{M}$ while the second is on Σ.) Then the equations on Σ are:

$$\mathrm{Ric}_g = u^{-1}D^2u + 2u^{-4}(\omega \otimes \omega - |\omega|^2g), \qquad (4.2)$$

$$\Delta u = -2u^{-3}|\omega|^2, \qquad (4.3)$$

$$d\omega = 0. \qquad (4.4)$$

The maximum principle applied to (4.3) immediately implies that if Σ is a closed 3-manifold, then (Σ, g) is flat and $u = \mathrm{const}$, and so $(\mathbf{M}, \mathbf{g})$ is a (space-like) isometric quotient of empty Minkowski space $(\mathbb{R}^4, \eta)$. Thus, we assume Σ is open, possibly with boundary.

Locally of course there are many solutions to the system (4.2)–(4.4); to obtain uniqueness, one needs to impose boundary conditions.

We consider first the global situation, and so assume that (Σ, g) is a complete, non-compact Riemannian 3-manifold. Boundary conditions are then at infinity, i.e., conditions on the asymptotic behavior of the metric. In this respect, one has the following classical result, cf. [37], [24].

Theorem 4.1. (Lichnerowicz). *The only complete, stationary vacuum space-time* $(\mathbf{M}, \mathbf{g})$ *which is asymptotically flat (AF) is empty Minkowski space-time* $(\mathbb{R}^4, \eta)$.

It is most always taken for granted that Σ should be AF. Stationary space-times are meant to model isolated physical systems, and the only physically reasonable models are AF, since in the far-field regime, general relativity should approximate Newtonian gravity. In fact, from this physical perspective, the Lichnerowicz theorem may be viewed as a triviality. Since there is no source for the gravitational field, it must be the empty Minkowski space-time.

However, mathematically, the Lichnerowicz theorem is not (so) trivial. More-over, the assumption that $(\mathbf{M}, \mathbf{g})$ is AF is contrary to the spirit of general relativity. Such a boundary condition is ad hoc, and its imposition is mathematically circular in a certain sense. A priori, there might well be complete stationary solutions for which the curvature does not decay anywhere to 0 at infinity. From this more gen-eral perspective, one should be able to *deduce* that the far-field regime of stationary space-times is necessarily AF and not have to assume this to begin with.

The following result from [6] clarifies this issue.

Theorem 4.2. (Generalized Lichnerowicz). *The only complete stationary vacuum space-time* $(\mathbf{M}, \mathbf{g})$ *is empty Minkowski space-time* $(\mathbb{R}^4, \eta)$, *or a discrete isometric quotient of it.*

The starting point of the proof of this result is to study first the moduli space of all complete stationary vacuum solutions. As noted above, any given solution may, a priori, have unbounded curvature, i.e., $|\operatorname{Ric}_g|$ may diverge to infinity on divergent sequences in Σ. Under such a condition, the first step is then to show, by taking suitable base points and rescalings, that one may obtain a new stationary vacuum solution, (i.e., a new point in the moduli space), with uniformly bounded curvature, and non-zero curvature at a base point. This step uses the Cheeger-Gromov theory, as described in §2–§3, and requires the special features of collapse in 3 dimensions.

The next step in the proof is to recast the problem in the Ernst formulation. Define the conformally related metric $\widetilde{g}$ by

$$\widetilde{g} = u^2 g. \tag{4.5}$$

A simple calculation shows that (4.2) becomes

$$\operatorname{Ric}_{\widetilde{g}} = 2(d \ln u)^2 + 2u^{-4}\omega^2 \geq 0. \tag{4.6}$$

Further, the system (4.2)–(4.4) becomes the Euler-Lagrange equations for an ef-fective 3-dimensional action given by

$$\mathcal{S}_{\text{eff}} = \int [R - \frac{1}{2}(\frac{|d\phi|^2 + |du^2|^2}{u^4})]dV.$$

Here ϕ is the twist potential, given by $d\phi = 2\omega$. (In general one must pass to the universal cover to obtain the existence of ϕ.)

This action is exactly 3-dimensional (Riemannian) gravity on $(\Sigma, \widetilde{g})$ coupled to a σ-model with target the hyperbolic plane $(H^2(-1), g_{-1})$. Thus, the Ernst map $E = (\phi, u^2)$ is a harmonic map

$$E : (\Sigma, \widetilde{g}) \to (H^2(-1), g_{-1}). \tag{4.7}$$

Now it is well known that harmonic maps $E : (M, g) \to (N, h)$ from Rie-mannian manifolds of non-negative Ricci curvature to manifolds of non-positive sectional curvature have strong rigidity properties, via the Bochner-Lichnerowicz formula,

$$\frac{1}{2}\Delta|DE|^2 = |D^2 E|^2 + \langle \text{Ric}_g, E^*(h) \rangle - \sum (E^* R_h)(e_i, e_j, e_j, e_i). \qquad (4.8)$$

By analyzing (4.8) carefully, one shows that E is a constant map, from which it follows easily that $(\mathbf{M}, \mathbf{g})$ is flat.

Remark 4.3.

(i) The same result and proof holds for stationary gravitational fields coupled to σ-models, whose target spaces are Riemannian manifolds of non-positive sectional curvature, i.e., $E : (\Sigma, \tilde{g}) \to (N, g_N)$ with $\text{Riem}_{g_N} \leq 0$.

(ii) Curiously, the Riemannian analogue of Theorem 4.2 remains an open problem. Thus, does there exist a complete non-flat Ricci-flat Riemannian 4-manifold which admits a free isometric S^1 action?

(iii) It is interesting to note that the analogue of Theorem 4.2 is false for stationary Einstein-Maxwell solutions. A counterexample is provided by the (static) Melvin magnetic universe [39], cf. also [28]. I am grateful to David Garfinkle for pointing this out to me. For the stationary Einstein-Maxwell system, the target space of the Ernst map is $SU(2,1)/S(U(1,1) \times U(1))$, $(SO(2,1)/SO(1,1)$ for static Einstein-Maxwell). Both of these target spaces have indefinite, (i.e., non-Riemannian), metrics.

The rigidity result Theorem 4.2 leads to a priori estimates on the geometry of general stationary solutions of the Einstein equations. Thus, if Σ is not complete, it follows that $\partial\Sigma \neq \emptyset$. Note that part of $\partial\Sigma$ may correspond to the horizon $H = \{u = 0\}$ where the Killing field vanishes. The following result is also from [6].

Theorem 4.4. (Curvature Estimate). *Let* $(\mathbf{M}, \mathbf{g})$ *be a stationary vacuum spacetime. Then there is a constant* $C < \infty$, *independent of* $(\mathbf{M}, \mathbf{g})$, *such that*

$$|\mathbf{R}|(x) \leq C/r^2[x], \qquad (4.9)$$

where $r[x] = \text{dist}_\Sigma(\pi(x), \partial\Sigma)$.

Here, the curvature norm $|\mathbf{R}|$ may be given by

$$|\mathbf{R}|^2 = |R_\Sigma|^2 + |d\ln u|^2 + |u^{-2}\omega|^2.$$

Note that Theorem 4.2 follows from Theorem 4.4 by letting $r \to \infty$. Conversely, it is a general principle for elliptic geometric variational problems that a global rigidity result as in Theorem 4.2 leads to a priori local estimates as in Theorem 4.4.

Remark 4.5.

(i) Using elliptic regularity, one also has higher-order bounds:

$$|\nabla^k \mathbf{R}|(x) \leq C_k/r^{2+k}[x]. \qquad (4.10)$$

(ii) A version of this result also holds for stationary space-times with energy-momentum tensor T. Thus, for example one has

$$|\mathbf{R}|(x) \leq C_\alpha \cdot |T|_{C^\alpha(B_{[x]}(1))}, \tag{4.11}$$

for any $\alpha > 0$, where $B_{[x]}(1)$ is the unit ball in (Σ, g) about $[x]$. The proof is the same as that of (4.9) given in [6].

Thus, one can use the Cheeger-Gromov theory to control the local behavior of stationary space-times, possibly with matter terms, away from any boundary.

The results above can in turn be applied to study the possible asymptotic behavior of general stationary or static vacuum space-times, without any a priori AF assumption. For example, (4.9) implies that the curvature decays at least quadratically in any end (E, g) of (Σ, g). For simplicity, we restrict here to static space-times.

Thus, let $(\mathbf{M}, \mathbf{g})$ be a static space-time with orbit space (Σ, g), with $\partial\Sigma \neq \emptyset$. Define $\partial\Sigma$ to be *pseudo-compact* if there exists $r_o > 0$ such that the level set $\{r = r_o\}$ in Σ is compact; recall that r is the distance function to the boundary $\partial\Sigma$. (There are numerous examples of static space-times for which $\partial\Sigma$ is non-compact, with $\partial\Sigma$ pseudo-compact.) Let $S(s) = r^{-1}(s) \subset \Sigma$. If E is an end of (Σ, g), define its mass m_E by

$$m_E = \lim_{s \to \infty} \frac{1}{4\pi} \int_{S(s)} \langle \nabla \ln u, \nabla r \rangle dA. \tag{4.12}$$

It is easily seen from the static vacuum equations that the integral is monotone non-increasing in s, and so the limit exists. The mass m_E coincides with the Komar mass in case E is AF. The following result is from [7].

Theorem 4.6. (Static Asymptotics). *Let $(\mathbf{M}, \mathbf{g}, u)$ be a static vacuum space-time with pseudo-compact boundary. Then $(\mathbf{M}, \mathbf{g})$ has a finite number of ends. Any end E on which*

$$\liminf_E u > 0, \tag{4.13}$$

is either:

$$AF$$

or

$$\text{small} \equiv_{def} \int_1^\infty [\text{area } S(r)]^{-1} dr < \infty. \tag{4.14}$$

Further, if $m_E \neq 0$ and $\sup_E u < \infty$, then E is AF.

This result is sharp in the sense that if any of the hypotheses are dropped, then the conclusion is false. For instance, if (4.13) fails, then there are examples of static vacuum solutions with ends neither small nor AF.

We note that when E is AF, the result implies it is AF in the strong sense that

$$|g - g_0| = \frac{2m}{r} + O(r^{-2}), \quad |R| = O(r^{-3}), \quad \text{and} \quad |u - 1| = \frac{m}{r} + O(r^{-2}). \tag{4.15}$$

More precise asymptotics can then be obtained by using standard elliptic estimates on the equations (4.2)–(4.4), or from [14]. Again, a version of Theorem 4.6 holds for static space-times with matter, cf. again [7] for further information.

The idea of the proof is to study the asymptotic behavior of an end E by blowing it down, as described in §1. Thus, for R large and any fixed k, consider the metric annuli $A(R, kR)$ about some base point $x_o \in (\Sigma, g)$ and consider the rescalings $g_R = R^{-2}g$. The annulus $A(R, kR)$ then becomes an annulus of the metric form $A(1, k)$ w.r.t. g_R. Further, the estimate (4.9) implies that the curvature of g_R in $A(1, k)$ is uniformly bounded. Thus, one may apply the Cheeger-Gromov theory as described in §2, §3, to a sequence $(A(1, k), g_{R_i})$, with $R_i \to \infty$. One proves that the convergence case gives rise to AF ends, while the collapse case gives rise to small ends.

Note that in the collapsing situation, one obtains an extra S^1 or T^2 symmetry when the collapse is unwrapped in covering spaces. Thus, the behavior in this case is described by axisymmetric static solutions, i.e., the Weyl metrics. Small ends typically have the same end structure as $\mathbb{R}^2 \times S^1$, where the S^1 factor has bounded length and so typically have at most quadratic growth for the area of geodesic spheres.

It is worth pointing out that there are static vacuum solutions, smooth up to a compact horizon, which have a single small end. This is the family of Myers metrics [40], or periodic Schwarzschild metrics, (discovered later and independently by Korotkin and Nicolai). The manifold Σ is topologically $(D^2 \times S^1) \setminus B^3$, so that $\partial\Sigma = S^2$ with a single end of the form $T^2 \times \mathbb{R}^+$. Metrically, the end is asymptotic to one of the (static) Kasner metrics. This is of course not a counterexample to the static black hole uniqueness theorem, since the end is not AF.

Note that since $\pi_1(\Sigma) = \mathbb{Z}$ here, one may take non-trivial covering spaces of the Myers metrics. This leads to static vacuum solutions with an arbitrary finite number, or even an infinite number, of black holes in static equilibrium. This situation is of course not possible in Newtonian gravity, and so is a highly non-linear effect of general relativity.

5. Lorentzian analogues and open problems

In this section, we discuss potential analogues of the results of §2 and §3 for Lorentzian metrics on 4-manifolds. The main interest is in space-times $(\mathbf{M}, \mathbf{g})$ for which one has control on the Ricci curvature of $\mathbf{g}$, or via the Einstein equations, control on the energy-momentum tensor T. In particular, the main focus will be on vacuum space-times, $\mathrm{Ric}_{\mathbf{g}} = 0$.

One would like to find conditions under which one can take limits of vacuum space-times. One natural reason for trying to do this is the following. There are now a number of situations where global stability results have been proved, namely: the global stability of Minkowski space-time [21], and of deSitter space-time [26], the global future stability of the Milne space-time [10], and the future $U(1)$ stability of certain Bianchi models [20]. These results are *openness* results, which state that the basic features of a given model, e.g., Minkowski, are preserved under suitably small perturbations of the initial data. It is then natural to consider what occurs when one tries to pass to limits of such perturbations.

The issue of being able to take limits is also closely related with the existence problem and singularity formation for the vacuum Einstein evolution equations. From this perspective, suppose one has an increasing sequence of domains $(\Omega_i, \mathbf{g_i})$, $\Omega_i \subset \Omega_{i+1}$ with $\mathbf{g_{i+1}}|_{\Omega_i} = \mathbf{g_i}$, which are evolutions of smooth Cauchy data on some fixed initial data set. If $\mathbf{M} = \cup \Omega_i$ is the maximal Cauchy development, then understanding $(\mathbf{M}, \mathbf{g})$ amounts to understanding the limiting behavior of $(\Omega_i, \mathbf{g_i})$.

There are two obvious but essential reasons why it is much more difficult to develop a Lorentzian analogue of the Cheeger-Gromov theory, in particular with bounds only on the Ricci curvature. The first is that the elliptic nature of the P.D.E. for Ricci curvature becomes hyperbolic for Lorentz metrics, and hyperbolic P.D.E. are much more difficult than elliptic P.D.E. The second is that the group of Euclidean rotations $O(4)$ is compact, while the group of proper Lorentz transformations $O(3, 1)$ is non-compact.

A: 1st Level problem

Consider first the problem of controlling the space-time metric $\mathbf{g}$ in terms of bounds, say L^∞, on the space-time curvature $\mathbf{R}$,

$$|\mathbf{R}|_{L^\infty} \leq K < \infty, \tag{5.1}$$

since already here there are significant issues.

First, the norm of curvature tensor $|\mathbf{R}|^2 = \mathbf{R}_{ijkl}\mathbf{R}^{ijkl}$ is no longer non-negative for Lorentz metrics, and so a bound on $|\mathbf{R}|^2$ does not imply a bound on all the components $\mathbf{R}_{ijkl}$. In fact, for a Ricci-flat 4-metric, there are exactly two scalar invariants of the curvature tensor:

$$\langle \mathbf{R}, \mathbf{R} \rangle = |\mathbf{R}|^2 = \mathbf{R}_{ijkl}\mathbf{R}^{ijkl} \text{ and } \langle \mathbf{R}, *\mathbf{R} \rangle = \mathbf{R}_{ijkl}(*\mathbf{R}^{ijkl}). \tag{5.2}$$

Both of these invariants can vanish identically on classes of Ricci-flat non-flat space-times; for instance this is the case for the class of plane-fronted gravitational waves, given by

$$\mathbf{g} = -dudv + (dx^2 + dy^2) - 2h(u, x, y)du^2,$$

$$\Delta_{(x,y)}h = 0,$$

cf. [15,§8] and references therein. Here, h is only required to harmonic in the variables (x, y), and is arbitrary in u. The class of such space-times is highly non-compact, and so one has no local control of the metric in any coordinate system under bounds on the quantities in (5.2).

Thus, one must turn to bounds on the components of $\mathbf{R}$ in some fixed coordinate system or framing. The most efficient way to do this is to choose a unit time-like vector $T = e_0$, say future directed, and extend it to an orthornormal frame $e_\alpha, 0 \le \alpha \le 3$. Since the space $T^\perp$ orthogonal to T is space-like and $O(3)$ is compact, the particular framing of $T^\perp$ is unimportant. One may then define the norm w.r.t. T by

$$|\mathbf{R}|_T^2 = \sum (\mathbf{R}_{ijkl})^2, \qquad (5.3)$$

where the components are w.r.t. the frame e_α. This is equivalent to taking the norm of $\mathbf{R}$ w.r.t. the Riemannian metric

$$g_E = \mathbf{g} + 2T \otimes T.$$

If, at a given point p, T lies within a compact subset W of the future interior null cone T_p^+, then the norms (5.3) are all equivalent, with constant depending only on W. Of course if D is a compact set in the space-time $(\mathbf{M}, \mathbf{g})$ and the vector field T is continuous in D, then T lies within a compact subset of T^+D, where T^+D is the bundle of future interior null cones in the tangent bundle TD.

It is quite straightforward to prove that if (M, g) is a smooth Riemannian manifold with an L^∞ bound on the full curvature, $|R| \le K$ then there are local coordinate systems in which the metric is $C^{1,\alpha}$ or $L^{2,p}$, with bounds depending only on K and a lower volume bound, cf. Remark 2.4(i).

However, this has been an open problem for Lorentzian metrics, apparently for some time, cf. [22],[47] for instance. The following result gives a solution to this problem.

To state the result, we need the following definition. Let Ω be a domain in a smooth Lorentz manifold $(\mathbf{M}, \mathbf{g})$, of arbitrary dimension $n + 1$. Then Ω is said to satisfy the **size conditions** if the following holds. There is a smooth time function t, with $T = \nabla t / |\nabla t|$ the associated unit time-like vector field on Ω, such that, for $S = S_0 = t^{-1}(0)$, the 1-cylinder

$$C_1 = B_p(1) \times [-1, 1] \subset\subset \Omega, \qquad (5.4)$$

i.e., C_1 has compact closure in Ω. Here $B_p(1)$ is the geodesic ball of radius 1 about p, w.r.t. the metric g induced on S and the product is identified with a subset of Ω by the flow of T.

It is essentially obvious that any point q in a Lorentz manifold has a neighborhood satisfying the size conditions, when the metric $\mathbf{g}$ is scaled up suitably.

Let $D = \Im T|_{C_1} \subset\subset T^+\Omega$.

Theorem 5.1. *Let Ω be a domain in a vacuum $(n + 1)$-dimensional space-time $(\mathbf{M}, \mathbf{g})$. Suppose Ω satisfies the size conditions, and that there exist constants $K < \infty$ and $v_o > 0$ such that*

$$|\mathbf{R}|_T \le K, \quad \mathrm{vol}_g\, B_p(\tfrac{1}{2}) \ge v_o. \qquad (5.5)$$

Then there exists $r_o > 0$, depending only on K, v_o and D, and coordinate charts on the r_o-cylinder

$$C_{r_o} = B_p(r_o) \times [-r_o, r_o],$$

such that the components of the metric $\mathbf{g}_{\alpha\beta}$ *are in* $C^{1,\alpha} \cap L^{2,p}$, *for any* $\alpha < 1$, $p < \infty$. *Further, there exists* R_o, *depending only on* K, v_o, D *and* p, *such that, on* C_{r_o},

$$\|\mathbf{g}_{\alpha\beta}\|_{L^{2,p}} \leq R_o. \qquad (5.6)$$

Here, the components $\mathbf{g}_{\alpha\beta}$ are the full space-time components of $\mathbf{g}$, and the estimate (5.6) gives bounds on both spatial and time derivatives of $\mathbf{g}$, up to order 2, in L^p, where L^p is measured on spatial slices of C_{r_o}.

This result is formulated in such a way that it is easy to pass to limits. Thus, if one has a sequence of smooth space-times $(\mathbf{M}_i, \mathbf{g}_i)$ satisfying the hypotheses of the Theorem, (with fixed constants K, v_o and uniformly compact domains D), then it follows that, in a subsequence, there is a limit $C^{1,\alpha} \cap L^{2,p}$ space-time $(\mathbf{M}_\infty, \mathbf{g}_\infty)$, defined at least on the r_o-cylinder C_{r_o}. Further, the convergence to the limit is $C^{1,\alpha}$ and weak $L^{2,p}$, and the estimate (5.6) holds on the limit.

We sketch some of the ideas of the proof; full details appear in [9]. First, one constructs a new local time function τ on small cylinders C_{r_1}, with $|\nabla \tau|^2 = -1$, so the flow of $\nabla \tau$ is by time-like geodesics. On the level sets Σ_τ of τ, one constructs spatially harmonic coordinates $\{x_i\}$, (w.r.t. the induced Riemannian metric). This gives a local coordinate system $(\tau, x_1, \ldots, x_n)$ on small cylinders about p. One then uses the transport or Raychaudhuri equation, together with the Bochner-Weitzenbock formula, (Simons' equation), and elliptic estimates to control $\mathbf{g}_{\alpha\beta}$.

The vacuum Einstein equations are needed in Theorem 5.1 only to prove the 2nd time derivatives of the shift $\partial_\tau \partial_\tau \mathbf{g}_{0\alpha}$ are in L^p, via use of the Bianchi identity. In place of vacuum space-times, it suffices to have a rather weak bound on the stress-energy tensor in the Einstein equations. All other bounds on $\mathbf{g}_{\alpha\beta}$ do not require the Einstein equations.

It would be interesting to apply this result, or variants of it, to obtain further information on the structure of the boundary of space-times.

If the volume bound on space-like hypersurfaces in (5.5) is dropped, then it is possible that space-like hypersurfaces may collapse with bounded curvature, as described in §3. Examples of this behavior occur on approach to Cauchy horizons, (as noted in §3 in connection with the Berger collapse and the Taub-NUT metric). More generally, Rendall [45] has proved the following interesting general result: if Σ is a *compact* Cauchy horizon in a smooth vacuum space-time in $3+1$ dimensions, then nearby space-like hypersurfaces collapse with bounded curvature on approach to Σ.

B: 2nd Level problem

While Theorem 5.1 represents a first step, one would like to do much better by replacing the bound on $|\mathbf{R}|_T$ by a bound on the Ricci curvature of $(\mathbf{M}, \mathbf{g})$, or assuming for instance the vacuum Einstein equations. Thus, one may ask if analogues of Theorems 2.2 or 2.3 hold in the Lorentzian setting.

The main ingredients in the proofs of these results are the splitting theorem – a geometric part – and the strong convergence to limits – an analytic part obtained from elliptic estimates for the Ricci curvature. Now one does have a direct analogue of the splitting theorem for vacuum space-times, (or more generally space-times satisfying the time-like convergence condition). Thus, by work of Eschenburg, Galloway and Newman, if $(\mathbf{M}, \mathbf{g})$ is a time-like geodesically complete, (or a globally hyperbolic), vacuum space-time which contains a time-like line, i.e., a complete time-like maximal geodesic, then $(\mathbf{M}, \mathbf{g})$ is flat, cf. [13] and references therein.

In analogy to the Riemannian case, define then the 1-cross $\mathrm{Cro}_1(x, T)$ of a Lorentzian 4-manifold $(\mathbf{M}, \mathbf{g})$ at x, in the direction of a unit time-like vector T, to be the length of the longest maximizing geodesic in the direction T, with center point x. For Ω a domain with compact closure in $\mathbf{M}$ and T a smooth unit time-like vector field, define

$$\mathrm{Cro}_1(\Omega, T) = \inf_{x \in \Omega} \mathrm{Cro}_1(x, T).$$

What is lacking is the regularity boost obtained from elliptic estimates. For space-times, the vacuum equations give a hyperbolic evolution equation, (in harmonic coordinates), for which one does not have a gain in derivatives. However, the smoothness of initial data is preserved under the evolution, until one hits the boundary of the maximal development.

Let $H^s = H^s(U)$ denote the Sobolev space of functions with s weak derivatives in $L^2(U)$, U a bounded domain in $\mathbb{R}^3$. For $s > 2.5$, (so that H^s embeds in C^1), and a space-like hypersurface $S \subset (\mathbf{M}, \mathbf{g})$, define the harmonic radius $\rho_s(x)$ of $x \in S$ in the same way as in Definition 2.1, where the components $\mathbf{g}_{\alpha\beta}$ and derivatives are in both space and time directions. For the following, we need only consider $s \in \mathbb{N}^+$, with s large, for instance, $s = 3$.

Now a well-known result of Choquet-Bruhat [19] states that the maximal vacuum H^s development of smooth (C^∞) initial data on S is the same for all s, provided $s > 2.5$. Thus, one does not have different developments of smooth initial data, depending on the degree of desired H^s regularity. Here, one may assume that S is compact, or work locally, within the domain of dependence of S. This qualitative result can be expressed as follows. Let S_t be space-like hypersurfaces obtained by evolution from initial data on $S = S_0$. If $x_t \in S_t$, then

$$\rho_s(x_t) \geq c_1 \Rightarrow \rho_{s+1}(x_t) \geq c_2, \tag{5.7}$$

where c_2 depends on c_1 and the (C^∞) initial data on S_0.

We raise the following problem of whether the qualitative statement (5.7) can be improved to a *quantitative* statement.

Regularity problem

Can the estimate (5.7) be improved to an estimate

$$\inf_{x_t \in S_t} \rho_{s+1}(x_t) \geq c_0 \cdot \inf_{x_t \in S_t} \rho_s(x_t), \tag{5.8}$$

where c_0 depends only on the initial data on S? One may assume, w.l.o.g, that $t \leq 1$.

The important point of (5.8) over (5.7) is that the estimate (5.8) is scale-invariant. Here, we recall that $\rho_s(x)$ measures the degree of concentration of derivatives of the metric in H^s, so that $\rho_s \to 0$ corresponds to blow-up of the metric in H^s locally.

If (5.8) holds, it serves as an analogue of the regularity boost. In such circumstances, one can imitate the proof of Theorems 2.2 or 2.3 to obtain similar results for sequences of space-times $(\mathbf{M}, \mathbf{g}_i)$.

In fact, the validity of (5.8) would have numerous interesting applications, even if it could be established under some further restrictions or assumptions.

Suppose next one drops any assumption on the 1-cross of $(\mathbf{M}, \mathbf{g})$ and maintains only a lower bound on the volumes of geodesic balls, as in (5.5), on space-like hypersurfaces. This leads directly to issues of singularity formation and the structure of the boundary of the vacuum space-time, where comparatively little is known mathematically.

A useful problem, certainly simple to state, is the following: for simplicity, we work in the context of compact, (i.e., closed, without boundary), Cauchy surfaces.

Sandwich problem

Let $(\mathbf{M}, \mathbf{g}_i)$ be a sequence of vacuum space-times, and let Σ_i^1, Σ_i^2 be two compact Cauchy surfaces in $\mathbf{M}$, with Σ_i^2 to the future of Σ_i^1 and with

$$1 \leq \mathrm{dist}_{\mathbf{g}}(x, \Sigma_i^1) \leq 10,$$

for all $x \in \Sigma_i^2$. Suppose the Cauchy data (g_i^j, K_i^j), $j = 1, 2$ on each Cauchy surface are uniformly bounded in H^s for some fixed $s > 2.5$, possibly large. Hence the data (g_i^j, K_i^j) converge, in a subsequence and weakly in H^s, to limit H^s Cauchy data g_∞^j, K_∞^j on Σ^j.

Do the vacuum space-times $A_i(1,2) \subset (M, g_i)$ between Σ^1 and Σ^2 converge, weakly in H^s, to a limit space time,

$$(A_i(1,2), g_i) \to (A_\infty, g_\infty)? \tag{5.9}$$

This question basically asks if a singularity can form between Σ_i^1 and Σ_i^2 in the limit. It is unknown even if there could be only a single singularity at an isolated point (event) $x_0 \in (A_\infty, g_\infty)$.

The existence of such a singularity may be related to the Choptuik solution. However, both the existence and the smoothness properties of the Choptuik solution have not been established well mathematically; cf. [34] for an interesting discussion.

Such a limit singularity would be naked in a strange way. It could be detected on Σ^2, since light rays from it propagate to Σ^2, but on Σ^2, no remnant of the singularity is detectable, since the data is smooth on Σ^2. Thus, the singularity is invisible to the future (or past) in a natural sense.

A resolution of this problem would be useful in understanding, for instance, limits of the asymptotically simple vacuum perturbations of deSitter space, given by Friedrich's theorem [26]. The sandwich problem above asks: suppose one has control on the space-time near past and future space-like infinity $\mathcal{I}^{\pm}$, does it follow that one has control in between?

Similar questions can be posed for non-compact Cauchy surfaces, and relate for instance to limits of the AF perturbations of Minkowski space given by Christodoulou-Klainerman, [21].

6. Future asymptotics and geometrization of 3-manifolds

In this section, we give some applications to the future asymptotic behavior of cosmological spaces times.

Let $(\mathbf{M}, \mathbf{g})$ be a vacuum cosmological space-time, i.e., $(\mathbf{M}, \mathbf{g})$ contains a compact Cauchy surface Σ of constant mean curvature (CMC). It is well known that Σ then embeds in a (local) foliation $\mathcal{F}$ by CMC Cauchy surfaces Σ_τ, all diffeomorphic to $\Sigma = \Sigma_1$, and parametrized by their mean curvature τ. The parameter τ thus serves as a time function, with respect to which one may describe the evolution of the space-time. We refer to the work of Bartnik [11], [12] and Gerhardt [29] for results on the existence of such foliations, and to the surveys by Marsden-Tipler [38] and Rendall [44] for an overview of this topic.

We assume throughout this section that Σ is of non-positive Yamabe type, i.e., Σ admits no metric of positive scalar curvature. It then follows from the Hamiltonian constraint equation that the mean curvature τ never achieves the value 0. Thus

$$\tau \in (-\infty, 0), \tag{6.1}$$

with τ increasing towards the future in $(\mathbf{M}, \mathbf{g})$. The sign of the mean curvature is chosen so that $\mathrm{vol}_{g_\tau} \Sigma_\tau$ is increasing with increasing τ, i.e., expanding towards the future. The foliated region $\mathbf{M}_{\mathcal{F}}$ is thus a subset of $\mathbf{M}$, although in general one cannot expect that $\mathbf{M} = \mathbf{M}_{\mathcal{F}}$ due to the formation of singularities.

Suppose that $(\mathbf{M}, \mathbf{g})$ is geodesically complete to the future of Σ, and that the future is foliated by CMC Cauchy surfaces, i.e., $\mathbf{M} = \mathbf{M}_{\mathcal{F}}$ to the future of Σ. These are of course strong assumptions, but are necessary if one wants to understand the future asymptotic behavior of $(\mathbf{M}, \mathbf{g})$ without the complicating issue of singularities.

The topology of Σ_τ is fixed, and so the metrics g_τ induced on Σ_τ by the ambient metric $\mathbf{g}$ give rise to a curve of Riemannian metrics on the fixed manifold Σ. In all known situations, one has $\mathrm{vol}_{g_\tau} \Sigma \to \infty$ as $\tau \to 0$, and the metrics g_τ become locally flat, due to the expansion, compare with the discussion in §1. It would be of interest to prove these statements in general, although it is hard to imagine situations where either one of them fails.

The local geometry of g_τ thus becomes trivial locally. This is of course not very interesting. As in §1 and §4, to study the asymptotic behavior, one should rescale by the distance to a fixed base point or space-like hypersurface. In this case, the distance is the time-like Lorentzian distance. Thus, for x to the future of $\Sigma = \Sigma_{-1}$, let $t(x) = \mathrm{dist}_{\mathbf{g}}(x, \Sigma)$ and let

$$t_\tau = t_{\max}(\tau) = \max\{t(x) : x \in \Sigma_\tau\} = \mathrm{dist}_{\mathbf{g}}(\Sigma_\tau, \Sigma). \tag{6.2}$$

It is natural to study the asymptotic behavior of the metrics

$$\bar{g}_\tau = t_\tau^{-2} g_\tau, \tag{6.3}$$

on Σ_τ. Observe that in the rescaled space-time $(\mathrm{M}, \bar{\mathbf{g}}_\tau)$, the distance of $(\Sigma_\tau, \bar{g}_\tau)$ to the "initial" singularity, (big bang), tends towards 1, as $\tau \to 0$. Any other essentially distinct scaling would have the property that the distance to the initial singularity tends towards 0 or ∞, and so is not particularly natural.

We need the following definition, closely related to Thurston's Geometrization Conjecture [46] on the structure of 3-manifolds.

Definition 6.1. Let Σ be a closed, oriented, connected 3-manifold, of non-positive Yamabe type. A *weak* geometrization of Σ is a decomposition

$$\Sigma = H \cup G, \tag{6.4}$$

where H is a finite collection of complete, connected hyperbolic manifolds, of finite volume, embedded in Σ, and G is a finite collection of connected graph manifolds, embedded in Σ. The union is along a finite collection of embedded tori $\mathcal{T} = \cup T_i = \partial H = \partial G$.

A *strong* geometrization of Σ is a weak geometrization as above, for which each torus $T_i \in \mathcal{T}$ is incompressible in Σ, i.e., the inclusion of T_i into Σ induces an injection of fundamental groups.

Of course it is possible that the collection $\mathcal{T}$ of tori dividing H and G is empty, in which case weak and strong geometrizations coincide. In such a situation, Σ is then either a closed hyperbolic manifold or a closed graph manifold. For a strong geometrization, the decomposition (6.4) is unique up to isotopy, but this is certainly not the case for a weak geometrization, cf. the end of §3.

In general, no fixed metric g on Σ will realize the decomposition (6.4), unless $\mathcal{T} = \emptyset$. This is because the complete hyperbolic metric on H does not extend to a metric on Σ. However, one can find sequences of metrics g_i on Σ which limit on a geometrization of Σ in the sense of (6.4). Thus, metrics g_i may be chosen to converge to the hyperbolic metric on larger and larger compact subsets of H, to be more and more collapsed with bounded curvature on G, and such that their behavior matches far down the collapsing hyperbolic cusps.

Next, to proceed further, we need to impose a rather strong curvature assumption on the ambient space-time curvature. Thus, suppose there is a constant $C < \infty$ such that, for x to the future of Σ,

$$|\mathbf{R}|(x) + t(x)|\nabla \mathbf{R}|(x) \le C \cdot t^{-2}(x). \tag{6.5}$$

Here, the curvature norm $|\mathbf{R}|$ may be given by $|\mathbf{R}|_T$ as in (5.3), where T is the unit normal to the foliation Σ_τ. Since $(\mathbf{M}, \mathbf{g})$ is vacuum, this is equivalent to

$|\mathbf{R}|^2 = |E|^2 + |B|^2$, where E, B is the electric/magnetic decomposition of $\mathbf{R}$, $E(X,Y) = \langle \mathbf{R}(X,T)T,Y \rangle$, $B(X,Y) = \langle (*\mathbf{R})(X,T)T,Y \rangle$ with X, Y tangent to the leaves. Similarly, $|\nabla \mathbf{R}|^2 = |\nabla E|^2 + |\nabla B|^2$.

The bound (6.5) is scale invariant, and analogous to the bound (4.9) or (4.10) for stationary space-times, (where it of course holds in general). The bound on $|\nabla \mathbf{R}|$ in (6.5) is needed only for technical reasons, (related to Cauchy stability), and may be removed in certain natural situations.

The discussion above leads to the following result from [8], to which we refer for further discussion and details.

Theorem 6.2. *Let* $(\mathbf{M}, \mathbf{g})$ *be a cosmological space-time of non-positive Yamabe type. Suppose that the curvature assumption* (6.5) *holds, and that* $\mathbf{M}_{\mathcal{F}} = \mathbf{M}$ *to the future of* Σ.

Then $(\mathbf{M}, \mathbf{g})$ *is future geodesically complete and, for any sequence* $\tau_i \to 0$, *the slices* $(\Sigma_{\tau_i}, \bar{g}_{\tau_i})$, *cf.* (6.3), *have a subsequence converging to a weak geometrization of* Σ, *in the sense following Definition* 6.1.

We indicate some of the basic ideas in the proof. The first step is to show that the bound (6.5) on the ambient curvature $\mathbf{R}$, in this rescaling, gives uniform bounds on the intrinsic and extrinsic curvature of the leaves Σ_τ. The proof of this is similar to the proof of Theorem 5.1.

Given this, one can then apply the Cheeger-Gromov theory, as described in §2–§3. Given any sequence $\tau_i \to 0$, there exist subsequences which either converge, collapse or form cusps. From the work in §3, one knows that the regions of $(\Sigma_{\tau_i}, \bar{g}_{\tau_i})$ which (fully) collapse, or which are sufficiently collapsed, are graph manifolds. This gives rise to the region G in (6.4). It remains to show that, for any fixed $\varepsilon > 0$, the ε-thick region Σ^ε of $(\Sigma_{\tau_i}, \bar{g}_{\tau_i})$ converges to a hyperbolic metric.

The main ingredient in this is the following volume monotonicity result:

$$\frac{\mathrm{vol}_{g_\tau} \Sigma_\tau}{t_\tau^3} \downarrow, \tag{6.6}$$

i.e., the ratio is monotone non-increasing in the distance t_τ. This result is analogous to the Fischer-Moncrief monotonicity of the reduced Hamiltonian along the CMC Einstein flow, cf. [25]. The monotonicity (6.6) is easy to prove, and is an analogue of the Bishop-Gromov volume monotonicity (2.12). It follows from an analysis of the Raychaudhuri equation, much as in the Penrose-Hawking singularity theorems, together with a standard maximum principle.

Moreover, the ratio in (6.6) is constant on some interval $[\tau_1, \tau_2]$ if and only if the annular region $\tau^{-1}(\tau_1, \tau_2)$ is a time annulus in a flat Lorentzian cone

$$\mathbf{g_o} = -dt^2 + t^2 g_{-1}, \tag{6.7}$$

where g_{-1} is a hyperbolic metric. Again, the ratio in (6.6) is scale invariant, and so

$$\frac{\mathrm{vol}_{g_\tau} \Sigma_\tau}{t_\tau^3} = \mathrm{vol}_{\bar{g}_\tau} \Sigma_\tau. \tag{6.8}$$

In the non-collapse situation, $\mathrm{vol}_{\bar{g}_\tau} \Sigma_\tau$ is uniformly bounded away from 0 as $\tau \to 0$, (i.e. $t_\tau \to \infty$), and hence converges to a non-zero limit. On approach to the $\tau = 0$ limit, the ratio (6.6) tends to a constant, and hence the corresponding limit manifolds are of the form (6.7). This implies that ε-thick regions converge to hyperbolic metrics, giving rise to the H factor in (6.4).

It would be of interest to construct large families of examples of vacuum space-times exhibiting the conclusions (and hypotheses) of Theorem 6.2.

Recent Note: (January, 04). The recent work of Grisha Perelman [49]–[51], currently under evaluation in the mathematics community, implies a solution of Thurston's Geometrization Conjecture, and hence in particular the Poincaré Conjecture.

Acknowledgments

I am grateful to many of the participants of the Cargèse meeting for their comments and suggestions, and in particular to Piotr Chruściel and Helmut Friedrich for organizing such a fine meeting.

References

[1] M. Anderson, Convergence and rigidity of manifolds under Ricci curvature bounds, Inventiones Math., **106**, (1990), 429–445.

[2] M. Anderson, Short geodesics and gravitational instantons, Jour. Diff. Geom., **31**, (1990), 265–275.

[3] M. Anderson and J. Cheeger, C^α compactness for manifolds with Ricci curvature and injectivity radius bounded below, Jour. Diff. Geom., **35**, (1992), 265–281.

[4] M. Anderson, Hausdorff perturbations of Ricci-flat metrics and the splitting theorem, Duke Math. Jour., **68**, (1992), 67–82.

[5] M. Anderson, Extrema of curvature functionals on the space of metrics on 3-manifolds, Calc. Var. & P.D.E., **5**, (1997), 199–269.

[6] M. Anderson, On stationary vacuum solutions to the Einstein equations, Ann. Henri Poincaré, **1**, (2000), 977–994.

[7] M. Anderson, On the structure of solutions to the static vacuum Einstein equations, Ann. Henri Poincaré, **1**, (2000), 995–1042.

[8] M. Anderson, On long-time evolution in general relativity and geometrization of 3-manifolds, Comm. Math. Phys., **222**, (2001), 533–567.

[9] M. Anderson, Regularity for Lorentz metrics under curvature bounds, Jour. Math. Physics, **44**, (2003), 2994–3012.

[10] L. Andersson and V. Moncrief, Future complete vacuum spacetimes, this volume and gr-qc/0303045.

[11] R. Bartnik, Regularity of variational maximal surfaces, Acta Mathematica, **161**, (1988), 145–181.

[12] R. Bartnik, Remarks on cosmological space-times and constant mean curvature surfaces, Comm. Math. Phys., **117**, (1988), 615–624.

[13] J. Beem, P. Ehrlich and K. Easley, Global Lorentzian Geometry, 2nd Edition, Marcel Dekker, New York, (1996).

[14] R. Beig and W. Simon, Proof of a multipole conjecture due to Geroch, Comm. Math. Phys., **78**, (1980), 75–82.

[15] J. Bičák, Selected solutions of Einstein's field equations: their role in General Relativity and Astrophysics, in Einstein's Field Equations and their Physical Applications, Lecture Notes in Physics, **540**, Springer Verlag, Berlin, (2000), 1–126.

[16] J. Cheeger, Finiteness theorems for Riemannian manifolds, Amer. Jour. Math., **92**, (1970), 61–75.

[17] J. Cheeger and D. Gromoll, The splitting theorem for manifolds of non-negative Ricci curvature, Jour. Diff. Geom., **6**, (1971), 119–128.

[18] J. Cheeger and M. Gromov, Collapsing Riemannian manifolds while keeping their curvature bounded, I, II, Jour. Diff. Geom., **23**, (1986), 309–346 and **32**, (1990), 269–298.

[19] Y. Choquet-Bruhat, Solutions C^∞ d'équations hyperboliques non-linéaires, C.R. Acad. Sci. Paris, Sér. A-B, **272**, (1971), A386–A388.

[20] Y. Choquet-Bruhat and V. Moncrief, Future global in time Einsteinian spacetimes with $U(1)$ isometry group, Ann. Henri Poincaré, **2**, (2001), 1007–1064.

[21] D. Christodoulou and S. Klainerman, The Global Nonlinear Stability of the Minkowski Space, Princeton Univ. Press, Princeton, (1993).

[22] C.J.S. Clarke, The Analysis of Space-Time Singularities, Camb. Lect. Notes in Physics, **1**, Camb. Univ. Press, (1993).

[23] D. Deturck and J. Kazdan, Some regularity theorems in Riemannian geometry, Ann. Sci. Ecole Norm. Sup, **14**, (1981), 249–260.

[24] A. Einstein and W. Pauli, Non-existence of regular stationary solutions of relativistic field equations, Annals of Math., ser. 2, **44**, (1943), 131–137.

[25] A. Fischer and V. Moncrief, The reduced Hamiltonian of general relativity and the σ-constant of conformal geometry, in Math. and Quantum Aspects of Relativity and Cosmology, S. Cotsakis and G.W. Gibbons, (Eds.), Springer Lecture Notes in Physics, **537**, (2000), 70–101.

[26] H. Friedrich, On the existence of n-geodesically complete or future complete solutions of Einstein's field equations with smooth asymptotic structure, Comm. Math. Phys., 107, (1986), 587–609.

[27] K. Fukaya, Collapsing Riemannian manifolds to ones of lower dimension, Jour. Diff. Geom., **25**, (1987), 139–156.

[28] D. Garfinkle and M. Melvin, Generalized magnetic universe solutions, Phys. Rev. **D50**, (1994), 3859–3866.

[29] C. Gerhardt, H-surfaces in Lorentzian manifolds, Comm. Math. Phys., **89**, (1983), 523–553.

[30] G.W. Gibbons and S.W. Hawking, Classification of gravitational instanton symmetries, Comm. Math. Phys., **66**, (1979), 291–310.

[31] M. Gromov, Structures Métriques Pour Les Variétés Riemanniennes, Cédic-Nathan, Paris, (1981).

[32] M. Gromov, Almost flat manifolds, Jour. Diff. Geom., **13**, (1978), 231–241.

[33] M. Gromov, Volume and bounded cohomology, Publ. Math. I.H.E.S., **56**, (1982), 5–99.

[34] C. Gundlach, Critical phenomena in gravitational collapse, Living Reviews in Relativity, **2:4**, (1999), www.livingreviews.org

[35] S. Hawking and G. Ellis, The Large Scale Structure of Space-Time, Cambridge Univ. Press, London, (1973).

[36] J. Jost and H. Karcher, Geometrische Methoden zur Gewinnung von a priori Schranken für harmonische Abbildungen, Manuscripta Math., 40, (1982), 21–71.

[37] A. Lichnerowicz, Théories Relativistes de la Gravitation et de l'Electromagnétisme, Masson, Paris, (1955).

[38] J. Marsden and F. Tipler, Maximal hypersurfaces and foliations of constant mean curvature in general relativity, Physics Reports, **66:3**, (1980), 109–139.

[39] M. Melvin, Pure magnetic and electric geons, Physics Letters, **8:1**, (1964), 65–68.

[40] R. Myers, Higher-dimensional black holes in compactified space-times, Phys. Rev. D., **35**, (1987), 455–466.

[41] P. Orlik, Seifert Manifolds, Springer Lecture Notes in Math., **291**, Springer Verlag, New York, (1972).

[42] R. Penrose, Any space-time has a plane-wave as a limit, in Differential Geometry and Relativity, Ed. Cahan and Flato, (1976), 271–275.

[43] P. Petersen, Riemannian Geometry, Graduate Texts in Mathematics, Springer Verlag, New York, (1998).

[44] A. Rendall, Constant mean curvature foliations in cosmological space-times, Helv. Physica Acta, **69:4**, (1996), 490–500.

[45] A. Rendall, Compact null hypersurfaces and collapsing Riemannian manifolds, Math. Nachrichten, **193**, (1998), 111–118.

[46] W. Thurston, Three-dimensional manifolds, Kleinian groups and hyperbolic geometry, Bull. Amer. Math. Soc., **6**, (1982), 357–381.

[47] F. Tipler, C. Clarke and G. Ellis, Singularities and horizons: a review article, in Gen. Rel. and Gravitation, vol. 2, A. Held, (Ed.), Plenum Press, New York, (1980), 87–206.

[48] F. Waldhausen, Eine Klasse von 3-dimensionalen Mannifaltigkeiten, I, II, Invent. Math. **3**, (1967), 308–333 and **4**, (1967), 87–117.

[49] G. Perelman, The entropy formula for the Ricci flow and its geometric applications, math.DG/0211159.

[50] G. Perelman, Ricci flow with surgery on three-manifolds, math.DG/0303109.

[51] G. Perelman, Finite extinction time for the solutions to the Ricci flow on certain three-manifolds, math.DG/0307245.

Michael T. Anderson
Department of Mathematics
S.U.N.Y. at Stony Brook
Stony Brook, N.Y. 11794-3651, USA
e-mail: anderson@math.sunysb.edu

Null Geometry and the Einstein Equations

Gregory J. Galloway

1. Introduction

This paper is based on a series of lectures given by the author at the Cargèse Summer School on Mathematical General Relativity and Global Properties of Solutions of Einstein's Equations, held in Corsica, July 29–August 10, 2002. The general aim of those lectures was to illustrate with some current examples how the methods of global Lorentzian geometry and causal theory may be used to obtain results about the global behavior of solutions to the Einstein equations. This, of course, is a long standing program, dating back to the singularity theorems of Hawking and Penrose [24]. Here we consider some properties of asymptotically de Sitter solutions to the Einstein equations with (by our sign conventions) positive cosmological constant, $\Lambda > 0$. We obtain, for example, some rather strong topological obstructions to the existence of such solutions, and, in another direction, present a uniqueness result for de Sitter space, associated with the occurrence of *eternal* observer horizons. As described later, these results have rather strong connections with Friedrich's results [11, 13] on the nonlinear stability of asymptotically simple solutions to the Einstein equations with $\Lambda > 0$; see also Friedrich's article elsewhere in this volume. The main theoretical tool from global Lorentzian geometry used to prove these results is the so-called null splitting theorem [16]. This theorem is discussed here, along with relevant background material.

The paper is divided into sections as follows. In Section 2 we present the basic elements of causal theory, emphasizing those parts of the subject needed for our work. In Section 3 we give a self-contained treatment of the geometry of smooth null hypersurfaces, and present a maximum principle for such hypersurfaces. In Section 4 we extend this maximum principle to the essential C^0 setting, and discuss the null splitting theorem. In Section 5 we consider the aforementioned applications.

Supported in part by NSF grant # DMS-0104042.

2. Elements of causal theory

Much of our work makes use of results from the causal theory of Lorentzian manifolds. In this section we recall some of the basic notions and notations of this subject, with an emphasis on what shall be needed later. There are many excellent treatments of causal theory, all varying somewhat in perspective and degree of rigor; see for example, [31, 24, 29, 33, 3].

Let M^{n+1} be a Lorentzian manifold, i.e., a smooth Hausdorff manifold equipped with a smooth metric $g = \langle\,,\,\rangle$ of Lorentz signature $(-+\cdots+)$. A vector $X \in T_pM$ is timelike (resp., null, causal, spacelike) provided $\langle X, X\rangle < 0$ (resp., $\langle X, X\rangle = 0, \leq 0, > 0$). This terminology extends to curves: A smooth curve $t \to \sigma(t)$ is timelike (resp., null, causal, spacelike) provided each of its velocity vectors $\sigma'(t)$ is timelike (resp., null, causal, spacelike). The causal character of curves extends in a natural way to piecewise smooth curves. For each $p \in M$, the set of null vectors at p-forms a double cone in T_pM. M is said to be time orientable provided the assignment of a future cone and past cone at each point of M can be made in a continuous manner over M. By a *spacetime*, we mean a connected, time oriented Lorentzian manifold M. Henceforth we restrict attention to spacetimes.

Let ∇ denote the Levi-Civita connection of M. Hence, for vector fields $X = X^a$ and $Y = Y^b$, $\nabla_X Y = X^a \nabla_a Y^b$ denotes the covariant derivative of Y with respect to X. For the most part we use index free notation. By definition, geodesics are curves $t \to \sigma(t)$ of zero covariant acceleration, $\nabla_{\sigma'}\sigma' = 0$.

The Riemann curvature tensor $(X, Y, Z) \to R(X, Y)Z$ is defined by,

$$R(X,Y)Z = \nabla_X \nabla_Y Z - \nabla_Y \nabla_X Z - \nabla_{[X,Y]} Z. \qquad (2.1)$$

The components of the curvature tensor are determined by the equation,

$$R(\partial_i, \partial_j)\partial_k = R^\ell{}_{kij}\partial_\ell.$$

The Ricci tensor and scalar curvature are obtained by tracing, $R_{ij} = R^\ell{}_{i\ell j}$ and $R = g^{ij}R_{ij}$.

We now introduce the notations for futures and pasts. $I^+(p)$ (resp., $J^+(p)$), the timelike (resp., causal) future of $p \in M$, is the set of points $q \in M$ for which there exists a future directed timelike (resp., causal) curve from p to q. Since small deformations of timelike curves remain timelike, the sets $I^+(p)$ are always open. However, the sets $J^+(p)$ need not in general be closed. To emphasize the particular spacetime involved, one sometimes writes $I^+(p, M)$, etc. More generally, for $A \subset M$, $I^+(A)$ (resp., $J^+(A)$), the timelike (resp., causal) future of A, is the set of points $q \in M$ for which there exists a future directed timelike (resp., causal) curve from a point $p \in A$ to q. Note, $I^+(A) = \cup_{p \in A} I^+(p)$, and hence is open.

By variational techniques one can establish the following fundamental causality result; cf. [29, p. 294].

Proposition 2.1. *If* $q \in J^+(p) \setminus I^+(p)$, *then any future directed causal curve from p to q must be a null geodesic (when suitably parametrized).*

The timelike and causal pasts $I^-(p)$, $J^-(p)$, $I^-(A)$, $J^-(A)$ are defined in a time dual manner.

Many causally defined sets of interest, for example, horizons of various sorts, arise essentially as *achronal boundaries*. By definition, an achronal boundary is a set of the form $\partial I^+(A)$ (or $\partial I^-(A)$), for some $A \subset M$. We wish to describe several important structural properties of achronal boundaries.

Proposition 2.2. *An achronal boundary $\partial I^+(A)$, if nonempty, is a closed achronal C^0 hypersurface in M.*

Recall, an achronal set is a subset of spacetime for which no two points can be joined by a timelike curve. We discuss briefly the proof of the proposition, beginning with the following simple lemma.

Lemma 2.3. *If $p \in \partial I^+(A)$ then $I^+(p) \subset I^+(A)$, and $I^-(p) \subset M \setminus \overline{I^+(A)}$.*

To prove the first part of the lemma, note that if $q \in I^+(p)$ then $p \in I^-(q)$, and hence $I^-(q)$ is a neighborhood of p. Since p is on the boundary of $I^+(A)$, it follows that $I^-(q) \cap I^+(A) \neq \emptyset$, and hence $q \in I^+(A)$. The second part of the lemma is proved similarly.

Since $I^+(A)$ is open, it does not meet its boundary. The first part of Lemma 2.3 then implies that $\partial I^+(A)$ is achronal. Lemma 2.3 also implies that $\partial I^+(A)$ is *edgeless*. The edge of an achronal set $S \subset M$ is the set of points $p \in \overline{S}$ such that every neighborhood U of p, contains a timelike curve from $I^-(p, U)$ to $I^+(p, U)$ that does *not* meet S. But Lemma 2.3 shows that for any $p \in \partial I^+(A)$, any timelike curve from $I^-(p)$ to $I^+(p)$ must meet $\partial I^+(A)$. The remainder of Proposition 2.2 then follows from the fact (cf. [29, p. 413]) that an achronal set without edge points is a C^0 hypersurface of M.

The next result shows that, in general, large portions of achronal boundaries are ruled by null geodesics.

Proposition 2.4. *Let $A \subset M$ be closed. Then each $p \in \partial I^+(A) \setminus A$ lies on a null geodesic contained in $\partial I^+(A)$, which either has a past end point on A, or else is past inextendible in M.*

We give a sketch of the proof. Choose a sequence of points $\{p_n\} \subset I^+(A)$ such that $p_n \to p$, and let γ_n be a past directed timelike curve from p_n to a point of A. By passing to a subsequence if necessary, $\{\gamma_n\}$ converges, in a suitable sense, to a past directed causal curve γ from p, which must be contained in $\partial I^+(A)$. We overlook here the technical difficulty that the limit curve γ need not be smooth; this, can be dealt with, however; see, [3, Sections 3.3, 14.1]. Since each segment of γ is both causal and achronal, it follows from Proposition 2.1 that γ is a null geodesic. Now, each γ_n is past inextendible in $M \setminus A$, and hence so is γ. Thus γ either has a past end point on A or is past inextendible in M.

Finally we make some remarks and recall some facts about global hyperbolicity. A spacetime M is *strongly causal* at $p \in M$ provided there are arbitrarily small neighborhoods U of p such that any causal curve γ which starts in, and leaves,

U never returns to U. M is *strongly causal* if it is strongly causal at each of its points. Thus, heuristically speaking, M is strongly provided there are no closed or "almost closed" causal curves in M.

A spacetime M is said to be *globally hyperbolic* provided (i) M is strongly causal and (ii) the sets $J^+(p) \cap J^-(q)$ are compact for all $p, q \in M$. The latter condition rules out the occurrence of naked singularities, and hence global hyperbolicity is closely related to the notion of cosmic censorship. Global hyperbolicity is also related to the existence of an ideal initial value hypersurface, that is to say a *Cauchy surface*, in spacetime. There are slight variations in the literature in the definition of a Cauchy surface. Here we adopt the following definition: A Cauchy surface for a spacetime M is an achronal C^0 hypersurface S of M which is met by every inextendible causal curve in M. We now recall several fundamental results.

Proposition 2.5. *M is globally hyperbolic if and only if M admits a Cauchy surface. If S is a Cauchy surface for M then M is homeomorphic to $\mathbb{R} \times S$.*

Along similar lines, one has that any two Cauchy surfaces in a given globally hyperbolic spacetime are homeomorphic. Hence, according to Proposition 2.5, any nontrivial topology in a globally hyperbolic spacetime must reside in its Cauchy surfaces. The following fact is often useful.

Proposition 2.6. *If S is a compact achronal hypersurface in a globally hyperbolic spacetime M then S must be a Cauchy surface for M.*

Cauchy surfaces can be characterized in terms of the domain of dependence. Let S be an achronal subset of M. The *future domain of dependence* of S is the set $D^+(S)$ consisting of all points $p \in M$ such that every past inexendendible causal curve from p meets S. Physically, $D^+(S)$ is the part of spacetime to the future of S that is predictable from S. The *future Cauchy horizon* of S, $H^+(S)$, is the future boundary of $D^+(S)$; formally, $H^+(S) = \overline{D^+(S)} \setminus I^-(D^+(S))$. Physically, $H^+(S)$ is the future limit of the part of spacetime predictable from S. The past domain of dependence $D^-(S)$ and past Cauchy horizon $H^-(S)$ are defined in a time-dual manner. Set $D(S) = D^+(S) \cup D^-(S)$ and $H(S) = H^+(S) \cup H^-(S)$; one has $\partial D(S) = H(S)$. Then, it is a basic fact that an achronal set S is a Cauchy surface for M iff $D(S) = M$ iff $H(S) = \emptyset$. Cauchy horizons have structural properties similar to achronal boundaries, as indicated in the following.

Proposition 2.7. *Let S be an achronal subset of a spacetime M. Then $H^+(S) \setminus$ edge S, if nonempty, is an achronal C^0 hypersurface of M ruled by null geodesics, each of which either is past inextendible M or has past end point on edge S.*

The proof of Proposition 2.7 is roughly similar to the proofs of Propositions 2.2 and 2.4. Proposition 2.6 can now be easily proved by showing, with the aid of Proposition 2.7, that $H(S) = \emptyset$.

We conclude this brief presentation with the following basic facts about global hyperbolicity.

Proposition 2.8. *Let M be a globally hyperbolic spacetime. Then,*

(i) *M is causally simple, i.e., the sets $J^\pm(A)$ are closed, for all compact $A \subset M$.*
(ii) *The sets $J^+(A) \cap J^-(B)$ are compact, for all compact $A, B \subset M$.*

3. The geometry of smooth null hypersurfaces

Here we review some aspects of the geometry of null hypersurfaces, along the lines developed in [16, 26], and present a maximum principle for such hypersurfaces.

Let (M^{n+1}, g) be a spacetime, with $n \geq 2$. A *(smooth) null hypersurface* in M is a smooth co-dimension one embedded submanifold S of M such that the pull-back of the metric g to S is degenerate. Because of the Lorentz signature of g, the null space of the pull-back is one-dimensional at each point of S. Hence, every null hypersurface S admits a smooth nonvanishing future directed null vector field $K \in \Gamma T S$ such that the normal space of K at $p \in S$ coincides with the tangent space of S at p, i.e., $K_p^\perp = T_p S$ for all $p \in S$. It follows, in particular, that tangent vectors to S not parallel to K are spacelike. Note also that the vector field K is unique up to a positive (pointwise) scale factor. The following fact is fundamental.

Proposition 3.1. *The integral curves of K when suitably parameterized, are null geodesics.*

Proof. It suffices to show that $\nabla_K K = \lambda K$. This will follow by showing that at each $p \in S$, $\nabla_K K \perp T_p S$, i.e., $\langle \nabla_K K, X \rangle = 0$ for all $X \in T_p S$. Extend $X \in T_p S$ by making it invariant under the flow generated by K, $[K, X] = \nabla_K X - \nabla_X K = 0$. X remains tangent to S, so along the flow line through p, $\langle K, X \rangle = 0$. Differentiating we obtain,

$$0 = K\langle K, X \rangle = \langle \nabla_K K, X \rangle + \langle K, \nabla_K X \rangle,$$

and hence,

$$\langle \nabla_K K, X \rangle = -\langle K, \nabla_X K \rangle = -\frac{1}{2} X \langle K, K \rangle = 0. \qquad \square$$

The integral curves of K are called the *null geodesic generators* of S.

Since K is orthogonal to S we can introduce the null Weingarten map and null second fundamental form of S with respect K in a manner roughly analogous to what is done for spacelike hypersurfaces or hypersurfaces in a Riemannian manifold. For technical reasons, one works "mod K", as described below.

We introduce the following equivalence relation on tangent vectors: For X, $X' \in T_p S$, $X' = X \mod K$ if and only if $X' - X = \lambda K$ for some $\lambda \in \mathbb{R}$. Let $\overline{X}$ denote the equivalence class of X. Let $T_p S/K = \{\overline{X} : X \in T_p S\}$, and $TS/K = \cup_{p \in S} T_p S/K$. TS/K, the mod K tangent bundle of S, is a smooth rank $n-1$ vector bundle over S. This vector bundle does not depend on the particular choice of null vector field K. There is a natural positive definite metric h on TS/K induced from $\langle , \rangle$: For each $p \in S$, define $h : T_p S/K \times T_p S/K \to \mathbb{R}$ by $h(\overline{X}, \overline{Y}) = \langle X, Y \rangle$. A simple computation shows that h is well defined.

The *null Weingarten map* $b = b_K$ of S with respect to K is, for each point $p \in S$, a linear map $b : T_p S/K \to T_p S/K$ defined by $b(\overline{X}) = \overline{\nabla_X K}$. It is easily checked that b is well defined. Note if $\widetilde{K} = fK$, $f \in C^\infty(S)$, is any other future directed null vector field tangent to S, then $\nabla_X \widetilde{K} = f\nabla_X K$ mod K. It follows that the Weingarten map b of S is unique up to positive scale factor and that b at a given point $p \in S$ depends only on the value of K at p.

A standard computation shows, $h(b(\overline{X}), \overline{Y}) = \langle \nabla_X K, Y \rangle = \langle X, \nabla_Y K \rangle = h(\overline{X}, b(\overline{Y}))$. Hence b is self-adjoint with respect to h. The *null second fundamental form* $B = B_K$ of S with respect to K is the bilinear form associated to b via h: For each $p \in S$, $B : T_p S/K \times T_p S/K \to \mathbb{R}$ is defined by $B(\overline{X}, \overline{Y}) = h(b(\overline{X}), \overline{Y}) = \langle \nabla_X K, Y \rangle$. Since b is self-adjoint, B is symmetric. We say that S is *totally geodesic* iff $B \equiv 0$. This has the usual geometric meaning: If S is totally geodesic then any geodesic in M starting tangent to S stays in S. This follows from the fact that, when S is totally geodesic, the restriction to S of the Levi-Civita connection of M defines an affine connection on S. Null hyperplanes in Minkowski space are totally geodesic, as is the event horizon in Schwarzschild spacetime.

The *null mean curvature* of S with respect to K is the smooth scalar field θ on S defined by, $\theta = \operatorname{tr} b$. Let Σ be the intersection of S with a hypersurface in M which is transverse to K near $p \in S$; Σ will be an $n - 1$-dimensional spacelike submanifold of M. Let $\{e_1, e_2, \ldots, e_{n-1}\}$ be an orthonormal basis for $T_p\Sigma$ in the induced metric. Then $\{\overline{e}_1, \overline{e}_2, \ldots, \overline{e}_{n-1}\}$ is an orthonormal basis for $T_p S/K$. Hence at p,

$$\theta = \operatorname{tr} b = \sum_{i=1}^{n-1} h(b(\overline{e}_i), \overline{e}_i) = \sum_{i=1}^{n-1} \langle \nabla_{e_i} K, e_i \rangle.$$
$$= \operatorname{div}_\Sigma K$$

Thus, the null mean curvature gives a measure of the divergence towards the future of the null generators of S. Note that if $\widetilde{K} = fK$ then $\widetilde{\theta} = f\theta$. Thus the null mean curvature inequalities $\theta \geq 0$, $\theta \leq 0$, etc., are invariant under positive rescaling of K. In Minkowski space, a future null cone $S = \partial I^+(p) \setminus \{p\}$ (resp., past null cone $S = \partial I^-(p) \setminus \{p\}$) has positive null mean curvature, $\theta > 0$ (resp., negative null mean curvature, $\theta < 0$).

The null second fundamental form of a null hypersurface obeys a well-defined comparison theory roughly similar to the comparison theory satisfied by the second fundamental forms of a family of parallel spacelike hypersurfaces (cf., Eschenburg [9], which we follow in spirit).

Let $\eta : (a, b) \to M$, $s \to \eta(s)$, be a future directed affinely parameterized null geodesic generator of S. For each $s \in (a, b)$, let

$$b(s) = b_{\eta'(s)} : T_{\eta(s)} S/\eta'(s) \to T_{\eta(s)} S/\eta'(s) \tag{3.2}$$

be the Weingarten map based at $\eta(s)$ with respect to the null vector $K = \eta'(s)$. The one parameter family of Weingarten maps $s \to b(s)$, obeys the following

Ricatti equation,

$$b' + b^2 + R = 0. \tag{3.3}$$

Here $'$ denotes covariant differentiation in the direction $\eta'(s)$: In general, if $Y = Y(s)$ is a vector field along η tangent to S, we define, $(\overline{Y})' = \overline{Y'}$. Then, if $X = X(s)$ is a vector field along η tangent to S, b' is defined by,

$$b'(\overline{X}) = b(\overline{X})' - b(\overline{X'}). \tag{3.4}$$

$R : T_{\eta(s)} S/\eta'(s) \to T_{\eta(s)} S/\eta'(s)$ is the curvature endomorphism defined by $R(\overline{X}) = \overline{R(X, \eta'(s))\eta'(s)}$.

We indicate the proof of Equation 3.3. Fix a point $p = \eta(s_0)$, $s_0 \in (a, b)$, on η. On a neighborhood U of p in S we can scale the null vector field K so that K is a geodesic vector field, $\nabla_K K = 0$, and so that K, restricted to η, is the velocity vector field to η, i.e., for each s near s_0, $K_{\eta(s)} = \eta'(s)$. Let $X \in T_p M$. Shrinking U if necessary, we can extend X to a smooth vector field on U so that $[X, K] = \nabla_X K - \nabla_K X = 0$. Then, $R(X, K)K = \nabla_X \nabla_K K - \nabla_K \nabla_X K - \nabla_{[X,K]} K = -\nabla_K \nabla_K X$. Hence along η we have, $X'' = -R(X, \eta')\eta'$ (which implies that X, restricted to η, is a Jacobi field along η). Thus, from Equation 3.4, at the point p we have,

$$
\begin{aligned}
b'(\overline{X}) &= \overline{\nabla_X K}' - b(\overline{\nabla_K X}) = \overline{\nabla_K X}' - b(\overline{\nabla_X K}) \\
&= \overline{X''} - b(b(\overline{X})) = -\overline{R(X, \eta')\eta'} - b^2(\overline{X}) \\
&= -R(\overline{X}) - b^2(\overline{X}),
\end{aligned}
$$

which establishes Equation 3.3.

By taking the trace of 3.3 we obtain the following formula for the derivative of the null mean curvature $\theta = \theta(s)$ along η,

$$\theta' = -\mathrm{Ric}(\eta', \eta') - \sigma^2 - \frac{1}{n-1}\theta^2, \tag{3.5}$$

where σ, the shear scalar, is the trace of the square of the trace free part of b. Equation 3.5 is the well-known Raychaudhuri equation (for an irrotational null geodesic congruence) of relativity. This equation shows how the Ricci curvature of spacetime influences the null mean curvature of a null hypersurface.

The following proposition is a standard application of the Raychaudhuri equation, a C^0 version of which will be needed later.

Proposition 3.2. *Let M be a spacetime which obeys the null energy condition (NEC), $\mathrm{Ric}(X, X) \geq 0$ for all null vectors X, and let S be a smooth null hypersurface in M. If the null generators of S are future geodesically complete then S has nonnegative null mean curvature, $\theta \geq 0$.*

Proof. Suppose $\theta < 0$ at $p \in S$. Let $s \to \eta(s)$ be the null generator of S passing through $p = \eta(0)$, affinely parametrized. Let $b(s) = b_{\eta'(s)}$, and take $\theta = \mathrm{tr}\, b$. By the invariance of sign under scaling, one has $\theta(0) < 0$. Raychaudhuri's equation and the NEC imply that $\theta = \theta(s)$ obeys the inequality,

$$\frac{d\theta}{ds} \leq -\frac{1}{n-1}\theta^2 , \tag{3.6}$$

and hence $\theta < 0$ for all $s > 0$. Dividing through by θ^2 then gives,

$$\frac{d}{ds}\left(\frac{1}{\theta}\right) \geq \frac{1}{n-1} , \tag{3.7}$$

which implies $1/\theta \to 0$, i.e., $\theta \to -\infty$ in finite affine parameter time, contradicting the smoothness of θ. ◻

Proposition 3.2 implies, under the given assumptions, that cross sections of S do not decrease in area as one moves towards the future. Proposition 3.2 is the most rudimentary form of Hawking's black hole area theorem. For a recent study of the area theorem, with a focus on issues of regularity, see [7].

3.1. Maximum principle for smooth null hypersurfaces

We present a maximum principle for smooth null hypersurfaces analogous to that for hypersurfaces in Riemannian manifolds and spacelike hypersurfaces in Lorentzian manifolds. Because of its natural invariance we restrict attention to the zero mean curvature case.

Theorem 3.3. *Let S_1 and S_2 be smooth null hypersurfaces in a spacetime M. Suppose,*

(1) S_1 and S_2 meet at $p \in M$ and S_2 lies to the future side of S_1 near p, and
(2) the null mean curvature scalars θ_1 of S_1, and θ_2 of S_2, satisfy, $\theta_2 \leq 0 \leq \theta_1$.
Then S_1 and S_2 coincide near p and this common null hypersurface has null mean curvature $\theta = 0$.

The heuristic here is that since the generators of S_1 are nonconverging, and the generators of S_2, which lie to the future of S_1 are nondiverging, the two sets of generators are forced to agree and form a nonexpanding congruence.

Proof. We give a sketch of the proof; for details, see [16]. (N.B. There is a bad typo in the statement of Theorem II.1 in [16], in which the mean curvature inequalities appear reversed.)

S_1 and S_2 have a common null direction at p. Let Q be a timelike hypersurface in M passing through p and transverse to this direction. By taking Q small enough, the intersections $\Sigma_1 = S_1 \cap Q$ and $\Sigma_2 = S_2 \cap Q$ will be smooth spacelike hypersurfaces of Q, with Σ_2 to the future side of Σ_1 near p.

Σ_1 and Σ_2 may be expressed as graphs over a fixed spacelike hypersurface V of Q (with respect to Gaussian normal coordinates), $\Sigma_1 = \text{graph}(u_1)$, $\Sigma_2 = \text{graph}(u_2)$. Let,

$$\theta(u_i) = \theta_i|_{\Sigma_i = \text{graph}(u_i)} , \quad i = 1, 2 .$$

By a computation,

$$\theta(u_i) = H(u_i) + \text{ lower-order terms},$$

where H is the mean curvature operator on spacelike graphs over V in Q. (The lower-order terms involve the second fundamental form of Q.) Thus θ is a second-order quasi-linear elliptic operator. In the present situation we have:

(i) $u_1 \leq u_2$, and $u_1(p) = u_2(p)$.
(ii) $\theta(u_2) \leq 0 \leq \theta(u_1)$.

A suitable version of the strong maximum principle then implies, $u_1 = u_2$. Thus, Σ_1 and Σ_2 agree near p. The normal geodesics to Σ_1 and Σ_2 in M will then also agree. This implies that S_1 and S_2 agree near p. □

4. C^0 null hypersurfaces and the null splitting theorem

The usefulness of the maximum principle for smooth null hypersurfaces presented in the previous section is limited by the fact that the most interesting null hypersurfaces arising in general relativity, e.g., horizons of various sorts, are C^0, but in general not C^1. Such hypersurfaces often arise as (the null portions of) achronal boundaries $\partial I^{\pm}(A)$. For example, (i) in black hole spacetimes, the (future) event horizon is defined as, $H = \partial I^-(\mathcal{I}^+)$, where $\mathcal{I}^+$ is *future null infinity* (roughly, the ideal boundary of future end points of null geodesics that escape to infinity), and (ii) the observer horizon of an observer (future inextendible timelike curve) γ is defined as $\partial I^-(\gamma)$. The aim of this section is to present a maximum principle for C^0 null hypersurfaces, similar in spirit to the maximum principle for C^0 spacelike hypersurfaces obtained in [2], and to describe how this is used to prove the null splitting theorem.

From the properties of achronal boundaries discussed in Section 2, a set of the form $S = \partial I^-(A) \setminus A$, with A closed, is an achronal C^0 hypersurface ruled by null geodesics which are future inextendible in S (in fact, which are future inextendible in the sub-spacetime $M \setminus A$). Though future inextendible in S, the null geodesics ruling S (i.e., the *null generators* of S) may have past end points on S. Consider, for example, the set $S = \partial I^-(A) \setminus A$, where A consists of two disjoint closed disks in the $t = 0$ slice of Minkowski 3-space. This surface, which represents the merger of two truncated cones, has a "crease", i.e., a curve of nondifferentiable points (corresponding to the intersection of the two cones) but which otherwise is a smooth null hypersurface. The null generators of S that reach the crease, leave S when extended to the past.

Sets of the form $\partial I^{\pm}(A) \setminus A$, A closed, are models for our notion of C^0 null hypersurfaces.

Definition 4.1. *A C^0 **future** (resp., **past**) null hypersurface is a locally achronal C^0 hypersurface ruled by null geodesics which are future (resp., past) inextendible in S.*

C^0 null hypersurfaces do not in general have null mean curvature in the classical sense, but may obey null mean curvature inequalities in the *support sense*, as described below.

Let S be a C^0 future null hypersurface, and let $p \in S$. A smooth null hypersurface W is said to be a past support hypersurface for S at p provided W passes through p and lies to the past of S near p. We note that any C^0 future null hypersurface S is supported from below at each point p by a smooth null hypersurface. Indeed, one may take $W = \partial I^-(q, U) \setminus \{q\}$, where U is a convex normal neighborhood of p, and $q \in U$ is on a null generator of S from p, slightly to the future of p. If S is actually smooth then, by an elementary comparison, $\theta_S(p) \geq \theta_W(p)$, provided the future directed null vector fields K_S and K_W used to define the null second fundamental forms on S and W, respectively, are scaled to agree at p. (Time-dual statement holds for C^0 past null hypersurfaces.) These considerations lead to the following definition.

Definition 4.2. *Let S be a C^0 future null hypersurface in M. We say that S has null mean curvature $\theta \geq 0$ **in the support sense** provided for each $p \in S$ and for each $\epsilon > 0$ there exists a smooth (at least C^2) null hypersurface $W_{p,\epsilon}$ such that,*

(1) *$W_{p,\epsilon}$ is a past support hypersurface for S at p, and*
(2) *the null mean curvature of $W_{p,\epsilon}$ at p satisfies $\theta_{p,\epsilon} \geq -\epsilon$.*

For this definition, it is assumed that the null vectors have been uniformly scaled, e.g., have unit length with respect to a fixed background Riemannian metric, otherwise the inequality in (2) would be meaningless. Note that if S is smooth and satisfies Definition 4.2 then $\theta_S \geq 0$ in the usual sense. If S is a C^0 past null hypersurface, one defines $\theta \leq 0$ *in the support sense* in an analogous manner in terms of *future* support hypersurfaces.

As a simple illustration of Definition 4.2, consider the future null cone $S = \partial I^+(p)$, where p is a point in Minkowski space. S is a C^0 future null hypersurface having mean curvature $\theta \geq 0$ in the support sense: One may use null hyperplanes, even at the vertex, as support hypersurfaces. A less trivial illustration is provided by the next proposition, which is a C^0 version of Proposition 3.2.

Proposition 4.1. *Let M be a spacetime which satisfies the NEC. Suppose S is a C^0 future null hypersurface in M whose null generators are future geodesically complete. Then S has null mean curvature $\theta \geq 0$ in the support sense.*

Proof. By restricting attention to a sufficiently small neighborhood of S, we may assume without loss of generality that S is globally achronal. Recall that if a null geodesic η contains a pair null conjugate points, then points of η are timelike related [29, p. 296]. Thus it follows that the null generators of S are free of null conjugate points.

Given $p \in S$, let $\eta : [0, \infty) \to S \subset M$, $s \to \eta(s)$, be a future directed affinely parametrized null geodesic generator of S from $p = \eta(0)$. For any $r > 0$, consider a small pencil of past directed null geodesics from $\eta(r)$ about η. This pencil, taken sufficiently small, will form a smooth (caustic free) null hypersurface $W_{p,r}$ containing $\eta([0,r))$. Moreover, since $W_{p,r} \subset J^-(S)$, it will be a lower support hypersurface for S at p.

Let $\theta = \theta(s)$, $0 \le s < r$, be the null mean curvature of $W_{p,r}$ along $\eta|_{[0,r)}$, where, as in the notation of Equation (3.2), $b(s) = b_{\eta'(s)}$. The differential inequality (3.6), which holds in the present situation, together with the initial condition $\theta(r) = -\infty$, implies,

$$\theta(0) \ge -\frac{n-1}{r}.$$

Since r can be taken arbitrarily large, the proposition follows. $\square$

Proposition 4.1 applies, in particular, to future event horizons in black hole spacetimes, in which the null generators are future complete. This fact provided the initial impetus for the development of a proof of the black hole area theorem which does not require the imposition of smoothness assumptions on the horizon; cf. [7].

We now present the maximum principle for C^0 null hypersurfaces.

Theorem 4.2. (Maximum Principle for C^0 null hypersurfaces.) *Let S_1 be a C^0 future null hypersurface and S_2 be C^0 past null hypersurface in a spacetime M. Suppose,*

(1) *S_1 and S_2 meet at $p \in M$, with S_2 lying to the future side of S_1 near p, and*

(2) *S_1 and S_2 have null mean curvatures satisfying, $\theta_2 \le 0 \le \theta_1$ in the support sense.*

Then S_1 and S_2 coincide near p and form a smooth null hypersurface with null mean curvature $\theta = 0$.

Comments on the proof: We first mention that, for simplicity, we have omitted a technical assumption from the statement of the theorem; there is a requirement that the second fundamental forms of the support hypersurfaces obey a local one-sided bound. In many geometric applications, this technical condition is automatically satisfied; cf., [16] for details. Although there are some significant technical issues, the proof proceeds more or less along the lines of the proof of Theorem 3.3.

One first observes that the point p is an interior point of a null generator common to both S_1 and S_2 near p. As before, one intersects S_1 and S_2 with a timelike hypersurface Q through p which is transverse to this generator. By taking Q small enough, the intersections $\Sigma_1 = S_1 \cap Q$ and $\Sigma_2 = S_2 \cap Q$ will be acausal C^0 hypersurfaces in Q passing through p, with Σ_2 to the future of Σ_1. One can again express Σ_1 and Σ_2 as graphs over a fixed smooth spacelike hypersurface $V \subset Q$ (with respect to Gaussian normal coordinates about V), $\Sigma_i = \text{graph}\, u_i$, $i = 1, 2$. The functions u_1 and u_2 are Lipschitz functions on V satisfying, $u_1 \le u_2$ and $u_1(p) = u_2(p)$. As in the proof of Theorem 3.3, let θ denote the null mean curvature operator. The null mean curvature assumption then implies that u_1 and u_2 satisfy the differential inequalities, $\theta(u_2) \le 0 \le \theta(u_1)$, in the support function sense. By the weak version of the strong maximum principle obtained in [2], which is a nonlinear generalization of Calabi's [5] weak version of the Hopf maximum principle, one concludes that u_1 and u_2 are smooth and agree near p. Thus, Σ_1 and Σ_2 are smooth spacelike hypersurfaces in Q which agree near p. One can then show that S_1 and S_2 are obtained locally by exponentiating normally out along a common smooth null orthogonal vector field along $\Sigma_1 = \Sigma_2$. The conclusion of Theorem 4.2 now follows. $\square$

4.1. The null splitting theorem

The main motivation for establishing a maximum principle for C^0 null hypersurfaces was the realization that such a result could be used to settle a problem that arose in the 80's concerning the occurrence of lines in spacetime. Recall, in a Riemannian manifold, a *line* is an inextendible geodesic, each segment of which has minimal length, while in a spacetime, a *timelike line* is an inextendible timelike geodesic, each segment of which has maximal length among causal curves joining its end points. The classical Cheeger-Gromoll splitting theorem [6] describes the rigidity of Riemannian manifolds of nonnegative Ricci curvature which contain a line. (Note that a complete Riemannian manifold with strictly positive Ricci curvature cannot contain any lines.) The standard Lorentzian splitting theorem [10, 15, 28], which is an exact Lorentzian analogue of the Cheeger-Gromoll splitting theorem, describes the rigidity of spacetimes obeying the strong energy condition, $\operatorname{Ric}(X, X) \geq 0$ for all timelike vectors X, which contain a timelike line. Yau [34] posed the problem of establishing a Lorentzian analogue of the Cheeger-Gromoll splitting theorem as an approach to establishing the rigidity of the Hawking-Penrose singularity theorems; see [3] for a more detailed discussion of these matters, as well as a nice presentation of the proof of the Lorentzian splitting theorem.

But here we are interested in null geometry. Motivated by the more standard cases discussed above, a *null line* in spacetime is defined to be an inextendible null geodesic which is globally achronal, *i.e.*, for which no two points can be joined by a timelike curve. (Hence, each segment of a null line is maximal with respect to the Lorentzian arc length functional.) We emphasize that the condition of being a null line is a global one. Although each sufficiently small segment of a null geodesic is achronal, this achronality need not hold in the large: Consider, for example a null geodesic winding around a flat spacetime cylinder (closed in space); eventually points on the null geodesic are timelike related. Null lines arise naturally in causal arguments; recall, for example, that sets of the form $\partial I^{\pm}(A) \setminus A$, A closed, are ruled by null geodesics which are necessarily achronal. Null lines have arisen, by various constructions, in the proofs of numerous results in general relativity; see, for example, [25, 18, 32, 22, 20]. All of the null geodesics in Minkowski space, de Sitter space and anti-de Sitter space are null lines. The null generators of the event horizon in extended Schwarzschild spacetime are null lines.

In analogy with the Lorentzian splitting theorem, one expects spacetimes which obey the NEC and contain a null line to exhibit some sort of *rigidity*, as suggested by the following considerations: The NEC tends to focus congruences of null geodesics, which can lead to the occurrence of null conjugate points. But a null geodesic containing a pair of null conjugate points cannot be achronal. Thus we expect that a spacetime which obeys the NEC and contains a complete null line should be special in some way. The question, which arose in the 80's, after the proof of the Lorentzian splitting theorem, as to what this rigidity should be, is addressed in the following theorem.

Theorem 4.3. *Let M be a null geodesically complete spacetime which obeys the NEC. If M admits a null line η then η is contained in a smooth properly embedded achronal* **totally geodesic** *null hypersurface S.*

The simplest illustration of Theorem 4.3 is Minkowski space: Each null line ℓ in Minkowski space is contained in a unique null hyperplane Π. Because of various analogies with the Lorentzian splitting theorem, we refer to the above theorem as the null splitting theorem. Here the "splitting" is understood as taking place in the null hypersurface S, rather than in the full spacetime.

Proof. The proof is an application of the maximum principle for C^0 null hypersurfaces. For simplicity we shall assume M is strongly causal; this however is not required; see [16] for details.

By way of motivation, note that the null plane Π in Minkowski space considered above can be realized as the limit of the future null cone $\partial I^+(x)$ as x goes to past null infinity along the null line ℓ. Π can also be realized as the limit of the past null cone $\partial I^-(x)$ as x goes to future null infinity along the null line ℓ. In fact, one sees that $\Pi = \partial I^+(\ell) = \partial I^-(\ell)$.

Thus, in the setting of Theorem 4.3, consider the achronal boundaries $S_+ = \partial I^+(\eta)$ and $S_- = \partial I^-(\eta)$. By results discussed in Section 2, S_+ and S_- are closed achronal C^0 hypersurfaces in M. Since η is achronal, it follows that S_+ and S_- both contain η. For simplicity, assume S_+ and S_- are connected (otherwise restrict attention to the component of each containing η). The proof then consists of showing that S_+ and S_- agree and form a smooth totally geodesic null hypersurface. (In a vague sense, this corresponds to showing, in the proof of the Lorentzian splitting theorem, that the level sets of the Busemann functions $b^\pm = 0$ associated to the timelike line coincide, which partially motivates thinking of Theorem 4.3 as a splitting theorem.)

We claim that S_- is a C^0 *future* null hypersurface whose generators are future complete. The assumption of strong causality implies that η is a closed subset of spacetime. Then, by (the time-dual of) Proposition 2.4, each point $p \in S_- \setminus \eta$ is on a null geodesic $\sigma \subset S_-$ which either is future inextendible in M or else has a future endpoint on η. In the latter case, σ meets η at an angle, and Proposition 2.1 then implies that there is a timelike curve from a point on σ to a point on η, violating the achronality of S_-. Thus, S_- is ruled by null geodesics which are future inextendible in M, and, hence by the completeness assumption, future complete, which establishes the claim. In a time dual manner, S_+ is a C^0 *past* null hypersurface whose generators are past complete.

Thus, by Proposition 4.1 and its time-dual, S_- and S_+ have null mean curvatures satisfying, $\theta_+ \leq 0 \leq \theta_-$, in the support sense. Let q be a point of intersection of S_+ and S_-. S_+ necessarily lies to the future side of S_- near q. We may now apply Theorem 4.2 to conclude that S_+ and S_- agree near q to form a smooth null hypersurface having null mean curvature $\theta = 0$. A fairly straightforward continuation argument shows that $S_+ = S_- = S$ is a smooth null hypersurface with $\theta = 0$. By setting $\theta = 0$ in the Raychaudhuri equation (3.5), and using the NEC, we see that the shear σ must vanish, and hence S is totally geodesic.   □

In the next section we consider some applications of Theorem 4.3.

5. Some global properties of asymptotically de Sitter spacetimes

In this section we present some global results for spacetimes M^{n+1} obeying the Einstein equations,

$$R_{ij} - \frac{1}{2}Rg_{ij} + \Lambda g_{ij} = 8\pi T_{ij}. \tag{5.8}$$

We will be concerned primarily with spacetimes that satisfy the null energy condition (NEC). In view of the Einstein equations, the NEC may be expressed in terms of the energy-momentum tensor $T = T_{ij}$, as the condition, $T(X, X) \geq 0$ for all null vectors X. (Note, in particular, that the NEC is insensitive to the sign of the cosmological constant.) In some situations we will specialize to the vacuum case, $T_{ij} = 0$, in which case the Einstein equations become,

$$R_{ij} = \lambda g_{ij} \tag{5.9}$$

where $\lambda = 2\Lambda/(n-1)$.

We mainly restrict attention to solutions of the Einstein equations with *positive* cosmological constant, $\Lambda > 0$. By our sign conventions, de Sitter space, which may be expressed in local coordinates as,

$$M = \mathbb{R} \times S^n, \qquad ds^2 = -dt^2 + \cosh^2 t \, d\omega^2 \tag{5.10}$$

is a vacuum solution of the Einstein equations with $\Lambda > 0$. (We have actually taken $\Lambda = n(n-1)/2$ in (5.10)). Thus, we will be typically dealing with spacetimes which behave asymptotically like de Sitter space. There has been increased interest in such spacetimes in recent years due, firstly, to observations concerning the rate of expansion of the universe, suggesting the presence of a positive cosmological constant in our universe, and, secondly, due to recent efforts to understand quantum gravity on de Sitter space via, for example, some de Sitter space version of the AdS/CFT correspondence (see [4] and references cited therein).

5.1. Asymptotically simple and de Sitter spacetimes

We use Penrose's notion of conformal infinity [30] to make precise what it means for spacetime to be asymptotically de Sitter. Recall, this notion is based on the way in which the standard Lorentzian space forms, Minkowski space, de Sitter space and anti-de Sitter space, conformally imbed into the Einstein static universe ($\mathbb{R} \times$

$S^n, -du^2 + d\omega^2$); see the article of Friedrich in this volume for further discussion. Under the transformation $u = \tan^{-1}(e^t) - \pi/4$, the metric (5.10) becomes

$$ds^2 = \frac{1}{\cos^2(2u)}(-du^2 + d\omega^2). \tag{5.11}$$

Thus, de Sitter space conformally imbeds onto the region $-\pi/4 < u < \pi/4$ in the Einstein static universe. Future conformal infinity $\mathfrak{I}^+$ (resp., past conformal infinity $\mathfrak{I}^-$) is represented by the *spacelike* slice $u = \pi/4$ (resp., $u = -\pi/4$). This serves to motivate the following definition.

Definition 5.1. *A spacetime* (M, g) *is* **asymptotically de Sitter** *provided there exists a spacetime-with-boundary* $(\tilde{M}, \tilde{g})$ *and a smooth function* Ω *on* $\tilde{M}$ *such that*

(a) *M is the interior of $\tilde{M}$; hence $\tilde{M} = M \cup \mathfrak{I}$, $\mathfrak{I} = \partial\tilde{M}$.*
(b) *$\tilde{g} = \Omega^2 g$, where (i) $\Omega > 0$ on M, and (ii) $\Omega = 0$, $d\Omega \neq 0$ along $\mathfrak{I}$.*
(c) *$\mathfrak{I}$ is spacelike.*

In general, $\mathfrak{I}$ decomposes into two disjoint sets, $\mathfrak{I} = \mathfrak{I}^+ \cup \mathfrak{I}^-$ where $\mathfrak{I}^+ \subset I^+(M, \tilde{M})$ and $\mathfrak{I}^- \subset I^-(M, \tilde{M})$. $\mathfrak{I}^+$ is future conformal infinity and $\mathfrak{I}^-$ is past conformal infinity. It is to be understood in the above definition that both $\mathfrak{I}^+$ and $\mathfrak{I}^-$ are nonempty. If a spacetime M obeys Definition 5.1 with $\mathfrak{I}^- = \emptyset$ (resp., $\mathfrak{I}^+ = \emptyset$), we will say that M is *future* (resp., *past*) asymptotically de Sitter. Expanding dust filled FRW models, which are solutions to the Einstein equations with $\Lambda > 0$, typically begin with a big bang singularity, cf. [8, Chpt. 23]. These cosmological models are future asymptotically de Sitter, but not past asymptotically de Sitter. We remark that no a priori assumption is made about the topology of $\mathfrak{I}^+$ and/or $\mathfrak{I}^-$.

Definition 5.2. *An asymptotically de Sitter spacetime is* **asymptotically simple** *provided each inextendible null geodesic in M has a future end point on $\mathfrak{I}^+$ and a past end point on $\mathfrak{I}^-$.*

Thus, spacetime is asymptotically simple provided each null geodesic extends to infinity both to the future and the past. Schwarzschild de Sitter spacetime (see, e.g., [23, 4]), which represents a Schwarzschild black hole in a de Sitter background, is an interesting example of a spacetime which is asymptotically de Sitter, but not asymptotically simple: Null geodesics entering the black hole cannot escape to infinity. In an obvious modification of the definition, we may also refer to spacetimes which are *future* (resp., *past*) asymptotically simple.

There are connections between asymptotic simplicity and the causal structure of spacetime, as illustrated in the next proposition.

Proposition 5.1. *Let M be a future asymptotically de Sitter spacetime, with future conformal infinity $\mathfrak{I}^+$.*

(i) *If M is future asymptotically simple then M is globally hyperbolic.*
(ii) *If M is globally hyperbolic and $\mathfrak{I}^+$ is compact then M is future asymptotically simple.*

In either case, the Cauchy surfaces of M are homeomorphic to $\mathfrak{I}^+$.

Proof. By extending $M \cup \mathfrak{J}^+$ a little beyond $\mathfrak{J}^+$, one may obtain a spacetime without boundary M' such that $\mathfrak{J}^+$ is achronal and has no future Cauchy horizon in M', $H^+(\mathfrak{J}^+, M') = \emptyset$.

Suppose M is future asymptotically simple. We claim that $\mathfrak{J}^+$ is a Cauchy surface for M'. Since, by construction, $H^+(\mathfrak{J}^+, M') = \emptyset$, we need only show that $H^-(\mathfrak{J}^+, M') = \emptyset$. If $H^-(\mathfrak{J}^+, M') \neq \emptyset$, then, by Proposition 2.7 and the fact that $\mathfrak{J}^+$ is edgeless, there exists a null geodesic η contained in $H^-(\mathfrak{J}^+, M')$ which is future inextendible in M'. By asymptotic simplicity, η must meet $\mathfrak{J}^+$, and hence enter $I^+(\mathfrak{J}^+, M')$. But this violates the achronality of $\mathfrak{J}^+$. Thus, M' is globally hyperbolic, from which it easily follows that M is, as well. This proves part (i).

Now assume M is globally hyperbolic and $\mathfrak{J}^+$ is compact. Since any Cauchy surface for M is clearly a Cauchy surface for M', M' is globally hyperbolic. Then, by Proposition 2.6, $\mathfrak{J}^+$ is a Cauchy surface for M'. Hence, any future inextendible null geodesic in M' starting in M meets $\mathfrak{J}^+$. It follows that M is future asymptotically simple, which proves part (ii). We leave the proof of the final statement to the reader.   □

5.2. A uniqueness theorem for de Sitter space

We present here a uniqueness theorem for de Sitter space associated with the occurrence of null lines. Every inextendible null geodesic in de Sitter space is a null line. This fact may be understood in terms of the causal structure of de Sitter space. The *observer horizon* of an observer (future inextendible timelike curve) γ is, by definition, the achronal boundary $\partial I^-(\gamma)$. The observer horizon describes the limit of the region of spacetime ultimately observable by γ. In a future asymptotically de Sitter spacetime, every observer has a nontrivial observer horizon, as follows from the fact that $\mathfrak{J}^+$ is spacelike. In de Sitter space, the observer horizon of every observer γ is *eternal*, i.e., extends from $\mathfrak{J}^+$ all the way back to $\mathfrak{J}^-$. If q is the future end point of γ on $\mathfrak{J}^+$, then the observer horizon $\partial I^-(\gamma)$ may be viewed as the past null cone from q, which, in de Sitter space, reconverges right on $\mathfrak{J}^-$ at a point q' "antipodal" to q. By properties of achronal boundaries, $\partial I^-(\gamma)$ is ruled by achronal null geodesics which, in de Sitter space, extend all the way from $\mathfrak{J}^-$ to $\mathfrak{J}^+$. Thus, to summarize, the observer horizon of every observer in de Sitter space is eternal and, as a consequence, is ruled by null lines.

We now consider the following rigidity result for asymptotically de Sitter spactimes, cf. [17].

Theorem 5.2. *Suppose M^4 is an asymptotically simple and de Sitter spacetime satisfying the vacuum Einstein equation (5.9), with $\lambda > 0$. If M contains a null line (i.e., if there is at least one eternal observer horizon) then M is isometric to de Sitter space.*

Theorem 5.2 may be interpreted in terms of the initial value problem for the vacuum Einstein equations, with $\lambda > 0$. According to the fundamental work of Friedrich [11], the set of asymptotically simple and de Sitter solutions to (5.9), with $\lambda > 0$, is open in the set of all maximal globally hyperbolic solutions with compact spatial sections. Thus, by Theorem 5.2, in conjunction with the work of Friedrich, a sufficiently small perturbation of the Cauchy data on a fixed Cauchy hypersurface in de Sitter space will in general destroy all the null lines of de Sitter space, i.e., the resulting spacetime that develops from the perturbed Cauchy data will not contain any null lines (or, equivalently, will not contain any eternal observer horizons). While one would expect many of the null lines to be destroyed, it is somewhat surprising that *none* of the null lines persist. The absence of null lines (or eternal observer horizons) implies, in particular, that the "past null cones" $\partial J^+(p)$ will be compact for all $p \in M$ sufficiently close to $\mathfrak{I}^+$. As all such sets in de Sitter space are noncompact, this further serves to illustrate the special nature of the causal structure of de Sitter space (see also [22, Corollary 1]). Finally, we remark, that a similar uniqueness result has also been obtained for Minkowski space, see [16, 17].

Proof. We present some comments on the proof; see [17] for further details. The main step is to show that M has constant curvature. Since M is Einstein, it is sufficient to show that M is conformally flat.

Let η be the assumed null line in M. By Theorem 4.3, η is contained in a smooth totally geodesic null hypersurface S in M. By asymptotic simplicity, η acquires a past end point p on $\mathfrak{I}^-$ and a future end point q on $\mathfrak{I}^+$. Let us focus attention on the situation near p. By the proof of Theorem 4.3, and the fact that p is the past end point of η, we have that,

$$S = \partial I^+(\eta) = \partial I^+(p, \tilde{M}) \cap M .$$

It follows that $N_p := S \cup \{p\}$ is a smooth null cone in $\tilde{M}$, generated by the future directed null geodesics emanating from p.

Since S is totally geodesic and the shear σ is a conformal invariant, the null generators of N_p have vanishing shear in the unphysical metric $\tilde{g}$. The trace free part of the Riccati equation (3.3) then implies (see [24, p. 88]) that the components of the conformal tensor suitably contracted in the direction of the null generators vanish,

$$\tilde{C}_{abcd}K^b K^d = 0 \quad \text{on} \quad S, \tag{5.12}$$

where K is a smooth tangent field to the null generators of S. An argument of Friedrich [12], in which N_p plays the role of an initial characteristic hypersurface, now implies that the conformal tensor of spacetime vanishes on the future domain of dependence of N_p,

$$C^i_{jkl} = 0 \quad \text{on} \quad D^+(N_p, \tilde{M}) \cap M . \tag{5.13}$$

Friedrich's argument makes use of the conformal field equations, specifically the divergencelessness of the rescaled conformal tensor,

$$\tilde{\nabla}_i d^i_{jkl} = 0, \qquad d^i_{jkl} = \Omega^{-1} C^i_{jkl} .$$

In a time-dual manner one obtains that C^i_{jkl} vanishes on $D^-(N_q, \tilde{M}) \cap M$. Since it can be shown that M is contained in $D^+(N_p, \tilde{M}) \cup D^-(N_q, \tilde{M})$, we conclude that M is conformally flat. Together with equation (5.9), this implies that M has constant (positive) curvature. Moreover, further global arguments show that M is geodesically complete and simply connected. It then follows from uniqueness results for Lorentzian space forms that M is isometric to de Sitter space.   □

As illustrated by Schwarzschild-de Sitter spacetime, the assumption of asymptotic simplicity cannot be dropped from Theorem 5.2. However, it appears that this assumption can be substantially weakened; the essential point is to assume the existence of a null line which extends from $\mathfrak{I}^-$ to $\mathfrak{I}^+$; the null lines in Schwarzschild-de Sitter spacetime do not have end points on $\mathfrak{I}$. It is also possible to weaken the vacuum assumption, for example, to allow for the possible presence of matter fields. These extensions of Theorem 5.2 are being considered in [19].

5.3. On the topology of asymptotically de Sitter spacetimes

The results of this subsection were obtained in joint work with Lars Andersson [1].

The result of Friedrich on the nonlinear stability of asymptotic simplicity mentioned in the previous subsection establishes the existence of an open set of solutions to the vacuum Einstein equations with compact Cauchy surfaces, which are asymptotically simple and de Sitter. One is naturally interested in the general features or properties of this class of solutions. Here we address the question of which Cauchy surface topologies are allowable within this class. Obviously, since de Sitter space is in this class, the spherical topology S^n is allowable. Moreover, since isometries of S^n extend in an obvious way to isometries of de Sitter space, any spherical space form S^n/Γ can be achieved. The next theorem shows that, at least in $3 + 1$ dimensions, these are all the topologies one can expect to get.

Theorem 5.3. *Let M^{n+1}, $n \geq 2$, be an asymptotically de Sitter spacetime (to both the past and future) satisfying the NEC. If M is asymptotically simple either to the past or future, then M is globally hyperbolic, and the Cauchy surfaces of M are compact with finite fundamental group.*

Remarks:
1. Theorem 5.3 implies that the universal cover S^* of S *finitely* covers S. Hence, S^* is compact and simply connected. In three spatial dimensions, this means that S^* is a homotopy 3-sphere (and in fact diffeomorphic to the 3-sphere if the Poincaré conjecture is valid). Thus, in $3 + 1$ dimensions, the Cauchy surfaces are homotopy 3-spheres, perhaps with identifications.

2. Theorem 5.3 may be reformulated as follows: If M is an asymptotically de Sitter spacetime obeying the NEC, having compact Cauchy surfaces with infinite fundamental group, then M cannot be asymptotically simple, either to the future or the past. This is well illustrated by Schwarzschild de Sitter spacetime, which has Cauchy surface topology $S^{n-1} \times S^1$. Formulated this way, Theorem 5.3 implies that, in the conformal framework of Friedrich [12], if one evolves, via the Einstein equations, suitable initial data on a compact $\mathfrak{I}^-$, with infinite fundamental group, something catastrophic must develop to the future, as the resulting maximal development cannot be asymptotically de Sitter to the future, i.e., cannot admit a

regular $\mathfrak{I}^+$. Presumably the resulting physical spacetime M is *globally* singular; it cannot simply develop a localized black hole, similar to that of Schwarzschild de Sitter spacetime. In any case, the time dual of Theorem 5.4 presented after the proof of Theorem 5.3 shows that some singularity (in the usual sense of causal incompleteness) must occur to the future.

Proof of Theorem 5.3: For the sake of definiteness, assume M is asymptotically simple to the future. That M is globally hyperbolic follows from Proposition 5.1. We show that the Cauchy surfaces of M are compact. One can extend M a little beyond $\mathfrak{I}^\pm$ to obtain a spacetime without boundary M' which contains $\tilde{M}$, such that any Cauchy surface for M is also a Cauchy surface for M'. Thus, it suffices to show the Cauchy surfaces of M' are compact.

Fix $p \in \mathfrak{I}^-$, and consider $\partial I^+(p, M')$. If $\partial I^+(p, M')$ is compact, then by Proposition 2.6, $\partial I^+(p, M')$ is a compact Cauchy surface, and we are done. If $\partial I^+(p, M')$ is noncompact then, by considering a sequence of points going to infinity in $\partial I^+(p, M')$, we can construct a null geodesic generator $\gamma \subset \partial I^+(p, M')$ which is future inextendible in M'. Since M is future asymptotically simple, γ will meet $\mathfrak{I}^+$ at q, say. Then, γ_0, the portion of γ between p and q is a null line in M. By Theorem 4.3, γ_0 is contained in a smooth totally geodesic null hypersurface S in M. By arguments like those of the preceeding subsection, the set $N = S \cup \{p, q\}$ forms a compact achronal hypersurface in M'; it represents a future null cone in M' emanating from the point p, and reconverging to a past null cone at q. (We do *not* use the fact here that S is totally geodesic.) By Proposition 2.6, N is a compact Cauchy surface for M'.

Thus, we are lead to the conclusion that the Cauchy surfaces of M', and hence the Cauchy surfaces of M, are compact. Now pass to the universal covering spacetime M^* of M. Since all of the hypotheses of Theorem 5.3 lift to M^*, the Cauchy surfaces of M^* are compact, as well. But since the Cauchy surfaces of M^* cover the Cauchy surfaces of M, and are simply connected, it follows that the universal covering of any Cauchy surface S for M is finite. This implies that the fundamental group of S is finite. $\qquad\square$

To conclude this subsection we present a singularity theorem for future asymptotically simple and de Sitter spacetimes.

Theorem 5.4. *Let M^{n+1}, $2 \leq n \leq 7$, be a future asymptotically simple and de Sitter spacetime with compact orientable Cauchy surfaces, which obeys the NEC. If the Cauchy surfaces of M have positive first Betti number, $b_1 > 0$, then M is past null geodesically incomplete.*

Note that if a Cauchy S contains a wormhole, i.e., has topology of the form $N \# (S^1 \times S^{n-1})$, then $b_1(S) > 0$. The theorem is somewhat reminiscent of previous results of Gannon [21], which show, in the asymptotically flat setting, how nontrivial spatial topology leads to the occurrence of singularities.

The proof of Theorem 5.4 is an application of the Penrose singularity theorem, stated below in a form convenient for our purposes.

Theorem 5.5. *Let M be a globally hyperbolic spacetime with noncompact Cauchy surfaces, satisfying the NEC. If M contains a past trapped surface, M is past null geodesically incomplete.*

Recall [24, 3, 29] that a past trapped surface is a compact co-dimension two spacelike submanifold W of M with the property that the two congruences of null normal geodesics issuing to the past from W have negative divergence along W.

Comments on the proof of Theorem 5.4: Since M is asymptotically de Sitter and $\mathfrak{I}^+$ is compact (see Proposition 5.1), one can find in the far future a smooth compact spacelike Cauchy surface Σ for M, with second fundamental form which is *positive definite* with respect to the *future* pointing normal. This means that Σ is *contracting* in all directions towards the *past*.

By Poincaré duality, and the fact that there is never any co-dimension one torsion, $b_1(\Sigma) > 0$ if and only if $H_{n-1}(\Sigma, \mathbb{Z}) \neq 0$. By well-known results of geometric measure theory (see [27, p. 51] for discussion; this is where the dimension assumption is used), every nontrivial class in $H_{n-1}(\Sigma, \mathbb{Z})$ has a least area representative which can be expressed as a sum of smooth, orientable, connected, compact, embedded minimal (mean curvature zero) hypersurfaces in Σ. Let W be such a hypersurface; note W is spacelike and has co-dimension two in M. As described in [14], since W is minimal in Σ, and Σ is contracting in all directions towards the past in M, W must be a past trapped surface in M.

Since W and Σ are orientable, W is two-sided in Σ. Moreover, since W represents a nontrivial element of $H_{n-1}(\Sigma, \mathbb{Z})$, W does not separate Σ, for otherwise it would bound in Σ. This implies that there is a loop in Σ with nontrivial intersection number with respect to W. There exists a covering space Σ^* of Σ in which this loop gets unraveled. Σ^* has a simple description in terms of cut-and-paste operations: By making a cut along Σ, we obtain a compact manifold Σ' with two boundary components, each isometric to W. Taking $\mathbb{Z}$ copies of Σ', and gluing these copies end-to-end we obtain the covering space Σ^* of Σ. In this covering, W is covered by $\mathbb{Z}$ copies of itself, each one separating Σ^*; let W_0 be one such copy. We know by global hyperbolicity that M is homeomorphic to $\mathbb{R} \times \Sigma$, and hence the fundamental groups of Σ and M are isomorphic. This implies that the covering spaces of M are in one-to-one correspondence with the covering spaces of Σ. In fact, there will exist a covering spacetime M^* of M in which Σ^* is a Cauchy surface for M^*. Thus, M^* is a spacetime obeying the NEC, which contains a noncompact Cauchy surface (namely Σ^*) and a past trapped surface (namely W_0). By the Penrose singularity theorem, M^* is past null geodesically incomplete, and hence so is M. $\qquad\square$

Acknowledgement

We wish to thank Piotr Chruściel and Helmut Friedrich for the invitation to speak at the aforementioned Cargése Summer School on Mathematical General Relativity. The School's recognition of the 50th anniversary of Professor Choquet-Bruhat's seminal work on the Einstein equations, made it, indeed, a special pleasure to participate.

References

[1] L. Andersson and G.J. Galloway, *DS/CFT and spacetime topology*, Adv. Theor. Math. Phys. **6** (2002) 307–327.

[2] L. Andersson, G.J. Galloway, and R. Howard, *A strong maximum principle for weak solutions of quasi-linear elliptic equations with applications to Lorentzian and Riemannian geometry*, Comm. Pure Appl. Math. **51** (1998) 581–624.

[3] J.K. Beem, P.E. Ehrlich, and K.L.Easley, *Global Lorentzian geometry*, 2 ed., Pure and Applied Mathematics, vol. 202, Marcel Dekker, New York, 1996.

[4] Raphael Bousso, *Adventures in de Sitter space*, preprint, hep-th/0205177.

[5] E. Calabi, *An extension of E. Hopf's maximum principle with an application to Riemannain geometry*, Duke Math. Jour. **25** (1958), 45–56.

[6] J. Cheeger and D. Gromoll, *The splitting theorem for manifolds of nonnegative Ricci curvature*, J. Diff. Geom. **6** (1971) 119–128

[7] P.T. Chruściel, E. Delay, G.J. Galloway, and R. Howard. *Regularity of horizons and the area theorem*. Annales H. Poincaré, **2** (2001) 109–178.

[8] Ray D'Inverno, *Introducing Einstein's relativity*, Oxford University Press, Oxford, 1992.

[9] J.-H. Eschenburg, *Comparison theorems and hypersurfaces*, Manuscripta Math. **59** (1987), no. 3, 295–323.

[10] _____ , *The splitting theorem for space-times with strong energy condition*, J. Diff. Geom. **27** (1988), 477–491.

[11] H. Friedrich, *On the existence of n-geodesically complete or future complete solutions of Einstein's field equations with smooth asymptotic structure.*, Comm. Math. Phys. **107** (1986) 587–609.

[12] _____ , *Existence and structure of past asymptotically simple solutions of Einstein's field equations with positive cosmological constant*, J. Geom. Phys. **3** (1986) 101–117.

[13] _____ , *On the global existence and the asymptotic behaviour of solutions to the Einstein-Maxwell-Yang-Mills equations*, J. Diff. Geom., **34** (1991) 275–345.

[14] G.J. Galloway, *Minimal surfaces, spatial topology and singularities in space-time*, J. Phys. A **16** (1983), no. 7, 1435–1439.

[15] _____ , *The Lorentzian splitting theorem without the completeness assumption*, J. Diff. Geom. **29** (1989), 272–389.

[16] _____ *Maximum Principles for null hypersurfaces and null splitting theorems*, Ann. Henri Poincaré **1** (2000) 543–567.

[17] _____ *Some global results for asymptotically simple spacetimes*, in: The conformal structure of space-times: Geometry, analysis, numerics, ed. by J. Frauendiener and H. Friedrich, Lecture Notes in Physics, vol. 604, pp 51–60, 2002, Springer Verlag.

[18] G. Galloway, K. Schleich, D. Witt, and E. Woolgar, *Topological Censorship and Higher Genus Black Holes*, Phys. Rev. D **60** (1999) 104039.

[19] G.J. Galloway and D. Solis, in preparation.

[20] G.J. Galloway, S. Surya and E. Woolgar, *A uniqueness theorem for the adS soliton*, Phys. Rev. Lett. **88** (2002) 101102.

[21] D. Gannon, *Singularities in nonsimply connected space-times*, J. Math. Phys., **16** (1975) 2364–2367.

[22] S. Gao and R.M. Wald, *Theorems on gravitational time delay and related issues*, Class. Quant. Grav. **17** (2000) 4999–5008.

[23] G. Gibbons and S.W. Hawking, *Cosmological event horizons, thermodynamics, and particle creation.*, Phys. Rev. D **15** (1977) 2738.

[24] S.W. Hawking and G.F.R. Ellis, *The large scale structure of space-time*, Cambridge University Press, Cambridge, 1973.

[25] S.W. Hawking and R. Penrose, *The singularities of gravitational collapse and cosmology*, Proc. Roy. Soc. Lond. **A314** (1970) 529–548.

[26] D.N. Kupeli, *On null submanifolds in spacetimes*, Geom. Dedicata **23** (1987), 33–51.

[27] H. Blaine Lawson, Jr., *Minimal varieties in real and complex geometry*, Les Presses de l'Université de Montréal, Montreal, Que., 1974, Séminaire de Mathématiques Supérieures, No. 57 (Été 1973).

[28] R.P.A.C. Newman, *A proof of the splitting conjecture of S.-T. Yau*, J. Diff. Geom. **31** (1990) 163–184.

[29] B. O'Neill, *Semi–Riemannian geometry*, Academic Press, New York, 1983.

[30] R. Penrose, *Zero rest-mass fields including gravitation: asymptotic behavior*, Proc. Roy. Soc. Lond. A **284** 159–203.

[31] _____ , *Techniques of differential topology in relativity*, SIAM, Philadelphia, 1972, (Regional Conf. Series in Appl. Math., vol. 7).

[32] R. Penrose, R.D. Sorkin and E. Woolgar, *A positive mass theorem based on the focusing and retardation of null geodesics*, preprint, gr-qc/9301015.

[33] R.M. Wald, *General relativity*, University of Chicago Press, Chicago, 1984.

[34] S.-T. Yau *Problem section*, in Ann. Math. Studies, vol. 102, ed. S.-T. Yau, Princeton, 1982, pp 669–706.

Gregory J. Galloway
Department of Mathematics
University of Miami
e-mail: galloway@math.miami.edu

Group Actions on Lorentz Spaces, Mathematical Aspects: A Survey

Thierry Barbot and Abdelghani Zeghib

Abstract. From a purely mathematical viewpoint, one can say that most recent works in Lorentz geometry, concern group actions on Lorentz manifolds. For instance, the three major themes: space form problem of Lorentz homogeneous spacetimes, the completeness problem, and the classification problem of large isometry groups of Lorentz manifolds, all deal with group actions. However, in the first two cases, actions are "zen" (e.g., proper), and in the last, the action is violent (i.e., with strong dynamics). We will survey recent progress in these themes, but we will focus attention essentially on the last one, that is, on Lorentz dynamics.

1. Introduction

Let Γ be a topological group acting continuously on a topological space X. Recall the notion of properness of such an action, it will be one key word in this text. So, the action is proper, if for any compact subset C of X, the set of return transformations $R_C = \{g \in \Gamma, gC \cap C \neq \emptyset\}$ is compact. In other words, if $x_i \in X$ converge, and $g_i x_i$ converge, then $\{g_i\}$ is confined in a compact subset of Γ.

1.1. Properness domain?

In general, it is not possible to define in a natural way a "maximal" properness domain D_Γ, i.e., an open Γ-invariant set where the action is proper, and such that on $L_\Gamma = X - D_\Gamma$, the dynamics is strong and far from being proper.

This explains why intermediate notions similar to properness are considered, as example, the notion of wandering...

There is however, at least one case, when this does work perfectly: X is the conformal sphere, and Γ is a discrete subgroup of the Möbius group of conformal transformations. In this case, X is a disjoint union $D_\Gamma \cup L_\Gamma$, where the action on

This work is partially supported by the ACI "Structures géométriques et Trous noirs".

D_Γ is proper, and in opposite the action on L_Γ is minimal: any orbit in L_Γ is dense (in L_Γ) (see for instance [79]).

Here, we will deal (even if not so stated) with properties of such a domain in a Lorentz situation (also, even if this is only roughly defined).

1.2. Causality domain?

In this paper, X will be a Lorentz manifold, and Γ acts always isometrically on it (expect the last part on conformal actions). We have a causality (partial) pre-order: $p < q$, which means, q is in the causal future of p.

Definition 1.1. *The action of Γ on X is causal, if for any $p \in X$, and $g \in \Gamma$, gp is not comparable to p, that is, neither $gp < p$, nor $p < gp$ are satisfied (for this notion one can take X just a pre-ordered space, not necessarily a Lorentz manifold).*

In fact, for $g \in \Gamma$, one can associate its causality set $C(g)$ which consists of points p such that p is not comparable to gp. The causality domain $C(\Gamma)$ of Γ equals $\cap_{g \in \Gamma} C(g)$. These notions seem to appear, for the first time, and as efficient tools in [17]. Therefore, the Γ-action is causal, if $C(\Gamma) = X$. If X is causal, that is, $<$ is actually an order, then the Γ-action is causal, iff, the quotient space X/Γ is causal.

As a dynamical condition, causality for actions is to be compared with properness. We think, from a purely mathematical point of view that it deserves to be considered (on general ordered spaces) for its own interest.

On the other hand, in physics, causality (or variants) is a realistic condition on a spacetime X.

1.3. Co-compactness

Mathematicians love compact manifolds! They are (unfortunately) never causal. For this reason, they are treated by physicists as non realistic. But, who knows, for instance, if causality like other physical notions are not violated, near the big bang? Why is non-causality not the right "physical" answer to the metaphysical question, what happened before the big bang?

Recall here that the Γ-action on X is co-compact if there is a compact set L in X whose iterates under Γ cover the whole X.

1.4. Content

We will essentially consider dynamics of group actions on Lorentz manifolds, with respect to the three notions above: properness, causality, cocompactness. They are the three key words which unify the content of the paper, even if they are not explicitly involved.

2. Geodesic flow

Let M be a Lorentz manifold. Its geodesic flow is a "local" flow on the tangent bundle TM. It is better to call it the geodesic vector field, since in general it is non-complete. For $c \in \mathbb{R}$, let $T_c M$ be the subset of vectors v_x, with $\| v_x \|^2 = c$. For example $T_0 M$ is the light cone bundle on M. Each level $T_c M$ is invariant under the geodesic vector field.

The Lorentz manifold M is complete if its geodesic vector field is complete (as a vector field).

In the analogous Riemannian case, only one $T_1 M$ ($c = 1$) is relevant. It has compact fibers, and hence this unit tangent bundle is compact when M is. In particular a compact Riemannian manifold is complete.

In the Lorentz case, one can fix -1, 0 and 1 as sufficient set of relevant values, that is one has to consider the three different dynamics of geodesic vector field on the bundles $T_{-1} M, T_0 M$, and $T_1 M$. None of them have compact fibers, hence these bundles are never compact (even if the basis M is compact). It is a purely Riemannian heritage to think that compactness automatically implies completeness! In fact, a priori, completeness is very special. Indeed, roughly speaking, one deals with quadratic like differential equations, which notoriously present, generically, explosions. In the sequel, we will recall completeness results which may be interpreted as a posteriori estimates.

Example 2.1. *The "simplest" non complete Lorentz metric is the Bohl metric on the torus. Endow $\mathbb{R}^2 - \{0\}$ with the metric $\frac{dx dy}{x^2 + y^2}$. Any line $\{x = Constant \neq 0\}$ is an isotropic non-complete geodesic.*

2.1. Completeness

Let us notice that for Lorentz manifolds there are many interesting notions of partial completeness: future (or past) completeness, lightlike (timelike, spacelike) completeness...

2.1.1. Projectivized geodesic foliation.
In the complete case, we have a geodesic flow which is a $\mathbb{R}$-action. It is therefore, a kind of a group action related to Lorentz geometry, which however will not be considered in our survey here. No systematic investigation of this dynamics, exist in the literature. Maybe, the mathematical (and psychological) difficulty comes from the noncompactness of the ambient manifold to this flow, even when the basis M is assumed compact. However, the projectivized tangent bundle $\mathbb{P}TM$ is compact in this case, and is endowed with a one-dimensional geodesic foliations. This seems to be a most tame object to study (see [29] for the 2-dimensional case).

2.2. b-Completeness

Usual non-completeness, means that some geodesic reaches "infinity" with finite energy. However, completeness does not prevent existence of non-geodesic curves reaches "infinity" by using only "finite energy".

The notion of b-completeness (b stands for bundle), implies a stronger "physical" completeness, which prohibits finite total curvature curves to reach infinity.

It also admits "nice" (at least coherently defined) completions and compactification (see [39, 76]). We also believe here that the extended group actions to the completed spaces, are interesting objects of study, although, we do not consider them here.

The definition goes as follows. It generalizes in fact to any manifold M endowed with a connection. Indeed, in this case the frame bundle P_M has an associated "canonical" parallelism (i.e., a trivialization of TP_M). Let's recall how to construct it. The connection induces a splitting of TP_M into horizontal and vertical bundles. The horizontal is tautologically parallelizable: it has a canonical frame field, obtained by identifying it with the tangent space of M. Any choice of a basis of the Lie algebra of the structural group (here $gl(n, \mathbb{R})$) determines a vertical parallelism. Therefore, we have a parallelism on P_M defined up to the choice of a basis of the Lie algebra. Any such parallelism determines a Riemannian metric, by letting it to be orthonormal. A change of the basis induces bi-Lipschitz equivalence between metrics. One says the connection is b-complete if (any) such a metric is complete.

Observe that a quotient of a b-complete manifold by an isometry group of the connection, acting properly discontinuously and freely, is b-complete.

2.3. "Bounded completeness"

One "dramatic" fashion which ensures completeness of a Lorentz compact manifold M, is to suppose that, each geodesic is bounded in TM, (by means of any fiberwise norm on TM, it does not matter since M is compact), or equivalently the geodesic is contained in a compact set of TM. Such a condition allows standard dynamical study of the geodesic flow. Obviously, there are weaker and also stronger (uniform) variants. It is worth investigating theses notions, and showing how much are different they are. Some examples of flat and anti de Sitter compact manifolds, given below are complete but not "boundly complete".

In contrast with b-completeness, a quotient of a boundly complete is not necessarily boundly complete.

2.4. Hopf-Rinow

The Lorentz Hopf-Rinow Theorem is false, in all its formulations. In particular Lorentz "geodesic connectedness" and completeness are different notions (see for instance [53]).

3. Completeness results

3.1. Geometric structures

Let (G, X) be a homogeneous "geometric structure", i.e., where X is a homogeneous space G/H of a Lie group G. A (G, X)-structure on M is an atlas on M with charts taking value in (open subsets of) X and such that chart transitions (defined on open subsets of X) are restriction of elements of G (seen as transformation of X). A manifold endowed with a (G, X)-structure is called (G, X)-manifold. If M is simply connected, the analytic continuation principle implies the existence of a local diffeomorphism $\mathcal{D} : M \to X$, which expresses in the chart of the structures as restrictions of elements of G. It is unique up to composition by automorphisms of X.

If M is not simply connected, then the developing map $\mathcal{D} : \tilde{M} \to X$, is defined on its universal cover. By the (essential) uniqueness of $\mathcal{D}$, there exists a holonomy homomorphism: $\rho : \pi_1(M) \to G$, such that $\mathcal{D}\gamma = \rho(\gamma)\mathcal{D}$, for any $\gamma \in \pi_1(M)$.

In the general case, $\mathcal{D}$ and ρ can be very pathological, for instance, $\mathcal{D}$ (resp. Γ, the image $\Gamma = \rho(\pi_1(M))$) is not necessarily a covering from $\tilde{M}$ to its image (resp. a discrete subgroup of G).

When there exists an open subset $\Omega \subset X$, on which Γ acts properly freely, and $\mathcal{D}$ is a covering from $\tilde{M}$ to Ω the structure is called Kleinian. This is the most regular property of the (G, X)-structure. For instance, we have:

Proposition 3.1. *Suppose M is compact.*

- *If the structure is Kleinian, then Ω is the unique maximal connected open set containing the developing image on which Γ acts properly.*
- *Assume that the G action preserves a complete connection on X. This induces in a natural way a connection on M. This connection is complete, iff, the (G, X)-structure is Kleinian with $\Omega = X$.*

3.2. Manifolds with constant (sectional) curvature

Fix a dimension n, and let $X(c)$ be the complete simply connected Lorentz space of constant curvature c. One can normalize c to be $-1, 0$ or 1. $X(0)$ is the Minkowski space Min_n, $X(1)$ is the de Sitter space dS_n defined as the set of vectors in Min_{n+1} with norm 1.

In order to define Anti de Sitter space AdS_n, consider $\mathbb{R}^{2,n-1}$, the linear space $\mathbb{R}^{1+n}$ equipped with a quadratic form of signature $(2, n-1)$: AdS_n is the the domain of $\mathbb{R}^{2,n-1}$ where the the quadratic form takes value -1, the Lorentzian metric being the restriction of the ambient quadratic form to the tangent spaces of this domain. Observe that AdS_n is not simply connected; $X(-1)$ is the universal covering of AdS_n (see [83] for details).

Remark 3.2. *In the sequel, we denote by AdS_n some cyclic quotient of $\widetilde{AdS_n}$: this*

A Lorentz manifold M^n of constant curvature c is modeled on $(G, X(c))$, where G is the isometry group of $X(c)$. Therefore, M is geodesically complete means that its universal cover $\tilde{M}$ is (globally) isometric to $X(c)$.

We believe the most revolutionary result in the subject is that proved by Y. Carrière in the flat case, and then adapted by B. Klingler to the general case (this result in the de Sitter case was also independently proved by M. Morrill in her thesis):

Theorem 3.3. [27, 56] *A compact Lorentz manifold of constant curvature is complete.*

In the flat case, the proof proceeds by checking the geodesic connectedness of $\tilde{M}$. In the non flat case, the universal space $X(c)$ itself is not geodesically connected. The point is to show that $\tilde{M}$ is "as geodesically connected as" $X(c)$ itself. . . . In all cases, the goal is achieved by a clever analysis of the dynamics of the holonomy group.

We stress out that this result is false for *locally homogeneous* Lorentz geometric structures, i.e., a Lorentz manifold M modeled on (G, X), where G acts transitively on X, and preserves a Lorentz metric on X, but not necessarily with constant curvature, may be non-complete. See for instance [54] for a construction of left invariant *non-complete* Lorentz metrics on the group $X = SL(2, \mathbb{R})$. Any (metric) quotient X/Γ, where Γ is a co-compact lattice of $SL(2, \mathbb{R})$ is a compact non-complete locally homogeneous Lorentz manifold. Observe here that non-completeness of M simply follows from that of X. In some sense, M is as complete as it could be! We dare ask:

Question 3.4. *If (G, X) is a homogeneous Lorentz space, is any compact (G, X)-manifold M Kleinian (in fact with $\Omega = X$)?*

3.2.1. Singular structures. As this will be done in the sequel, the next step towards understanding compact manifolds of constant curvature, is to consider the holonomy group (or equivalently because of completeness, the fundamental group). The de Sitter case is "hyper-rigid" due to the so-called Calabi-Markus phenomenon [25] which states that only finite groups can act properly on the (full) de Sitter space dS_n (assume here $n \geq 3$ to avoid complication with the non-simply connected dS_2). Therefore, there is no compact Lorentz manifold of positive constant curvature!

Here, we want to emphasize the importance of compactness. Indeed, let M^n be a compact (Riemannian) hyperbolic (i.e., of constant curvature -1) manifold. Let $x \to H(x)$ be a hyperplane field on M. Lift it as a hyperplane field $x \to \tilde{H}(x)$ on the hyperbolic space $\mathbb{H}^n$. To a tangent hyperplane of $\mathbb{H}^n$, corresponds a geodesic hyperplane which is interpreted as a point of dS_n. Thus, we get a mapping $\mathcal{D}_H : \mathbb{H}^n \to dS_n$. It is equivariant with respect to the $\pi_1(M)$ action (seen as a subgroup of $O(1, n)$) on both $\mathbb{H}^n$ and dS_n. The previous results implies in particular that $\mathcal{D}_H$ can never be the developing map of a $(O(1, n), dS_n)$-structure, that is $\mathcal{D}_H$ can not be a local diffeomorphism. In the generic case, $\mathcal{D}_H$ will have "tame" singularities. Therefore, we get a singular de Sitter structure on M.

However, we will see in Section 5.2.3 how this construction provide fair regular geometric structures in interesting (non-compact) cases.

3.3. Completeness in presence of Killing fields

Riemannian homogeneous manifolds are complete, even if they are non-compact. This is false in the Lorentz case, as mentioned in the case of $SL(2, \mathbb{R})$ above. Marsden proved the first general completeness theorem, for homogeneous and compact pseudo-Riemannian manifolds [63]. In fact, he proved "boundly completeness". For instance, in the Lorentz case, from an everywhere *timelike* Killing field, one construct a Clairaut first integral of the geodesic flow, with compact levels. Homogeneity does not lead to existence existence of such Killing fields, but a one concludes after a little bit work (see [53] for related results).

Remark 3.5. *It seems interesting to get extension of Marsden's Theorem to other classes of (non pseudo-Riemannian) connections.*

4. The π_1-action, algebraic classification

At this stage we know that a compact Lorentz manifold of constant curvature is a quotient $M = X/\Gamma$, where X is the Minkowski space or the anti de Sitter space (the case of de Sitter space was excluded above 3.2.1), and Γ is a discrete group of X acting properly co-compactly and freely. The question is to "classify" such Γ. In the Euclidean case, the similar question is the classical classification of crystallographic groups, for which some aspects still remain fascinating problems for geometers. However, the first step of the classification was the celebrated Bieberbach Theorem (for crystallographic groups). The fundamental question that we will ask in our Lorentz case is in fact in the same vein as Bieberbach Theorem[1].

4.1. Bieberbach rigidity

Consider the general case of a homogeneous space I/H, quotient of a *connected* Lie group I by a *connected* Lie subgroup H. One central problem about homogeneous (non-Riemannian) spaces, is the study of discrete subgroups $\Gamma \subset I$ acting properly co-compactly and freely on I/H (so that $M = \Gamma \setminus I/H$ is a compact manifold).

One may start considering a radically simpler and soft problem which is, first, to find a connected Lie subgroup $G \subset I$ acting co-compactly (or say, transitively) and *properly* on I/H, and next to find a co-compact lattice Γ in G.

One says that I/H satisfies a " Bieberbach rigidity" if all its compact quotients are of this type (say, up to finite covers to avoid obvious trivial counterexamples).

One says to have a "unique Bieberbach rigidity", if up to conjugacy, there is only one group G as above (for all Γ's).

4.1.1. Flat manifolds. As example, after many works during the last decade, the structure of compact flat Lorentz manifolds, was elucidated, as in the following Theorem:

Theorem 4.1. ([44], [48], [51],...) *Let $M = Min_n/\Gamma$ be a compact Lorentz flat manifold. Then there is a solvable group G acting isometrically and simply transitively on the Minkowski space Min_n and a lattice Γ in G such that up to finite covers, $M = Min_n/\Gamma$ $(= G/\Gamma)$.*

---

[1] As a matter of fact; Bieberbach's Theorem, as formulated in [28], is a fundamental ingredient of the study in [17].

Actually this structure theorem for the fundamental groups of compact Min-kowskian manifolds, was proved before the completeness Theorem, that is we only deal at this time with complete manifolds.

In other words, the result says that M is a quotient I/Γ, where I is a solv-able group endowed with a Lorentz left invariant metric which is complete and flat. There are many examples of such solvable Lie groups, see [43] for concrete constructions in the case of the 3-dimensional Heisenberg and SOL groups and [51] and [52] for a general study.

4.1.2. Anti de Sitter manifolds. A compact anti de Sitter manifold must have odd dimension, since according to Gauss-Bonnet formula, for even-dimensional anti de Sitter manifolds, the Euler number equals the volume, up to a non-trivial multiplicative constant. But, any compact Lorentz manifold has a vanishing Euler number, since it possesses a direction field.

Conversely, for any odd dimension, there are closed anti de Sitter manifolds. This was mentioned for the first time by R. Kulkarni [59]. They are just obtained by taking $G = U(1,d) \subset O(2,2d)$, in the introduction of the Bieberbach rigidity above. Indeed, one verifies that $U(1,d)$ acts isometrically on AdS_{2d+1} (this is the meaning of the inclusion $U(1,d) \subset O(2,2d)$) transitively and properly, the isotropy group being $U(d)$. As said above, any co-compact lattice in $U(1,d)$ gives rise to a compact anti de Sitter manifold of dimension $2d+1$. To fix ideas, let us introduce as in [60] the next terminology:

Definition 4.2. *An anti de Sitter manifold of dimension $2d+1$ is called standard (resp. special standard) if up to conjugacy, its holonomy group Γ is contained in $U(1,d)$ (resp. $SU(1,d)$).*

The special anti de Sitter manifolds have the following Riemannian descrip-tion. Let $\mathbb{H}_{\mathbb{C}}^d$ be the hyperbolic complex space of (complex) dimension d. It is nothing but the homogeneous space $= U(1,d)/S^1 \times U(d)$. Therefore, AdS_{2d+1} is a circle fiber bundle over $\mathbb{H}_{\mathbb{C}}^d$. In fact, AdS_{2d+1} is the circle bundle associated to the canonical line bundle (in the complex meaning) of $\mathbb{H}_{\mathbb{C}}^d$.

4.1.3. Anti de Sitter, Dimension > 3. The results of [84] leads one to hope that a unique Bieberbach rigidity phenomenon holds, for compact anti de Sitter manifolds of dimension ≥ 5. In other words, we dare ask:

Conjecture 4.3. *Up to finite coverings, every compact anti de Sitter manifold of dimension ≥ 5, is standard.*

4.1.4. Anti de Sitter, dimension $= 3$. For $d = 1$, the circle fibration $AdS_3 \to \mathbb{H}_{\mathbb{C}}^1$ is just the usual fibration over the hyperbolic plane $\mathbb{H}^2$ of its unit tangent space. Also, (an index 2 quotient of) AdS_3 is identified to the group $PSL(2,\mathbb{R})$ and $O(2,2)$ is identified to $PSL(2,\mathbb{R}) \times PSL(2,\mathbb{R})$. The action of this last group on $X = PSL(2,\mathbb{R})$ (seen at the same time as a group and the anti de Sitter space) is given by $(g,h).x = gxh^{-1}$.

Up to switch of factors, a special standard quotient is such that

$$\Gamma \subset PSL(2, \mathbb{R}) \times \{1\},$$

and special means that $\Gamma \subset SL(2, \mathbb{R}) \times S^1$

In their pioneering work, Kulkarni and Raymond [60] showed that any co-compact holonomy group $\Gamma \subset PSL(2, \mathbb{R}) \times PSL(2, \mathbb{R})$, is (up to switch of factors) a graph: there is a closed hyperbolic surface S, and two homomorphisms $\rho_L, \rho_R :$ $\pi_1(S) \to PSL(2, \mathbb{R})$, such that ρ_L is fuchsian (discrete and injective), and such that Γ is the image of $\rho_L \times \rho_R$:

$$\Gamma = \{(\rho_L(\gamma), \rho_R(\gamma) \in PSL(2, \mathbb{R}) \times PSL(2, \mathbb{R})), \gamma \in \pi_1(S)\}.$$

The non-standard case corresponds to the fact that the image of ρ_R is not contained in a circle (in $PSL(2, \mathbb{R})$). The first examples were observed by Ghys and Goldman [47, 49], by just taking ρ_R small enough. F. Salein showed in particular that ρ_R can be "very" big, that is to say not homotopic to the trivial representation, or equivalently, with non-vanishing Euler number. For instance, let $f : S \to S'$ be a (non-trivial) ramified covering, where S' is another hyperbolic surface, which is holomorphic, when S is endowed with the structure given from ρ_L. Let ρ_L the homomorphism induced by f, and Γ determined by (ρ_L, ρ_R). F. Salein [73] proved that Γ acts *properly* freely and co-compactly on AdS_3. (Exercise: all these closed manifolds are complete and b-complete. For which holonomy Γ is the manifold boundedly complete?)

4.2. Margulis spacetimes

In the flat case, according to Theorem 4.1, compact complete flat manifolds have (virtually) solvable fundamental groups. The difficulty to find non-solvable proper actions on Min led J. Milnor to wonder: *does the free group admit a proper action on Min_3?* In [66], when Milnor addressed this question, he suggested a way to produce an example. In [62], following this (partial) hint, Margulis answers affirmatively to the question. Since [62], proper quotients of Minkowski by free groups of isometries are called *Margulis spacetimes*.

Afterwards, T. Drumm introduced the notion of *crooked planes*, giving a more intuitive geometric vision on these spacetimes ([36]), and extended considerably the list of Margulis's spacetimes by proving that *every discrete free subgroup of $SO_0(1, 2)$ is the linear part of the holonomy of a Margulis spacetime* ([37]).

In his work [62], G. Margulis associates to every hyperbolic element g of $\text{Isom}(Min_3)$ a real number $\alpha(g)$ – the so-called *Margulis invariant* – and proved that a discrete purely hyperbolic subgroup of $\text{Isom}(Min_3)$ can act properly on Minkowski space only if all the Margulis invariants have the same sign (see the good survey [1] giving in particular a lucid account of why this sign condition is necessary). On the other hand, it is not true that the positivity of all Margulis invariants ensure the properness of the action. The correct reverse statement was recently proved by Goldman, Labourie and Margulis in a forthcoming preprint "Proper affine actions and geodesic flows of hyperbolic surfaces": once the linear part $F \subset SO_0(1,2)$ is fixed, we consider the geodesic flow on the surface $F\backslash \mathbb{H}^2$, and denote by $\mathcal{P}(F)$ the space of all invariant probability measures for this geodesic flow. For any affine deformation $\rho : F \to \text{Isom}(Min_3)$, and for any closed orbit c of the geodesic flow, define $A(c) = \alpha(\rho(g))/l(c)$ where g is the element of F corresponding to c, $\alpha(\rho(g))$ is the Margulis invariant of $\rho(g)$, and $l(c)$ the period of c (the length of the corresponding closed geodesic). Closed orbits can be considered as particular elements of $\mathcal{P}(F)$ (a kind of Dirac measures), and the map A extends uniquely to a continuous map $A : \mathcal{P}(F) \to \mathbb{R}$. Then, the affine deformation ρ is proper (i.e., $\rho(\Gamma)$ acts properly on Mink$_3$ if and only if $A(m)$ vanishes nowhere.

Let us also mention [38], and, more recently, [30], where it is proven that Margulis invariant characterize completely the free group of isometries in the purely hyperbolic case. The survey [32] covers the topics discussed in this section.

In Section 5.6, we will add some comments on Margulis spacetimes.

Remark 4.4. *Fundamental domains for Lorentz groups in the anti de Sitter case, and in general, polyhedra in Lorentz manifolds, were recently investigated by for A. Pratoussevitch and J.M. Schlenker, respectively* [71, 77].

5. Global hyperbolicity

Let M be an (open) Lorentz manifold with a closed (compact without boundary) spacelike Cauchy hypersurface. By general theory [33], there is an abstract maximal globally hyperbolic (abbreviation MGH) extension $\bar{M}$ of M.

Assume now that M has a (G,X)-structure, hence so does $\bar{M}$ (by analyticity). The question is how to describe this geometric structure of $\bar{M}$ (say by means of that of M)?

The most natural case is that of 3-dimensional manifolds of constant curvature, since they are the only solutions of the vacuum Einstein equations (with cosmological constant) in dimension $2 + 1$. One may then ask the general question for all constant curvature manifolds of any dimension.

The first work in this direction is [65], even if the results there are not stated in the terminology presented here. This celebrated preprint completely solves the negative curvature case in dimension $2 + 1$.

K. Scannell, Mess student, solves the positive curvature case in any dimension in his thesis [75], where he established a natural $1 - 1$ correspondence between $n + 1$-dimensional spatially closed MGH de-Sitter spacetimes with flat conformal Riemannian structures on closed n-manifolds.

Remark 5.1. *We have to pay attention to the meaning of "flat conformal Riemannian structure". Here, we mean a (G,X) structure where X is the conformal sphere and G the Möbius group. This is a quite lazy convention that we maintain from* [75]: *it is maybe more usual to define flat conformal structures as the conformal classes of Riemannian metrics which can be written everywhere locally as scalar multiples of local flat Riemannian metrics.*

According to Liouville's Theorem, in dimension $n \geq 3$, these two notions coincide, but this is dramatically false in dimension 2. Nevertheless, their is a huge mathematical literature on Möbius structures on closed manifolds, particularly in the 2-dimensional case, which is fairly well understood.

Finally, and maybe quite surprisingly, the flat case has been systematically studied only recently by one of the authors ([17]). However, some fundamental observations appeared in [65] where the $2 + 1$-dimensional case is treated, and in [9], the classification is performed in any dimension, but in a particular case, assuming that the spacelike Cauchy hypersurface admits some Riemannian metric with negative constant curvature. We should also indicate [74], specifying the possible geometric character in the Thurston's terminology of spacelike hypersurfaces of MGH flat $3 + 1$-spacetimes.

5.1. The Anti de Sitter case

5.1.1. The 3-dimensional case. We present here briefly Mess results, and address some questions. We consider as in 4.1.4 the model $PSL(2, \mathbb{R})$, endowed with the Lorentzian metric defined by its Killing form, and admitting as isometry group the product $PSL(2, \mathbb{R}) \times PSL(2, \mathbb{R})$ acting by left and right translations.

Let S be a closed surface, and Γ be its fundamental group. Denote by $Teich(\Gamma)$ the Teichmüller space of S: we consider it as the space of discrete injective representations of Γ in $PSL(2, \mathbb{R})$ with cocompact image, modulo inner automorphisms of $PSL(2, \mathbb{R})$. Any pair (ρ_L, ρ_R) of elements of $Teich(\Gamma)$ thus defines a representation $\rho : \Gamma \to Iso(AdS_3)$, well defined up to conjugacy. The causality domain of $\rho(\Gamma)$ in AdS_3, as defined in Section 1.2, is a (convex) domain $C(\rho)$. Actually, the action of $\rho(\Gamma)$ on $C(\rho)$ is proper, and the quotient manifold $M(\Gamma) = \rho(\Gamma) \backslash C(\rho)$ is MGH, admitting a Cauchy surface homeomorphic to S. Moreover, Mess proved that any spatially closed GH spacetime locally modelled on AdS_3 can be isometrically embedded in such a $M(\Gamma)$, in particular, they are Kleinian.

In a more precise way, Mess did not consider the case where S is a torus. But this case is quite simple: it is the Torus Universe as presented in [26]. In particular, the left and right holonomy groups are necessarily contained in 1-parameter hyperbolic subgroups of $PSL(2, \mathbb{R})$ (see [20]).

Mess results involves actually many interesting geometric and physically relevant features as real trees, earthquakes, and so on, that we cannot pretend to develop here further.

5.1.2. Higher dimensions. Many propositions in [65] still apply in higher dimensions: for example, MGH spacetimes with constant negative curvature are still Kleinian, and thus uniquely defined by their holonomy groups. However, the complete classification remains an open question. A natural way to define GH *AdS* spacetimes in dimension $n + 1$ is the following: let Γ be a cocompact lattice of $SO_0(1, n)$, the identity component of $O(1, n)$. The quotient space $S = \Gamma \backslash \mathbb{H}^n$ is a closed hyperbolic manifold. Let ρ_0 be the composition of the inclusion $\Gamma \subset$

$SO_0(1, n)$ with the natural embedding $SO_0(1, n) \subset SO_0(2, n)$. Since $SO_0(2, n)$ is the isometry group of AdS, we can define the causality domain $C(\rho_0)$, which actually has a simple geometrical description. The quotients $M(\rho_0) = \rho_0(\Gamma) \backslash C(\rho_0)$ are then MGH spacetimes with Cauchy hypersurfaces homeomorphic to S. These spacetimes are *static,* meaning that they admit a timelike vector field with integrable orthogonal plane fields.

This construction still applies for small deformations $\rho : \Gamma \to SO_0(2, n)$ of ρ_0, providing nonstatic MGH spacetimes $M(\rho)$. But some rigidity aspect appears here: it is not a really trivial task to exhibit such a deformation; the only known procedure is the deformation along codimension 1 geodesic hypersurfaces of S given in [55]. Anyway, these deformations exist, and a first question is the following: is the space of holonomy representations of spatially closed AdS spacetimes with Cauchy hypersurface homeomorphic to S, connected?

There is another natural question: what are the possible topologies for closed Cauchy hypersurfaces of AdS GH spacetimes? Are they necessarily homeomorphic (up to finite coverings and products by flat tori) to hyperbolic manifolds or to the product of two hyperbolic manifolds?

5.2. The de Sitter case

We now present Scannell's work ([75]) relating spatially closed MGH $n + 1$-spacetimes locally modelled on dS with Cauchy hypersurface S with flat conformal structures on S. Unfortunately, the description is quite delicate, since, in general, these spacetimes are *not* Kleinian.

Actually, up to time reversing isometries, these spacetimes are all geodesically complete in the future (it is a nontrivial by-product of Scannell's classification), so we will restrict to this case for our description. The fundamental group of S is still denoted Γ.

5.2.1. A quick presentation of dS. We use the projective model of dS. More precisely, it will be more convenient to lift the usual projective model in the sphere, the double covering of the projective space. Here, the projective model is the space of vectors $(x_0, x_1, \ldots, x_{n+1})$ in $Mink_{n+2}$ with positive norm contained in the sphere S of equation $\sum x_i^2 = 1$. In order to distinguish this model from the usual definition of de Sitter space given in 3.2, we denote it by ds. The boundary of ds is $x_0^2 = x_1^2 + \cdots + x_{n+1}^2$: this is the union of two n-dimensional subspheres ∂ds_+ (where $x_0 > 0$) and ∂ds_- (where $x_0 < 0$). Each $\partial ds_\pm$ is also the boundary of a copy of the Klein model of the hyperbolic space. Geodesics are intersections of round circles in S with ds; a geodesic is spacelike if this circle avoids $\partial ds_\pm$, lightlike if it is tangent to $\partial ds_\pm$, and timelike if the circle intersects each $\partial ds_\pm$ at exactly two points.

The choice of a chronological orientation of ds is equivalent to the choice of one of the $\partial ds_\pm$ – let say, ∂ds_+ – as the ideal boundary *in the future.* Now, for any point x in ds, the timelike future-complete geodesic rays starting from x hit ∂ds_+ in a bounded region, which, when we identify $\partial ds_+ \approx \partial \mathbb{H}^{n+1}$ with the conformal sphere $\mathbb{S}^n \approx \partial\mathbb{H}^{n+1}$, is an open n-ball $B(x)$. Conversely, every round n-ball B is the "visible domain of infinity" $B(x)$ of an unique element x of ds.

5.2.2. Scannell's results. We first consider the Kleinian case: consider a Kleinian flat conformal structure on S, i.e., a simply connected domain D of $\partial ds_+ \approx \mathbb{S}^n$ and a representation $\rho : \Gamma \to \text{Conf}(\partial ds_+) = SO_0(1, n)$ with image preserving D, and acting freely and properly discontinuously on it, such that S is homeomorphic to the quotient space $\rho(\Gamma)\backslash D$. Then, the points x of ds such that $B(x)$ is included in D form an open domain Ω (in ds) on which $\rho(\Gamma) \subset SO_0(1, n)$ acts freely and properly discontinuously. Therefore, the quotient space $M(\rho) = \rho(\Gamma)\backslash\Omega$ is well defined. Observe that it is obviously geodesically complete in the future.

When D is the entire sphere ∂ds_+, we obtain by this construction (finite quotients of) the entire de Sitter space. When D is the complement of a single point x_0 of ∂ds_+, we obtain the parabolic case: Γ is then necessarily abelian (up to finite index) and only contains parabolic elements of $SO_0(1, n)$; the n-dimensional spheres in S tangent to ∂ds_+ at x_0 foliates Ω and they induce on the quotient n-dimensional tori which are Cauchy hypersurfaces

These two cases being from now excluded, the boundary $\partial\Omega$ in ds is a null-hypersurface. For any point x in Ω, we can define its "maximal time of existence" $\tau(x)$: this is the supremum of total proper times of timelike curves starting from $\partial\Omega$ and ending at x. Then, τ is a $\rho(\Gamma)$-invariant C^1-function. It induces on the quotient a "cosmological time" function whose levels sets are spacelike hypersurfaces which are Cauchy hypersurfaces, proving that $M(\Gamma)$ is globally hyperbolic.

In general, a conformal structure on S is given precisely by:

- a local homeomorphism $\overline{\mathcal{D}} : \tilde{S} \to \partial ds_+$,
- a morphism $\rho : \Gamma \to SO_0(1, n)$ such that $\rho(\gamma) \circ \overline{\mathcal{D}} = \overline{\mathcal{D}} \circ \rho(\gamma)$.

To such a data, associate the space $\mathcal{M}$ formed by closed subsets $\tilde{B}$ of $\tilde{S}$ on which $\overline{\mathcal{D}}$ restrict as a homeomorphism with image a closed round n-ball. Equip $\mathcal{M}$ with its obvious topology: it is a n-manifold admitting a natural action of Γ which is properly discontinuous. Denote by M the quotient space. There is a natural map $\mathcal{D} : \mathcal{M} \to ds$: define $\mathcal{D}(\tilde{B})$ as the unique point x in ds for which $B(x) = \overline{\mathcal{D}}(\tilde{B})$. This map is a local homeomorphism, respecting the respective Γ-actions. It follows that $\mathcal{D}$ is the developing map of a de Sitter structure on M. K. Scannell proved that the cosmological time function on M is well defined, admitting as level sets Cauchy hypersurfaces for M. Moreover, this construction provides all spatially closed MGH de Sitter spacetimes.

To be complete, we must give a flavor of the non-Kleinian case, this particular flavor arising from the "pathologies" of flat conformal structures: restrict to $n = 2$, and take as surface S the 2-torus. The most obvious flat conformal structures on S are the conformal class of flat metrics on the torus: i.e., the quotient of $\mathbb{C}$ by lattices. But it is far from exhausting the entire list of flat conformal structures on the torus! Indeed, consider the torus as the quotient of $\mathbb{C}$ by a lattice Λ. Take as developing map the exponential $\exp : \mathbb{C} \to \mathbb{C}^*$. The translations by elements of Λ

correspond to homotheties by elements of $\exp(\Lambda)$. This simple family of examples prove that:

- the holonomy can be noninjective: it happens when Λ contains a integer multiple of $2i\pi$,
- the developing map can be non-injective,
- two non-isometric MGH spacetimes can admit the same holonomy.

Actually, "worse" situations arise: when the surface S has higher genus, any irreducible representation $\Gamma \to SL(2, \mathbb{C})$ is the holonomy of one (or maybe an infinite number of) flat conformal structure on S (see [45]). For example, the holonomy group can be dense on $SL(2, \mathbb{C})$.

Remark 5.2. *Anyway, we have to indicate here that we gave above the complete description of conformal (Möbius) structures on the 2-torus, i.e., of globally hyperbolic dS_3-spacetimes admitting a Cauchy surface homeomorphic to the 2-torus.*

5.2.3. The dual Riemannian version: hyperbolic ends.

There is a dual Riemannian version, which in the 2-dimensional case is much more popular in the mathematical community, although a bit more delicate to define than the Lorentzian version: the theory of hyperbolic ends (see [80, 61]).

First of all, we must state clearly that this construction does not apply in the particular case of finite quotients of dS-spacetimes: indeed, finite subgroups of $SO(1, n)$ never act freely on the hyperbolic space!

Constructing directly from the conformal structure on S the hyperbolic structure on $S \times \mathbb{R}$ is a delicate task, but is easier when the Lorentzian version is available: indeed, the construction of the hyperbolic end from the globally hyperbolic is essentially given by the (inverse of) process described at Section 3.2.1 above. More precisely: the dS-spacetime M associated to the given conformal structure is foliated by the level sets of the function τ defined above. The tangent planes of these spacelike hypersurfaces form a field $y \mapsto L(y)$ of spacelike hyperplanes on M. But the pairs (y, L) where y is an element of the spacelike hyperplane L of dS correspond bijectively and naturally with the pairs (x, H) where x is an element of $\mathbb{H}$ and H a totally geodesic hyperplane containing x: take H as the geodesic hypersurface admitting $\partial B(y)$ as boundary at infinity, and x as the intersection point between H and the great circle in S orthogonal to L at y. This procedure composed with the field $y \mapsto L(y)$ produces then a Γ-equivariant map $\mathcal{D}$ from the universal covering $\widetilde{M}$ to $\mathbb{H}$. Moreover, Scannell proved that the level sets of τ are strictly convex. *In fine*, this strict convexity property ensures that $\mathcal{D}$ is a local homeomorphism. It can thus be interpreted as the developing map of a hyperbolic structure on M: this is precisely the associated hyperbolic end M_{hyp}.

The procedure described above actually defines for every strictly convex spacelike hypersurface Σ in M a map f from Σ into M_{hyp}, with image a strictly convex hypersurface. There is an inverse procedure defining from a strictly convex hypersurface in M_{hyp} a spacelike hypersurface in M. We will use this remark later;

we will actually need there the following fact: the eigenvalues of the second fundamental form[2] of $f(\Sigma)$ at $f(x)$ are the inverses of the eigenvalues of the second fundamental form of Σ at x.

5.3. The flat case

The most obvious examples of MGH flat manifolds are quotients of Min_{n+1} by abelian discrete groups of rank n of spacelike translations; we call these examples *translation spacetimes*. They admit as Cauchy hypersurfaces flat tori.

The second natural family of examples are what we call *Misner spacetimes*. Let v_1, v_2 be two lightlike vectors in Min_{n+1}, and denote by $v_i^\perp$ their orthogonal. Let L be the 1-parameter group formed by pure (i.e., without elliptic part) loxodromic isometries (or "boosts") preserving the directions v_i.

Vectors v in Min_{n+1} for which the Minkowski scalar products $\langle v \mid v_i \rangle$ are negative form a domain which has two connected components $\Omega^\pm$, one being geodesically complete in the future and the other, in the past. These domains are "quarter of space". Isometries of Min_n preserving $\Omega^\pm$ form a group which is a compact extension of the abelian group A of dimension n: elements of A are compositions of loxodromic elements in L with translations by vectors in $v_1^\perp \cap v_2^\perp$. This group A acts properly discontinuously on $\Omega^\pm$, and the orbits of this action are product of hyperbolae with euclidean spaces: these orbits are spacelike hypersurfaces isometric to the euclidean space.

For any lattice Λ of A, the quotient spaces $\Lambda \backslash \Omega^\pm$ are globally hyperbolic spacetimes geodesically complete in the future or in the past, and the orbits of A project in these quotients as toroidal Cauchy hypersurfaces.

The last family are the so-called *standard spacetimes* that G. Mess has already described in [65]: the simplest members of this family of examples are constructed from cocompact lattices of $SO_0(1,n)$. Let Γ be such a lattice. The set of timelike vectors of Min_n admits two connected components $\Omega^\pm$ respectively geodesically complete in the future and in the past. The action of Γ on $\Omega^\pm$ is free and properly discontinuous: we denote by $M^\pm(\Gamma)$ the quotient manifold.

Every level set $\{Q = -t^2\} \cap \Omega^\pm$ is Γ-invariant; it induces in $M^\pm(\Gamma)$ a hypersurface with induced metric of constant sectional curvature $-\frac{1}{t^2}$. Since $\{Q = -1\}$ is the usual representant of the hyperbolic space, the flat Lorentzian metric on $M^\pm(\Gamma)$ admits the warped product form $-dt^2 + t^2 g_0$, where g_0 is the hyperbolic metric on $\Gamma \backslash \mathbb{H}^n$. We call these examples *radiant standard spacetimes*. Observe that $M^+(\Gamma)$ (resp. $M^-(\Gamma)$) is geodesically complete in the future (resp. in the past), and that there is a time reversing isometry between them.

New examples are obtained by adding translation parts (see [65] or [9]): any representation of a Γ in Isom(Min_n) admitting as linear part an embedding onto a cocompact lattice of $SO(1,n)$ is the holonomy group of a flat spatially closed globally hyperbolic spacetime. Actually, there are two such globally hyperbolic

---

[2]Maybe physicists are more acquainted with the notion of shape operator: it has the same eigenvalues than the second fundamental form.

spacetimes, one being geodesically complete in the future, and the other, geodesically complete in the past. Moreover, they are Kleinian, more precisely, they are quotients by the holonomy group of convex domains of Min_n. Of course, if M is a flat GH spacetime with Cauchy hypersurface S, and N a flat euclidean torus, then the product $M \times N$, equipped with the product metric, is still a flat GH manifold, with Cauchy hypersurface $S \times N$.

In [17], we prove that any flat globally hyperbolic spacetime admitting a closed Cauchy hypersurface is finitely covered by a globally hyperbolic spacetime which can be isometrically embedded in a translation spacetime, in a Misner spacetime or in a (twisted) product of a standard spacetime by an euclidean torus.

A key point of this analysis, entering in the spirit of the present survey, is that standard spacetimes are precisely the quotients by the holonomy group of the causality domain defined in Section 1.2. In some way, we can say that the properness of the action is ensured by the causality restriction. This viewpoint is developed in [18].

Remark 5.3. *Actually, motivated by some examples appearing in the literature, we do not reduce in [17] to the case where the Cauchy hypersurface are closed, but we also consider the case where the Cauchy hypersurfaces are complete for their induced metric. This case is considerably more difficult: for example, it is not clear that any subgroup of $Isom(Min_n)$ admitting as linear part a Kleinian group, i.e., a discrete subgroup of $SO(1,n)$, is the holonomy group of globally hyperbolic spacetime. We proved that this statement is true for* convex cocompact *Kleinian groups, i.e., geometrically finite (admitting a finite sided polyhedral fundamental domain) and containing only hyperbolic elements[3]. This problem when the Kleinian group is geometrically finite is an open interesting question.*

5.4. Absolute time, CMC foliations

Everybody knows that from the general relativity point of view, there is no natural global time function on spacetime. Globally hyperbolic spacetimes do have many time functions –, i.e., strictly increasing along timelike paths – but *a priori*, none of them has a preferred status. However, some of them have a special interest, at least from the mathematical point of view. We present here two of them.

5.4.1. Cosmological time function.
We have already mentioned previously this function when discussing de Sitter spacetimes. The cosmological time function is defined in any spacetime as follows: $\tau(x)$ is the supremum of the proper times of future oriented timelike curves ending at x. In all examples of globally hyperbolic spacetimes discussed in this section, except for the cases of translation spacetimes, finite quotients of de Sitter spacetimes and parabolic de Sitter spacetimes, this function, if the chronological orientation of spacetime is well chosen, has only finite values and is C^1. This function provides quick proofs for the global hyperbolicity of these examples. Moreover, it is a "gauge-invariant" intrinsic feature of spacetime. On the other hand, it has a poor level of differentiability: even if in all

---

[3]For $n = 2$, a Kleinian group is geometrically finite if and only if it is finitely generated.

circumstances considered here it is C^2 almost everywhere (for the Lebesgue measure), in general, it is *not* C^2 everywhere – for example, in the case of standard spacetimes, it is C^2 only in the radiant case.

Spacetimes with regular cosmological time function are defined and studied in [11]. We also mention [21], where the cosmological function of $2+1$-dimensional standard spacetimes is geometrically studied, discussing the link between these notions and measured foliations, real trees, etc. In particular, there is a remarkable discussion on the striking fact that standard spacetimes geometrically realize the well-known and extensively studied correspondence between measured geodesic laminations and measured foliations. There is also the description of an interesting (mainly suggested) gluing operation obtaining from two standard spacetimes a globally hyperbolic spacetime with constant curvature -1, and relying on the cosmological function.

Let's also mention the very up-to-date [22], where the geometrical notion of *geodesic stratification* is associated to flat higher-dimensional standard spacetimes.

5.4.2. CMC foliations. The constant mean curvature of a spacelike hypersurface at a point is the trace of the shape operator at this point. A CMC hypersurface is a spacelike hypersurface with constant mean curvature. It is well known that Einstein's equations has a considerably most tractable form when considered in the neighborhood of a CMC hypersurface. Of course, this viewpoint is perfectly suited for the local study of Einstein equations, since local pieces of CMC hypersurfaces always exist; but the global existence of CMC hypersurfaces is a strong hypothesis. Let's mention here the survey [72] presenting similar questions for nonvacuum Einstein equations under strong or weak conditions.

We will call CMC foliation a foliation of spacetimes admitting as leaves CMC hypersurfaces. A function f is a CMC time function if:

- it is a time function, i.e., is increasing along timelike curves oriented towards the future,
- its value at a point x is the mean curvature of the level set $f^{-1}(f(x))$.

If we reverse the time orientation, we change the chronological orientations of timelike curves, but we also change the sign of shape operator: the CMC time function changes its sign, but is still a CMC time function.

Actually, the maximum principle of CMC hypersurfaces implies the uniqueness of CMC time function on a given spacetime – but not its existence! More precisely, for right conventions of sign on the shape operator, if S' is a spacelike hypersurface contained in the future of another spacelike hypersurface S, then, at every common point of tangency, the mean curvature of S is greater than the mean curvature of S'. Therefore, in a spacetime admitting a CMC time function τ, every closed CMC hypersurface is a fiber of τ. Indeed, its constant mean curvature value has to be greater (respectively less) than the maximal (respectively minimal) value of τ on itself.

Thus, a CMC time function, when it exists, is an intrinsic feature of the spacetime.

In [9], L. Andersson proved that every flat standard spacetime admits a CMC time function. The proof follows from the observation, already pointed out in [65], that these standard spacetimes can always be considered as small deformations of radiant standard spacetimes. In the radiant case, the CMC time function is obvious – and coincide with the CT function – and the level sets of this radiant CT function persist in the deformations as spacelike with controlled mean curvature. This control on the curvature enables the successful use of barrier methods of [46].

We should mention that CMC time functions of future (resp. past) geodesically complete standard spacetimes take value in $]-\infty, 0[$ (resp. $]0, +\infty[$)).

Observe that according to [17], the same conclusion holds more generally for every flat MGH spacetime with closed Cauchy hypersurfaces, except for translations spacetimes.

The authors, jointly with F. Béguin, established the existence of CMC time functions for locally AdS_3 globally hyperbolic spacetimes with closed spacelike surfaces. ([19, 20]). The proof is similar in many points with the Andersson's proof in the flat case.

As a comment for the genus 1 case[4] (the Torus Universe), we only mention that the proof is quite easy, the fibers of the CMC time function being the orbits of a 2-dimensional abelian Lie group.

The proof for the higher genus case is based as in the flat case on the exhibition of barriers to which are applied Gerhard's criteria [46]. But there is a fundamental difference: the AdS_3-spacetimes admitting obvious CMC functions are the static ones, but it is not possible to consider general AdS_3-spacetimes as small deformations of the static ones. The construction of barriers is thus undertaken with another method: basically, these barriers are constructed as smooth approximations of level sets of the CT function. More precisely, some level sets of the CT time function are strictly convex, other level sets are strictly concave; and in [20] we approximate these C^1 spacelike surfaces by smooth spacelike surfaces which are still respectively (strictly) convex and concave. In particular, their constant mean curvature values are respectively negative and positive everywhere. According to [46], the existence of a maximal hypersurface –, i.e., with null mean curvature – follows. The proof is then achieved thanks to the main result of [12]: in dimension $2+1$, the existence of a single CMC hypersurface ensures the existence of a CMC time function. This CMC time function in this context is a surjection onto $\mathbb{R}$.

The proof should presumably extend to higher dimensions, but it requires approximation of convex (or concave) hypersurfaces by smooth convex (concave) hypersurfaces – this is not so easy a task – and, more seriously, to supply an alternative to the use of [12] which has been established only in dimension $2+1$. Anyway, this draft of proof would be successful only through a better knowledge of the topological type of GH AdS-spacetimes.

---

[4]The genus 0 does not occur: the sphere cannot be a spacelike surface in a AdS_3 spacetime.

Finally, the existence of CMC time functions in globally hyperbolic spacetimes locally modelled on dS_3 with closed surfaces has not been published or announced anywhere. Here we give a sketchy proof, using the dual theory of hyperbolic ends.

First of all, we must point out that de Sitter spacetime itself does *not* admit CMC time functions! Of course, it admits (many) CMC foliations, but the leaves of these foliations are totally geodesic, and thus, this CMC foliation violates the increasing hypothesis for CMC time functions. More generally, this observation applies for finite quotients of de Sitter space, i.e., the elliptic case.

The same phenomena applies for the parabolic case: the n-dimensional spheres we mentioned earlier as Cauchy surfaces have all constant mean curvature -2, thus the argument above apply here also, proving that parabolic spacetimes do not admit CMC time functions.

However:

Theorem 5.4. *Cauchy-closed MGH dS_3-spacetimes which are not elliptic or parabolic all admit CMC time functions.*

Remark 5.5. *In this context, the CMC time function takes value in $]-\infty, -2[$ when the time orientation is selected so that the spacetime is future geodesically complete.*

Proof. Let S be a Cauchy surface of the spacetime M under study. Since M is nonelliptic, S is not a sphere. Moreover, since M is not parabolic, and if S has genus 1, it follows from remark 5.2 that the image of the developing map of the flat conformal structure is the complement of 2 points. These points are the extremities of a geodesic in $\mathbb{H}^3$ that we will use later; let's denote it by c.

In [61], it is proved that, hyperbolic ends always admit a *smooth* foliation by closed surfaces with constant *scalar* curvature. More precisely, the constant scalar curvature values of leaves vary between 0 and 1, and the principal eigenvalues are negative (here, the reader must trust our unexpressed sign conventions). Actually, [61] only deals with the case where S has genus bigger than 2. In the case of the 2-torus, the fibers of the distance function to the geodesic c defined above are the leaves of the required foliation.

This foliation provides a dual foliation in M (cf. Section 5.2.3). Moreover, the leaves of this foliation have constant scalar curvature too! Last but not least, the scalar curvatures of the leaves increases in time between 1 and $+\infty$. If L_t denote the leaf of this foliation with constant scalar curvature t, at any point x of L_t, and the principal eigenvalues λ and μ are both negative. and satisfy of course $\lambda\mu = t$. Hence, once of them is less than $-\sqrt{t}$, and the same is true for the mean curvature value $\lambda + \mu$. This is uniform on x: *the mean curvature of L_t is everywhere less than $-\sqrt{t}$.* On the other hand, if L_t^s denote the image of L_t under the Gauss flow which pushes every point during the proper time s along the normal of L_t, the principal eigenvalues of L_t^s converges uniformly to -1. Hence, for $t > 4$, and for s sufficiently big the mean curvature of L_t^s is everywhere greater that the maximal mean curvature value on L_t. Therefore, L_t and L_t^s form a pair of barriers to which [46] can be applied: M contains a CMC hypersurface. We then conclude as in the AdS_3 case by application of [12].   □

Remark 5.6. *In* [10], *answering a question in* [21], *L. Andersson proves that in standard spacetimes, the cosmological time functions and the CMC time functions have the same asymptotic properties.*

5.5. BTZ Black holes and wormholes

BTZ black holes was defined in [15, 16]: these are $2 + 1$-dimensional spacetimes locally modeled on AdS_3 presenting common features with realistic Schwarzschid (for the non-rotating case) and Kerr black holes (for the rotating case). Loosely speaking, they admit a natural conformal boundary at infinity and there is an open domain – the "black hole" – of spacetime which cannot be "seen" from this conformal boundary. In other words, future oriented lightlike rays starting from the "black hole" domain are incomplete, whereas outside the "black hole", there are complete lightlike rays reaching the conformal boundary.

Our quick presentation is very poor, however, these geometric objects have been extensively studied from the beginning of the 90's as toy models to which questions about the interaction of quantum phenomena and gravity can be tested. Let us mention [26] or [82] as cautions for the physical interest of these examples. And let's mention as good texts presenting BTZ black holes: [7, 6, 24, 8].

From (our) mathematical point of view, these black holes, and their multi-connected versions called "multiblack holes" and "wormholes" are the natural generalizations of globally hyperbolic geometric spacetimes. It has been understood from the beginning that all of them can be defined a the quotient by a discrete group Γ of isometries of a domain $\Omega(\Gamma)$ of AdS_3, the domains being defined in some way as the causality domain of the discrete group. Actually, there is some nontrivial result here: these domains of causality is not precisely $C(\Gamma)$. Elements of Γ are exponentials of elements $X(\gamma)$ of the Lie algebra $sl(2, \mathbb{R}) \times sl(2, \mathbb{R})$, and the domain $\Omega(\Gamma)$ is defined as the open domain where all the right invariant vector fields defined by the $X(\gamma)$ are spacelike. As a matter of fact, the action of Γ is proper and causal. In the rotating cyclic case, $C(\Gamma) \neq \Omega(\Gamma)$. But, as it is proved in [6] in a particular case, and that we will generalize in a forthcoming paper, the equality $C(\Gamma) = \Omega(\Gamma)$ holds as soon as the group Γ is not virtually cyclic.

Let us also mention that the so-called *angular momenta* of BTZ (multi) black holes (which is a physical notion) is the perfect analog in the AdS context of the Margulis invariant discussed in 4.2. It is an interesting challenge to evaluate how far this analogy can be continued. In particular, the validity of the results of [38, 30] in the BTZ context would be very interesting.

5.6. Causal properties of Margulis spacetimes

In his book, S. Carlip, refering to Margulis spacetimes, wrote: "The resulting geometries are fairly bizarre...., and they could potentially serve as counterexamples for a number of plausible claims about $(2 + 1)$-dimensional gravity" ([26], p. 11).

Here, we mainly want to stress out that globally hyperbolic parts of Margulis spacetimes are fairly well identified: according to [17], any finitely generated purely hyperbolic subgroup Γ of $\text{Isom}(Min_3)$ is the holonomy group of two MGH spacetimes. More precisely, there are two (disjoint) Γ-invariant convex open domains $C^\pm$ in Min_3 whose quotients $U^\pm$ are MGH spacetimes containing every globally hyperbolic spacetimes with holonomy group Γ. This is true in particular when Γ acts properly, i.e., when $M = \Gamma \backslash Min_3$ is a Margulis spacetime. In other words, M contains disjoint isometric copies of $U^\pm$, and the complement of these two subdomains is a region N where causality properties are dramatically violated. Let's now state some properties of M:

- for every timelike geodesic $t \mapsto \gamma(t)$, there are two values $t_\pm$ such that for $t < t_-$ (resp. $t > t_+$) $\gamma(t)$ is in U^- (resp. U^+),
- every point in the interior of N belongs to a closed timelike curve (CTC), and CTC are all contained in N,
- $\partial U^\pm$ are null-surfaces covered by future or past (depending on the sign $\pm$) null geodesic rays,
- M does not contained closed lightlike geodesics,
- (Penrose boundary) M can be naturally "completed" by future and past conformal ideal boundaries $\mathcal{J}^+$, $\mathcal{J}^-$, each of them being a finite number of annular components.

Naïvely, one is tempted at first glance to consider N as a black hole since, at least, it could be the feeling of observers inside U^-: they cannot observe any singularity, since their past cone is complete, enjoy the nice sensation to be part of a globally hyperbolic spacetime, but all of them are promised to a dramatic issue: the entrance in N.

On the other hand, the convexity of C^- implies that the volume of the intersection between its boundary and spacelike planes of Min_3 decreases when the plane is moved in the future: this is not compatible with the increasing entropy property of black holes. Anyway, this phenomena does not contradict Hawking's Theorem, since for this theorem, the black hole is the region which cannot be observed from the future conformal boundary $\mathcal{J}^+$. Here, the conformal boundary of C^- is contained in $\mathcal{J}^-$. Thus, to follow the classical treatment of black holes, we must be concerned with C^+: then, the singularity region N is observed by everybody: this is a naked singularity, that all relativists reject as physically unrealistic.

6. Isometric actions

Now we investigate isometric actions on Lorentz manifolds. (To begin with, let us mention that, even if in setting problems the compactness of the Lorentz manifold is not needed, the whole results of the present section, except Paragraph 6.7, concern compact manifolds.)

Question 6.1. *Let (M, g) be a Lorentz manifold, and $G = Isom(M, g)$ its isometry group. When is the action of G on M essential?*

Firstly, by "essential", we mean that it is really a (pure) Lorentz isometry group, i.e., it cannot preserve a Riemannian metric on M?

Here, one may ask, why in this notion, we are comparing Lorentz metrics with Riemannian ones, and not with any other structures. The point is that the comparison is from a dynamical point of view. A Riemannian metric is in a dynamical sense a structure of lower order: Riemannian isometries are equicontinuous, and therefore have no "chaotic" dynamics. In contrast (as it is well known, and we will recall below), Lorentz isometries can be, for instance of Anosov type (this is reminiscent to the revolutionary facts of special Relativity, asserting possibility of contraction of local time, and dilation of lengths). It is thus natural to call inessential a Lorentz metric having an isometry group which coincides with that of (an auxiliary) Riemannian metric.

In fact, one can try to generalize a notion of "essentiality" to other geometric structures. This is not so easy to formulate, but the idea is to find a good notion of dynamical hierarchy between geometric structures, that is to decide how stronger is the dynamics (i.e., isometry group) generated by someone with respect to the other. As a paradigmatic example, a conformal pseudo-Riemannian structure is essential, when its conformal group does not preserve a pseudo-Riemannian metric (in the same conformal class) (see §7.)

Now, it is known that preserving a Riemannian metric is equivalent to acting properly (all objects are smooth). Therefore, the question becomes

When is the action of G on M non proper?

If furthermore the manifold is compact, then the G-action is proper iff G is compact, hence our question becomes:

When is the isometry group of a compact Lorentz manifold non-compact?

6.0.1. Sub-question: Lorentz homogeneous spaces.
Here we specialize the question to the homogeneous case. Therefore, the question is to classify with the following (algebraic) conditions:

– $M = G/H$ (G a Lie group, and H a closed subgroup of G).

(To simplify, We suppose that G acts faithfully on M, i.e., we cannot simplify G/H to a smaller G'/H'.)

– The left G-action $((g, xH) \in G \times M \to (gx)H \in M)$ preserves a Lorentz metric.

– The isotropy group H is not compact (this means non-properness).

If M is compact, the last condition becomes: G is not compact.

6.0.2. Super-question: stable properness. Let $\mathrm{Diff}^k(M)$ be the group of diffeomorphisms of class k of M. It acts on $Lor^{k-1}(M)$, the space of C^{k-1} Lorentz metrics on M. Endow $Lor^{k-1}(M)$ with the Banach or Fréchet topology (Fréchet for $k = \infty$). For the sake of simplicity, we will not note k, and assume that M is compact.

- It is known that $\mathrm{Diff}(M)$ acts properly on $Riem(M)$, the space of Riemannian metrics on M. In particular, the quotient $Riem(M)/Diff(M)$ is Hausdorff, it is the modular space of M.
- Notice that, any function on $Riem(M)/Diff(M)$ is a Riemannian invariant: e.g., volume, diameter...

Super-question 6.2. *When is the $Diff(M)$-action on $Lor(M)$ proper?*

For $g \in Lor(M)$, $\mathrm{Stabilizer}(g) = \mathrm{Isom}(g)$. If the $\mathrm{Diff}(M)$-action is proper then, $\forall\, g \in Lor(M)$, $\mathrm{Isom}(g)$ is compact, that is the super-question is stronger than the question! We quote from [35] that the difficulty in the global studying of Lorentz manifolds lies in the fact that $Lor(M)/Diff(M)$ is not Hausdorff.

6.0.3. Some motivations. The question we are asking is reminiscent to the (former) Lichnérowitch conjecture, "Conformal groups of Riemannian manifolds", solved by Ferrand and Obata [40, 70]. It starts by the observation that, although $\mathrm{Conf}(M, g)$, the conformal group of a compact Riemannian manifold (M, g), is not, a priori, compact, the only known examples for which the group is indeed non-compact are the Euclidean spheres. The result, which was actually proved, in its final form by J. Ferrand, confirms this fact: only the usual spheres have non-compact conformal group (among compact Riemannian manifolds).

Ferrand-Obata Theorem and our present question are in fact particular cases of a rigidity phenomenon in geometric dynamics (see for instance [35]).

Our sub-question concerns classification of a small class in the wide world of compact homogeneous spaces. The homogeneous Riemannian problem, is "trivial": take $M = G/H$, where G is any compact Lie group and H is a closed subgroup of it. In contrast, we know very little information about general compact non-Riemannian homogeneous spaces. The interest of the Lorentz case (that is, our sub-question) is that it seems to be the easiest non-Riemannian homogeneous problem.

The case where H is discrete is special. Indeed, in this case, G covers G/H, and one can pull back the G-invariant geometric structure on G itself. Therefore, the nature of this geometric structure can be seen on G, in fact at its Lie algebra level $\mathcal{G}$.

Fact 6.3. *The problem of closed homogeneous Lorentz manifolds with discrete isotropy is equivalent to find a co-compact lattice H in a Lie group G, and a Lorentz scalar product on $\mathcal{G}$ preserved by the adjoint action of H.*

Proof. Left translate on G the Lorentz scalar product on $\mathcal{G}$ which is $Ad(H)$-invariant. The Lorentz metric on G is: G-left-invariant, and H-right invariant. Therefore, it passes to a G-invariant Lorentz metric on G/H.

In particular, if $\mathcal{G}$ admits a bi-invariant (i.e., $Ad(G)$-invariant) Lorentz scalar product, then any quotient G/H, where H is discrete is a homogeneous Lorentz manifold.   □

6.1. Examples: Lie algebra with bi-invariant Lorentz metrics

We will start here by giving examples of compact homogeneous Lorentz manifolds. Obviously, the only interesting non-trivial cases are when the isometry group is neither compact nor abelian.

6.1.1. "Baby" example: $PSL(2, \mathbb{R})$.
The Killing form of $PSL(2, \mathbb{R})$ is non-degenerate (as the "simplest" semi-simple Lie group). Since the dimension is 3, the Killing form has a type $- + +$ or $- - +$. Anyway, up to a change of sign it is Lorentzian. It is bi-invariant as for any group. Therefore $PSL(2, \mathbb{R})$ has a bi-invariant Lorentz metric, and any compact quotient $M = PSL(2, \mathbb{R})/H$ where H is a co-compact lattice (a surface group) is a compact homogeneous Lorentz manifold. In fact, the isometry group of M is essentially (i.e., up to finite index) $PSL(2, \mathbb{R})$. Actually, as already discussed 4.1.4, these examples are locally modelled on AdS_3.

6.2. Oscillator groups

The oscillator groups (sometimes called warped Heisenberg groups as in [85]) is a family of "sympathetic groups": they are *solvable* but look like $SL(2, \mathbb{R})$ (we use the adjective "sympathetic" for these guys because we find they are so, and also this adjective is used in some literature to refer to groups which enjoy many properties of semi-simple Lie groups). They admit bi-invariant Lorentz metrics (which are of course different from the Killing form, which is degenerate for these groups). Also, they do have co-compact lattices (co-compact is superfluous, since any lattice in a solvable Lie group is co-compact). Let us anticipate here and say that an oscillator group has essentially one lattice (all of them are commensurable).

6.2.1. The simplest example.
The semi-direct product $G = S^1 \ltimes Heis$.

Recall the definition of $Heis$, the Heisenberg group of dimension 3:

$$Heis = \left\{ \begin{pmatrix} 1 & x & z \\ 0 & 1 & y \\ 0 & 0 & 1 \end{pmatrix}, x, y, z \in \mathbb{R} \right\}.$$

$Heis$ is characterized essentially, by the existence of a non-split exact sequence: $1 \to \mathbb{R} \to Heis \to \mathbb{R}^2 \to 1$.

The circle S^1 acts automorphically on $Heis$, where the action is trivial on the center $\mathbb{R}$, and it is by rotation on $\mathbb{R}^2$.

The simplest oscillator group is the semi-direct product $G = S^1 \ltimes Heis$.

G can also be characterized as a non-trivial central extension of Ec, the group of Euclidean isometries of the plane: $S^1 : 1 \to \mathbb{R} \to G \to Ec \to 1$.

6.2.2. Generalization: "canonical" oscillator groups. Recall the construction of Heisenberg algebras $\mathcal{HE}_d$ (dim $= 2d+1$). Consider $\mathbb{R} \oplus \mathbb{C}^d$, with basis $Z, e_1, \ldots, e_d$ The only non-vanishing brackets are: $[e_k, ie_k] = Z$ (here $i = \sqrt{-1}$). Equivalently, $[X, Y] = \omega(X, Y)Z$, where ω is the symplectic form $\omega(X, Y) = \langle X, iY \rangle_0$, where, $\langle, \rangle_0$ is the Hermitian product.

- Canonical oscillator algebras are obtained by adding an exterior element t, such that: $[t, e_k] = ie_k, [t, ie_k] = -e_k$, and $[t, Z] = 0$. Denote by $\mathcal{HE}_d^t$ the resulting Lie algebra.
- Define on it a scalar product $\langle, \rangle$ as follows. Endow $\mathbb{C}^d$ with its Hermitian structure $\langle, \rangle_0$. Decree: $\mathbb{C}^d$ orthogonal to $Span\{t, Z\}$, $\langle t, t \rangle = \langle Z, Z \rangle = 0$ and $\langle t, Z \rangle = 1$.
- It turns out that $\langle, \rangle$ is a $Ad(\mathcal{HE}_d^t)$-invariant Lorentz scalar product. In other words, for every u in $\mathcal{HE}_d^t$, ad_u is antisymmetric with respect to $\langle, \rangle$. (Exercise: why this does not work for the Heisenberg algebras themselves?)
- Consider $\tilde{G} = \tilde{H}e_d^t$ the simply connected Lie group generated by $\mathcal{HE}_d^t$.
- $\tilde{H}e_d^t$ is a semi-direct product of $\mathbb{R}$ by He_d: the action of $\mathbb{R}$ on the center is trivial, and its action on $\mathbb{C}^d$ is via multiplication by $\exp is$.
- This is in fact an action of S^1. Consider then the semi-direct product $G = He_d^t = \tilde{H}e_d^t/\mathbb{Z} = S^1 \ltimes He_d$ (here $\mathbb{Z}$ is simply the subgroup of integers of $\mathbb{R}$)
- Any lattice in the Heisenberg group He_d is also a lattice in He_d^t (since He_d is co-compact in He_d^t). As example of lattice in He_1, we have:

$$Heis_\mathbb{Z} = \left\{ \begin{pmatrix} 1 & x & z \\ 0 & 1 & y \\ 0 & 0 & 1 \end{pmatrix}, x, y, z \in \mathbb{Z} \right\}.$$

6.2.3. General construction of oscillator groups. The most general oscillator groups are defined as above as semi-direct product $\mathbb{R} \ltimes He_d$, where $\mathbb{R}$ acts on $\mathbb{C}^d$ via a homomorphism $s \to \exp(2\pi sA) \in U(d)$ such that: $\exp(2\pi A) = 1$, and A diagonalizable with eigenvalues say, $\lambda_1, \ldots, \lambda_d \in \mathbb{Z}$ having the same sign. This last condition on signs guarantees that the obtained Lie algebra admits a bi-invariant Lorentz metric. The arithmetic (or say quantum) condition, λ_i integers, implies that the $\mathbb{R}$-action factors via an action of S^1. An oscillator group is any such semi-direct product $G = S^1 \ltimes He_d$. It enjoys the same property as that of canonical oscillator groups, i.e., when the matrix A is scalar.

6.2.4. Further remarks.

1. Remember that the Lorentz scalar product was defined, among other conditions, by the fact that $\langle t, t \rangle = 0$. In fact, one can take $\langle t, t \rangle = $ Constant $\neq 0$, and multiply the other given products by any constant ($\neq 0$), and gets another bi-invariant Lorentz metric.

2. However, up to automorphism, there exists only one bi-invariant Lorentz metric on a oscillator algebra. In particular a metric is isometric to any multiple of itself. This follows from existence of homotheties. (This is true for $\mathbb{R}^n$ but not for $PSL(2, \mathbb{R})$.)

3. Oscillator groups are (locally) symmetric Lorentz spaces of non reductive type, that is they have non-reductive holonomy. This means they have a codimension 1 *parallel* foliation which has no supplementary parallel direction field. This foliation is nothing but determined by translates of the Heisenberg group (on the left or the right, it is the same thing since Heisenberg group is normal in the oscillator group).
4. The Ricci curvature of an oscillator group equals its Killing form (up to constant).

6.2.5. Historical comments.

1. Actually, it is the He_1^t, the 4-dimensional canonical oscillator example (the simplest example) which was named in the literature as the oscillator group [78], see justification below. It is also known as the diamond group in Representation Theory.
2. The bi-invariant Lorentz metrics were known to Medina-Revoy, and "partially" to Zimmer [89] and Gromov [50].
   This seems folkloric in relativistic literature: some gravitational plane waves spacetimes...
   Also, Witten and Nappi [69] used the oscillator group to built "a WZW model based on a non semi-simple group".

6.2.6. Justification of the name "oscillator".

The Lie algebra $\mathcal{H}\mathcal{E}_1^t$ has the following representation in the algebra of operators of the Hilbert space $E = L^2(\mathbb{R})$:

$$Z \to 1, X \to q, Y \to p, t \to p^2 + q^2$$

where the operators q and p are given by:

$$q(f) = xf, f \in L^2(\mathbb{R}), p(f) = \frac{\partial f}{\partial x}$$

(1 is the Identity (operator), q the position, p the momentum, and $p^2 + q^2$ the energy).

To show that this gives a homomorphism, one verifies in particular: $[q, p] = 1$, which is the Heisenberg uncertainty principle.

Finally, $p^2 + q^2$ is the energy of the harmonic oscillator, which explains the origin of the terminology, that is this representation gives a quantification of the harmonic oscillator.

6.3. Other examples: discrete isometry groups, general constructions

6.3.1. Discrete case. So far, our examples have the form $M = G/H$, where G is a Lie group group, which is non-compact, and (implicitly) connected. However, it might happen that G is not the full isometry group of M. However, this fact causes no loss for us. The true difficulty, is when a homogeneous Lorentz space has the form G/H, where G is *compact* (and connected), so at first glance, M looks like inessential, but it might happen that Isom(M) is not compact. For instance, G could be the identity component of the isometry group, which has "a discrete part" Isom$(M)/G$ non-compact. Let's give an example illustrating this phenomenon (which might be the general one?).

On $\mathbb{R}^n$, consider a lorentz scalar product g, and let $O(g)$ be its orthogonal group. The essential point is that we consider $\mathbb{R}^n$ together with its lattice $\mathbb{Z}^n$. Hence, $O(g)$ is isomorphic to $O(1, n-1)$, but in general, $O(g, \mathbb{Z}) = O(g) \cap GL(n, \mathbb{Z})$ is not isomorphic to $O(1, n-1; \mathbb{Z}) = O(1, n-1) \cap GL(n, \mathbb{Z})$.

Consider the flat torus $(T^n, g) = (\mathbb{R}^n, g)/\mathbb{Z}^n$. Then, Isom $(T^n, g) = T^n \rtimes O(g, \mathbb{Z})$. Therefore, the identity component is compact, and the "discrete part" is $O(g, \mathbb{Z})$.

Let us consider the simplest example, apparently firstly observed by Avez: let A be a hyperbolic element of $SL(2, \mathbb{Z})$ hyperbolic, that is, with real eigenvalues of norm $\neq 1$, e.g.,

$$A = \begin{pmatrix} 2 & 1 \\ 1 & 1 \end{pmatrix}.$$

Let $\{\omega^s, \omega^u\}$ be a diagonalisation basis for the dual of A: ω^s and ω^u are linear forms on $\mathbb{R}^2$ defined up to scalars. Finally, $g = c\omega^u \omega^s$, where c is an arbitrary constant $(\neq 0)$. Then, A preserves g. In fact, up to a finite index, Isom$(T^2, g) = T^2 \rtimes \mathbb{Z}$ ($\mathbb{Z}$ is generated by A).

In dimension > 2, the discrete part $O(g, \mathbb{Z})$ may be much bigger. Indeed, by a Harisch-Chandra Borel Theorem [23], if g is rational (that is up to a multiplicative constant, the coefficients of g in the canonical basis, are all rational), then $O(g, \mathbb{Z})$ is a lattice in $O(g)$ (in particular $O(g, \mathbb{Z})$ is isomorphic to the fundamental group of a *finite volume* hyperbolic manifold, which can be compact).

One remarks here that a non co-compact lattice in $O(g)$, or say $O(1, n-1)$ since we are allowed to identify them at this stage, contains hyperbolic and parabolic elements (in the sense of $O(1, n-1)$). We then see, from a dynamical point of view, that Lorentz isometries may have, Anosov (in dimension 2), partially hyperbolic, and also horocycle-like behaviors. This contrasts with Riemannian isometries which are built up from "blocks", on which dynamics is equivalent to a translation on a torus.

6.3.2. Warped products. Let M be an essential Lorentz manifold. Observe that its (direct) product by a Riemannian manifold is also essential. If both are homogeneous, then the same is true for the product.

In fact, local products may also preserve essentiallity. To define them (in our Lorentz context), let $M = \tilde{N}$ be Lorentz and $\tilde{L}$ be Riemannian. Consider $M = \tilde{N} \times \tilde{L}/\Gamma$, where Γ is a subgroup of $Isom(\tilde{N}) \times Isom(\tilde{L})$. Assume Γ is *non split* (or in other terms irreducible), that is, M is not a product. If the centralizer of Γ acts non-properly on $\tilde{N} \times \tilde{L}$, then M is essential.

We recall now another way to preserve essentiallity, the warped product construction. Let (N, g) be Lorentz, and (L, h) Riemannian and $w : L \to R^+$ a (warping) function. The warped product $M = L \times_w N$, is the topological product $L \times N$, endowed with the metric $h \oplus wg$.

The crucial property for us here is that, if $f : N \rightarrow N$, is an isometry then, its trivial extension: $\bar{f} : (x, y) \in L \times N \rightarrow (x, f(y)) \in L \times N$, is an isometry of $L \times_w N$

In particular, as above, in the class of essential Lorentz manifolds , one can perform warped products by (any) Riemannian manifolds. Similarly to local products, one can define local warped products, which may preserve essentiallity.

6.3.3. "Counter-examples". Consider the "hyperbolic torus" T_A^3, where $A \in SL(2, \mathbb{Z})$ is hyperbolic (see §6.3.1). So, T_A^3 is the suspension of A seen as a diffeomorphism (of Anosov type) of the 2-torus T^2. As was said in §6.3.1, A preserves a Lorentz metric g on T^2. Endow $T^2 \times \mathbb{R}$ with the flat product metric $g \oplus dt^2$. The mapping $\phi : (x, t) \rightarrow (Ax, t + 1)$ is isometric. Therefore, T_A^3 inherits a flat Lorentz metric, with the suspension flow acting isometrically. It is an Anosov flow, and hence T_A^3 is in particular essential. Nevertheless, T_A^3 is not Lorentz homogeneous. Its isometry group is (up to a finite cover) generated by the suspension flow. On the other hand this is a sol-manifold: $T_A^3 = SOL/\mathbb{Z} \times_A \mathbb{Z}^2$. Summarizing, T_A^3 is an essential Lorentz manifold, topologically homogeneous (i.e., the isometry group admits a dense orbit), but not Lorentz homogeneous.

This "contrasting" fact is also valid for 3-Nil manifolds, i.e., compact quotients of the (3-dimensional) Heisenberg group (these manifolds are however not essential, since their isometry group are cyclic).

6.4. Classification of Killing algebras

We have the following result which answers, at least at the Lie algebra level our question 6.1. It was proved independently in its final form in [3] and [85] (at the same month!). Partial steps were done in [50] and [89].

Before announcing it, recall that the Lie algebra of a group acting isometrically on a compact Riemannian manifold is a sum of an abelian Lie algebra with a semi-simple Lie algebra of *compact* type (i.e., the Lie algebra of a compact semi-simple Lie group).

The result says that, in the Lorentz case, the new factor that might occur, is a subalgebra of $\mathcal{S}$, where $\mathcal{S}$ is the Lie algebra of $SL(2, \mathbb{R})$ or an oscillator group.

Theorem 6.4. *Let G be a connected Lie group acting isometrically on a compact Lorentz manifold M. Then, up to compact objects, G is a subgroup of $PSL(2, \mathbb{R})$ or of an oscillator group.*

More precisely, the Lie algebra $\mathcal{G}$ is isomorphic to a direct algebra sum:

$$\mathcal{K} + \mathbb{R}^k + \mathcal{S},$$

where $\mathcal{K}$ is the Lie algebra of a compact semi-simple Lie group, $k \geq 0$ is an integer and $\mathcal{S}$ is a subalgebra of:

- *$sl(2, \mathbb{R})$.*
- *an oscillator algebra.*

Furthermore, the group S associated to $\mathcal{S}$ acts on M locally freely, i.e., stabilizer in S are discrete

Corollary 6.5. *The stabilizer of any point of M is* "almost discrete" *: its connected component is compact.*

Remark 6.6. *The Corollary is far from being a priori obvious. It is false for non-compact homogeneous spacetimes, and for general homogeneous pseudo-Riemannian manifolds, even compact.*

6.4.1. The full Killing algebra. We dealt above with a group acting on M, i.e., a subgroup of $\mathrm{Isom}(M)$. One may ask about the structure of the full $\mathrm{Isom}(M)$ itself. In other words, can any group as described above (at the Lie algebra level) be *exactly* the (full) isometry group of some compact Lorentz manifold? For example, can the S-factor for the full $\mathrm{Isom}(M)$ be the affine group $\mathrm{Aff}(\mathbb{R})$, or in contrary, once $\mathrm{Aff}(\mathbb{R})$ acts isometrically, then does its action automatically extend (isometrically) to $SL(2,\mathbb{R})$ (always at a Lie algebra level)? The answer to this precise example is that extension indeed exists. In general, the answer was given in [86] and [4] independently (in the same season!)

By the Killing algebra of M we mean the Lie algebra of $\mathrm{Isom}(M)$.

Theorem 6.7. [86, 4] *The Killing Lie algebra of a compact Lorentz manifold is isomorphic to a direct sum*

$$\mathcal{K} + \mathbb{R}^k + \mathcal{S},$$

where $\mathcal{K}$ is the Lie algebra of a compact semi-simple Lie group, $k \geq 0$ is an integer and $\mathcal{S}$ is trivial or isomorphic to:

- *a Heisenberg algebra,*
- *an oscillator algebra, or*
- *$sl(2,\mathbb{R})$.*

Conversely, any such algebra is isomorphic to the Lie algebra of the isometry group of some compact Lorentz manifold.

6.5. Sub-question: Homogeneous case

6.5.1. Algebraic classification.

Theorem 6.8. [3, 4, 85, 86] *Let $M = G/H$ be a compact homogeneous Lorentz manifold. Then, up to compact objects: G is $SL(2,\mathbb{R})$ or an oscillator group.*

More precisely, there is a subgroup $S \subset G$, such that:

- *S is normal, and the Lie algebra of S is a factor in $\mathcal{G}$*
- *S is co-compact in G (i.e., G/S is compact)*
- *S is isomorphic to $PSL_k(2,\mathbb{R})$ the k-folded cover of $PSL(2,\mathbb{R})$, or*
- *S is an oscillator group*
- *S acts on M locally freely, that is H is "almost discrete", in the sense that its identity component is compact.*

6.5.2. Geometric classification.

Theorem 6.9. [85] *Let $M = G/H$ be a compact homogeneous Lorentz manifold. Then, up to compact objects, it is isometric to S/H, where H is a co-compact lattice (in particular discrete) in S, where S is $PSL(2, \mathbb{R})$ or an oscillator group.*

- *Roughly, M is a "local product" modeled on $S \times \tilde{L}$, where $\tilde{L}$ is a homogeneous Riemannian manifold*
- *The case $S = PSL_k(2, \mathbb{R})$ (due to Gromov [50]):*
  - *$M = S \times \tilde{L}/H$:*
  - *$\tilde{L}$ is a compact homogeneous Riemannian manifold*
  - *There is H_0 a lattice in S, such that H is the graph of a homomorphism $\rho : H_0 \to Isom(\tilde{L})$*
  - *The centralizer of $\rho(H_0)$ acts transitively on $\tilde{L}$.*
  - *The metric on $S \times \tilde{L}$ equals: $c.Killing \otimes r_{\tilde{L}}$, for some constant c, where $r_{\tilde{L}}$ denotes the Riemannian metric of $\tilde{L}$.*
  - *Conversely, with these data, one constructs a compact homogeneous space-time.*

- *In the case where S is an oscillator group, the geometric description is a little bit complicated: the (local) product structure can be somehow "twisted" (see [85]).*

6.6. Super-question

Remember that the super-question concerns the action of $\mathrm{Diff}(M)$ on $Lor(M)$, the space of Lorentz metrics on M. Consider two converging sequences, g_n and h_n of Lorentz metrics (in the C^2 topology). Suppose they are isometric, that is, there is a sequence of diffeomorphisms ϕ_n, such that $h_n = (\phi_n)_* g_n$. Properness, means that, after passing to a subsequence, the sequence ϕ_n must converge (the limit will be an isometry between the two limit metrics).

6.6.1. Main ingredient: actions of discrete groups.
So far, only actions of connected Lie groups were considered. However, as was seen in the case of flat tori 6.3.1, the essentiality may come from the discrete part. It seems that the only works which deal with Lorentz isometries, without, a priori, connectedness hypothesis, are firstly that of D'Ambra [34] (where actually connectedness is proved at an intermediate step), and [87, 88], which investigate dynamics of Lorentz isometries in a systematic way. In fact, also sequences of isometries were considered there (seen as generalized dynamical systems). It was also observed that the approach can be adapted to sequences of isometries between two different Lorentz metrics, or even sequences of Lorentz metrics, which is exactly the situation you meet, as above, when dealing with properness. The philosophy of this work is to see how such sequences of isometries degenerate. One consider their graphs in $M \times M$, which are totally geodesic (and isotropic) for the product metric. The limits are geodesic laminations in a suitable space. It turns out however, that, by projecting on M, one gets a codimension one foliation, with geodesic and lightlike leaves (the metric

on them is degenerate). At least in dimension 3, one knows many obstructions to the existence of such foliations. As a corollary, we get:

Theorem 6.10. [88] *For M the 3-sphere, the Diff(M)-action on Lor(M) is proper.*

6.6.2. Case of compact surfaces. P. Mounoud pushed forward the analogy between the case of isometries of a fixed metric, and that between sequences of metrics (as in the definition of properness). He applied this for Lorentz compact surfaces which must be (topologically) a Klein Bottle or a torus. New ideas are needed here since lightlike (in fact isotropic) foliations always exist. He firstly proved:

Theorem 6.11. [68] *For M = Klein bottle, the Diff(M)-action on Lor(M) is proper.*

Let M be now a 2-torus, and $\mathcal{F}$ the space of flat Lorentz metrics on it. Any such a metric is linear on the universal cover $\mathbb{R}^2$ (up to a diffeomorphism). One observed that the $\mathrm{Diff}_0(M)$-action on $\mathcal{F}$ is proper, where $\mathrm{Diff}_0(M)$ is the group of diffeomorphisms isotopic to the identity. The quotient is the Lorentz Teichmüller space of the 2-torus. It is identified to the de Sitter space:

$$dS_2 = SL(2, \mathbb{R})/\{ \begin{pmatrix} e^t & 0 \\ 0 & e^{-t} \end{pmatrix}, t \in \mathbb{R}\}.$$

The action of $\mathrm{Diff}(M)$ on $\mathcal{F}/Diff_0(M)$ is identified to the action of $SL(2, \mathbb{Z})$ on dS_2. This action is dual to the action of $\{ \begin{pmatrix} e^t & 0 \\ 0 & e^{-t} \end{pmatrix}, t \in \mathbb{R}\}$ (the geodesic flow) on the unit tangent bundle $SL(2, \mathbb{R})/SL(2, \mathbb{Z})$ of the modular surface $\mathbb{H}^2/SL(2, \mathbb{Z})$. Therefore the $\mathrm{Diff}(M)$-action on $\mathcal{F}$ is in particular ergodic.

The main result of [68] is that in contrast:

Theorem 6.12. *The Diff(M)-action on Lor(M) $- \mathcal{F}$ (the space of non-flat metrics) is proper.*

Therefore, the $\mathrm{Diff}(M)$ dynamics on $Lor(M)$ is fuchsian-like: strong (ergodic...) on its "limit set" $\mathcal{F}$, and proper on its "discontinuity domain" $Lor(M) - \mathcal{F}$.

Among other beautiful ideas, the proof uses an amazing lemma, which states that, if a Lorentz metric on the torus has its curvature constant along one isotropic foliation, then this metric is flat.

6.7. Non-compact manifolds

We consider here our question (or its variants, sub, or super) in the case of non-compact Lorentz manifolds. Despite its natural both mathematical and physical interest (only non-compact spacetimes are realistic in physics), the question here is far from being elucidated, and sufficiently investigated. Maybe, the reason is that no prompt answer seems to be available.

From a relativistic point of view, one observes that only few classical exact solutions have essential isometry groups. One may dare ask:

Question 6.13. *(for relativists) Classify physical solution (i.e., an exact solution, a spacetime with a natural energy-momentum tensor, a spacetime satisfying suitable causality conditions) having an essential isometry group.*

As example, spaces of constant curvature, dS_n, Min_n and AdS_n have essential groups. They are homogeneous, with non-compact isotropy.

As celebrated exact solutions, *pp*-waves have essential isometry groups. Our question above asks for a rigidity of essential "physical" spacetimes, that is they must belong to a (small) list to be founded and enumerated.

From a purely mathematical point of view, it seems that Nadine Kowalsky was the first to consider this problem, in her thesis (supervised by Zimmer). As the general question looks too waste to be systematically investigated, she made an algebraic hypothesis on the acting group. It was with a little bit variation of this same hypothesis that other authors contribute. So, N. Kowalsky asked, and solved the following:

Question 6.14. *When a simple Lie group acts isometrically non-properly on a Lorentz manifold?*

Observe that the de Sitter and de Sitter spaces, are examples of such spaces. Their isometry groups are $O(1,n)$ and $O(2,n)$ (they are simple except for $O(2,2)$).

The answer of Kowalsky is that these are the only examples, at an algebraic level.

Theorem 6.15. [57] *If a simple Lie group acts isometrically non-properly on a Lorentz manifold, then it is isomorphic to $O(1,n)$ or $O(2,n)$ (for some n).*

This was the principal result of [57], and was also announced in [58], together with another announcements of results. Unfortunately, Kowalsky died prematurely, before publishing half of the announced results. One of the announced results is on a geometric description of the Lorentz manifolds as in the theorem above. Here again dS, and AdS appear (essentially) as unique examples.

Theorem 6.16. *If a simple Lie group G acts isometrically non-properly on a Lorentz manifold M, then M is a warped product of dS_n or AdS_n with some Riemannian metric (here we must assume G not locally isomorphic to $SL(2, \mathbb{R})$ in order to avoid consideration of local warped products...).*

D. Witte [81] proved this result, assuming Theorem 6.15 and that the action is transitive. Let us observe that, even with these assumptions, the result is by no means obvious! In [13], the authors introduce a new geometric approach allowing a unified proof of both the previous two theorems. They also consider some generalizations. Let us notice here that S. Adams was the first and principal "investigator" on Kowalsky's heritage. In particular, he relaxed in many ways the algebraic condition (simplicity) on the Lie group (the conclusions are different). He also yields another proof of Theorem 6.15 (see for instance [2]). Notice however that all approaches, except in [13], are deeply algebraic.

6.8. Idea of proof of Theorem 6.4

Consider the L^2 bilinear form on the Lie algebra $\mathcal{G}$:

$$\kappa(X, Y) = \int_M \langle X(x), Y(x) \rangle_x \, dx$$

where X, Y are Killing fields and $\langle, \rangle$ is the Lorentz metric.

6.8.1. Steps.

1) κ is a *bi-invariant* quadratic form on $\mathcal{G}$. This is a general fact: an action of $\mathcal{G}$ means, a homomorphism (of Lie brackets) $X \in \mathcal{G} \to \bar{X} \in$ Vector-fields on M.

   So, to $Y \in \mathcal{G}$, is associated:
   - ϕ^t a one-parameter subgroup of G, and
   - $\bar{\phi}^t$ a one parameter group of diffeomorphisms on M.

   Naturally:

   $$\bar{\phi}^t_* \bar{X} = \overline{Ad(\phi^t)X}$$

   It then follows, if G preserves a volume dx and a q-covariant tensor T, then, the formula:

   $$\kappa^T(X_1, \ldots, X_q) = \int_M T(\overline{X_1}(x), \ldots, \overline{X_q}(x)) dx$$

   determines a *bi-invariant* q-tensor on $\mathcal{G}$.

2) However, κ might be trivial! For instance, if $G = SO(n), n > 2$, let T be any left invariant quadratic form (degenerate or not, positive or not...). Then necessarily, κ^T is a multiple of the Killing form, by simplicity. In particular, it may happen that $\kappa^T = 0$ for T Lorentz.

3) The point is thus to show that, in our situation, κ is sufficiently non-trivial.... A major step in the proof will be to show that κ satisfies a condition (*), which roughly speaking means that κ is between being a Lorentz and a Euclidean scalar product!

4) Theorems 6.4 follows from an "Algebraic Lemma" classifying" those Lie algebras admitting Ad-invariant scalar product satisfying (*). This classification is similar (but of higher-order difficulty) to the lemma saying that a Lie algebra with an Ad-invariant positive scalar product is a sum of an abelian algebra and compact one.

6.8.2. Condition (*).
Behind the condition (*) is the following:

Lemma 6.17. *(Fundamental non-degeneracy Fact) Let M be a compact Lorentz manifold, $\phi^t \subset Isom(M)$ a one parameter group with infinitesimal generator (a Killing field) X. Suppose ϕ^t is non-precompact (i.e., non-equicontinuous, or equivalently the closure of $\{\phi^t, t \in \mathbb{R}\}$ in $Isom(M)$ is not compact). Then, X is everywhere non-timelike: $\langle X(x), X(x) \rangle \geq 0, \forall x$.*

Corollary 6.18. *(Condition (*)): Let $\mathcal{P}$ a linear subspace of $\mathcal{G}$ containing a dense set of non-precompact Killing fields. Then, $\kappa|\mathcal{P} \geq 0$, and $\dim Ker(\kappa|\mathcal{P}) \leq 1$.*

6.8.3. Proof of Lemma 6.17. The proof of the Fundamental non-degeneracy Fact is based on two uniformity facts:

Fact 6.19. *Let* $\{\phi^t\} \subset Isom(M)$ *be a one parameter group of isometries. If for some* $t_i \to \infty$, $\{\phi^{t_i}\}$ *is precompact (i.e., equicontinuous), then* $\{\phi^t\}$ *is precompact.*

Sketch. Let $L = \overline{\{\phi^t\}} \subset \mathrm{Isom}(M)$. Then, L is an abelian Lie group, and hence it is a cylinder $T^k \times \mathbb{R}^d$ (where T^k is a torus).

But, L has a dense one parameter group ($\{\phi^t\}$ itself), i.e., a dense geodesic (when L is seen as a flat Euclidean cylinder). It then follows that $L = T^k$, or $L = \mathbb{R}$. Now, if there $\exists \{\phi^{t_i}\}$ equicontinuous, then $L \neq \mathbb{R}$, and hence $L = T^k$, i.e., $\{\phi^t\}$ is equicontinuous. $\qquad\square$

Fact 6.20. *If for some* $x_i \in M, t_i \to \infty$, $\{D_{x_i}\phi^t\}$ *is equicontinuous (i.e.,* $\parallel D_{x_i}\phi^t \parallel$ *and* $\parallel (D_{x_i}\phi^t)^{-1} \parallel$ *bounded), then* $\{\phi^{t_i}\}$ *is equicontinuous (and therefore by the fact above* $\{\phi^t\}$ *is equicontinuous)*

Sketch. By definition of its Lie group structure, $\mathrm{Isom}(M)$ acts properly (and freely) on the frame bundle $P(M)$. $\qquad\square$

These two facts are true in affine dynamics, i.e., for $\{\phi^t\}$ preserving any linear connection.

We need a third fact special to the Lorentz case:

Fact 6.21. *If a Killing field* X *is somewhere timelike (i.e.,* $\langle X(x_0), X(x_0)\rangle < 0$*), then it generates an equicontinuous flow* $\{\phi^t\}$.

Sketch. Let U be a neighborhood of x_0 where X is timelike. By Poincaré recurrence Lemma, there exist x_i near x_0, $t_i \to \infty$, such that $\phi^{t_i}x_i$ is near x_0.

Now, near x_0, $D_{x_i}\phi^{t_i}$ behave as Riemannian isometries, and are then equicontinuous. Apply the second fact, and then the first one to deduce that $\{\phi^t\}$ is equicontinuous. $\qquad\square$

7. Conformal actions

We will be very succinct at this §. We essentially take the opportunity to mention recent works on the domain, and quote references for detailed and complete exposition.

As for Riemannian geometry, one defines, conformal Lorentz manifolds, conformal actions, conformally flat structures.... The conformal group is essential when it cannot be reduced to the isometry group of some Lorentz metric in the conformal class. The "vague" conjecture is that it is possible, and anyway interesting, to classify essential conformal Lorentz manifolds. To be precise one may ask a kind of Lorentz conformal Lichnérowicz conjecture (see [35]). Recall for this that the universal substratum of conformal Lorentz geometry, is the (static) Einstein cosmos Ein_n. Its conformal structure is obtained (up to a 2 folded cover) as follows. Consider on $\mathbb{R}^{2+n}$ a quadratic form of type $--+,\ldots+$, and let $C^{2,n}$

be its isotropic cone. Then, Ein_n is the projectivization of $C^{2,n}$, endowed with its natural conformal Lorentz structure. Its conformal group is $O(2, n)$, which acts essentially.

C. Frances [41] exhibited a huge class of conformally flat manifolds, which are essential. As amazing fact, the Einstein cosmos itself as a topological manifold, which is $S^1 \times S^n$, has a big Teichmüller space of conformally flat (Lorentz) structures, some of which are essential.

Unrelated to essentiallity, C. Frances [42] also studied closed conformally flat Lorentz manifolds, for themselves. Their holonomy groups are in some sense the Lorentz parallel of Kleinian groups, i.e., discrete groups of the Möbius group. As the usual (Riemannian) sphere is the boundary at infinity of the hyperbolic space, the Einstein cosmos is the conformal boundary. Therefore, we find ourselves here in the heart of an "equivariant" AdS/CFT correspondence.

Coming back to essentiallity, the geometrical ingredient of the Lorentzian Lichnérowicz's conjecture, still stands up. That is, it seems that a compact essential Lorentz manifold is conformally flat. The non-compact case is false (see for instance [5]). This contrasts with the Riemannian non-compact case, since (even if Lichnérowicz did not dare to ask it) the Euclidean space is the unique essential Riemannian manifold.

Acknowledgments

We would like to thank the referee for his valuable remarks and suggestions, and the editors, Piotr Chrusciel and Helmut Friedrich, for firstly inviting us to the Cargèse school, and then giving us the opportunity to write this paper.

References

[1] H. Abels, *Properly discontinuous groups of affine transformations, A survey,* Geometriae Dedicata **87** (2001), 309–333.

[2] S. Adams, *Dynamics on Lorentz manifolds,* World Scientific Publishing Co., Inc., River Edge, NJ.

[3] S. Adams, G. Stuck, *The isometry group of a compact Lorentz manifold, I,* Invent. Math. **129** (1997), 239–261.

[4] S. Adams, G. Stuck, *The isometry group of a compact Lorentz manifold, II,* Invent. Math. **129** (1997), 263–287.

[5] D. Alekseevski, *Self-similar Lorentzian manifolds,* Ann. Global Anal. Geom. **3** (1985), no. 1, 59–84.

[6] S. Aminneborg, I. Bengtsson, S. Holst, *A Spinning Anti-de Sitter Wormhole,* Class. Quant. Grav. **16** (1999) 363–382, gr-qc/9805028.

[7] S. Aminneborg, I. Bengtsson, D. Brill, S. Holst, P. Peldan, *Black Holes and Wormholes in 2+1 Dimensions,* Class. Quant. Grav. **15** (1998) 627–644, gr-qc/9707036.

[8] S. Aminneborg, I. Bengtsson, S. Holst, P. Peldan, *Making Anti-de Sitter Black Holes,* Class. Quant. Grav. **13** (1996), 2707–2714, gr-qc/9604005.

[9] L. Andersson, *Constant mean curvature foliations of flat space-times*, Comm. Anal. Geom. **10** (2002), no. 5, 1125–1150.

[10] L. Andersson, *Constant mean curvature foliations of simplicial flat space-times*, math.DG/0307338.

[11] L. Andersson, G.J. Galloway, R. Howard *The cosmological time function*, Classical Quantum Gravity **15** (1998), 309–322.

[12] L. Andersson, V. Moncrief, A. Tromba *On the global evolution problem in 2 + 1 gravity*, J. Geom. Phys. **23** (1997), no. 3–4, 191–205.

[13] A. Arouche, M. Deffaf, Y. Raffed, *A geometric approach of groups actions on Lorentz non-compact manifolds*, To appear.

[14] U. Bader, A. Nevo, *Conformal actions of simple Lie groups on compact pseudo-Riemannian manifolds*, J. Differential Geom. **60** (2002), no. 3, 355–387.

[15] M. Banados, M. Henneaux, C. Teitelboim, J. Zanelli, *Geometry of the 2 + 1 Black Hole*, Phys. Rev. D **48** (1993) 1506–1525, gr-qc/9302012.

[16] M. Banados, C. Teitelboim, J. Zanelli, *The Black Hole in Three-Dimensional Space Time*, Phys. Rev. Lett. **69** (1992) 1849–1851, hep-th/9204099.

[17] T. Barbot, *Flat globally hyperbolic spacetimes*, accepted for publication in Journal of Geometry and Physics, math. GT/0402257.

[18] T. Barbot, *Limit sets of discrete Lorentzian groups*, in preparation.

[19] T. Barbot, F. Béguin, A. Zeghib, *Feuilletages des espaces temps globalement hyperboliques par des hypersurfaces à courbure moyenne constante*, C.R. Acad. Sci. Paris, Ser. I **336** (3) (2003), 245–250.

[20] T. Barbot, F. Béguin, A. Zeghib, *CMC foliations on globally hyperbolic spacetimes*, in preparation.

[21] R. Benedetti, E. Guadagnini, *Cosmological time in (2 + 1)-gravity*, Nuclear Phys. B **613** (2001), no. 1–2, 330–352.

[22] F. Bonsante, *Flat Spacetimes with Compact Hyperbolic Cauchy Surfaces*, math.DG/0311019.

[23] A. Borel, Harish-Chandra, *Arithmetic subgroups of algebraic groups*, Ann. of Math. (2) **75** (1962), 485–535.

[24] D. Brill, *Multi-Black-Hole Geometries in (2+1)-Dimensional Gravity*, Phys.Rev. D **53** (1996) 4133–4176, gr-qc/9511022.

[25] E. Calabi., L. Markus, *Relativistic space forms*, Ann. Math. **75** (1962), 63–76

[26] S. Carlip, *Quantum gravity in 2 + 1 dimensions*, Cambridge Monographs on Math. Phys. (1998), Cambridge University Press.

[27] Y. Carrière, *Autour de la conjecture de L. Markus sur les variétés affines*, Invent. Math. **95** (1989), 615–628

[28] Y. Carrière, F. Dal'bo, *Généralisations du 1er Théorème de Bieberbach sur les groupes cristalographiques*, Enseignement Math. **35** (1989), 245–262

[29] Y. Carrière, L. Rozoy, *Complétude des métriques lorentziennes de T^2 et difféomorphismes du cercle*, Bol. Soc. Brasil. Mat. (N.S.) **25** no. 2 (1994), 223–235.

[30] V. Charette, T.A. Drumm, *Strong marked isospectrality of affine Lorentzian groups*, math.DG/0310464.

[31] V. Charette, T.A. Drumm, D. Brill, *Closed time-like curves in flat Lorentz space-times.* J. Geom. Phys. **46** (2003), no. 3–4, 394–408.

[32] V. Charette, T. Drumm, W. Goldman, M. Morrill, *Complete flat affine and Lorentzian manifolds,* Geometriae Dedicata **97** (2003), 187–198.

[33] Y. Choquet-Bruhat, R. Geroch, *Global aspects of the Cauchy problem in general relativity,* Comm. Math. Phys. **14**, 1969 329–335.

[34] G. D'Ambra, *Isometry groups of Lorentz manifolds,* Invent. Math. **92** (1988), 555–565.

[35] G. D'Ambra and M. Gromov, *Lectures on transformation groups: geometry and dynamics, Surveys in Differential Geometry,* (Supplement to the Journal of Differential Geometry), 1 (1991) 19–111.

[36] T.A. Drumm, *Fundamental polyhedra for Margulis space-times,* Topology **31** (4) (1992), 677–683.

[37] T.A. Drumm, *Linear holonomy of Margulis space-times,* J. Differential Geom. **38** (3) (1993), 679–690.

[38] T.A. Drumm, W. Goldman, *Isospectrality of flat Lorentz 3-manifolds,* J. Diff. Geom. **58** (2001), 457–465.

[39] G. Ellis and S. Hawking Hawking, *The large scale structure of space-time,* Cambridge Monographs on Mathematical Physics, No. 1. Cambridge University Press, London-New York, 1973.

[40] J. Ferrand, *The action of conformal transformations on a Riemannian manifold,* Math. Ann. **304** (1996) 277–291.

[41] C. Frances, Thesis, ENS-Lyon 2002, www.umpa.ens-lyon.fr/~cfrances/these2-frances.pdf

[42] C. Frances, *Sur les groupes kleiniens lorentziens,* www.umpa.ens-lyon.fr/~cfrances/kleinlorentz.pdf

[43] D. Fried, W. Goldman, *Three-dimensional affine crystallographic groups,* Adv. Math. **47** (1983), 1–49.

[44] D. Fried, W. Goldman, M. Hirsch, *Affine manifolds with nilpotent holonomy,* Comment. Math. Helv. **56** (1981), 487–523.

[45] D. Gallo, M. Kapovitch, A. Marden, *The monodromy groups of Schwarzian equations on closed Riemann surfaces,* Ann. Math. **151** (2000), 625–704.

[46] C. Gerhardt, *H-surfaces in Lorentzian manifolds,* Comm. Math. Phys. **89** (1983), no. 4, 523–553.

[47] E. Ghys, *Flots d'Anosov dont les feuilletages stables et instables sont différentiables,* Ann. Sc. Ec. Norm. Sup., **20** (1987) 251–270.

[48] W. Goldman, Y. Kamishima, *The fundamental group of a compact Lorentz space is virtually polycyclic,* J. Differential. Geom. **19** (1984), 233–240.

[49] W. Goldman, *Nonstandard Lorentz space forms,* J. Differential. Geom. **21** (1985), 301–308.

[50] M. Gromov, *Rigid transformation groups,* "Géométrie différentielle ", D. Bernard et Choquet-Bruhat. Ed. Travaux en cours 33. Paris. Hermann (1988).

[51] F. Grunewald, G. Margulis, *Transitive and quasi-transitive actions of affine groups preserving a generalized Lorentz structure,* J. Geom. Phys. **5** (1988), 493–530.

[52] M. Guediri, *Compact flat spacetimes*, J. Diff. Geom. Appli., to appear. available at: www.mpim-bonn.mpg.de/html/preprints/preprints.html.

[53] M. Guediri, *On the geodesic connectedness of simply connected Lorentz surfaces*, Ann. Fac. Sci. Toulouse Math. (6) 6 (1997), no. 3, 499–510.

[54] M. Guediri, J. Lafontaine, *Sur la complétude des variétés pseudo-riemanniennes*, J. Geom. Phys. **15** (1995), no. 2, 150–158.

[55] D. Johnson, J. Millson, *Deformation spaces associated to compact hyperbolic manifolds*, in "Discrete Groups in geometry and analysis" (New Haven), 48–106, Progr. Math. **67** (1987).

[56] B. Klingler, *Complétude des variétés lorentziennes à courbure constante*, Math. Ann. **306** (1996), 353–370.

[57] N. Kowalsky, *Noncompact simple automorphism groups of Lorentz manifolds*, Ann. Math. **144** (1997), 611–640.

[58] N. Kowalsky, *Actions of non-compact simple groups of Lorentz manifolds*,C. R. Acad. Sci. Paris Sér. I Math. **321** (1995), no. 5, 595–599.

[59] R. Kulkarni, *Proper actions and pseudo-Riemannian space forms*, Adv. Math. **40** (1981), 10–51.

[60] R. Kulkarni, F. Raymond, *3-dimensional Lorentz space-forms and Seifert fiber spaces*, J. Diff. Geom. **21** (1985), 231–268.

[61] F. Labourie, *Problème de Minkowski, et surfaces à courbure constante dans les variétés hyperboliques*, Bull. Soc. Math. Fr. **119** (1991), 307–325.

[62] G. Margulis, *Free properly discontinuous groups of affine transformations*, Dokl. Akad. Nauk. SSSR **272** (1983), 937–940.

[63] J. Marsden, *On completeness of homogeneous pseudo-Riemannian manifolds*, Ind. Univ. Math. J, Vol **22** (1973) 1065–1066.

[64] A. Medina, Ph. Revoy, *Les groupes oscillateurs et leurs réseaux*, Manuscripta. Math. **52** (1985), 81–95.

[65] G. Mess, *Lorentz spacetimes of constant curvature*, preprint IHES/M/90/28 (1990).

[66] J. Milnor, *On fundamental groups of complete affinely flat manifolds*, Adv. Math. **25** (1977), 178–187.

[67] M. Morrill, UCLA thesis (1996).

[68] P. Mounoud, *Dynamical properties of the space of Lorentzian metrics*, Comment. Math. Helv. **78** (2003), no. 3, 463–485.

[69] C. Nappi, E. Witten, *Wess-Zumino-Witten model based on a nonsemisimple group* Phys. Rev. Lett. **71** (1993), no. 23, 3751–3753.

[70] M. Obata, Morio *The conjectures on conformal transformations of Riemannian manifolds* J. Differential Geometry **6** (1971/72), 247–258.

[71] A. Pratoussevitch, *Fundamental domains in Lorentzian geometry*, available at: www.math.uni-bonn.de/people/anna/publications.html.en.

[72] A.D. Rendall, *Constant mean curvature foliations in cosmological spacetimes*, Helv. Phys. Acta **69** (1996), no. 4, 490–500, gr-qc/9606049.

[73] F. Salein, *Variétés anti-de Sitter de dimension 3 exotiques*, Ann. Inst. Fourier **50** (2000), no. 1, 257–284.

[74] K. Scannell, *3-manifolds which are spacelike slices of flat spacetimes*, Classical Quantum Gravity **18** (2001), no. 9, 1691–1701.

[75] K. Scannell, *Flat conformal structures and the classification of de Sitter manifolds*, Comm. Anal. Geom. **7** (1999), no. 2, 325–345.

[76] B. Schmidt, *The local b-completeness of space-times*, Comm. Math. Phys. **29** (1973), 49–54.

[77] J.M. Schenker, *Convex polyhedra in Lorentzian space-forms*, Asian J. Math. **5** (2001), no. 2, 327–363.

[78] R. F. Streater, *The representations of the oscillator group*, Commun. Math. Phys. **4** (1967), 217–236.

[79] W. Thurston, *Three-dimensional geometry and topology*, Vol. 1. Edited by Silvio Levy. Princeton Mathematical Series, 35. Princeton University Press, Princeton, NJ.

[80] W. Thurston, in: D.B.A. Epstein (Ed.), London Math. Soc. Lecture Notes, vol 111, Cambridge University Press, 1987.

[81] D. Witte, *Homogeneous Lorentz manifolds with simple isometry group*, Beiträge Algebra Geom. **42**, no. 2 (2001) 451–461.

[82] E. Witten, *2 + 1-dimensional gravity as an exactly soluble system*, Nucl. Phys. B **311** (1988), 46–78.

[83] J. Wolf, *Spaces of constant curvature*, New York, McGraw-Hill, 1967

[84] A. Zeghib, *On closed anti de Sitter spacetimes*, Math. Ann., **310** (1998) 695–716.

[85] A. Zeghib, *Sur les espaces-temps homogènes*, Geometry and Topology Monographs **1** (1998) 531–556: http://www.maths.warwick.ac.uk/gt/GTMon1/paper26.abs.html

[86] A. Zeghib, *The identity component of the isometry group of a compact Lorentz manifold*, Duke Math. J., **92** (1998) 321–333.

[87] A. Zeghib, *Isometry groups and geodesic foliations of Lorentz manifolds. Part I: Foundations of Lorentz dynamics*. GAFA, **9** (1999) 775–822.

[88] A. Zeghib, *Isometry groups and geodesic foliations of Lorentz manifolds. Part II: Geometry of analytic Lorentz manifolds with large isometry groups*. GAFA, **9** (1999) 823–854.

[89] R. Zimmer, *On the automorphism group of a compact Lorentz manifold and other geometric manifolds*, Invent. Math. **83** (1986) 411–426.

Thierry Barbot and Abdelghani Zeghib
CNRS, UMPA
École Normale Supérieure de Lyon
46, allée d'Italie
F-69364 Lyon cedex 07, France
e-mail: {tbarbot, Zeghib}@umpa.ens-lyon.fr
URL: http://umpa.ens-lyon.fr/

Gauge, Diffeomorphisms, Initial-Value Formulation, Etc.

Robert Geroch

Abstract. We introduce a large class of systems of partial differential equations on a base manifold M, a class that, arguably, includes most systems of physical interest. We then give general definitions – applicable to any system of equations in this class – of "having the diffeomorphisms on M as a gauge group", and, for such a system, of "having an initial-value formulation, up to this gauge". These definitions, being algebraic in the coefficients of the partial differential equations, are relatively easy to check in practice. The Einstein system, of course, satisfies our definitions.

1. Introduction

Two features of Einstein's equation play a fundamental role in the structure of the general theory of relativity.

The first of these is that Einstein's equation manifests a certain gauge freedom, the corresponding gauge transformations being associated with diffeomorphisms on the underlying space-time manifold. This freedom is perhaps *the* striking feature of the relativity theory. It impacts the way we understand and address a variety of topics within the theory, e.g., gravitational energy [17] [2] [8], and gravitational radiation [13] [6]. Indeed, a substantial shift in attitude toward this diffeomorphism freedom has taken place over the years. This gauge freedom was, at the beginning, regarded as a curious and novel feature of the theory. But it is more common now to argue [12] that diffeomorphism gauge is a necessary feature of *any* viable physical theory based on partial differential equations on a manifold.

The second feature of Einstein's equation is that it manifests, up to this gauge freedom, a well-posed initial-value formulation [3] [5]. Having an initial-value formulation is by now so much a part of our thinking that it is difficult to imagine doing physics without it. For example, one might be interested in whether a physical system is stable in time, or whether it can send signals superluminally. But both of these notions depend directly on having a description of the system that involves an initial-value formulation. Stability in time, for example, deals with the evolution of small perturbations of the system, while inherent in "evolution" is an initial-value formulation.

These two features are, clearly, intertwined. For example, Einstein's equation as it stands does *not* admit an initial-value formulation in the traditional sense, precisely because the gauge freedom prohibits it. Given the central role these two features play in the structure of the theory, then, it would be of some interest to understand better how they operate and how they interact. What is the mechanism by which general relativity manifests these features? For example, it is not obvious, merely by examining the Einstein system of partial differential equations, that it has an initial-value formulation up to diffeomorphism gauge. One route to understanding these issues would be to formulate a general characterization of "having gauge freedom associated with diffeomorphisms", and "having, up to that freedom, a well-posed initial-value formulation". That is, one would like to formulate precise definitions of these notions, definitions that are applicable, say, to virtually any system of partial differential equations on a manifold. We here propose such definitions.

Given such definitions, we may ask what other systems of partial differential equations satisfy them; and whether any of those systems are likely to underlie viable physical theories. We have been able to find just a few other classes systems of partial differential equations on a manifold having diffeomorphism gauge freedom combined with an initial-value formulation. One is that for special relativity – regarding the Minkowski metric as dynamic. The other is a certain class of systems involving a preferred vector field. None of the latter seem likely to underlie interesting physical theories. What other systems of partial differential equations are there that manifest an initial-value formulation up to diffeomorphism freedom? Can all such systems be "found"? Do any appear to represent viable physical theories?

In Section 2, we review briefly systems of partial differential equations, and the conditions under which such a system admits an initial-value formulation. The key idea is to generate a universal framework, into which the partial differential equations describing physical systems fit. In Section 3, we introduce the notion of gauge for a system of partial differential equations; and, in particular, that of diffeomorphism gauge freedom. It turns out that "gauge" can be defined using only the partial differential equation itself, and not any physical interpretation of that equation. In Section 4, we introduce our definition of a system's having an initial-value formulation up to diffeomorphism gauge. This definition has the important feature that it is essentially algebraic, i.e., to decide whether a given partial differential equation satisfies it involves, essentially, algebraic manipulations with the coefficients in that equation. Examples and a few other issues are discussed briefly in the Conclusion, Section 5. Appendix A is a plea for settlement of an important open question involving the existence of an initial-value formulation for general hyperbolic systems with constraints.

2. Partial differential equations

It might seem at first thought that the task of this section – to set up a universal framework for the physically interesting partial differential equations – is nearly an impossible one. There is clearly an enormous variety of possible systems of partial differential equations in this World: How, given this apparent diversity, will we ever get sufficient control over these equations to be able to analyze the character of the the "general" one? It turns out, however, that this problem is more apparent than real. There exists a general formulation of the subject of partial differential equations – a formulation that, on the one hand, is systematic, and, on the other, is sufficiently broad to include virtually all equations of physical interest. The key idea of this formulation is to restrict consideration to a certain class of systems of partial differential equations – namely to systems that are first order (i.e., involve only first derivatives of the fields) and quasilinear (i.e., are linear in those first derivatives). This class is much broader than it might appear at first. For example, higher-order systems are cast into this form by introducing new fields to represent the lower derivatives. This class, it turns out, admits a systematic treatment, and at the same time appears to be adequate for the description of physical phenomena[1]. This general formulation of partial differential equations for physics (which is discussed in more detail, with many examples, in [7]) is summarized below.

Let there be given a fibre bundle, consisting of some base manifold M, some bundle manifold $\mathcal{B}$, and some smooth projection mapping $\mathcal{B} \xrightarrow{\pi} M$. Typically, M will be the 4-dimensional manifold of space-time events (but it could be any smooth manifold). By the *fibre* over a point x of M, we mean the set of all points b of $\mathcal{B}$ such that $\pi(b) = x$. Think of the fibre over $x \in M$ as "the set of possible field-values at x". Then $\mathcal{B}$ is interpreted as the set of "all possible choices of field-values at all points of M", and π as the mapping that assigns, to each such choice, the underlying point of M. Thus, point b of $\mathcal{B}$ could be written as $b = (x, \phi)$, with $x \in M$ and ϕ in the fibre over x. The action of the projection mapping would then be given by $\pi(x, \phi) = x$. Typically, the fibre over a point $x \in M$ will be some collection of tensors or other geometrical objects (such as derivative operators, spinors, etc.) at x, possibly subject to various symmetries or other algebraic conditions; whence $\mathcal{B}$ will be the manifold of all such collections of objects at all points of M. A tangent vector at a point of $\mathcal{B}$ is said to be *vertical* if it is tangent to the fibre at that point (or, what is the same thing, if its image under the projection π vanishes). Thus, a vertical vector represents an "infinitesimal change in the field-values" at a fixed point of M. We shall adopt the convention that all "algebraic constraints" on our fields have been incorporated already at the level of the construction of the bundle manifold $\mathcal{B}$, and thus that no such constraints

---

[1] The present class is, essentially, what is called the *symmetric hyperbolic* systems of partial differential equations. There exists at least one other, still larger, class, the *strongly hyperbolic* systems [15]. Apparently, neither the status of the initial-value formulation in this larger class, nor the role of this class in physics, has been settled decisively.

are to be imposed as additional conditions on $\mathcal{B}$. While we shall think of $\mathcal{B}$ as representing "field values", it can in general be any smooth manifold, subject only to the local-product condition in the definition of a fibre bundle[2].

For Maxwell theory, for example,

  i) M is the four-dimensional space-time manifold;
 ii) $\mathcal{B}$ is the ten-manifold consisting of pairs (x, F_{ab}), where $x \in M$ and F_{ab} is a skew tensor at x;
iii) the projection π sends (x, F_{ab}) to x;
 iv) the fibre over $x \in M$ is the six-manifold of skew F_{ab} at x; and
  v) a vertical vector at point (x, F_{ab}) of $\mathcal{B}$ may be represented by a skew tensor[3] δF_{ab} at x.

For general relativity, M is again the four-manifold of space-time events, while the fibre over $x \in M$ consists of pairs (g_{ab}, ∇_a), where g_{ab} is a Lorentz-signature metric and ∇_a a derivative operator[4], at x. Thus, the dimension of the fibres in this case is 50 ($= 10 + 40$); while the dimension of the manifold $\mathcal{B}$ is 54. A vertical vector at point (x, g_{ab}, ∇_a) of $\mathcal{B}$ may be represented by pair of tensors, $(\delta g_{ab}, s^m{}_{ab})$, with δg_{ab} (representing the g_{ab}-component of the vector) and $s^m{}_{ab}$ (representing the ∇_a-component of the vector[5]) each symmetric the indices "a" and "b".

Returning to the general case, by a *cross-section* of such a bundle we mean a smooth mapping $M \overset{\phi}{\to} \mathcal{B}$ such that $\pi \circ \phi$ is the identity map on M. In other words, a cross-section ϕ assigns, to each point x of M, some point, $\phi(x)$, of the fibre over x. Think of a given cross-section as representing a particular choice of a "field" (of the type represented by the bundle) over M. Thus, for Maxwell theory, a cross-section of the bundle is represented by a smooth skew tensor field F_{ab} on the space-time manifold M; for general relativity, by smooth fields g_{ab} and ∇_a on M.

Our partial differential equation will be an equation on such a cross-section map ϕ, linear in its first derivative. In order to write out this equation, we must

---

[2]Recall that this condition requires, essentially, that, locally in M, $\mathcal{B}$ can be written as a product, $M \times F$, of M with some other fixed manifold F, in such a way that the projection mapping π becomes the projection to the M-factor in this product. This condition guarantees, e.g., that, locally, all the fibres of the bundle are diffeomorphic with this fixed manifold F, and so with each other. We shall take the term "fibre bundle", to mean such a smooth mapping of manifolds, $\mathcal{B} \overset{\pi}{\to} M$, subject only to this local-product condition. That is, we shall not require (as is sometimes done [14]) that there also be given a group action on the bundle manifold.

[3]This δF_{ab}, which we may think of as a "small change in F_{ab}" is, more precisely, the tangent vector, at F_{ab}, to some curve in the manifold of skew tensors at x.

[4]A derivative operator at a point x of M could be defined, for example, as a map from smooth covector fields on M to second-rank covariant tensors at x, subject to additivity, the Leibnitz rule, and consistency with the exterior derivative. We shall take all derivative operators to be torsion-free.

[5]Recall that the difference of two derivative operators, ∇_a and $\tilde{\nabla}_a$ on a manifold is represented by a tensor $C^m{}_{ab} = C^m{}_{(ab)}$, which is defined by the property that, for any smooth covector field k_a, $\nabla_a k_b - \tilde{\nabla}_a k_b = -C^m{}_{ab} k_m$. Thus, a tangent vector in the space of derivative operators at a point gives rise naturally, i.e., without any "reference" derivative operator $\tilde{\nabla}_a$, to such a tensor $s^m{}_{ab}$.

introduce two new smooth fields, $k^{Aa}{}_\alpha$ and j^A, on $\mathcal{B}$. Being fields on $\mathcal{B}$, these depend of course on the point $b = (x, \phi)$ of $\mathcal{B}$, i.e., they depend on a choice of "point x of the base manifold, as well as field-value ϕ at that point". The index "α" on $k^{Aa}{}_\alpha$ is a tensor index in $\mathcal{B}$ at the point, $b \in \mathcal{B}$, at which this field is evaluated; the index "a" is a tensor index in M at the corresponding point, $\pi(b)$, of the base manifold. The index "A", of both $k^{Aa}{}_\alpha$ and j^A, lies in some new vector space (which will turn out to be the vector space of equations). Finally, our partial differential equation, on a cross-section ϕ, is

$$k^{Aa}{}_\alpha (\nabla \phi)_a{}^\alpha = j^A. \tag{1}$$

This equation is to be imposed at each point $x \in M$, with the fields k and j evaluated at $\phi(x) \in \mathcal{B}$, i.e., on the cross-section. Here, $(\nabla\phi)_a{}^\alpha$ denotes the derivative of the map ϕ (i.e., a map from tangent vectors in M at x to tangent vectors in $\mathcal{B}$ at $\phi(x)$). The index "A" in Eqn. (1) is free, i.e., Eqn. (1) can be reduced, by taking components, to a number of scalar equations equal to the dimension of the vector space in which "A" lies. We demand that $\nu_A k^{Aa}{}_\alpha = 0$ only when $\nu_A = 0$, i.e., that every the equation of the system (1) really be "differential".

For Maxwell theory, Eqn. (1) is Maxwell's equations,

$$\nabla_{[a} F_{bc]} = 0, \tag{2}$$

$$\nabla^a F_{ab} = 0. \tag{3}$$

The index "A" in this case stands for three antisymmetric M-tensor indices, together with a single M-tensor index (this being the index-structure of Eqns. (2)–(3)). Thus, "A" in this example lies in a vector space of dimension 8 (= 4 + 4). For general relativity (say, with vanishing sources), Eqn. (1) is

$$\nabla_a g_{bc} = 0, \tag{4}$$

$$R_{ab(c}{}^m g_{d)m} = 0, \tag{5}$$

$$R_{amb}{}^m = 0, \tag{6}$$

where we have defined the Riemann tensor $R_{abc}{}^d$ (the "derivative of the derivative operator") by the requirement that $\nabla_{[a} \nabla_{b]} k_c = 1/2 R_{abc}{}^d k_d$ for every covector field k_d on M. Note that these are indeed first-order, quasilinear equations in the fields g_{ab} and ∇_a. Note also that we include in this system *all* first-order equations, i.e., we include (5), even though it is the curl of (4). The equation-index "A" in this example lies in a vector space of dimension 110 (= 40 + 60 + 10, where these three terms correspond, respectively, to the three equations (4)–(6)).

Returning to the general case, we are concerned with the issue of when the system (1) has an initial-value formulation. To this end, we consider a submanifold, T, of M of codimension one, together with a cross-section ϕ_0, of $\mathcal{B}$ over T[6]. Think of the cross section ϕ_0 as the "initial data" at the "time" represented by T. When

---

[6]That is, ϕ_0 is a smooth map from T to $\mathcal{B}$ such that $\pi \circ \phi_0$ is the identity on T.

do these data give rise to (i.e., are the restrictions to T of) some solution, ϕ, of (1), and when is that solution unique? In order to answer these questions, we require two further notions.

Fix a bundle, $\mathcal{B} \xrightarrow{\pi} M$, and a system (1) of partial differential equations on that bundle. By a *constraint* of this system, at a point of $\mathcal{B}$, we mean a tensor $c^a{}_A$ at that point satisfying $c^{(a}{}_A k^{|A|b)}{}_{\alpha'} = 0$ there. Here, and hereafter, a prime on a Greek subscript means "applied only to vertical vectors". [These primes will appear frequently, for it is usually only the "vertical parts" of things that are of interest.] Note that the constraints at each point of $\mathcal{B}$ form a vector space. Each constraint, it turns out, has two distinct facets. It is one of the nice features of this subject that these two, apparently quite dissimilar, facets coalesce into the simple, geometrical, definition above.

As to the first facet: Each constraint gives rise to an integrability condition. Fix any constraint field, $c^a{}_A$, and any cross-section, ϕ. Contract both sides of Eqn. (1) with $c^b{}_A$, and apply to both sides of the result some derivative operator, $\tilde{\nabla}_b$, on M. Then, by the defining equation for a constraint, terms involving second derivatives of ϕ disappear, leaving an algebraic equation (indeed, a polynomial of degree at most two) in the first derivative, $(\nabla\phi)_a{}^\alpha$, of ϕ. The given constraint field is said to be *integrable* if this equation is an algebraic consequence of Eqn. (1), i.e., if the difference of its two sides can be written as the product of two factors: one some expression (at most linear in field-derivatives), and the other the difference of the two sides of (1). Note that integrability of a fixed constraint field is independent of the choice, above, of the derivative operator $\tilde{\nabla}$, for a change in this choice is compensated for by a change in the first factor in the product above. Note also that any linear combination, with coefficients functions on $\mathcal{B}$, of integrable constraint fields is again an integrable constraint field. We say that the constraints of the system are *integrable* if every such constraint field is. Failure of integrability of the constraints may be interpreted as meaning that "not all of the partial differential equations appropriate to the given system have been included in (1)"[7].

As to the second facet, each constraint gives rise to a consistency condition on initial data. Fix a constraint field, $c^a{}_A$, and a solution ϕ of Eqn. (1); as well as some submanifold T of M of codimension one. Then, at each point of T, we have

$$n_m c^m{}_A k^{Aa}{}_\alpha (\nabla\phi)_a{}^\alpha = n_m c^m{}_A j^A, \tag{7}$$

where n_m is the normal to T at that point. But, by virtue of the definition of a constraint, the index "a" in the expression $n_m c^m{}_A k^{Aa}{}_\alpha$ is tangent to T (for the

---

[7]On discovering that integrability fails for some given system of partial differential equations, one might contemplate constructing a new system, all of whose constraints *are* integrable, in the following manner. First augment the given system, in some manner, by additional quasilinear equations so chosen that they cause the original integrability conditions to be satisfied. This new system, so constructed, may now give rise to new constraints, and so to additional integrability conditions. If so, repeat the first step above, adding as necessary still more quasilinear equations; and continue in this way. In many cases, such a procedure terminates in an integrable system. However, there is, as far as I am aware, no simple criterion for when this will happen.

contraction of this expression with the normal, n_a, vanishes). So, Eqn. (7) takes the derivative of ϕ only in directions tangent to T, and so it refers only to the value of ϕ on T, i.e., only to the initial data induced on T from ϕ. In short, Eqn. (7) represents a consistency condition on initial data. If any consistency condition obtained in this way were not satisfied, for given initial data on T, then we would have have no hope of finding, for those data, a corresponding solution of Eqn. (1). Integrability of the constraints guarantees, in essence, that these consistency conditions "evolve in time". We say that the constraints of the system (1) are *complete* if, in a certain sense, *every* consistency condition on initial data on T arises in the manner of Eqn. (7). The definition[8], in more detail, is the following. We demand that, for every point b of $\mathcal{B}$, there is an open set of covectors n_a at the corresponding base point, $\pi(b)$, such that the following holds: Given any n_a lying in that open set, and any ν_A, such that $n_a \nu_A k^{Aa}{}_{\alpha'} = 0$, then there exists a constraint $c^m{}_A$ of the system such that $\nu_A = c^m{}_A n_m$. Here, the n_a represents a normal to a candidate initial surface (and the open set a restriction on the allowed surface-normals), while the ν_A represents the selection, from (1), of a particular consistency condition. The definition then requires that every such consistency condition arise, via (7), from some constraint $c^m{}_A$.

Integrability and completeness, as defined above, hold, as far as I am aware, for every system of partial differential equations of physical interest.

Consider, for example, the Maxwell system, with field F_{ab}, and equations (2)–(3). Then initial data consist of the specification of a skew field F_{ab} over the 3-submanifold T of space-time, where, of course, these tensor indices lie within the full 4-dimensional space-time M. A constraint, $c^m{}_A$, is represented by a pair of tensors, (c^{mabc}, cg^{ma}), where c^{mabc} is totally antisymmetric, and c is a number. Here, the equation-index "A" is represented by an antisymmetric triple of indices, "abc", together with a single index, "a". The vector space of constraints at each point, then, is 2-dimensional. The corresponding integrability conditions, obtained by taking the curl of (2) and the divergence of (3) are of course identities. So, the constraints are integrable. The corresponding consistency conditions (7) become, in this example, the familiar conditions $\nabla \cdot B = 0$ and $\nabla \cdot E = 0$ on the induced initial data on a space-like T. This system, we claim, is complete. Indeed, let the open set of n_a be that consisting of the timelike covectors. Completeness then becomes the following assertion: Let, for fixed timelike n_a (i.e., for fixed n_a in the open set), $\nu^{abc} = \nu^{[abc]}$ and ν^a (i.e., ν_A) be such that

$$\nu^{abc} n_a \delta F_{bc} + \nu^a n^b \delta F_{ab} = 0 \tag{8}$$

for every $\delta F_{ab} = \delta F_{[ab]}$ (i.e., such that $n_a \nu_A k^{aA}{}_\alpha = 0$). Then, for some $c^{abcd} = c^{[abcd]}$ and some number c, (i.e., for some $c^a{}_A$) we have $\nu^{bcd} = n_a c^{abcd}$ and $\nu^b =$

---

[8]This definition replaces a more awkward one, involving dimensions of various vector spaces, that was given earlier [7]. While the two definitions are equivalent in the presence of a hyperbolization (defined below), the present condition is, in the general case, much more convenient than the earlier one. For example, the Einstein system satisfies the present definition of completeness, but not the earlier one.

$cg^{ab}n_a$ (i.e., we have $\nu_A = n_a c^a{}_A$). But, as is easily checked, this assertion is true. Note that the open set of n_a in this example depends only on the base point, and not the point of the fibre.

Consider, as a second example, the Einstein system, with fields (g_{ab}, ∇_a), and equations (4)–(6). The initial data in this case consist of the specification of fields g_{ab} and ∇_a on a 3-manifold T.

Note that our initial data for the Einstein system consist of the *entire*, four-dimensional, spacetime metric g_{ab} as well as the *entire* derivative operator ∇_a, at points of T. These choices stand in contrast to those for the usual initial-value formulation, in which the data consist of only part of g_{ab} (namely, the induced metric q_{ab} on T) and only part of ∇_a (namely, the extrinsic curvature p_{ab} of T). This is not a difference of substance: (g_{ab}, ∇_a) determine (q_{ab}, p_{ab}) uniquely, while (q_{ab}, p_{ab}) determine (g_{ab}, ∇_a) uniquely up to diffeomorphisms in M leaving T pointwise invariant. Choosing for the initial data (q_{ab}, p_{ab}) is more convenient when one wishes to consider an initial-data set as an entity in its own right; while choosing (g_{ab}, ∇_a) allows the Einstein system to fit into the general framework for the initial-value formulation of systems of partial differential equations on a manifold.

For the Einstein system, the general constraint, $c^m{}_A$, at a point is represented by three tensors, $c^{mabc} = c^{[ma](bc)}$, $c^{mabcd} = c^{[mab](cd)}$, and $c^{mab} = g^{m(a}c^{b)}$. Here, the equation-index "A" is represented by the three index-combinations "abc", "$abcd$", and "ab", with appropriate symmetries, corresponding to the three equations, (4)–(6), of the Einstein system. The dimension of the vector space of constraints at a point is 104 ($= 60+40+4$). The corresponding integrability conditions correspond to taking a curl of Eqn. (4) and to applying the Bianchi identities to Eqns. (5)–(6). These constraints are integrable (the integrability condition for (4) being (5), and those for (5)–(6) being "identities"). Furthermore, we claim, this system is complete. Indeed, let the open set of n_a consist of the timelike covectors. [Note that this set, in contrast to the Maxwell case, *depends* on fibre-point.] Completeness then becomes the following assertion: Let, for fixed timelike n_a, $\nu^{abc} = \nu^{a(bc)}$, $\nu^{abcd} = \nu^{[ab](cd)}$ and $\nu^{ab} = \nu^{(ab)}$ (i.e., ν_A) be such that

$$\nu^{abc} n_a \delta g_{bc} + \nu^{abcd} n_a s^p{}_{bc} g_{dp} + \nu^{ab} n_{[a} s^m{}_{m]b} = 0 \qquad (9)$$

for every $\delta g_{bc} = \delta g_{(bc)}$ and $s^m{}_{ab} = s^m{}_{(ab)}$ (i.e., such that $n_a \nu_A k^{aA}{}_\alpha = 0$). Then, for some $c^{mabc} = c^{[ma](bc)}$, $c^{mabcd} = c^{m[ab](cd)}$ and $c^{mab} = g^{m(a}c^{b)}$ (i.e., for some $c^a{}_A$) we have $\nu^{abc} = n_m c^{mabc}$, $\nu^{abcd} = n_m c^{mabcd}$ and $\nu^{ab} = n_m c^{mab}$ (i.e., we have $\nu_A = n_a c^a{}_A$). This assertion, again, is true.

The consistency conditions, (7), for the constraint tensors $c^{mabcd} = c^{[mab](cd)}$ and $c^{mab} = g^{m(a}c^{b)}$ become, when expressed in terms of the usual initial data, (q_{ab}, p_{ab}), the usual [12] Einstein constraint equations: $\mathcal{R} - p_{ab}p^{ab} + (p^m{}_m)^2$, $D_a(p^{ab} - p^m{}_m q^{ab}) = 0$. It is, of course, of interest to ask how many and of what character are the solutions of these constraint equations. But these questions are not germane to the issues with which we are dealing here, namely, how to characterize in a general way the presence of diffeomorphism gauge and the existence of an initial-value formulation up to that gauge.

We return now to the general case. The key to achieving an initial-value formulation for the system (1) is an object called a *hyperbolization* , a field $h_{\beta A}$ on the bundle manifold $\mathcal{B}$ having the properties described below. Fix any point, (x, ϕ), of $\mathcal{B}$, and consider, for any covector n_m at $x \in M$ and any two vertical vectors, $\delta\phi^\alpha, \delta'\phi^\alpha$, at $(x, \phi) \in \mathcal{B}$, the expression

$$n_m h_{\beta A} k^{Am}{}_\alpha \delta\phi^\alpha \delta'\phi^\beta. \tag{10}$$

This expression is a bilinear form in $\delta\phi^\alpha$ and $\delta'\phi^\alpha$. We demand, in order that this $h_{\beta A}$ be a hyperbolization, that this expression be symmetric under interchange of $\delta\phi^\alpha, \delta'\phi^\alpha$ for every n_m, and positive-definite (i.e., positive whenever $\delta'\phi^\beta = \delta\phi^\beta \neq 0$) for every n_m lying in some open set. Generally speaking, the most direct way to specify a hyperbolization for a system of partial differential equations is simply to give this bilinear expression. Such an expression indeed defines a hyperbolization provided it is symmetric and positive-definite, as described above; and furthermore, that it is of the form (10), i.e., that it is some multiple of the result of replacing, in the left side of Eqn. (1), "$(\nabla\phi)_a{}^\alpha$" by "$n_a\delta\phi^\alpha$". For example, the Maxwell system possess a hyperbolization. Fix any timelike vector u^a, and consider the quadratic form given by

$$u_a n_b(\delta F^{am} \delta' F^b{}_m - 1/4 g^{ab} \delta F_{mn} \delta' F^{mn}). \tag{11}$$

This quadratic form is indeed symmetric under interchange of δF_{ab} and $\delta' F_{ab}$, and that it is indeed positive-definite, for n_a timelike with $u^a n_a > 0$. Furthermore, this quadratic form does indeed arise as a linear combination of $n_{[a}\delta F_{bc]}$ and $n_a\delta F^{ab}$ (this fact being, e.g., the essence of the proof that conservation of the standard electromagnetic stress-energy follows from Maxwell's equations). Note that each choice of timelike u_a gives rise to some particular hyperbolization, i.e., that there are many hyperbolizations in this Maxwell example. More generally, consider any system of equations on fields, for which there is a symmetric stress-energy tensor that i) is a quadratic algebraic function of all the fields; ii) is, by virtue of the field equations, conserved; and ii) satisfies a suitable energy condition. Then that system of equations admits a hyperbolization via this stress-energy, just as for the Maxwell case. The Einstein system admits no hyperbolization, a feature that, as we shall see later, is closely related to the invariance of this system under diffeomorphisms.

A key assertion of this subject is to the effect that a system of partial differential equations, provided it satisfies certain conditions, must have an initial-value formulation. Consider the system (1), and let us suppose that

  i) the constraints of this system are integrable and complete,
 ii) this system admits some hyperbolization field, $h_{\alpha A}$, on $\mathcal{B}$, and
iii) the open sets of covectors n_a for completeness and for the hyperbolization have, at each point of $\mathcal{B}$, nonempty intersection.

Next, let there be given initial data, i.e., a cross section ϕ_0 over some submanifold T of M of codimension one. Let this initial-data set, (T, ϕ_o),

  i) satisfy the consistency conditions, (7), and
  ii) be such that, at each point of T, the normal, n_a, to T at that point lies within the two open sets above.

Note that the second supposition, which is essentially the requirement that the data set (T, ϕ_o) be "non-characteristic", in general involves *both* T and the data ϕ_o thereon. Now consider, under the suppositions above, the following assertion: There exists one and only one solution of the partial differential equation, (1), defined in a neighborhood of T, that manifests the given initial data. This assertion is discussed in more detail in Appendix A. It is, unfortunately, not a theorem, because there is a gap in the proof. Closing this gap (possibly with the introduction of some benign further hypotheses) is, in my opinion, an important open question in the subject of partial differential equations.

It is not hard to see intuitively that the assertion above is reasonable. Let the system (1) satisfy the three conditions above, and fix n_a, lying in the two open sets above. Now consider $n_a k^{Aa}{}_\alpha$, regarded as a linear mapping from the vector space of equations to the vector space of covectors in field-space. The domain of this mapping is the vector space of equations; the kernel of this mapping is (by completeness) the vector space of consistency conditions (7); and the range of this mapping is (by existence of a hyperbolization) the vector space of vertical vectors. We now have, using an elementary fact about linear mappings, the following relation: (dimension of space of equations) - (dimension of space of consistency conditions) = (dimension of space of fields). But this relation asserts that the number of equations in (1) involving "time-derivatives" (the left side) is equal to the number of field-components. In other words, this relation guarantees that Eqn. (1) can be solved for the "time-derivatives" of all fields, in terms of the values and space-derivatives of those fields. Integrability guarantees, by a similar argument, that the time-derivatives of the consistency expressions can be expressed as a linear combination of the values and space-derivatives of those expressions. In short, completeness, integrability and the existence of a hyperbolization, taken together, guarantee that the system (1) has a "naive" initial-value formulation.

3. Gauge

A key concept is the notion of a gauge transformation. We begin with the "infinitesimal" ones.

Fix a first-order, quasilinear system of partial differential equations, (1). By a *gauge vector field* for this system, we mean a smooth vector field, ξ^α, on the bundle manifold $\mathcal{B}$ that

  i) preserves the fibres, i.e., has the property that, for any two points b, b' of the bundle manifold lying in the same fibre, we have $(\nabla\pi)_\alpha{}^a(\xi^\alpha)|_b = (\nabla\pi)_\alpha{}^a(\xi^\alpha)|_{b'}$; and

ii) preserves the system (1), i.e., satisfies the equations

$$\mathcal{L}_\xi k^{Aa}{}_\alpha = S^A{}_B k^{Ba}{}_\alpha + \gamma^{aA}{}_m (\nabla \pi)_\alpha{}^m, \tag{12}$$

$$\mathcal{L}_\xi j^A = S^A{}_B j^B + \gamma^{mA}{}_m, \tag{13}$$

for some fields $S^A{}_B$ and $\gamma^{aA}{}_m$ on $\mathcal{B}$.

The first condition, preservation of the fibres, is precisely the requirement that there exist some vector field in the base manifold (the *drop* of ξ) having ξ^α as its lift. So, for example, every vertical vector field on $\mathcal{B}$ is fibre-preserving (for it is a lift of the zero vector field on the base manifold). In the second condition, the Lie derivative operator, $\mathcal{L}_\xi$ is defined as follows: Consider the one-parameter family of diffeomorphisms on the bundle manifold generated by ξ^α. Take the image of the field ($k^{Aa}{}_\alpha$ or j^A) under that family of diffeomorphisms (noting that this is well defined, in the case of $k^{Aa}{}_\alpha$, by condition i)). Finally, take the parameter-derivative of this family at parameter-value zero. The arbitrary fields $S^A{}_B$ and $\gamma^{aA}{}_m$ in Eqns. (12)–(13) reflect a certain freedom in how our basic differential equation (1) is represented in terms of (k, j). Indeed, replacement of $(k^{Aa}{}_\alpha, j^A)$ by $(W^A{}_B k^{Ba}{}_\alpha, W^A{}_B j^B)$, for $W^A{}_B$ an arbitrary invertible tensor field, results in an identical system of equations. This is the origin of the $S^A{}_B$ in (12)–(13). Furthermore, adding to $k^{Aa}{}_\alpha$ and j^A the fields $\gamma^{aA}{}_m (\nabla \pi)_\alpha{}^m$ and $\gamma^{mA}{}_m$, respectively, where $\gamma^{aA}{}_m$ is an arbitrary field on $\mathcal{B}$, also results in an identical system of equations[9]. This is the origin of the $\gamma^{aA}{}_m$ in (12)–(13).

These gauge vector fields represent, of course, "infinitesimal gauge transformations". We may also define, similarly, a (full) *gauge transformation* as a diffeomorphism, Ψ, on $\mathcal{B}$ that sends fibres to fibres, and the fields $k^{Aa}{}_\alpha, j^A$ to an equivalent pair. It is immediate that, for Ψ a gauge transformation, there exists a unique diffeomorphism, ψ, (the *drop* of Ψ) on M such that $\pi \circ \Psi = \psi \circ \pi$. These two versions of gauge are indeed related as we would expect: A vector field on $\mathcal{B}$ generates (locally) gauge transformations if and only if it is a gauge vector field. The gauge vector fields are generally simpler to work with computationally, while the full gauge transformations are easier to think about.

Every gauge transformation sends every solution cross-section of (1) to another solution cross-section (clearly, since a diffeomorphism on $\mathcal{B}$, in order to be a gauge transformation, must preserve everything involved in (1))[10]. The Lie bracket of two gauge vector fields is, of course, a gauge vector field, and so the gauge vector fields form a Lie algebra. Furthermore, the drop of the bracket is the bracket of the drops. That is, "drop" is a homomorphism from the Lie algebra of gauge vector fields to the Lie algebra of smooth vector fields on M. Similarly, the gauge

---

[9]To see this, use the relation $(\nabla \phi)_a{}^\alpha (\nabla \pi)_\alpha{}^b = \delta^b{}_a$, which is precisely the derivative of the equation $\pi \circ \phi = $ (identity on M).

[10]The converse of this assertion – that a fibre-preserving diffeomorphism on $\mathcal{B}$ that sends solution cross-sections to solution cross-sections must be a gauge transformation – also holds, under the additional condition that there are "sufficiently many" solutions of (1). What is required, in more detail, is that, given any tensor $\mu_a{}^\alpha$ at any point of $\mathcal{B}$ such that $k^{aA}{}_\alpha \mu_a{}^\alpha = j^A$ at that point, then there exists some solution cross-section through that point, with $(\nabla \phi)_a{}^\alpha = \mu_a{}^\alpha$ there.

transformations form, under composition, a group; and "drop" is a homomorphism from this group to the group of diffeomorphisms on the manifold M.

The traditional picture of a gauge transformation is of a change in the mathematical objects used to describe a physical system, which, however, reflects no change in the physical system itself. According to this picture, then, the notion of a gauge transformation is intimately connected with the physical interpretation that is attached to the mathematical objects. This picture stands in sharp contrast to the present formulation: Our definition of "gauge transformation" refers only to mathematical objects – specifically, only to the partial differential equation (1) – making no direct reference to any physical interpretation of the field ϕ or of the equation itself. How is it that this information – about what constitutes a gauge transformation – gets transferred from the physics to the mathematics? The idea is that the "physical interpretation" is already inherent, in some sense, in the partial differential equation itself. This interpretation, after all, is merely a compilation of the various physical effects that the fields produce; and these effects, in turn, are described entirely by the equations those fields satisfy (possibly augmented by additional fields that describe the various measuring instruments). This remark will be illustrated in the examples below.

In the Maxwell case, with zero sources in flat space-time, the gauge transformations consist precisely of duality rotations and scaling transformations on the Maxwell field (replacement F_{ab} by any linear combination, with constant coefficients, of itself and its dual), together with the replacement of F_{ab} by the result of applying to that field any Poincaré transformation. That is, the gauge group is a Lie group, of dimension twelve. The corresponding gauge vector fields are vertical for the duality rotations and scaling, but not for the Poincaré transformations. If we replace this system by one in a curved background space-time (say, having no symmetries), then the Poincaré group of gauge transformations disappears. If we further replace this Maxwell system by one with a fixed charge-current source, then we no longer have duality or scaling as gauge transformations. Quite generally, gauge transformations lose their character when there is turned on an interaction that breaks the corresponding gauge-symmetry. Suppose next that the charge-current source was not fixed, but instead was expressed in terms of additional fields, where these were included within the system, (1), of partial differential equations. In this case, we would expect that the scalings would reappear as gauge transformations. Note that in order to recover the "standard" gauge transformations in the Maxwell case, it is necessary to introduce an additional field, A_a, (thus enlarging $\mathcal{B}$ to a fourteen-manifold), and an additional equation,

$$\nabla_{[a} A_{b]} = F_{ab}. \tag{14}$$

For this enlarged system, the gauge transformations include adding to A_a the gradient of any smooth scalar field on M. The corresponding gauge vector fields are all vertical. Note that it not possible, within the present framework, to admit only the field A_a, and not F_{ab}, for we are demanding right from the beginning that our system be first order.

For the Einstein system with zero sources, the gauge transformations consist [1] of scaling transformations (multiplication of g_{ab} by a constant factor, keeping ∇_a fixed), and the diffeomorphisms on M (which we shall consider in more detail shortly). If sources are included in Einstein's equation – and if those sources are represented by fields, which are included within the bundle manifold $\mathcal{B}$ and on which suitable equations are imposed – then these gauge transformations above will generally remain such. Suppose that we wished to restrict consideration to a particular class of solutions of Einstein's equation – say, those having a Killing field. What are the gauge transformations in this case? We are not permitted simply to impose "having a Killing field" on top of Eqns. (4)–(6), for the general rule is that *all* information is to be encoded, once and for all, into the single system, (1), of partial differential equations. So, we might proceed as follows. First, introduce a new bundle, with fibre over $x \in M$ consisting of quadruples, $(g_{ab}, \nabla_a, \zeta^a, \zeta_a{}^b)$, with g_{ab} Lorentz-signature and the combination $g_{ma}\zeta_b{}^m$ symmetric. [Thus, the fibres in this case have dimension 64 $(= 10 + 40 + 4 + 10)$.] Let the equations on these fields consist of all those already given for the Einstein system, (4)–(6), together with two new equations: $\nabla_a \zeta^b = \zeta_a{}^b$ and $\nabla_a \zeta_b{}^m + R_{sab}{}^m \zeta^s = 0$. Thus, the new field ζ^a represents the Killing field, and $\zeta_a{}^b$ its derivative (the latter being necessary to retain the first-order character). Note that this new system has, for its gauge transformations, not only the g-scalings and diffeomorphisms above, but also ζ-scalings. An alternative treatment of the Einstein system with Killing field is the following. Fix, once and for all, a (say, nowhere vanishing) vector field ζ^a on M. Let the bundle $\mathcal{B}$ consist of all (g_{ab}, ∇_a), with g_{ab} Lorentz signature, such that $g_{m(a}\nabla_{b)}\zeta^m = 0$. Note that the latter is an *algebraic* condition on the fields. The fibres in this case have dimension forty. Let the equations be the usual ones for the Einstein system, (4)–(6). For this system, the gauge transformations include, not all M-diffeomorphisms, but rather only those that are ζ^a-preserving.

The examples above – of various Maxwell and Einstein systems – are sufficiently familiar that it was possible to determine their gauge transformations by inspection. But what about more complicated systems? Is there some simple, general procedure that, applied to any system, (1), of partial differential equations, will yield the complete class of gauge vector fields for that system? None, apparently, is known; but here is a possible line that might yield such a procedure. Fix the system, (1). First, find all fields ξ^α and $S^A{}_B$ such that the equation

$$\mathcal{L}_\xi k^{Aa}{}_{\alpha'} = S^A{}_B k^{Ba}{}_{\alpha'} \tag{15}$$

holds. Eqn. (15) is precisely Eqn. (12), restricted to vertical vectors. But this equation has the great advantage over (12) that it acts within each fibre separately: It is virtually "algebraic". Next, compute the $\gamma^{aA}{}_m$ (which, by (15), must exist; and which must be unique) such that Eqn. (12) holds. And then, finally, demand (as a condition on the original choices of ξ^α and $S^A{}_B$) that this $\gamma^{aA}{}_\alpha$ also satisfy Eqn. (13). It might be of interest to see if there could be generated, along this line, some simple procedure for finding the gauge vector fields.

Up to this point, we have been dealing with "gauge" in a very general context. We now wish to consider a special case of particular interest: The gauge of diffeomorphisms. The system (1) of partial differential equations will be said to admit *diffeomorphism gauge* if every smooth vector field ξ^a on M admits a lifting to some gauge vector field, ξ^α, on $\mathcal{B}$. In terms of the full gauge transformations, the condition is that every diffeomorphism on M lifts to a gauge transformation on $\mathcal{B}$. In the presence of diffeomorphism gauge, every gauge vector field can be written as the sum of one of the ξ^α above and a vertical gauge vector field. It is apparently not known whether, for a system with diffeomorphism gauge, one can always assign, to each vector field on ξ^a on M, a specific lift such that the resulting gauge vector fields themselves form a Lie algebra. Note that, in the presence of diffeomorphism gauge, the Lie algebra of gauge vector fields is always infinite-dimensional.

The Einstein system, of course, admits diffeomorphism gauge. The Maxwell system in Minkowski space-time does not. However, if we modify the latter by including the metric and derivative operator among the fields, and including Eqn. (4) and the vanishing of the Riemann tensor (i.e., the "equations of special relativity") among the equations, then we again recover a system having diffeomorphism gauge.

Diffeomorphism gauge is particularly tractable in the case in which the gauge vector field ξ^α on $\mathcal{B}$ arises from the vector field ξ^a on M by means of a differential operator. Fix any derivative operator $\tilde{\nabla}_a$ on M, and consider the equation

$$\xi^\alpha = \delta^{\alpha a \cdots c}{}_m \tilde{\nabla}_a \cdots \tilde{\nabla}_c \xi^m + \cdots . \tag{16}$$

Here, the right side represents a general linear combination of derivatives of ξ^a up to the nth (with $n \geq 1$). We have written out only the highest-order term, with the remaining terms (with orders $n-1$ down to zero) indicated by dots. The coefficient, $\delta^{\alpha a \cdots c}{}_m$, of this highest-order term is a natural tensor, independent of the choice of derivative operator $\tilde{\nabla}_a$ used on the right in (16). By contrast, the coefficients of the lower-order terms do depend on this choice. In (16), the coefficients depend on point of the bundle manifold (and so in particular $\delta^{\alpha a \cdots c}{}_m$ is a field on $\mathcal{B}$). The right side of Eqn. (16) is thus a function on $\mathcal{B}$, where, for $b \in \mathcal{B}$, the expressions $\tilde{\nabla}_a \cdots \tilde{\nabla}_c \xi^m$ are to be evaluated at $\pi(b) \in M$. Thus, this right side indeed defines a vector field, ξ^α, on $\mathcal{B}$. We next demand that, for every smooth field ξ^a on M, the field ξ^α on $\mathcal{B}$ given by (16) be a lift of ξ^a. This implies that the index "α" of $\delta^{\alpha a \cdots c}{}_m$ is vertical. [To see this, apply $(\nabla \pi)_\alpha{}^s$ to both sides of Eqn. (16), using $n \geq 1$ and noting that the left side then involves no derivatives of ξ^a.] Finally, we impose on the ξ^α of (16) the condition that it be a gauge vector field. We claim: If, for every smooth field ξ^a on M, the field ξ^α on $\mathcal{B}$ given by (16) is a gauge vector field, then the coefficient $\delta^{\alpha a \cdots b}{}_m$ in (16) must satisfy the following two conditions:

i) For every tensor field $L_{a \cdots c}{}^m$ on M, the vertical vector field given by $\kappa^\alpha = \delta^{\alpha a \cdots c}{}_m L_{a \cdots c}{}^m$ satisfies

$$\mathcal{L}_\kappa k^{Aa}{}_{\alpha'} - [k^{Ac}{}_{\alpha'} L_c{}^m] = S^A{}_B k^{Ba}{}_{\alpha'}, \tag{17}$$

for some $S^A{}_B$; and

ii) $$k^{A(d}{}_\alpha \delta^{|\alpha|a \cdots c)}{}_m = 0. \tag{18}$$

In condition i), the term in square brackets in (17) is to be included *only* for the case $n = 1$. [In fact, this term only makes sense for $n = 1$.] Condition i) follows from Eqn. (15), retaining in that equation only the part of highest order (namely, n) in ξ^a-derivatives. Since the vertical part is being taken in (15), i.e., since the index α is primed there, only the vertical derivative of ξ^α is taken in that equation, and so no further derivatives of ξ^a are introduced there. The square-bracketed term appears in (17) because the Lie derivative on the left in (15) involves a term $k^{Am}{}_\alpha \tilde{\nabla}_m \xi^\alpha$, which must be retained when and only when $n = 1$. Note that the left side of (17) is effectively the Lie derivative within the fibre, since κ^α is vertical. For condition ii), first take that part of Eqn. (12) of highest order (namely, $(n+1)$) in ξ^a-derivatives. There results $\gamma^{dA}{}_m = k^{Ad}{}_\alpha \delta^{\alpha a \cdots c}{}_s \tilde{\nabla}_m \tilde{\nabla}_a \cdots \tilde{\nabla}_c \xi^s$. [The highest-order term on the left of (12) is that arising from $k^{Aa}{}_\alpha \tilde{\nabla}_\beta \xi^\alpha$.] Now substitute this $\gamma^{aA}{}_m$ into Eqn. (13), and again take the part highest-order (namely, $(n+1)$) in ξ^m-derivatives. [Neither the left side of (13), nor the first term on the right, contribute at all, for they both involve terms of order at most n. Thus, it is only the $\gamma^{aA}{}_m$-term that contributes at this order.]

I do not believe that there are any further simple conditions on $\delta^{\alpha a \cdots c}{}_m$ that follow from the demand that ξ^α be a gauge vector field[11]. On the other hand, it appears that any sort of converse of the above (i.e., any result to the effect that every $\delta^{\alpha a \cdots c}{}_m$ satisfying the two conditions above, together, possibly, with some further conditions, leads, via (16) to a gauge vector field) is likely to be extremely complicated. The problem, of course, is that the demand that the ξ^α given by (16) be a gauge vector field imposes conditions on all the lower-order terms on the right. These terms will already be awkward (since they will be derivative-operator dependent), and the conditions that must be imposed on them will surely be complicated.

I am not aware of any example of a system, (1), of first order, quasilinear partial differential equations, having diffeomorphism gauge freedom, for which the corresponding gauge vector fields, ξ^α, cannot be expressed as the result of applying a suitable differential operator to ξ^a, in the manner of (16).

Let us now return to the example of general relativity. Recall that the fields in this case consist of a Lorentz-signature metric, g_{ab}, together with a derivative operator, ∇_a; and that the equations are (4)–(6). Fix a smooth vector field, ξ^a, on M. Then the corresponding gauge vector field is given by

$$\xi^\alpha = (2g_{m(b}\nabla_{a)}\xi^m, \nabla_{(a}\nabla_{b)}\xi^c) + \cdots . \tag{19}$$

The two components of the first term on the right reflect the behavior of the space-time metric and of the derivative operator, respectively, under diffeomorphisms. This term is vertical (as, as we have seen, it must be). The remaining terms on

---

[11] A further condition we might reasonably impose (but won't, because it is not needed for what follows) is that the differential operator in (16) commute with taking the Lie bracket, i.e., that $[\xi^\alpha(\xi^a), \xi^\beta(\xi'^b)] = \xi^\gamma([\xi^a, \xi'^b])$. Note that the drop of this equation is automatic. In the case $n > 1$, this equation implies: For any two tensor fields, $L_{a \cdots c}{}^m$ and $L'_{a \cdots c}{}^m$ on M, the vertical vector fields that result from contracting these with $\delta^{\alpha a \cdots c}{}_m$ commute.

the right are lower order[12] in derivatives of ξ^a, and are not all vertical. Thus, the order of the differential operator relating ξ^α and ξ^a in (16), is, in this example, $n = 2$; and the tensor $\delta^{\alpha b \cdots c}{}_m$ in (16) is given by

$$\delta^{\alpha a \cdots b}{}_m = (0, \delta^{(a}{}_{(r} \delta^{b)}{}_{s)} \delta^p{}_m). \tag{20}$$

The two expressions on the right in this equation correspond, respectively, to the g-component and the ∇-component inherent in the index "α". One can check (with a little algebra) that this $\delta^{\alpha a b}{}_m$ does indeed satisfy Eqn. (18).

It appears to be difficult to find a tractable criterion that decides, given a general system (1) of partial differential equations, whether or not that system has diffeomorphism gauge freedom. For common examples (such as Einstein system, above), all the fields are "geometrical objects", i.e., are fields on which the action of diffeomorphisms has been pre-specified. In this case, there is a simple intuitive criterion: Diffeomorphisms act as gauge when and only when all fields are "dynamical", i.e., all are included within the fibres of the bundle $\mathcal{B}$. But in a general case – with the bundle $\mathcal{B}$ and the system (1) specified in some less concrete way – the situation may not be so clear. In this connection, it would be useful, at least as a first step, to be able to find, given the general system (1), those tensors $\delta^{\alpha b \cdots c}{}_m$ satisfying the two conditions given above. From simple examples, these conditions appear to be rather stringent, i.e., there doesn't appear to be an excessive number of solutions. Eqn. (18) is purely algebraic, and so it may be possible to "solve" it. Note that every solution, $\delta^{\alpha b \cdots c}{}_m$, of Eqn. (18) gives rise to a whole class of solutions namely, those given by $\delta'^{\alpha a \cdots e}{}_m = \delta^{\alpha(a \cdots c}{}_n \lambda^{d \cdots e)n}{}_m$, where $\lambda^{d \cdots en}{}_m = \lambda^{(d \cdots e)n}{}_m$, but is otherwise arbitrary. In this way, one can easily raise (or, at minimum, keep the same) the order of δ. It is typically the lowest-order solutions of Eqn. (18) that are of interest. With respect to condition i), fix any point of the base manifold M; and restrict to order $n > 1$. Then the vertical vector fields κ^α on the fibre over this point, satisfying $\mathcal{L}_\kappa k^{Aa}{}_{\alpha'} = S^A{}_B k^{aA}{}_{\alpha'}$ for some $S^A{}_B$, form a vector space. [Essentially, this equation asserts that the Lie derivative, by κ^α, of each of a certain collection of covector fields in the fibre, yields again a certain linear combination of those covectors.] This vector space is "usually" finite-dimensional. It is possible that finite-dimensionality follows already from the conditions of the following section. In any case, fixing now a point of M, $\delta^{\alpha a \cdots c}{}_m$ in the fibre over this point can now be regarded as a linear map from the vector space of tensors $S_{a \cdots c}{}^m$ in M at this point to the vector space such vector fields κ^α on the fibre. It would be of interest to determine whether these remarks actually lead to some effective procedure for finding, given the system (1) of partial differential equations, the $\delta^{\alpha a \cdots c}{}_m$ for that system.

---

[12]There is, unfortunately, a confusing technical issue regarding (19). The term given explicitly on the right is *not* a vector field on $\mathcal{B}$, for it involves not only a point, (x, g_{ab}, ∇_a), of $\mathcal{B}$, but also the "derivative of ∇_a" there. For this term to make sense requires, if you like, a cross-section. However, this dependence on the derivative of ∇_a is lower order in ξ^a, and is compensated for by the remaining terms in (19) (which are not vector fields either). The final result is, indeed, a vector field on $\mathcal{B}$.

As an example of the usefulness of the differential operator, (16), we now use it to prove the following: Given any system (1) of equations that admits diffeomorphism gauge freedom via (16), then that system cannot have a hyperbolization. Intuitively, we might have expected such a result, for the diffeomorphism gauge freedom could be invoked to change any given solution in a region away from some initial surface T while leaving that solution unchanged on T itself. This possibility would seem to be inconsistent with the presence of an initial-value formulation.

The proof of this result is quite simple. Let, $\delta^{aa\cdots c}{}_m \neq 0$ yield the diffeomorphism gauge freedom, via (16); and suppose, for contradiction, that there also exists a hyperbolization, $h_{A\alpha}$. Let, at a point, n_a be such that the quadratic form (10) is positive-definite. Consider the expression

$$n_r \cdots n_s \delta^{\alpha r \cdots s}{}_m [n_a h_{A\alpha} k^{Aa}{}_\beta] n_p \cdots n_q \delta^{\beta p \cdots q}{}_n. \tag{21}$$

On the one hand, this expression must vanish, by (18), since the n's enforce symmetrization over the contravariant indices $a, p, \cdots q$. But, on the other hand, the tensor in square brackets is positive-definite. We conclude that $n_p \cdots n_q \delta^{\beta p \cdots q}{}_n = 0$. But this must hold for every n_a in an open set, and so we have that $\delta^{\beta p \cdots q}{}_n = 0$, a contradiction.

So, diffeomorphism gauge freedom indeed precludes an initial-value formulation. But other types of gauge freedom need not. For example, the system of Maxwell's equations with zero sources, (2)–(3), admits gauge transformations associated with duality and scaling, and yet this system has an initial-value formulation.

4. Initial-value formulation for systems with gauge

Consider a first-order, quasilinear system, (1), of partial differential equations. Let the constraints of this system be integrable and complete. Suppose further that this system manifests some nonzero group of gauge transformations. Then this system as it stands may have no initial-value formulation, for, while we have required integrability and completeness, we have not required the existence of a hyperbolization. Indeed, as we have just seen, a large degree of gauge freedom (such as that for diffeomorphism gauge) is typically incompatible with a hyperbolization. Nevertheless, it may be possible, for certain such systems, to recover an "effective" initial-value formulation. Consider initial data for such a system, consisting of a submanifold T of M of codimension one and a cross-section, ϕ_o, of the bundle over T, subject to the consistency conditions, (7), on those data that flow from the constraints. Roughly speaking, this system has an initial-value formulation "up to gauge" provided the following holds: Given such initial data, subject to some further inequality to the effect that (T, ϕ_o) is "non-characteristic", then there exists a solution of the system (1) manifesting that initial data, and that solution is unique up to gauge transformations[13]. Clearly, this is the closest one could reasonably expect to an initial-value formulation, in the presence of gauge freedom.

Gauge conditions

There is a standard technique for demonstrating that certain classes of systems of partial differential equations have an initial-value formulation up to gauge, in the sense described above. This technique involves introducing what are called "gauge conditions" – certain additional equations imposed on the fields of the system. These additional equations may be purely algebraic (i.e., requiring passage to a subbundle of the original bundle B), purely differential (i.e., requiring a simple enlargement of the system (1)) or some combination of the two. The idea is to choose these equations such that, at least locally, they have the following two properties:

i) Given any solution of the original system, (1), these additional equations can always be achieved by some suitable gauge transformation; and

ii) the system that results from combining the original system, (1), with these additional equations *does* have an initial-value formulation in the traditional sense, i.e., that described in Section 2.

Note that these additional equations cannot themselves be gauge-invariant: Indeed, if they were, then *neither* of the two conditions above could hold. The gauge transformation whose existence is demanded by property i) need not be unique. The initial-value formulation demanded by property ii) can be achieved in a variety of ways. For example, the additional equations can, among other things, create or destroy constraints, turn some of the original differential equations in (1) into identities, or cause the appearance of a hyperbolization when there was none before. Clearly, if we can manage to find gauge conditions having the two properties above, then our system does have, in the sense described above, an initial-value formulation up to gauge. We should emphasize, however, that such gauge conditions are by no means necessary. There could well be a system of partial differential equations that has an initial-value formulation up to gauge, and yet for which there exist no suitable gauge conditions whatever. Here are two examples of gauge conditions.

Consider the Maxwell system, with fields F_{ab}, A_a, and equations (2), (3), and (14). The gauge transformations of interest here are given by addition to A_a of the gradient of a smooth scalar field on M. Consider now the gauge condition given by

$$\nabla^a A_a = 0. \tag{22}$$

This equation satisfies the two properties given above. Indeed, given any solution of (2), (3), and (14), the gauge condition (22) can always by achieved by some

---

[13]In more detail, we require, for existence, that there exist a solution in some neighborhood of T. For uniqueness, we require that, given two such solutions in neighborhoods, there exists a gauge transformation that leaves the initial data (T, ϕ_o) invariant (and so, in particular, that leaves the submanifold T pointwise invariant), and that sends the first solution to one that coincides, in some neighborhood of T, with the second.

transformation. [The gauge scalar field must be chosen to satisfy the wave equation with a suitable source.] Furthermore, the system that results from combining Eqns. (2), (3), and (14) with (22) does indeed have an initial-value formulation. [The vector potential, A_a, now satisfies the wave equation.] So, Eqn. (22) is a suitable gauge condition: It yields, for this Maxwell system, an initial-value formulation up to gauge. This gauge condition is purely differential. The gauge transformation to achieve (22) is never unique. Inclusion of Eqn. (22) with Eqns. (2), (3), and (14) neither creates nor destroys constraints; nor does it render any of the original equations identities. However, (22) does give rise to a hyperbolization where there was none before. Eqn. (22) is, of course, the familiar Lorentz gauge condition in Maxwell theory.

As a second example, consider the Einstein system, given by (4)–(6). The gauge transformations of interest here are the M-diffeomorphisms. Fix any totally symmetric tensor field $W^{abc} = W^{(abc)}$ on M such that at each point of M there exists a covector n_a (and, hence, an open set of n_a) such that the combination $n_a W^{abc}$ is positive-definite. Now consider the following set of gauge conditions:

$$W^{abc} g_{bc} = 0, \tag{23}$$

$$(\nabla_m W^{abc}) g_{bc} = 0. \tag{24}$$

Note that these equations are purely *algebraic* in the fields: Eqn. (23) is algebraic in the metric g_{ab} alone; while (24) is algebraic in (g_{ab}, ∇_a). Eqn. (24) is, of course, merely the result of combining Eqns. (23) and (4). To understand what Eqn (23) means, consider a particular choice of W^{abc}, given by $W^{abc} = u^{(a} h^{bc)}$, where u^a is any nowhere-vanishing vector field, and h^{bc} any positive-definite metric. This choice indeed satisfies the positive-definiteness condition above, e.g., for $n_a = u^b h_{ab}$, where h_{ab} denotes the inverse of h^{ab}. For this particular W^{abc}, Eqn. (23) requires that $g_{ab} u^b$ be a certain multiple of the fixed vector $u^b h_{ab}$. Think of u^b as a "time-direction". Then Eqn. (23) fixes the "time-time" and "time-space" components of the metric g_{ab}. Thus, in this special case, the gauge condition (23) is a version of the familiar lapse-shift gauge in the traditional treatment [12], [5] of the initial-value formulation for general relativity.

There are many other possible choices for the W^{abc} above. Another, for instance, is $W^{abc} = u^{(a} h^{bc)}$, where h^{ab} is Lorentz-signature and u^a is h-timelike. But note that we cannot choose $h^{bc} = g^{bc}$, for we did not permit W^{abc} to depend on the space-time metric. In fact, what follows is actually true for a class of gauge conditions more general than that given above. This class is described as follows. Fix any vector-valued function, v^a, which depends, at each point of M, algebraically on the value there of the space-time metric g_{ab} and of some additional tensor fields. Let this function be such that $\partial(v^a)/\partial(g_{bc}) = W^{abc}$ is totally symmetric, and satisfies the positive-definiteness condition above. The new gauge conditions are now those that result from replacing Eqn. (23) by the equation $v^a = 0$, and retaining Eqn. (24) as given. This is indeed a generalization, for the original formulation (23)–(24), arises as the special case $v^a = W^{abc} g_{bc}$.

We turn now to the issue of whether, for the Einstein system, Eqns. (23)–(24) qualify as a set of gauge conditions, in the sense above.

Can Eqns. (23)–(24) be achieved, at least locally, via a gauge transformation? Fix a solution, (g_{ab}, ∇_a), of the Einstein system. Then, by virtue of Eqn. (4) of that system, it suffices to achieve only (23), for Eqn. (24) then follows. The statement that a diffeomorphism, ψ, on M achieve (23) is

$$W^{abc}(\nabla\psi)_b{}^{d'}(\nabla\psi)_c{}^{e'}g_{d'e'} = 0, \tag{25}$$

where primed indices refer to the image point, $x' = \psi(x)$, and $g_{d'e'}$ denotes the metric evaluated at this point. We may understand this equation in the following manner. Consider the fibre bundle over M whose fibre, over each point $x \in M$, is a copy of M itself. Then the diffeomorphism ψ can be regarded as a cross-section of this bundle; whence (25) becomes a first-order equation on this cross-section. But this equation is not even quasilinear! Consider, however, its linearized version. In this version, the diffeomorphism ψ is replaced by its generator, vector field ξ^a on M; and Eqn. (25) is replaced by an equation on this ξ^a, with principal part $W^{abc}\nabla_b(g_{cd}\xi^d)$. But this linearized equation *does* admit an initial-value formulation: Its only constraint is zero, and it admits a hyperbolization, by virtue of precisely the conditions imposed above on W^{abc}. Thus, while we cannot guarantee solutions of the full system (25), we can guarantee solutions of its linearized version. This behavior – a first-order system of partial differential equations that is not even quasilinear, whose corresponding linearized system is not only quasilinear but actually has an initial-value formulation – seems surprising. It seems likely that, at least in the present case, the initial-value formulation for the linearized system will imply appropriate solutions also of Eqn. (25). It would be of interest to try to prove this – either for the present case, or more generally. A possible method might be first to choose a diffeomorphism ψ that makes the right side of Eqn. (25) "small", and then perform a sequence of corrections via generators ξ^a of infinitesimal diffeomorphisms.

Note that for the Einstein system, in contrast to the Maxwell example above, we only expect to be able to satisfy the gauge conditions via a gauge transformation when the original cross-section ϕ satisfies the field equations, (4)–(6).

Does the system whose fields consist of (g_{ab}, ∇_a) subject to the algebraic conditions (23)–(24), and whose equations consist of (4)–(6), have an initial-value formulation? The constraints of this system, as is not hard to check, are integrable and complete. But does this system admit a hyperbolization?

In order to answer this question, let us return briefly to the original Einstein system, i.e., that without (23)–(24). For this system, the dimension of the space of equations (4)–(6) is 110 (= 40 + 60 + 10), while the dimension of the space of consistency conditions (7) is 64 (= 30 + 30 + 4). This leaves 46 (= 110 − 64) dynamical equations. But the space of independent variables has dimension 50 (= 10+40). This excess – four more independent variables than equations – reflects the diffeomorphism gauge freedom inherent in the Einstein system. Let us now see how this arithmetic is affected by the imposition of the gauge equations (23)–(24).

The dimension of the space of equations is reduced, by the gauge equations, by 40 (= 16 (contraction of (4) with W^{bcd}) +24 (contraction of (5) with W^{cdn})). The dimension of the space of consistency conditions is reduced by 24 (= 12 (from (4)) +12 (from (5))). Hence the dynamical equations are reduced in dimension by 16 (= 40 − 24). But the independent variables are reduced in dimension by 20 (= 4 (from (23)) +16 (from (24))). Thus, imposition of the gauge equations (23)–(24) reduces dimension of the dependent variables by four more than it reduces that of the dynamical equations. In other words, in the Einstein system, supplemented with the algebraic equations (23)–(24), the number of dependent variables precisely matches the number of dynamical equations. What this means is that, for this system, there are equations that give the "time-derivatives" of all the fields, in terms of their values and space-derivatives. Naively, then, one would expect an initial-value formulation for this system.

But is there really such a formulation, i.e., does there actually exist a hyperbolization for this system? This is, apparently, an open question. But it seems extremely likely that there exists none. Here is a possible line for a proof. Suppose for a moment that we replace Eqns. (5)–(6) by the single equation $R_{abc}{}^d = 0$ That is, we replace the Einstein system by that of "special relativity". This replacement merely adds to the Einstein system some additional constraints, from which it follows that every hyperbolization (if any) of the Einstein system must also be a hyperbolization of this new system. In this new system, the constraints continue to be integrable and complete, and, just as above, there is a matching between dimensions of the space of dynamical equations and dependent variables. However, in this case, there *is* a hyperbolization. The corresponding quadratic form is

$$n_a[W^{abc}g_{bm}g_{cn}s^m{}_{rs}s'^n{}_{pq} + u^a\delta g_{rs}\delta' g_{pq}]\, V^{rspq} \tag{26}$$

where $V^{rspq} = V^{(pq)(rs)}$ is any tensor positive-definite in its two symmetric index-pairs and u^a is any timelike vector. It now suffices to prove two assertions:

i) that the expression (26) represents the *only* hyperbolization for this new system, and

ii) that the expression (26) is not a hyperbolization of the constrained Einstein system.

Both of these assertions seem plausible. It would be of interest to prove, by this method or otherwise, that this constrained Einstein system, (4)–(6), (23)–(24), has no initial-value formulation.

These two examples illustrate the point that imposing gauge conditions on systems of partial differential equations with gauge freedom is potentially a complicated business. The basic problem is that there is so much variety: The gauge conditions themselves can be algebraic or differential; these conditions can be achieved by a gauge transformation in a variety of ways; and these conditions can restore to the system an initial-value formulation through a variety of mechanisms. It appears to be necessary to deal with systems of equations on a case by case basis.

We shall now introduce an alternative method for determining that a given system of first-order, quasilinear partial differential equations has an initial-value formulation up to gauge. Our method is systematic, relatively simple and manifestly gauge-invariant.

Fix a first-order, quasilinear system, (1) of partial differential equations, and let this system have some gauge group. The idea is to introduce certain additional geometrical structure on the bundle manifold $\mathcal{B}$, this structure consisting, at each point of $\mathcal{B}$, of a vertical flat σ, i.e., of a subspace of the vector space of vertical vectors at that point of $\mathcal{B}$. We demand that this field of flats be smooth in its dependence on point of $\mathcal{B}$. We further demand that this field be integrable, so the integral surfaces of σ give a foliation of each fibre by submanifolds; and each of these submanifolds has, for its tangent space at each of its points, precisely the flat σ at that point. In all the examples of which I am aware, these flats are also invariant under the action of gauge transformations on the system (1), although this invariance is not actually needed in what follows.

Think of the flat σ, at a fixed point of $\mathcal{B}$, as representing "physical directions" in field-space. Then the vertical vectors not lying in this flat are to be thought of as having "unphysical" (i.e., gauge) components. A crucial point here is that we do not attempt to single out any particular complementary subspace, i.e., any space of specifically "unphysical directions". In the case in which the gauge group consists only of the identity (as well as in certain other cases with small gauge group), the flat σ, at each point of $\mathcal{B}$, will consist of *all* vertical vectors at that point, whence the integral surfaces of the flats will be precisely the fibres of the fibre bundle $\mathcal{B}$.

Existence

Recall that a first-order, quasilinear system of partial differential equations has (subject to some resolution of the gap discussed in Appendix A) an initial-value formulation provided the following four conditions are satisfied:

  i) its constraints are integrable,
 ii) its constraints are complete,
iii) it admits a hyperbolization, and
 iv) there exists some nonempty open set of covectors n_a that suffices for both completeness and the hyperbolization.

The idea is to demonstrate existence of an initial-value formulation up to gauge (subject to the same gap) by suitably modifying, taking into account the flats σ, these four conditions.

The first condition, integrability of the constraints, remains unchanged. For the second condition, recall, from Section 2, that completeness means that, for every n_a in some open set of covectors, whenever ν_A is such that $\nu_A n_a k^{Aa}{}_\alpha$ annihilates all vertical vectors, then $\nu_A = c_A{}^a n_a$ for some constraint $c_A{}^a$. We now replace this second condition with what we call σ-*completeness*: For every such n_a, whenever $\nu_A n_a k^{Aa}{}_\alpha$ annihilates all vectors *in the flat σ*, then $\nu_A = c_A{}^a n_a$ for some constraint $c_A{}^a$. Note that σ-complete implies complete; and, furthermore, that σ-completeness follows from completeness, together with the property: If $\nu_A n_a k^{Aa}{}_\alpha$ annihilates all vectors in σ, then it annihilates all vertical vectors. The condition of σ-completeness means physically that none of the equations of the system serve to restrict the "time-derivatives" *only* of unphysical degrees of freedom. The third condition is modified to the following. We demand that there exist, at each point of the bundle manifold $\mathcal{B}$, a σ-*hyperbolization*, i.e., a tensor $h_{A\beta}$ at that point such that the combination $n_a h_{A\beta} k^{Aa}{}_\alpha$, applied only to vectors in σ, is symmetric, and, for all n_a lying in some open set, is positive-definite. This is precisely the same as the original definition of a hyperbolization, except that now the quadratic form is restricted to σ, i.e., to "physical degrees of freedom". In particular, every hyperbolization is a σ-hyperbolization. Finally, the fourth condition remains unchanged.

To summarize, what we have done here is expand the notion of completeness, and contract that of a hyperbolization, to take account of the flats σ.

We shall show shortly that, under these four conditions, as broadened in the paragraph above, existence holds for the system (1) of partial differential equations. But first, we give two examples.

Consider first the Maxwell system with vector potential – the system with equations (2), (3), and (14) and gauge the addition to A_a of the gradient of a scalar field – in a general curved spacetime. Now fix any smooth, nowhere vanishing timelike vector field u^a on the manifold M. Let, at each point of $\mathcal{B}$, the flat σ at that point consist of all vertical tangent vectors, $(\delta F_{ab}, \delta A_a)$, at that point satisfying $u^a \delta A_a = 0$. Thus, the flats in this case are nine-dimensional (in ten-dimensional fibres). These flats are smooth and integrable, the integral surfaces being those of constant $u^a A_a$. Furthermore, these flats are gauge-invariant, for the action of a gauge transformation, within a fixed fibre, is to send the point (F_{ab}, A_a) to the point $(F_{ab}, A_a + w_a)$, where w_a is some fixed vector (which, however, is different for different gauge transformations). But clearly these leave the integral surfaces surfaces above, and so the flats σ, invariant.

We now claim that this Maxwell system, with these flats σ, satisfies the four conditions given above. Indeed, the first condition, integrability, was already shown in Section 2; as was completeness of the system (2)–(3). Hence, the second condition, σ-completeness, (with the open set of covectors n_a consisting of the timelike ones) becomes: Given any tensor $\nu^{ab} = \nu^{[ab]}$ and any timelike n_a, such that $\nu^{ab} n_a \delta A_b = 0$ for all δA_b satisfying $u^b \delta A_b = 0$, then we have $\nu^{ab} n_a = 0$. But this assertion is true, for the vanishing of $\nu^{ab} n_a \delta A_b$ for all δA_b with $u^b \delta A_b = 0$ implies that $\nu^{ab} n_a$ is a multiple of u^b; whence, contracting with n_b and using the timelike character, that $\nu^{ab} n_a = 0$. For the third condition, existence of a σ-hyperbolization, consider the quadratic form given by the usual Maxwell one, (11), on the Maxwell field, plus the additional term $n_{[a} \delta A_{b]} u^a g^{bc} \delta' A_c$ This expression i) arises from (14) (for it is the contraction of $n_{[a} \delta A_{b]}$ with something); and, with δA_b and $\delta' A_c$ restricted by $u^b \delta A_b = u^c \delta' A_c = 0$, is ii) symmetric; and iii) positive-definite, whenever $u^a n_a > 0$. The fourth condition is immediate (the appropriate open set of n_a consisting of the timelike ones with $u^a n_a > 0$).

Consider, as a second example, the Einstein system, (4)–(6) with diffeomorphism gauge. Recall that, at a general point, (g_{ab}, ∇_a), of the fibre over base point $x \in M$, vertical tangent vectors are represented by $(\delta g_{ab}, s^m{}_{ab})$, where $s^m{}_{ab} = s^m{}_{(ab)}$ represents the "change in the derivative operator". Let the flat σ at that point consist of those vertical vectors satisfying $g^{ab} s^m{}_{ab} = 0$. Thus, the flats in this case are forty-six-dimensional (in fifty-dimensional fibres). These flats are smooth and integrable. To see what the integral surfaces are, fix any derivative operator, $\tilde{\nabla}_a$, at x, and represent ∇_a there by the tensor, $C^m{}_{ab}$, expressing the difference between ∇_a and $\tilde{\nabla})_a$. The integral surfaces of the flats σ within this fibre are now given by the surfaces of constant $g^{ab} C^m{}_{ab}$. [Note that these surfaces are independent of the choice of the auxiliary operator $\tilde{\nabla}_a$. However, which constant vector represents which surface will, of course, depend on this choice.] That these flats are gauge-invariant follows from the fact that they are defined without reference to any "external objects".

We now claim that this Einstein system, with these flats σ, satisfies the four conditions given above. The first condition, integrability, was already shown in Section 2; as was completeness of the system (4)–(6). Hence, the second condition, σ-completeness, requires (with the open set of covectors n_a consisting of the timelike ones): Given any tensors $\nu^{abcd} = \nu^{[ab](cd)}$ and $\nu^{ab} = \nu^{(ab)}$ and any timelike n_a such that

$$\nu^{abcd} n_a s^m{}_{bc} g_{md} + 1/2 \nu^{ab}(n_m s^m{}_{ab} - n_b s^m{}_{am}) = 0 \qquad (27)$$

for all $s^m{}_{ab} = s^m{}_{(ab)}$ with $g^{ab} s^m{}_{ab} = 0$, then Eqn. (27) holds for *all* $s^m{}_{ab} = s^m{}_{(ab)}$. [Here, the left side of Eqn. (27) is the equation that results from replacing the derivative of ∇ by $n_a s^b{}_{cd}$ in Eqns. (5) and (6), contracting with ν^{abcd} and ν^{ab}, respectively, and adding.] To see that this assertion is true, note that the hypothesis implies that $n_r[\nu^{r(ab)s} g_{sm} + 1/2 \delta^r{}_m \nu^{ab} - 1/2 \delta^{(a}{}_m \nu^{b)r}]$ is a multiple of g^{ab}. Contracting with $n_a n_b$ and using the timelike character, we conclude that that multiple is zero. But this is precisely the statement that Eqn. (27) holds for all $s^m{}_{ab} = s^m{}_{(ab)}$. For the third condition, the existence of a σ-hyperbolization, we proceed as follows. Let $s^m{}_{ab} = s^m{}_{(ab)}$ and $s'^m{}_{ab} = s'^m{}_{(ab)}$, be any two tensors, and n_a and u^a any two vectors. We then have the following identity, whose proof is straightforward but a little tedious:

$$s'_{(cd)n} u^n \Sigma(ab)[-2n_{[m} s^m{}_{a]b} + 4n_{[q} s^m{}_{b](a} g_{p)m} g^{pq}] - s'_{(cd)}{}^m u^n [s_{(ab)[m} n_{n]}] \qquad (28)$$

$$= s_{(ab)m} s'_{(cd)n}[-(u^p n_p) g^{mn} + 2u^{(m} n^{n)}] - n_{(a} s_{b)m}{}^m s'_{(cd)n} u^n, \qquad (29)$$

where we have freely raised and lowered indices with the metric g_{ab}, and where the symbol "$\Sigma(ab)$" is the instruction to symmetrize over the indices a and b. Now, the expressions in square brackets on the left involve only the combinations that result from the equations, (5)–(6), of the Einstein system by replacing the derivative of ∇ by $n_a s^b{}_{cd}$. The last term on the right vanishes for $s^m{}_{ab}$ lying in the flat σ, for this implies $s_{bm}{}^m = 0$. Now fix timelike u^a, and let n_a also be timelike, with $u^a n_a > 0$.

Then the right side, contracted with $(g^{ac}+2u^a u^c/(u^m u_m))(g^{bd}+2u^b u^d/(u^m u_m))$, is the required quadratic form for the σ-hyperbolization. Finally, the fourth condition is again immediate, with, again, the open set of n_a consisting of the timelike ones with $u^a n_a > 0$. Thus, we have verified that these flats σ for the Einstein system satisfy our four conditions above. It is perhaps a reasonable conjecture that these flats are the unique ones for the Einstein system satisfying our conditions. It would be interesting to prove this.

We now return to the general case. Why do we not not further require that the hyperbolization $h_{A\alpha}$ of the third condition also be invariant under gauge transformations? We would if we could, but, unfortunately, such invariance cannot be achieved in examples (e.g., the Einstein system). Note that the conditions above refer *only* to the flats σ, and not at all to gauge transformations. The gauge transformations enter at this stage only indirectly, through our inability to find a hyperbolization that is symmetric when applied to *all* vertical vectors. Note, incidentally, σ-completeness favors larger flats σ, while existence of a σ-hyperbolization favors smaller. The final flats represent a compromise between these competing demands.

The condition of σ-completeness has several immediate, and very useful, consequences. Let us adopt the convention, for purposes of this paragraph that a σ, appended to any covariant Greek index, means "applied only to vectors in the flat σ". Then σ-completeness implies in particular: $n_a \nu_A k^{Aa}{}_{\alpha\sigma} = 0$ implies $n_a \nu_A k^{Aa}{}_{\alpha'} = 0$. [Recall that a prime on a covariant Greek index means "applied only to vertical vectors".] The first consequence is this: Whenever $\nu_A k^{Aa}{}_{\alpha\sigma} = 0$, then $\nu_A = 0$. [To see this, note that the hypothesis implies $\nu_A n_a k^{Aa}{}_{\alpha\sigma} = 0$ for every n_a; and so, by σ-completeness, that $\nu_A n_a k^{Aa}{}_{\alpha'} = 0$ for every n_a in some open set; and so, that $\nu_A k^{Aa}{}_{\alpha'} = 0$; which in turn implies $\nu_A = 0$.] What this means is that no equation of the system has the property that, when restricted to physical degrees of freedom, it becomes purely algebraic. We turn next to the constraints. First, note that, quite generally, every constraint of the system (1), i.e., every $c^a{}_A$ satisfying $c^{(a}{}_A k^{|A|b)}{}_{\alpha'} = 0$, is also automatically a σ-*constraint*, i.e., also satisfies $c^{(a}{}_A k^{|A|b)}{}_{\alpha\sigma} = 0$. But σ-completeness implies the reverse, i.e., that *every* σ-constraint arises in this manner from some constraint of the original system. [Indeed, let $\hat{c}^{(a}{}_A k^{|A|b)}{}_{\alpha\sigma} = 0$. Contract with $n_a n_b$ to obtain $(n_a \hat{c}^a{}_A) n_b k^{Ab}{}_{\alpha\sigma} = 0$. Choosing n_a to lie in the open set, this condition now implies, by σ-completeness, that $(n_a \hat{c}^a{}_A) n_b k^{Ab}{}_{\alpha'} = 0$; which, since this now holds for all n_a in an open set, implies $(\hat{c}^{(a}{}_A) k^{|A|b)}{}_{\alpha'} = 0$.] In addition, again from σ-completeness, no nonzero constraint gives rise to the zero σ-constraint. To summarize, σ-completeness implies that the equations of the original system (1), as well as its constraints, go over, essentially unchanged in number and character, when restricted to the physical degrees of freedom.

Fix a first-order, quasilinear system of partial differential equations, together with a smooth, integrable field of flats σ, satisfying the four conditions (integrability, σ-completeness and existence of a σ-hyperbolization, for a common open set of n_a) above. There is a key result to the effect that this system must, for suitable initial data, manifest existence of solutions. This result, in more detail, is the following. Let (T, ϕ_o) be initial data for this system, so T is a submanifold of the base manifold M of codimension one, and ϕ_o is a cross-section of the bundle $\mathcal{B}$ over T. Let these data satisfy all the consistency conditions (7) arising from the constraints of the system. Let, furthermore, these data be non-characteristic, in the sense that the normal n_a to T at each point lies within the open set specified in the fourth condition. Then there exists a solution ϕ of the system (1), defined in a neighborhood of T, that reduces to the given data, ϕ_o, on T. The proof of this result will emerge from the discussion that follows.

Consider, in this system, a submanifold, $\tilde{\mathcal{B}}$, of $\mathcal{B}$ having the following two properties:

i) $\tilde{\mathcal{B}}$ meets each fibre of $\mathcal{B}$ in a single integral surface of σ, and

ii) the projection mapping $\tilde{\mathcal{B}} \xrightarrow{\pi} M$ is a surjection onto M, i.e., every tangent vector at every point of M is the image, under the derivative of this map, of some tangent vector at some point of $\tilde{\mathcal{B}}$.

We remark that there are many such submanifolds. Indeed, fixing any cross-section, ϕ, of $\mathcal{B}$, the union of all the σ-integral surfaces that meet $\phi[M]$ is such a submanifold (and, in fact, the unique one containing the image of this cross-section). Furthermore, every such submanifold $\tilde{\mathcal{B}}$ arises in this way. Next, note that every such submanifold $\tilde{\mathcal{B}}$ is itself also a bundle over the same base manifold M. In this bundle, the fibre over $x \in M$ is the corresponding integral surface of σ in the $\mathcal{B}$-fibre over x. Note further that every cross-section of $\tilde{\mathcal{B}}$ is automatically also a cross-section of $\mathcal{B}$; and that every cross-section of $\mathcal{B}$ – provided it lies within the submanifold $\tilde{\mathcal{B}}$ – is also a cross-section of the bundle $\tilde{\mathcal{B}}$.

We now introduce a certain partial differential equation based on this subbundle $\tilde{\mathcal{B}} \xrightarrow{\pi} M$, namely, that which results from simply restricting our original equation, (1), to $\tilde{\mathcal{B}}$. That is, we regard a cross-section $\tilde{\phi}$, of $\tilde{\mathcal{B}}$ as a cross-section also of the original bundle $\mathcal{B}$, and, as such, impose on it Eqn. (1). The result is a first-order, quasilinear system of partial differential equations, based on the bundle $\tilde{\mathcal{B}}$. The "k" for this new system is simply the restriction of the original $k^{Aa}{}_\alpha$ to σ, and the "j" the original j^A; where both of these are now evaluated on $\tilde{\mathcal{B}}$ (since that is where the cross-section $\tilde{\phi}$ lives). We note that, by the discussion of σ-completeness above, none of the partial differential equations, (1), of the original system are "lost" (i.e., become algebraic equations) on restriction to $\tilde{\mathcal{B}}$. Hence, every solution of the restricted equation, in $\tilde{\mathcal{B}}$, is also, when regarded as a cross-section in $\mathcal{B}$, a solution of (1). We now claim that this new system of partial differential equations satisfies the conditions of Section 2 for having an initial-value formulation. Indeed, this is immediate. Integrability passes from the original system, (1), to the new one because the two systems have the same constraints; σ-completeness and the σ-hyperbolization yield completeness and a hyperbolization for the new system.

Now consider our given initial data (T, ϕ_o), satisfying the consistency conditions (7), and such that the normal, n_a, to T, at each of its points, lies in the open set of covectors given above. Extend this cross-section ϕ_o over T to *any* cross-section over the entire base manifold M; and then expand this cross-section to a submanifold $\tilde{\mathcal{B}}$ of $\mathcal{B}$, in the manner described above. There results a subbundle, $\tilde{\mathcal{B}} \xrightarrow{\pi} M$, of our original bundle, $\mathcal{B} \xrightarrow{\pi} M$, and, as we have seen above, this subbundle inherits, from (1), its own first-order, quasilinear partial differential equation. Now, the original initial data, (T, ϕ_o), for (1) is (since $\phi_0[T] \subset \tilde{\mathcal{B}}$, by construction) also initial data for this new system. Furthermore, these initial data on $\tilde{\mathcal{B}}$ also satisfy the consistency conditions there (since the original initial data satisfied the consistency conditions for (1), and since the two systems have precisely the same constraints). Hence, since the equation in $\tilde{\mathcal{B}}$ has an initial-value formulation, there exists a unique solution, $\tilde{\phi}$, of the $\tilde{\mathcal{B}}$-equation manifesting these initial data. But, as we remarked above, this $\tilde{\phi}$ is also a solution of the original system, (1), in $\mathcal{B}$. In short, the original initial data, (T, ϕ_o), in $\mathcal{B}$ admit a solution of (1).

This completes the proof of existence of solutions for systems of partial differential equations satisfying the four conditions above. Note that we do not in general have uniqueness, because we had the freedom, in the argument above, in the choice of the subbundle $\tilde{\mathcal{B}}$. Note also that the treatment above nowhere made any mention of gauge transformations or gauge conditions.

Uniqueness

Under certain further conditions, the solution obtained above will be unique, up to gauge. To make things more explicit, let us restrict consideration, for the moment, to the case in which the gauge is that of diffeomorphisms. What follows can be generalized to other gauge groups. In any case, we have, under this assumption, the tensor $\delta^{\alpha a \cdots c}{}_m$ (with n indices $a \cdots c$) of Eqn. (16), which describes the lifting of any smooth vector field ξ^a on M to a gauge vector field ξ^α on $\mathcal{B}$.

The idea is to consider two extensions of the given cross-section ϕ_o over T to subbundles $\tilde{\mathcal{B}}$ of $\mathcal{B}$. We must show that there exists a gauge transformation sends one such subbundle to the other. In more detail, we shall rewrite "sends one such subbundle to the other" as a system of partial differential equations on the gauge transformation itself. This system will turn out to be n^{th} order, with principal part $\delta^{\alpha a \cdots c}{}_m$. We will then demand that this system have an initial-value formulation. This system, when rendered first-order, will automatically have its constraints integrable (as systems usually do!). But we shall have to impose on it (as appropriate conditions on $\delta^{\alpha a \cdots c}{}_m$) completeness and existence of a hyperbolization. These conditions, in more detail, are the following.

First, we shall demand that, for every covector n_a in our open set, we have: Given any ν_α that annihilates all vectors in σ and satisfies $\nu_\alpha \delta^{\alpha a \cdots c}{}_m n_a \cdots n_c = 0$, then ν_α annihilates all vertical vectors. This is a kind of completeness, but now referring to the action of the gauge group[14]. It not only has a form similar to σ-completeness, but also has similar consequences. The following, for example, are

---

[14] In fact, it appears that, in practice, this condition may provide the simplest route to "finding", appropriate flats σ, given only some partial differential equation (1), and the action on the bundle $\mathcal{B}$ of the diffeomorphisms.

direct consequences of this condition (with proofs completely analogous to those for σ-completeness):

i) For any ν_α that annihilates all vectors in σ and satisfies $\nu_\alpha \delta^{\alpha a \cdots c}{}_m = 0$, we have that this ν_α annihilates all vertical vectors;

ii) For $c^a{}_\alpha$ annihilating all vectors in σ, and satisfying $c^{(a}{}_\alpha \delta^{|\alpha|c \cdots d)}{}_m = 0$, then $c^a{}_\alpha$ annihilates all vertical vectors.

Second, we shall demand that there exist a tensor $H_n{}^{d \cdots e}{}_\alpha = H_n{}^{(d \cdots e)}{}_\alpha$, vanishing when contracted with any $\mu^\alpha \in \sigma$, and a tensor

$$S_n{}^{d \cdots e a b \cdots c}{}_m = S_n{}^{(d \cdots e) a (b \cdots c)}{}_m,$$

with symmetry

$$S_n{}^{d \cdots e (a b \cdots c)}{}_m = 0,$$

such that the combination

$$H_n{}^{d \cdots e}{}_\alpha \delta^{\alpha a b \cdots c}{}_m + S_n{}^{d \cdots e a b \cdots c}{}_m \tag{30}$$

is symmetric under interchange of the index-pairs "$n \, d \cdots e$" and "$m \, b \cdots c$", and, contracted with any covector n_a lying in our open set, is positive-definite in these index-pairs. This H will turn out to be an effective hyperbolization for our system of equations on the gauge transformation.

As we shall see shortly, these two conditions taken together guarantee, in an appropriate sense, uniqueness of solutions up to gauge.

Let us return to the example of the Einstein system. Recall that the tensor $\delta^{\alpha a \cdots b}{}_m$ in this case has rank two, and is given by Eqn. (20). The flats are given by tangent vectors, $(\delta g_{ab}, s^m{}_{ab})$, satisfying $g^{ab} s^m{}_{ab} = 0$; and the open set of n_a is that consisting of the timelike covectors. This system, we claim, satisfies the two conditions above. The first asserts, in this example, the following. For every timelike n_a, if tensor $\nu^{ab}{}_m$ (i.e., ν_α) is such that $\nu^{ab}{}_m$ is a multiple of g^{ab} (i.e., $\nu_{\alpha^\sigma} = 0$) and $\nu^{ab}{}_m n_a n_b = 0$ (i.e., $\nu_\alpha \delta^{\alpha ab}{}_m n_a n_b = 0$), then $\nu^{ab}{}_m = 0$. But this assertion, clearly, is true. For the second condition, consider the choices

$$H_n{}^{drs}{}_m = -(g_{mn} + u_m u_n/(u^q u_q)) g^{rs} u^d, \tag{31}$$

$$S_n{}^{dab}{}_m = (g_{mn} + u_m u_n/(u^q u_a)) u^{[a} g^{b]d}, \tag{32}$$

where the index combination "mrs" in $H_n{}^{drs}{}_m$ stands for "α". These two clearly satisfy the symmetries given above. Furthermore, the quadratic form given by (30) becomes $(g_{mn} + u_m u_n/(u^q u_q))[u^a g^{db} - 2g^{a(d} u^{b)}]$, which has the requisite symmetry and positive-definite properties. Thus, the Einstein system, (4)–(6), satisfies our two conditions.

The key result is that the two conditions above imply uniqueness, up to gauge, of our solution of (1) with the given initial data (T, ϕ_o). The method, as discussed earlier, is to show that given any two extensions of the given cross-section, ϕ_o, over T to a subbundles, $\tilde{B}$ and $\tilde{B}'$, of B, then these two subbundles are gauge-related.

It is convenient to begin with the "infinitesimal argument", which is somewhat easier conceptually. Fix any cross-section, ϕ, of the original bundle B, and any vertical field, μ^α, defined at points of $\phi[M]$. Consider now the following partial differential equation

$$[\delta^{\alpha a \cdots c}{}_m \tilde\nabla_a \cdots \tilde\nabla_c \xi^m + \cdots] - (\nabla\phi)_a{}^\alpha \xi^a - \mu^\alpha \in \sigma \tag{33}$$

The term in square brackets on the left in Eqn. (33) is precisely the expression, (16) for the gauge vector field defined by ξ^a. [Here, $\tilde\nabla_a$ denote some arbitrary, fixed derivative operator on the manifold M.] The second term on the left in (33), which is just the lift of ξ^a to the cross-section, serves to make the sum of the first two terms vertical. That is, Eqn. (33) is the assertion that the difference between the vertical vector given by the first two terms on the left and the given vertical vector μ^α lies in the flat σ. This is a first-order, quasilinear system of partial differential equations on the vector field ξ^a on M (as is seen, e.g., by contracting (33) with covectors in the fibre orthogonal to σ). The geometrical meaning of this equation is the following. Think of the given cross-section, ϕ, as defining some subbundle, $\tilde{B}$, of the original bundle B. Think of the vertical vector μ^α as representing the vertical "connecting vector" from $\tilde{B}$ to some nearby subbundle $\tilde{B}'$. Then this μ^α is meaningful only up to addition of vertical vectors tangent to $\tilde{B}'$, i.e., only up to vectors lying in the flat σ. [In other words, if we change μ^α by addition to it of a vector in σ, then this new μ^α connects to the same nearby subbundle $\tilde{B}'$.] Eqn. (33), then, is the assertion that the vertical part of the gauge vector field generated by the vector field ξ^a on M send $\tilde{B}$ to $\tilde{B}'$.

Eqn. (33) is a system of nth order, quasilinear partial differential equation. [The "vector space of equations" is the quotient space of the vector space of vertical vectors by the vector subspace σ.] We now claim: By virtue of the two conditions above, this system has an initial-value formulation. The first step is to convert (33) to a first-order system. To this end, we introduce, in addition to the original vector field ξ^m, tensor fields $\xi_a{}^m, \xi_{ab}{}^m, \cdots \xi_{a\cdots c}{}^m$, each totally symmetric in its covariant indices, and having covariant indices ranging in number from from one to $(n-1)$. On these fields, we now impose the following three sets of equations:

$$\tilde\nabla_a \xi^m = \xi_a{}^m, \ldots, \tilde\nabla_{(a}\xi_{b\cdots c)}{}^m = \xi_{a\cdots c}{}^m, \tag{34}$$

$$\tilde\nabla_{[a}\xi_{b]}{}^m = \cdots, \quad \ldots, \quad \tilde\nabla_{[a}\xi_{b]\cdots c} = \cdots, \tag{35}$$

$$[\delta^{\alpha a \cdots c}{}_m \tilde\nabla_a \xi_{b\cdots c}{}^m + \cdots] - (\nabla\phi)_a{}^\alpha \xi^a - \mu^\alpha \in \sigma. \tag{36}$$

Eqn. (34) consists of $(n-1)$ equations, which allow us to "interpret" each of $\xi_a{}^m, \ldots, \xi_{a\cdots c}{}^m$ as the symmetrized derivative of its predecessor, and so, ultimately, as a symmetrized derivative of ξ^m. Eqn. (35), consists of the $(n-1)$ equations resulting from taking the curl of each equation in (34). Thus, Eqns. (35) are the integrability conditions for Eqns. (34). Finally, Eqn. (36) is precisely Eqn. (33), rewritten in first-order form.

Note that (34)–(36) is indeed a first-order, quasilinear system of partial differential equations, on the n tensor fields $\xi^m, \xi_a{}^m, \ldots, \xi_{a\cdots c}{}^m$. In the case of Eqn. (36), the equation lies in the vector space given by the quotient of the space of vertical vectors by the subspace consisting of those vectors lying in flat σ. This system, we now claim, satisfies the conditions, of Section 2, for having an initial-value formulation. To see this, consider first the equations consisting of all $(n-1)$ in (34) and all but the last in (35). These $(2n-3)$ equations express the derivatives of each of the the first $(n-1)$ of our fields (i.e., all but the last one, the $\xi_{a\cdots c}{}^m$ appearing in (36)) algebraically in terms of these fields. This system has an obvious initial-value formulation. [The constraints correspond to taking the curls of (34) and of (35), the corresponding integrability conditions being (35) and identities, respectively. A hyperbolization is immediate.] The critical field is the final one, $\xi_{a\cdots c}{}^m$, and the equations on it are the last equation in (35), and Eqn. (36). The first of these equations has a constraint corresponding to taking its curl. The corresponding integrability condition is an identity (by the choice of right-hand side of the last equation in (35)). A constraint for (36) is represented by a tensor $c^a{}_\alpha$ such that $c^a{}_{\alpha\sigma}=0$, and such that $c^{(a}{}_\alpha\delta^{|\alpha|b\cdots c)}{}_m=0$. But our first condition above, completeness, implies that the only such $c^a{}_\alpha$ is zero. Thus, the equations on $\xi_{a\cdots c}{}^m$ are integrable and complete. But these equations also have a hyperbolization, namely the H of (30), for our second condition is precisely the statement that this H has the requisite properties. Thus, we have shown that the system (34)–(36) satisfies the conditions of Section 2 for an initial-value formulation.

Next, let there be given initial data, (T,ϕ_o), satisfying the consistency conditions (7) and with the normal n_a to T lying, at each of its points, in our open set. Let the given vertical field μ^α be chosen to vanish on (T,ϕ_o). Now choose, on T, initial data for the system (34)–(36) with all the fields $\xi^m,\cdots,\xi_{a\cdots c}{}^m$ vanishing there. These data clearly satisfy all the consistency conditions for the system (34)–(36). So, there exists a solution of this system. That is, there exists a vector field ξ^m on M whose corresponding gauge vector field, ξ^α, coincides with the given vertical μ^α on the subbundle $\tilde{B}$, and leaves invariant the given initial data, (T,ϕ_o). We have shown, then, uniqueness of solutions up to "infinitesimal" gauge.

There follows similarly a corresponding result for the full gauge group. Consider the fibre bundle over M, whose fibre, over each point x of M, is a copy of M itself. Then a cross-section of this bundle is precisely a smooth map ψ from M to M. Now demand that the gauge transformation $B \xrightarrow{\Psi} B$ arising from this ψ send the submanifold $\tilde{B}$ to $\tilde{B}'$. This is an nth order, quasilinear differential equation on the cross-section ψ, an equation having precisely the principal part of Eqn. (33). Just as with (33), we convert this to a first-order partial differential equation (so the fields will now be "point of M", together with certain tensors that can be interpreted as the first $(n-1)$ derivatives of the smooth map ψ). These equations will have precisely the character of the system (34)–(36), and for the same reason will have an initial-value formulation. We conclude that there exists an M-diffeomorphism whose corresponding gauge transformation sends the submanifold $\tilde{B}$ to $\tilde{B}'$ – and, therefore, a diffeomorphism that sends the one solution ϕ of (1) to the other solution ϕ'.

This completes the demonstration of uniqueness of solutions of (1) up to diffeomorphism gauge, under the two conditions listed above. In particular solutions of the Einstein system are unique up to the diffeomorphism gauge. Note that even though the diffeomorphism gauge group is infinite-dimensional, all manifolds in the present treatment are finite-dimensional. We also remark that the gauge transformation generated by the argument above is unique.

Finally, we note that similar considerations apply to certain other gauge groups. The crucial property we needed, above, is that the gauge transformations be represented by cross-sections of a suitable bundle, and that the requirement that a gauge transformation send one submanifold $\tilde{B}$ to another be a system of quasilinear partial differential equations on cross-sections of that bundle. An example is the Maxwell system considered earlier, with flats σ given by $u^a \delta A_a = 0$. In this case, the corresponding bundle has fibre, over each point x of M, consisting of the reals (the possible values of the gauge-potential). The conditions analogous to those above again hold; and again we conclude uniqueness of solutions, up to gauge transformations.

5. Conclusion

Fix any first-order, quasilinear system, (1), of partial differential equations, and let there act on that system some group of gauge transformations. We have shown that, under certain conditions, this system manifests an initial-value formulation up to gauge, i.e., that, given suitable initial data, there exist solutions of the system manifesting those initial data, and those solutions are unique, up to gauge transformations. These "certain conditions" are relatively simple: They may be checked by carrying out algebraic manipulations involving the coefficients of the differential equation. The key idea is to introduce a certain field, σ, of flats within the bundle manifold, where these flats represent "physical directions" in field space. To prove existence, we effectively restrict the original equation, (1), to the "physical degrees of freedom". Then the conditions in this case require that the equation, so restricted, manifest completeness of its constraints, and a hyperbolization. To prove uniqueness, we consider two solutions of the system, and write out the partial differential on a gauge transformation that guarantees that that gauge transformation send one of those solutions to the other. The conditions in this case are those that guarantee that this partial differential equation have an initial-value formulation. In particular, the Einstein system satisfies the "certain conditions" above, and so we conclude as a special case what is well known [5], [3]: The Einstein system manifests an initial-value formulation up to gauge.

In the example of the Einstein system, the appropriate flats σ carry information similar to that used in the standard harmonic gauge condition in relativity. However, we use these flats in quite a different way. We do not directly impose any gauge conditions on the fields, and, in particular, we impose on initial data no further constraints, other than those that follow from the original Einstein system (4)–(6). In the Maxwell case, on the other hand, the flats carry information similar to that of the Coulomb gauge. It is curious that these two gauge conditions –

harmonic for general relativity and Coulomb for Maxwell – are here placed on a similar footing.

Are there, besides the Einstein system, other examples of systems of partial differential equations that have an initial-value formulation up to the gauge freedom of diffeomorphisms, in the sense of Section 4? I am aware of, essentially, just two.

One such system, it turns out, is that of special relativity. The fields are the same as those for the Einstein system: a Lorentz-signature metric g_{ab} and a derivative operator ∇_a. However, the equations for this system consist of (4) together with $R_{abc}{}^d = 0$ (the latter replacing (5)–(6) for the Einstein system). That is, this system is merely the Einstein system, augmented with some additional constraint equations. Note that here the Minkowski metric is taken to be "dynamical", resulting in a system subject to the gauge freedom of diffeomorphisms. The M-diffeomorphisms act on (g_{ab}, ∇_a) in the same manner as in the Einstein system, and so in particular we have the same tensor $\delta^{\alpha a b}{}_m$ of Eqn. (20). For the flats σ, choose precisely the same ones as for the Einstein system. We now claim that this system – special relativity – satisfies all the conditions given in Section 4. For the σ-hyperbolization $h_{A\alpha}$ and the $H_n{}^{dab}{}_m$ choose precisely the same objects as for the Einstein system. The demonstrations that the various conditions of Section 4 are satisfied is virtually identical to the corresponding demonstrations for the Einstein system.

Here is a second example. Let the fields, on the base manifold M, consist of a nowhere vanishing vector field u^a together with a tensor field $\alpha^b{}_{cd}$. Let the equation be $\mathcal{L}_u \alpha^b{}_{cd} = 0$, i.e., the requirement that the Lie derivative of $\alpha^b{}_{cd}$ by u^a vanish. This system has an action of M-diffeomorphisms as gauge: These diffeomorphisms act on $(u^a, \alpha^b{}_{cd})$ in the usual way, and this action clearly preserves the equation of the system. Note that the order of this action (i.e., the number of ξ-derivatives that appear in (16)) is one, in contrast to the order, two, for the Einstein system. We now claim that this system satisfies all the conditions of Section 4. First note that the only constraint of this system is zero. Let, at each point of the bundle manifold $\mathcal{B}$, the flat σ consist of those vertical tangent vectors, $(\delta u^a, \delta \alpha^b{}_{cd})$, with $\delta u^a = 0$. This field of flats is clearly smooth and integrable. Let the open set of covectors n_a, at each point of $\mathcal{B}$ consist of those with $n_a u^a > 0$. Then σ-completeness becomes: Let n_a satisfy $n_a u^a > 0$, and let $\nu_b{}^{cd}$ be any tensor such that the expression

$$\nu_b{}^{cd}[u^m n_m \delta\alpha^b{}_{cd} + \alpha^m{}_{cd}\delta u^b n_m - \alpha^b{}_{md}\delta u^m n_c - \alpha^b{}_{cm}\delta u^m n_d] \tag{37}$$

vanishes for all $(\delta u^a, \delta\alpha^b{}_{cd})$ with $\delta u^a = 0$. Then this expression vanishes for *all* $(\delta u^a, \delta\alpha^b{}_{cd})$. But this assertion is true (as follows immediately from the fact that the hypothesis implies, using $u^m n_m > 0$, that $\nu_b{}^{cd} = 0$). For the σ-hyperbolization, fix any positive-definite metric p_{ab}, with inverse p^{ab}, and let $h_{A\alpha}$ be represented by the tensor $h_m{}^{rs}{}_b{}^{cd} = p_{mb}p^{rc}p^{sd}$. The completeness condition for the gauge transformation reads: If $(\nu_a, \nu_b{}^{cd})$ (i.e., ν_α) is such that $\nu_b{}^{cd} = 0$ (i.e., $\nu_{\alpha^\sigma} = 0$) and such that $(n_s u^s)\nu_m = 0$ (i.e., $n_s \nu_\alpha \delta^{\alpha s}{}_m = 0$), then $(\nu_a, \nu_b{}^{cd}) = 0$. This is immediate. Finally, for the hyperbolization for the gauge transformation, choose $H_{n\alpha}$

such that $H_{na}\delta\phi^\alpha = p_{na}\delta u^a$, and $S_n{}^a{}_m = 0$. Then the corresponding quadratic form, (30), is given by $u^a p_{nm}$; and this form indeed has the required symmetry and positive-definiteness.

Thus, this system – with fields u^a and $\alpha^b{}_{cd}$ and equation $\mathcal{L}_u \alpha^b{}_{cd} = 0$ – has an initial-value formulation, up to the gauge freedom of diffeomorphisms. Of course, this conclusion was already clear geometrically: Use the diffeomorphisms to "fix" the field u^a, whence the equation, $\mathcal{L}_u \alpha^b{}_{cd} = 0$, serves to fix the field $\alpha^b{}_{cd}$ along the u-trajectories. There are similar, more general, examples in which the field $\alpha_b{}^{cd}$ is replaced by a tensor field with general index structure, or, even more generally, by any finite collection of tensor fields, each with some index structure. Then the single equation $\mathcal{L}_u \alpha^b{}_{cd} = 0$ must be replaced by the equations that specify that each of these fields has vanishing Lie-derivative by u^a. Still more generally, there could be imposed various algebraic or differential equations (such as, e.g., the vanishing of a contraction, or of the exterior derivative of a form) on the fields; and the equations could demand that the u-Lie derivative of each field, be, not zero, but rather some some expression algebraic in the fields.

Further examples can be obtained from these by combining systems. Thus, for example, the Einstein-perfect fluid system[15], as well as the special-relativity-Maxwell system, also represent systems having an initial-value formulation up to diffeomorphism gauge. Still more examples could, presumably, be constructed by taking the derivative system [9] of these examples.

The examples described above are the only ones of which I am aware that have an initial value formulation up to the gauge freedom of diffeomorphisms. Is there any hope of finding *all* first-order, quasilinear systems of partial differential equations having this property? While such a classification of systems of partial differential equations is certainly an ambitious program, it does not appear to be an impossible one, for the various conditions (Section 4) that must be imposed seem tractable. It would be particular interesting to determine if there are examples having diffeomorphism order (the number of ξ-derivatives in Eqn. (16)) exceeding two. Such a classification – particularly if it is found that there exist relatively few systems with an initial-value formulation up to diffeomorphism gauge – might be of physical interest. After all, one could argue that the system of partial differential equations appropriate for any viable physical theory on a manifold should manifest an initial-value formulation and diffeomorphism gauge. Thus, such a classification would represent the beginnings of a classification of allowed physical theories – at least, of physical theories described by partial differential equations on a manifold.

---

[15]In order that the perfect-fluid system have an initial-value formulation, we must avoid "fluid boundaries". This is accomplished by imposing a suitable restriction on the fibres of the bundle $\mathcal{B}$: We demand that the velocity field u^a be unit timelike, and that the mass density be strictly positive. This restriction may be both unnecessary and unphysical.

Appendix A – Initial-value formulation

Fix a of first-order, quasilinear system, (1), of partial differential equations. Let this system have constraints that are integrable and complete; and let it also admit a hyperbolization, h_{Aa}. Let T be a submanifold of M of codimension one, and ϕ_0 a cross-section over T, such that

  i) this (T, ϕ_0) satisfies all the consistency conditions, (7), arising from (1), and
  ii) the normal n_a of T at each of its points lies in the open sets of covectors associated with completeness and with the hyperbolization. [That is, let (T, ϕ_0) be "non-characteristic".]

We sketch a partial proof of the following assertion: Under these conditions, there exists, in some neighborhood of T, one and only one smooth solution ϕ of the system, (1), satisfying $\phi|_T = \phi_0$. What follows is only a "partial proof" because, as we shall see, it contains a gap. This appendix is an open invitation to fill that gap.

As we shall remark later, various specific systems of physical interest (Maxwell; Einstein, suitably formulated; etc) *do* have an initial-value formulation. We are here asking, more generally, whether this behavior, known already for these specific examples, holds for certain general systems of partial differential equations. Consider first the system of equations given by

$$h_{A\beta} k^{Aa}{}_\alpha (\nabla \phi)_a{}^\alpha = h_{A\beta} j^A, \tag{38}$$

i.e., the subsystem of (1) that results from contracting it with $h_{A\beta}$. This system also has a hyperbolization (and, indeed, it is already in "symmetric-hyperbolic form"), and its only constraint is zero. There is a standard theorem [4], [11], [10], [16], [7], whose proof uses an energy argument, to the effect that *this* system, by virtue of its having a hyperbolization and zero constraints, admits one and only one smooth solution, ϕ, in some neighborhood of T, manifesting the given initial data (T, ϕ_o). Thus (since every solution of (1) is certainly a solution of (38)), we have shown uniqueness of solutions of (1). There remains only existence, and the plan for this is to show that the solution ϕ, just obtained, of the subsystem (38) in fact satisfies the *full* system, (1), of partial differential equations. To this end, consider the field I^A, defined in a neighborhood of T by

$$I^A = k^{Aa}{}_\alpha (\nabla \phi)_a{}^\alpha - j^A, \tag{39}$$

where, on the right, we have substituted our solution ϕ of (38). What we must show, to prove existence, is $I^A = 0$. To this end, we note that this field I^A has the following three properties. First, this I^A satisfies the algebraic (linear) system

$$h_{A\alpha} I^A = 0 \tag{40}$$

everywhere in our neighborhood of T. This is exactly (38). Second, this I^A vanishes on the submanifold T itself. To see this, first note that, by the consistency conditions (7), we have that $n_a c^a{}_A I^A = 0$ on T, for every constraint $c^a{}_A$ of our system. But, by completeness, the $h_{A\alpha}$ and $n_a c^a{}_A$ (as $c^a{}_A$ runs through the constraints) together span the equation covector space. The result now follows from

(40). The third property is that this I^A satisfies a linear system of partial differential equations of the form

$$c^a{}_A \nabla_a I^A = W^A{}_B I^B. \tag{41}$$

Here, $c^a{}_A$ denotes any constraint of the original system, (1), and there is one equation in (41) for each such constraint. Indeed, fixing the constraint $c^a{}_A$, that this equation hold, for some field $W^A{}_B$ on M, follows immediately from integrability of that constraint.

Now consider the following first-order, quasilinear (in fact, linear) system of partial differential equations: The field is I^A subject to (40), and the system of equations is (41). One solution of this system is the I^A given by (39), and this particular solution has $I^A = 0$ on T. So, in order to prove that the I^A of (39) vanishes in a neighborhood of T, we need only show uniqueness for this system, (40)–(41), with given initial values on T.

We first remark that, by completeness of (1), the system above takes "evolution form", i.e., it expresses the derivative, off T, of every component of I^A in terms of the value of I^A on T and its derivative there within T. But this property alone is not, apparently, sufficient to establish uniqueness. Uniqueness would follow, from the standard theorems of existence and uniqueness discussed earlier, if we could find a hyperbolization for the system (40)–(41) (for this system has only the zero constraint). In fact, for every physical example of which I am aware, this system does admit a hyperbolization. Thus, for all these physical examples, the system (1) does indeed manifest existence and uniqueness. But, unfortunately, there has not been given, as far as I am aware, any general proof of the existence of a hyperbolization for (40)–(41). Such a proof could, of course, make use of the assumed integrability and completeness of the constraints, as well as the existence of a hyperbolization, for (1). It would also be acceptable to impose, on the original system (1), suitable additional, mild hypotheses. Of course, it would also suffice to prove uniqueness of the system (40)–(41) by some other means, i.e., by some method that does not invoke the standard theorems of existence and uniqueness. This, then, is the gap we mentioned earlier. It would, I believe, be of some interest to fill it.

Consider, as an example, the Maxwell system, (2)–(3). A hyperbolization for this system is, as we have seen, determined by a timelike vector field u^a on space-time. Eqn. (38) reads in this case

$$u^c \nabla_{[a} F_{bc]} = 0, \quad u_{[a} \nabla^m F_{b]m} = 0. \tag{42}$$

The I^A of (39) is given by

$$I_{abc} = \nabla_{[a} F_{bc]}, \quad I_b = \nabla^a F_{ab}. \tag{43}$$

The algebraic conditions, (40), on the I^A become

$$u^c I_{abc} = 0, \quad u_{[a} I_{b]} = 0. \tag{44}$$

Finally, the equations, (41) on the I^A become

$$\nabla_{[d} I_{abc]} = 0, \quad \nabla^b I_b = 0. \tag{45}$$

So, our question becomes in this case: Does uniqueness hold for the system (45) of partial differential equations, on fields consisting of (I_{abc}, I_b) subject to (44)? The answer to this question is yes. Consider, e.g., the field I_b. Then the first equation in (44) is precisely the statement that $I_b = \gamma u_b$ for some function γ, whence (45) becomes $u^b \nabla_b \gamma = -\gamma \nabla_b u^b$. But, clearly, this partial differential equation on function γ on M satisfies uniqueness. [In fact, it not only does this system have a hyperbolization, and so an initial-value formulation; but it is actually an ordinary differential equation.] Dually for I_{abc}.

So, by the argument above, the Maxwell system admits an initial-value formulation. This same method – checking directly and explicitly that the system (40), (41) satisfies uniqueness – works for other systems (1) of partial differential equations of physical interest. In particular, such an argument works for the Einstein system, reduced, as in Section 4, to a hyperbolic system. What is now needed, is to replace this piecemeal approach by a general theorem.

Acknowledgment

I would like to thank Oscar Reula, Gabriel Nagy, David Garfinkle, and Maciej Dunajski for discussions.

References

[1] I.M. Anderson, C.G. Torre, *Classification of Local Generalized Symmetries for the Vacuum Einstein Equations*, Comm. Math. Phys. 176, 479 (1996).

[2] R. Arnowitt, S. Deser, C.W. Misner, *The Dynamics of General Relativity*, in *Gravitation: An Introduction to Current Research*, L. Witten, ed., (Wiley, New York, 1962).

[3] Y. Choques-Bruhat, *Théorème d'existence Pour Certain Systèmes d'équations aux Dérivées Partielles Non Linéaires*, Acta Math. 88, 141 (1952).

[4] K.O. Friedrichs, *Symmetric Hyperbolic Linear Differential Equations*, Comm. Pure and Appl. Math. 7, 345 (1954).

[5] H. Friedrich, A. Rendall, *The Cauchy Problem for the Einstein Equations*, in *Einstein's Field Equations and their Physical Interpretation*, B.G. Schmidt, ed., (Springer, Berlin, 2000). Available as gr-qc/0002074.

[6] R. Geroch, *Asymptotic Structure of Space-Time*, in *Asymptotic Structure of Space-Time*, F.P. Esposito, L. Witten, eds. (Plenum, New York, 1977).

[7] R. Geroch, *Partial Differential Equations of Physics*, in *General Relativity (Proceedings of the Forty-Sixth Scottish Summer School in Physics)*, G.S. Hall, J.R. Pulham, Ed. (SUSSP Publ., Edinburgh, 1996). Available as gr-qc/9602055.

[8] R. Geroch, S.M. Perng, *Total Mass-Momentum of Arbitrary Initial-Data Sets in General Relativity*, J. Math. Phys. 35, 4157 (1994).

[9] R. Geroch, G. Nagy, O. Reula, *Relativistic Lagrange Formulation*, J. Math. Phys. 42, 3789 (2001). Available as gr-qc/0102071.

[10] F. Johns, *Partial Differential Equations*, (Springer-Verlag, New York, 1982).

[11] P.D. Lax, *On Cauchy's Problem for Hyperbolic Equations and the Differentiability of Solutions of Elliptic Equations*, Comm. Pure and Appl. Math. 8, 615 (1955).

[12] C.W. Misner, K.S. Thorne, J.A., *Gravitation*, (Freeman, San Francisco, 1973).

[13] R. Penrose, *Asymptotically Flat Spacetimes*, Proc. R. Soc. A284, 159 (1965).

[14] N. Steenrod, *The Topology of Fibre Bundles*, (Princeton Univ. Press, Princeton, 1954).

[15] M.E. Taylor *Pseudodifferential Operators*, (Princeton Univ. Press, Princeton, 1981).

[16] M.E. Taylor *Partial Differential Equations, Vol. III Nonlinear Equations*, (Springer, Berlin, 1996).

[17] A. Trautman, *Conservation Laws in General Relativity*, in *Gravitation: An Introduction to Current Research*, L. Witten, Ed. (Wiley, New York, 1962).

Robert Geroch
Enrico Fermi Institute
5640 Ellis Ave
Chicago, IL 60464, USA
e-mail: `geroch@midway.uchicago.edu`

Index

Abbott-Deser mass, 56
ADM
  dynamical equations, 11
  Hamiltonian, 10
  mass, 42
Anti de Sitter space, 405
asymptotic simplicity, 140
asymptotically
  simple spacetimes, 33
  static, 174
asymptotically de Sitter, 393
attractor, 91
AVTD, 95

b-Completeness, 403
Bartnik's quasi-local mass, 63
Bel–Robinson energy, 314
Bel–Robinson tensor, 314
Bianchi spacetimes, 84
Bieberbach rigidity, 407
BKL, 108
boundary problem, 220
Bounded completeness, 404
BTZ black hole, 420

Carrières theorem, 405
Cauchy
  development, 74
  formulation, 208
  generalized surface, 332
  horizon, 110
  surface, 110
causal theory, 379
causality domain, 402
characteristic, 169
  formulation, 208
Choptuik solution, 371
circular fibers, 251
closed timelike curves, 332
CMC, 79
CMC time function, 417
CMCSH gauge, 80, 305

cn-gauge, 148
Codazzi equation, 5
Codazzi tensor, 309
collapse of metrics, 357
complete, 447
Completeness, 403
conformal
  boundary, 141
  constraints, 140
  field equations, 132
    general, 134
    metric, 138
    reduced, 138
  Gauss gauge, 134
  geodesics, 133
constant mean curvature, 79, 305
  conformal construction, 17
  existence, 22
  non-existence, 22, 32
constraint, 446
  equations, 6
    linearized, 11
    preserved under ADM evolution, 12
  Hamiltonian, 259, 264
  integrable, 446
  momentum, 259
  operators, 11
  σ-, 465
convergence of metrics, 349
Convex cocompact Kleinian groups, 416
cosmic censorship, 77
Cosmological time function, 416
Cotton tensor, 150
critical collapse, 245
critical sets, 170
cross-section, 444
curvature estimate, 364
cylinder at space-like infinity, 169

d-jet, 174
de Sitter space, 405

decay estimate, 295
diffeomorphism gauge, 454
dischronal region, 336
discrete-energy, 221
domain of dependence, 110
dominant energy condition, 238
drop, 451

Einstein
  -Hilbert Lagrangian, 10
  equations
    reduced, 7
  evolution equations, 78
Einstein cosmos, 434
elliptic-hyperbolic system, 80
ellipticity constant, 305
energy
  first, 265
    corrected, 282
  second, 267
    corrected, 289
  total corrected, 294
essential, 422
eternal observer horizon, 394
extrinsic curvature, 4

Fermi-propagated, 304
Ferrand-Obata theorem, 423
fibre, 443
first energy, 265
formation of cusps, 360
formulations issue, 218
free conformal data, 14
Fuchsian, 95, 102
  methods, 247

G_2 spacetime, 92
galaxies, 231
Galloway–Schleich–Witt–Woolgar
  inequality, 59
gauge transformation, 451
gauge vector field, 450
Gauss equation, 5
generalized Kasner exponents, 241
geodesically complete, 324
geometric boundary conditions, 28
Geometric structure, 404
geometrization conjecture, 373
Global hyperbolicity, 410
global Lorentzian geometry, 379
globally hyperbolic, 110
gluing constraint solutions, 30
Gowdy spacetime, 92

graph manifold, 358
gravitational degrees of freedom, 22, 28

Hamiltonian constraint, 259, 264
Hamiltonian phase space, 12, 13
harmonic
  coordinate condition, 80
  coordinates, 305
  radius, 351, 370
  time gauge, 82
Hawking mass, 48
Heisenberg group, 424
hyperbolic cone space-time, 306
Hyperbolic ends, 414
hyperbolic manifold, 373
hyperbolization, 449

inner mass, 63
inverse mean curvature flow, 28, 50
Isometric actions, 421
isotropic singularity, 247

Kasner
  billiard, 89
  circle, 88
  relations, 88
Killing
  algebra, 429
  initial data, 33
  symmetry, 251
Kleinian, 405

lapse, 4
Lichnerowicz equation, 15, 31
Lichnerowicz theorem, 362
linearization stability, 12
linearized constraint operator, 33
Lorentz homogeneous spaces, 422

Mainardi equation, 5
Margulis spacetime, 409
mass
  Abbott-Deser, 56
  ADM, 42
  Bartnik's, 63
  Hawking, 48
  inner, 63
  quasi-local, 62
  Trautman-Bondi, 56
maximum principle for null hypersurfaces,
  389
Minkowski space, 405
Misner space, 336
Mixmaster, 89

momentum constraint, 259

non linear stability theorem, 251
normal expansion, 163
null
    condition, 83
    lines, 336
    quasi-spherical coordinates, 30
    splitting theorem, 390
null mean curvature, 384
numerical-relativity, 205

optically stable, 343
Oscillator groups, 424

Penrose conjecture, 28
Penrose Inequality, 40
polarized, 95, 101
Positive mass theorem, 44
prescribed mean curvature, 80

quasi-local mass, 28, 62
quasi-spherical equation, 27
quasi-spherical foliations, 28

radiation field, 141
regularity condition, 178
Riemannian Penrose Inequality, 46
rigidity (of hyperbolic
    cone space-times), 310

σ-completeness, 462
σ-constraint, 465
σ-hyperbolization, 463
s-jet, 174
Sachs peeling, 142
scri, 33
second variation
    Lorentzian, 10
    Riemannian, 26
shift vector, 4
spacetime, 380
    G_2, 92
    globally hyperbolic, 331
    Gowdy, 92
    U(1) symmetric, 99
spatially compact, 239
spatially homogeneous, 236
spherically symmetric, 243
splitting theorem, 353, 370
stable properness, 423
static, 243
static vacuum field equations, 149
strong energy condition, 74

strongly hyperbolic, 443
sub- and super-solutions, 18
summation by parts, 221
symmetric hyperbolic, 138, 443

tangent vector
    vertical, 443
Taub-NUT, 90
Teichmüller parameters, 260
thick-thin decomposition, 360
thin sandwich, 23
    conformal, 24
    conjecture, 23
    equations, 23
    reduced equations, 24
Thurston geometries, 85
time function, 4
time nonorientable, 332
total characteristic, 169
transport equations, 169
Trautman-Bondi mass, 56
TT-tensor, 307
Twist potential, 255

U(1) symmetric spacetime, 99
universe-from-nothing, 331

vacuum data set, 73
vacuum extension, 73
Vlasov equation, 231

Warped products, 427
Wave map, 258
wave map, 101
    equation, 83
    gauge, 8
well posed, 82
Weyl connection, 131
Weyl field, 311
wormhole
    construction, 32

Yamabe invariant of RP^3, 55
Yamabe theorem, 18

Editors-in-Chief:
A. Boutet de Monvel, Université Paris VII
Denis Diderot, France
G. Kaiser, The Virginia Center for Signals and Waves, Glen Allen, USA

Progress in Mathematical Physics is a book series encompassing all areas of theoretical and mathematical physics. It is intended for mathematicians, physicists, and other scientists, as well as graduate students in the above related areas. This distinguished collection of books includes authored monographs and textbooks, the latter primarily at the senior undergraduate and graduate levels. Edited collections of articles on important research developments or expositions of particular subject areas may also be included.

PMP 38: Dalibard, J. / Duplantier, B. / Rivasseau, V.
Poincaré Seminar 2003. Bose-Einstein Condensation – Entropy (2004)

The Poincaré Seminar is held twice a year at the Institute Henri Poincaré in Paris. The goal of this seminar is to provide up-to-date information about general topics of great interest in physics. Both the theoretical and experimental results are covered, with some historical background. Particular care is devoted to the pedagogical nature of the presentation.
This volume contains the lectures of the third and fourth Poincaré Seminar, both held in 2003. The third one is devoted to Bose-Einstein Condensation: it covers the physics of superfluid liquid helium as well as the recently discovered atomic Bose-Einstein condensates. Major experimental results are presented, together with relevant theoretical approaches and remaining open questions. The fourth one concentrates on Entropy, giving a comprehensive account of the history and various realizations of this concept, from thermodynamics to black holes, and including theoretical and experimental discussions of the corresponding fluctuations for mesoscopic systems near equilibrium.

Contributing authors: Roger Balian, Sébasten Balibar, Claude Cohen-Tannoudji, Jean Dalibard, Thibault Damour, Olivier Darrigol, Christian Maes, Philippe Nozières, Félix Ritort, Christophe Salomon, Gora V. Shlyapnikov

ISBN 3-7643-7106-4

PMP 37: Nikiforov, A.F. / Novikov, V. / Uvarov, V.B.
Quantum-Statistical Models of Hot Dense Matters. Methods for Computation Opacity and Equation of State (2004)
ISBN 3-7643-2183-0

PMP 36: Chruściński, D. / Jamiołkowski, A.
Geometric Phases in Classical and Quantum Mechanics (2004)
ISBN 0-8176-4282-X

PMP 35: Sommerfeld, A.
Mathematical Theory of Diffraction. Translated by R.J. Nagem, M. Zampolli and G. Sandri (2004)
ISBN 0-8176-3604-8

PMP 34: Abłamowicz, R. (Ed.)
Clifford Algebras. Applications to Mathematics, Physics and Engineering (2004)
ISBN 0-8176-3525-4

PMP 33: Hehl, W. / Obukhov, Y.N.
Foundations of Classical Electrodynamics. Charge, Flux, and Metric (2003)
ISBN 0-8176-4222-6